Smart Protective Coatings for Corrosion Control

Smart Protective Coatings for Corrosion Control

Lingwei Ma and Dawei Zhang

WILEY-VCH

Authors

Prof. Lingwei Ma
University of Science and
Technology Beijing
Beijing, 100083
China

Prof. Dawei Zhang
University of Science and
Technology Beijing
Beijing, 100083
China

Cover Image: © Oleg Breslavtsev/Getty Images

Library of Congress Card No.: applied for

British Library Cataloguing-in-Publication Data
A catalogue record for this book is available from the British Library.

Bibliographic information published by the Deutsche Nationalbibliothek The Deutsche Nationalbibliothek lists this publication in the Deutsche Nationalbibliografie; detailed bibliographic data are available on the Internet at <http://dnb.d-nb.de>.

Print ISBN: 978-3-527-35386-6
ePDF ISBN: 978-3-527-84714-3
ePub ISBN: 978-3-527-84713-6
oBook ISBN: 978-3-527-84715-0

Typesetting Straive, Chennai, India

Contents

Foreword

Corrosion represents natural and gradual deterioration of materials (usually metals) by chemical or electrochemical interaction with their environment. This process jeopardizes functional and structural integrity in equipment and infrastructure, shortens their operational lifespan, and ultimately may trigger catastrophic failures and environmental contamination, imposing grave threats to public safety and property. It is universally acknowledged that corrosion is a global issue, affecting our buildings, roads, machinery, bridges, vessels, and any locale where materials are exposed to industrial or natural corrosive environments, including atmospheric moisture, salts, alkalis, acids, and diverse chemical agents. Authoritative data indicate that the economic losses due to corrosion amount to 3–5% of the gross domestic product (GDP) in many industrialized countries. Like earthquakes, fires, and pollution, corrosion poses a serious hazard, rendering the onus of corrosion mitigation paramount.

To minimize the functional and structural impact caused by corrosion, the application of corrosion protection coatings emerges as one of the most effective methods to inhibit material corrosion degradation and reduce economic losses. Such coatings can slow down the rate of material corrosion under exposure to aggressive environmental conditions, thereby extending the service life of the object. In ancient times, substances such as egg whites, tree gums, and asphalt served as coating materials, providing a physicochemical barrier between the environment and the underlying metal. Nowadays, engineers can integrate functionality and aesthetics into coating designs using polymer technology and nanotechnology. In modern society, nearly all man-made items employ protective or decorative coatings, with the global coatings industry generating approximately one trillion US dollars in annual production. With continuous industrial development and the need for applications exposed to extreme conditions, the performance specifications for corrosion protection coatings have escalated, leading to the emergence of smart coatings distinguished by eco-friendliness, self-cleaning, self-healing, corrosion sensing, and enhanced durability. Combining new smart coating technologies with traditional methodologies, or using smart coatings to replace traditional ones, has emerged as one of the focal points of research topics in preventing metal corrosion.

This book comprehensively and systematically introduces the latest advancements and future prospects of smart coatings. It elaborates on the raw materials, preparation methods, corrosion prevention principles, and monitoring technologies used for each category of smart coatings. It is a valuable resource summarizing the currently achieved performance level and further development status of smart coating technologies. The smart coating technologies and corrosion monitoring methods presented in this book will have a favorable effect on improving the service life of coatings in complex and aggressive service environments. This publication is a timely compendium, serving as a pivotal reference for scholars and engineers across a wide variety of backgrounds, providing theoretical support for the innovation of new coatings in the corrosion protection engineering field, and acting as a technical guide for professionals in industry. It guides the conceptualization and design of state-of-the-art technologies for intelligent corrosion control through theoretical insights and practical examples, which has profound implications for the transformation toward a sustainable industry and circular economy. I also hope that this book will inspire a broader audience to engage with the development of corrosion protection technology, enhance comprehension and awareness of corrosion issues and corrosion protection strategies in various industries, and contribute to addressing the global material and energy transition challenges.

TU Delft, The Netherlands
29 February 2024

Arjan Mol

Preface

Corrosion refers to the degradation and deterioration of metals owing to chemical, electrochemical, or other reactions at the surface or interface. This process damages engineering structures, which shortens their service life, increases the maintenance cost, and causes accidents that threaten the environment and public safety. In terms of magnitude, it was estimated that almost one ton of steel is converted to rust every minute around the world. Each year, up to one-third of global steel production is scrapped owing to corrosion, approximately one-third of which is irrecoverable. The annual economic loss due to corrosion far exceeds that due to natural disasters and other accidents: the global economic loss due to corrosion amounts to approximately US$4 trillion per year, with China's annual economic loss from corrosion estimated at 3 trillion RMB.

Several methods and materials have been developed to prevent corrosion. Corrosion protection coatings account for approximately two-thirds of all anticorrosion measures. A dense covering layer effectively isolates the metal substrate from corrosive media, thus preventing or slowing down electrochemical reactions on the substrate. However, the coating can become damaged or cracked in harsh environments during transportation and service. Unless it is effectively repaired, such damage will seriously weaken adhesion between the coating and metal substrate, and the barrier properties of the coating will also decline to some degree. A new research focus is smart coatings, defined as coatings that change their properties in response to environmental stimuli. Smart coatings can detect corrosion on metal surfaces at the earliest stage and enable preventive measures. The development of smart coatings for intelligent corrosion control has been expanding in recent years, with continuous application of new knowledge gained from different fields.

This book introduces the latest research on smart coatings for corrosion protection, including self-healing coatings, self-reporting coatings, superhydrophobic coatings, weathering-resistant coatings, and corrosion monitoring techniques based on big data. The book reviews the major processes and strategies for preventing corrosion; techniques for the synthesis and application of smart coatings and multifunctional coatings; and current and future trends in coatings for marine corrosion protection, nuclear corrosion protection, oil/gas corrosion protection, and related fields.

The chapters of this book are organized to provide a comprehensive review of the fabrication, properties, mechanism, and applications of different types of smart coatings. Chapter 1 explores the fundamentals of smart coatings and their corrosion protection capabilities. After a brief introduction to corrosion protection coatings and smart design approaches for intelligent corrosion control, the processes of coating degradation and metal corrosion are discussed in detail so that nonexperts can gain a basic understanding of the corrosion protection techniques. The concepts of self-healing coatings, self-reporting coatings, and superhydrophobic coatings are also explained, and the application fields of smart coatings are surveyed. Chapters 2–5 introduce different types of self-healing coatings, including thermally activated, photothermally activated, pH- and redox-activated, and ion exchange-activated ones. Specifically, Chapter 2 discusses two thermally activated self-healing coating systems, covering aspects such as coating preparation, filler preparation, surface characterization, macroscopic and microscopic electrochemical properties, and self-healing performance. In one system, $Ce(NO_3)_3$-ethylene-vinyl acetate (EVA) microspheres are added to a shape memory epoxy resin to recover both the corrosion protection properties and adhesion strength. In another coating system containing poly (ε-caprolactone) microspheres loaded with 8-Hydroxyquinoline, the good corrosion protection performance is due to the triple actions of (1) the shape memory effect triggered by thermal healing of the coating, (2) the filling effect of molten microspheres, and (3) the action of corrosion inhibitors. Chapter 3 describes the development of a photothermally activated self-healing coating. Photothermal conversion species such as graphene oxide, titanium nitride, and Fe_3O_4 are incorporated into micro/nanocontainers and loaded into the coatings. Effects from the type, amount, and dispersion of the photothermal filler on the corrosion resistance, photothermal conversion performance, and self-healing performance are analyzed. Chapter 4 focuses on two types of pH-activated and one type of redox-activated self-healing coatings. Silica nanoparticles and PDMAEMA macromolecules are pH-sensitive materials, and redox-responsive polyaniline can accelerate the release of corrosion inhibitors and the formation of a stable rust layer. Chapter 5 describes the application of layered double hydroxide (LDH) to ion exchange-activated self-healing coatings. An LDH nanocontainer containing sustained-release metavanadate corrosion inhibitors was successfully loaded onto graphene oxide surface, which was added to waterborne epoxy resin to achieve the self-healing effect.

In Chapters 6–8, pH-, metal ion-, and mechanically responsive self-reporting coatings are introduced. The self-reporting coatings can autonomously indicate coating damage and metal corrosion in the early stages. They also allow the observation of corrosion initiation over time. Chapter 6 covers several self-reporting coatings based on different types of corrosion indicators (phenolphthalein, sulfosalicylic acid-modified carbon dots, and phenanthroline) that either emit fluorescence or display color changes due to pH variation or metallic cations within local defects. The corrosion-sensing performance and barrier property of these self-reporting coatings are investigated, and their future development is also discussed. Chapter 7 describes a novel coating with dual self-reporting and self-healing functions.

2′,7′-Dichlorofluorescein (DCF) is loaded into PU/UF microcapsules, which are then dispersed in epoxy to produce a mechanically responsive self-reporting coating. DCF reacts with the coating matrix containing residual amino groups to produce a prominent red color. For self-healing performance, the high mobility of the polymer network endows a shape memory effect, allowing the coating to recover its original shape and barrier properties. Chapter 8 presents the design of an active protective epoxy coating with weathering-resistant, self-reporting, and self-healing properties. The mesoporous silica nanocontainers loaded with natural polyphenols (tannic acid and tea polyphenol) can alleviate coating degradation by scavenging radicals generated during UV irradiation. Once the coating is damaged, the polyphenols are released to react with Fe^{3+} ions, forming a black-colored chelate that indicates the onset of corrosion. This chelate also inhibits corrosion propagation and therefore offers significant self-healing effect. Chapter 9 discusses different types of superhydrophobic coatings with improved anti-icing, anticorrosion, mechanical stabilization, and self-healing properties. The superhydrophobic property of the coating is achieved by adding CNT–SiO_2 hybrids into the coating, and the self-healing property is achieved by adjusting the proportion of the epoxy resin and curing agent with a flexible chain segment. The damaged coating can heal under thermal or electrical heating by activating the shape memory effect. The stable three-dimensional network structure and high proportion of flexible chain segments endow the coating with high mechanical durability. Chapter 10 applies atmospheric corrosion monitor (ACM) technique to investigate the self-healing performance of coatings under three corrosion conditions: room temperature immersion, alternating wet–dry, and outdoor atmospheric exposure. The coating containing zinc phosphate exhibited lower ACM corrosion current values and corrosion charge values compared with the black epoxy coating in all three corrosion conditions, indicating better corrosion resistance, and self-healing properties. By using the ACM technique, a new approach was developed to assess the failure behavior of coatings in practical application environment. Finally, Chapter 11 overviews future developments of self-healing, self-reporting, superhydrophobic, and other novel coatings as well as their various industrial application scenarios. The application of atmospheric corrosion-monitoring technology to smart coating design is discussed in detail.

This book is largely driven by the requirements of practical applications; therefore, engineers and researchers can choose the type of smart coating most suitable for their application scenarios. This book discusses new synthesis methods to improve the self-healing efficiency and extend the service life of coatings, new ideas such as combining self-healing and self-reporting properties, and new techniques to study localized microscale electrochemical corrosion behavior.

By presenting the latest research progresses in smart protection coatings for intelligent corrosion control, the book should be useful to scientists, researchers, and students from diverse backgrounds. Beyond smart coatings and corrosion protection, it is also relevant to other fields such as materials science, chemical/industrial/manufacturing engineering, and nanotechnology. The book strongly emphasizes the characterization techniques, particularly new techniques for coatings. Note that some of the explanations and expressions are based on the authors' own views and expertise in the field. Readers are also encouraged to take a detailed look at the cited references to further understand the subjects covered in the book.

The authors are grateful to many individuals who contributed to the manuscript, artwork, and typesetting. The authors thank the editorial crew of Wiley Press for accepting the proposal and providing the opportunity to write this book. They are also grateful for the technical support from National Materials Corrosion and Protection Data Center, Liaoning Academy of Materials, and Prof. Xiaogang Li. The authors also thank Dr. Hongchang Qian, Dr. Yao Huang, Dr. Jinke Wang, Dr. Chenhao Ren, Dr. Fan Zhang, Dr. Tong Liu, Dr. Shanghao Wu, Dr. Panjun Wang, Dr. Yue Wang, Dr. Xin Guo, Dr. Zhibin Chen, Dr. Di Xu, Dr. Leping Deng, Yajie Wang, Hao Yang, Xin Wang, Xuanbo Wang, Bing Zhao, Miao Zhao, Xiaolun Ding, Zongbao Li, Weiting Chen, Zhuoyao Chen, Dongfang Jia, and Xueqing Yang for providing assistance with the project. Although the cover lists two authors, this book is really a product of cumulative efforts from numerous scientists and engineers across generations.

February, 2024

Lingwei Ma
Dawei Zhang
Beijing, China

1

Fundamentals of Smart Corrosion Protection Coatings

1.1 Introduction of Corrosion Protection Coatings

Metal materials are ubiquitous due to their good mechanical and processing properties, which have been widely applied in industrial production and daily life [1, 2]. However, they are vulnerable to corrosion under the influence of chloride ions, water, oxygen, atmospheric pollutants, etc., causing both economic loss and safety issues [3–5]. According to a survey by the Chinese Academy of Engineering in 2017, the annual cost of corrosion in China is more than 2100 billion yuan, accounting for 3.34% of the gross national product [6].

Many methods and materials have been developed to retard corrosion, with organic coating being one of the most widely used [7–9]. The principle of organic coating is to create a dense covering layer that can effectively isolate the metal substrate from corrosive media, preventing or slowing down the electrochemical reactions on the underlying metal substrate [10, 11]. Corrosion protection coatings account for approximately two-thirds of all anticorrosion measures. However, these coatings can become damaged or cracked by the harsh environment during transportation and service. Unless effectively repaired in time, such damage will seriously weaken the adhesion between the coating and the metal substrate, and the barrier properties of the coating will also decline to some degree.

Organic coating primers provide corrosion protection in multilayer systems because they have high adhesion strength with the metal substrate, good barrier properties, and can be loaded with large amounts of pigments and fillers. Chromate is often used as an anti-rust pigment in the organic primer layer due to its excellent corrosion inhibition properties. In recent years, more nontoxic and harmless new corrosion inhibitors have been developed to improve the corrosion resistance of coatings. The limited solubility of the corrosion inhibitor in the coating substrate causes the corrosion inhibitor to be exhausted in a certain period, thus reducing the corrosion protection effectiveness. The disadvantages of directly adding the inhibitor as a pigment can now be overcome by using microcapsules or nanocontainers that store both organic and inorganic inhibitors without any negative impact on the organic material matrix. The top paint should not only have good weather resistance and aging resistance but also must have the necessary decorative effects. It can provide a protective layer for the primer layer. Therefore, the top paint

Smart Protective Coatings for Corrosion Control, First Edition. Lingwei Ma and Dawei Zhang.

used outdoors must be selected with excellent weather resistance, such as alkyd, polyester, fluorocarbon, polysiloxane, polyacrylate, and polyurethane polymers.

1.1.1 Mechanisms of Corrosion Protection Coatings

The corrosion of metal materials involves three basic processes that occur in parallel: metal dissolution, the reduction of the cathode depolarizing agent, and the conduction of electronic and ionic currents in the cathode and anode [12]. As long as one of the processes is blocked, the metal corrosion rate decreases and the corrosion is suppressed. The protection mechanism of organic coatings is achieved by inhibiting one or more of the abovementioned steps. Some generally recognized anticorrosion mechanisms of coatings are listed below [13]:

(1) Shielding effect: The corrosion reaction of metals requires the presence of water, oxygen, and ions. When these corrosive media penetrate the coating to the coating–metal interface and accumulate at a certain concentration, metal corrosion occurs. The significance of the existence of organic coatings is to effectively prevent or slow down the direct contact between water, oxygen, and the metal matrix, thereby preventing or slowing down the occurrence of corrosion, which requires the coating to have good water resistance, oxygen permeability resistance, and wet adhesion [14]. When there is less water and oxygen on the substrate surface, the anode and cathode reactions of corrosion are slow, and the corrosion current is reduced. At the same time, flake pigments, such as mica powder and peeling scales, can be added to the coating. These flake pigments can cut off the pinhole channels in the coating and shield the diffusion of water, oxygen, and ions to the metal substrate [15]. Coatings can effectively block the mutual diffusion of corrosive ions, effectively preventing the generation of corrosion products by their moist-resistant adhesion.

(2) Cathodic protection: When added to the coating, a large amount of metal powder can function as a sacrificial anode in the corrosion process, while the metal substrate acts as the cathode. The metal powder contained within the coating corrodes as a sacrificial anode in the corrosion process of the organic coating/metal system, thus delaying the initiation and development of matrix metal corrosion and effectively protecting the cathode metal [16, 17]. For example, Zn powder can be used for this purpose because Zn has a more negative electrode potential than common metals, such as iron, steel, and aluminum alloys. Moreover, the corrosion products of Zn (the basic zinc chloride and zinc carbonate) can fill pores in the coating, tighten the coating layer, and further reduce the corrosion rate.

(3) Coating adhesion: Coating adhesion is an important characteristic that affects the performance of the corrosion resistance [18]. When the adhesion becomes weaker at the coating–metal interface, more corrosive media can leak into the area, and the coating may even fall off. Therefore, stronger coating adhesion (especially after encountering water) can significantly enhance the corrosion resistance of metals.

1.1.2 Failure Type and Failure Mechanism of Corrosion Protection Coatings

Serious coating defects, such as peeling, cracks, loss of mechanical strength, and foaming, occur between the coating and metal interface, which not only affect the aesthetics but also compromise the protection [19]. Figure 1.1 shows the microstructure of corrosion products formed at defects in organic coatings in different systems. The four samples were corroded to different degrees owing to surface defects.

The failure of the organic coating itself does not involve electrochemical corrosion or produce electrical current; rather, it occurs through aging caused by physical and chemical changes. To fundamentally strengthen the protective performance of organic coatings, accurately predict the coating life, and reduce the occurrence of accidents, it is important to find out the failure mechanism of the coating. The microscopic failure mechanism of organic coatings can be divided into four categories:

(1) Aging: Physical aging refers to the deterioration of a coating's protective properties. Swelling, infiltration of corrosive media, and cracking may occur under the action of environmental media and external stress during the coating's service [20]. In chemical aging, macromolecules in the chain or network structure are gradually degraded into smaller units, cross-linked, or undergo other chemical reactions under the actions of light, heat, acid, alkali, and oxygen during the coating's service. Aging facilitates the penetration of external corrosive media and degrades the chemical and physical properties of the coating.

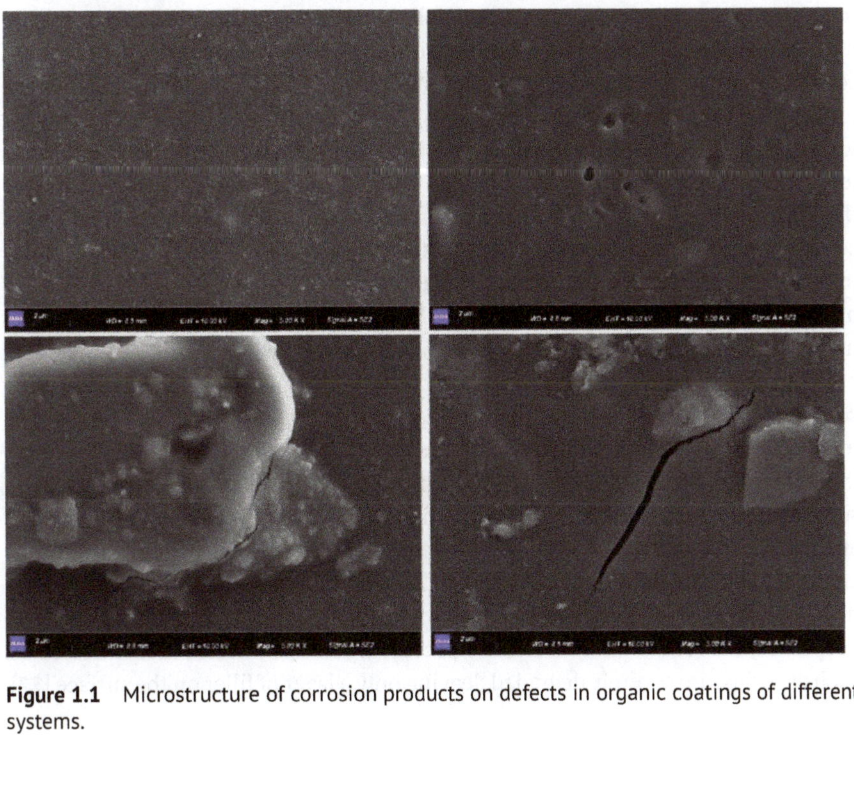

Figure 1.1 Microstructure of corrosion products on defects in organic coatings of different systems.

(2) Hydrolysis: When exposed to external corrosive media, organic coatings expand, causing bubbles or layer damage due to decomposition reactions. Hydrolysis is a typical decomposition reaction [21, 22]. The water resistance of organic coatings depends strongly on their molecular structures.
(3) Photo-oxygen degradation: The organic coating in outdoor service is subjected to the double action of sunlight and oxygen, and the organic coating will undergo photo-oxygen aging. Photons with enough energy from light will trigger a light reaction, causing polymer excitation or bond breakage, and the second energy of the light wave will be absorbed.
(4) Thermal failure: When the coating is exposed to high temperatures for a long time, decomposition, aging, and weight loss can easily occur [23]. These changes are mainly caused by the destruction of molecular chains and the cross-linking of film-forming materials. After the destruction of the molecular chains, the length and molecular weight of the polymers decrease, the film becomes soft and thin, and its resistance to permeability decreases. On the other hand, molecular cross-linking makes the polymer less elastic, which also tends to diminish the coating's protective effects, especially under the combined actions of environmental media and stress.

Common failure modes of coatings include peeling, foaming, burst holes, cracking, and pulverization. (i) Peeling: When adhesion between the coating and metal substrate falls below the tensile strength of the coating, the coating peels from the substrate or separates into layers [24]. This phenomenon is mainly due to poor pretreatment of the metal substrate, mixed pollutants between the primer layer, middle layer and top coating, and other reasons. (ii) Foaming: Foaming is the main form of coating failure and is more likely to occur for coatings in marine environments due to the presence of abundant corrosive ions [25]. It usually starts with water reacting with hydrophilic groups and soluble salts in the coating, forming channels for ions to diffuse to the metal surface. Corrosion reactions on the metal surface enhance the alkali content in the cathode region and reduce the adhesion strength of the coating. With the gradual accumulation of corrosion products, the coating expands in volume, and bubbles are formed. (iii) Burst holes: Foaming is the formation of bubbles on the coating surface. If the bubble breaks without the coating supplement and flattens, this phenomenon is called burst hole, and pinholes are small burst holes. When the temperature rises, the solution volatilized by the underlying coating cannot penetrate into the air through the high-viscosity surface layer, resulting in bubbles within the coating [26]. As the temperature continues to increase, the bubbles expand and exert pressure on the coating. When the expansion pressure exceeds the binding force of the surface coating film, the coating film breaks and forms a burst hole. (iv) Cracking: Cracks include both visible cracks and fine cracks that are generally invisible to the naked eye. They are mainly caused by a mismatch between the coating resin and additive fillers. Stress causes existing cracks to deepen. (v) Pulverization: Pulverization occurs when cross-linked macromolecules, such as resins, are severely degraded, leaving only a layer of filler on the surface [27].

This phenomenon is caused by ultraviolet (UV) radiation from the sun, while moisture and air pollutants also participate in the formation of powder. The presence of pigments in the coating also affects pulverization. Flaky aluminum powder and micro filler can reflect sunlight and thus retard pulverization.

1.2 Smart Self-healing Coatings

At present, artificial repair or replacement is the main treatment method for damaged coating, which is complicated and costly. Due to the limitations of site construction conditions, the surface state of repaired coatings is difficult to meet the requirements of related performance parameters. Therefore, the development of anticorrosion coatings with self-healing functions to realize active repair after coating damage is of great significance to extend the service life of coatings and reduce maintenance costs. Smart self-healing coatings have emerged, which can effectively repair the damage with no or little external intervention [28, 29]. Due to the great economic value and good development prospects in improving corrosion protection performance, many domestic and international scholars have focused on self-healing anticorrosive coatings in recent years [10, 30, 31]. These coatings can be classified into intrinsic and extrinsic types. Intrinsic self-healing coatings can be repaired by restoring internal chemical bonds and/or physical conformations in the polymer network [32]. Extrinsic self-healing coating generally loads healing agents or corrosion inhibitors into the coating and can recover the corrosion protection performance based on the adsorption of these species [33]. Compared to intrinsic self-healing coatings, extrinsic ones are more widely used in production and daily life because of their simple preparation and more options for loading species. Microencapsulation technology is often used to enhance their stability to avoid the premature release of film-forming substances and corrosion inhibitors and the occurrence of side reactions.

1.2.1 Intrinsic Self-healing Coating

Instead of film-forming substances or corrosion inhibitors used in extrinsic self-healing, intrinsic self-healing occurs by restoring internal chemical bonds and/or physical conformations in the polymer network, with little relationship with the metal substrate. Thermal energy and light radiation may provide sufficient activation energy for the formation or breakage of chemical bonds. Thermal stimulation can effectively improve the fluidity of polymers, allowing the atoms to closely approach each other and restore broken bonds. Thermal energy may be supplied artificially (e.g. using a heat gun) or from the environment (such as heating by sunlight). Ultraviolet, near-infrared, and sunlight have all been used as light stimuli. By adding corrosion-sensitive components, the stimulated response in the coating can be effectively promoted to repair the coating defect. Thus, the coating can be positioned and healed in a more timely and efficient manner.

1.2.1.1 Self-healing Based on Dynamic Bonds

The first type of intrinsic self-healing mechanism relies on dynamic bonding, such as hydrogen bonds, dynamic covalent bonds, and ionic interactions [34]. For instance, a thermoreversible bond breaks above a certain temperature, which allows the polymer chains to gather, cross-link at the defect, and heal the defect [35]. The main merit of this healing system is that no other healing agent and the healing process can theoretically be repeated endlessly.

A typical reaction used for dynamic bond repair is the Diels–Alder reaction (DA), which is achieved through a [4+2] cycloaddition between conjugated dienes and dienophiles (such as furans and maleimides) [36]. Wouters et al. [37] and Tian et al. [38] showed that by fabricating furfuryl glycidyl ether and *N*,*N*-diglycidyl furfuryl amine and doping reversible furan and maleimide bonds in epoxy coatings, self-healing can be achieved under heating. Wang et al. [39] synthesized UV-curable polyurethane based on a thermoreversible DA structure by copolymerizing an isocyanate monomer with a polycross-ester diol (Figure 1.2). The thermal characteristics of these materials were verified using differential scanning calorimetry and dynamic mechanical analysis. Surface scratching and subsequent healing tests have shown that the scratches on the polyurethane film surface can be repeatedly healed in as little as 100 seconds under heating conditions at 120 °C.

Light can also be used to trigger the formation of dynamic bonds [40, 41]. For instance, UV-sensitive self-healing polymers have been designed based on dynamic covalent bonds (such as S—S bonds) and cross-linking reactions (such as the cycloaddition of cinnamyl or anthracene derivatives). Compared with heat-induced self-healing, light-induced self-healing can be triggered instantly, remotely, and on demand [42]. It is also easier to limit stimulation to a specific area, preventing possible side reactions and degradation during the healing process and preserving the integrity of the coating. For coatings used in outdoor practical environments, sunlight-triggered self-healing is a convenient way. However, in addition to the need for photosensitive bonds, the coating must be sufficiently stable to withstand long-term sunlight exposure and maintain its overall chemical composition and performance. Moreover, light stimulation can be combined with other self-healing mechanisms, such as the controlled release of corrosion inhibitors in micro-/nanocontainers and damage repair by monomer polymerization [43, 44].

1.2.1.2 Self-healing Based on Shape Memory Effect

Shape memory materials, including shape memory alloys (SMAs) and shape memory polymers (SMPs), are novel self-healing materials that can recover from a temporarily fixed deformation state to the original state under external stimuli [42]. For SMAs, such as NiTi and Cu–Zn–Al, their shape memory effect is controlled by transitions among three phases: twinned martensite, detwinned martensite, and austenite [45]. SMA fibers can assist microencapsulated composites containing polymerizable restorers in achieving complete closure of cracks [46]. For SMPs, the viscoelastic transition occurs mainly through the polymer chains during the thermal transition temperature (T_{trans}) cycle, and therefore, the glass transition temperature (T_g) and melting temperature (T_m) are important factors (Figure 1.3).

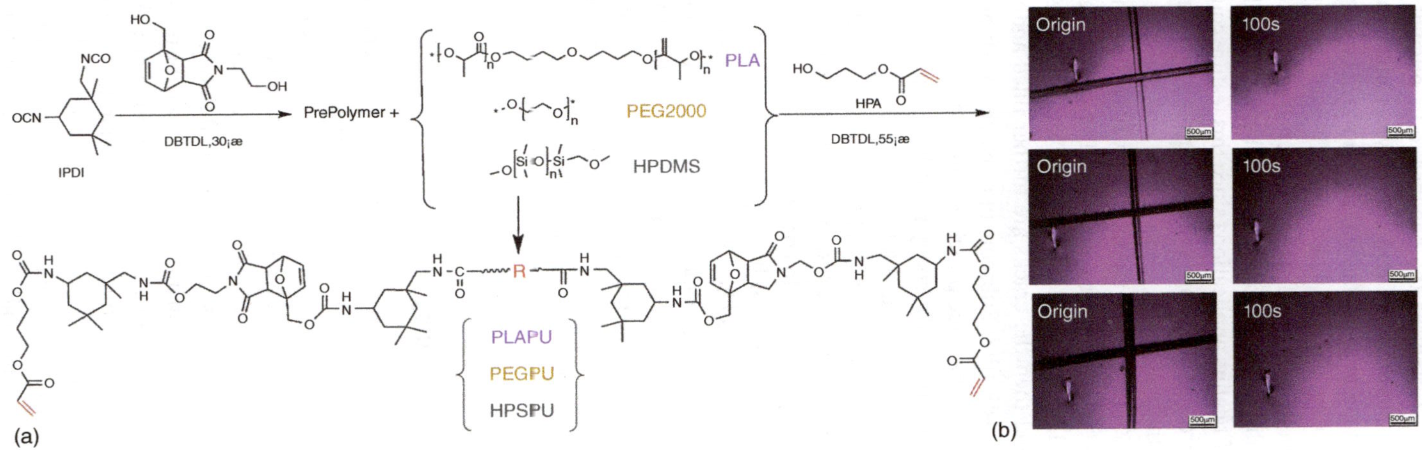

Figure 1.2 (a) Synthesis of UV-curable polyurethane coating with Diels–Alder structure, (b) optical microscope images of polylactide diol polyurethane coating with artificial scratches in three different areas and healed at 120 °C. Source: Wang et al. [39]/with permission of Elsevier.

For example, an SMP deformed at $T < T_g$ is "frozen" because the polymer chains are locked. Heating above T_g returns the polymer to its original shape, defined by chemical or physical cross-linking, and the polymer chains regain mobility and return to the random coiling state, where the entropy is most stabilized [48]. Recently, a novel self-healing coating based on SMP was developed to prevent corrosion. Compared with other types of self-healing coatings, SMP coatings have a strong strain-recovery ability and can effectively close larger physical damages. This is very meaningful in self-healing applications since this feature could significantly reduce the amount of corrosion inhibitors and film-forming agents required for the damaged parts of the coating.

Li and coworkers [49, 50] designed a clever strategy of sealing-and-then-healing to recover physical cracks in a polymer composite. The composite consisted of a shape memory matrix and a polyester thermoplastic filler. The shape memory effect narrows the scratches, and then the cracks are further filled and sealed by melting the polyester thermoplastic filler. Subsequent research on self-healing corrosion protection coatings employed the same principle. Luo and Mather [51] synthesized self-healing coatings by injecting SMP epoxy resin into a polycaprolactone (PCL) fiber scaffold (Figure 1.3). When the temperature reached 80 °C, the shape memory effect of the resin was triggered (above the T_g of epoxy) to close scratches in the coating, and the PCL fiber melted and flowed to seal the defect (above the T_m of PCL). Linear scanning voltammetry was used to confirm that the shielding properties of the coating recovered well after thermal healing. However, this process could not restore coating adhesion near the scratches. Wang et al. [47] developed a similar but more effective approach by applying carnauba wax microcapsules as restoratives in SMP coatings. After heating for 60 minutes, the effects of molten SMP and wax on the self-healing performance were determined. The best self-healing was observed at 15 wt% carnauba wax microcapsules. When this ratio was increased to 20 wt%, the molten wax oozed from the coating substrate after heating, which made the coating surface more hydrophobic and reduced the adhesion strength at the coating–steel interface.

1.2.2 Extrinsic Self-healing Coating

One of the most direct ways to realize extrinsic self-healing is to embrace the coating with a healing agent, mostly stored in micro- or nanocapsules [52, 53]. When subjected to mechanical impact, the coating breaks, causing the capsules to break and release the healing agent, which can polymerize to form a protective film to repair the coating and restore its barrier function and integrity. In addition, there is a class of extrinsic self-healing coatings that use the corrosion inhibition effect of inhibitors [54]. In this repair process, the corrosion inhibitor oozes from the micro- or nanocapsules to form a complex with metal ions through adsorption or chemical reaction that inhibits the corrosion behavior of the metal substrate. Such coatings often need to interact with the specific environmental factors in which they are located. Compared with intrinsic self-healing mechanisms, these extrinsic self-healing coatings have more extensive applications in aerospace, construction, medical, and electronic fields.

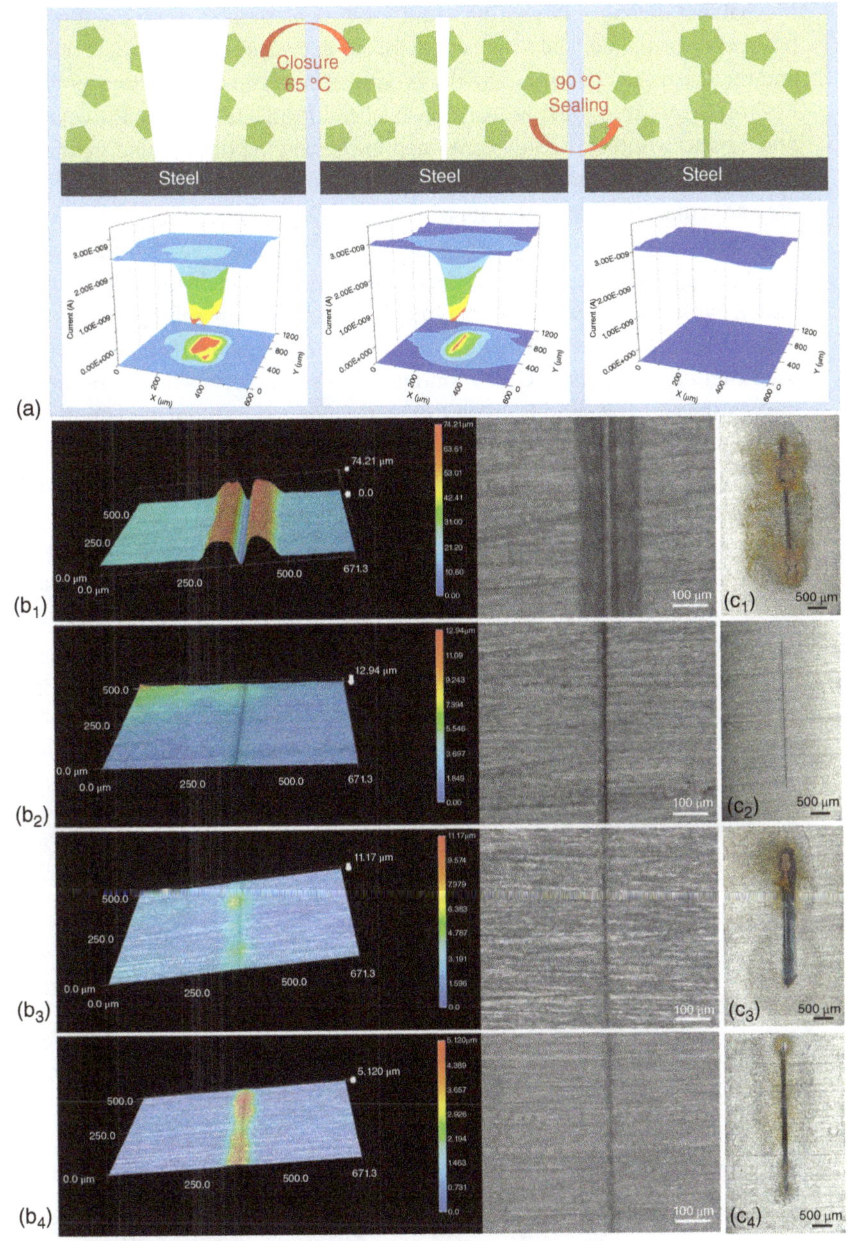

Figure 1.3 (a) Optical images during self-healing of damaged coating: (b_1) original defect, (b_2) after the first healing step at 65 °C, (b_3) after the second healing step at 90 °C, and (b_4) wax-free coating after two-step healing. Optical images of samples after seven days of immersion in 3.5 wt% NaCl solution: (c_1) scratched SMC coating, (c_2) healed SMC coating, (c_3) scratched wax-free coating, and (c_4) healed wax-free coating. Source: Wang et al. [47]/with permission of Elsevier.

1.2.2.1 Self-healing Coating Based on Defect Filling

Mechanical damages in coating require the construction of a new film to restore barrier performance. The film used to fill the defects must possess a certain strength and thickness [15, 55]. In self-healing coatings based on defect filling, polymerizable healing agents are spontaneously released and form a film under the catalytic effect of substances (either already contained in the coating or provided by the corrosive environment). The proper encapsulation of healing agents can extend their service life and reduce their interactions with the coating matrix.

Because of the fast curing rate, good fluidity, and high reactivity of dicyclopentadiene (DCPD), White et al. [56] designed urea–formaldehyde resin microcapsules coated with DCPD, which is a typical first-generation self-healing system. The authors prepared urea–formaldehyde resin microcapsules (>200 μm in size) and used them to store the DCPD monomer. These microcapsules were dispersed on the coating substrate and isolated from a Grubbs catalyst that could polymerize the monomers. When the coating was damaged, the monomer was released from the microcapsules and underwent catalytic polymerization to bond with the fracture interface, thus restoring the protective performance of the coating.

To rebond the crack interface, the healing agents must have an ideal chemical composition and provide sufficient bonding strength between the coating and metal interface. Epoxy resin is regarded as an excellent choice owing to its high adhesion strength and wide use [57]. Attaei et al. [58] developed a variety of microcapsules and selectively enclosed isophorone diisocyanate (IPDI) (Figure 1.4), hexamethylene diisocyanate, and the less toxic hexamethylene diisocyanate trimer as single-component healing agents in the coating. The microcapsules at the

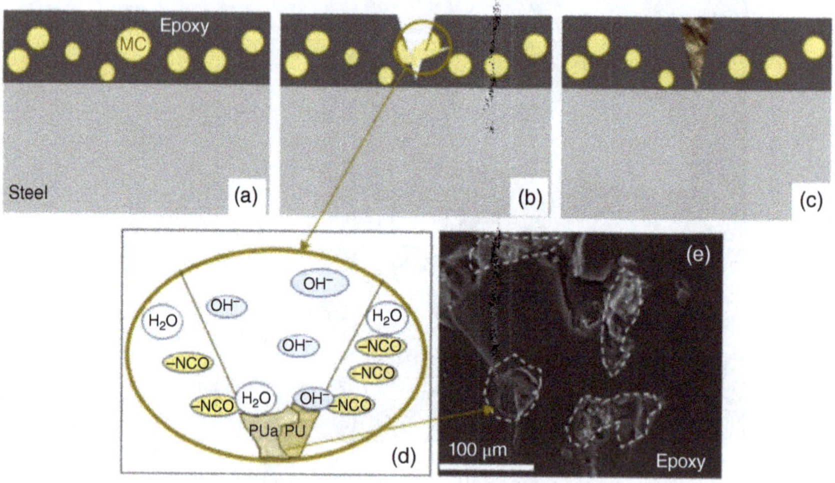

Figure 1.4 Diagram of self-healing process of microcapsule coating containing isophorone diisocyanate as a healing agent: (a) intact coating with added microcapsules, (b) damaged coating containing a scratch, (c) filling of the healing agent after polymerization, (d) self-healing reaction, (e) SEM micrographs of artificially damaged epoxy coating after immersion in 0.005 mol l^{-1} NaCl for seven days. Dashed lines indicate polymer formed by the reaction of IPDI released from microcapsules. Source: Attaei et al. [58]/with permission of Elsevier.

coating–damage interface expand and break due to water infiltration, prompting the healing agent to polymerize and fill the damage. However, because of the limited storage capacity of the microcapsules and reaction losses at the damaged area, the self-healing effect can only be effective for cracks with a width below 30 μm [59]. In actual applications, moisture from the environment inevitably penetrates the whole coating and reacts with the healing agent, and there may not be sufficient healing agent by the time the coating becomes damaged, which reduces the self-healing performance [60].

Among existing self-healing coatings based on defect filling by healing agents, the healing ability is directly affected by the mechanical and chemical properties, size, and quantity of microcapsules [61]. Ideally, the microcapsule wall should have sufficient stiffness; meanwhile, it must also be brittle to ensure that the microcapsules are broken when coating healing is needed [62]. Good compatibility and strong adhesion between the microcapsule wall and the coating substrate help to minimize water seepage through the capsule–coating interface. To provide durable self-healing capacity, the capsule material should also be stable against water, oxygen, and other corrosive media in different environments. Currently, the size of microcapsules used for storing polymerizable healing agents (often ranging from tens to hundreds of microns) is relatively large, which leads to difficulties in the practical application of corrosion protection coatings with thin thickness. Recently, oil-loaded smaller capsules (hundreds of nanometers in size) based on the sol–gel technology have been successfully applied to thinner coatings [63]. To ensure complete self-healing performance, a sufficient number of microcapsules have to be broken to fill the damaged area completely; otherwise, the damaged coating will have difficulty achieving the expected self-healing effect. To seal the coating defects effectively, the microcapsules were evenly mixed with the primer and coexisted close to the metal substrate. Although this design somewhat reduced the adhesion between the primer and metal, the researchers controlled the loading of capsules to reach a good compromise between the coating's adhesion strength, barrier properties, and self-healing performance [64].

1.2.2.2 Self-healing Coating Based on Corrosion Inhibitors

Another method to achieve extrinsic self-healing is to embed corrosion inhibitors in the coating [65]. Corrosion inhibitors can be classified into three categories according to their mechanisms: oxide film-forming, precipitation film-forming, and adsorption film-forming inhibitors [66]. Generally, there are many types of inorganic corrosion inhibitors used in self-healing coatings, including nitrite, molybdate, phosphate, and rare-earth salts [17]. Organic corrosion inhibitors contain adsorption centers (electron-rich regions) with an affinity to the metal surface [67]. The typical adsorption center is a relatively unstable unsaturated bond, such as C=C (alkenes), C≡C (alkyne), C=O (carbonyl), C=N (imine), and C=S (carbonyl-sulfur) [68]. The selection of corrosion inhibitors depends on the metal substrate that needs protection. For aluminum alloys, benzotriazole, mercaptobenzothiazole, and 8-hydroxyquinoline have good effects [69]. For steels, the choice also depends on the solution conditions. For instance, precipitation film-forming inhibitors in neutral solutions include zinc salts, cerium salts, and phosphate salts that can be applied [70]. Most adsorption film-forming inhibitors are organic compounds and are typically used in industrial acid pickling processes, such as ionic liquids, Schiff base derivatives, amine-based organic molecules, and imidazoles [71]. In the protection mechanism of film-forming inhibitors, the corrosion inhibitor released from the coating can quickly inhibit the anodic dissolution and cathodic reduction reactions at the defects [72]. In the anodic dissolution inhibition mechanism, inhibitors are adsorbed onto the metal surface and form a dense passivation film, whereas the cathodic reaction may be inhibited by the deposition of oxides and hydroxides by the corrosion inhibitors [73]. In addition, mixed-type corrosion inhibitors function via a combination of chemical reactions, physical adsorption, and/or complexation [74]. The majority of studies on self-healing coatings are based on various corrosion inhibition mechanisms because the design concept is simple and the inhibition action is direct.

Corrosion inhibitors can be mixed directly with the coating matrix to induce self-healing. For instance, cerium nitrate and phosphate have been directly incorporated into organic coatings to achieve the self-healing effect of coatings on zinc, steel, and aluminum alloy surfaces [75–77]. Lithium salts have also been added to organic coatings as corrosion inhibitors to protect aluminum alloys. The lithium salt created an alkaline environment in the striated area of the coating and induced the formation of three layers of alumina/alumina hydroxide films (including dense, porous, and columnar layers). Compatibility has always been the most important consideration for corrosion inhibitors and coatings, and it has an important impact on the self-healing process. The corrosion inhibitors must be properly leached from the coating into the damaged area. Undesirable side reactions between the corrosion inhibitors and coating should be avoided to ensure good corrosion inhibition efficiency of inhibitors, and agglomeration of the inhibitor should also be prevented because it creates more channels for water penetration, which is fatal for the barrier properties of coatings.

In view of these factors, there is increasing attention on the use of microcontainers or nanocontainers to load corrosion inhibitors and ensure their release in a sustained or controlled manner. Using organic microcapsules, corrosion inhibitors can be released either with or without breaking the capsules [78]. For nanocontainers, the commonly used ones include nanoparticles, nanotubes, and porous nanofibers [79]. These have great advantages in thin organic coatings or sol–gel layers and can be evenly distributed in the coating matrix because such particles are sufficiently small and disperse easily. However, organic nanoparticles are often unsuitable for applications because of their poor stability, low mechanical properties, and vulnerability to solvents or heat during service. Inorganic nanoparticles with more stable chemical properties are often used instead. In self-healing coating applications, the micro- or nanocontainers usually release their content in response to specific conditions according to their own characteristics, surface grafting, modification, or intelligent functional group modification, as illustrated in Figure 1.5 [80].

Self-healing will be unsatisfactory if corrosion inhibitors cannot completely cover the exposed metal substrate in the damaged area. Therefore, the inhibitor is mainly

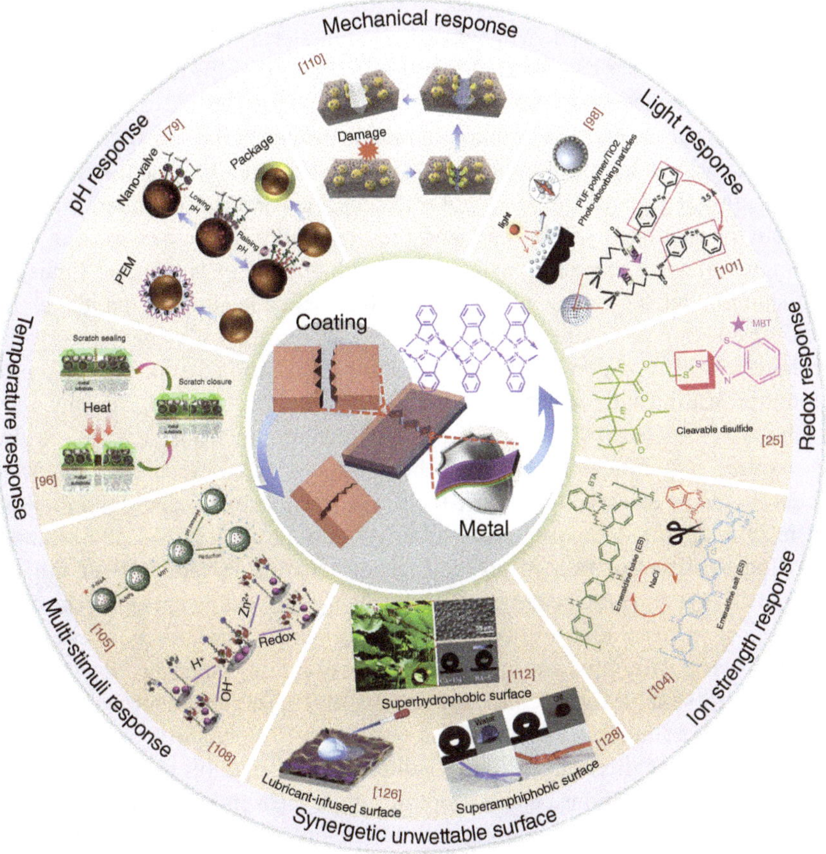

Figure 1.5 Release of corrosion inhibitors from containers in response to different stimuli. Source: Fu et al. [80]/with permission of Elsevier.

doped into the primer or pretreatment layer to be as close as possible to the metal substrate [81]. It should be noted that excessive use of corrosion inhibitors (or their containers) should be avoided to maintain overall good barrier properties and adhesion to the metal substrate. Containers with a high loading capacity for corrosion inhibitors and corrosion inhibitors with high performance are the best choices [82]. Once these conditions are satisfied, the next priority is evenly dispersing the corrosion inhibitor or container throughout the coating.

To achieve long-term restoration capacity, the corrosion inhibitors should be carefully screened with consideration of the types of corrosive media at the damage site and the electrochemical reactions during corrosion. Second, the compatibility between the container and coating should be optimized by design. Third, the environmental stability of the container should also be considered. The self-healing properties of a coating can only be confirmed after prolonged simulation or corrosion tests in an actual service environment. Although studies have demonstrated the release of corrosion inhibitors from the container in response to external stimuli, a better understanding of their release from the coated substrate or container and transport to the damaged site will help achieve long-term self-healing.

Self-healing coatings based on extrinsic healing mechanisms have great potential in industrial applications, and the healing of coating defects can be achieved without external energy. The addition of microcapsules to the substrate can play a toughening role, delay the expansion of fatigue cracks, and increase the fatigue life of the coating. However, the variety of microcapsules or nanocontainers currently available for coating is still limited, and the preparation process of most of them is complex and cumbersome, which is difficult to use in large-scale industrial production. After the release of corrosion inhibitors or healing agents, the empty microcapsules or nanocontainers provide a new infiltration route for corrosive substances. The corrosion resistance and durability of the coating depend on the amount of healing agent/corrosion inhibitor carried. If the coating has only a few healing cycles, the corrosion resistance in harsh environments is far from meeting the required requirements. Therefore, the volume of microcapsules or nanocontainers should be enlarged to improve their loading capacity of corrosion inhibitors and healing agents. Moreover, these additives should be designed and arranged according to the most needed positions in the coating to reduce their random distribution. For intrinsic coatings, the dependence on healing agents or corrosion inhibitors is reduced, and in theory, the coating can be repaired countless times, with the potential for rapid and large-scale repair of coating damages. However, this coating is only suitable for specific resin systems, and it has some drawbacks, such as environmental pollution for some coating systems, poor mechanical properties, and weak fracture resistance. The healing condition usually requires high temperatures or other severe conditions. For the optimization measure, multiphase polymer systems can be developed by binding multiple chemical bonds to achieve a balance between high mechanical properties and high self-healing properties of coatings.

1.2.3 Characterization Methods of Self-healing Properties

The characterization methods of self-healing properties in coatings are broadly categorized into visualization and quantitative techniques. Visualization methods encompass optical microscopy (OM), scanning electron microscopy (SEM), atomic force microscopy (AFM), confocal laser scanning microscopy (CLSM), and other methods. OM, being straightforward to operate and cost-effective, is widely accessible in most laboratories. It is used to monitor variations in the size of coating damage, release of healing agents, and peeling phenomena in coatings. However, its magnification capability is limited to about 1500 times, with a restricted depth of field. SEM, on the other hand, offers magnifications up to 100 000 times, enabling detailed investigation of surface topography and morphology at the coating damage area and on the underlying metal substrate. SEM, coupled with energy dispersive spectrum (EDS) analysis, provides insights into the morphologies and composition of microcapsules, nanocontainers, and nanofibers, their distribution in coating systems, and the release of healing agents and corrosion inhibitors. It also facilitates the comparison of corroded morphologies between traditional and self-healing coatings to analyze the self-healing mechanism. SEM is instrumental in studying nanoparticle migration and intrinsic healing of minor cracks. AFM, a nanotechnology tool, is used for detailed analysis of metal substrates and coating surfaces due to its excellent resolution, minimal sample preparation needs, and the ability to preserve the original sample state. It can in situ monitor scratch closure, changes in Young's modulus, and adhesion strength of coatings in response to external stimuli, such as pH, heat, and water. CLSM, offering higher resolution than OM, does not require probe contact with the coating surface, thus preventing sample damage and preserving sample integrity during imaging. It is useful for studying the self-healing performance of cracks on coating surfaces. Localized electrochemical measurements, including scanning vibrating electrode technique (SVET), scanning kelvin probe (SKP), localized electrochemical impedance spectroscopy (LEIS), and scanning electrochemical microscopy (SECM), can also be applied for analyzing the self-healing behaviors of coatings. SVET is utilized to study potential gradients generated by reaction currents on sample surfaces in solution environments, pinpointing sites of anodic oxidation and cathodic reduction reactions at the micron scale on corroded sample surfaces. The technique involves a probe vibrating at a fixed frequency, driven by a piezoelectric unit, and an AC signal monitored through a phase-locked amplifier to enhance the signal-to-noise ratio. SVET is extensively used in corrosion systems, including corrosion inhibitors and self-healing coatings. SKP can acquire surface morphology and work function distribution of samples in atmospheric environments or with a single microdrop without direct contact. It aids in studying coating peeling kinetics, such as accelerated peeling rates in polypyrrole membranes due to small cation penetration. LEIS, which measures the ratio between applied AC voltage and the tested AC current, provides high-precision information, such as pitting or scratches. Although a powerful tool for studying under-coating reactions, the distance between the probe and sample is crucial for result resolution. SECM, combining high electrochemical sensitivity and spatial resolution, analyzes electrochemical reactions in different coating systems, sensitively detecting localized oxygen reduction, metal ion oxidation processes, and pH changes. These advanced characterization methods are pivotal in the ongoing development and refinement of self-healing coatings.

For quantitative analysis, common methods include electrochemical impedance spectroscopy (EIS), potentiodynamic polarization tests, oxygen or water permeability tests, and water contact angle (WCA) measurements. In EIS studies, variations in impedance modulus, phase angle, and fitting parameters are indicative of a coating's self-healing properties. The efficiency of self-healing can be gauged using the values of charge transfer resistances (R_{ct}) of both the damaged ("blank") coating and the healed one. This technique offers a deep insight into the electrical characteristics of coatings and their changes post-healing. Potentiodynamic polarization testing, as an electrochemical approach, involves scanning the potential of the working electrode while concurrently measuring the current density. This method is particularly suitable for examining the release of healing agents and evaluating the effectiveness of corrosion inhibitors in protecting the underlying material. Oxygen and water permeability tests are critical for assessing the barrier properties of polymer films. In the context of self-healing coatings, these tests can be used to gauge the extent of healing, especially for coatings relying on intrinsic healing mechanisms. Additionally, measuring the WCA on both damaged and healed coating surfaces is a straightforward approach to observe changes in surface structure and composition. This method provides a visual and quantitative assessment of the coating's hydrophobic or hydrophilic nature before and after healing. These methods collectively contribute significantly to the development and industrial application of self-healing coatings by elucidating the mechanisms of self-repair and providing both qualitative and quantitative means of assessment. However, it is important to note that, despite the availability of various evaluation techniques, there is still a lack of standardized and universally established methods for assessing self-healing performance in coatings. Therefore, the introduction of relevant standards is imperative, as it will facilitate the comparison of self-healing performances across different research works and aid in the advancement of this field. Establishing these standards will provide a consistent framework for evaluating and comparing the efficacy of various self-healing coatings, thereby promoting their development and application in various industries.

1.3 Smart Self-reporting Coatings

Organic coatings can effectively extend the service life of metal substrates. However, even nano- or micron-scale damage to the coating surface would destroy the integrity of the coating matrix. If not detected in time and firmly repaired, the spread of corrosion can lead to catastrophic consequences and pose serious economic and safety problems. Therefore, there is an urgent need to develop self-reporting coatings that can detect coating damage at an early stage of corrosion so that maintenance can be carried out before the materials show serious corrosion or aging performance [83].

Self-reporting coatings are primarily divided into pH-sensitive type, metal-ion-sensitive type, and aggregation-induced luminescence material (AIE)-fluorescent type [84]. Zhang and Frankel [85] developed a new self-reporting coating on an aluminum alloy. Two pH indicators, phenolphthalein and bromothymol blue, were directly added to the acrylic resin. After immersion in the NaCl solution for eight days, the resin with phenolphthalein indicator showed pink spots while that with bromothymol blue indicator turned overall dark orange in color. Between these two, phenolphthalein performed better for early self-reporting in the coating. Dhole et al. [86] introduced modified 5-acrylamide-1,10-phenanthroline into acrylic polymer segments. After reacting with ferrous ions, the polymer changed its color from light yellow to bright red, as a sign of corrosion in steel substrates.

Self-reporting functional materials can be prepared by directly dispersing or modifying responsive dye molecules or fluorescent molecules into the coating matrix. However, these small molecules or their functional groups are prone to react with active groups in the coating matrix, resulting in photobleaching or degradation under UV light. To better protect the dye or fluorescent molecules, microcapsules or nanocontainers have been used to prevent their direct contact

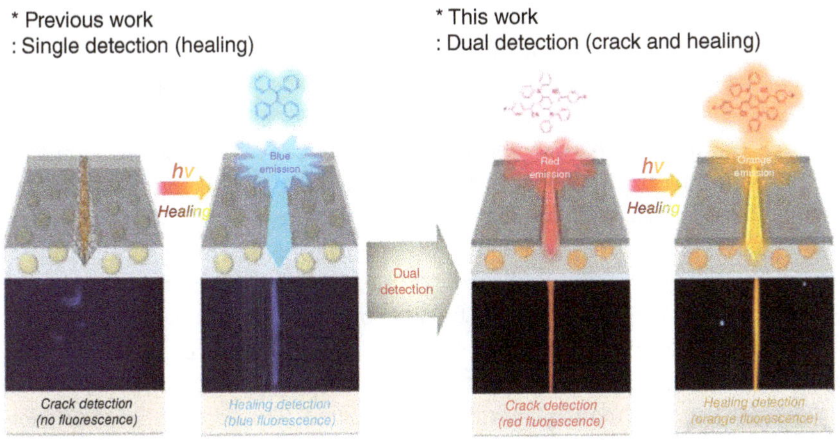

Figure 1.6 Schematic diagram of AIE-based self-reporting and self-healing mechanism. Source: Song et al. [88]/with permission of Elsevier.

with the coating matrix [21]. Song et al. [87, 88] conducted extensive research on self-reporting coatings based on the aggregation of fluorescence quenching (ACQ) and AIE (Figure 1.6). In their early research, AIE and ACQ fluorescent agents and liquid healing agents were encapsulated in urea–formaldehyde resin microcapsules. When microcapsules in the coating were destroyed, the liquid healing agents flowed out from the collapsed microcapsule, and the ACQ dye showed yellow fluorescence signals. Under UV irradiation, the fluorescence of the ACQ dye was inhibited by photoradical polymerization, and the blue fluorescence of the AIE dye was significantly enhanced. However, these microcapsules loaded with two dyes still have some limitations, such as adjusting the loading ratio owing to the different solubilities of the ACQ and AIE dyes. Further optimization is needed for the AIE- and ACQ-based self-reporting coatings.

1.4 Superhydrophobic Coatings

Inspired from "biological bionics," many researchers have studied superhydrophobic coating surfaces, which have great potential in self-cleaning, antifog, antifreeze, low-viscosity, reducing resistance, and other applications. Superhydrophobic surfaces are also useful for minimizing corrosion in atmospheric and marine conditions, soils, and even industrial environments [89, 90]. Therefore, the development of superhydrophobic coatings with comprehensive properties, such as self-cleaning, corrosion resistance, and wear resistance, is an important research direction.

An increasing number of superhydrophobic surfaces with microscale roughness have been used to protect metals from corrosion [91]. However, there has been no thorough discussion on how these surfaces improve the physical barrier effect and inhibit corrosion initiation. It is generally believed that the micro- and nanostructures of superhydrophobic surfaces can trap a large amount of air and greatly increase the contact angle of water droplets on the superhydrophobic surface [92]. The water droplets have difficulty staying on the substrate, and it also reduces the contact time between the water droplet and the surface. When superhydrophobic surfaces are immersed in water, the trapped air forms a gas film, which somewhat enhances the physical barrier effect against water and other corrosive media. A biomimetic superhydrophobic surface has also been applied to prevent ice formation. The formation of an ice layer on a surface is very complicated and involves both the environment and the phase transition of water. By reducing the contact area between water droplets and substrate, the superhydrophobic surface lowers the heat transfer rate and delays the freezing of water droplets. During icing, the formation of heterogeneous ice nuclei is delayed, the generation of surface fog/frost becomes slower, and the transformation of water droplets from the Cassie to Wenzel state on the superhydrophobic surface is prevented. The probability of droplets sliding on the surface also increases, and the adhesion strength at the ice–solid interface decreases, both of which are beneficial for achieving the anti-icing effect.

Making the surface more hydrophobic often reduces the corrosion rate of metals, mainly because the hydrophobic coating limits the interaction of water or other corrosive particles with the substrate material [93]. Organic anticorrosive coatings can effectively prevent the diffusion of water molecules and extend the time required for water to reach the metal surface. As mentioned above, hydrophobicity depends mainly on the inherent chemical properties of the materials and the surface microstructure. The most obvious way to improve surface hydrophobicity is by using low-surface-energy materials. For instance, the surface energy of silicone resins can be as low as 22 mN m^{-1}, whereas that of fluorinated resins is even lower (10 mN m^{-1}). Nevertheless, the WCA of coatings prepared using these materials does not exceed 120°. On the other hand, a designed rough surface could increase the WCA to above 150° to realize a superhydrophobic state [94].

As mentioned above, a microstructured superhydrophobic surface can trap a thin film of air, which significantly reduces the contact between water droplets and the surface [95]. Therefore, such surfaces are expected to reduce corrosion by preventing the generation of electrolyte films. These properties of the superhydrophobic surface will facilitate the future development of anticorrosion coatings in rainy environments. However, the water resistance of a superhydrophobic surface also depends on the droplet size [96]. Ultrafine droplets in the actual environment, such as water vapor, can directly condense onto the surface microstructures, precluding the formation of a protective gas film. To solve this problem effectively, it is recommended to use materials with finer dimensions or higher aspect ratios to prepare the superhydrophobic surface. The mechanical durability of the microstructure should also be considered in the manufacturing process.

A superhydrophobic surface prepared by Zhang and coworkers has prevented atmospheric corrosion caused by deliquesced salt [97, 98]. Such a surface makes it difficult for a deliquesced salt solution to diffuse. Instead, the salt solution remains a spherical droplet until the salt is completely dissolved. Salt particles on an inclined and highly lubricated substrate surface are extremely easy to fall off owing to the presence of the surface air film, thus reducing the corrosion of the substrate by

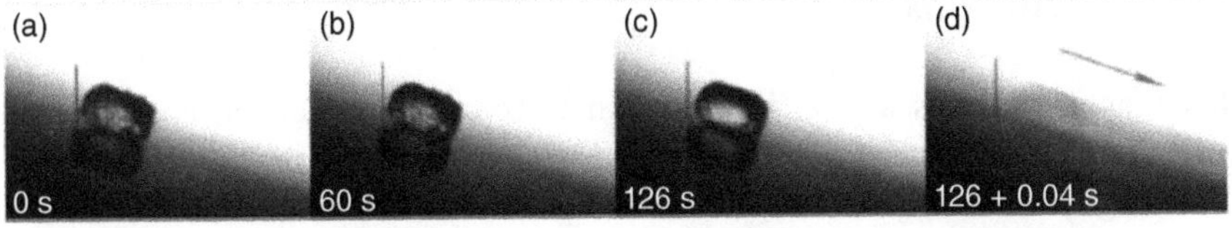

Figure 1.7 Time-lapse photographs of a NaCl particle dissolving under atmospheric environment at RH = 90% on an inclined superhydrophobic surface: (a) 0 s, (b) 60 s, (c) 126 s, and (d) 126 + 0.04 s. Source: Lu et al. [97]/with permission of Elsevier.

the dissolved salt (Figure 1.7). However, this effect depends on the microstructure of the superhydrophobic coating. For example, the brine in a deliquesced salt solution will roll off a surface with smaller microstructures (~5 μm); however, it penetrates another one with larger microstructures (~30 μm) [99]. In addition, fine salt particles deposited in the marine environment can easily enter the surface microstructure. Therefore, the microscale structure of superhydrophobic coatings should be fully considered when the application environment contains brine droplets or salt particles.

In superhydrophobic surfaces immersed in water, the air layer in the microstructure can form a barrier that somewhat protects the surface from the surrounding moisture, enhancing the overall corrosion resistance [100]. The superior corrosion inhibition performance of the superhydrophobic surface is also directly reflected by the corrosion current density from the polarization curve. After prolonged soaking, water inevitably penetrates the air layer. In other words, the Cassie contact may change to a Wenzel contact, and the corrosion resistance of the superhydrophobic surface decreases. In a previous study, Wang and coworkers eliminated the residual air captured on the surface of superhydrophobic copper by changing the solvent from ethanol to water [101]. Unlike the conventional superhydrophobic surface, the degassed surface formed a Wenzel contact with the surrounding water. In the absence of the air barrier, both the potentiodynamic polarization curve and EIS data confirmed a dramatic reduction in corrosion resistance.

The protective effect of a superhydrophobic surface in an immersed environment largely depends on the stability of the air film in water, which also rapidly decays at a higher hydraulic pressure, high flow rate, or high salinity of the surrounding fluid [102, 103]. To achieve hydrophobicity, a highly porous microstructure with a large surface area is required. However, the microstructure also renders the material more vulnerable to a corrosive environment. By comparing the hydrophobic and superhydrophobic surfaces of TiO_2/ZnO structures, Yu and Tian [104] found that the superhydrophobic TiO_2/ZnO layer has high porosity, good self-cleaning properties, and good corrosion resistance in an atmospheric environment. Nevertheless, a porous superhydrophobic surface is more susceptible to the intrusion of corrosive media during immersion, making this surface less resistant to corrosion than a hydrophobic one. Further studies are required to fully elucidate the relationship among the microstructure, surface roughness, and corrosion resistance of the generated superhydrophobic surfaces [105].

1.5 Conclusions

This chapter provides a brief introduction to corrosion protection coatings and smart design approaches for intelligent corrosion control. The degradation process of coatings and the corrosion process of metals are discussed in detail so that the reader can gain a basic understanding of the corrosion protection techniques. The concepts of self-healing coatings, self-reporting coatings, and superhydrophobic coatings are also explained. Compared to traditional corrosion protection coatings, smart coatings can ensure the lifetime safety of coatings, greatly reducing the repair costs for damaged coatings and sending warnings of coating damage at a very early stage. Therefore, smart coatings with self-healing, self-reporting, and superhydrophobic functions have broad application prospects in concrete, marine shipment, automotives, 3C products (computer, communication, and consumer electronics), and biomedicine. The following chapters will discuss in detail the different types of smart coatings to inspire further development in this area.

References

1 Sathiyamoorthi, P. and Kim, H.S. (2022). High-entropy alloys with heterogeneous microstructure: processing and mechanical properties. *Progress in Materials Science* 123: 100709. https://doi.org/10.1016/j.pmatsci.2020.100709.

2 Srivastava, M., Rathee, S., Patel, V. et al. (2022). A review of various materials for additive manufacturing: recent trends and processing issues. *Journal of Materials Research and Technology* 21: 2612–2641. https://doi.org/10.1016/j.jmrt.2022.10.015.

3 Harb, S.V., Trentin, A., de Souza, T.A.C. et al. (2020). Effective corrosion protection by eco-friendly self-healing PMMA-cerium oxide coatings. *Chemical Engineering Journal* 383: 123219. https://doi.org/10.1016/j.cej.2019.123219.

4 Lazorenko, G., Kasprzhitskii, A., and Nazdracheva, T. (2021). Anti-corrosion coatings for protection of steel railway structures exposed to atmospheric environments: a review. *Construction and Building Materials* 288: 123115. https://doi.org/10.1016/j.conbuildmat.2021.123115.

5 Ye, Y., Chen, H., Zou, Y. et al. (2020). Corrosion protective mechanism of smart graphene-based self-healing coating on carbon steel. *Corrosion Science* 174: 108825. https://doi.org/10.1016/j.corsci.2020.108825.

6 Hou, B., Li, X., Ma, X. et al. (2017). The cost of corrosion in China. *npj Materials Degradation* 1 (1): 4. https://doi.org/10.1038/s41529-017-0005-2.

7 Yan, H., Zhang, L., Li, H. et al. (2020). Towards high-performance additive of Ti_3C_2/graphene hybrid with a novel wrapping structure in epoxy coating. *Carbon* 157: 217–233. https://doi.org/10.1016/j.carbon.2019.10.034.

8 Wu, Y., He, Y., Zhou, T. et al. (2020). Synergistic functionalization of h-BN by mechanical exfoliation and PEI chemical modification for enhancing the corrosion resistance of waterborne epoxy coating. *Progress in Organic Coatings* 142: 105541. https://doi.org/10.1016/j.porgcoat.2020.105541.

9 Ramezanzadeh, M., Tati, A., Bahlakeh, G., and Ramezanzadeh, B. (2021). Construction of an epoxy composite coating with exceptional thermo-mechanical properties using Zr-based NH_2-UiO-66 metal-organic framework (MOF): experimental and DFT-D theoretical explorations. *Chemical Engineering Journal* 408: 127366. https://doi.org/10.1016/j.cej.2020.127366.

10 Cui, G., Bi, Z., Wang, S. et al. (2020). A comprehensive review on smart anti-corrosive coatings. *Progress in Organic Coatings* 148: 105821. https://doi.org/10.1016/j.porgcoat.2020.105821.

11 Zhang, R., Yu, X., Yang, Q. et al. (2021). The role of graphene in anti-corrosion coatings: a review. *Construction and Building Materials* 294: 123613. https://doi.org/10.1016/j.conbuildmat.2021.123613.

12 Gong, K., Wu, M., and Liu, G. (2020). Stress corrosion cracking behavior of rusted X100 steel under the combined action of Cl^- and HSO_3^- in a wet–dry cycle environment. *Corrosion Science* 165: 108382. https://doi.org/10.1016/j.corsci.2019.108382.

13 Le Pen, C., Lacabanne, C., and Pébère, N. (2000). Structure of waterborne coatings by electrochemical impedance spectroscopy and a thermostimulated current method: influence of fillers. *Progress in Organic Coatings* 39 (2–4): 167–175. https://doi.org/10.1016/S0300-9440(00)00148-X.

14 Zhang, Y., Li, T.T., Shiu, B.C. et al. (2022). Multifunctional sodium Alginate@ urushiol fiber with targeted antibacterial, acid corrosion resistance and flame retardant properties for personal protection based on wet spinning. *Applied Surface Science* 584: 152573. https://doi.org/10.1016/j.apsusc.2022.152573.

15 Ren, S., Cui, M., Liu, C., and Wang, L. (2022). A comprehensive review on ultrathin, multi-functionalized, and smart graphene and graphene-based composite protective coatings. *Corrosion Science* 212: 110939. https://doi.org/10.1016/j.corsci.2022.110939.

16 Ma, Q., Wang, L., Sun, W. et al. (2022). Effect of chemical conversion induced by self-corrosion of zinc powders on enhancing corrosion protection performance of zinc-rich coatings. *Corrosion Science* 194: 109942. https://doi.org/10.1016/j.corsci.2021.109942.

17 Wang, A., De Silva, K., Jones, M. et al. (2023). Anticorrosive coating systems for marine propellers. *Progress in Organic Coatings* 183: 107768. https://doi.org/10.1016/j.porgcoat.2023.107768.

18 Liu, T., Zhao, H., Zhang, D. et al. (2021). Ultrafast and high-efficient self-healing epoxy coatings with active multiple hydrogen bonds for corrosion protection. *Corrosion Science* 187: 109485. https://doi.org/10.1016/j.corsci.2021.109485.

19 Dong, H., Qiao, Y., Peng, S. et al. (2023). 2D material/epoxy composite coatings, a perspective from the regulation of 2D materials. *Progress in Organic Coatings* 183: 107817. https://doi.org/10.1016/j.porgcoat.2023.107817.

20 Heimann, R.B. (2021). Magnesium alloys for biomedical application: advanced corrosion control through surface coating. *Surface and Coatings Technology* 405: 126521. https://doi.org/10.1016/j.surfcoat.2020.126521.

21 Xiao, H., Wang, Y., Gu, L. et al. (2023). Smart sensing coatings for early warning of degradations: a review. *Progress in Organic Coatings* 177: 107418. https://doi.org/10.1016/j.porgcoat.2023.107418.

22 Zhang, Z., Zhong, X., Li, L. et al. (2023). Unmasking the delamination mechanisms of a defective coating under the co-existence of alternating stress and corrosion. *Progress in Organic Coatings* 180: 107560. https://doi.org/10.1016/j.porgcoat.2023.107560.

23 Olad, A., Maryami, F., Mirmohseni, A., and Shayegani-Akmal, A.A. (2021). Potential of slippery liquid infused porous surface coatings as flashover inhibitors on porcelain insulators in icing, contaminated, and harsh environments. *Progress in Organic Coatings* 151: 106082. https://doi.org/10.1016/j.porgcoat.2020.106082.

24 Wagle, P.G., Tamboli, S.S., and More, A.P. (2021). Peelable coatings: a review. *Progress in Organic Coatings* 150: 106005. https://doi.org/10.1016/j.porgcoat.2020.106005.

25 Shi, Y., Chen, C., Li, Y., and Zhao, W. (2023). Achieving dual functional corrosion resistance for epoxy coatings under alternating hydrostatic pressure via constructing P-phenylenediamine/$Ti_3C_2T_x$ hybrids. *Carbon* 201: 1048–1060. https://doi.org/10.1016/j.carbon.2022.09.089.

26 Czachor-Jadacka, D., Pilch-Pitera, B., Kisiel, M., and Thomas, J. (2023). Polyurethane powder coatings with low curing temperature: research on the effect of chemical structure of crosslinking agent on the properties of coatings. *Progress in Organic Coatings* 182: 107662. https://doi.org/10.1016/j.porgcoat.2023.107662.

27 Chen, Y., Liu, R., and Luo, J. (2023). Improvement of anti-aging property of UV-curable coatings with silica-coated TiO_2. *Progress in Organic Coatings* 179: 107479. https://doi.org/10.1016/j.porgcoat.2023.107479.

28 Udoh, I.I., Shi, H., Daniel, E.F. et al. (2022). Active anticorrosion and self-healing coatings: a review with focus on multi-action smart coating strategies. *Journal of Materials Science and Technology* 116: 224–237. https://doi.org/10.1016/j.jmst.2021.11.042.

29 Liu, S., Li, Z., Yu, Q. et al. (2021). Dual self-healing composite coating on magnesium alloys for corrosion protection. *Chemical Engineering Journal* 424: 130551. https://doi.org/10.1016/j.cej.2021.130551.

30 Wang, L., Li, S., and Fu, J. (2023). Self-healing anti-corrosion coatings based on micron-nano containers with different structural morphologies. *Progress in Organic Coatings* 175: 107381. https://doi.org/10.1016/j.porgcoat.2022.107381.

31 Ye, K., Bi, Z., Cui, G. et al. (2020). External self-healing coatings in anticorrosion applications: a review. *Corrosion* 76 (3): 279–298. https://doi.org/10.5006/3430.

32 Wang, S. and Urban, M.W. (2020). Self-healing polymers. *Nature Reviews Materials* 5 (8): 562–583. https://doi.org/10.1038/s41578-020-0202-4.

33 Ouarga, A., Lebaz, N., Tarhini, M. et al. (2022). Towards smart self-healing coatings: advances in micro/nano-encapsulation processes as carriers for anti-

corrosion coatings development. *Journal of Molecular Liquids* 354: 118862. https://doi.org/10.1016/j.molliq.2022.118862.

34 Roy, N., Bruchmann, B., and Lehn, J.M. (2015). DYNAMERS: dynamic polymers as self-healing materials. *Chemical Society Reviews* 44 (11): 3786–3807. https://doi.org/10.1039/C5CS00194C.

35 García, S.J., Fischer, H.R., and Van Der Zwaag, S. (2011). A critical appraisal of the potential of self healing polymeric coatings. *Progress in Organic Coatings* 72 (3): 211–221. https://doi.org/10.1016/j.porgcoat.2011.06.016.

36 Ratwani, C.R., Kamali, A.R., and Abdelkader, A.M. (2023). Self-healing by Diels–Alder cycloaddition in advanced functional polymers: a review. *Progress in Materials Science* 131: 101001. https://doi.org/10.1016/j.pmatsci.2022.101001.

37 Wouters, M., Craenmehr, E., Tempelaars, K. et al. (2009). Preparation and properties of a novel remendable coating concept. *Progress in Organic Coatings* 64 (2–3): 156–162. https://doi.org/10.1016/j.porgcoat.2008.09.023.

38 Tian, Q., Rong, M.Z., Zhang, M.Q., and Yuan, Y.C. (2010). Synthesis and characterization of epoxy with improved thermal remendability based on Diels–Alder reaction. *Polymer International* 59 (10): 1339–1345. https://doi.org/10.1002/pi.2872.

39 Wang, Z., Yang, H., Fairbanks, B.D. et al. (2019). Fast self-healing engineered by UV-curable polyurethane contained Diels–Alder structure. *Progress in Organic Coatings* 131: 131–136. https://doi.org/10.1016/j.porgcoat.2019.02.021.

40 Huang, S., Kong, X., Xiong, Y. et al. (2020). An overview of dynamic covalent bonds in polymer material and their applications. *European Polymer Journal* 141: 110094. https://doi.org/10.1016/j.eurpolymj.2020.110094.

41 Li, B., Cao, P.F., Saito, T., and Sokolov, A.P. (2022). Intrinsically self-healing polymers: from mechanistic insight to current challenges. *Chemical Reviews* 123 (2): 701–735. https://doi.org/10.1021/acs.chemrev.2c00575.

42 Herath, M., Epaarachchi, J., Islam, M. et al. (2020). Light activated shape memory polymers and composites: a review. *European Polymer Journal* 136: 109912. https://doi.org/10.1016/j.eurpolymj.2020.109912.

43 Chen, Z., Scharnagl, N., Zheludkevich, M.L. et al. (2023). Micro/nanocontainer-based intelligent coatings: synthesis, performance and applications – a review. *Chemical Engineering Journal* 451: 138582. https://doi.org/10.1016/j.cej.2022.138582.

44 Yimyai, T., Crespy, D., and Rohwerder, M. (2023). Corrosion-responsive self-healing coatings. *Advanced Materials* 47 (35): 2300101. https://doi.org/10.1002/adma.202300101.

45 Kumari, S. and Abhishek, K. (2021). Study of machinability aspects of shape memory alloys: a critical review. *Materials Today Proceedings* 44: 1336–1343. https://doi.org/10.1016/j.matpr.2020.11.390.

46 Kirkby, E.L., Michaud, V.J., Månson, J.A. et al. (2009). Performance of self-healing epoxy with microencapsulated healing agent and shape memory alloy wires. *Polymer* 50 (23): 5533–5538. https://doi.org/10.1016/j.polymer.2009.05.014.

47 Wang, L., Deng, L., Zhang, D. et al. (2016). Shape memory composite (SMC) self-healing coatings for corrosion protection. *Progress in Organic Coatings* 97: 261–268. https://doi.org/10.1016/j.porgcoat.2016.04.041.

48 Zhao, Q., Qi, H.J., and Xie, T. (2015). Recent progress in shape memory polymer: new behavior, enabling materials, and mechanistic understanding. *Progress in Polymer Science* 49: 79–120. https://doi.org/10.1016/j.progpolymsci.2015.04.001.

49 Li, G. and Nettles, D. (2010). Thermomechanical characterization of a shape memory polymer based self-repairing syntactic foam. *Polymer* 51 (3): 755–762. https://doi.org/10.1016/j.polymer.2009.12.002.

50 Nji, J. and Li, G. (2012). Damage healing ability of a shape-memory-polymer-based particulate composite with small thermoplastic contents. *Smart Materials and Structures* 21 (2): 025011. https://doi.org/10.1088/0964-1726/21/2/025011.

51 Luo, X. and Mather, P.T. (2013). Shape memory assisted self-healing coating. *ACS Macro Letters* 2 (2): 152–156. https://doi.org/10.1021/mz400017x.

52 Gong, H., Yu, C., Zhang, L. et al. (2020). Intelligent lubricating materials: a review. *Composites Part B: Engineering* 202: 108450. https://doi.org/10.1016/j.compositesb.2020.108450.

53 Zehra, S., Mobin, M., and Aslam, R. (2023). Nanocontainers: a comprehensive review on their application in the stimuli-responsive smart functional coatings. *Progress in Organic Coatings* 176: 107389. https://doi.org/10.1016/j.porgcoat.2022.107389.

54 Wang, Y., Wang, J., Huang, L. et al. (2023). Photothermally activated self-healing coatings for corrosion protection: a review. *Progress in Organic Coatings* 185: 107886. https://doi.org/10.1016/j.porgcoat.2023.107886.

55 Nazari, M.H., Zhang, Y., Mahmoodi, A. et al. (2022). Nanocomposite organic coatings for corrosion protection of metals: a review of recent advances. *Progress in Organic Coatings* 162: 106573. https://doi.org/10.1016/j.porgcoat.2021.106573.

56 White, S.R., Sottos, N.R., Geubelle, P.H. et al. (2001). Autonomic healing of polymer composites. *Nature* 409 (6822): 794–797. https://doi.org/10.1038/35057232.

57 Ahmadi, Z. (2019). Nanostructured epoxy adhesives: a review. *Progress in Organic Coatings* 135: 449–453. https://doi.org/10.1016/j.porgcoat.2019.06.028.

58 Attaei, M., Calado, L.M., Taryba, M.G. et al. (2020). Autonomous self-healing in epoxy coatings provided by high efficiency isophorone diisocyanate (IPDI) microcapsules for protection of carbon steel. *Progress in Organic Coatings* 139: 105445. https://doi.org/10.1016/j.porgcoat.2019.105445.

59 Wang, W., Xu, L., Li, X. et al. (2014). Self-healing mechanisms of water triggered smart coating in seawater. *Journal of Materials Chemistry A* 2 (6): 1914–1921. https://doi.org/10.1039/C3TA13389C.

60 Zhang, D., Peng, F., and Liu, X. (2021). Protection of magnesium alloys: from physical barrier coating to smart self-healing coating. *Journal of Alloys and Compounds* 853: 157010. https://doi.org/10.1016/j.jallcom.2020.157010.

61 Tomić, N.Z., Mustapha, A.N., AlMheiri, M. et al. (2022). Advanced application of drying oils in smart self-healing coatings for corrosion protection: feasibility for industrial application. *Progress in Organic Coatings* 172: 107070. https://doi.org/10.1016/j.porgcoat.2022.107070.

62 Wu, M., Johannesson, B., and Geiker, M. (2012). A review: Self-healing in cementitious materials and engineered cementitious composite as a self-healing material. *Construction and Building Materials* 28 (1): 571–583. https://doi.org/10.1016/j.conbuildmat.2011.08.086.

63 Zhang, F., Ju, P., Pan, M. et al. (2018). Self-healing mechanisms in smart protective coatings: a review. *Corrosion Science* 144: 74–88. https://doi.org/10.1016/j.corsci.2018.08.005.

64 Kumar, A., Stephenson, L.D., and Murray, J.N. (2006). Self-healing coatings for steel. *Progress in Organic Coatings* 55 (3): 244–253. https://doi.org/10.1016/j.porgcoat.2005.11.010.

65 Sun, W., Yang, Y., Yang, Z. et al. (2021). Review on the corrosion-promotion activity of graphene and its inhibition. *Journal of Materials Science and Technology* 91: 278–306. https://doi.org/10.1016/j.jmst.2020.12.084.

66 Sheetal, Batra, R., Singh, A.K. et al. (2023). Advancement of corrosion inhibitor system through N-heterocyclic compounds: a review. *Corrosion Engineering, Science and Technology* 58 (1): 73–101. https://doi.org/10.1080/1478422X.2022.2137979.

67 Mamand, D.M. and Qadr, H.M. (2023). Corrosion inhibition efficiency and quantum chemical studies of some organic compounds: theoretical evaluation. *Corrosion Reviews* 41 (4): 427–441. https://doi.org/10.1515/corrrev-2022-0085.

68 Berdimurodov, E., Kholikov, A., Akbarov, K. et al. (2022). Novel gossypol–indole modification as a green corrosion inhibitor for low–carbon steel in aggressive alkaline–saline solution. *Colloids and Surfaces A: Physicochemical and Engineering Aspects* 637: 128207. https://doi.org/10.1016/j.colsurfa.2021.128207.

69 Shchukina, E., Wang, H., and Shchukin, D.G. (2019). Nanocontainer-based self-healing coatings: current progress and future perspectives. *Chemical Communications* 55 (27): 3859–3867. https://doi.org/10.1039/C8CC09982K.

70 Gao, F., Mu, J., Bi, Z. et al. (2021). Recent advances of polyaniline composites in anticorrosive coatings: a review. *Progress in Organic Coatings* 151: 106071. https://doi.org/10.1016/j.porgcoat.2020.106071.

71 Aslam, R., Mobin, M., Zehra, S., and Aslam, J. (2022). A comprehensive review of corrosion inhibitors employed to mitigate stainless steel corrosion in different environments. *Journal of Molecular Liquids* 364: 119992. https://doi.org/10.1016/j.molliq.2022.119992.

72 Chauhan, D.S., Quraishi, M.A., Ansari, K.R., and Saleh, T.A. (2020). Graphene and graphene oxide as new class of materials for corrosion control and protection: present status and future scenario. *Progress in Organic Coatings* 147: 105741. https://doi.org/10.1016/j.porgcoat.2020.105741.

73 Shinato, K.W., Zewde, A.A., and Jin, Y. (2020). Corrosion protection of copper and copper alloys in different corrosive medium using environmentally friendly

corrosion inhibitors. *Corrosion Reviews* 38 (2): 101–109. https://doi.org/10.1515/corrrev-2019-0105.

74 Kobzar, Y.L. and Fatyeyeva, K. (2021). Ionic liquids as green and sustainable steel corrosion inhibitors: recent developments. *Chemical Engineering Journal* 425: 131480. https://doi.org/10.1016/j.cej.2021.131480.

75 Montemor, M.F., Pinto, R., and Ferreira, M.G.S. (2009). Chemical composition and corrosion protection of silane films modified with CeO_2 nanoparticles. *Electrochimica Acta* 54 (22): 5179–5189. https://doi.org/10.1016/j.electacta.2009.01.053.

76 Carneiro, J., Tedim, J., Fernandes, S.C. et al. (2012). Chitosan-based self-healing protective coatings doped with cerium nitrate for corrosion protection of aluminum alloy 2024. *Progress in Organic Coatings* 75 (1–2): 8–13. https://doi.org/10.1016/j.porgcoat.2012.02.012.

77 Mirzakhanzadeh, Z., Kosari, A., Moayed, M.H. et al. (2018). Enhanced corrosion protection of mild steel by the synergetic effect of zinc aluminum polyphosphate and 2-mercaptobenzimidazole inhibitors incorporated in epoxy-polyamide coatings. *Corrosion Science* 138: 372–379. https://doi.org/10.1016/j.corsci.2018.04.040.

78 Chen, S., Huang, Z., Yuan, M. et al. (2022). Trigger and response mechanisms for controlled release of corrosion inhibitors from micro/nanocontainers interpreted using endogenous and exogenous stimuli: a review. *Journal of Materials Science and Technology* 125: 67–80. https://doi.org/10.1016/j.jmst.2022.02.037.

79 Habib, S., Shakoor, R.A., and Kahraman, R. (2021). A focused review on smart carriers tailored for corrosion protection: developments, applications, and challenges. *Progress in Organic Coatings* 154: 106218. https://doi.org/10.1016/j.porgcoat.2021.106218.

80 Fu, X., Tian, L., Fan, Y. et al. (2021). Stimuli-responsive self-healing anticorrosion coatings: from single triggering behavior to synergetic multiple protections. *Materials Today Chemistry* 22: 100575. https://doi.org/10.1016/j.mtchem.2021.100575.

81 Bijapur, K., Molahalli, V., Shetty, A. et al. (2023). Recent trends and progress in corrosion inhibitors and electrochemical evaluation. *Applied Sciences* 13 (18): 10107. https://doi.org/10.3390/app131810107.

82 Cui, G., Zhang, C., Wang, A. et al. (2021). Research progress on self-healing polymer/graphene anticorrosion coatings. *Progress in Organic Coatings* 155: 106231. https://doi.org/10.1016/j.porgcoat.2021.106231.

83 Niu, Y., Wu, J., Kang, Y. et al. (2023). Encapsulation technique and application progress of mechanical stimuli-responsive microcapsules. *Progress in Organic Coatings* 176: 107390. https://doi.org/10.1016/j.porgcoat.2022.107390.

84 Chen, C., Yu, M., Tong, J. et al. (2022). A review of fluorescence based corrosion detection of metals. *Corrosion Communications* 6: 1–15. https://doi.org/10.1016/j.corcom.2021.11.009.

85 Zhang, J. and Frankel, G.S. (1999). Corrosion-sensing behavior of an acrylic-based coating system. *Corrosion* 55 (10): 957–967. https://doi.org/10.5006/1.3283932.

86 Dhole, G.S., Gunasekaran, G., Ghorpade, T., and Vinjamur, M. (2017). Smart acrylic coatings for corrosion detection. *Progress in Organic Coatings* 110: 140–149. https://doi.org/10.1016/j.porgcoat.2017.04.048.

87 Song, Y.K., Lee, T.H., Kim, J.C. et al. (2019). Dual monitoring of cracking and healing in self-healing coatings using microcapsules loaded with two fluorescent dyes. *Molecules* 24 (9): 1679. https://doi.org/10.3390/molecules24091679.

88 Song, Y.K., Lee, T.H., Lee, K.C. et al. (2020). Coating that self-reports cracking and healing using microcapsules loaded with a single AIE fluorophore. *Applied Surface Science* 511: 145556. https://doi.org/10.1016/j.apsusc.2020.145556.

89 Farag, A.A., Mohamed, E.A., and Toghan, A. (2023). The new trends in corrosion control using superhydrophobic surfaces: a review. *Corrosion Reviews* 41 (1): 21–37. https://doi.org/10.1515/corrrev-2022-0020.

90 Liravi, M., Pakzad, H., Moosavi, A., and Nouri-Borujerdi, A. (2020). A comprehensive review on recent advances in superhydrophobic surfaces and their applications for drag reduction. *Progress in Organic Coatings* 140: 105537. https://doi.org/10.1016/j.porgcoat.2019.105537.

91 Yang, Q., Cao, J., Ding, R. et al. (2023). The synthesis and mechanism of superhydrophobic coatings with multifunctional properties on aluminum alloys surface: a review. *Progress in Organic Coatings* 184: 107875. https://doi.org/10.1016/j.porgcoat.2023.107875.

92 Hooda, A., Goyat, M.S., Pandey, J.K. et al. (2020). A review on fundamentals, constraints and fabrication techniques of superhydrophobic coatings. *Progress in Organic Coatings* 142: 105557. https://doi.org/10.1016/j.porgcoat.2020.105557.

93 Vazirinasab, E., Jafari, R., and Momen, G. (2018). Application of superhydrophobic coatings as a corrosion barrier: a review. *Surface and Coatings Technology* 341: 40–56. https://doi.org/10.1016/j.surfcoat.2017.11.053.

94 Moghadam, S.G., Parsimehr, H., and Ehsani, A. (2021). Multifunctional superhydrophobic surfaces. *Advances in Colloid and Interface Science* 290: 102397. https://doi.org/10.1016/j.cis.2021.102397.

95 Huang, W., Huang, J., Guo, Z., and Liu, W. (2022). Icephobic/anti-icing properties of superhydrophobic surfaces. *Advances in Colloid and Interface Science* 304: 102658. https://doi.org/10.1016/j.cis.2022.102658.

96 Huang, W., Tang, X., Qiu, Z. et al. (2020). Cellulose-based superhydrophobic surface decorated with functional groups showing distinct wetting abilities to manipulate water harvesting. *ACS Applied Materials & Interfaces* 12 (36): 40968–40978. https://doi.org/10.1021/acsami.0c12504.

97 Lu, Z., Wang, P., and Zhang, D. (2015). Super-hydrophobic film fabricated on aluminium surface as a barrier to atmospheric corrosion in a marine environment. *Corrosion Science* 91: 287–296. https://doi.org/10.1016/j.corsci.2014.11.029.

98 Wang, P., Zhang, D., and Lu, Z. (2015). Advantage of super-hydrophobic surface as a barrier against atmospheric corrosion induced by salt deliquescence. *Corrosion Science* 90: 23–32. https://doi.org/10.1016/j.corsci.2014.09.001.

99 Zhang, D., Li, H., Chen, X. et al. (2014). Effect of surface microstructures on hydrophobicity and barrier property of anticorrosive coatings prepared by soft lithography. *Advances in Materials Science and Engineering* 2014: 342184. https://doi.org/10.1155/2014/342184.

100 George, J.S., Hoang, A.T., Kalarikkal, N. et al. (2022). Recent advances in bio-inspired multifunctional coatings for corrosion protection. *Progress in Organic Coatings* 168: 106858. https://doi.org/10.1016/j.porgcoat.2022.106858.

101 Fan, Y., Chen, Z., Liang, J. et al. (2014). Preparation of superhydrophobic films on copper substrate for corrosion protection. *Surface and Coatings Technology* 244: 1–8. https://doi.org/10.1016/j.surfcoat.2014.01.005.

102 Zhao, Y., Xu, Z., Wang, X., and Lin, T. (2012). Photoreactive azido-containing silica nanoparticle/polycation multilayers: durable superhydrophobic coating on cotton fabrics. *Langmuir* 28 (15): 6328–6335. https://doi.org/10.1021/la300281q.

103 Wu, D.Y., Meure, S., and Solomon, D. (2008). Self-healing polymeric materials: a review of recent developments. *Progress in Polymer Science* 33 (5): 479–522. https://doi.org/10.1016/j.progpolymsci.2008.02.001.

104 Yu, D. and Tian, J. (2014). Superhydrophobicity: is it really better than hydrophobicity on anti-corrosion? *Colloids and Surfaces A: Physicochemical and Engineering Aspects* 445: 75–78. https://doi.org/10.1016/j.colsurfa.2014.01.016.

105 Chen, Y., Ao, J., Zhang, J. et al. (2023). Bioinspired superhydrophobic surfaces, inhibiting or promoting microbial contamination? *Materials Today* 67: 468–494. https://doi.org/10.1016/j.mattod.2023.06.006.

2

Development of Thermally Activated Self-healing Coating

2.1 Introduction

Heat serves as the most widely employed stimulus for initiating the self-healing behavior of polymer coatings [1–4]. Thermal heating can be applied in both intrinsic and extrinsic healing approaches. In the intrinsic healing process, thermal heating can accelerate the movement of the polymer matrix to reform bonds such as hydrogen bonding [5–9], dynamic covalent bonds [10–14], and coordination bonds [15–18]. In the extrinsic healing process, thermal heating fosters the release of corrosion inhibitors or healing agents, leading to the formation of a protective layer on the exposed metal surface [19–24]. This chapter introduces several thermally activated self-healing coating systems, encompassing aspects such as coating preparation, filler preparation, surface characterization, macroscopic and microscopic electrochemical properties, and self-healing performance. Furthermore, the advantages and disadvantages of thermally activated self-healing coatings are thoroughly discussed.

2.2 Thermally Activated Self-healing Coatings to Simultaneously Recover the Corrosion-resistance and Adhesion Strength

While numerous studies have concentrated on designing self-healing coatings, the primary emphasis remains on restoring the barrier properties of coatings to impede electrochemical reactions at scratches, often overlooking the loss of adhesion strength (AS) in these exposed coating areas [25–27]. It is imperative for researchers to consider both corrosion protection performance and AS when designing coating systems. Coatings with excellent adhesion can prevent accelerated corrosion, whereas poor adhesion can lead to rapid coating failure and a significant reduction in durability [28–31].

To enhance the AS of coating systems, Wang et al. [32] grafted dopamine (DA) onto the surface of metal–organic frameworks (MOFs). By incorporating this dopamine-grafted organic framework (DA-MOFs) into water-based epoxy coatings, they effectively reinforced the adhesion between the coating and the substrate. This

Smart Protective Coatings for Corrosion Control, First Edition. Lingwei Ma and Dawei Zhang.

is attributed to the addition of DA-MOFs reducing the number of micropores formed on the coating surface during evaporation, thereby effectively improving the coating's adhesion to the metal substrate.

This section introduces a self-healing coating capable of simultaneously restoring the barrier properties and AS of coatings. Leveraging the shape memory effect of the epoxy coating, the corrosion-inhibiting action of $Ce(NO_3)_3$, and the melting of ethylene-vinyl acetate (EVA) microcapsules, the self-healing ability of the system is substantially enhanced. EVA, a commonly used hot-melt adhesive, acts both as a carrier for corrosion inhibitors and as a healing agent [33, 34]. When heated to a specific temperature, the shape recovery of the epoxy matrix and the melting of the EVA microcapsules are triggered. Consequently, the scratch narrows, allowing the healing agent to flow into the scratched area, sealing the scratches, and restoring the barrier properties of the coating.

Moreover, this process effectively reinstates the interfacial bonding between the coating and the metal substrate. The morphologies of the coating scratches before and after self-healing are analyzed to elucidate the healing mechanism. Macro- and micro-electrochemical tests are conducted to assess the protective performance of the self-healing coatings. Additionally, AS values of the scratched coating before and after healing are analyzed. Molecular dynamics simulation is employed to reveal the interaction modes of EVA and $Ce(NO_3)_3$, and the mechanism for adhesion restoration.

2.2.1 Materials and Methods

2.2.1.1 Materials

The materials and reagents used in this section are shown in Table 2.1.

2.2.1.2 Preparation of Dispersant Cerium Oleate

EVA microcapsules loaded with $Ce(NO_3)_3$ were prepared using a water–oil–water solvent evaporation method. A 1 ml 0.1 g ml^{-1} $Ce(NO_3)_3$ aqueous solution was

Table 2.1 Main experimental materials and reagents.

Materials	Purity	Company
Carbon steel	Q235	Hengde Steel
Ethylene-vinyl acetate		DuPont
$Ce(NO_3)_3 \cdot 6H_2O$	99.95%	Shanghai Macklin
Sodium oleate	97%	Sinpharm
Sodium dodecyl sulfate (SDS)	≥99%	Aladdin
Trichloromethane	≥99%	Sinpharm
Bisphenol A diglycidyl ether (BADGE)	≥85%	Aladdin
Neopentyl glycol diglycidyl ether (NGDE)	≥40%	Aladdin
Jeffamine D230	Mn ~ 230	Aladdin

added to a 10 ml trichloromethane solution containing 0.3 g EVA and 0.025 g CeOA, homogenized under sonication. The ultrasonic emulsion was then added to a 100 ml SDS aqueous solution (0.5 wt%) and stirred for five hours until the organic solvent completely evaporated. The resulting $Ce(NO_3)_3$-EVA microcapsules were filtered, washed with deionized water, and freeze-dried for 24 hours. For comparison, CeOA-EVA microcapsules containing only CeOA dispersant were prepared using a similar method (referred to as EVA microcapsules).

2.2.1.3 Preparation of $Ce(NO_3)_3$-EVA Microcapsules

EVA microcapsules loaded with $Ce(NO_3)_3$ were prepared using a water–oil–water solvent evaporation method [35–37]. A 1 ml 0.1 g ml^{-1} $Ce(NO_3)_3$ aqueous solution was added to a 10 ml trichloromethane solution containing 0.3 g EVA and 0.025 g CeOA, homogenized under sonication. The ultrasonic emulsion was then added to a 100 ml SDS aqueous solution (0.5 wt%) and stirred for five hours until the organic solvent completely evaporated. The resulting $Ce(NO_3)_3$-EVA microcapsules were filtered, washed with deionized water, and freeze-dried for 24 hours. For comparison, CeOA-EVA microcapsules containing only CeOA dispersant were prepared using a similar method (referred to as EVA microcapsules).

2.2.1.4 Characterization of Microcapsule

The morphology and size distribution of the microcapsules were analyzed by scanning electron microscopy (SEM, Quanta 250, FEI). Fourier transform infrared spectroscopy (FTIR, VERTEX 70, Bruker) was used to study the chemical structures of the microcapsules. To determine the loading ratio of $Ce(NO_3)_3$ in the microspheres, the Ce content in the $Ce(NO_3)_3$-EVA microcapsules was analyzed by inductively coupled plasma emission spectrometry (ICP-OES, 730, Agilent).

2.2.1.5 Preparation of Coatings

The coating substrate was a Q235 steel sheet, and surface treatment was carried out before use to remove the oxides and oils attached to the metal surface. The specific coating configuration process involved mixing 2.86 g BADGE, 1.82 g NGDE, and 0.68 g $Ce(NO_3)_3$-EVA microcapsules under magnetic stirring. Subsequently, 1.94 g D230 was added to the mixture solution and stirred evenly, with the fraction of $Ce(NO_3)_3$-EVA microcapsules in the coating being 10 wt%. The mixed coating solution was evenly applied to the Q235 matrix to obtain the $Ce(NO_3)_3$-EVA microcapsule coating. As controls, a blank epoxy coating and an epoxy coating with 10 wt% EVA microspheres were prepared using the same method. Additionally, a coating containing 10 wt% $Ce(NO_3)_3$-EVA microcapsules cured at 55 °C for 24 hours were prepared to study the effect of curing temperature on the coating properties. The film thickness of all coatings after curing was approximately 80 μm.

2.2.1.6 Self-healing Tests

The glass transition temperature (T_g) of the coating and the melting temperature (T_m) of the microspheres were characterized by differential scanning calorimetry (DSC, DSC25, TA). The surface of the coating was intentionally damaged with a scalpel, creating a scratch of ~40 μm wide on the coated sample. To test the self-healing performance, the coating was heated in an oven at 90 °C for 30 minutes to heal scratches. The morphological changes of the coatings were characterized by SEM and CLSM.

Electrochemical impedance spectroscopy (EIS, PARSTAT 2273, AMETEK) was used to test the recovery of corrosion protection ability. The experimental data were analyzed using ZSimpwin software. The morphology of the coating after immersion in a 3.5 wt% NaCl solution for 24 hours was characterized by SEM, and the elements in the scratched area were analyzed by EDS (EDAX, HORIBA).

Scanning electrochemical microscopy (SECM, 920D, CHI) was employed to analyze the in situ oxygen reduction process in the scratch area. Oxygen reduction currents at different time periods during the immersion of a 0.1 mol l^{-1} NaCl solution were recorded.

The coating AS was determined by a pull-off test [38].

2.2.1.7 Molecular Dynamic Simulation

Molecular dynamics simulations were executed using the Visualizer and Forcite modules within Materials Studio 2020 software to investigate the adhesion behavior among the metal substrate, coating, and microspheres. Crystallographic planes, specifically (100) in FeO, (110) in Fe_2O_3, and (111) in CeO_2, were chosen to examine the bonding interactions with EVA on the oxide substrates [39, 40]. Subsequently, periodic boxes with a vacuum space of 30 Å were constructed, and the geometric dimensions of the slab models were set as $30.63 \times 30.63 \times 40.83$ Å^3 for FeO, $29.59 \times 21.68 \times 42.59$ Å^3 for Fe_2O_3, and $30.61 \times 30.61 \times 42.50$ Å^3 for CeO_2. For the upper absorbed EVA, an EVA fragment (vinyl:acetateethylene = 1 : 4) was established to analyze the interaction between EVA molecules and metal oxide substrates in localized areas. All initially established slab models underwent geometric optimization, with the bottom five layers of metal oxide substrate atoms constrained. Subsequent molecular dynamics simulations were conducted at 298 K in the NVT canonical ensemble, employing the COMPASS force field, with a simulation time of 1000 ps and a step time of 1 fs. Based on the optimized model, a separate molecular dynamics simulation was carried out at 298 K, NVT ensemble, and COMPASS force field, with a simulation time of 1000 ps and a step length of 1 fs. The system's energy was calculated using single-point energy based on the balanced structure of molecular dynamics.

2.2.2 Characterization of Microcapsules

Figure 2.1a,b display SEM images of EVA and $Ce(NO_3)_3$-EVA microspheres, respectively, both exhibiting diameters ranging from 5 to 15 μm. Figure 2.1c depicts the corresponding FT-IR spectra of the microspheres. In the EVA microspheres, the ester bond appeared at 1740 cm^{-1}, while the stretching vibration peaks at 2853 and 2923 cm^{-1} represented $-CH_3$ and $-CH_2$, respectively. Given the water solubility of $Ce(NO_3)_3$, the centrifugal washing process effectively removed the $Ce(NO_3)_3$

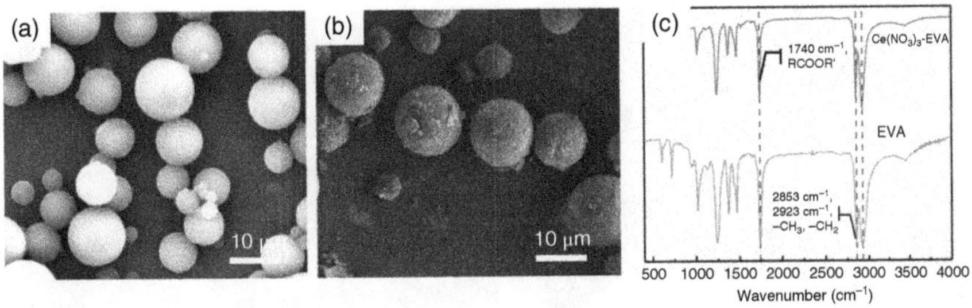

Figure 2.1 SEM images of (a) EVA and (b) $Ce(NO_3)_3$-EVA microcapsules; (c) corresponding FT-IR spectra.

adsorbed on the surface of $Ce(NO_3)_3$-EVA, resulting in the absence of the N–O peak in its FT-IR spectrum.

2.2.3 Thermal Properties and Self-healing Behavior of Coatings

To ascertain the self-healing temperature, DSC curves of $Ce(NO_3)_3$-EVA microcapsules, a blank epoxy coating, and an epoxy coating containing $Ce(NO_3)_3$-EVA microcapsules were examined. Figure 2.2 shows that the $Ce(NO_3)_3$-EVA microcapsule exhibited two endothermic peaks at approximately 44 and 82 °C, corresponding to the T_g and T_m of EVA, respectively. The T_g of both the blank epoxy coating and the $Ce(NO_3)_3$-EVA microcapsule coating fell between 45 and 50 °C. Consequently, setting the self-healing temperature at 90 °C was deemed appropriate, as it exceeded the T_m of EVA, activating the shape memory effect of the epoxy coating and facilitating the melting of EVA microcapsules to fill the coating scratches.

Surface morphologies of the scratched coatings before and after heating were characterized in Figure 2.3. Following 30 minutes of heating at 90, all coating scratches exhibited a significant reduction in width, which was attributed to the shape recovery of the epoxy matrix. Figure 2.3b,c illustrate that the scratches on coatings with the addition of EVA and $Ce(NO_3)_3$-EVA microcapsules (cured at 75 °C) were nearly

Figure 2.2 DSC curves of EVA microcapsules, epoxy coating, and epoxy coating containing $Ce(NO_3)_3$-EVA microcapsules.

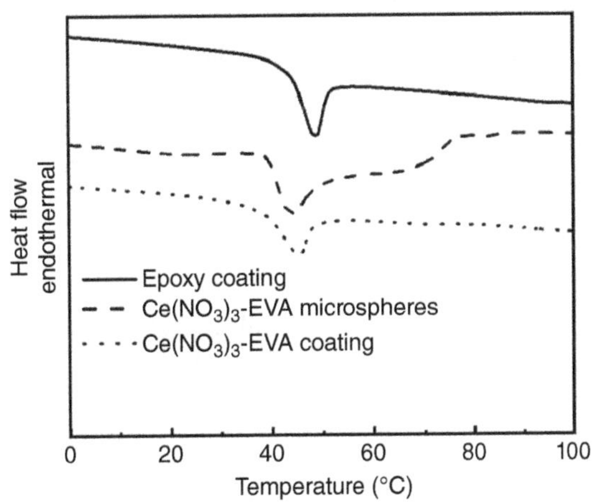

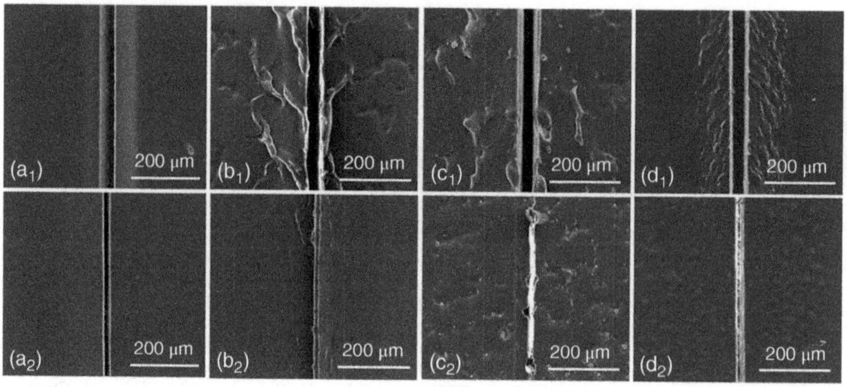

Figure 2.3 SEM images of (a$_1$) scratched and (a$_2$) healed blank epoxy coating; (b$_1$) scratched and (b$_2$) healed coating containing EVA microcapsules; (c$_1$) scratched and (c$_2$) healed coating containing $Ce(NO_3)_3$-EVA microcapsules cured at 75 °C; (d$_1$) scratched and (d$_2$) healed coating containing $Ce(NO_3)_3$-EVA microcapsules cured at 55 °C.

completely filled with molten microcapsules. In Figure 2.3d, representing the surface morphology of the coating containing $Ce(NO_3)_3$-EVA microcapsules cured at 55 °C, some filaments were observed bonding the scratching interface, indicating a less effective filling effect.

2.2.4 Healing Mechanisms of Coatings at Different Curing Temperatures

To elucidate the impact of curing temperature on self-healing properties, Figure 2.4a,b present optical microscope images of coatings cured at 55 and 75 °C with added $Ce(NO_3)_3$-EVA microcapsules, respectively. Figure 2.4a suggests that microcapsules in the coating cured at 55 °C remained well-dispersed in the epoxy resin, maintaining a good separation. Conversely, at the higher curing temperature of 75 °C (Figure 2.4b), microcapsules began to partially melt, forming a semi-continuous network. Graphical representations of the self-healing mechanism for coatings cured at 55 and 75 °C are shown in Figure 2.4c,d, respectively. In Figure 2.4c, only the microcapsules near the scratch could melt upon heating, yet they could not completely fill the scratch, as illustrated in Figure 2.4d. At 75 °C, microcapsules evenly bonded to create a network. Both exposed and connected microcapsules melted, flowing to fill the scratch. Consequently, coatings cured at 75 °C, with interconnected microcapsules, exhibited superior self-healing properties compared to those cured at 55 °C.

2.2.5 Electrochemical Analysis of the Coatings

The self-healing effects of coatings with different microcapsules and curing temperatures were investigated further using EIS measurements [41–43]. Figure 2.5 illustrates the EIS results for the healed epoxy coating, coating with EVA microcapsules, coating with the addition of $Ce(NO_3)_3$-EVA microcapsules cured at 75 °C,

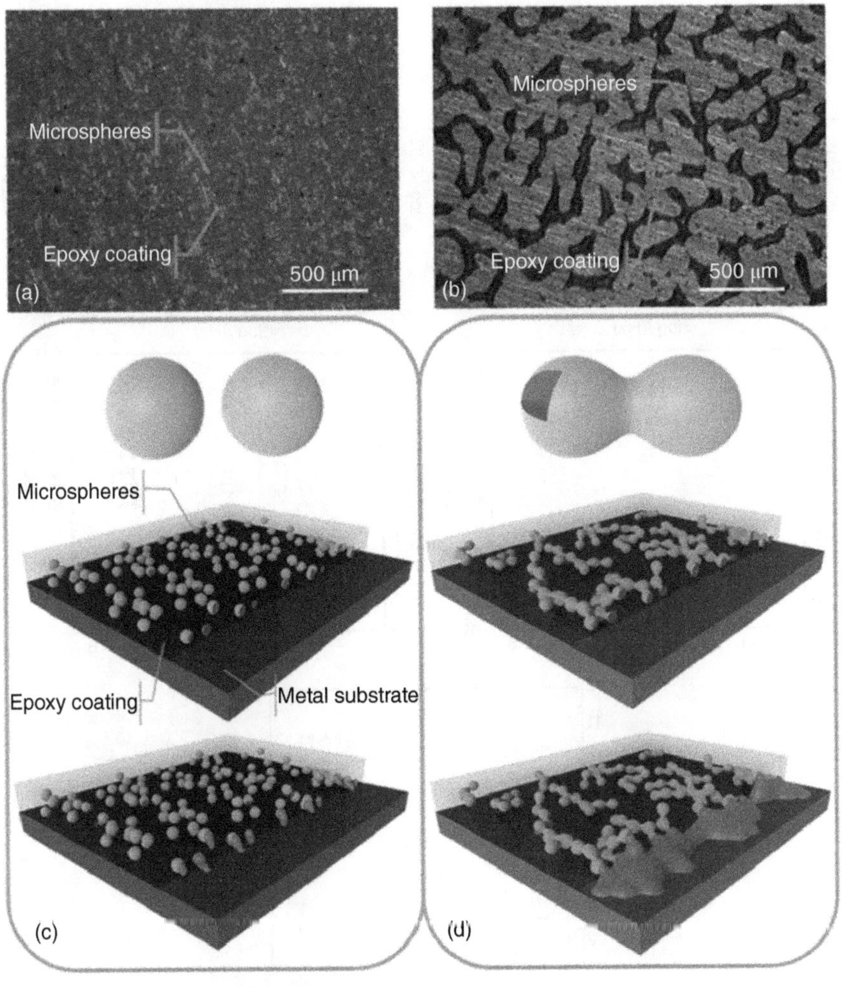

Figure 2.4 Optical images of coatings containing $Ce(NO_3)_3$-EVA microcapsules cured at (a) 55 and (b) 75 °C; the schematic illustration of the self-healing mechanisms of the coatings cured at (c) 55 and (d) 75 °C. Source: Ren et al. [3]/with permission of Elsevier.

and coating with the addition of $Ce(NO_3)_3$-EVA microcapsules cured at 55 °C after immersion in 3.5 wt% NaCl solution for 0.5, 2, 4, 6, and 24 hours, respectively. It can be observed from Figure 2.5a that the $|Z|_{0.01Hz}$ value of the healed blank epoxy coating remained consistently low throughout the immersion period. It decreased from $6.0 \times 10^5\ \Omega\,cm^2$ (0.5 hours) at the initial stage to $3.3 \times 10^5\ \Omega\,cm^2$ (24 hours). The phase angle at high frequency was very low, indicating incomplete repair of scratches and the inability to prevent the penetration of corrosive media. As depicted in Figure 2.5b, the healed coating containing EVA microcapsules exhibited good barrier performance during the initial 0.5 hours of immersion. The Bode figure showed a low-frequency impedance value of $4.2 \times 10^{10}\ \Omega\,cm^2$, maintaining a high phase angle value at high frequency. The $|Z|_{0.01Hz}$ value exhibited a continuous

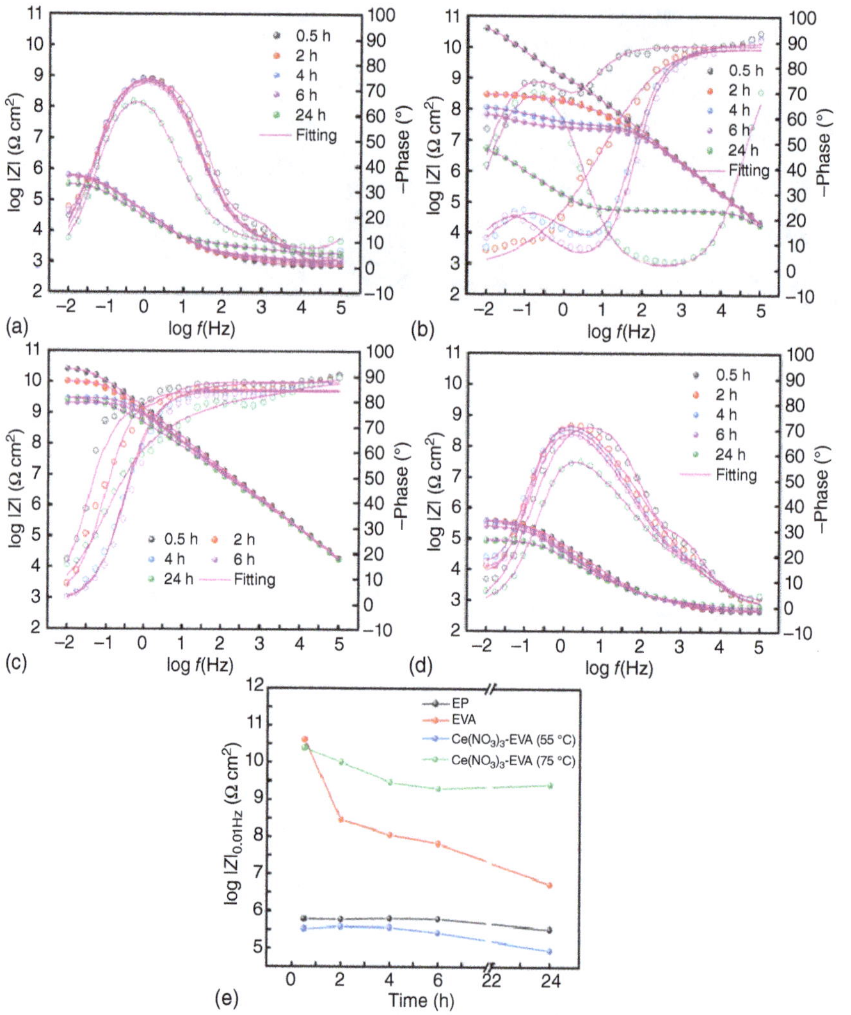

Figure 2.5 Bode plots of healed coatings during immersion in 3.5 wt% NaCl solutions for 24 hours: (a) blank epoxy coating, (b) coating containing EVA microcapsules, (c) coating containing $Ce(NO_3)_3$-EVA microcapsules cured at 75 °C, and (d) coating containing $Ce(NO_3)_3$-EVA microcapsules cured at 55 °C; (e) the $|Z|_{0.01Hz}$ values of different healed coatings. Source: Ren et al. [3]/with permission of Elsevier.

decreasing trend over the extension of immersion time, reaching $5.4 \times 10^6\ \Omega\,cm^2$ after 24 hours. Simultaneously, a second-time constant appeared for the EVA microcapsule coating after four hours of immersion, indicating corrosion of the metal substrate; this became more evident with longer immersion times. Although the EVA microcapsule coating could largely fill the scratch, there were still some gaps in the healed scratch area, making it difficult to block the penetration of corrosive media during prolonged immersion. For the coating with $Ce(NO_3)_3$-EVA microcapsules cured at 75 °C (Figure 2.5c), the $|Z|_{0.01Hz}$ value after 0.5 hours was $2.5 \times 10^{10}\ \Omega\,cm^2$, remaining high throughout the immersion process. After 24 hours, the $|Z|_{0.01Hz}$ value decreased to $2.6 \times 10^9\ \Omega\,cm^2$, exhibiting capacitance behavior at high frequency and resistance behavior at low frequency in the corrosive solution. Compared with the blank epoxy coating and EVA microcapsule coating, the corrosion inhibitor $Ce(NO_3)_3$ released from the $Ce(NO_3)_3$-EVA microcapsules inhibited corrosion in the gap, maintaining the corrosion protection performance of the healed coating at a higher level. Figure 2.5d shows the EIS results of the coating with $Ce(NO_3)_3$-EVA microcapsules cured at 55 °C. Because the scratch could not be completely closed, the $|Z|_{0.01Hz}$ value consistently remained low, allowing corrosive medium access to the metal substrate, and causing corrosive activities. Its $|Z|_{0.01Hz}$ value decreased from $3.3 \times 10^5\ \Omega\,cm^2$ (0.5 hours) at the beginning of immersion to $8.8 \times 10^4\ \Omega\,cm^2$ (24 hours). Figure 2.5e depicts the variation trend of the $|Z|_{0.01Hz}$ value of the different healed coatings with the extension of immersion time. The protective properties of the coating with $Ce(NO_3)_3$-EVA microcapsules cured at 75 °C maintained high stability during the immersion process, showing the best self-healing properties.

The equivalent circuit in Figure 2.6a was employed to fit the EIS results of the coating with $Ce(NO_3)_3$-EVA microcapsules (cured at 75 °C). The EIS results of the healed blank epoxy coating and the coating with $Ce(NO_3)_3$-EVA microcapsules (cured at 55 °C) were fitted based on the equivalent circuit in Figure 2.6b. Here, R_s represents the solution resistance, CPE_c and R_{po} represent the coating capacitance and pore resistance, respectively, CPE_{dl} represents the double-layer capacitance, and R_{ct} represents the charge transfer resistance. Acknowledging the nonuniformity of the surface and the nonideal capacitance characteristics, the pure capacitor was replaced by a constant phase element (CPE).

R_{ct} is commonly used to gauge the strength of corrosion resistance – the higher the R_{ct} value, the more challenging the electron transfer, and the better the corrosion resistance. The fitted R_{ct} values are depicted in Figure 2.6c. Throughout the immersion in 3.5 wt% NaCl solution, the R_{ct} values of the blank epoxy coating, EVA microcapsule coating, and 55 °C cured $Ce(NO_3)_3$-EVA microcapsule coating

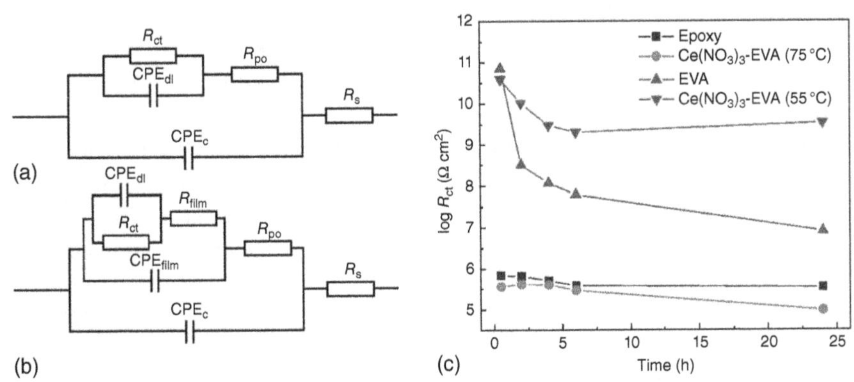

Figure 2.6 (a,b) Equivalent circuits of coatings immersed in 3.5 wt% NaCl solution; (c) variation of R_{ct} values.

exhibited a decreasing trend, while the R_{ct} values of the 75 °C cured $Ce(NO_3)_3$-EVA microcapsule coating remained relatively stable. The order of R_{ct} values was as follows – $Ce(NO_3)_3$-EVA microcapsule coating (55 °C) < blank epoxy coating < EVA microcapsule coating < $Ce(NO_3)_3$-EVA microcapsule coating (75 °C), indicating that the 75 °C cured $Ce(NO_3)_3$-EVA microcapsule coating possessed the best corrosion resistance. This is because the shape memory effect of the epoxy matrix was triggered during the heating process, causing the EVA microcapsules to melt and fill the scratches. Simultaneously, $Ce(NO_3)_3$ in the microspheres released Ce^{3+} to react with the OH^- generated in the corroded cathode region, forming a cerium oxide film on the metal matrix to effectively suppress local corrosion behavior. These three healing mechanisms worked in tandem to enhance the anticorrosion properties of the scratched coating.

2.2.6 Multiple Self-healing Mechanisms

By maintaining the platinum microelectrode potential at −0.6 V versus AgCl, the alteration in dissolved oxygen concentration at the scratch can be monitored, providing insights into the local corrosion activity of the metal. Figure 2.7 illustrates the SECM mapping results for the scratched blank epoxy coating, the scratched $Ce(NO_3)_3$-EVA microcapsule coating, the healed EVA microcapsule coating, and the healed $Ce(NO_3)_3$-EVA microcapsule coating. For the blank epoxy coating (Figure 2.7a), at 0.5 hours, the platinum microelectrode detected a very low oxygen reduction current due to oxygen competition between the microelectrode tip and the cathodic corrosion reaction on the exposed metal matrix. This indicated that oxygen was consumed by the cathodic reaction on the metal matrix. After four and eight hours, the oxygen reduction current slightly increased above the scratch area, possibly caused by the formation of corrosion products that partially inhibited the corrosion activity on the metal matrix. In the case of unhealed coatings containing $Ce(NO_3)_3$-EVA microcapsules (Figure 2.7b), the initially low oxygen reduction current increased with immersion time, attributed to the release of Ce^{3+} in $Ce(NO_3)_3$, acting as a corrosion inhibitor. Figure 2.7c depicts the SECM image of the coating containing EVA microcapsules. At 0.5 hours, the coating exhibited good barrier properties, but gradual NaCl solution penetration led to corrosion activities, and the oxygen reduction current detected on the probe began to decrease at four and eight hours. As shown in Figure 2.7d, the healed $Ce(NO_3)_3$-EVA microcapsule coating displayed a consistently high oxygen reduction current throughout the immersion process, indicating weak corrosion activity. The exceptional self-healing effect of the coating was attributed to the triple action of the shape memory effect, melting filling effect of the microcapsules, and corrosion inhibition effect of $Ce(NO_3)_3$.

To assess the release and adsorption of the corrosion inhibitor, $Ce(NO_3)_3$, Figure 2.8 presents the SEM images and EDS spectra of the scratched epoxy coating and scratched $Ce(NO_3)_3$-EVA microcapsule coating after immersion in 3.5 wt% NaCl solution for 24 hours. A multitude of corrosion products were observed on the blank epoxy coating. Conversely, on the scratched $Ce(NO_3)_3$-EVA microcapsule

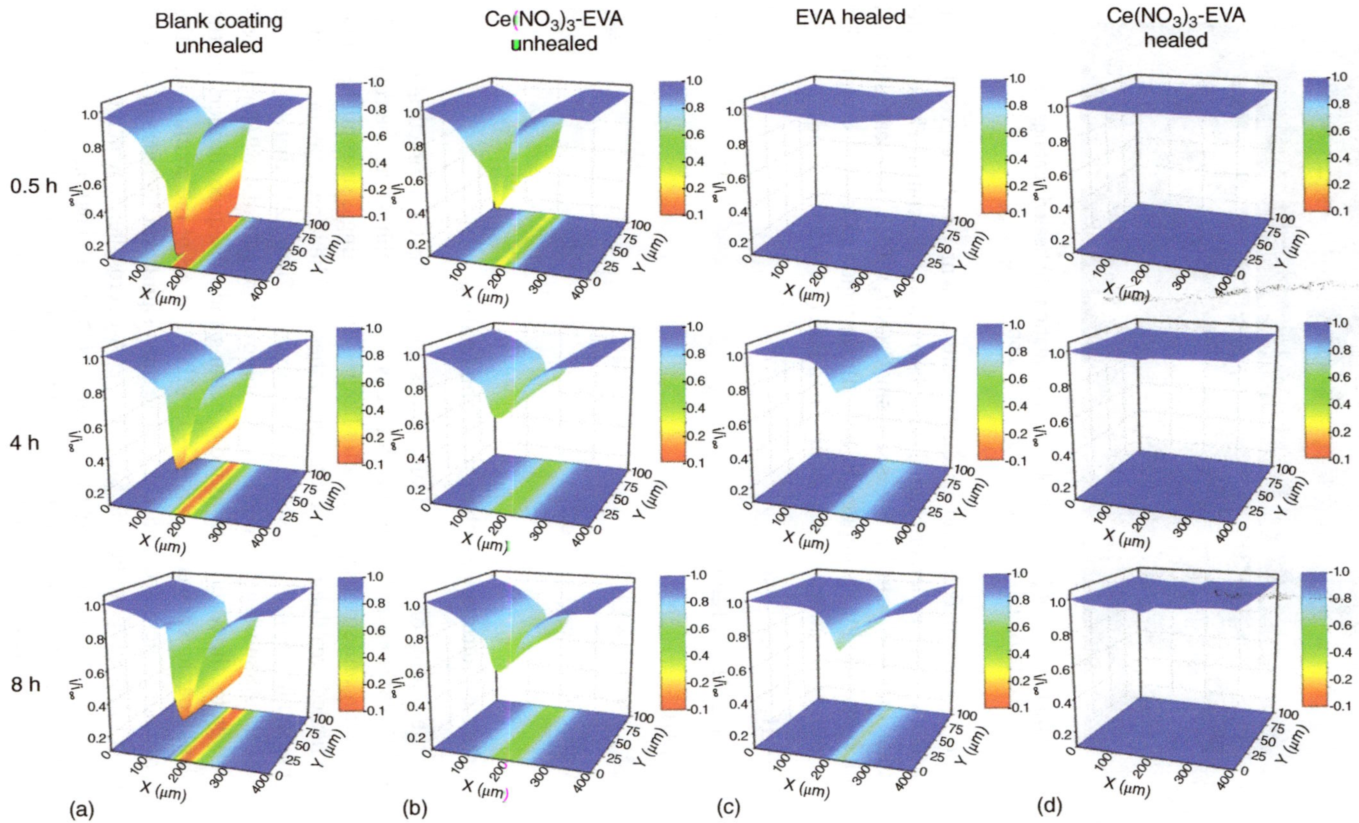

Figure 2.7 SECM mapping of (a) scratched blank epoxy coating, (b) scratched coatings containing $Ce(NO_3)_3$-EVA microcapsules, (c) healed coatings containing EVA microcapsules, (d) healed coatings containing $Ce(NO_3)_3$-EVA microcapsules during immersion in 3.5 wt% NaCl solution. Source: Ren et al. [3]/with permission of Elsevier.

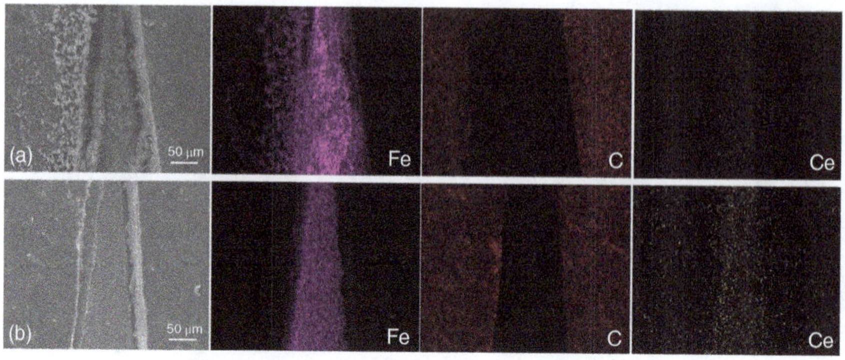

Figure 2.8 EDS mapping of (a) healed blank epoxy coating, (b) healed coatings containing $Ce(NO_3)_3$-EVA microcapsules after 24 hours of immersion.

coating, no significant corrosion products were evident on the exposed metal matrix, and a strong Ce element signal was observed at the scratch. These results demonstrate that Ce^{3+} can be released from $Ce(NO_3)_3$-EVA microcapsules and react with hydroxide ions generated in the cathode region to form a cerium oxide protective film on the metal matrix, as shown in Eqs. (2.1)–(2.5) [44–46].

$$O_2 + H_2O + 4e^- \rightarrow 4OH^- \tag{2.1}$$

$$Fe^{2+} + 2OH^- \rightarrow Fe(OH)_2 \tag{2.2}$$

$$Fe(OH)_2 \rightarrow Fe(OH)_3 \rightarrow Fe_2O_3 \tag{2.3}$$

$$Ce^{3+} + 3OH^- \rightarrow Ce(OH)_3 \tag{2.4}$$

$$2Ce(OH)_3 \rightarrow Ce_2O_3 \rightarrow CeO_2 \tag{2.5}$$

2.2.7 Adhesion Recovery Mechanism of the Coating

The AS of the coating on the metal substrate is a key indicator in practical service processes [47, 48]. Generally, the scratch area is a weak point where the coating is prone to layering and stripping. Even after healing, the coating is susceptible to failure in the healed area due to weak interface adhesion. The three types of coatings were immersed in 3.5 wt% NaCl solution. As shown in Figure 2.9, after 48 hours of immersion, the AS substantially decreased to approximately 3.8–4.4 MPa for all coatings. However, for the coatings subjected to heating and subsequent immersion, the healed blank coating showed an AS of ~4.1 MPa. The AS of the coating with the EVA microspheres increased to 6.8 MPa. The most significant increase in AS was observed on the coating containing $Ce(NO_3)_3$-EVA microspheres, which increased to ~8.1 MPa, approximately 80% of the AS value of the intact coating.

To further understand the effect of EVA on AS recovery, molecular dynamics simulations were used to evaluate the interactions between EVA and typical corrosion products, such as FeO and Fe_2O_3, and the corrosion inhibition film, CeO_2. The results in Figure 2.10 showed that EVA fragments tended to be adsorbed

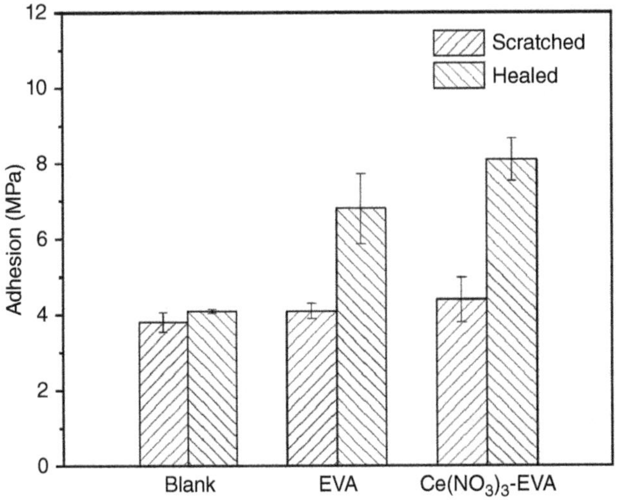

Figure 2.9 Adhesion strengths of (left) scratched blank epoxy coating, coating containing EVA microcapsules, coating containing $Ce(NO_3)_3$-EVA microcapsules after immersion in 3.5 wt% NaCl solution for 48 hours, and (right) heated to heal before measuring the adhesion strength.

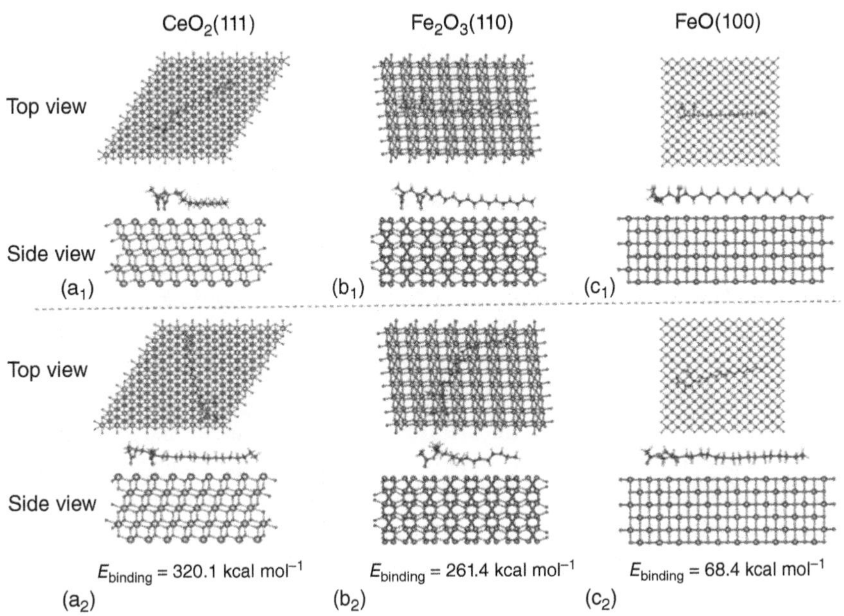

Figure 2.10 Initial (a_1,b_1,c_1) and final (a_2,b_2,c_2) structure of EVA fragment over (a) CeO_2(111), (b) Fe_2O_3(110), and (c) FeO(100) substrates. Source: Ren et al. [3]/with permission of Elsevier.

Table 2.2 Interaction and binding energy between EVA and different surfaces.

Oxide	Interaction energy ($kcal\ mol^{-1}$)	Binding energy ($kcal\ mol^{-1}$)
CeO_2	−320.1	320.1
Fe_2O_3	−261.4	261.4
FeO	−68.4	68.4

horizontally on the surface of the oxides. Additionally, the adsorption of EVA on the surface of $Fe_2O_3(110)$ is curved, whereas on the surface of FeO(100) and $CeO_2(111)$, it is linear. During dynamic simulation, the oxygen atom in the EVA copolymer gradually approached the surface of the metal oxide, and strong interactions occurred between the oxygen atom and the surface of the metal oxide, which is conducive to improving the adhesion performance. The binding energy ($E_{binding}$) represents the adsorption strength of EVA on the oxide surfaces of Fe and Ce. The positive binding energy indicates that the EVA copolymer will form a stable adsorption layer on the corresponding oxide surface.

$$E_{interaction} = E_{total} - (E_{surface} + E_{EVA}) \tag{2.6}$$

$$E_{binding} = -E_{interaction} \tag{2.7}$$

where E_{total} is the total energy of the final equilibrium system and $E_{surface}$ and E_{EVA} are the energies of the oxide surface and isolated EVA molecules, respectively.

Table 2.2 shows that the binding energy of EVA on the surface of CeO_2 is higher than that on the oxide surface of Fe, indicating that the EVA copolymer has stronger adhesion on the surface of CeO_2. This also explains the improved AS of the healed coating with the addition of $Ce(NO_3)_3$-EVA microcapsules.

Figure 2.11 illustrates the AS recovery mechanisms of different coatings. When a coating is damaged, corrosion products can accumulate near the coating scratch, creating a gap that leads to the delamination of the coating at the metal interface. When EVA microcapsules are added to the coating (Figure 2.11b), the molten EVA flows into the scratch defect and tightly bonds the polymer coating and metal substrate on both sides. This protects the metal substrate from the external environment and prevents coating delamination and loss of adhesion. With the release of the corrosion inhibitor $Ce(NO_3)_3$ into the EVA microcapsules, a CeO_2 film can be formed to inhibit the corrosion reaction, prevent the accumulation of corrosion products, and avoid coating delamination. Meanwhile, EVA exhibits a strong binding force on the corrosion inhibition film CeO_2, contributing to the further recovery of the coating adhesion.

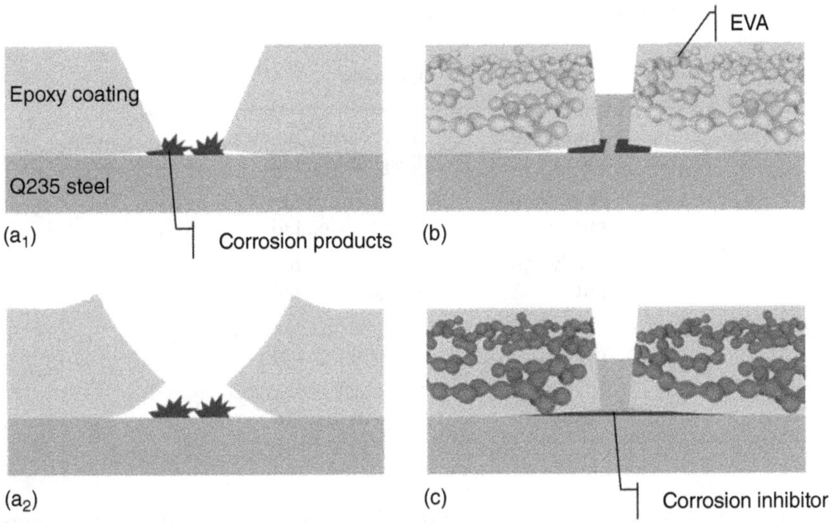

Figure 2.11 Schematic illustration of the healing mechanism of coating adhesion. (a_1,a_2) Blank epoxy coating; (b) the coatings containing EVA microcapsules; (c) the coatings containing $Ce(NO_3)_3$-EVA microcapsules.

2.3 Thermally Activated Self-healing Coatings Based on Dual Actions

Polymer microspheres are new functional materials with excellent properties that exhibit volume, magnetic, and surface effects, as well as biocompatibility, functional groupings, and more [49–59]. Polymer microspheres are widely used in biomedicine for the measurement of standard information, analytical chemistry, colloidal science, and chromatographic separation [60–70]. In recent years, with the introduction of various functional groups of monodispersed polymer microspheres, they have shown good application prospects in many high-tech fields, such as bioengineering, electronics, microelectronics, and high-performance liquid chromatography [71–79].

Preparation methods for microspheres mainly include physical and mechanical types [80–82], including phase separation [83–89] and chemical methods (e.g. emulsion crosslinking [90–93] and monomer polymerization [94, 95]). In this section, a single emulsification–solvent volatilization method is adopted, noting that oil/water type (O/W-type) emulsification and solvent volatilization are common methods for preparing drug-loaded hydrophobic microspheres. Although the 8 hydroxyquinoline (8HQ) corrosion inhibitor is difficult to dissolve in water, it is suitable for O/W-type emulsification and solvent volatilization. This method also has the advantages of a short drying time and good microsphere dispersion.

Table 2.3 Main experimental materials and reagents.

Materials	Purity	Company
n-Decylamine	Analytical purity	Sinopharm Chemical Reagent Co., Ltd
NaCl	Analytical purity	Sinopharm Chemical Reagent Co., Ltd
Ethanol	Analytical purity	Sinopharm Chemical Reagent Co., Ltd
8-Hydroxyquinoline	Analytical purity	Sinopharm Chemical Reagent Co., Ltd
Acetone	Analytical purity	Sinopharm Chemical Reagent Co., Ltd

To prepare corrosion protection coatings with dual self-healing effects using thermal stimuli, poly(ε-caprolactone) microspheres loaded with 8HQ (PCL/8HQ) were incorporated in a coating matrix with a shape memory effect to ensure the combination of dual self-healing mechanisms. It could better realize the corrosion-resistant effect of coatings on the metal substrate [96].

2.3.1 Materials and Methods

2.3.1.1 Materials

The materials and reagents used in this section are shown in Table 2.3:

2.3.1.2 Preparation of PCL/8HQ Microcapsules

8HQ was used as the corrosion inhibitor, and PCL was used as the main material to prepare the microspheres. First, 0.6 g PCL and 0.4 g 8HQ were added to a 30 ml dichloromethane solution, and 0.1 g surfactant polyvinyl alcohol (PVA) was added to 200 ml of deionized water. The two solutions were then mixed for five minutes, and the dichloromethane oil solution was evenly broken down into small droplets and suspended in an aqueous solution containing PVA. A rotor was placed in the mixed solution, and the beaker was placed on a magnetic mixer. The solution was stirred again until the dichloromethane solvent was completely volatilized, and the mixed solution was placed into six centrifugal tubes and spun. The supernatant was poured out, and an equal amount of deionized water was added for centrifugal cleaning. Afterwards, the products were frozen in a vacuum freezer and dried for 24 hours. Finally, dry PCL/8HQ microspheres were obtained. For comparison, pure 0.5 g PCL microspheres were prepared by dissolution in a 15 ml methylene chloride solution, and the same process was administered.

2.3.1.3 Preparation of Coatings

(1) Preparation of shape memory coatings: Bisphenol A diglycidyl ether was heated at 80 °C for two minutes to a molten state, and a D230 epoxy curing agent with

n-decyl amine was added to the molten epoxy resin and mixed well. The three materials were in equimolar proportions. The obtained coating was uniformly applied to a metal substrate and cured at 55 °C for 24 hours, followed by room temperature curing for another 24 hours. Finally, a shape memory coating was obtained.

(2) Preparation of self-healing coatings containing microspheres: The bisphenol A diglycidyl ether and D230 were mixed and heated at 35–45 °C for ~5 minutes, and the PCL/8HQ microspheres or pure PCL microspheres were poured into the molten epoxy resin mixture and fully mixed. Finally, *n*-decylamine was added to obtain a homogeneous mixture. The microsphere content of the coatings was 10 wt%, and the coating was uniformly applied to the substrate material using a scraper bar, followed by high-temperature curing at 55 °C for 24 hours and room temperature curing for another 24 hours.

2.3.1.4 Morphology Observation

The microstructures of the prepared microspheres and coating surfaces were observed using a FEl Quanta 250 scanning electron microscope (SEM). As the microspheres and coatings are polymer materials with no conductivity, it was necessary to spray gold onto them to increase their conductivity. A KEYENCE VK-X2000 series laser confocal microscope was used to observe the morphology of the damaged coatings. A laser was used as the scanning light source, and 3D images capable of observing the depth of damage were obtained by rapid side-to-side scanning. In addition, the morphology of the damaged coating after thermal healing was observed and compared.

2.3.1.5 Corrosion Inhibitor Loading and Releasing Test

A HITACHI U-3900H ultraviolet spectrophotometer was used to measure the loading of the corrosion inhibitor onto the PCL/8HQ microspheres. By measuring the UV absorption spectra of a series of known corrosion inhibitors and selecting the characteristic wavelength spectra in the strong UV absorption band, a linear regression equation was used to quantify the relationship between the absorption value and substance concentration, which was used to determine the corrosion inhibitor loading rate and release amount from the microspheres.

2.3.1.6 Focused Ion Beam Microscopy

To obtain a smooth cross-sectional morphology of the coating, a focused ion beam with a strong current was used to peel the surface and complete the micron processing. The experimental instrument was a LYRA3-focused ion beam microscope (TESCAN).

2.3.1.7 Electrochemical Measurement

(1) Macroscopic electrochemical testing: EIS was used to evaluate the macroscopic corrosion characteristics of the coatings (PARSTAT-2273 electrochemical workstation, Princeton Company). The test electrolyte was a 3.5 wt% NaCl solution [97–99].

(2) Microscopic electrochemical testing: An M370 electrochemical workstation (Princeton Applied Research) was used in this study. The probe diameter was 10 m, the tip voltage was −0.75 V (SCE), and the scanning rate was 50 ms^{-1}. The experiment adopted a four-point working system, and the experimental model used was the redox competition model. The SECM consisted of a Pt probe, a 370 scan control unit, an M236A potentiostat, and a phase-locked amplifier. The *x*, *y*, and *z* axes had an independent linear coding scale with a resolution of 100 nm, and the stepper motor had a resolution of 8 nm.

In most corrosive environments, dissolved oxygen is directly involved in the corrosion reaction via a reduction at the cathode. The SECM electrode probe directly detects the presence of dissolved oxygen by setting the potential of the probe in the region of oxygen diffusion potential. When the coating sample in this test was immersed in the electrolyte solution, a corrosion reaction began, kicking off the coating defect. The semi-reaction of a corroded anode requires dissolved metal Fe ions to emerge from the metal substrate. At the same time, the cathodic half-reaction of corrosion involves the depletion of dissolved oxygen from the solution. Therefore, when the SECM electrode probe swept over the coating defect, oxygen diffusion was confirmed. At this point, a competitive mechanism was established between the electrode probe and the metal substrate, both of which consumed oxygen in the solution.

2.3.2 Characterization of PCL/8HQ Microcapsules

2.3.2.1 SEM Images of PCL/8HQ Microcapsules

The SEM images of the PCL/8HQ and PCL microspheres are shown in Figure 2.12. Uniform microspheres with a diameter of ~5 μm were prepared by O/W emulsification and solvent volatilization. The microspheres were smooth and uniform in size. Compared with the PCL microspheres, the PCL/8HQ microspheres showed little adhesion amongst themselves because there was a trace of 8HQ dissolved in the water. Hence, the surfactant could not completely separate the oil phase and the water phase near the traces. However, a small number of microspheres adhered during freeze-drying [100].

2.3.2.2 Loading Rate of 8HQ in PCL/8HQ Microcapsules

To measure the 8HQ content in the PCL/8HQ microspheres, the UV absorption spectrum was measured using a spectrophotometer, and a quantitative analysis was carried out. First, a standard UV absorption curve for 8HQ in dichloromethane was obtained from a series of concentrations. Figure 2.13 shows that the UV absorption curve of the 8HQ fell in the wavelength range of 190–450 nm, and the maximum absorption wavelength of 309 nm was selected to obtain the relationship between the absorption peak and the 8HQ concentration in dichloromethane. A standard curve of the concentration absorption peak at 309 nm was obtained to detect the actual concentration of 8HQ in a dichloromethane solution with an unknown 8HQ concentration.

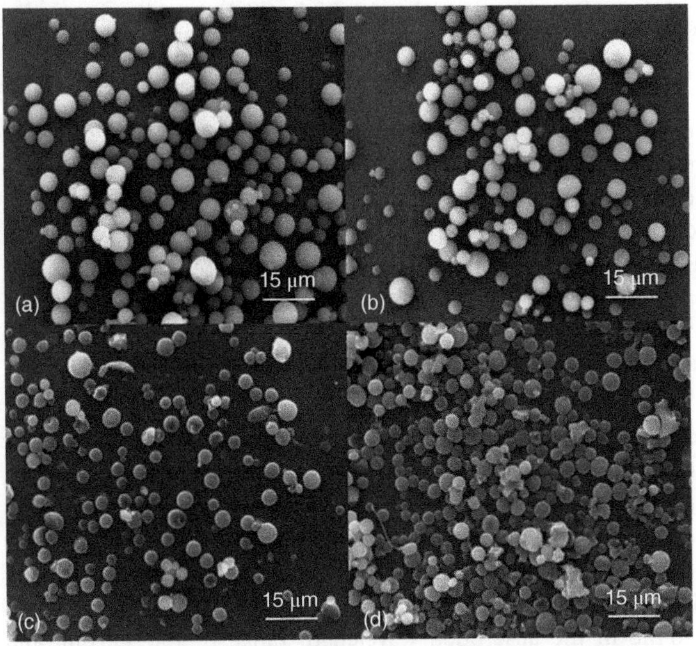

Figure 2.12 (a,b) Pure PCL microcapsules, and (c,d) PCL/8HQ microcapsules.

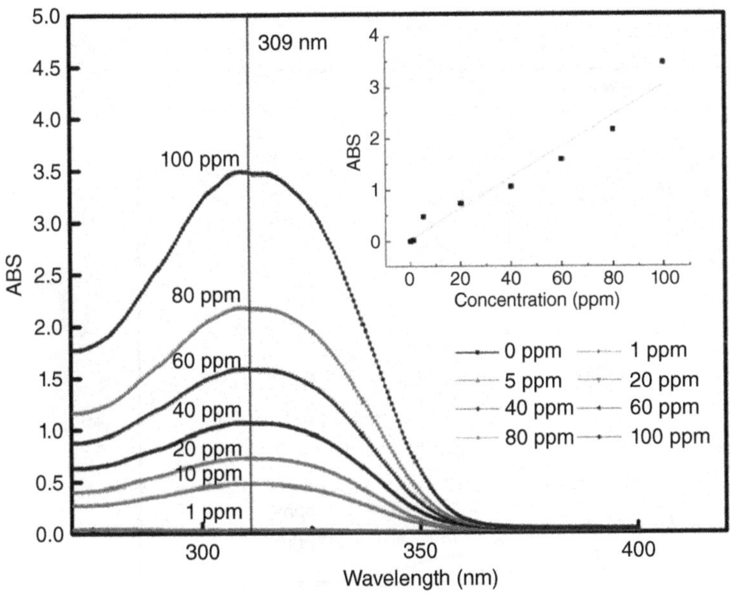

Figure 2.13 UV absorption spectra and standard curves of 8HQ in dichloromethane solution.

$$\text{Microcapsule loading rate (\%)} = \frac{\text{weight of inhibitors in microcapsule}}{\text{weight of microcapsules}} \times 100\% \tag{2.8}$$

Then, 20 mg of PCL/8HQ microspheres were dissolved in a 100 ml dichloromethane solvent at room temperature, and the UV absorbance spectra were determined under the same conditions. The quality of the corrosion inhibitor in the solution was calculated based on the linear relationship between the UV absorption intensity of the 8HQ concentration. The 8HQ loading rate in the microspheres was also calculated, as was the loading rate of the corrosion inhibitor. According to the literature, a corrosion inhibitor loading rate of 20% is desirable so that composite microspheres will exhibit a sustained release effect of corrosion inhibitors.

2.3.2.3 Corrosion Inhibitor Release Process from Microcapsules

The simulated environment in this experiment was seawater; therefore, the service environment of the microspheres included a 3.5 wt% NaCl solution [101–103]. The standard curve of the corrosion inhibitor 8HQ in an NaCl solution was measured as follows. The 8HQ corrosion inhibitors at different concentrations were accurately weighed and added to the 3.5 wt% NaCl solution. The UV absorption peak was observed by scanning in the absorption wavelength range of 190–450 nm, and the UV absorption wavelength of 8HQ at 239 nm was observed and recorded. Figure 2.14 illustrates the UV absorption spectrum and standard curves of 8HQ in the NaCl solution. The data obtained from these measurements exhibited good linearity and high reliability.

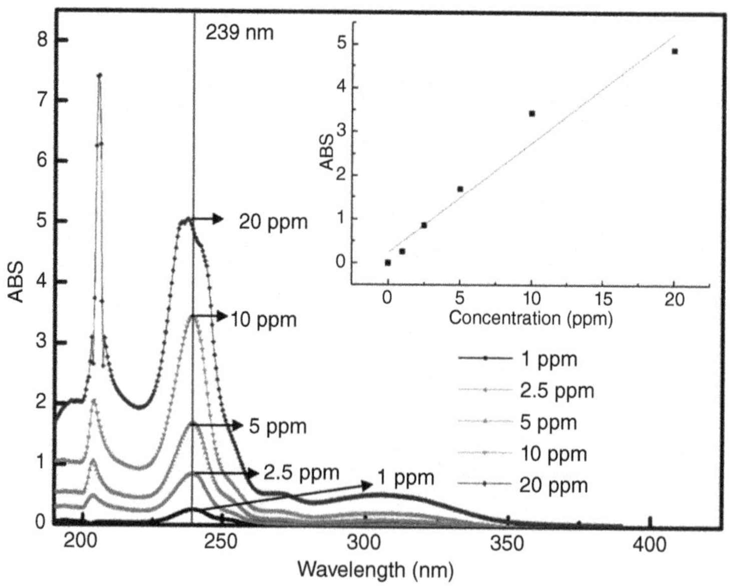

Figure 2.14 UV absorption spectra and standard curves of 8HQ in 3.5 wt% NaCl solution.

A dialysis bag wrapped with PCL/8HQ microspheres was used to study the release of the corrosion inhibitor. Prior to the experiment, a small section of the dialysis bag was cut and soaked in deionized water for 24 hours until it was completely soft. Then, 0.2 g of the PCL/8HQ microspheres were weighted, and according to the loading rate of 8HQ in the microspheres, the mass of 8HQ contained in 0.2 g of PCL/8HQ microspheres should be 0.04 g. The PCL/8HQ microspheres were then placed in the dialysis bag, tightened on both sides and soaked in a sealed 500 ml 3.5 wt% NaCl solution. The solution was shuttled to and from a colorimetric dish at regular intervals to measure the concentration of 8HQ using a UV spectrophotometer. After the measurement, the solution in the colorimetric dish was poured back into the beaker to reduce the error, and the 500 ml beaker was sealed at any time to reduce the error caused by evaporation.

Figure 2.15 shows the release process of PCL/8HQ microspheres containing corrosion inhibitor 8HQ in 3.5 wt% NaCl solution over time. It can be seen that in the first six hours, 8HQ in the microspheres was rapidly released into the NaCl solution, and 92% of the corrosion inhibitors were released from the microspheres. Afterward, the release rate slowed greatly, and the amount of corrosion inhibitor released from the microspheres moved from 92% to 95% until 48 hours had passed. This shows that the corrosion inhibitors in the microspheres were quickly released to protect the metal substrate and prevent corrosion reactions.

2.3.2.4 Corrosion Inhibitor Release Process from Coatings

To simulate the release of the 8HQ corrosion inhibitor into the coating after damage, a coating was cut through the middle, and some of the microspheres contained in the coating were dissected. The exposed cross-section was immersed in a 3.5 wt% NaCl solution to simulate the exposure of microcracks and electrolytes after damage (see Figure 2.16). It is obvious that the inhibitor was released rapidly during the first

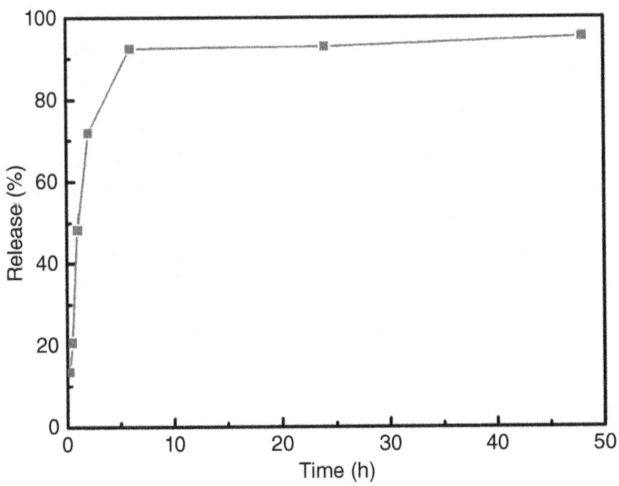

Figure 2.15 Release of corrosion inhibitors from PCL/8HQ microcapsules in 3.5 wt% NaCl solution.

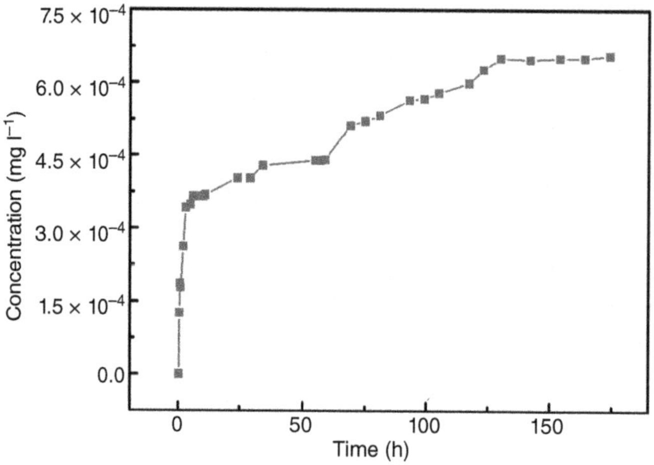

Figure 2.16 Release of corrosion inhibitors from coating with PCL/8HQ microcapsules in 3.5 wt% NaCl solution.

six hours, followed by a great slowing, which is consistent with the release process detected earlier. The release rate slowed after 6 hours and was nearly stable after 42 hours. With the increase in immersion time at the coating cross-section, the electrolyte slowly entered the entire coating, infiltrated other microspheres within the coating, and slowly released the corrosion inhibitor, which continuously protected the metal substrate [104, 105].

In this section, PCL/8HQ microspheres were successfully prepared by the O/W emulsification–solvent volatilization method. The content and mass fraction of the inhibitor reached 20%. In addition, through the corrosion inhibitor release experiment, it is concluded that the inhibitor in the mixed microspheres or the inhibitor buried in the microspheres in the coating can be released rapidly within the first six hours and continue to release with the increase of immersing time.

2.3.3 Characterization of Self-healing Coatings

2.3.3.1 The Influence of Microcapsule Content on the Coating Property

To observe the shape memory effect of the coating and the healing effect of the melting microspheres, a series of scratches of equal width (~8 mm long and 50 μm wide) were prepared on the surface of the coating using a sharp knife. Coatings with microsphere contents of 0, 5, 10, and 15 wt% were scratched in turn, and the damaged coatings were placed in the oven and heated at 80 °C for 30 minutes for healing. After removal, the healing status of the coatings was observed using a laser confocal microscope and SEM observation.

Figures 2.17 and 2.18 show the laser confocal 3D scans of the coatings with different microsphere contents before and after healing, respectively. The width of the scratch before healing was about 50 μm. After heating, all gaps narrowed, and the width of the defect was reduced to about 5 μm. The shape memory effect triggered by heating caused significant coating defects but restored the barrier effect of the

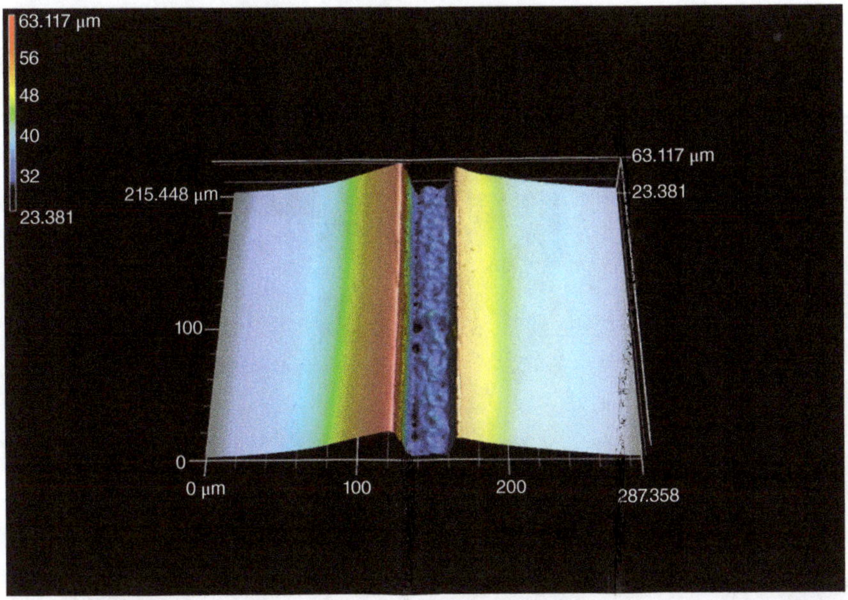

Figure 2.17 Laser confocal 3D scan of the coating after scratching.

coating. However, the laser confocal microscope did not show the effects of different microsphere contents on the healing of coating defects. SEM was then used to compare the images before and after the scratch defect healing. Figure 2.19 shows the SEM image of the coating scratches after healing. The results indicate that the scratch size was significantly reduced after heating, and the scratch width of the coating without microspheres was reduced by the shape memory effect, whereas a bridge of the filling material at the gap was clearly observed in the coating with microspheres. Compared with the blank coating, the gap closure effect was better for the coatings with microspheres because the PCL microspheres melted at 60 °C, filling the scratch defect and improving the repair performance. The combination of the shape memory effect and the melting of microspheres is thus shown to heal coating defects and protect the metal substrate. Coatings with different microsphere contents were compared, but the influence of different microsphere contents on the coating healing effect could not be clearly detected. Hence, FIB was used to examine the cross-sections of the coating. As shown in Figure 2.20, the shape memory of the coating significantly reduced the defects at the scratch, and microsphere melting filled the defects at the gap to achieve a good healing effect.

To further determine the optimal microsphere content required for the healing effect of self-healing coatings, the coatings containing different contents of microspheres were marked with defects of the same size and heated. The healed coating defects were immersed in a 3.5 wt% NaCl solution and EIS testing were performed to observe the barrier performance [106–111]. Figure 2.21 shows the EIS results of the self-healing coatings with different pure PCL microsphere contents after defect healing. The results showed that the self-healing coatings with microspheres

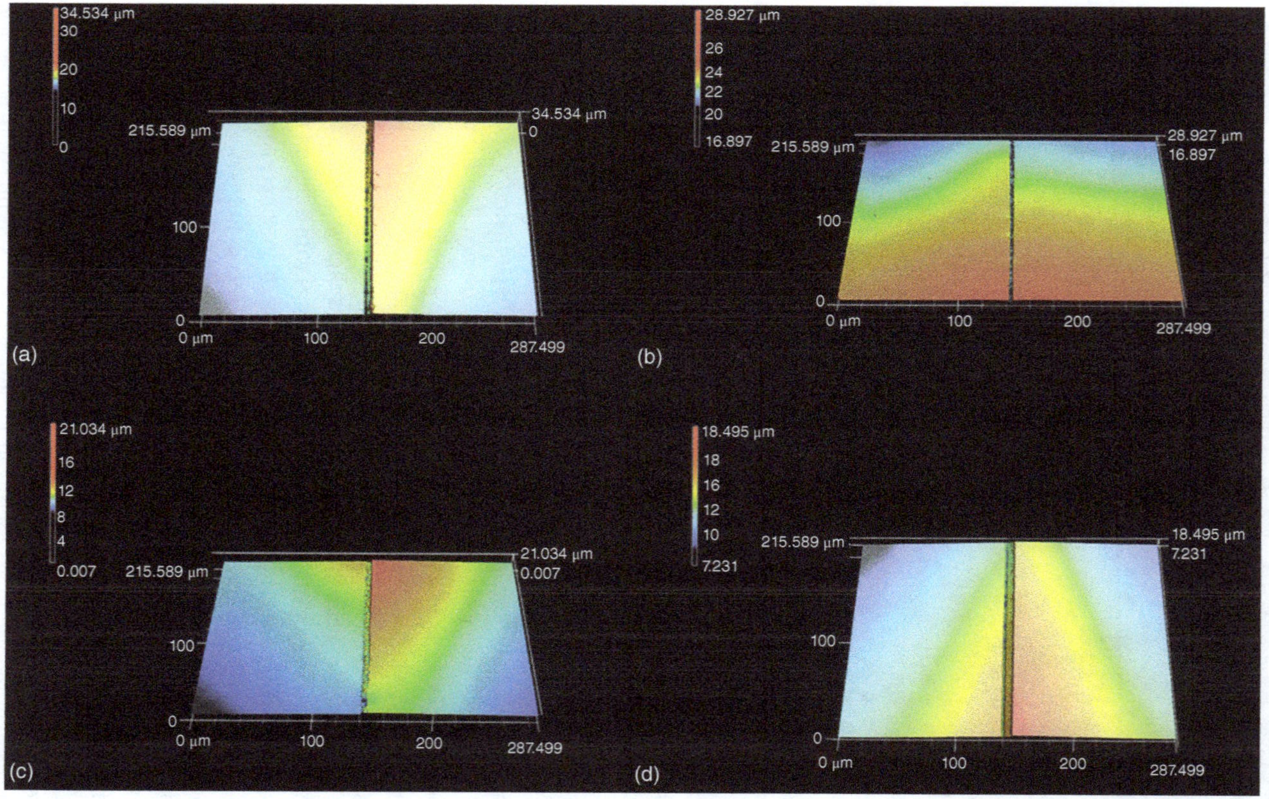

Figure 2.18 Confocal laser 3D scanning images of coatings with (a) 0, (b) 5, (c) 10, and (d) 15 wt% microcapsule content after scratch healing.

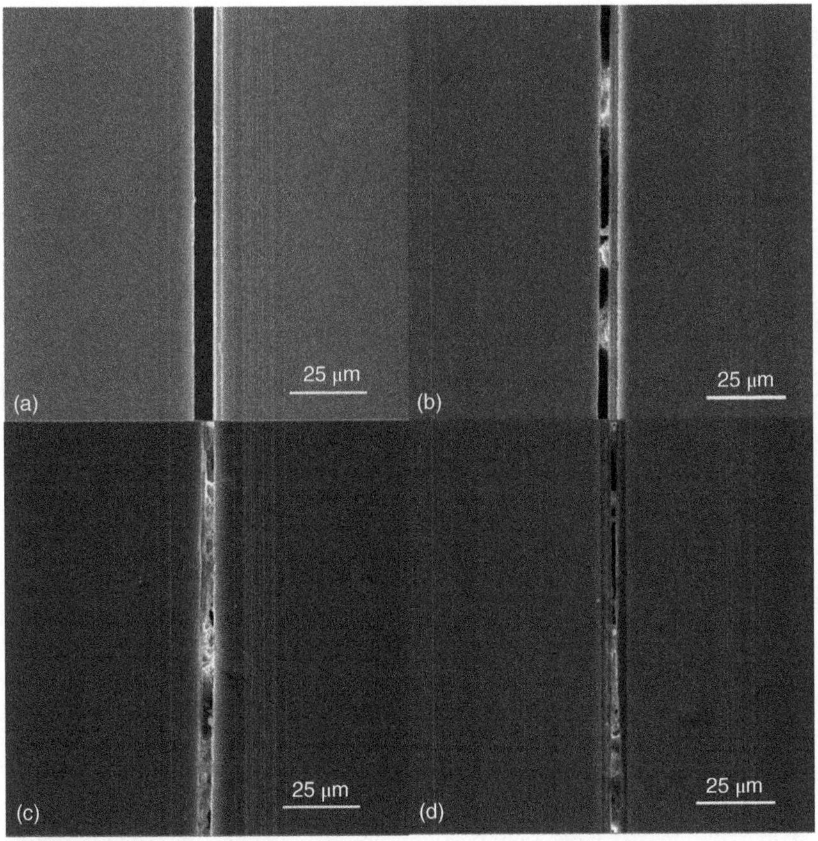

Figure 2.19 SEM images of coatings with (a) 0, (b) 5, (c) 10, and (d) 15 wt% microcapsule content after scratch healing. Source: Huang et al. [96]/with permission of American Chemical Society.

exhibited excellent corrosion resistance after healing. The $|Z|_{0.01\mathrm{Hz}}$ value exceeded $1 \times 10^{11}\ \Omega\,\mathrm{cm}^2$, whereas the barrier effect of the blank coating did not recover. After healing, the $|Z|_{0.01\mathrm{Hz}}$ was $\sim 5 \times 10^7\ \Omega\,\mathrm{cm}^2$ at the beginning of the immersion, which lost the protective performance against the metal substrate. The $|Z|_{0.01\mathrm{Hz}}$ of coatings with different microsphere contents decreased with the immersion time, and the impedance modulus of the coating with 10 wt% microspheres decreased the most slowly. After three days, the $|Z|_{0.01\mathrm{Hz}}$ of coatings with 5, 10, and 15 wt% microspheres were 3×10^7, 8×10^8, and $4 \times 10^5\ \Omega\,\mathrm{cm}^2$, respectively. It can be seen that the coating with 10 wt% microspheres restored the corrosion-resistant property of the coating at the highest level.

The macroscopic electrochemical test showed that when the coating was completely healed, the Bode diagram of the EIS produced a straight line, and the $|Z|_{0.01\ \mathrm{Hz}}$ values exceeded $10^{10}\ \Omega\,\mathrm{cm}^2$. Hence, the shape memory effect caused the scratch to close to ∼5 μm, which greatly reduced the metal surface exposure to the electrolyte solution and improved the corrosion resistance of the coating to a certain extent.

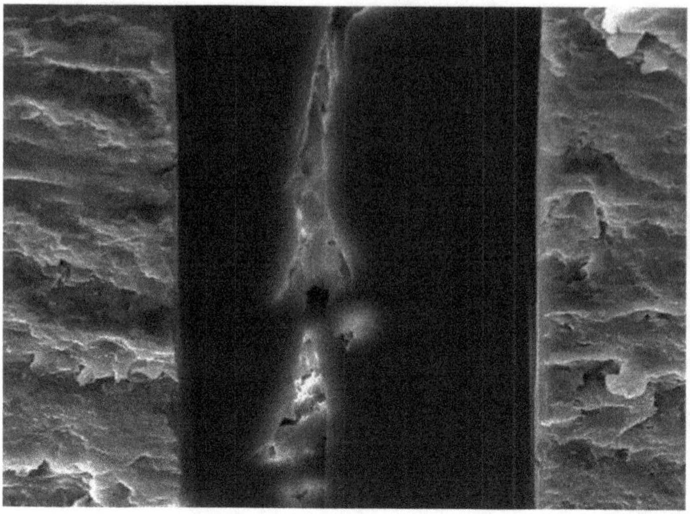

Figure 2.20 FIB image of cross-section after restoration by adding 10 wt% PCL microcapsule in the coating. Source: Huang et al. [96]/with permission of American Chemical Society.

However, due to the lack of a bridge at both ends of the scratch, the scratch could not be completely healed, and the barrier properties were not fully restored.

For the self-healing coating prepared in this experiment, a filling of PCL microspheres was added, and when the heating temperature reached the T_m of the microspheres, they melted and filled the gaps left by the shape memory effect. Hence, the self-healing coatings recovered their barrier properties after heating. The EIS results also showed that the effect of adding different amounts of PCL microspheres on the filling of the coating scratch produced different results. When microspheres were added at 0–10 wt%, the barrier effect after coating healing increased with the mass of the microspheres; however, when the mass fraction of the microspheres added to the coating reached 15%, the micro defects in the coating increased. That is, although the melted microspheres could fill the micro-cracks, the increasing interface gap between the microspheres and the coating led to many tiny channels forming in the coating, which made the electrolyte solution more permeable to the coating and hence reduced the barrier properties. As a result, 10 wt% of microspheres was selected as the benchmark for further investigations into dual action self-healing performance.

2.3.3.2 Macroscopic Electrochemical Tests of Self-healing Coatings

(1) EIS results for scratched coatings with 10 wt% PCL or PCL/8HQ microcapsules. Figure 2.22 shows the EIS results for scratched coatings with 10 wt% pure PCL microspheres and PCL/8HQ microspheres during immersion in a 3.5 wt% NaCl solution. For the scratched coating with 10 wt% pure PCL microspheres, Figure 2.22a shows that at the beginning of immersion, the $|Z|_{0.01Hz}$ value was $3.0\times10^5\ \Omega\,cm^2$, and it was reduced to $7.1\times10^4\ \Omega\,cm^2$ after 72 hours.

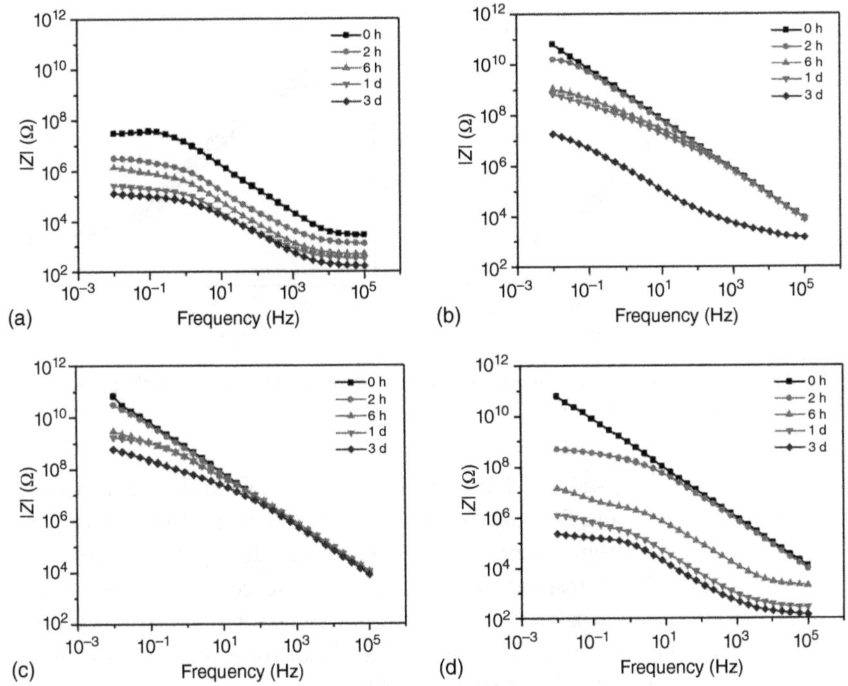

Figure 2.21 (a–d) EIS Bode diagrams of the healed coating with microcapsule content of 0, 5, 10, and 15 wt%.

The rapid decline in $|Z|_{0.01Hz}$ indicated that the electrolyte solution and the corrosive medium reached the metal substrate through the coating defects. The impedance of the coating tended to stabilize with an increase in the immersion time. Meanwhile, the metal substrate formed a corrosion product layer to gradually suppress the diffusion of corrosive media, thus playing a role in slowing the corrosion process. The $|Z|_{0.01Hz}$ of the coating increased slowly in the next three days, and it reached 1.0×10^5 Ω·cm^2 after seven days. Figure 2.22b shows the EIS results for scratched coatings with 10 wt% PCL/8HQ microspheres after immersion in a 3.5 wt% NaCl solution. Because the coating was not thermally healed, the metal substrate exposed to the electrolyte solution was large, and its $|Z|_{0.01Hz}$ was reduced from 6.9×10^5 Ω cm^2 at the beginning of immersion to 2.4×10^5 Ω cm^2 after 120 hours. However, compared to the coating with pure PCL microspheres, the $|Z|_{0.01Hz}$ value of the coating with PCL/8HQ microspheres decreased extremely slowly. The corrosion inhibitor was diffused into the solution and was adsorbed onto the metal substrate to form a chelate layer so that the corrosion reaction was prevented. However, because the corrosion inhibition behavior was not ideal, the $|Z|_{0.01Hz}$ value of the coating decreased; nonetheless, it was higher than that of the microspheres without the corrosion inhibitor. After 120 hours, both the corrosion inhibitor layer and the corrosion product film were generated to provide corrosion protection for the metal substrate, and in the next three days, the $|Z|_{0.01Hz}$

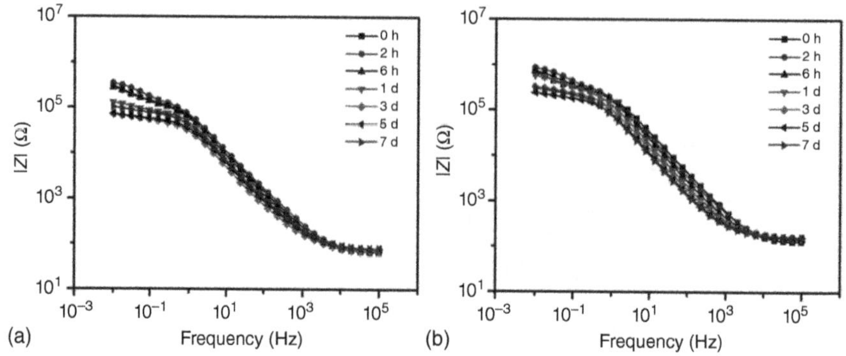

Figure 2.22 Bode plots of scratched coatings after immersion (a) with 10 wt% pure PCL microcapsules; (b) with 10 wt% PCL/8HQ microcapsules.

value of the coating stabilized at approximately $3.0 \times 10^5\ \Omega\,cm^2$. The coatings containing 10 wt% pure PCL microspheres and 10 wt% PCL/8HQ microspheres exhibited two different electrochemical processes during immersion. The 10 wt% PCL/8HQ microsphere coating reduced the corrosion rate because of the addition of corrosion inhibitors within the coating, which blocked the contact between corrosive media and the metal substrate, possibly slowing the processes involved in the interfacial corrosion reaction.

(2) EIS results for healed coatings with 10 wt% pure PCL or PCL/8HQ microspheres. Figure 2.23 presents the EIS results for scratched coatings with 10 wt% pure PCL microspheres and PCL/8HQ microspheres after thermal healing and during immersion in a 3.5 wt% NaCl solution. As shown in Figure 2.23a, for the healed coating with 10 wt% PCL, the $|Z|_{0.01Hz}$ value reached $6.9 \times 10^{10}\ \Omega\,cm^2$, which restored the barrier effect of coating matrix. However, the $|Z|_{0.01Hz}$ value decreased rapidly over time, reaching $3.0 \times 10^9\ \Omega\,cm^2$ after six hours. The area of the metal substrate exposed to the electrolyte solution after coating healing was extremely small; thus, the coating exhibited good corrosion resistance at zero hour. Subsequently, the corrosive medium rapidly penetrated the substrate through the coating defects, and the $|Z|_{0.01Hz}$ of the coating decreased. After five days, two electrochemical reaction processes were observed in the EIS results. The $|Z|_{0.01Hz}$ of the coating reached $1.0 \times 10^7\ \Omega\,cm^2$ on day 9. Figure 2.23b shows the EIS results for the healed coatings with 10 wt% PCL/8HQ microspheres after immersion. A good barrier effect was observed with $|Z|_{0.01Hz} = 1.2 \times 10^{11}\ \Omega\,cm^2$. The $|Z|_{0.01Hz}$ of the coating decreased slowly but was still far higher than that of the coating without the corrosion inhibitor. After nine days, the $|Z|_{0.01Hz}$ value had declined to $2.0 \times 10^9\ \Omega\,cm^2$. This indicates that the healing function of the coating and corrosion inhibitor plays a strong role in preventing the corrosion of the metal substrate. Physical corrosion protection due to shape memory and microsphere melting, as well as chemical corrosion protection from corrosion inhibitors, function together to optimize the self-healing performance.

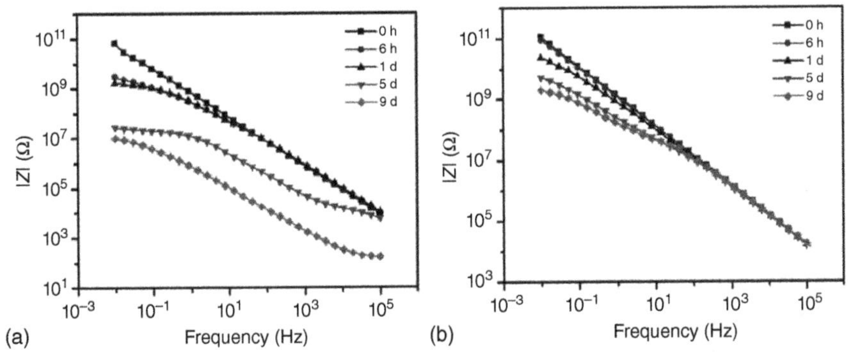

Figure 2.23 Bode plots of healed coatings after immersion with (a) 10 wt% pure PCL microcapsules; (b) 10 wt% PCL/8HQ microcapsules.

2.3.3.3 Microscopic Electrochemical Tests of Self-healing Coatings

Although the macroscopic electrochemical method can be used to examine the corrosion protection effect, it reflects the overall barrier performance of the entire immersion area, including the intact coating and defect area. To better observe the shape memory effect in localized coating defects and the corrosion inhibition effect of the inhibitors, microscopic electrochemical methods were employed to clarify the corrosion reactions at the defects, along with the self-healing effect of the coatings [112–116].

(1) SECM results for scratched coating with 10 wt% pure PCL microspheres.
Figure 2.24a shows the SECM results for the scratched coating with 10 wt% pure PCL microspheres before heating. Because the coating was not healed, the current value at the defect area was less than 2.0×10^{-10} A, indicating a severe corrosion reaction at the defect that consumed oxygen in the solution. After immersion in a 3.5 wt% NaCl solution for eight hours, the corrosion current was essentially unchanged. After 24 hours, the severe corrosion region at the center of the defect began to spread to the surrounding area, and the detected current in the surrounding area decreased. Because the coating absorbed water and dissolved after immersion, the corrosion current detected on the coating surface decreased. After 48 hours, the severe corrosion area at the center of the defect was enlarged, and the current of the coating in the area surrounding the defect fluctuated, indicating that the coating lost its barrier properties after the immersion and absorption of water.

(2) SECM results for scratched coating with 10 wt% PCL/8HQ microspheres.
Figure 2.24b shows the SECM test results for the scratched coatings with 10 wt% PCL/8HQ microspheres before heating. Initially, the metal substrate at the defect was heavily exposed because of the lack of healing, and a severe corrosion reaction occurred at the defect. After eight hours, the corrosion inhibitor 8HQ in the coating was released. The corrosion reaction occurring at the center of the defect generated an $Al(8\text{-}HQ)_3$ chelate that covered the exposed metal substrate and began to slowly prevent the diffusion of corrosive media. It can also be seen from the figure that the current in the central region of the defect

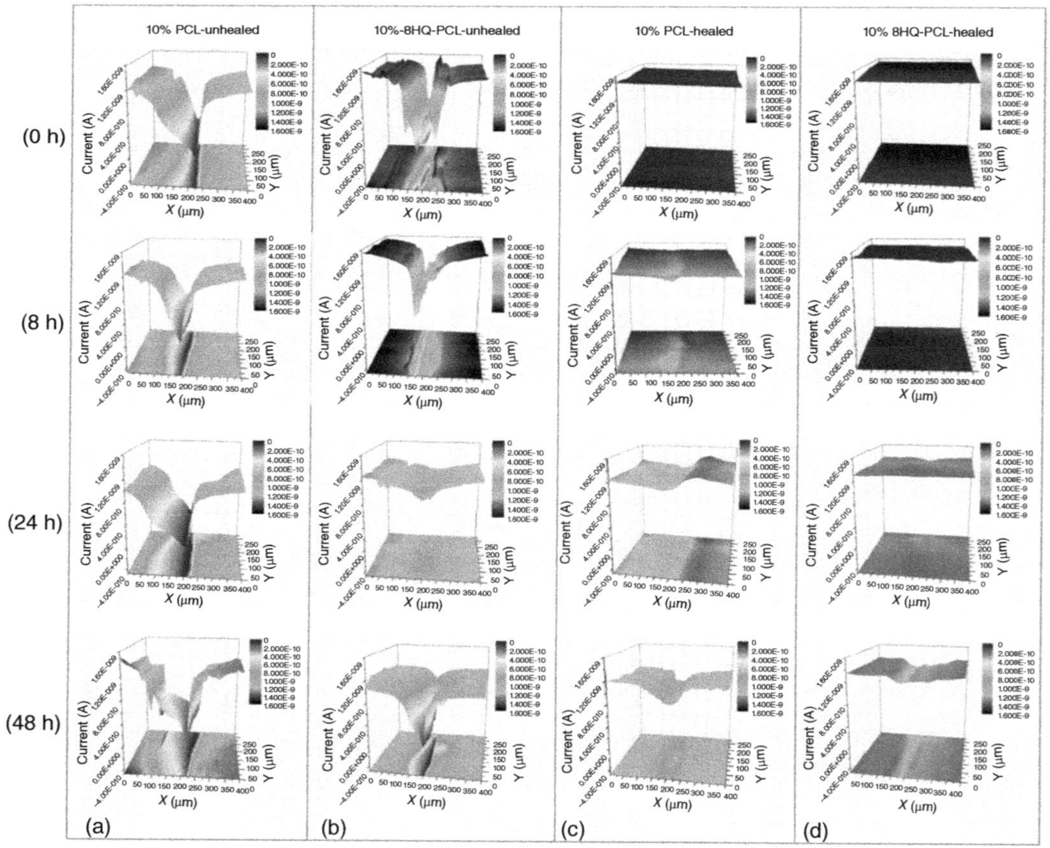

Figure 2.24 SECM results of (a) scratched coatings with 10 wt% PCL; (b) scratched coatings with 10 wt% PCL/8HQ; (c) healed coatings with 10 wt% PCL; (d) healed coatings with 10 wt% PCL/8HQ. Source: Huang et al. [96]/with permission of American Chemical Society.

increased from 2.0×10^{-10} to 8.0×10^{-10} A. After immersion of the coating defects for up to 24 hours, the current in the central region of the coating defect increased to 1×10^{-9} A, which was on the same order of magnitude as the probing current around the defect, and the corrosion reaction at the surface defect was reduced with low oxygen consumption. The corrosion inhibitor added to the coating provided sufficient protection to the metal substrate, thereby inhibiting corrosion reaction. However, when the immersion time reached 48 hours, the corrosion condition was similar to that at the beginning of the corrosion reaction, which can be explained by the corrosive ions breaking through the chelate film so that the corrosion of the metal substrate continued. The addition of 8HQ to the coating can inhibit the corrosion reaction; however, when the coating scratch was large, the corrosion inhibitor did not provide sufficient protection for the metal substrate.

(3) SECM results for healed coating with 10 wt% pure PCL microspheres.
Figure 2.24c presents the SECM test results for the scratched coating with 10 wt% PCL microspheres after thermal healing and during immersion in a 3.5 wt% NaCl solution. Heating led to extensive defect closure, reducing the area of the exposed metal substrate and slowing the corrosion expansion. As shown in the figure, the coating scratches were completely closed after the thermal healing, and the currents in the coating defects and their surroundings were maintained at a level far higher than that for the same scratched coating immersed for zero hour in Figure 2.24a. This indicated that the self-healing effect of the coating defects significantly restored the barrier properties and prevented corrosion reactions. When the immersion time reached eight hours, no significant corrosion reactions occurred. After 24 hours of immersion, the current fluctuated, and the current in the central area was lower than that around the defect. After 48 hours, corrosion occurred at the coating defects. This indicates that after thermal healing, the coating defects can be closed to partially restore the barrier effect of the coating.

(4) SECM results for healed coating with 10 wt% PCL/8HQ microspheres.
Figure 2.24d shows the SECM results for the scratched coating with 10 wt% PCL/8HQ microspheres after thermal healing. As shown, the scratched coating was closed to a great extent after healing, reducing the probability of a corrosion reaction. After one hour, the current in the defective part of the coating was equal to the surrounding current, which was similar to that in the case of the coating with 10 wt% pure PCL microspheres, indicating that the coating inhibited the corrosion reaction after thermal healing. When the coating was immersed for eight hours, the current at the coating defects fluctuated slightly, but the change did not exceed 0.2×10^{-10} A, indicating that although the coating underwent slight water absorption and swelling, no obvious corrosion reaction occurred. Compared with the healed coatings with 10 wt% pure PCL microspheres, the current in the defect central region and the surrounding current changed significantly, indicating that the addition of corrosion inhibitors delayed the corrosion reactions. When the coating absorbed water, 8HQ was released from the coating, inhibiting the corrosion of the metal substrate.

After 24 hours, obvious corrosion reactions occurred, and the current in the coating defect area was significantly lower than the surrounding current. As the immersion time increased, the protective effects of the coating weakened. However, compared with the results for the healed coating with 10 wt% pure PCL microspheres after immersion, the coatings with 8HQ significantly delayed the corrosion reaction. This is because, during the water absorption process of the coating, the corrosion inhibitors in the coating were released, reacting with the metal substrate and suppressing the corrosion reaction.

Comparing the SECM results for the four coatings revealed that the added corrosion inhibitors can react with metal ions to form chelates that cover the metal substrate surface, preventing further diffusion of the corrosion medium and delaying the corrosion reaction. However, because the amount of corrosion inhibitor in the coating was limited, the adsorbed corrosion inhibitor layer broke over time, and the corrosion reaction continued. Thermal healing can significantly close the coating defects and reduce the exposure area of the metal substrate in the solution, reducing the probability of a corrosion reaction; however, it cannot achieve the optimal corrosion protection effect. The thermally activated shape memory effect of the coating resulted in the best healing effect when combined with the corrosion inhibition effect of the corrosion inhibitors, providing long-lasting corrosion protection for the metal substrate.

2.3.3.4 Characterization of Scratched Coatings After Immersion

To study the adhesion of corrosion inhibitor on the metal substrate at the coating defect, the experiments were carried out using EDS energy spectrometry to determine the surface products near the coating defect after immersion [117–119]. Meanwhile, SEM was used to observe the morphological changes in the coating defects after immersion.

Figure 2.25 shows the morphologies and elemental distributions of the scratched coatings with 10 wt% PCL/8HQ microspheres after 0, 8, 24, and 48 hours of immersion in a 3.5 wt% NaCl solution. The central region of the scratch was selected for elemental mapping. As shown in the SEM images, no filler material was present inside the scratch before immersion. Because the substrate at the center of the scratched coating was selected, the aluminum contents were evenly and densely spread over the entire area, and nitrogen and carbon contents were almost absent, indicating that the scratch contained no substances other than the substrate metal. When the immersion time reached eight hours, loose granular material began to grow on both sides of the coating scratches. These results indicated that the nitrogen and carbon contents of the metal substrate began to increase. After 24 hours of immersion, the scratches in the coating were filled with loose granular fillers, and the nitrogen and carbon contents continued to increase. After 48 hours, the coating defects were completely filled, and aluminum, nitrogen, and carbon were densely and uniformly distributed in the area. Because the corrosion inhibitor 8HQ contains nitrogen, the appearance of fillers at the defects of the coating after immersion may have resulted from a reaction between the released 8HQ and the metal aluminum ions to form a chelate layer.

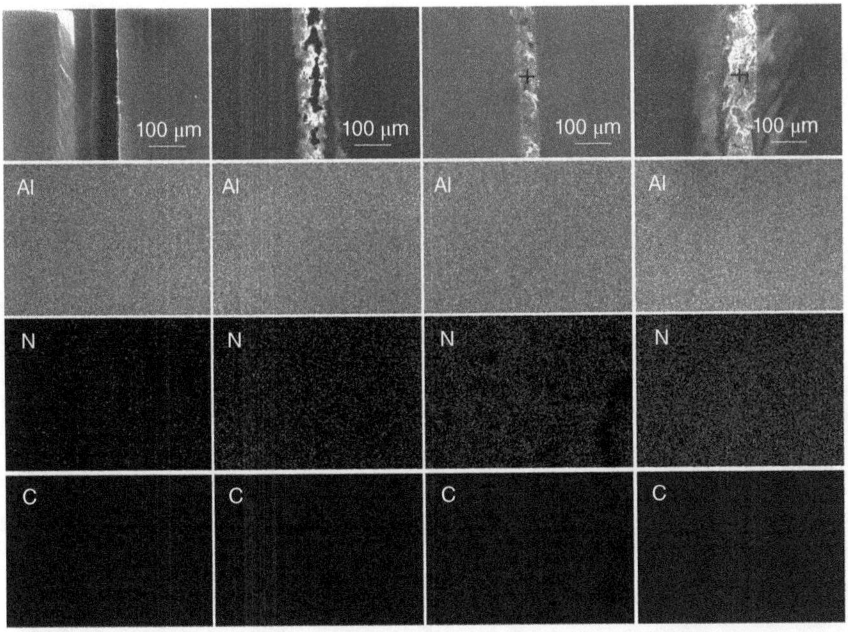

Figure 2.25 SEM morphology and element distribution of the scratched coatings with 10 wt% PCL/8HQ microcapsules after immersion in 3.5 wt% NaCl solution for 0, 8, 24, and 48 hours. Source: Huang et al. [96]/with permission of American Chemical Society.

Figure 2.26 shows the SEM images and elemental distribution of the scratched coatings with 10 wt% PCL microspheres after immersion. The width of the coating scratches was approximately 70 μm, and the center of the scratches was selected for elemental mapping. Before immersion, aluminum was uniformly and densely distributed in the scratched area, and nitrogen and carbon elements were almost absent, indicating that the metal substrate was directly exposed and that there was no other material covering it. Similar to the scratched coatings without immersion, there was no significant increase in the nitrogen or carbon content. This indicated that the coating with pure PCL microspheres did not protect the metal substrate, because of the absence of corrosion inhibitors. The corrosion of the immersed aluminum substrate occurred when it was in direct contact with the electrolyte solution.

To verify the adsorption of the corrosion inhibitor onto the metal substrate, the scratched coating was subjected to FT-IR spectroscopy after 48 hours of immersion. From top to bottom, the FT-IR spectra in Figure 2.27 correspond to 8HQ, the substance produced within the scratches of the scratched coatings with 10 wt% PCL/8HQ microspheres, the substance produced within the scratches of scratched coatings with 10 wt% PCL microspheres in 3.5 wt% NaCl after immersion for 48 hours, and the coating with 10 wt% PCL/8HQ microspheres. After immersion, an Al-8HQ in-plane bending vibration absorption peak was observed near 1380 cm^{-1} in the FT-IR spectrum of the scratched coating containing corrosion inhibitors, indicating that the generated substance was an Al-8HQ composite. The obvious absorption peaks near 1508 and 1471 cm^{-1} were caused by the stretching vibration

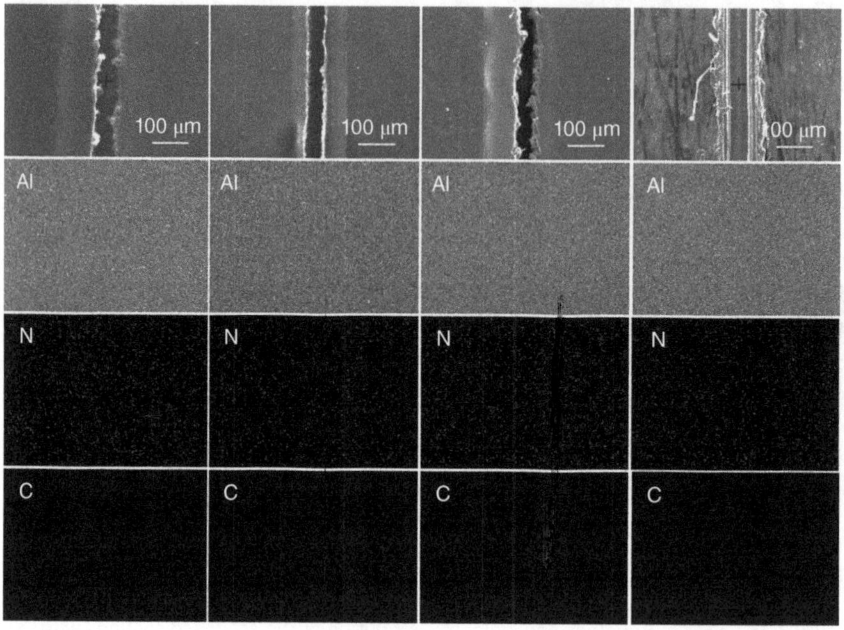

Figure 2.26 The SEM image and the element distribution of the scratched coatings with 10 wt% pure PCL microcapsules after immersion in 3.5 wt% NaCl solution for 0, 8, 24, and 48 hours. Source: Huang et al. [96]/American Chemical Society.

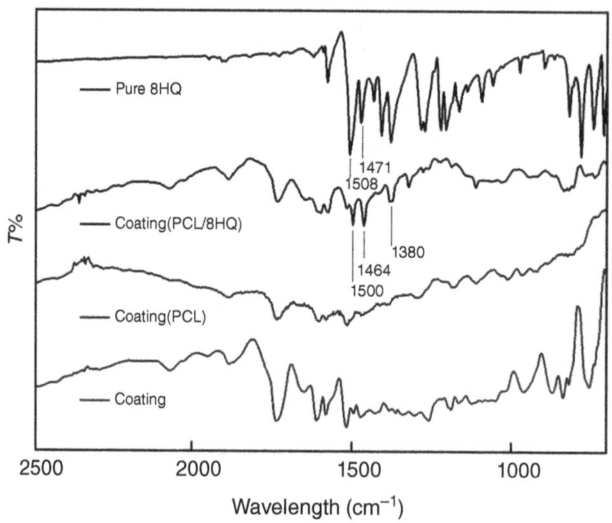

Figure 2.27 FTIR spectra of blank epoxy coating (named as Coating), coating containing PCL (named as Coating(PCL)), coating containing PCL/8HQ microspheres (named as Coating(PCL/8HQ)) and pure 8HQ material.

of the C=C bond in the benzene ring. For the other three FT-IR spectra, no obvious absorption peaks were observed at the corresponding positions of pure 8HQ. In addition, the other two spectra did not exhibit obvious absorption peaks, indicating that the corrosion inhibitor 8HQ was released to generate the Al-8HQ composite. The resulting chelate covered the bare metal surface, preventing the spread of corrosive substances and protecting the metal substrate.

2.4 Chapter Summary

In this chapter, two types of thermally activated self-healing coatings were discussed. For the epoxy coating containing $Ce(NO_3)_3$-EVA microspheres, the microspheres were added to the shape memory epoxy resin to recover both the corrosion protection properties and the AS. The effect of the curing temperature (55 and 75 °C) on the self-healing performance of the coating was investigated. When cured at 75 °C, $Ce(NO_3)_3$-EVA microcapsules uniformly adhered to each other and formed a bonding network. In addition to the microcapsules near the scratch, the adjacent ones were heated, causing them to melt and flow, filling the scratch. For the 55 °C curing condition, the $Ce(NO_3)_3$-EVA microcapsules were individually dispersed, and only the microcapsules near the scratch could be heated to melt and flow out, which was not conducive to the filling of the scratch. The shape memory effect of the coating matrix, the melt filling of EVA microspheres, and the $Ce(NO_3)_3$ corrosion inhibition can function together to enhance the protective performance of the coatings and significantly inhibit the corrosion behavior. The results of an AS test and molecular dynamic simulations indicated that EVA, as a hot-melt adhesive, can restore the adhesion of the healed coating on the metal surface and that the release of $Ce(NO_3)_3$ can form a CeO_2 film layer with a corrosion inhibition effect in the scratched area, which prevents delamination of the coating interface due to the accumulation of corrosion products. Meanwhile, because EVA has a high binding energy with CeO_2, it promotes further recovery of the coating adhesion.

For the PCL/8HQ microsphere-containing coating, the shape memory effect of the coating matrix can close the coating scratch to a large extent but cannot close it completely, which cannot provide effective corrosion protection for the metal substrate. The shape memory effect triggered by the thermal healing of the coating, the filling effect of molten microspheres, and the corrosion inhibition effect of the inhibitors all contributed to the good corrosion protection of the triple-action self-healing coating, which is characterized by intrinsic and extrinsic self-healing based on macroscopic- and microscopic-scale electrochemical measurement methods.

Thermal healing not only increases the temperature of the polymer matrix to make it higher than the T_g or T_m of the polymer, realizing the closure of the defect area and melting of coating fillers but also promotes the movement of dynamic reversible covalent bonds or noncovalent bonds to re-bond the polymer matrix. However, the self-healing mechanism based on traditional heating stimulation often requires heating the entire polymer for a long time at a relatively short distance, which can cause new damage to other intact parts of the coating. Therefore, the heating temperature and heating range should be controlled to avoid an excessive impact on the intact coating. In addition, the dynamic reversible covalent bonds or noncovalent bonds in coatings are often inversely proportional to the corrosion resistance; that is, coatings with better intrinsic self-healing performance, mechanical properties of the resin, and barrier properties are often undesirable. Thus, there is considerable demand for the development of new coating systems to realize the synergistic improvement of the intrinsic self-healing ability, barrier, and mechanical properties of coatings. Advanced heating methods must be designed to achieve self-healing in outdoor environments.

References

1 Fortunato, G., Marroccoli, V., Corsini, F. et al. (2020). A facile approach to durable, transparent and self-healing coatings with enhanced hardness based on Diels–Alder polymer networks. *Progress in Organic Coatings* 147: 105840. https://doi.org/10.1016/j.porgcoat.2020.105840.

2 González-García, Y., Mol, J.M.C., Muselle, T. et al. (2011). A combined mechanical, microscopic and local electrochemical evaluation of self-healing properties of shape-memory polyurethane coatings. *Electrochimica Acta* 56: 9619–9626. https://doi.org/10.1016/j.electacta.2011.03.081.

3 Ren, C., Huang, Y., Hao, W. et al. (2022). Multi-action self-healing coatings with simultaneous recovery of corrosion resistance and adhesion strength. *Journal of Materials Science and Technology* 101: 18–27. https://doi.org/10.1016/j.jmst.2021.05.070.

4 Zhang, F., Ju, P., Pan, M. et al. (2018). Self-healing mechanisms in smart protective coatings: a review. *Corrosion Science* 144: 74–88. https://doi.org/10.1016/j.corsci.2018.08.005.

5 Cordier, P., Tournilhac, F., Soulié-Ziakovic, C. et al. (2008). Self-healing and thermoreversible rubber from supramolecular assembly. *Nature* 451: 977–980. https://doi.org/10.1038/nature06669.

6 Hofmeier, H., Hoogenboom, R., Wouters, M.E.L. et al. (2005). High molecular weight supramolecular polymers containing both terpyridine metal complexes and ureidopyrimidinone quadruple hydrogen-bonding units in the main chain. *Journal of the American Chemical Society* 127: 2913–2921. https://doi.org/10.1021/ja042919e.

7 Liu, T., Zhao, H., Zhang, D. et al. (2021). Ultrafast and high-efficient self-healing epoxy coatings with active multiple hydrogen bonds for corrosion protection. *Corrosion Science* 187: 109485. https://doi.org/10.1016/j.corsci.2021.109485.

8 Wang, T., Feng, H., Wang, W. et al. (2022). Interfacial controllable heterojunctions nanosheets as photothermal catalyzer for cyclic photothermal self-healing of polydimethylsiloxane coating. *Composites Part B: Engineering* 240: 110002. https://doi.org/10.1016/j.compositesb.2022.110002.

9 Yu, C., Salzano de Luna, M., Russo, A. et al. (2021). Role of diisocyanate structure on self-healing and anticorrosion properties of waterborne polyurethane coatings. *Advanced Materials Interfaces* 8: 2100117. https://doi.org/10.1002/admi.202100117.

10 Chen, K., Zhu, H., Zhang, Z. et al. (2023). Self-healing polyurethane coatings based on dynamic chemical bond synergy under conditions of photothermal response. *Chemical Engineering Journal* 474: 145811. https://doi.org/10.1016/j.cej.2023.145811.

11 Yu, C., Salzano de Luna, M., Marotta, A. et al. (2021). NIR light-triggered self-healing waterborne polyurethane coatings with polydopamine-coated reduced graphene oxide nanoparticles. *Progress in Organic Coatings* 161: 106499. https://doi.org/10.1016/j.porgcoat.2021.106499.

12 Ha, Y.-M., Kim, Y.N., and Jung, Y.C. (2021). Rapid and local self-healing ability of polyurethane nanocomposites using photothermal polydopamine-coated graphene oxide triggered by near-infrared laser. *Polymers* 13: https://doi.org/10.3390/polym13081274.

13 Zhu, K., Li, Z., Cheng, F. et al. (2021). Preparation of durable superhydrophobic composite coatings with photothermal conversion precisely targeted configuration self-healability and great degradability. *Composites Science and Technology* 213: 108926. https://doi.org/10.1016/j.compscitech.2021.108926.

14 Kang, Q., Wang, W., Wang, T. et al. (2023). Self-healing protective coating with dual reversible bonds triggered by photothermal responsive $Co_9S_8@Bi_2S_3$ hollow core-shell nanofillers. *Progress in Organic Coatings* 185: 107891. https://doi.org/10.1016/j.porgcoat.2023.107891.

15 Hofmeier, H. and Schubert, U.S. (2003). Supramolecular branching and crosslinking of terpyridine-modified copolymers: complexation and decomplexation studies in diluted solution. *Macromolecular Chemistry and Physics* 204: 1391–1397. https://doi.org/10.1002/macp.200350003.

16 Pollino, J.M., Nair, K.P., Stubbs, L.P. et al. (2004). Cross-linked and functionalized 'universal polymer backbones' via simple, rapid, and orthogonal multi-site self-assembly. *Tetrahedron* 60: 7205–7215. https://doi.org/10.1016/j.tet.2004.05.055.

17 Li, B., Xue, S., Mu, P. et al. (2022). Robust self-healing graphene oxide-based superhydrophobic coatings for efficient corrosion protection of magnesium alloys. *ACS Applied Materials & Interfaces* 14: 30192–30204. https://doi.org/10.1021/acsami.2c06447.

18 Wang, T., Wang, W., Feng, H. et al. (2022). Photothermal nanofiller-based polydimethylsiloxane anticorrosion coating with multiple cyclic self-healing and long-term self-healing performance. *Chemical Engineering Journal* 446: 137077. https://doi.org/10.1016/j.cej.2022.137077.

19 Li, S., Xu, Y., Xiang, F. et al. (2023). Enhanced corrosion resistance of self-healing waterborne polyurethane coating based on tannic acid modified cerium-montmorillonites composite fillers. *Progress in Organic Coatings* 178: 107454. https://doi.org/10.1016/j.porgcoat.2023.107454.

20 Mirzakhanzadeh, Z., Kosari, A., Moayed, M.H. et al. (2018). Enhanced corrosion protection of mild steel by the synergetic effect of zinc aluminum polyphosphate and 2-mercaptobenzimidazole inhibitors incorporated in epoxy-polyamide coatings. *Corrosion Science* 138: 372–379. https://doi.org/10.1016/j.corsci.2018.04.040.

21 Carneiro, J., Tedim, J., Fernandes, S.C.M. et al. (2012). Chitosan-based self-healing protective coatings doped with cerium nitrate for corrosion protection of aluminum alloy 2024. *Progress in Organic Coatings* 75: 8–13. https://doi.org/10.1016/j.porgcoat.2012.02.012.

22 Mardel, J., Garcia, S.J., Corrigan, P.A. et al. (2011). The characterisation and performance of $Ce(dbp)_3$-inhibited epoxy coatings. *Progress in Organic Coatings* 70: 91–101. https://doi.org/10.1016/j.porgcoat.2010.10.009.

23 Hughes, A., Laird, J., Ryan, C. et al. (2017). Particle characterisation and depletion of Li_2CO_3 inhibitor in a polyurethane coating. *Coatings* 7: https://doi.org/10.3390/coatings7070106.

24 Wang, P., He, B., Wang, B. et al. (2023). MXene/metal–organic framework based composite coating with photothermal self-healing performances for antifouling application. *Chemical Engineering Journal* 474: 145835. https://doi.org/10.1016/j.cej.2023.145835.

25 Wang, J., Ma, L., Huang, Y. et al. (2022). Photothermally activated self-healing protective coating based on the "close and seal" dual-action mechanisms. *Composites Part B: Engineering* 231: 109574. https://doi.org/10.1016/j.compositesb.2021.109574.

26 Huang, Y., Liu, T., Ma, L. et al. (2022). Saline-responsive triple-action self-healing coating for intelligent corrosion control. *Materials & Design* 214: 110381. https://doi.org/10.1016/j.matdes.2022.110381.

27 Huang, Y., Wang, P., Tan, W. et al. (2022). Photothermal and pH dual-responsive self-healing coating for smart corrosion protection. *Journal of Materials Science and Technology* 107: 34–42. https://doi.org/10.1016/j.jmst.2021.08.044.

28 Fang, H., Wang, C., Zhou, S. et al. (2020). Enhanced adhesion and anticorrosion of silk fibroin coated biodegradable Mg–Zn–Ca alloy via a two-step plasma activation. *Corrosion Science* 168: 108466. https://doi.org/10.1016/j.corsci.2020.108466.

29 Wang, C., Fang, H., Hang, C. et al. (2020). Fabrication and characterization of silk fibroin coating on APTES pretreated Mg–Zn–Ca alloy. *Materials Science and Engineering: C* 110: 110742. https://doi.org/10.1016/j.msec.2020.110742.

30 Fang, H., Wang, C., Zhou, S. et al. (2021). Exploration of the enhanced performances for silk fibroin/sodium alginate composite coatings on biodegradable Mg–Zn–Ca alloy. *Journal of Magnesium and Alloys* 9: 1578–1594. https://doi.org/10.1016/j.jma.2020.08.017.

31 Li, H., Si, S., Yang, K. et al. (2023). Hexafluoroisopropanol based silk fibroin coatings on AZ31 biometals with enhanced adhesion, corrosion resistance and biocompatibility. *Progress in Organic Coatings* 184: 107881. https://doi.org/10.1016/j.porgcoat.2023.107881.

32 Wang, N., Zhang, Y., Chen, J. et al. (2017). Dopamine modified metal-organic frameworks on anti-corrosion properties of waterborne epoxy coatings. *Progress in Organic Coatings* 109: 126–134. https://doi.org/10.1016/j.porgcoat.2017.04.024.

33 Tomić, N.Z., Marinković, A.D., Veljović, Đ. et al. (2018). A new approach to compatibilization study of EVA/PMMA polymer blend used as an optical fibers adhesive: mechanical, morphological and thermal properties. *International Journal of Adhesion and Adhesives* 81: 11–20. https://doi.org/10.1016/j.ijadhadh.2017.11.002.

34 He, X.-r., Zhang, R., Chen, Q. et al. (2014). Different surface functionalized nano-Fe_3O_4 particles for EVA composite adhesives. *International Journal of Adhesion and Adhesives* 50: 128–135. https://doi.org/10.1016/j.ijadhadh.2014.01.023.

35 Kimura, I., Sekine, T., and Endo, Y. (2021). Production of calcium magnesium phosphate microspheres in a water-in-oil-in-water dispersion. *Colloids and Surfaces A: Physicochemical and Engineering Aspects* 613: 126089. https://doi.org/10.1016/j.colsurfa.2020.126089.

36 Li, D., Li, X., Chen, C. et al. (2018). Monodisperse water-in-oil-in-water emulsions generation for synthesising alginate hydrogel microspheres via locally hydrophobic modification to PMMA microchannels. *Sensors and Actuators B: Chemical* 255: 1048–1056. https://doi.org/10.1016/j.snb.2017.08.152.

37 Jie, L., Hongyan, L., and Shujun, W. (2011). Fabrication of ABS/PC alloy hollow microspheres via water/oil/water emulsion solvent evaporation. *Materials Letters* 65: 2696–2699. https://doi.org/10.1016/j.matlet.2011.06.016.

38 Qian, H., Xu, D., Du, C. et al. (2017). Dual-action smart coatings with a self-healing superhydrophobic surface and anti-corrosion properties. *Journal of Materials Chemistry A* 5: 2355–2364. https://doi.org/10.1039/C6TA10903A.

39 Bahlakeh, G., Ramezanzadeh, B., and Ramezanzadeh, M. (2017). Cerium oxide nanoparticles influences on the binding and corrosion protection characteristics of a melamine-cured polyester resin on mild steel: an experimental, density functional theory and molecular dynamics simulation study. *Corrosion Science* 118: 69–83. https://doi.org/10.1016/j.corsci.2017.01.021.

40 Bahlakeh, G., Ghaffari, M., Saeb, M.R. et al. (2016). A close-up of the effect of iron oxide type on the interfacial interaction between epoxy and carbon steel: combined molecular dynamics simulations and quantum mechanics. *The Journal of Physical Chemistry C* 120: 11014–11026. https://doi.org/10.1021/acs.jpcc.6b03133.

41 Jiao, Z.-J., Yu, C., Wang, X.-M. et al. (2023). Corrosion resistance enhanced by an atomic layer deposited Al_2O_3/micro-arc oxidation coating on magnesium alloy AZ31. *Ceramics International* https://doi.org/10.1016/j.ceramint.2023.11.309.

42 Aslam, M.S., Ahmed, N., and Khan, Z. (2024). High performance aluminized Monel coatings with prolonged corrosion resistance in saline environment. *Materials Chemistry and Physics* 313: 128695. https://doi.org/10.1016/j.matchemphys.2023.128695.

43 Guo, C., Li, Y., Qi, C. et al. (2024). Preparation and corrosion resistance characterization of MAO coating on AZ31B magnesium alloy formed in the mixed silicate and phosphate electrolytes with pectin as an additive. *Surface and Coatings Technology* 476: 130209. https://doi.org/10.1016/j.surfcoat.2023.130209.

44 Decroly, A. and Petitjean, J.-P. (2005). Study of the deposition of cerium oxide by conversion on to aluminium alloys. *Surface and Coatings Technology* 194: 1–9. https://doi.org/10.1016/j.surfcoat.2004.05.012.

45 Li, F.-B., Newman, R.C., and Thompson, G.E. (1997). In situ atomic force microscopy studies of electrodeposition mechanism of cerium oxide films: nucleation and growth out of a gel mass precursor. *Electrochimica Acta* 42: 2455–2464. https://doi.org/10.1016/S0013-4686(96)00433-1.

46 Davenport, A.J., Isaacs, H.S., and Kendig, M.W. (1991). XANES investigation of the role of cerium compounds as corrosion inhibitors for aluminum. *Corrosion Science* 32: 653–663. https://doi.org/10.1016/0010-938X(91)90113-4.

47 Afshari, M., Eivaz Mohammadloo, A., Roshan, S. et al. (2023). Mixed simultaneous barrier and passivation mechanisms by Cloisite 30B and cerium conversion coating to reinforce the epoxy coating's adhesion and anti-corrosion properties. *Colloids and Surfaces A: Physicochemical and Engineering Aspects* 675: 132048. https://doi.org/10.1016/j.colsurfa.2023.132048.

48 Zhao, Y., Li, Y., Yin, T. et al. (2023). Investigation of the post-treatment process for improving corrosion behaviors and adhesion strength of plasma electrolytic oxidation (PEO) coatings on pure copper. *Surface and Coatings Technology* 472: 129873. https://doi.org/10.1016/j.surfcoat.2023.129873.

49 Yang, H., Kang, W., Yu, Y. et al. (2017). A new approach to evaluate the particle growth and sedimentation of dispersed polymer microsphere profile control system based on multiple light scattering. *Powder Technology* 315: 477–485. https://doi.org/10.1016/j.powtec.2017.04.001.

50 Yuan, D., Chen, L., Yuan, L. et al. (2016). Superparamagnetic polymer composite microspheres supported Schiff base palladium complex: an efficient and reusable catalyst for the Suzuki coupling reactions. *Chemical Engineering Journal* 287: 241–251. https://doi.org/10.1016/j.cej.2015.10.081.

51 Dai, C., Liu, Y., Zou, C. et al. (2017). Investigation on matching relationship between dispersed particle gel (DPG) and reservoir pore-throats for in-depth profile control. *Fuel* 207: 109–120. https://doi.org/10.1016/j.fuel.2017.06.076.

52 Yang, H., Kang, W., Tang, X. et al. (2018). Gel kinetic characteristics and creep behavior of polymer microspheres based on bulk gel. *Journal of Dispersion Science and Technology* 39: 1808–1819. https://doi.org/10.1080/01932691.2018.1462192.

53 Yang, H., Zhou, B., Zhu, T. et al. (2021). Conformance control mechanism of low elastic polymer microspheres in porous medium. *Journal of Petroleum Science and Engineering* 196: 107708. https://doi.org/10.1016/j.petrol.2020.107708.

54 Wang, H., Lin, M., Dong, Z. et al. (2020). Study on the association behavior of synthesized hydrophobically associating polymer microspheres. *Colloids and Surfaces A: Physicochemical and Engineering Aspects* 598: 124829. https://doi.org/10.1016/j.colsurfa.2020.124829.

55 Yao, C., Lei, G., Li, L. et al. (2012). Selectivity of pore-scale elastic microspheres as a novel profile control and oil displacement agent. *Energy & Fuels* 26: 5092–5101. https://doi.org/10.1021/ef300689c.

56 Yang, H., Shen, J., Liu, D. et al. (2023). Stabilization mechanism of viscoelastic polymer microspheres on N_2 foam system. *Journal of Molecular Liquids* 392: 123444. https://doi.org/10.1016/j.molliq.2023.123444.

57 Kim, J., Lee, S., Park, J. et al. (2023). Preparation and properties of antibacterial polymer microspheres and their films having renewable cardanol moiety. *European Polymer Journal* 194: 112140. https://doi.org/10.1016/j.eurpolymj.2023.112140.

58 Liu, Y., Qiu, X., Zhang, M. et al. (2024). Boronate affinity-based surface molecularly imprinted polymer microspheres using polyethyleneimine/dopamine coating for efficient selective recognition and separation of Ginsenoside Rb1. *Reactive and Functional Polymers* 194: 105780. https://doi.org/10.1016/j.reactfunctpolym.2023.105780.

59 Wang, Z., Li, Z., Yan, R. et al. (2023). Facile fabrication of hollow molecularly imprinted polymer microspheres via pickering emulsion polymerization stabilized with TiO_2 nanoparticles. *Arabian Journal of Chemistry* 16: 105304. https://doi.org/10.1016/j.arabjc.2023.105304.

60 Wawrzeńczyk, J., Molendowska, A., and Kłak, A. (2016). Effect of ground granulated blast furnace slag and polymer microspheres on impermeability and freeze-thaw resistance of concrete. *Procedia Engineering* 161: 79–84. https://doi.org/10.1016/j.proeng.2016.08.501.

61 Islam, M.S., Yeum, J.H., and Das, A.K. (2012). Synthesis of poly(vinyl acetate-methyl methacrylate) copolymer microspheres using suspension polymerization. *Journal of Colloid and Interface Science* 368: 400–405. https://doi.org/10.1016/j.jcis.2011.11.002.

62 Halaciuga, I., Njagi, J.I., Redford, K. et al. (2012). Deposition of continuous nickel shells on polymer microspheres. *Journal of Colloid and Interface Science* 383: 215–221. https://doi.org/10.1016/j.jcis.2012.05.055.

63 Wang, Z.-G., Cheng, G., Liu, Y.-L. et al. (2014). Novel core–shell cerium(IV)-immobilized magnetic polymeric microspheres for selective enrichment and rapid separation of phosphopeptides. *Journal of Colloid and Interface Science* 417: 217–226. https://doi.org/10.1016/j.jcis.2013.11.004.

64 Zhao, G., Li, J., Gu, C. et al. (2018). Dispersed particle gel-strengthened polymer/surfactant as a novel combination flooding system for enhanced oil recovery. *Energy & Fuels* 32: 11317–11327. https://doi.org/10.1021/acs.energyfuels.8b02720.

65 Zhang, K., Wu, W., Meng, H. et al. (2009). Pickering emulsion polymerization: preparation of polystyrene/nano-SiO_2 composite microspheres with core-shell structure. *Powder Technology* 190: 393–400. https://doi.org/10.1016/j.powtec.2008.08.022.

66 Kang, W.-L., Hu, L.-L., Zhang, X.-F. et al. (2015). Preparation and performance of fluorescent polyacrylamide microspheres as a profile control and tracer agent. *Petroleum Science* 12: 483–491. https://doi.org/10.1007/s12182-015-0042-9.

67 Shagymgereyeva, S., Sarsenbekuly, B., Kang, W. et al. (2024). Advances of polymer microspheres and its applications for enhanced oil recovery. *Colloids and Surfaces B: Biointerfaces* 233: 113622. https://doi.org/10.1016/j.colsurfb.2023.113622.

68 Chon, C.H., Kim, J.H., On, H. et al. (2023). A microfluidic application for mass production of drug-loaded polymeric microspheres for a long-acting injectable with IVL-DrugFluidic, a novel microfluidic microsphere manufacturing platform technology. *OpenNano* 12: 100153. https://doi.org/10.1016/j.onano.2023.100153.

69 Yi, Y., Gong, J., Shi, K. et al. (2023). Isolation of antibody by polymer microspheres embedded with *E. coli* displaying IgG-binding domain. *Journal of Chromatography B* 1228: 123825. https://doi.org/10.1016/j.jchromb.2023.123825.

70 Liu, X., Li, T., Hu, C. et al. (2023). Using novel polymer microspheres as an efficient UV filter and blue light converter for constructing multifunctional transparent polyvinyl alcohol films. *Colloids and Surfaces A: Physicochemical and Engineering Aspects* 666: 131236. https://doi.org/10.1016/j.colsurfa.2023.131236.

71 Zhang, S., Tang, R., Wang, D. et al. (2021). Fabrication of highly crosslinked and monodispersed silicon-containing polymeric microspheres via photo-initiated polymerization and their application in capillary liquid chromatography. *Journal of Chromatography A* 1659: 462643. https://doi.org/10.1016/j.chroma.2021.462643.

72 Okubo, M., Iwasaki, Y., and Yamamoto, Y. (1992). Preparation of micron-size monodisperse polymer microspheres having cationic groups. *Colloid and Polymer Science* 270: 733–737. https://doi.org/10.1007/BF00776143.

73 Guo, F., Ding, Y., Wang, Y. et al. (2021). Functional monodisperse microspheres fabricated by solvothermal precipitation co-polymerization. *Chinese Journal of Chemical Engineering* 34: 323–331. https://doi.org/10.1016/j.cjche.2020.09.036.

74 Liang, R., Hu, Y., and Li, G. (2020). Monodisperse pillar[5]arene-based polymeric sub-microsphere for on-line extraction coupling with high-performance liquid chromatography to determine antioxidants in the migration of food contact materials. *Journal of Chromatography A* 1625: 461276. https://doi.org/10.1016/j.chroma.2020.461276.

75 Li, L., Liu, X., Xu, J. et al. (2019). Microfluidic preparation of thiol-containing monodisperse polymer microspheres and their adsorption of Pb^{2+} in water. *Chemical Engineering Journal* 375: 122012. https://doi.org/10.1016/j.cej.2019.122012.

76 Huang, H., Ding, Y., Chen, X. et al. (2016). Synthesis of monodisperse micron-sized poly(divinylbenzene) microspheres by solvothermal precipitation polymerization. *Chemical Engineering Journal* 289: 135–141. https://doi.org/10.1016/j.cej.2015.12.092.

77 Kim, B., Lee, Y.-H., Han, S.-J. et al. (2007). Electrorheological properties of carbon nanotubes-coated monodisperse polymeric microspheres. *Colloids and Surfaces A: Physicochemical and Engineering Aspects* 298: 245–251. https://doi.org/10.1016/j.colsurfa.2006.11.003.

78 Huang, Y., Yuan, J., Tang, J. et al. (2019). One step synthesis of monodisperse thiol-ene clickable polymer microspheres and application on biological functionalization. *European Polymer Journal* 110: 22–30. https://doi.org/10.1016/j.eurpolymj.2018.11.018.

79 Bai, F., Yang, X., Li, R. et al. (2006). Monodisperse hydrophilic polymer microspheres having carboxylic acid groups prepared by distillation precipitation polymerization. *Polymer* 47: 5775–5784. https://doi.org/10.1016/j.polymer.2006.06.014.

80 Li, C., Guo, X., and Frankel, G.S. (2021). Corrosion inhibition of AA2024-T3 by a coating containing dual-pH sensitive, corrosion inhibitor loaded microspheres. *Corrosion Science* 192: 109835. https://doi.org/10.1016/j.corsci.2021.109835.

81 Li, W., Chen, S., Zhang, L. et al. (2020). Inhalable functional mixed-polymer microspheres to enhance doxorubicin release behavior for lung cancer treatment. *Colloids and Surfaces B: Biointerfaces* 196: 111350. https://doi.org/10.1016/j.colsurfb.2020.111350.

82 Cao, Y., Yuan, X., Wang, X. et al. (2021). Synthesis and controlled release kinetics of pH-sensitive hollow polyaniline microspheres encapsuled with the corrosion inhibitor. *Journal of Molecular Liquids* 342: 117497. https://doi.org/10.1016/j.molliq.2021.117497.

83 Li, L., Yao, J., Xiao, P. et al. (2013). One-step fabrication of ZIF-8/polymer composite spheres by a phase inversion method for gas adsorption. *Colloid and Polymer Science* 291: 2711–2717. https://doi.org/10.1007/s00396-013-3024-8.

84 Zeng, Y., Wang, K., Yao, J. et al. (2014). Hollow carbon beads fabricated by phase inversion method for efficient oil sorption. *Carbon* 69: 25–31. https://doi.org/10.1016/j.carbon.2013.11.036.

85 Liu, M. and Xue, D. (2007). Effect of heating rate on the crystal composition of ferroelectric lithium niobate crystallites. *Journal of Alloys and Compounds* 427: 256–259. https://doi.org/10.1016/j.jallcom.2006.02.065.

86 Chen, F., Luo, G.S., and Wang, Y.J. (2004). Preparation of polysulfone-coated chitosan beads by emulsion phase-inversion method. *Journal of Applied Polymer Science* 91: 3542–3548. https://doi.org/10.1002/app.13577.

87 Wang, P., Ma, J., Shi, F. et al. (2013). Behaviors and effects of differing dimensional nanomaterials in water filtration membranes through the classical phase inversion process: a review. *Industrial & Engineering Chemistry Research* 52: 10355–10363. https://doi.org/10.1021/ie303289k.

88 Astakhov, E.Y., Zhironkin, S.F., Kolganov, I.M. et al. (2011). Study of the formation of the porous structure of membranes during phase separation of a poly(ester sulfone) solution. *Polymer Science, Series A* 53: 613–620. https://doi.org/10.1134/S0965545X11070017.

89 Duraikkannu, S.L., Castro-Muñoz, R., and Figoli, A. (2021). A review on phase-inversion technique-based polymer microsphere fabrication. *Colloid and Interface Science Communications* 40: 100329. https://doi.org/10.1016/j.colcom.2020.100329.

90 Li, W., Tao, J., Chen, Y. et al. (2023). Porous microspheres with corrosion sensing and active protecting abilities towards intelligent self-reporting and anti-corrosion coating. *Progress in Organic Coatings* 178: 107468. https://doi.org/10.1016/j.porgcoat.2023.107468.

91 Zhao, D., Wang, M., Xu, Y. et al. (2014). The fabrication and corrosion resistance of benzotriazole-loaded raspberry-like hollow polymeric microspheres. *Surface and Coatings Technology* 238: 15–26. https://doi.org/10.1016/j.surfcoat.2013.10.031.

92 Yu, T., Zhao, Y., Zheng, P. et al. (2021). Ultra-durable superhydrophobic surfaces from 3D self-similar network via co-spraying of polymer microspheres and nanoparticles. *Chemical Engineering Journal* 410: 128314. https://doi.org/10.1016/j.cej.2020.128314.

93 Liu, Y., Tang, P., Pu, H. et al. (2020). Study on the synthesis and drug-loading optimization of beta-cyclodextrin polymer microspheres containing ornidazole. *Journal of Drug Delivery Science and Technology* 58: 101836. https://doi.org/10.1016/j.jddst.2020.101836.

94 Bao, J., Zhang, H., Zhao, X. et al. (2018). Biomass polymeric microspheres containing aldehyde groups: immobilizing and controlled-releasing amino acids as green metal corrosion inhibitor. *Chemical Engineering Journal* 341: 146–156. https://doi.org/10.1016/j.cej.2018.02.047.

95 Chen, J., Shi, D., Wu, Q. et al. (2023). Magnetically-separable quasi-homogeneous catalyst: brush-type ionic liquid polymer coated magnetic polymer microspheres for tandem reactions to produce 4H-pyrans/biodiesel. *Colloids and Surfaces A: Physicochemical and Engineering Aspects* 665: 131209. https://doi.org/10.1016/j.colsurfa.2023.131209.

96 Huang, Y., Deng, L., Ju, P. et al. (2018). Triple-action self-healing protective coatings based on shape memory polymers containing dual-function microspheres. *ACS Applied Materials & Interfaces* 10 (27): 23369–23379. https://doi.org/10.1021/acsami.8b06985.

97 Sun, C., Liu, C., Wang, Y. et al. (2024). Corrosion sensing and self-healing composite coatings on magnesium alloy AZ31 enabled by stimuli-responsive microcapsules loaded with phenolphthalein and epoxy resin. *Progress in Organic Coatings* 186: 108043. https://doi.org/10.1016/j.porgcoat.2023.108043.

98 Li, J., Qian, S., Bian, D. et al. (2023). Effect of heating self-healing and construction of composite structure on corrosion resistance of superhydrophobic phosphate ceramic coating in long-term water environment. *Ceramics International* https://doi.org/10.1016/j.ceramint.2023.11.177.

99 Ren, A., Kang, M., Fu, X. et al. (2024). Preparation of Ni-B composite coating doped with hydrothermally modified synthetic WC@MoS_2 powders via ultrasonic-assisted jet electrodeposition for corrosion protection. *Surfaces and Interfaces* 44: 103674. https://doi.org/10.1016/j.surfin.2023.103674.

100 Marhamati, F., Ghaderi, M., Haddadi, S.A. et al. (2023). Nitrogen-doped carbon hollow sphere capsules for a novel hybrid inhibitor based on lanthanum cations and 8-hydroxyquinoline: synthesis, characterization, and self-healing properties in epoxy coating. *Colloids and Surfaces A: Physicochemical and Engineering Aspects* 675: 132065. https://doi.org/10.1016/j.colsurfa.2023.132065.

101 Chen, Q., Xu, Y., Ma, A. et al. (2023). Study of the passivation and repassivation behavior of pure titanium in 3.5 wt% NaCl solution and 6 M HNO_3 solution. *Corrosion Science* 224: 111538. https://doi.org/10.1016/j.corsci.2023.111538.

102 Song, Q.N., Wang, Y., Wu, Y.Q. et al. (2023). Effect of pre-corrosion on the cavitation erosion performance of two aluminum bronzes in 3.5% NaCl solution. *Materials Today Communications* 37: 107265. https://doi.org/10.1016/j.mtcomm.2023.107265.

103 Zhang, D., Yan, Z., Gao, L. et al. (2024). Corrosion behavior of AA5052 aluminum alloy in the presence of heavy metal ions in 3.5% NaCl solution under negative pressure. *Desalination* 570: 117082. https://doi.org/10.1016/j.desal.2023.117082.

104 Wang, W., Yi, P., Mo, J. et al. (2023). Acid resistance and failure mechanism of zinc-rich primer/micaceous iron oxide/polyurethane composite coating immersed in 5 wt% sulfuric acid. *International Journal of Electrochemical Science* 18: 100205. https://doi.org/10.1016/j.ijoes.2023.100205.

105 Sørensen, P.A., Dam-Johansen, K., Weinell, C.E. et al. (2010). Cathodic delamination of seawater-immersed anticorrosive coatings: mapping of parameters affecting the rate. *Progress in Organic Coatings* 68: 283–292. https://doi.org/10.1016/j.porgcoat.2010.03.012.

106 He, Y. and Zhou, B. (2021). The influence of mica iron oxide pigments on epoxy coating properties prepared on AISI 1045 steel rebar immersed in 3.5 wt% NaCl solution. *International Journal of Electrochemical Science* 16: 210726. https://doi.org/10.20964/2021.07.41.

107 Wei, Z., Hong, S., Wei, Z. et al. (2022). Comparison on long-term corrosion performance of WC-CoCr and Al_2O_3-TiO_2 ceramic coatings in sulphide-containing 3.5 wt% NaCl solution. *International Journal of Refractory Metals and Hard Materials* 107: 105906. https://doi.org/10.1016/j.ijrmhm.2022.105906.

108 Bai, W., Zhang, X., He, Z. et al. (2023). Intelligent anti-corrosion coating with multiple protections using active nanocontainers of ZnAl LDH equipped with ZIF-8 encapsulated environment-friendly corrosion inhibitors. *Progress in Organic Coatings* 185: 107940. https://doi.org/10.1016/j.porgcoat.2023.107940.

109 Chengwen, S., Cheng, W., and Dejun, K. (2024). Tribo–corrosive performance of laser cladded AlFeCoCr–*x*TiC coatings in 3.5% NaCl solution. *Materials Letters* 356: 135577. https://doi.org/10.1016/j.matlet.2023.135577.

110 Li, J., Wang, L., Bai, H. et al. (2023). Development of an eco-friendly waterborne polyurethane/catecholamine/sol-gel composite coating for achieving long-lasting corrosion protection on Mg alloy AZ31. *Progress in Organic Coatings* 183: 107732. https://doi.org/10.1016/j.porgcoat.2023.107732.

111 Cheng, J., Wu, Y., Zhu, S. et al. (2023). The coupling effect of cavitation-erosion and corrosion for HVOF sprayed Cu-based medium-entropy alloy coating in 3.5 wt.% NaCl solution. *Journal of Materials Research and Technology* 25: 2936–2947. https://doi.org/10.1016/j.jmrt.2023.06.109.

112 Feng, J., Wang, Y., Lin, X. et al. (2023). SECM in situ investigation of corrosion and self-healing behavior of trivalent chromium conversion coating on the zinc.

Surface and Coatings Technology 459: 129411. https://doi.org/10.1016/j.surfcoat.2023.129411.

113 Trinh, D., Vosgien-Lacombre, C., Bouvet, G. et al. (2020). Use of ionic liquids in SECM experiments to distinguish effects of temperature and water in organic coating swelling. *Progress in Organic Coatings* 139: 105438. https://doi.org/10.1016/j.porgcoat.2019.105438.

114 Hou, Z.-z., Yue, Q.-x., Lv, J. et al. (2022). Research on the anti-corrosion performance of organosilane modified Basalt scales epoxy coating by EIS and SECM. *International Journal of Electrochemical Science* 17: 220327. https://doi.org/10.20964/2022.03.12.

115 Xia, D.-H., Wang, J., Wu, Z. et al. (2019). Sensing corrosion within an artificial defect in organic coating using SECM. *Sensors and Actuators B: Chemical* 280: 235–242. https://doi.org/10.1016/j.snb.2018.10.051.

116 González-García, Y., García, S.J., Hughes, A.E. et al. (2011). A combined redox-competition and negative-feedback SECM study of self-healing anti-corrosive coatings. *Electrochemistry Communications* 13: 1094–1097. https://doi.org/10.1016/j.elecom.2011.07.009.

117 Galai, M., Dahmani, K., Kharbouch, O. et al. (2024). Surface analysis and interface properties of a newly synthesized quinoline-derivative corrosion inhibitor for mild steel in acid pickling bath: mechanistic exploration through electrochemical, XPS, AFM, contact angle, SEM/EDS, and computational studies. *Journal of Physics and Chemistry of Solids* 184: 111681. https://doi.org/10.1016/j.jpcs.2023.111681.

118 Fakhry, H., El Faydy, M., Benhiba, F. et al. (2021). A newly synthesized quinoline derivative as corrosion inhibitor for mild steel in molar acid medium: characterization (SEM/EDS), experimental and theoretical approach. *Colloids and Surfaces A: Physicochemical and Engineering Aspects* 610: 125746. https://doi.org/10.1016/j.colsurfa.2020.125746.

119 Garcia, S.J., Markley, T.A., Mol, J.M.C. et al. (2013). Unravelling the corrosion inhibition mechanisms of bi-functional inhibitors by EIS and SEM–EDS. *Corrosion Science* 69: 346–358. https://doi.org/10.1016/j.corsci.2012.12.018.

3

Development of Photothermally Activated Self-healing Coating

3.1 Introduction

Photothermal heating is a convenient method for healing the localized damage on coatings. Self-healing coatings triggered by photothermal effects have the advantages of high efficiency, high precision, and long-distance repair [1–3]. The development of self-healing coating based on the photothermal effect is of great significance for smart coatings with high precision and long-distance healing ability [4–6]. In this chapter, the photothermal conversion species used in self-healing coatings, such as graphene oxide (GO), titanium nitride, and Fe_3O_4 are introduced. These species are incorporated into micro/nanocontainers and loaded into the coating matrix. The photothermal conversion performance of different coatings has been investigated, and the coating preparation, filler preparation, surface characterization, macroscopic and microscopic electrochemical properties, and self-healing performance were studied. The advantages and disadvantages of the photothermally activated self-healing coatings are also discussed.

3.2 Self-healing Coating Based on GO–CeO_2 Photothermal Fillers

As a two-dimensional single-layer carbon material, graphene has excellent properties, including a large specific surface area, high mechanical strength, good toughness, and good barrier properties [7–9]. As such, it can improve the mechanical and anticorrosion properties of coatings [10–12]. The addition of graphene with heat absorption properties to thermally stimulated self-healing coatings can improve their heat absorption ability, and promote the chain diffusion and re-entanglement of the coating material, thereby promoting the healing of coating defects [13, 14].

However, the strong van der Waals forces, π–π double bonds, and high specific surface area of graphene sheets increases their polymerization and agglomeration into the polymer matrix [15–17]. Fortunately, GO can effectively address this problem. In particular, GO has several hydrophilic groups in its sheet structure, which can be easily ionized to form a negative electric repulsion, increasing its solubility and dispersion in water [18–20]. Although the unique structure of graphene sheets

Smart Protective Coatings for Corrosion Control, First Edition. Lingwei Ma and Dawei Zhang.

promotes van der Waals forces between graphene sheets and reduces their agglomeration, the six-membered ring C structure in GO accounts for a large proportion and poor dispersion in polymer resins. Moreover, several O-containing functional groups on the GO surface make it easy to be modified, thereby improving the dispersion of GO in coating materials, such as epoxy resins, and realizing the functionalization of GO [21, 22].

In this section, nano-cerium oxide (CeO_2) particles were used as corrosion inhibitors to disperse GO in a shape-memory coating with a photothermal self-healing effect. The dispersion, mechanical properties, barrier properties, and photothermal self-healing effects of GO loaded with nano-CeO_2 particles in the coating were characterized and analyzed. The healing status of the damaged coatings after near-infrared (NIR) light irradiation was studied, providing new data for the exploration and development of self-healing coatings for corrosion and protection applications.

(1) GO loaded with a nano-CeO_2 corrosion inhibitor (GO–CeO_2) was prepared, and its morphology and composition were characterized. The effects of different pH values, titration rates, and reaction times on the morphology of GO–CeO_2 and size of the CeO_2 nanoparticles were studied. Finally, the release of the corrosion inhibitor from the prepared GO–CeO_2 was analyzed under different conditions.
(2) Blank epoxy resin and epoxy coatings with GO and GO–CeO_2 additives were prepared. The dispersion of GO–CeO_2 in the coating was better than that of the GO-loaded coating. The mechanical properties, such as hardness and wear resistance, were tested for all the three coatings. The corrosion resistance of the coatings was tested by electrochemical impedance spectroscopy (EIS). The endothermic and self-healing effects of the three coatings under the same photothermal conditions were compared, and their self-healing properties were characterized and analyzed.

3.2.1 Materials and Methods

3.2.1.1 Materials

The materials and regents used in this section are shown in Table 3.1.

3.2.1.2 Instruments

The instruments used in this section are shown in Table 3.2.

3.2.1.3 Preparation of GO–CeO_2

First, 50 mg GO powder was dispersed in 100 ml deionized (DI) water under stirring and placed in an ultrasonic dispersion instrument at a power of 100 W for 30 minutes to obtain a uniformly dispersed GO suspension [23]. Meanwhile, 100 mg $Ce(NO_3)_3 \cdot 6H_2O$ was dissolved in 100 ml DI water under stirring until precipitation was not observed in the solution. The GO suspension and $Ce(NO_3)_3$ solution were then mixed, and the pH was titrated to 3 using a dilute HCl solution. In an acidic environment, the Ce^{3+} ions can form cation–π bonds with the O functional groups

Table 3.1 Experimental materials and reagents.

Materials	Purity	Company
$Ce(NO_3)_3{\cdot}6H_2O$	AR	Shandong Xiya
Graphene oxide (GO)	99.0%	Suzhou Tanfeng
Hydrogen peroxide (H_2O_2)	AR	Sinopharm
NaOH	AR	Aladdin
HCl	AR	Sinopharm
Ethyl alcohol (C_2H_5OH)	AR	Sinopharm
Bisphenol A diglycidyl ether (DGEBA)	AR	Aladdin
Neopentyl glycol diglycidyl ether (NGDE)	AR	Aladdin
Jeffamine D230 (D230)	AR	Aladdin
N,*N*-Dimethylformamide (DMF)	AR	Aladdin
Sodium chloride (NaCl)	AR	Sinopharm
Acetone (CH_3COCH_3)	AR	Sinopharm
AA2024-T3 aluminum alloy	$50 \times 60 \times 0.5\ mm^3$	Merchant

Table 3.2 The instruments and equipment.

Instrument	Production type	Manufacturer
Scanning electron microscopy (SEM)	FE-SEM G300	Merlin, ZEISS
Transmission electron microscopy (TEM)	JEM-2100F	JEM-2100F, JEOL
Oil bath with magnetic stirrer	ZNCL-GS	Jieang
pH meter	PHS-3C	Leici
Centrifuge	Allegra X-30	Beckman
Freeze dryer	Lab-1D-50	Boyikang
X-ray diffraction (XRD)	Ultima IV	Rigaku
X-ray photoelectron spectroscopy (XPS)	Escalab 250Xi	Thermo Scientific
Confocal laser scanning microscopy (CLSM)	VK-X2000 series	KEYENCE
Nano-indentation tester	U9820A Nano Indenter G200	Agilent
Simultaneous thermal analyzer	STA449F3	Netzsch
Electrochemical workstation	PARSTAT-2273	Princeton
Salt spraying tester	CK/YWX-90C	Marit
Inductive coupled plasma emission spectrometer (ICP)	Gilent 7700	Agilent
Air blowing thermostatic oven	303-3S	Baihui

on the GO surface. In addition, the negatively charged carboxyl groups on the surface of the GO sheet can combine with the Ce^{3+} ions via electrostatic adsorption [24]. The mixture was placed in an ultrasonic dispersion instrument for ultrasonic shocking for 30 minutes to form a uniform GO–Ce^{3+} suspension. Subsequently, 50 mg of a 33 wt% H_2O_2 solution was added to the suspension and subjected to ultrasound for 15 minutes to complete the oxidation of the Ce^{3+} ion to form the GO–Ce^{4+} suspension [25]. Then, 0.5 wt% NaOH solution was added slowly to adjust the pH value of suspension to 9 to complete the replacement process of OH^- ion and NO_3^- ion, forming the GO–$Ce(OH)_2^{2+}$ suspension [26]. The mixture was then transferred to a Teflon stainless-steel autoclave, and the hydrothermal reaction was carried out at 180 °C for two hours. The obtained GO–CeO_2 composite was centrifuged and washed thrice with a mixture of 1 : 3 anhydrous ethanol and DI water. Finally, GO–CeO_2 was freeze-dried in a freeze dryer for 24 hours.

3.2.1.4 Morphological and Structural Characterization of GO–CeO_2

The morphology, elemental composition, and structure of GO and GO–CeO_2 were characterized by SEM, TEM, XRD, and XPS.

3.2.1.5 Surface Treatment of the Substrates

A 2024-T3 aluminum alloy was selected as the substrate. Before applying the coating, the alloy sheet was polished until it exhibited a metallic luster and cleaned with acetone to remove oil. Subsequently, it was completely immersed in anhydrous ethanol and ultrasonically cleaned for five minutes to remove surface impurities. After three cleaning cycles, the cleaned AA2024-T3 sheets were completely immersed in anhydrous ethanol to prevent surface oxidation [27, 28].

3.2.1.6 Preparation of Composite Coatings

(1) Preparation of blank epoxy coating: DGEBA was heated at 45–50 °C for 15 minutes to melt it. NGDE was added to the molten DGEBA and evenly mixed. Subsequently, D230 was added to the mixture, and an evenly mixed liquid was obtained after shaking and stirring. DGEBA, NGDE, and D230 were mixed at a molar ratio of 1 : 1 : 1. Finally, the epoxy coating was evenly coated on the metal substrate with a rod applicator and cured in the oven at 80 °C for 24 hours to obtain the coating sample.
(2) Preparation of epoxy coatings containing GO and GO–CeO_2: GO and GO–CeO_2 sheets were separately added to dimethylformamide (DMF) and ultrasonically dispersed to obtain uniformly dispersed solutions. Subsequently, DGEBA was heated at 50 °C for 15 minutes to melt it, and then added into the fully dispersed GO and GO–CeO_2 solution. The mixture was placed in an ultrasonic cleaning instrument for 15 minutes for an even dispersion. DMF was evaporated from the mixture using a rotary evaporator. Subsequently, the mixture was combined with NGDE and D230 to obtain a coating solution. DGEBA, NGDE, and D230 were used at a molar ratio of 1 : 1 : 1, and 1 wt% GO and 1 wt% GO–CeO_2 were added to the coating, respectively. Finally, the epoxy coating was evenly coated on the AA2024-T3 substrate with a rod applicator and cured in the oven at 80 °C for

24 hours to obtain the coating samples. The obtained composite coatings were referred to as GO/epoxy and GO–CeO_2/epoxy coatings.

3.2.1.7 Characterization of Coatings

The compositions and structures of the coatings were measured by XRD. The thermal properties of the coatings were evaluated using differential scanning calorimetry (DSC) under N_2 protection. The coating morphologies were measured using confocal laser scanning microscopy and SEM. Hardness and elastic modulus were measured using a nanoindentation tester.

3.2.1.8 Photothermal Conversion of GO–CeO_2/Epoxy Coating

The coated specimen was placed on a white polytetrafluoroethylene board with a thermometer installed at the center of the back surface of each specimen. An NIR lamp with a power of 300 W was placed at 200 mm above the sample. The temperature changes were recorded for 30 minutes at intervals of 2 seconds. After photothermal self-healing, the coatings were placed in a salt spray at an angle of 30° ± 5° from the vertical surface. The concentration of the NaCl solution was 5 ± 0.5 wt%. The EIS data of the coatings were tested under different test times.

3.2.2 Characterization and Property of GO–CeO_2 Composites

The combination of inorganic nanomaterials and GO improves the mechanical properties of GO and increases the layer spacing of GO sheets to improve their dispersion [29]. Herein, the GO–CeO_2 filler was prepared by a hydrothermal method. The precursor chemical substances for the preparation of CeO_2 nanoparticles and a well-dispersed GO aqueous solution react at high temperature and pressure to achieve the successful loading of CeO_2 nanoparticles on the GO sheet. During the synthesis of GO–CeO_2, the desired product was obtained by the combination of Ce^{3+} ions with GO, the oxidation of Ce^{3+} to Ce^{4+}, and the formation of $Ce(OH)_2{}^{2+}$. The preparation process is shown in Figure 3.1.

3.2.2.1 Morphological and Structural Characterization of GO–CeO_2

SEM images of the GO and GO–CeO_2 sheets are shown in Figure 3.2. The GO sheet exhibits a lamellar structure with minimal wrinkles, which is a typical microstructural characteristic of graphene. After the hydrothermal reaction, the surface roughness of the GO–CeO_2 sheet exceeded that of pure GO sheet, and its thickness was significantly increased owing to the successful loading of the CeO_2 layer. In addition, the two-dimensional structure of the GO and GO–CeO_2 sheets were ~20 μm, indicating that the combination of the GO sheet and CeO_2 nanoparticles did not affect the size of GO sheet [30, 31].

Figure 3.3 exhibits a partially enlarged drawing of a GO–CeO_2 sheet. The surface of GO–CeO_2 sheet was covered with several small particles. In addition, the EDS mapping results in Figure 3.3b show that the high Ce content on the surface of the GO–CeO_2 sheet, verifying the successful synthesis and loading of the CeO_2 particles [32, 33].

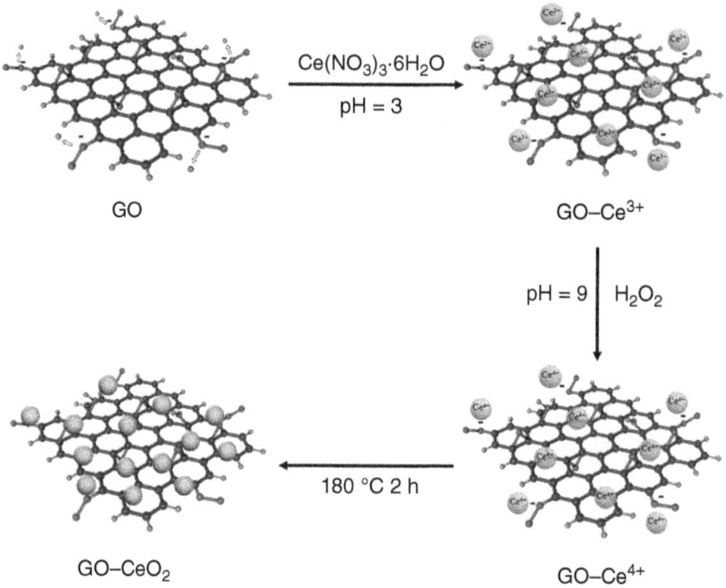

Figure 3.1 The schematic diagram of GO–CeO_2 preparation.

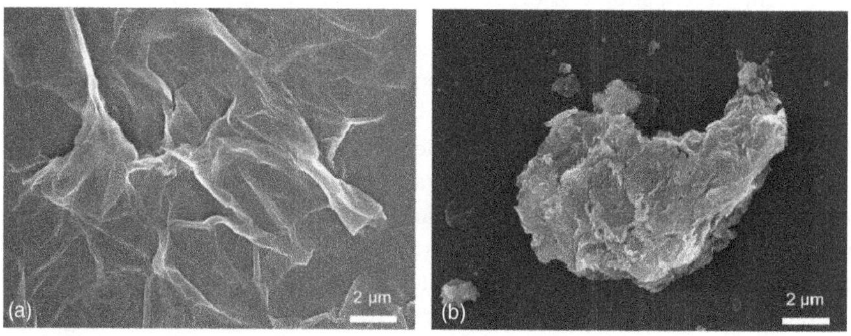

Figure 3.2 SEM images of (a) GO sheet and (b) GO–CeO_2 sheet.

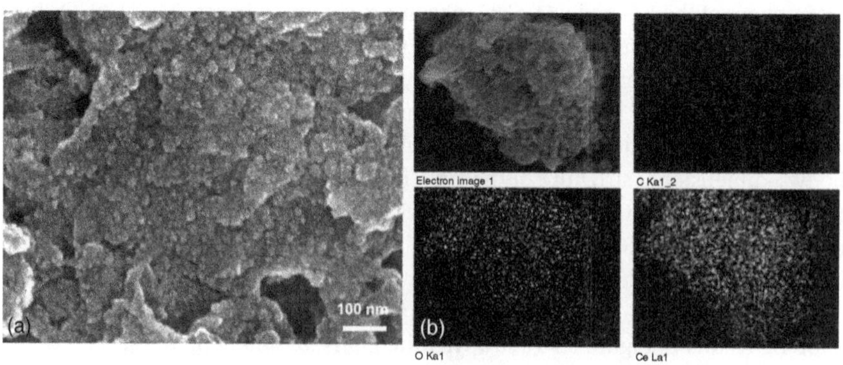

Figure 3.3 (a) The enlarged SEM images and (b) EDS mapping results of GO–CeO_2 sheet.

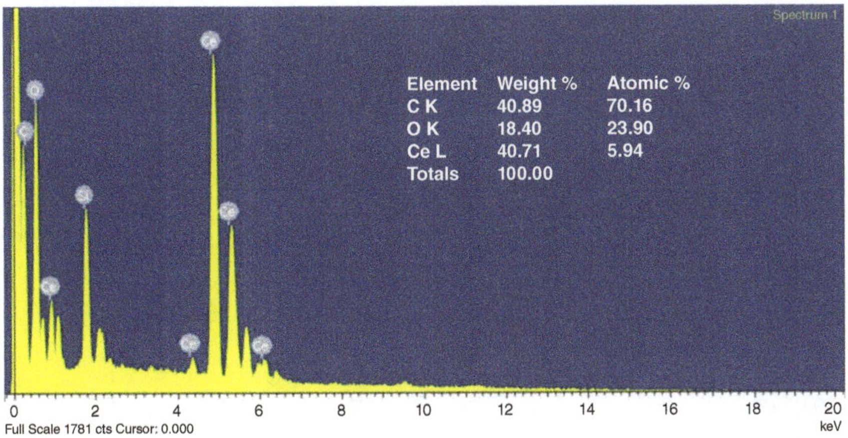

Figure 3.4 The element content analysis of GO–CeO_2 sheet.

Figure 3.4 shows the elemental composition of GO–CeO_2. C, O, and Ce have atomic contents of 69.22%, 24.68%, and 6.10%, respectively. The high Ce content confirmed the successful formation of CeO_2 on the GO sheet.

As shown in Figure 3.5, the TEM images of the GO–CeO_2 sheet provided high-resolution morphological information on the lamellar structure of GO and size of the CeO_2 particles. Several CeO_2 nanoparticles with diameters of ~10 nm was well distributed on the GO sheet, which can be explained by the chemical combination between the O functional groups of the GO sheet and CeO_2 particles. In addition, the interplanar spacing of approximately 0.3 nm was attributed to the (111) plane of CeO_2 [34, 35].

The crystalline structures of GO and GO–CeO_2 sheets were evaluated by XRD. As shown in Figure 3.6a, a sharp diffraction peak is observed at 2θ of ~11.3°, corresponding to the (002) plane of GO [36]. Moreover, a weak characteristic peak is noted at 2θ of ~42.3°, representing a short-range ordered arrangement of stacked GO sheets [37]. In contrast, four characteristic peaks at 2θ of approximately 28.5°, 33.1°,

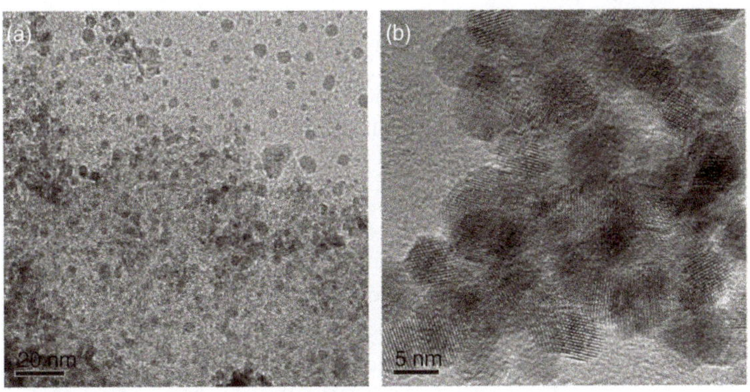

Figure 3.5 TEM images of GO–CeO_2 sheet.

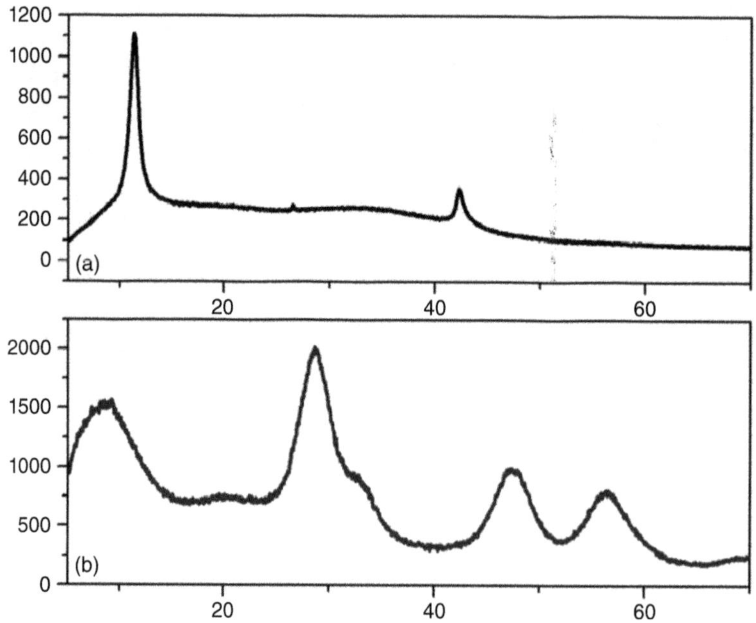

Figure 3.6 XRD patterns of (a) GO and (b) GO–CeO_2 sheet [12].

47.5°, and 56.3°, which are consistent with the characteristic peaks of the (111), (200), (220), and (311) crystal planes of CeO_2 in the Joint Committee on Powder Diffraction Standard (JCPDS) card [38], respectively, are noted in Figure 3.6b. This proved that the nanoparticles on the surface of the GO sheets were CeO_2. The characteristic peak of GO at 2θ of ~11.6° shifted to ~8.4°, indicating that the presence of CeO_2 nanoparticles increased the spacing between GO sheets [39, 40].

The chemical compositions and oxidation states of Ce in GO–CeO_2 were analyzed using XPS. Figure 3.7 shows the overall XPS survey of the GO–CeO_2 composite. The characteristic peak of Ce confirmed the presence of Ce on the GO surface [41, 42]. High-resolution Ce 3d was analyzed at 875–925 eV and its energy segment was decomposed into eight peaks, as shown in Figure 3.8. The eight peaks were divided into u and v groups, representing the $3d_{3/2}$ and $3d_{5/2}$ orbits, respectively [43, 44]. Peaks u, u″, u‴, v, v″, and v‴ represent Ce^{4+} ions, whereas u′ and v′ represent the Ce^{3+} ions. The u′ and v′ peaks have lower intensity than the other peaks, indicating the existence of Ce in the quaternary valence state in GO–CeO_2 composites.

3.2.2.2 Dispersion Properties of GO–CeO_2

As shown in Figure 3.9a, GO and GO–CeO_2 exhibit good dispersion in anhydrous ethanol after ultrasonic dispersion; no substance exhibits agglomeration or stratification in anhydrous ethanol. After standing for 12 hours, an obvious delamination phenomenon was noted in the anhydrous ethanol solution of GO, owing to its high specific surface area and interlayer forces; thus, GO was completely deposited at the bottom of the solution (Figure 3.9b). The GO–CeO_2 in ethanol solution did not

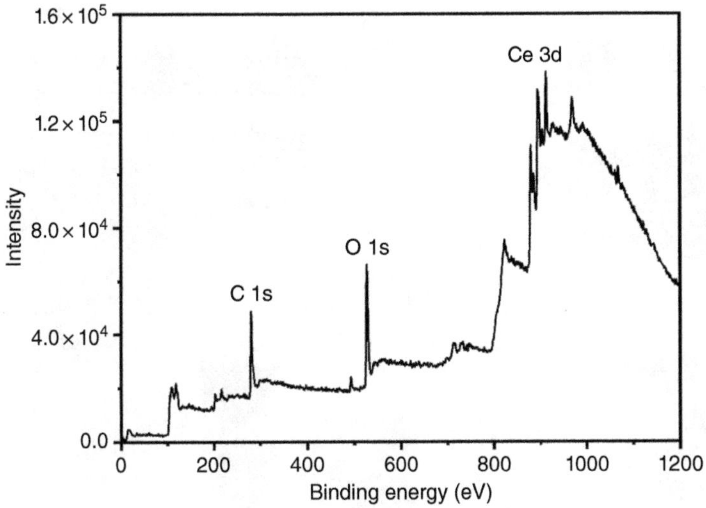

Figure 3.7 Overall XPS survey of GO–CeO_2 sheet. Source: Ma et al. [12]/with permission of Springer Nature.

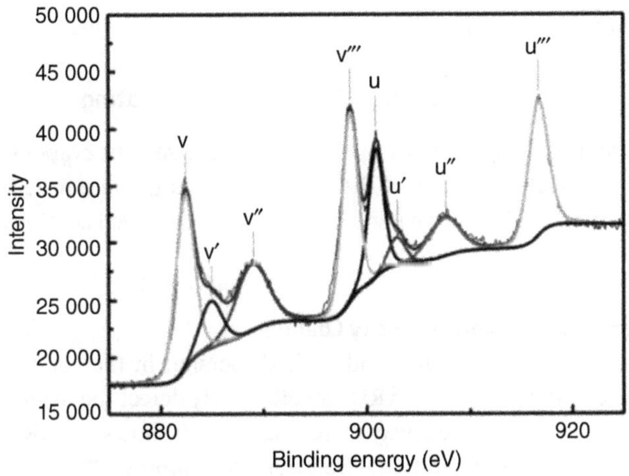

Figure 3.8 XPS spectra of Ce $3d_{5/2}$ and Ce $3d_{3/2}$ of GO–CeO_2 sheet. Source: Ma et al. [12]/with permission of Springer Nature.

exhibit a delamination phenomenon. After 48 hours of standing, the GO–CeO_2 in ethanol solution did not stratify, and GO–CeO_2 composite was uniformly dispersed in anhydrous ethanol. This is mainly attributed to the loading of CeO_2 nanoparticles on the GO surface, neutralizing the negative charge of the GO sheet and eliminating the original strong electrostatic attraction between the GO sheets. Therefore, the agglomeration between the GO sheets significantly improved without electrostatic attraction [45].

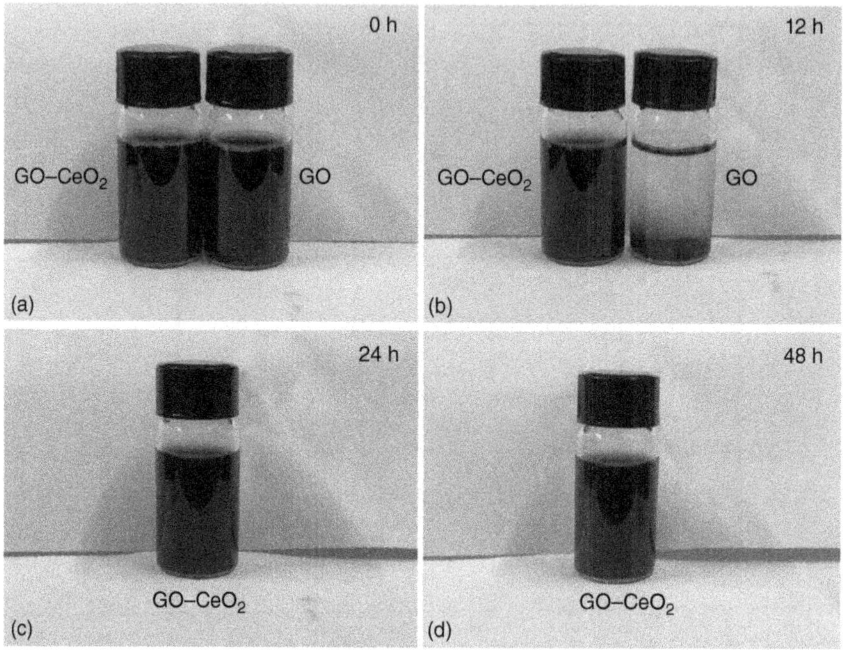

Figure 3.9 Dispersion performance of GO and GO-CeO_2 sheets after different time.

3.2.3 Characterization and Properties of the GO-CeO_2/Epoxy Coating

GO–CeO_2 composites were introduced into the epoxy resin by the solvent evaporation method, and the mechanical properties of the prepared coatings were tested. The shape-memory recovery and self-healing performance of the damaged coatings with different fillers were analyzed and compared.

3.2.3.1 Dispersion Properties of GO-CeO_2 in Epoxy Coating

The degree of stripping of GO and GO–CeO_2 and their dispersion in the coating have critical effects on the coating properties. XRD can effectively detect the degree of stripping of GO and GO–CeO_2 from the epoxy resin. Figure 3.10 shows the XRD patterns of the blank epoxy, GO/epoxy, and GO–CeO_2/epoxy coatings. The three coatings only have a wide peak at approximately 18.5° in the 2θ range from 10° to 30°, which is a characteristic peak of cured amorphous epoxy resin [46]. The characteristic peaks of GO and GO–CeO_2 were located at 2θ of 11.3° and 8.4°, respectively, whereas that characteristic peak of graphite was at 2θ of 26.6°, indicating the absence of GO reduction. No other characteristic peaks were observed in the XRD spectra, except for an obvious peak. This indicates the complete stripping of GO and GO–CeO_2 in the coating and the absence of GO reduction during solvent evaporation [47].

Although GO exhibits a high degree of stripping in epoxy resins, it is prone to agglomeration in uncured epoxy coatings under the van der Waals and Coulomb forces. The optical stereoscopic microscopy images of the coatings were obtained to

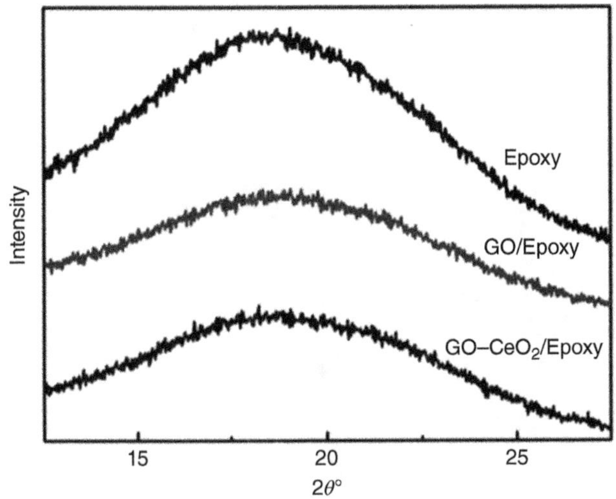

Figure 3.10 XRD patterns of three coatings.

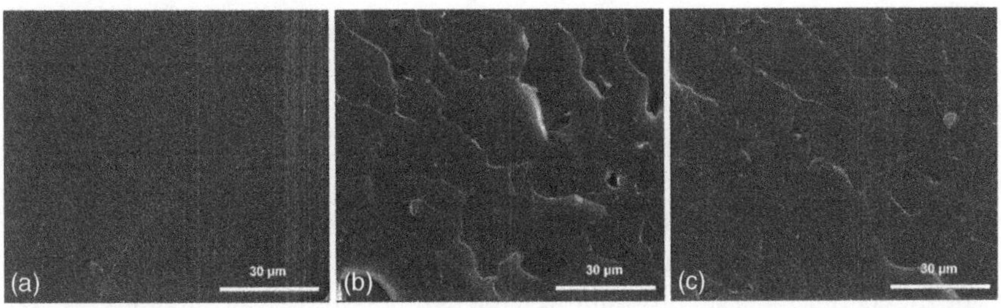

Figure 3.11 Surface morphology of different coatings: (a) blank epoxy, (b) GO/epoxy, (c) GO–CeO_2/epoxy.

evaluate the dispersion of GO and GO–CeO_2 in the epoxy resin coating. As shown in Figure 3.11b, several spots of different sizes were distributed throughout the GO/epoxy coating, indicating the serious agglomeration of a large amount of GO in the epoxy resin. In contrast, the GO–CeO_2/epoxy coating had minimal GO agglomeration spots, indicating the uniform dispersion of stripped GO sheets within the coating with a small number of black clumps inevitably scattered throughout the coating [48] (Figure 3.11c). The results indicated the better dispersion performance of the GO–CeO_2 sheets in the polymer than that of GO. This could be attributed to the increased spacing between the GO layers owing to the steric hindrance of CeO_2 on the GO sheets, greatly weakening the mutual attraction between the GO sheets and inhibiting the agglomeration of GO [49, 50].

3.2.3.2 Mechanical Properties of GO–CeO_2/Epoxy Coating

The effect of adding GO and GO–CeO_2 sheet on the mechanical properties of the epoxy matrix was investigated by obtaining the elastic modulus and hardness of

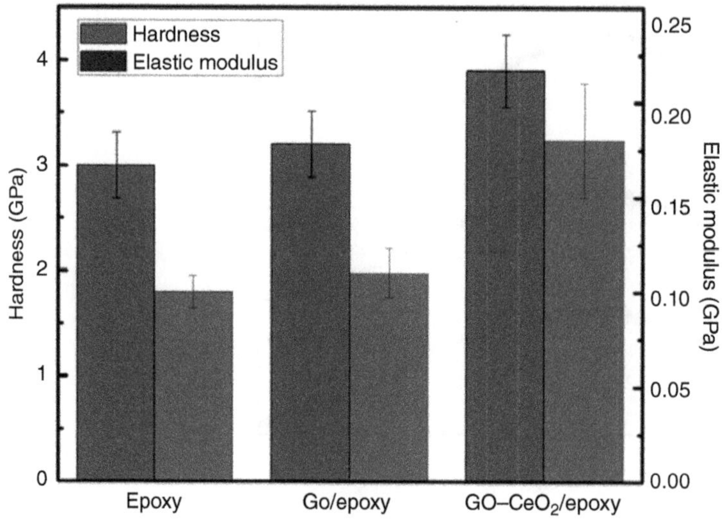

Figure 3.12 Hardness and elastic modulus of different coatings.

the blank epoxy, GO/epoxy, and GO–CeO_2/epoxy coatings through nanoindentation measurements. As shown in Figure 3.12, the hardness values of the blank epoxy, GO/epoxy, and GO–CeO_2/epoxy coatings are 3.0, 3.2, and 3.7 GPa, respectively. The elastic modulus of the blank epoxy, GO/epoxy, and GO–CeO_2/epoxy coatings are 0.10, 0.11, and 0.18 GPa, respectively. Although various studies have noted that the introduction of GO significantly improves the mechanical properties of coatings, the unsatisfactory dispersion of GO in the coating, caused by the aggregation of GO inside the coating, only slightly improved the hardness and elastic modulus [51, 52]. The notable improvements in the mechanical properties of GO–CeO_2/epoxy coating can be ascribed to the highly dispersed GO–CeO_2 composite material compensating for common defects, such as holes, inside the coating, realizing the reconnection between the coating holes and matrix.

3.2.3.3 Photothermal Conversion Properties of GO–CeO_2/Epoxy Coating

Figure 3.13 shows the DSC curves of the blank epoxy and GO–CeO_2/epoxy coatings. For the blank epoxy coating, an obvious endothermic peak is noted at ~42 °C, corresponding to its T_g. In comparison, the T_g of the GO–CeO_2/epoxy coating slightly shifted to ~45 °C. The slight increase in T_g is mainly ascribed to the hindered flow between the chain segments of the GO–CeO_2 fillers in the epoxy coating, inhibiting the flow within the chain segments. The photothermal self-healing effect of the GO–CeO_2/epoxy coating was achieved by ensuring that the temperature at the scratch area of the coating should exceed the T_g of the GO–CeO_2/epoxy coating [53].

The photothermal conversion capacity of the coating was assessed by placing an NIR lamp (250 W) at 25 cm above the coating to irradiate the coating surface and taping a thermocouple to the back of the aluminum alloy sheet to monitor the heating rate and temperature change of the coating surface. Figure 3.14 shows the temperature change curve of the GO–CeO_2/epoxy composite coating after NIR irradiation

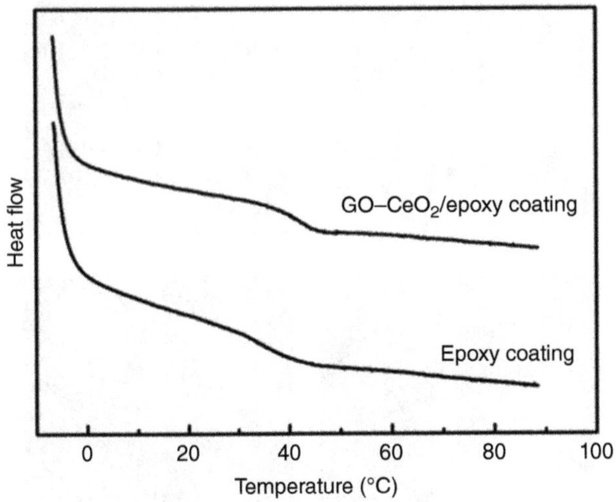

Figure 3.13 DSC curves of different coatings.

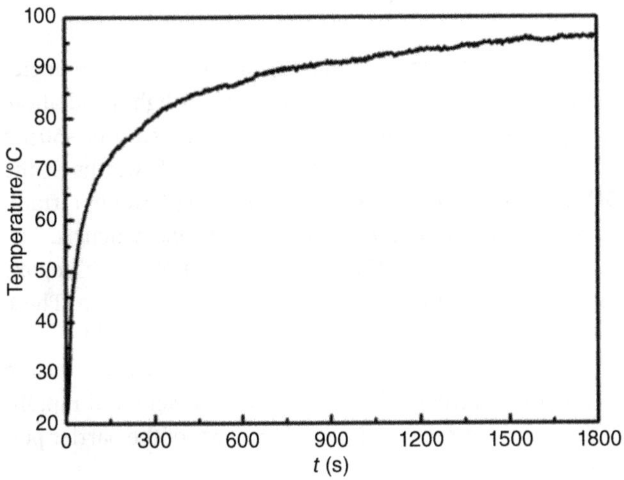

Figure 3.14 Surface temperature of GO–CeO_2/Epoxy coating upon exposure to NIR light for 1800 seconds.

for 1800 seconds. After irradiation for 30 minutes, the epoxy coating temperature of the GO–CeO_2/epoxy composite coating exceeded 90 °C, indicating good photothermal conversion characteristics of the GO–CeO_2/epoxy composite coating to realize the shape-memory recovery of the damaged coating area [54].

3.2.3.4 Self-healing Properties of the GO–CeO_2/Epoxy Coating

The self-healing performance was observed by creating a scratch with a length of ~5 mm and width of ~50 μm on the blank epoxy, GO/epoxy, and GO–CeO_2/epoxy coatings. Subsequently, the coatings were irradiated with an NIR light for 15 minutes to trigger a photothermal process. The scratch widths before and after repair were

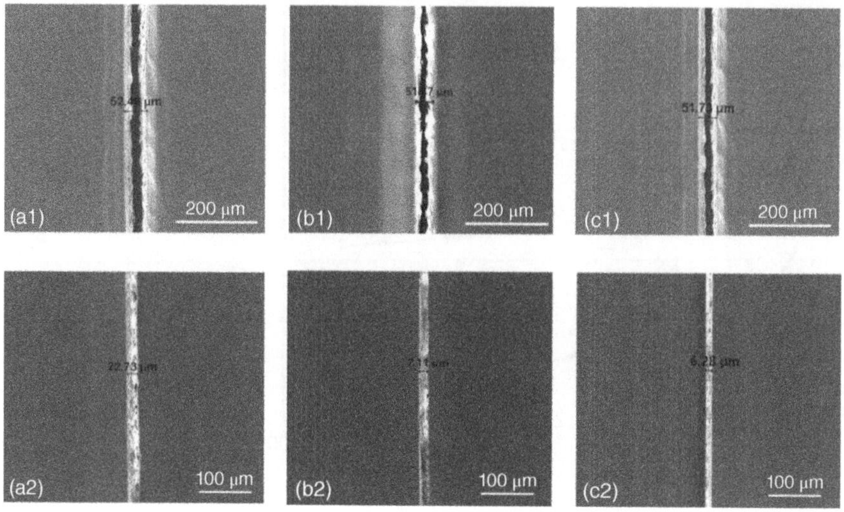

Figure 3.15 The shape recovery performance of different coatings within 15 minutes of NIR irradiation: (a_1,a_2) epoxy; (b_1,b_2) GO/epoxy; (c_1,c_2) GO-CeO_2/epoxy.

observed and compared using SEM. Figure 3.15 shows the surface morphologies of the scratch areas of the different coatings before and after NIR light irradiation. As shown in Figure 3.15a_1–c_1, the scratch widths of the three coatings are all ~50 μm before NIR irradiation. After 15 minutes of irradiation, the scratch widths of the GO/epoxy coating and GO–CeO_2/epoxy decreased, indicating that NIR light irradiation successfully stimulated the shape-memory recovery of the epoxy matrix. The scratch width of blank epoxy was more than 20 μm, whereas that of the GO/epoxy and GO–CeO_2/epoxy coating decreased to only ~6 μm after NIR irradiation. These results demonstrate the ability of GO to improve the heat absorption ability of the coating through its excellent photothermal conversion capacity; thus, the diffusion and re-entanglement of the epoxy matrix chain segments can be achieved rapidly, promoting the healing of the coating defects and partial recovery of the barrier performance of the coating [55].

The healing of the defects in the coating morphology does not indicate that the coating has fully recovered its barrier properties. Nonetheless, the recovery of the self-healing properties should successfully prevent the penetration of the corrosive medium into the coating and prevent corrosive electrochemical reactions. Therefore, electrochemical tests on the coating after photothermal healing should be conducted to evaluate its barrier properties. Figure 3.16a,b show the EIS test results for the blank epoxy coatings [56, 57]. As shown in Figure 3.16a, the low-frequency impedance modulus at 0.01 Hz ($|Z|_{0.01\text{Hz}}$) of the coating gradually decreases from 6.2×10^6 to 5.4×10^6 Ω cm^2 during the five-day salt spray test. This is ascribed to the complete loss of protective effect of the coating on the metal substrate at the beginning of the salt spray test; thus, the corrosive medium contacted the metal substrate. Figure 3.16c,d show the EIS test results of the GO–CeO_2/epoxy coating after photothermal self-healing. As shown in Figure 3.16c, the $|Z|_{0.01\text{Hz}}$ value of

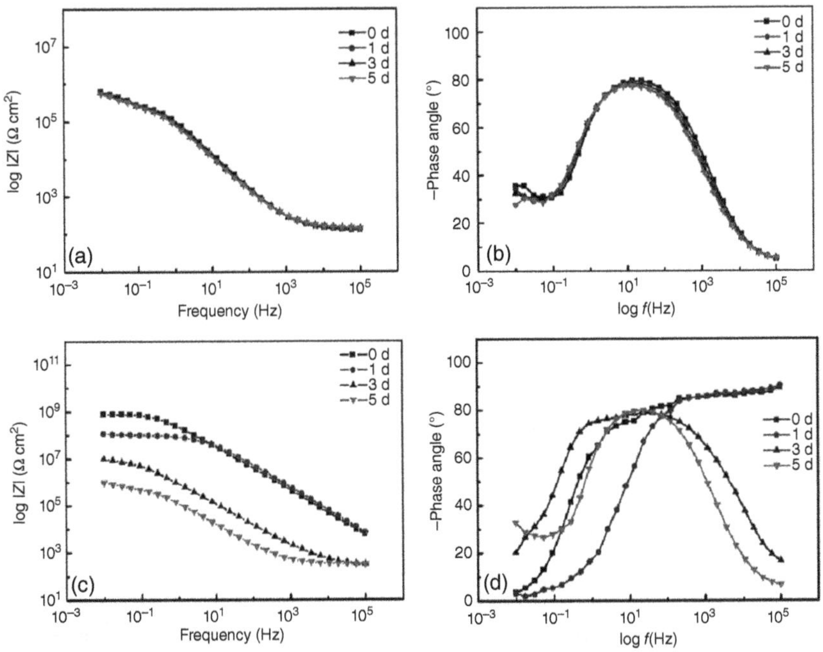

Figure 3.16 (a,b) Bode diagram of blank epoxy coating, and (c,d) Bode diagram of GO–CeO_2/epoxy coating after photothermally self-healing.

the GO–CeO_2/epoxy coating is considerably higher than that of the blank coating owing to the closure of the coating scratch due to the shape-memory effect. Further, the release of the corrosion inhibitor CeO_2 contributes to the self-healing effect [58, 59]. After salt spraying for three and five days, corrosive media gradually penetrated the coating because the coating scratches were not completely closed, thereby decreasing the $|Z|_{0.01Hz}$ values of the coating. Therefore, the self-healing effect of the GO–CeO_2/epoxy coating requires further optimization.

3.3 Self-healing Coating Based on TiN–BTA@MSN Photothermal Filler

Titanium nitride nanoparticles (TiN NPs) exhibit an excellent photothermal response owing to their strong plasmonic properties [60, 61]. In addition, TiN NPs have several other advantages, such as low price, high melting point, and good chemical stability [62, 63], and are ideal photothermal response materials for constructing photothermal self-healing coating systems [64, 65]. To fully exploit the plasmonic properties of TiN NPs and maximize their photothermal effects, we proposed and synthesized a novel photothermal composite filler with TiN NPs as the core and mesoporous silica nanoparticles as the shell, loaded with benzotriazole (BTA) molecules. The photothermal response of the TiN NPs core was expected to realize the dual self-healing effect of accelerating the release of BTA molecules

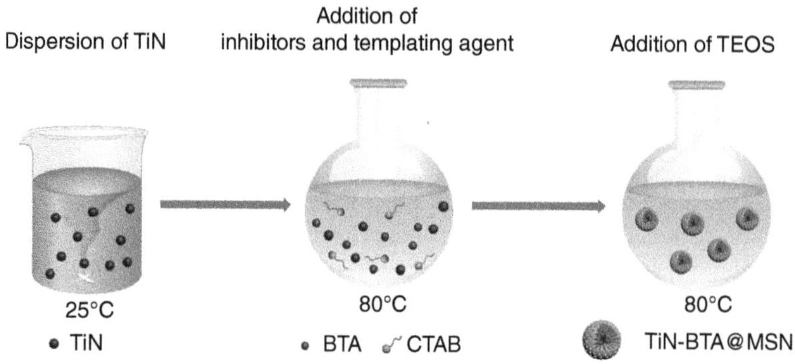

Figure 3.17 Schematic diagram of TiN-BTA@MSN preparation.

and physically closing the shape-memory coating matrix, thus improving the active protection of the coating [66].

The traditional method for the synthesis of mesoporous silica nanocontainers (MSN) often requires multiple experimental steps: (i) preparation of templated silica nanoparticles, (ii) removal of the template agent by ion exchange or high-temperature calcination, and (iii) loading of the corrosion inhibitors via immersion at low atmospheric pressure [67, 68]. This method is cumbersome, time-consuming, and labor-intensive. In this study, a one-pot sample preparation method was used to prepare MSN with TiN NPs loaded with BTA molecules (denoted as TiN-BTA@MSN). Compared to the traditional preparation method, the preparation time of the TiN-BTA@MSN filler was effectively shortened, and the corrosion inhibitor loading rate of the filler was similar to that of the traditional preparation method [69, 70]. The schematic diagram of TiN-BTA@MSN preparation is shown in Figure 3.17.

To study the effects of the photothermal action on the shape-memory recovery of the coating matrix and the accelerated release of corrosion inhibitors from TiN-BTA@MSN, different amounts of TiN-BTA@MSN were added to the SMP epoxy matrix and NIR light irradiation was used to trigger the photothermal conversion in the damaged coating area. The dual self-healing mechanism of the photothermal self-healing coating based on TiN-BTA@MSN (epoxy/TBS) was clarified. Finally, the long-term corrosion protection of the epoxy/TBS coatings was tested and analyzed.

3.3.1 Materials and Methods

3.3.1.1 Materials

The materials and regents used in this section are shown in Table 3.3.

3.3.1.2 Instruments

The instruments used in this section are shown in Table 3.4.

Table 3.3 Experimental materials and reagents.

Materials	Purity	Company
Titanium nitride nanoparticles (TiN NPs)	99.9%	Sigma-Aldrich
Tetraethyl orthosilicate (TEOS)	99.0%	Sinopharm
Cetyltrimethylammonium bromide (CTAB)	99.0%	Sinopharm
Benzotriazole (BTA)	AR	Sigma-Aldrich
Concentrated hydrochloric acid (HCl)	AR	Sinopharm
AA2024-T3 alloy	$40 \times 30 \times 5\ mm^3$	Merchant

Table 3.4 Instruments and equipment.

Instrument	Production type	Manufacturer
Scanning electron microscopy (SEM)	FE-SEM G300	Merlin, ZEISS
Transmission electron microscopy (TEM)	JEM-2100F	JEM-2100F, JEOL
X-ray diffraction (XRD)	D/Max-2500	Rigaku
UV/Vis/NIR spectrometer	Lambda 750	PerkinElmer
Physisorption analyzer	ASAP2460	Micromeritics
Fourier transform infrared spectrometer	Vector 22	Bruker
Thermogravimetry-differential thermal analysis	STA449F5	Netzsch
Thermocouple thermometer	TES-1310	TES
Infrared camera	FLIRONEPRO	FLIR
Electronic balance	ME201	Mettler Toledo
Cell breaker	S-450D	SONICS
pH meter	PHS-3C	Leici
Oil bath with magnetic stirrer	ZNCL-GS	Jieang
Centrifuge	Allegra X-30	Beckman
Electrochemical workstation	PARSTAT-2273	Princeton
Air blowing thermostatic oven	303-3S	Baihui
Confocal laser scanning microscopy (CLSM)	VK-X2000 series	KEYENCE
Simultaneous thermal analyzer	STA449F3	Netzsch
Raman spectroscope	Lab-RAM	Horiba
Microelectrochemical station	920D	CH Instruments

3.3.1.3 Preparation of TiN-BTA@MSN

A photothermal composite filler (TiN-BTA@MSN) was prepared via a hydrothermal method using CTAB as the template agent, BTA as the corrosion inhibitor, and TEOS as the silicon source [71, 72]. First, 0.2 g of TiN NPs were added to 240 ml of water and thoroughly dispersed using a cell breaker. The dispersion was mixed with 1.4 g of NaOH, 0.8 g BTA, and 0.5 g CTAB, transferred to a three-necked flask, and heated at 80 °C under magnetic stirring. Then, 2.5 ml of TEOS was added to the mixture and allowed to react for two hours for sufficient hydrolysis and polycondensation. Finally, the TiN-BTA@MSN was obtained by centrifugation (8000 r min^{-1}, five minutes), washed with DI water, and dried overnight at 70 °C. For comparison, a sample without TiN NPs was prepared using the same procedure and is denoted as BTA@MSN. In addition, BTA and CTAB molecules were washed from TiN-BTA@MSN with hydrochloric acid and ethanol (mass ratio 1 : 9), and dried at 70 °C for 24 hours, and denoted as TiN@MSN.

3.3.1.4 Morphology and Structure Characterization of TiN-BTA@MSN

The morphologies of the nanoparticles were characterized using SEM and TEM. An XRD pattern shows the crystal phase of TiN NPs. The chemical composition of TiN-BTA@MSNs was determined using FTIR spectroscopy. The plasmonic properties of TiN-BTA@MSN and BTA@MSN were evaluated using UV–Vis spectroscopy.

3.3.1.5 Photothermal Conversion Performance of TiN-BTA@MSN

The photothermal conversion performances of TiN-BTA@MSN and BTA@MSN were evaluated using a thermocouple thermometer combined with an infrared camera. First, 0.01 g of TiN-BTA@MSN and BTA@MSN were separately dispersed in 100 ml of water. A NIR laser (2.5 W cm^{-2}) was used to irradiate the TiN-BTA@MSN and BTA@MSN dispersion solutions for 300 seconds. Time-dependent temperature evolution was recorded using a thermocouple thermometer and camera [73, 74].

3.3.1.6 Inhibitor Release Behavior of TiN-BTA@MSN

First, the loading rate of BTA in TiN-BTA@MSN was determined using UV/Vis/NIR spectrometry [75, 76]. The UV/Vis/NIR spectra of BTA at different concentrations were examined and a standard linear relationship was constructed. The UV/Vis/NIR spectra of the collected residual solutions were measured to determine the number of unloaded BTA molecules during the synthesis process of TiN-BTA@MSN. Then, the loading ratio of BTA molecules (η_{BTA}) was calculated by the following equation:

$$\eta_{\mathrm{BTA}}\,(\mathrm{wt\%}) = \frac{m_{\mathrm{BTA,Pristine}} - m_{\mathrm{BTA,Unloaded}}}{m_{\mathrm{TiN-BTA@MSN,Synthesized}}} \tag{3.1}$$

To study the inhibitor release performance of TiN-BTA@MSN, NaCl solutions with different pH values were prepared and NIR irradiation was applied to test their effect on the inhibitor release behavior. 0.1 g of TiN-BTA@MSN was dispersed in NaCl solution (12 ml) with a mass fraction of 3.5 wt% (pH = 4.0, 7.0, 11.0) and added into a dialysis bag (molecular weight cut off 500). The dialysis bag was immersed in 100 ml of NaCl solution, keeping the pH inside and outside the dialysis bag consistent, and the solution was stirred at 1000 rpm. At regular intervals, the solution in the beaker was absorbed using a handheld pipette gun, and the UV/Vis/NIR spectra were measured. According to the characteristic UV absorption peak of BTA at different pH values, the corresponding peak value was recorded, and the corresponding solution was poured back after the examination. To analyze the effect of light on the inhibitor release behavior, the sample was irradiated by a NIR laser for five minutes at certain time intervals, and then the UV/Vis/NIR spectra were measured. The released amounts of BTA were calculated using the standard linear relationship of BTA in different pH environments [77].

3.3.1.7 Corrosion Inhibition Properties of TiN-BTA@MSN

The corrosion inhibition properties of TiN-BTA@MSN on the AA2024-T3 substrate were evaluated using EIS measurements. The 3.5 wt% NaCl solution with and without TiN-BTA@MSN addition was used as an electrolyte. After the open circuit potential was stabilized, EIS measurements were conducted. The effect of NIR irradiation on the corrosion inhibition of TiN-BTA@MSN was investigated after 30 minutes of illumination.

3.3.1.8 Preparation of Photothermal Composite Coating

An AA2024-T3 aluminum alloy was used as the substrate. The aluminum sheet was abraded with sandpaper until it exhibited a metallic luster. The polished metal substrate was ultrasonicated in anhydrous ethanol for 10 minutes, cleaned with acetone, dried, and stored in a dryer for experimental application.

A shape-memory epoxy matrix was constructed by mixing DGEBA, NGDE, and D230 in a molar ratio of 1 : 1 : 1. First, DGEBA and NGDE were thoroughly mixed, and then the TiN-BTA@MSN with mass fractions of 1%, 2%, and 3% were added, respectively. The mixtures were fully stirred, followed by the addition of curing agent D230 and then stirred for an additional 10 minutes. The mixture was dripped onto a metal substrate until it was completely coated, using a rotating coater for 30 seconds at a speed of 300 rpm. The coated substrates were cured for 24 hours at room temperature and then dried for 12 hours in a 55 °C drying oven. The coatings containing 1, 2, and 3 wt% TiN-BTA@MSN were prepared and named as epoxy/TBS-1, epoxy/TBS-2, and epoxy/TBS-3, respectively. A coating without TiN-BTA@MSN was prepared using the same preparation steps as those for the control sample (epoxy). The coating containing ~1.7 wt% BTA@MSN was prepared as another control sample (epoxy/BS). All coating thicknesses were ~80 μm.

3.3.1.9 Photothermal Conversion Property of Epoxy/TBS Coating

To characterize the photothermal conversion performance of the epoxy/TBS coating, the temperature probe of the thermocouple thermometer was fixed to the back of the metal substrate. Then, the surface position of the coating corresponding to the probe was illuminated by an 808-nm laser source, and the real-time evolution of the coating surface temperature was recorded using a thermometer.

Furthermore, the photothermal self-healing properties of the different coatings were determined based on the photothermal performance of TiN-BTA@MSN. An artificial scratch defect of ~5 mm length and 80 μm width was made on different coated samples with a scalpel, before the scratch was irradiated for 30 seconds. A laser confocal microscope was used to observe the morphological changes in the scratched areas of the coatings before and after illumination. Thermal infrared images of different coatings after 30 seconds of laser irradiation were captured using an infrared imager to explore the influence of the TiN-BTA@MSN content on the shape-memory performance of the coating.

3.3.1.10 Corrosion Protection of Epoxy/TBS Coating

The corrosion protection performances of the intact and scratched coatings were evaluated using EIS measurements. To explore the long-term protective abilities of the coating samples with different filler contents, a long-term immersion test was performed. After immersion for 30, 90, and 270 days, EIS tests were performed to explore the variation in the electrochemical properties of the different coating samples during immersion. Equivalent circuit fitting of the EIS data was performed using the ZSimpWin software [78, 79].

3.3.2 Preparation and Characterization of TiN-BTA@MSN

3.3.2.1 Morphology and Structure of TiN-BTA@MSN

Figure 3.18a–d respectively show the SEM image, TEM image, and XRD pattern of the TiN NPs. SEM and TEM microscopy revealed that the diameter of the TiN NPs was approximately 20–30 nm, showing a uniform size and well-dispersed nanoparticle morphology. The crystal structures of the TiN NPs were determined using XRD. All XRD peaks corresponded to different crystal faces of the TiN NPs, indicating that the TiN NPs possessed a face-centered cubic structure [80, 81]. Figure 3.19 shows an SEM image of TiN-BTA@MSN. The TiN-BTA@MSN prepared by the one-pot method presented a uniformly dispersed near-spherical morphology with a narrow diameter of 150–200 nm.

As shown in Figure 3.20, TiN-BTA@MSN displayed a near-spherical morphology with a obvious core-shell structure. In addition, the EDS mapping results demonstrate the elemental distribution of TiN-BTA@MSN. The distributions of Ti and N mainly correspond to TiN NPs. The distribution of C corresponds to the corrosion inhibitor BTA and template agent CTAB. The distributions of Si and O correspond to the mesoporous silica shells. By comparing the distributions of N and Ti, it was

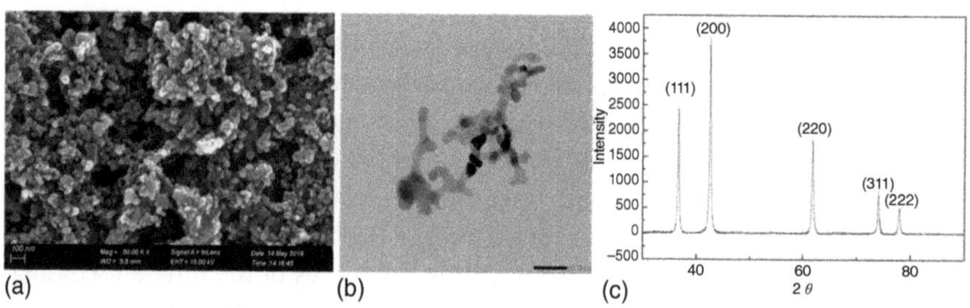

Figure 3.18 (a) SEM, (b) TEM image, (c) XRD pattern of TiN NPs.

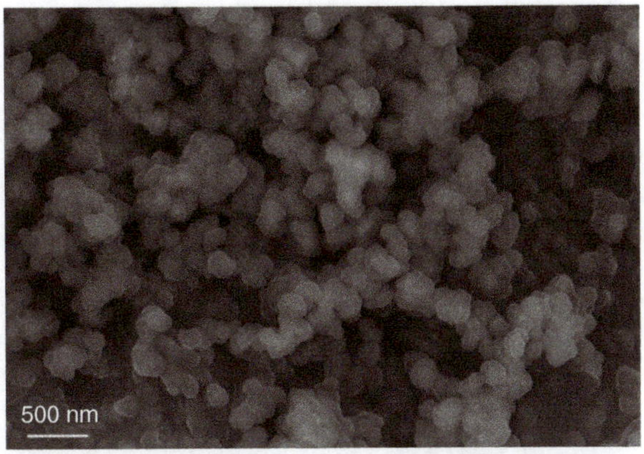

Figure 3.19 SEM image of TiN-BTA@MSN.

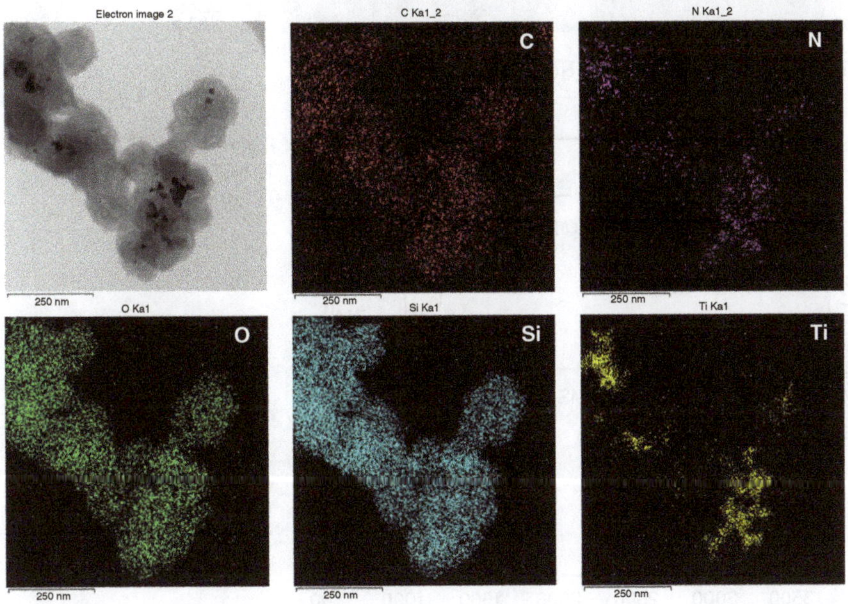

Figure 3.20 TEM image and EDS mapping of TiN-BTA@MSN.

found that N still had some additional distribution in the silica shell region compared to the position distribution of Ti. These N atoms originated from the BTA and CTAB, indicating that BTA and CTAB was successfully loaded into the mesoporous silica framework [82, 83].

Figure 3.21 shows the morphologies of the TiN@MSN samples. TiN@MSN was found to possess a mesoporous structure, which facilitated the efficient loading of the BTA corrosion inhibitor [84, 85].

To further determine the loading of the BTA and CTAB molecules and TiN NPs into the mesoporous silica structure, the chemical compositions of TiN-BTA@MSN and TiN@MSN were qualitatively characterized using FTIR analysis. Figure 3.22 shows the FTIR spectra of the TiN-BTA@MSN and TiN@MSN samples. Both the

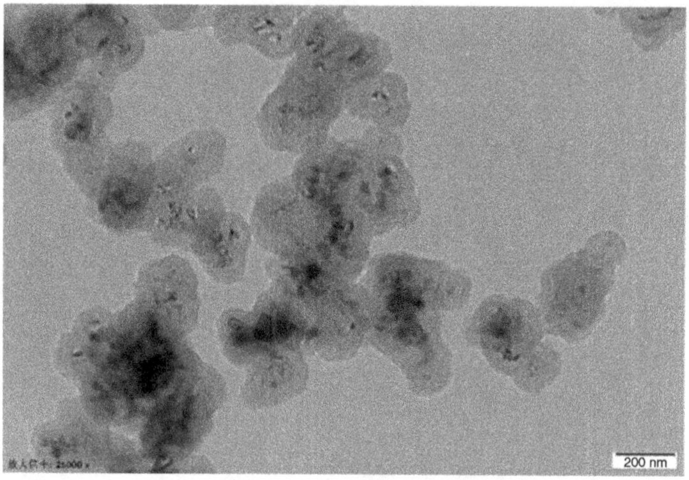

Figure 3.21 TEM image of TiN@MSN.

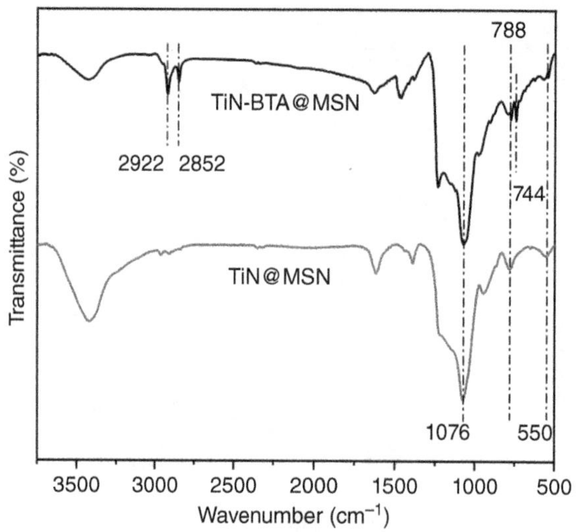

Figure 3.22 FTIR spectra of TiN-BTA@MSN and TiN@MSN.

TiN-BTA@MSN and TiN@MSN samples showed absorption peaks at 788 and 1076 cm^{-1} [86], corresponding to Si–O–Si symmetric stretching and asymmetric vibration, respectively, indicating the formation of a mesoporous silica structure. The 550 cm^{-1} peak corresponds to the Ti—N bond of TiN [87, 88]. Compared to TiN@MSN, TiN-BTA@MSN displayed additional characteristic peaks at 744, 2852, and 2922 cm^{-1}, which were attributed to the bending vibration in the C–H plane of the benzene ring of BTA and the stretching vibration of the C—H bond of CTAB [89, 90]. These results indicate that BTA was successfully loaded into the mesoporous silica. It was also demonstrated that the BTA molecules were entirely released after washing, which was conducive to the determination of the mesoporous structure of TiN-BTA@MSN.

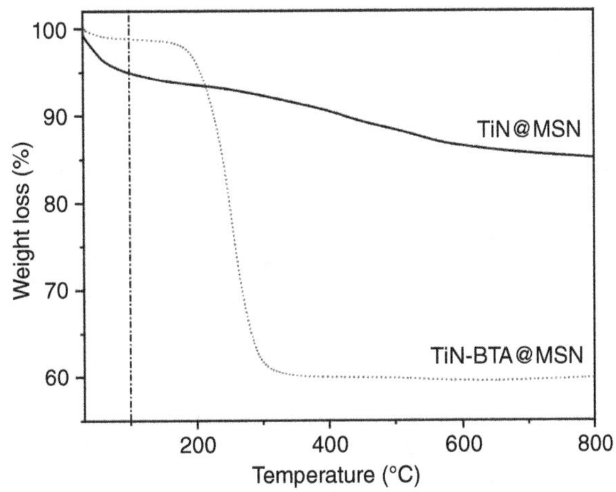

Figure 3.23 TGA curves of TiN-BTA@MSN and TiN@MSN.

Figure 3.23 shows the TGA curves of TiN-BTA@MSN and TiN@MSN. The weight loss over the range from room temperature to 100 °C is primarily caused by the evaporation of water. In addition, TiN@MSN displayed a weight loss of ~10% from 100 to 800 °C, while the weight loss of TiN-BTA@MSN was ~40%. This difference is mostly due to the effective loading of BTA and CTAB molecules in TiN-BTA@MSN [91].

The effective loading of the corrosion inhibitor BTA benefitted from the mesoporous structure of the silica framework. N_2 adsorption–desorption experiments were performed to investigate the porous structure. Figure 3.24 shows the N_2 isothermal adsorption–desorption curves and pore sizes of TiN-BTA@MSN and TiN@MSN. TiN-BTA@MSN displayed a very low N_2 adsorption capacity and a

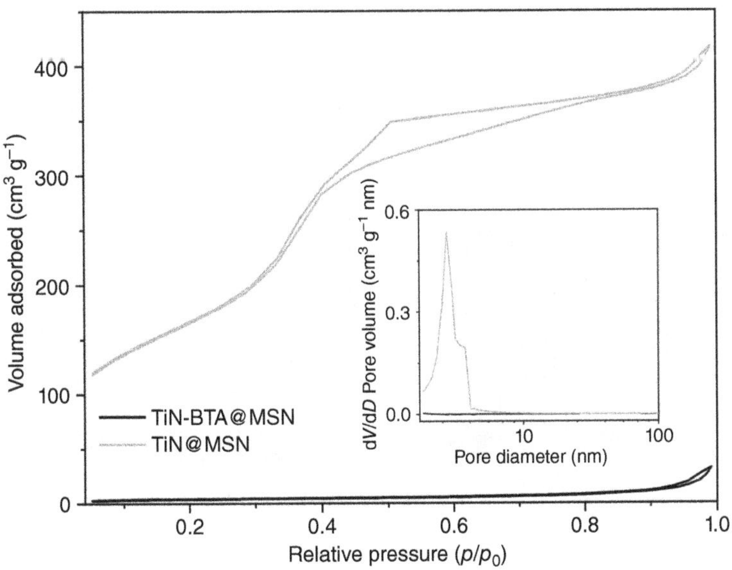

Figure 3.24 BET results of TiN-BTA@MSN and TiN@MSN.

Brunauer–Emmett–Teller (BET) specific surface area of 13.6 $m^2\ g^{-1}$, indicating that the porous structure was filled. The BET specific surface area of TiN@MSN rapidly increased to 619.6 $m^2\ g^{-1}$, and the pore volume was 0.65 $cm^3\ g^{-1}$. The pore size exhibited a narrow distribution of ~3.3 nm. This indicates that the prepared fillers have a rich porous structure, thus ensuring efficient loading of BTA molecules in the mesoporous silica pores [92, 93].

In summary, the one-step synthesized TiN-BTA@MSN possessed abundant pores so that the corrosion inhibitor, BTA, could be successfully loaded. The photothermal conversion properties of TiN-BTA@MSN are further characterized in Section 3.3.2.2.

3.3.2.2 Photothermal Conversion Properties of TiN-BTA@MSN

The photothermal conversion capability of TiN-BTA@MSN was evaluated because of the excellent light absorption properties of the TiN core in the NIR wavelength range. As shown in Figure 3.25a, the temperature of the TiN-BTA@MSN dispersion increased rapidly within 200 seconds of illumination, rising nearly 45 °C and gradually becoming stable, while the temperature of the BTA@MSN dispersion rose only ~7 °C after 300 seconds of NIR irradiation. In addition, some water droplets formed by water vapor appeared on the test tube wall of the TiN-BTA@MSN sample, but almost no change was observed in the test tube wall of the BTA@MSN sample. The infrared thermal images in Figure 3.25b also demonstrate the photothermal properties of TiN-BTA@MSN. Almost no significant temperature shift was detected in the BTA@MSN dispersion after 300 seconds of laser irradiation. In contrast, the infrared thermal image of TiN-BTA@MSN showed a gradual increase in temperature, indicating that the TiN core caused the TiN-BTA@MSN dispersion to produce a significant thermogenic effect [94].

3.3.2.3 Inhibitor Release Behavior of TiN-BTA@MSN

First, different amounts of BTA were dissolved into 3.5 wt% NaCl solutions at different pH values. Figure 3.26 shows the UV/Vis/NIR spectra of BTA solutions at different concentrations. An obvious absorption peak at 260 nm was observed for BTA

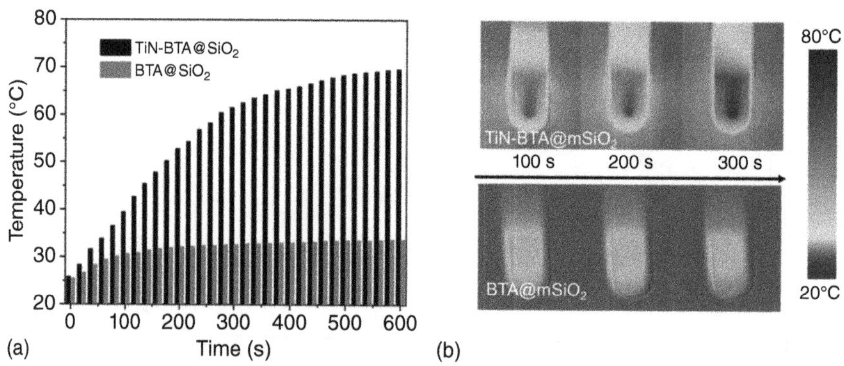

Figure 3.25 (a) Temperature variations of solutions containing TiN-BTA@MSN and BTA@MSN under laser irradiation; (b) the infrared thermal images of BTA@MSN and TiN-BTA@MSN dispersion after NIR illumination.

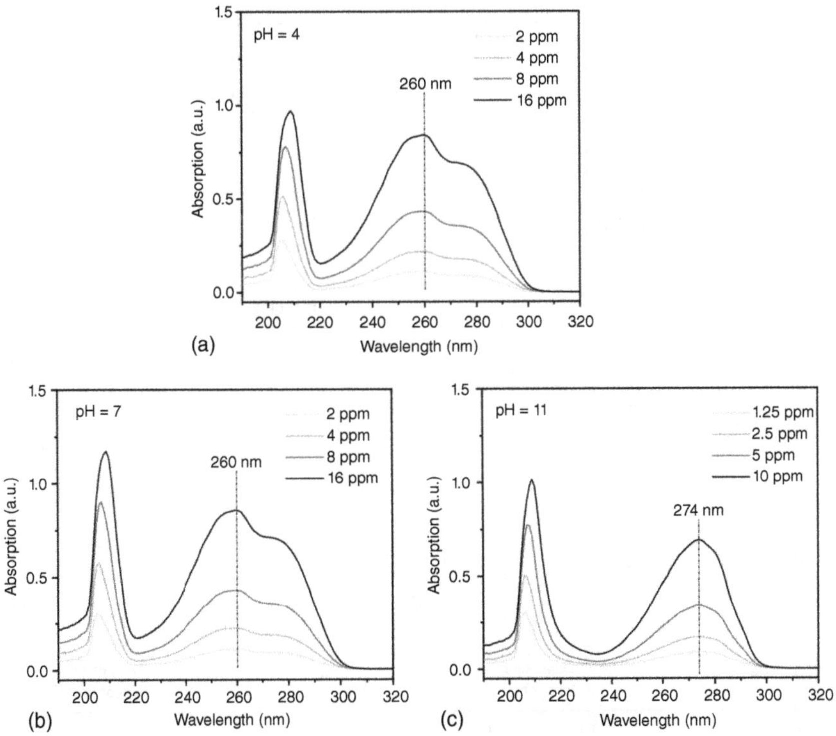

Figure 3.26 UV/Vis/NIR spectra of BTA in 3.5 wt% NaCl solution at different pH values: (a) pH = 4; (b) pH = 7; (c) pH = 11.

solutions at pH 4 and pH 7. At pH 11, the position of the absorption peak red-shifted to 274 nm, indicating an increase in the degree of BTA deprotonation in an alkaline environment [95, 96]. The concentration of the BTA solution and the corresponding characteristic absorption peak intensities are plotted in Figure 3.27, exhibiting a linear relationship. Therefore, the amount of BTA released could be calculated [97].

As shown in Figure 3.28, when the BTA molecule release experiment was performed for 60, 120, 180, and 240 minutes; 5 minutes of light irradiation was applied to the dialysis bag in the experimental group, while no light irradiation was applied to the control sample. Compared to the control sample, the release rate of BTA molecules in the experimental group significantly increased after intermittent irradiation. After 300 minutes of release, the inhibitor release amount in the experimental group was approximately 30% higher than that in the control group. This was because the plasmonic resonance of the TiN NPs core of TiN-BTA@MSN was stimulated by NIR laser irradiation and generated heat, which effectively promoted the diffusion and release of corrosion inhibitors in the TiN-BTA@MSN pores [98].

As shown in Figure 3.29, the experimental group was exposed to light irradiation for five minutes at each test point in the release experiment, whereas the control group was not exposed to light irradiation. In the control group, the release behavior of the corrosion inhibitor remained nearly constant under different pH conditions.

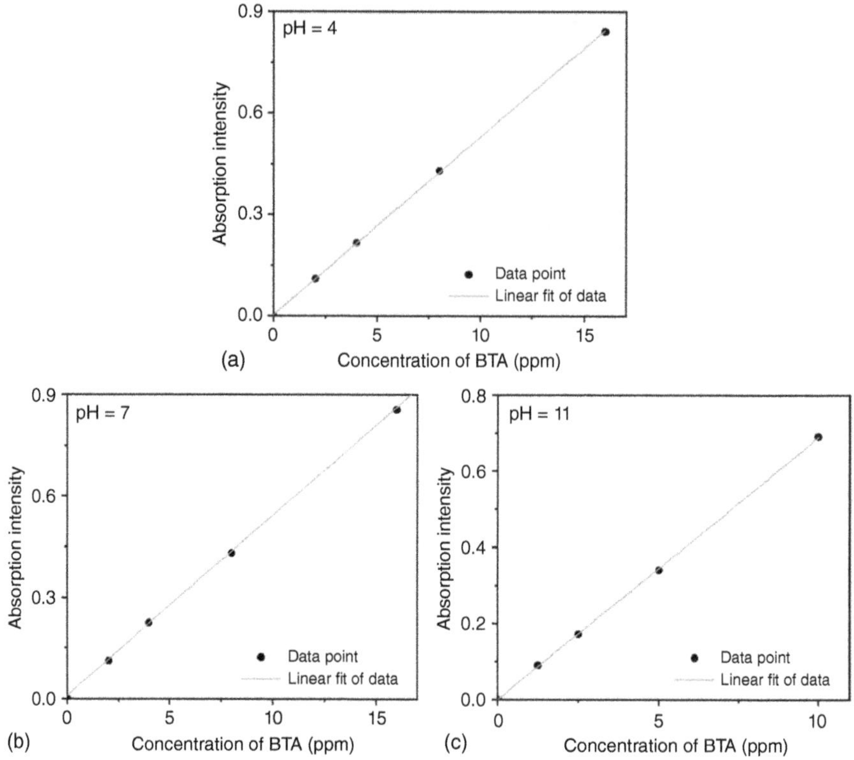

Figure 3.27 Linear relationship between the concentrations of BTA and the characteristic UV–Vis peak intensities: (a) pH = 4; (b) pH = 7; (c) pH = 11.

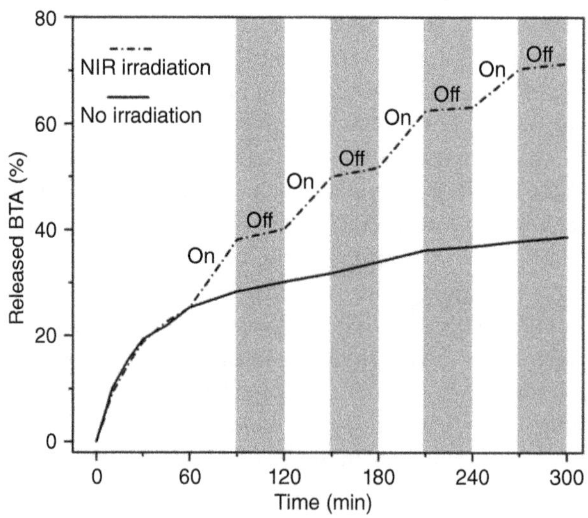

Figure 3.28 Release rate of BTA from TiN-BTA@MSN in NaCl solution with/without intermittent NIR irradiation.

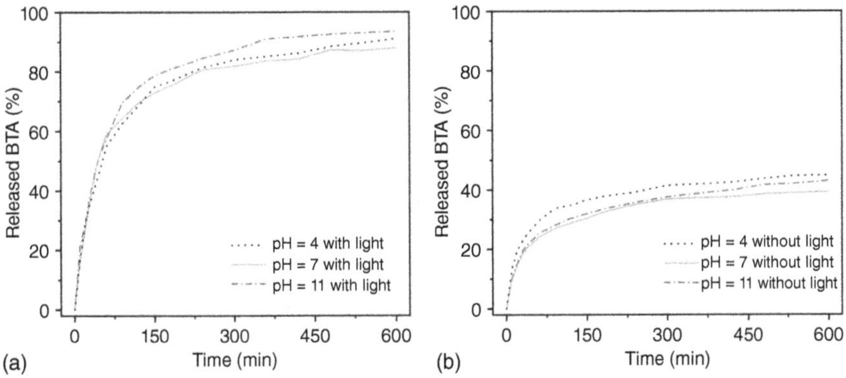

Figure 3.29 Release profiles of BTA from TiN-BTA@MSN under laser irradiation at different pH conditions.

Compared to the control group, the release rate of BTA molecules in the experimental group significantly increased and reached a plateau within ~400 minutes, which further proved that light promoted the release of corrosion inhibitor molecules [99, 100]. In addition, there was little difference in the release of BTA molecules in different pH environments after light application, indicating that the photothermal promotion of BTA release was not affected by pH.

3.3.2.4 Corrosion Inhibition Property of TiN-BTA@MSN

Owing to the excellent photothermal conversion performance of TiN-BTA@MSN and the facilitated release performance of the BTA inhibitor, the photothermal effect of TiN-BTA@MSN on the corrosion inhibition of the aluminum alloy substrate was further explored. Figure 3.30 presents the Bode diagrams of the AA2024-T3 substrate immersed in different solutions. The samples immersed in a 3.5 wt% NaCl solution with and without TiN-BTA@MSN ($3.0\,g\,l^{-1}$) were denoted as Particles and Al, respectively. The alloy sample immersed in a NaCl solution with $3.0\,g\,l^{-1}$ of TiN-BTA@MSN and illuminated by NIR light for 30 minutes was denoted as Particles-light.

For the aluminum alloy substrate, the $|Z|_{0.01Hz}$ value was low and decreased significantly as the immersion time was extended to 24 hours. In contrast, for the Particles sample, $|Z|_{0.01Hz}$ was higher and increased slightly with time. Moreover, the frequency range of the phase angle was also large because BTA was released from TiN-BTA@MSN and adsorbed onto the alloy surface. $|Z|_{0.01Hz}$ was highest for the Particles-light sample because the photothermal effect of TiN can significantly accelerate the release of BTA, so that more BTA can be adsorbed onto the alloy surface earlier. In summary, the excellent photothermal conversion performance of TiN-BTA@MSN promoted the release of inhibitors and inhibited the corrosion of the metal surface [101].

In this section, a simple one-step synthesis method is used to successfully prepare a photothermal composite filler (TiN-BTA@MSN), based on the plasmonic properties of TiN TPs and loaded with the corrosion inhibitor BTA. Its excellent photothermal

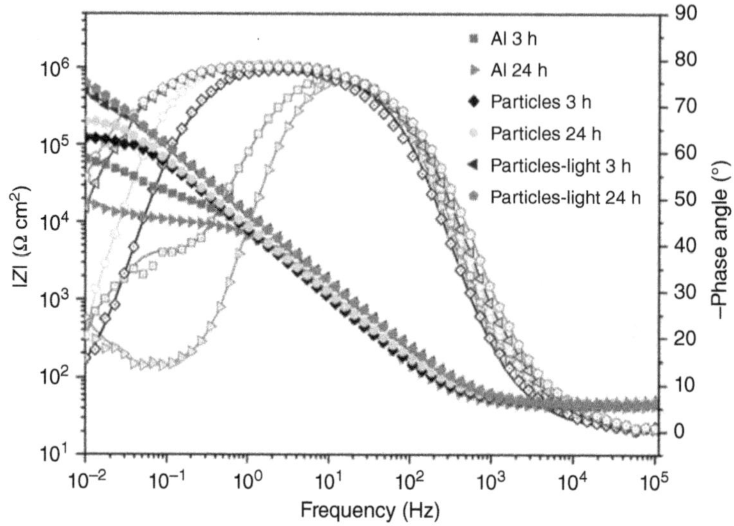

Figure 3.30 Bode plots of AA2024-T3 immersed in different solutions. Source: Adapted from Ma et al. [66]/Elsevier.

conversion ability is beneficial for constructing self-healing coating systems. The prepared TiN-BTA@MSN had a diameter of 150–200 nm, and its rich mesoporous structure endowed the composite filler with a high corrosion inhibitor loading rate (~16.5 wt%). At the same time, TiN-BTA@MSN possess excellent light absorption characteristics in a wide wavelength range of NIR light and its aqueous solution sample can be heated up to 70 °C after being irradiated by an 808 nm NIR laser for 300 seconds.

By studying the release behavior of the BTA inhibitor, it was found that the application of light played an essential role in promoting the release of the inhibitor and that the pH environment did not affect the photothermal promotion mechanism of inhibitor release from TiN-BTA@MSN. The results show that TiN-BTA@MSN can rapidly produce heat under the stimulation of NIR light and considerably promote the release of corrosion inhibitor molecules. Characterization of the corrosion-inhibition properties of TiN-BTA@MSN after photothermal release revealed that the promotion of BTA release caused by the photothermal effect was of great significance for improving the response rate of metal corrosion inhibition. Therefore, the addition of an appropriate amount of TiN-BTA@MSN to the coating matrix will help achieve a rapid release of the corrosion inhibitor and subsequent adsorption of the damaged area by light stimulation and improve the active protection of the coating.

3.3.3 Preparation and Characterization of Epoxy/TBS Coatings

In Section 3.3.3, a composite filler, TiN-BTA@MSN, with excellent photothermal conversion ability was prepared, and the experimental results proved that TiN-BTA@MSN can promote the rapid release of corrosion inhibitors through the photothermal effect. In this section, different amounts of TiN-BTA@MSN were added to the SMP epoxy matrix, and NIR irradiation was used to trigger the photothermal conversion in the damaged coating area to explore the effects of photothermal action on the shape-memory recovery of the coating matrix and the accelerated release of corrosion inhibitors from TiN-BTA@MSN. The dual self-healing mechanism of the photothermal self-healing coating based on TiN-BTA@MSN (epoxy/TBS) was clarified.

3.3.3.1 Photothermal Conversion Properties of Epoxy/TBS Coating

First, DSC measurements were performed on different coating samples. As shown in Figure 3.31, the DSC curve of blank epoxy possessed a clear endothermic peak at ~46 °C, corresponding to its T_g. The T_g of a coating refers to the relaxation phenomenon in which the amorphous part of the polymer changes from a glassy state (frozen state) to a highly elastic state (thawed state) [102, 103]. When the temperature is higher than T_g, the molecular segments of the polymer begin to flow, leading to the shape-memory recovery. The DSC curves of the epoxy/TBS-1, epoxy/TBS-2, and epoxy/TBS-3 coatings were almost unchanged, indicating that the presence of nanofillers barely affects the thermal properties of coatings.

Subsequently, the photothermal conversion performance of the composite coatings was characterized. As shown in Figure 3.32, the temperatures of the epoxy/TBS-1, epoxy/TBS-2, and epoxy/TBS-3 samples rapidly increased within 30 seconds and reached a balanced state of heat generation and dissipation at 180 seconds. As the amount of TiN-BTA@MSN increased, the temperatures of the coatings gradually increased. The approximate temperatures of epoxy/TBS-1, epoxy/TBS-2, and epoxy/TBS-3 coatings were 35, 55, and 60 °C, respectively, after 180 seconds of NIR laser irradiation. In comparison, the temperature of blank epoxy did not change much, only rising by ~7 °C. Based on the above analysis, the temperatures of the epoxy/TBS-2 and epoxy/TBS-3 samples were higher than the T_g of the epoxy matrix within 30 seconds of laser irradiation, demonstrating the capability of the coatings to trigger the shape-memory effect [104].

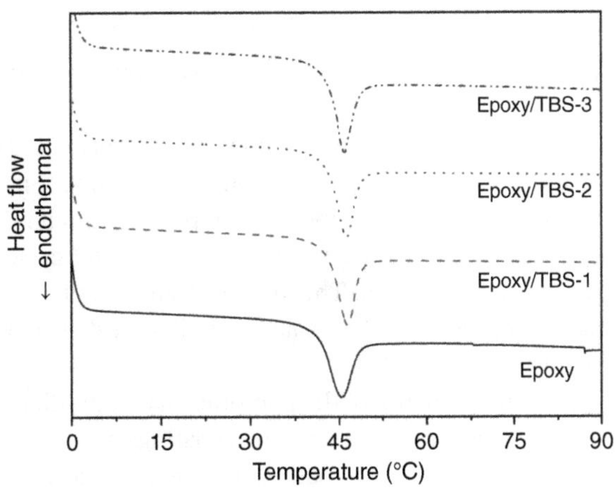

Figure 3.31 DSC curves of different coatings.

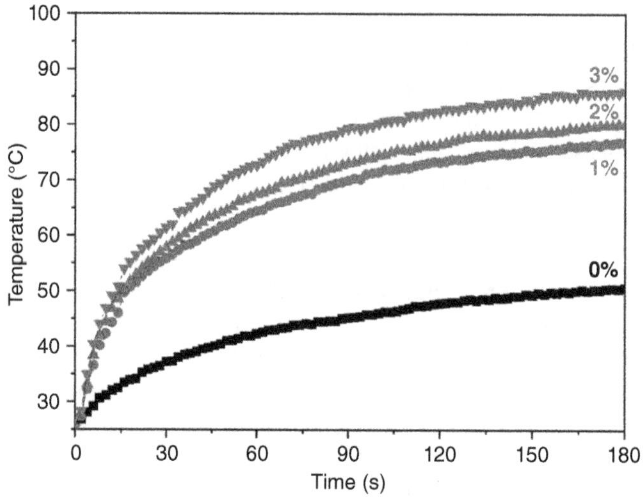

Figure 3.32 Surface temperatures of different coatings upon NIR exposure.

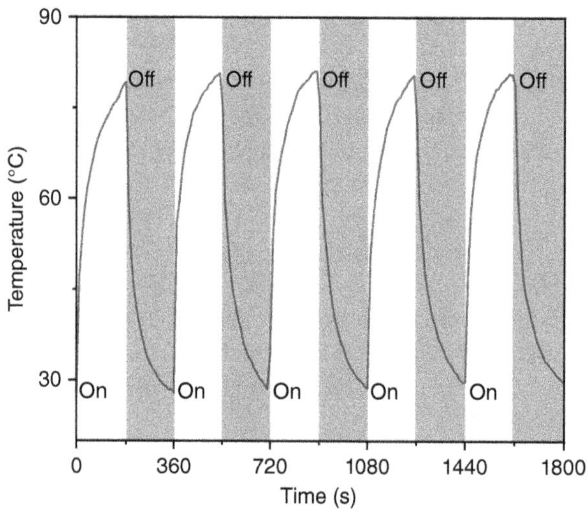

Figure 3.33 Temperature variations of Epoxy/TBS-2 coating within five light on/off cycles.

NIR laser irradiation was applied at intervals of 180 seconds to test the repeatability of the photothermal conversion of the coating (Figure 3.33). Under illumination, the coating temperature increased rapidly; without NIR light illumination, the coating temperature decreased rapidly. Five on–off cycles showed a reasonably stable heat-production effect. This demonstrates that the coating has a stable photothermal performance and that laser irradiation does not damage the structure of the coating matrix [105].

Next, the influence of the photothermal nanofiller TiN-BTA@MSN on the shape-memory recovery of the scratched coatings was explored. Figure 3.34 shows optical images of the different coatings before and after 30 seconds NIR laser irradiation.

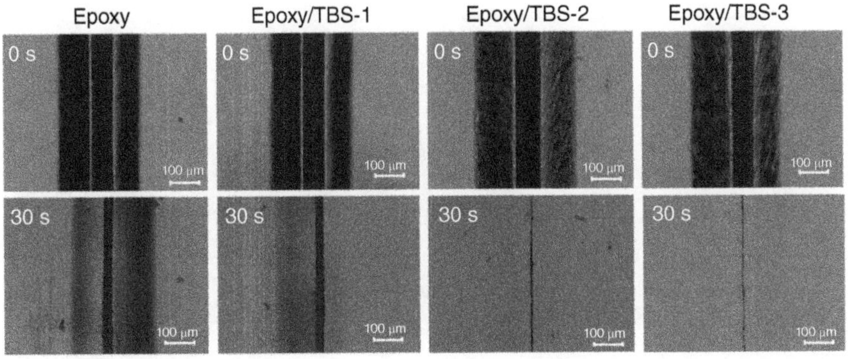

Figure 3.34 Shape recovery performance of different coatings within 30 seconds of NIR laser irradiation [66].

After 30 seconds, different samples exhibited different temperature changes. For the blank epoxy coating, the scratch size showed limited reduction because the surface temperature of ~30 °C was not significantly higher than the T_g value. For the epoxy/TBS-1 coating, the scratch was not closed, and the gap was still obvious. For the epoxy/TBS-2 and epoxy/TBS-3 coatings, the scratches were nearly closed, leaving only gaps of a few micrometers. When the surface temperature exceeded the T_g of the epoxy matrix, the molecular segments at the scratches began to flow effectively, resulting in scratch closure.

As shown in Figure 3.35, the blank epoxy and epoxy/TBS-1 coatings exhibited relatively low surface temperatures, whereas epoxy/TBS-2 and epoxy/TBS-3 showed higher surface temperatures, indicating that the epoxy coatings had excellent photothermal conversion capability when the amount of added TiN-BTA@MSN was 2 wt% or greater. Simultaneously, the temperature of the area directly irradiated by the laser light source was the highest. Comparatively, the temperature of the surrounding area was much lower, indicating that self-healing of the damaged coating area can be accurately achieved using NIR laser irradiation without affecting the undamaged areas of the coatings. Thus, the accuracy of the coating healing can be guaranteed.

3.3.3.2 Corrosion Protection Property of Intact Coatings

Based on the previous results, the addition of TiN-BTA@MSN enabled the epoxy coating to undergo photothermal conversion in a relatively short time to achieve a

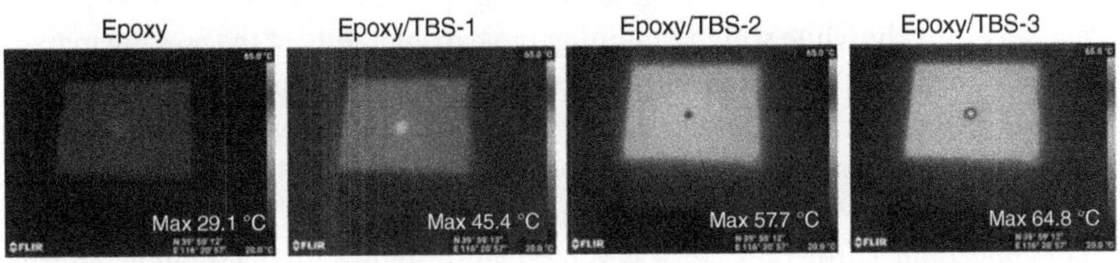

Figure 3.35 Shape recovery performance of different coatings within 30 seconds of NIR laser irradiation [66].

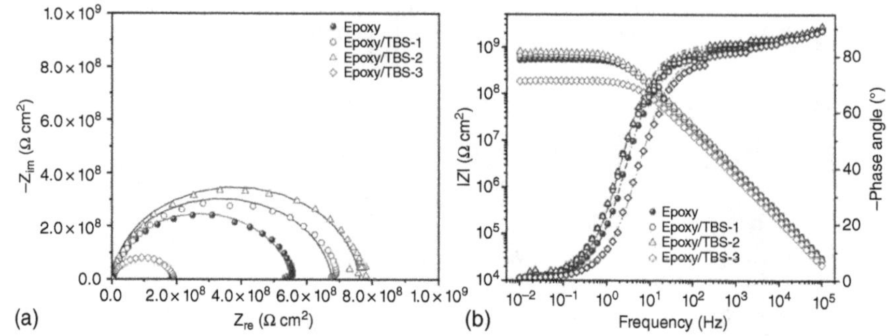

Figure 3.36 (a) Nyquist diagrams and (b) Bode plots of different coatings after immersion in 3.5 wt% NaCl solutions for two days [66].

shape-memory effect. EIS measurements were conducted on different intact coatings to explore the influence of the amount of TiN-BTA@MSN on the corrosion resistance of coatings. Figure 3.36 shows the Nyquist and Bode diagrams of different intact coatings after immersion in 3.5 wt% NaCl solution for two days. The Nyquist diagram of the coating exhibits a semicircular shape. The impedance in the Bode diagram shows a plateau shape in the low-frequency region, and there is only one-time constant in the phase angle plot, indicating that the coating has good barrier properties [106]. It is worth noting that epoxy/TBS-2 has the largest arc radius in the Nyquist diagram and $|Z|_{0.01\text{Hz}}$ value in the Bode plot. By comparing the different coatings, it was found that the barrier effect of the coating could be improved to a certain extent when the added amount of TiN-BTA@MSN was 1 and 2 wt%. The corrosion protection performance of the coatings decreased significantly when the amount of additive was further increased to 3 wt%. The barrier property of the epoxy/TBS-2 coating was better than that of the epoxy/TBS-1 coating, which can be attributed to the fact that the epoxy/TBS-2 coating had sufficient corrosion inhibitors to enhance the corrosion resistance of the epoxy matrix. The deterioration in the barrier properties of epoxy/TBS-3 can be explained by the agglomeration and uneven dispersion of excess TiN-BTA@MSN, resulting in more diffusion channels within the coating, making it easier for the corrosive medium to reach the metal surface [107].

3.3.3.3 Self-healing Property of Scratched Coatings

Based on the discussion of the photothermal and protective properties of the different coatings, it can be concluded that when the amount of photothermal composite filler is 2 wt%, the coating can achieve excellent photothermal conversion performance very quickly, while still maintaining the barrier ability of the coating matrix. Therefore, an epoxy/TBS-2 coating with excellent anti-corrosion properties and photothermal conversion effects was selected to assess the self-healing properties and mechanisms of the coatings. To further investigate the self-healing properties of the blank epoxy and epoxy/TBS-2 coatings, the epoxy and epoxy/TBS-2 were subjected to scratching, and a drop of water was subsequently applied to the scratch and irradiated with a NIR laser for 30 seconds. The scratched area of the coating sample was then immersed in a NaCl solution for 15 days. Figure 3.37 shows the Bode diagram

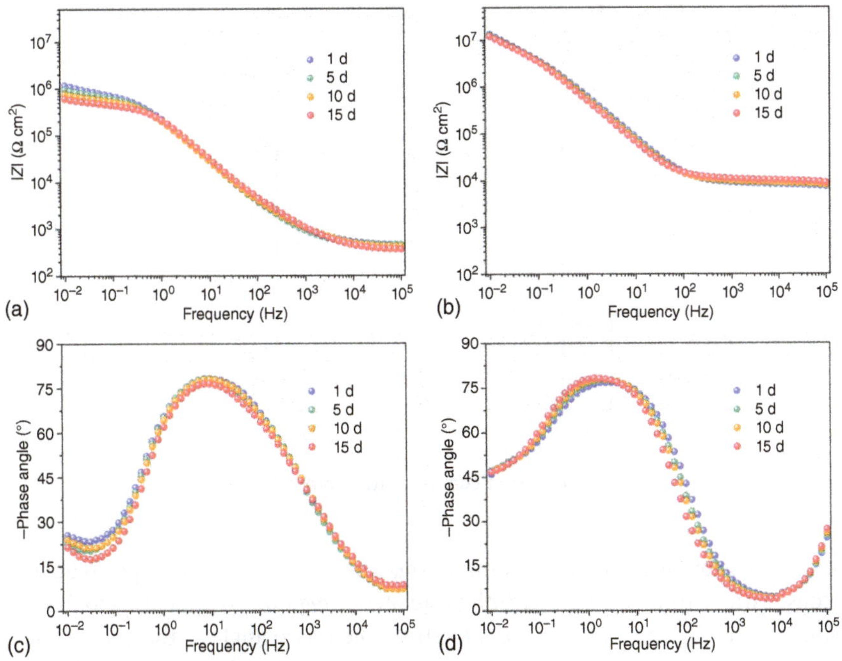

Figure 3.37 Bode-impedance plots of the scratched (a) blank epoxy and (b) epoxy/TBS-2 coating during immersion in 3.5 wt% NaCl solution; Bode-phase angle plots of the scratched (c) blank epoxy and (d) epoxy/TBS-2 coating during immersion in 3.5 wt% NaCl solution. Source: Adapted from Ma et al. [66]/Elsevier.

of the coating sample after immersion in the NaCl solution for different numbers of days.

First, the $|Z|_{0.01Hz}$ value of epoxy/TBS-2 coating ($1.4\times10^7\,\Omega\,cm^2$) was nearly 1 order of magnitude higher than that of the blank epoxy ($1.2\times10^6\,\Omega\,cm^2$) after one day of immersion. Then the $|Z|_{0.01Hz}$ value of the blank epoxy decreased gradually to $6.1\times10^5\,\Omega\,cm^2$ after 15 days of immersion. This is because the corrosion medium (Cl^-, H_2O, O_2, etc.) continues to invade and reach the metal surface, leading to gradual intensification of the corrosion phenomenon in the damaged region. Conversely, the $|Z|_{0.01Hz}$ value of epoxy/TBS-2 coating was maintained at $\sim 1.2\times10^7\,\Omega\,cm^2$ after immersing for 15 days. This self-healing effect can be explained in two ways. On the one hand, the coating scratch was physically closed owing to the photothermal conversion of the epoxy/TBS-2 coating. On the other hand, the photothermal conversion promoted the release of the BTA inhibitor from the TiN-BTA@MSN, forming a corrosion inhibitor film on the metal surface in the damaged area of the coating to inhibit the corrosion effectively.

In addition, it can be observed from Figure 3.38c that the blank epoxy coating shows three-time constants: in the low-frequency region corresponding to the corrosion activity on the metal substrate, in the mid-frequency region corresponding to the corrosion inhibitor layer on the surface of the aluminum alloy substrate, and in the high-frequency region corresponding to the coating barrier properties [108]. The

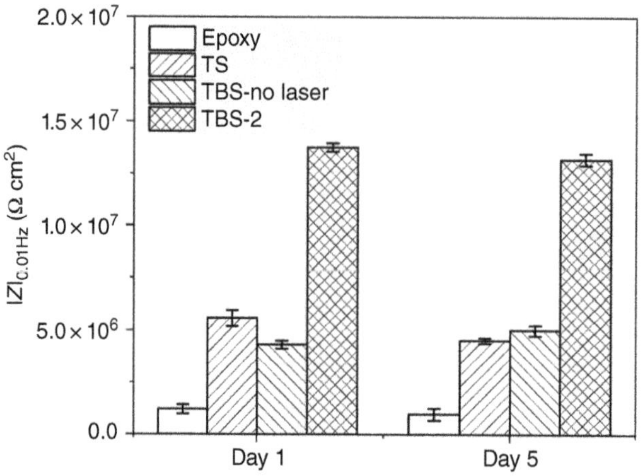

Figure 3.38 $|Z|_{0.01\mathrm{Hz}}$ values of different coatings after immersion.

epoxy/TBS-2 coating exhibited two-time constants in the mid- and high-frequency regions, with no time constant in the low-frequency region corresponding to the corrosion activity of the metal surface. Moreover, the phase angle of the epoxy/TBS-2 coating at a high frequency (10^5 Hz) was higher than that of the blank coating, indicating that the epoxy/TBS-2 had good corrosion resistance [109].

Based on the above analysis, the epoxy/TBS-2 epoxy coating exhibits good self-healing properties owing to the photothermal conversion effect of TiN-BTA@MSN. However, the roles of shape-memory recovery and corrosion inhibitor release in the self-healing process remain unclear. EIS was performed on the different coatings to further clarify the self-healing mechanism. The blank epoxy, epoxy/TS, and epoxy/TBS coatings were irradiated with NIR light, whereas the epoxy/TBS-NL coating was not irradiated. The epoxy/TBS coating represents the dual-healing mechanism of the shape-memory recovery effect and the accelerated release of the corrosion inhibitor owing to the photothermal effect, the epoxy/TS coating represents the shape-memory recovery mechanism, and the epoxy/TBS-NL coating represents the mechanism of corrosion inhibitor release.

Figure 3.38 shows the $|Z|_{0.01\mathrm{Hz}}$ values of the different coatings after immersion in 3.5 wt% NaCl solution for one and five days. After one day, the blank epoxy coating had the lowest $|Z|_{0.01\mathrm{Hz}}$ value, and the epoxy/TBS coating had the highest $|Z|_{0.01\mathrm{Hz}}$ value. The $|Z|_{0.01\mathrm{Hz}}$ values of epoxy/TS and epoxy/TBS-NL coatings were both $\sim 5 \times 10^6\ \Omega\,\mathrm{cm}^2$. These results demonstrate that a single self-healing mechanism can slow down the corrosion process of the metal substrate; however, the epoxy/TBS coating with a dual self-healing mechanism was superior to that with a single self-healing mechanism. After five days, the $|Z|_{0.01\mathrm{Hz}}$ values of the epoxy and epoxy/TS coatings decreased, which can be explained by the continuous invasion of corrosive media and the development of the corrosion reaction. However, the $|Z|_{0.01\mathrm{Hz}}$ value of the epoxy/TBS-NL coating showed a slight increase owing to the release and adsorption of the BTA inhibitor, which suppressed the corrosion

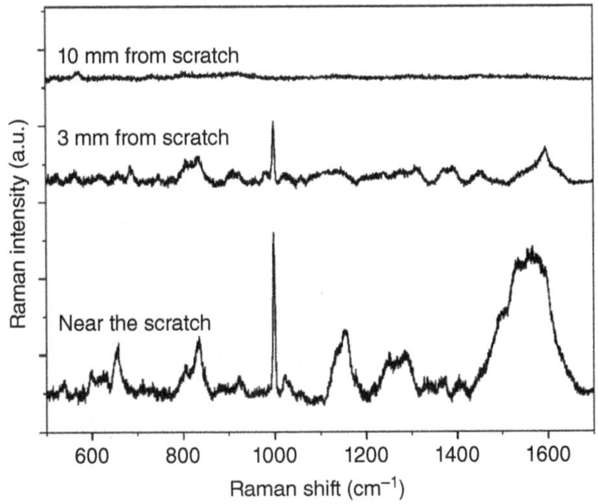

Figure 3.39 Raman spectra of BTA molecules at different detection positions.

process. The $|Z|_{0.01\text{Hz}}$ of the epoxy/TBS coating remained almost unchanged owing to the synergistic effect of shape-memory recovery and corrosion inhibitor release.

To determine the adsorption and corrosion inhibition of BTA molecules on the metal substrate in the scratched area, the coating was stripped from the metal substrate after photothermal healing and 24-hours immersion, and the metal substrate near the damaged area of the coating was examined by Raman spectroscopy [110]. As shown in Figure 3.39, clear Raman signals of BTA were observed at the detection points located on the damaged area of coating, with characteristic peaks located at 1010, 1146, 1292, 1449, and 1595 cm^{-1}. The peak at 1010 cm^{-1} corresponds to the benzene ring skeleton and CH bending. The peak at 1146 cm^{-1} was attributed to the bending vibration in the NH plane. The peak at 1292 cm^{-1} corresponds to the NH bending and CH bending of bone elongation. The peaks at 1449 and 1595 cm^{-1} correspond to benzene ring stretching [111, 112]. In comparison, the BTA Raman signal at the detection point ~3 mm out of the scratch was much weaker, and almost no Raman signal was detected far from the scratch.

3.4 Self-healing Coating Based on PCL@TiN Photothermal Fillers

In real applications of corrosion protection coatings, coating damages or defects inevitably occur, seriously weakening the service life of coatings. Because the sizes of damaged areas differ greatly, a single self-healing method cannot be expected to achieve an ideal healing effect. However, by using a shape-memory polymer as the coating matrix, the damaged region can be reduced, resulting in a substantial reduction in defect size and damage. However, the mechanism of a self-healing shape-memory solution does not achieve complete recovery of the barrier effect

[113, 114]. After a certain period, the corrosion activity in the remaining scratched area continues to intensify, leading to gradual failure of the protection system. To improve the self-healing performance, the remaining defects should be filled with a filler having a low T_m value so that corrosion medium intrusion can be further suppressed.

In this section, a novel composite filler was prepared by combining TiN NPs and Poly (ε-caprolactone) (PCL) as the photothermal substance and the low-T_m material, respectively [115]. First, PCL microspheres of appropriate sizes were obtained via the solvent evaporation method. Then, TiN NPs were deposited on the surface of the PCL via chemical modifications and electrostatic interactions to obtain photothermal fillers (PCL@TiN). Owing to the plasmonic properties of TiN, the PCL@TiN composite microspheres exhibit excellent photothermal properties as well. The photothermal effect of TiN can trigger not only a shape-memory effect of the coating matrix but also the melting of PCL microspheres to further close and seal the coating defect.

3.4.1 Materials and Methods

3.4.1.1 Materials

The materials and regents used in this section are shown in Table 3.5.

3.4.1.2 Instruments

The instruments used in this section are shown in Table 3.6.

3.4.1.3 Preparation of PCL Microspheres

First, 0.6 g of PCL and 1.0 g of PVA were separately dissolved in 30 ml of CH_2Cl_2 and 200 ml of DI water to form oil phase and water phase, respectively. They were then mixed and homogenized via sonication. Subsequently, the O/W emulsion was magnetically stirred until the CH_2Cl_2 completely evaporated. PCL microspheres were then obtained by centrifugation at 5000 rpm for 3 minutes, washed with water three times, and freeze-dried for 24 hours [116].

3.4.1.4 Preparation of PCL@TiN Composite Microspheres

Figure 3.40 shows a schematic of the PCL@TiN preparation. First, the PCL microspheres were treated with an alkaline solution to obtain carboxyl groups (–COOH)

Table 3.5 Experimental materials and reagents.

Materials	Purity	Company
TiN NPs	99.9%	Aladdin
Poly(ε-caprolactone) (PCL, ~45 000 g mol^{-1})	99.0%	Sigma-Aldrich
(3-Aminopropyl) triethoxysilane (APTES)	AR	Aladdin
Polyvinyl alcohol (PVA, ~27 000 g mol^{-1})	AR	Xiya
Dichloromethane (CH_2Cl_2)	AR	Sinopharm

Table 3.6 Instruments and equipment.

Instrument	Production type	Manufacturer
Scanning electron microscopy (SEM)	FE-SEM G300	Merlin, ZEISS
Transmission electron microscopy (TEM)	JEM-2100F	JEM-2100F, JEOL
UV/Vis/NIR spectrometer	UH-5300	Hitachi
Thermocouple thermometer	TES-1310	TES
Infrared camera	FLIRONEPRO	FLIR
Electronic balance	ME204	Mettler Toledo
Electrokinetic analyzer	Zetasizer Nano	Malvern
Centrifuge	Allegra X-30	Beckman
Air blowing thermostatic oven	303-3S	Baihui
Simultaneous thermal analyzer	STA449F3	Netzsch
UV/Vis spectrometer	LAMBDA 950	PerkinElmer
Electrochemical workstation	PARSTAT-2273	Princeton
Atmospheric corrosion monitors (ACM)		Qianlang
Focused ion beam equipment (FIB)	LYRA 3	TESCAN
Microelectrochemical station	920D	CH Instruments

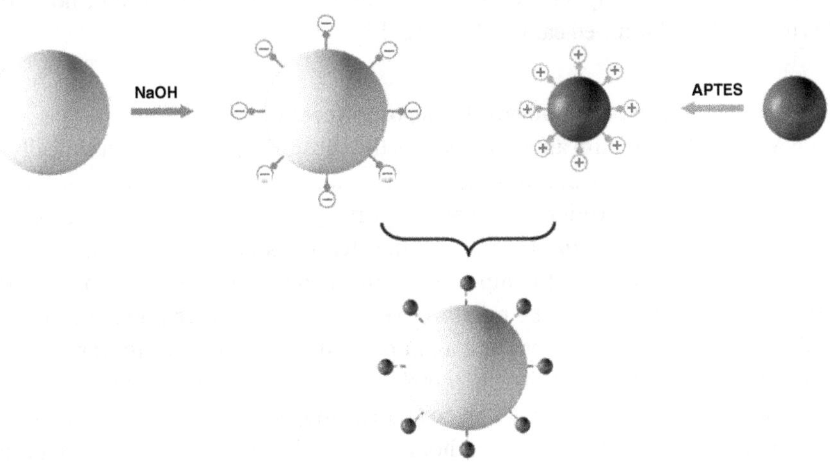

Figure 3.40 Schematic diagram of PCL@TiN preparation.

on their surfaces [117, 118]. That is, 2 g of NaOH was dissolved in a 100 ml mixture of DI water and ethanol. Next, 1 g of the PCL microspheres was fully dispersed into the mixture, followed by ultrasonic treatment for five minutes. The products (PCL–COOH) were centrifuged at 5000 rpm for 3 minutes, washed with DI water three times, and freeze-dried for 24 hours. Next, 2 ml of APTES was dissolved in 100 ml of a mixture of DI water and ethanol. TiN NPs (0.2 g) were dispersed into the mixture and stirred for four hours [119]. The products (TiN–NH_2) were then centrifuged at 8000 rpm for five minutes, washed three times with ethanol, and dried in an oven.

Finally, the PCL–COOH and TiN–NH_2 were coupled to prepare PCL@TiN microspheres via dispersion in DI water and ethanol, respectively [120]. The two suspensions were mixed and stirred at 800 rpm for eight hours. Three ratios of TiN–NH_2 to PCL–COOH were produced at 1 : 10, 2 : 10, and 3 : 10. The PCL@TiN products were then centrifuged, washed, and freeze-dried.

3.4.1.5 Morphological and Structural Characterization of PCL@TiN

The morphology and elemental composition of the TiN NPs, PCL, and PCL@TiN were characterized using SEM and TEM observation. Their zeta potentials were characterized using an electrokinetic analyzer. The light absorption capacity of PCL@TiN was then evaluated using UV–Vis spectroscopy.

3.4.1.6 Photothermal Conversion Performance of PCL@TiN

The photothermal conversion performance of the PCL@TiN composite microspheres was evaluated using a thermocouple and an infrared camera. First, 0.05 g of PCL and PCL@TiN microspheres were separately dispersed in 100 ml of DI water. To evaluate the photothermal conversion, an 808 nm NIR laser was used as the light source to irradiate the PCL and PCL@TiN dispersion solutions for 400 seconds. A time-dependent temperature evolution was recorded using the thermocouple thermometer and infrared camera [121, 122].

3.4.1.7 Preparation of Photothermal Composite Coating

An AA2024-T3 aluminum alloy was used as the substrate, polished ultrasonically in anhydrous ethanol for 10 minutes, cleaned with acetone, dried, and stored in a dryer for experimental application. A shape-memory epoxy matrix was constructed by mixing DGEBA, NGDE, and D230. First, the DGEBA and NGDE were thoroughly mixed, and then PCL@TiN composite microspheres with mass fractions of 5%, 10%, and 15% were added. The mixtures were thoroughly stirred, followed by the addition of a D230 curing agent and additional stirring. The mixture was dropped onto a metal substrate until it was completely coated, and a rotating coater was used for 30 seconds at 300 rpm. The coated substrates were cured for 24 hours at room temperature and dried for 12 hours in a 50 °C oven. Coatings containing 0%, 5%, 10%, and 15% PCL@TiN microspheres were prepared and labeled Epoxy, Epoxy/PT-5, Epoxy/PT-10, and Epoxy/PT-15, respectively. Atmospheric corrosion monitors coated with Epoxy and Epoxy/PT-10 coatings were prepared to monitor the galvanic corrosion currents.

3.4.1.8 Characterization of Epoxy/PT Coating

The thermal properties of the PCL microspheres and different coatings were evaluated via DSC from −10 to 110 °C under a nitrogen atmosphere.

3.4.1.9 Photothermal Conversion Property of Epoxy/PT Coating

To characterize the photothermal effects of the Epoxy/PT coating, the coating surface was illuminated using an NIR light source with different power densities, and the real-time evolution of the coating surface temperature was recorded using a thermometer. Furthermore, photothermal self-healing of different coatings was performed. Artificial scratches were made on the surfaces of coated samples using a scalpel, and the coating scratches were then irradiated with a laser at 3.0 W cm^{-2} for 30 seconds. The morphologies of the coatings were then characterized using CLSM and SEM equipment.

3.4.1.10 Corrosion Resistance and Self-healing Property of Epoxy/PT Coating

The corrosion resistance and self-healing properties of both intact and scratched coatings were evaluated using EIS. SECM was conducted on the scratched areas to monitor the local corrosion activity [123, 124]. A conventional four-electrode system was used, and the distance between the probe tip and the coating surface was 25 μm. The tip potential was −0.75 V versus SCE.

ACM sensors coated with Epoxy and Epoxy/PT-10 coatings were also used to evaluate variations in corrosive currents. As shown in Figure 3.41, the ACM sensor comprises seven Al/graphite galvanic couples [125–127]. To monitor its self-healing behavior, an artificial scratch was made on the coating surface across three pairs of Al/graphite electrode couples, and a water droplet of 6 μl was dropped on the scratched regions. NIR laser illumination was then employed to evaporate the water from the coating surface until complete evaporation. To simulate a practical environment, coated ACM sensors were used for outdoor exposure tests.

3.4.2 Preparation and Characterization of PCL@TiN

3.4.2.1 Morphology and Structure Characterization of PCL@TiN

A zeta potential test was performed to demonstrate the formation of the PCL@TiN microspheres. Figure 3.42 shows the zeta potentials of the different samples. The pristine TiN NPs exhibited a negative ζ potential of −15.7 mV, which increased to

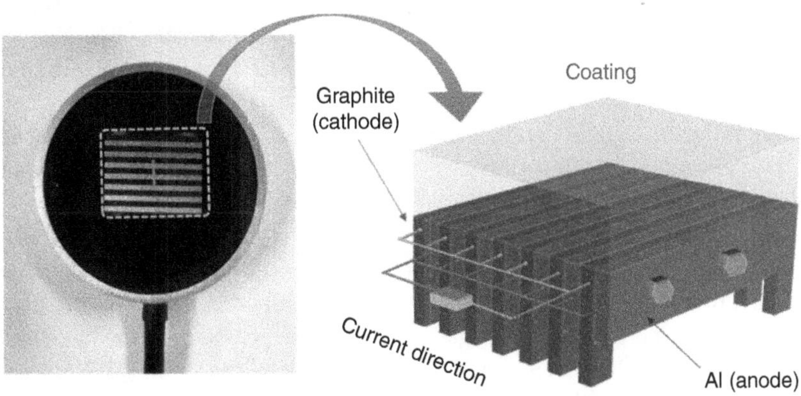

Figure 3.41 Al/graphite-type ACM sensor and its schematic diagram.

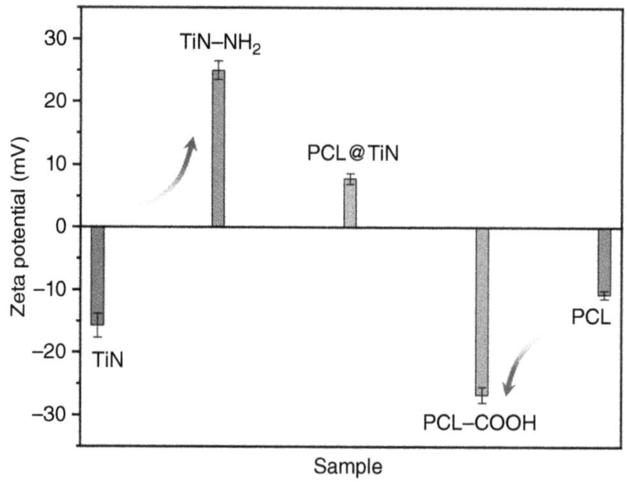

Figure 3.42 Zeta potentials of TiN, TiN−NH_2, PCL, PCL−COOH, and PCL@TiN.

+25.0 mV after the modification of -NH_2 groups [128]. In contrast, the ζ potential of the blank PCL microspheres further decreased from −10.7 to −26.8 mV due to the formation of -COOH groups on the microspheres [129]. After coupling PCL–COOH and TiN–NH_2, the interaction between the negatively charged PCL–COOH and the positively charged TiN–NH_2 lead to a potential of +7.8 mV for the PCL@TiN composites. The zeta potential results confirmed the formation of PCL@TiN.

Figure 3.43 shows the SEM images of the TiN NPs, PCL, PCL–COOH microspheres, and PCL@TiN microspheres. It is obvious that the TiN NPs possessed an average size of ~30 nm, and the PCL microspheres presented a uniformly dispersed spherical morphology with diameter distributions between 2 and 10 μm. After modification with the -COOH groups, the sizes of the PCL microspheres exhibited few changes. For the prepared PCL@TiN composite microspheres, their surfaces were completely covered with TiN NPs [130].

To assess the effects of the amount of TiN NPs on the morphology of PCL@TiN microspheres, microspheres with different weight ratios of TiN–NH_2 NPs to PCL–COOH microspheres (1,10 and 3,10) were prepared. Figure 3.44 shows that the surfaces of the PCL microspheres were insufficiently covered by TiN NPs when the weight ratio was 1 : 10. At a ratio of 3 : 10, the number of TiN NPs was excessive. Thus, a weight ratio of 2 : 10 was determined to be optimal for the preparation of PCL@TiN microspheres.

The PCL microspheres were also mixed with TiN NPs. Figure 3.45 shows that this direct mixing did not create a hierarchical structure due to insufficient surface functional groups and identical surface electrical properties. That is, the zeta potentials of PCL and TiN were all negative. In contrast, after surface modification, PCL–COOH and TiN–NH_2 were successfully combined to create PCL@TiN microspheres via the electrostatic interactions between the -COOH and -NH_2 groups (see Figure 3.45).

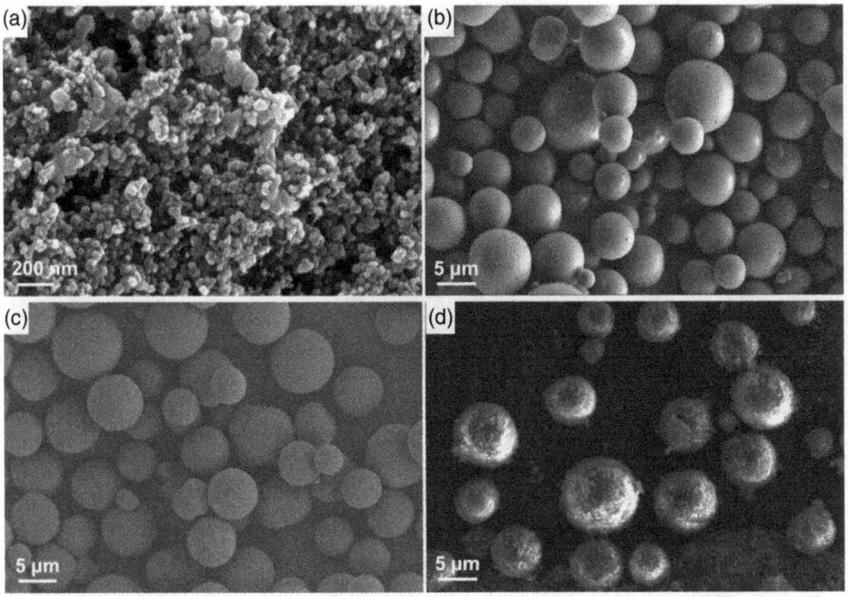

Figure 3.43 SEM images of (a) TiN NPs, (b) PCL microspheres, (c) PCL–COOH microspheres, and (d) PCL@TiN microspheres.

Figure 3.44 SEM images of PCL@TiN microspheres prepared with different weight ratio between TiN–NH_2 NPs and PCL–COOH: (a) 1 : 10, (b) 3 : 10.

3.4.2.2 Photothermal Conversion Properties of PCL@TiN

Figure 3.46 shows that the dispersion solution of PCL microspheres ($0.5\,g\,l^{-1}$) had a very low light absorption in the wavelength from 300 to 1100 nm. In contrast, the PCL@TiN microsphere suspension exhibited a strong light absorption capacity from 500 to 1100 nm [131].

Owing to their intense light absorption performance, the PCL@TiN microspheres were expected to achieve a desirable photothermal effect. As shown in Figure 3.47, the temperature variation of the PCL@TiN dispersion solution under 808 nm irradiation was recorded to evaluate the photothermal effects. The temperature of the PCL@TiN suspension increased rapidly within 80 seconds of illumination

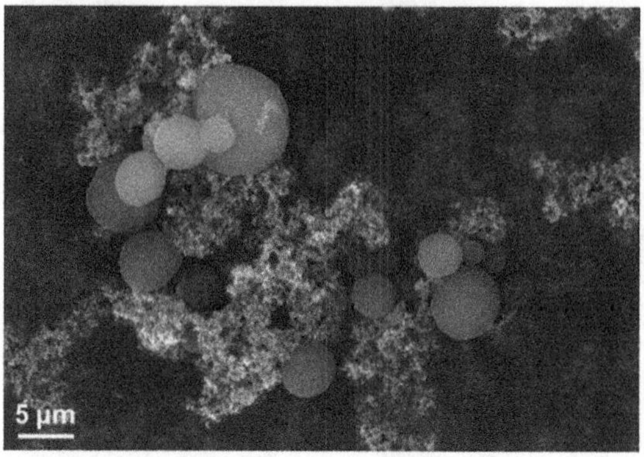

Figure 3.45 SEM images of the direct mixture of TiN NPs and PCL microspheres.

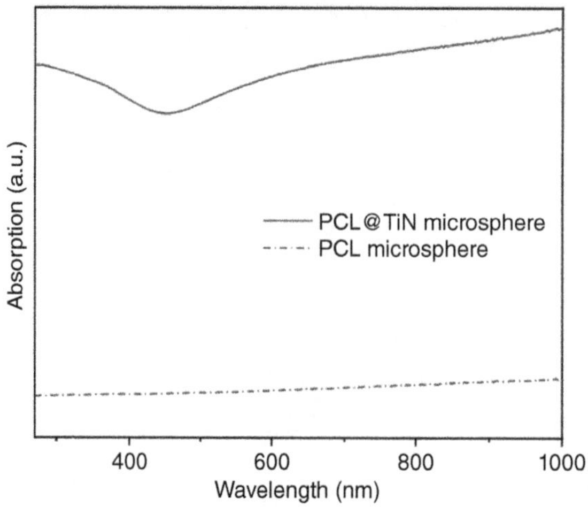

Figure 3.46 Absorption spectra of PCL dispersion and PCL@TiN dispersion.

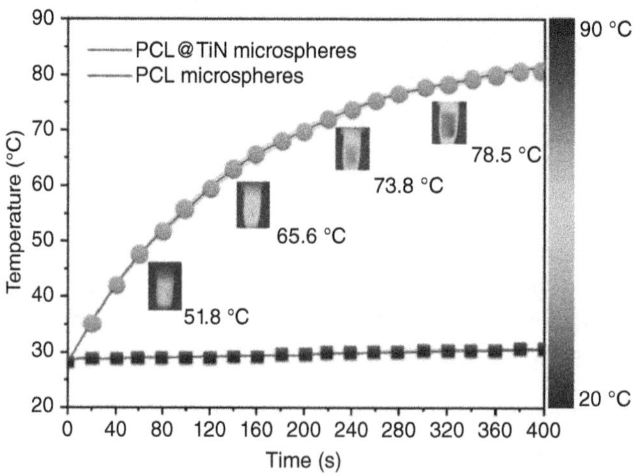

Figure 3.47 Temperature variations and infrared thermal images of PCL and PCL@TiN suspension during NIR irradiation [115].

Figure 3.48 SEM images of PCL@TiN composite microspheres after 30 seconds of NIR light irradiation.

from room temperature to 51.8 °C. After 400 seconds, the temperature of PCL@TiN suspension reached a plateau at ~80 °C, whereas the temperature of the PCL dispersion rose only 2 °C after the same period with the same irradiation. Infrared thermal images confirmed the photothermal properties of the PCL@TiN microspheres.

Figure 3.48 presents a SEM image of the PCL@TiN configuration after laser irradiation for 30 seconds. The microspheres completely melted due to the photothermal effects of TiN, indicating that the PCL@TiN microspheres could melt and seal the scratches during self-healing.

3.4.3 Preparation and Characterization of Epoxy/PT Photothermal Coatings

From the previous experimental results, PCL@TiN composite microspheres with excellent photothermal properties were prepared. After 30 seconds of NIR irradiation, the PCL@TiN microspheres would fully melt. Although the PCL@TiN microspheres have excellent photothermal conversion properties, the photothermal conversion capacity and the heat generation properties of the shape-memory polymer coating after incorporating PCL@TiN composite microspheres still need to be further investigated.

Hence, different amounts of composite microspheres were added to the shape-memory coating matrix, and NIR light was used to trigger photothermal conversions in the damaged coating area to explore the effects of photothermal action on the shape-memory effect of coating and the melting of PCL microspheres, clarifying the dual self-healing mechanism of the photothermal PCL@TiN (Epoxy/PT) coating. Moreover, the influence of PCL@TiN addition on the water evaporation performance of the coating was then evaluated, including an application of the Epoxy/PT coating to ACM sensors in an outdoor environment.

3.4.3.1 Characterization of Epoxy/PT Coating

First, DSC measurements were performed on the PCL microspheres and different coating samples [132]. Figure 3.49 shows that the DSC curves of PCL microspheres and blank epoxy coating possessed obvious endothermic peaks at ~45 and ~62 °C, respectively. The DSC curves of Epoxy/PT-5, Epoxy/PT-10, and Epoxy/PT-55 coatings all exhibited two endothermic peaks at ~45 and ~64 °C, which can be attributed

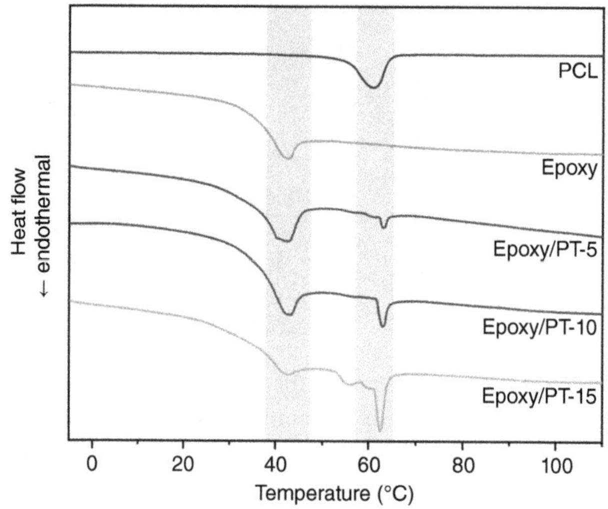

Figure 3.49 DSC curves of PCL and different coatings.

to the T_g of the epoxy matrix and the T_m of the PCL. As such, the shape recovery of the epoxy matrix and the melting of PCL microspheres for self-healing could be triggered when the surface temperature exceeds 64 °C [133].

3.4.3.2 Photothermal Conversion Property of Epoxy/PT Coating

NIR laser irradiation at $3.0\,W\,cm^{-2}$ was applied at 120 seconds intervals to test the stability of the photothermal conversion performance of the Epoxy/PT-10 coating (see Figure 3.50). When the light source was on, the coating temperature increased dramatically to ~90 °C. When the light source was turned off, the coating rapidly decreased to room temperature. From the shapes of the respective cycle curves, it

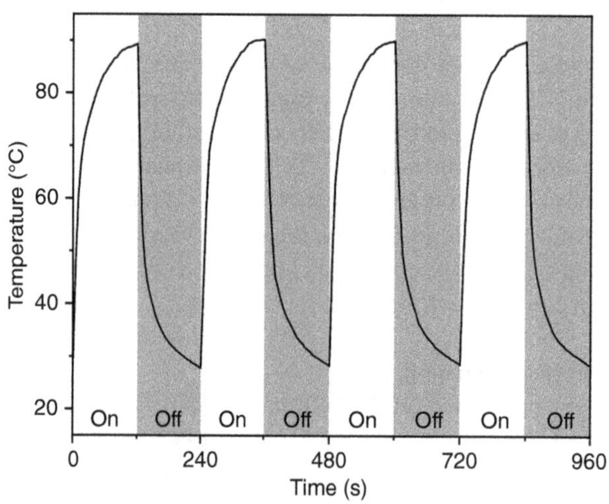

Figure 3.50 Temperature variations of Epoxy/PT-10 coating within four on/off cycles.

can be seen that the four on/off cycles showed reasonably stable heat production effects. These results demonstrate that the coating has a stable photothermal effect for achieving a desirable dual-action self-healing property and that laser irradiation does not damage the structure of the coating matrix [134].

The influence of photothermal nanofiller (PCL@TiN) content on the shape-memory recovery of the scratched coatings was further evaluated. Figure 3.51 presents optical images of the different coatings before and after 30 seconds of NIR irradiation. After 30 seconds, the scratched areas of the samples exhibited different recovery results. For the epoxy coating, the initial scratch size (~70 μm) exhibited little change. For the Epoxy/PT-5 coating, the gap was still obvious. For Epoxy/PT-10 and Epoxy/PT-15 coatings, the scratches were nearly closed, leaving a gap of only a few microns. These results demonstrate that the Epoxy/PT-10 and Epoxy/PT-15 coatings could generate sufficient heat to induce the evident shape-memory recovery.

In addition to surface morphology, the cross-sectional morphology of the Epoxy/PT-10 coating was examined after healing. As shown in Figure 3.52, the pristine scratch of ~70 μm decreased greatly to less than 10 μm, indicating photothermally triggered shape-memory recovery. In addition, the remaining scratch was filled with melted PCL@TiN microspheres.

3.4.3.3 Self-healing Properties of Scratched Coatings

To further investigate the self-healing properties, an artificial scratch was made on the coating surface, which was then exposed to NIR laser irradiation for 30 seconds. The scratched area of the coating was immersed in 3.5 wt% NaCl solution for seven days. Figure 3.53 shows the Nyquist diagrams and Bode impedance plots of the different coatings.

The initial $|Z|_{0.01\text{Hz}}$ value of Epoxy/PT-10 coating was $2.6 \times 10^8\ \Omega\,\text{cm}^2$, which was much higher than that of the blank epoxy ($4.8 \times 10^6\ \Omega\,\text{cm}^2$). With an increasing immersion time, the $|Z|_{0.01\text{Hz}}$ value of the blank epoxy decreased to $1.3 \times 10^6\ \Omega\,\text{cm}^2$ after seven days. These results can be explained by the continuous corrosion activity of wide-open scratches. The $|Z|_{0.01\text{Hz}}$ value of the Epoxy/PT-10 coating held steady at

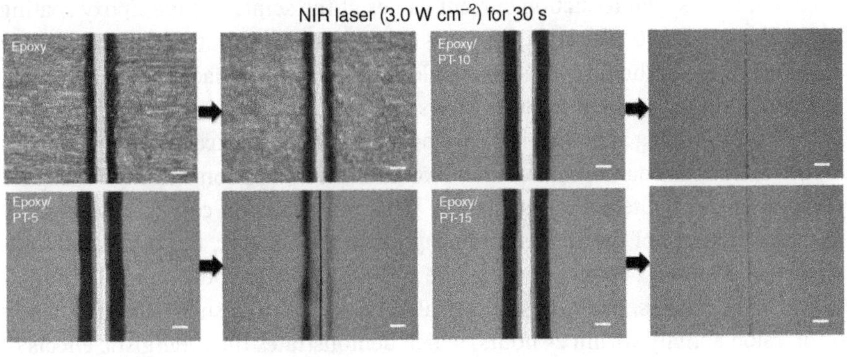

Figure 3.51 Shape recovery performance of different coatings within 30 seconds of NIR laser irradiation, scale bar: 100 μm [115].

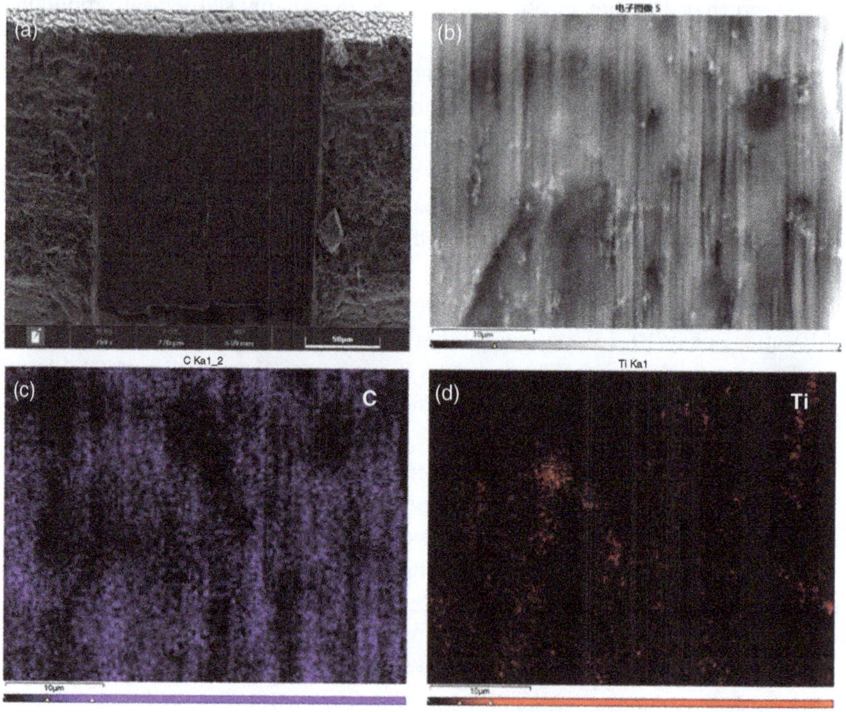

Figure 3.52 (a) Cross-section image of the healed Epoxy/PT-10 coating, (b–d) SEM and EDS result of PCL@TiN microsphere on the cross-section of Epoxy/PT-10 coating.

$\sim 2.0 \times 10^8\ \Omega\ cm^2$ after seven days of immersion. On one hand, the scratch was physically closed due to its narrow shape. On the other hand, the photothermal conversion triggered the melting of PCL in the PCL@TiN microspheres filled the damaged area, effectively suppressing the invasion of the corrosion medium [135].

To further clarify the self-healing mechanism, SECM measurements were performed on different coatings. Figure 3.54 presents these results for different samples. As shown, the oxygen reduction current values at the scratch of the Epoxy coating were as low as $\sim 2 \times 10^{-10}$ A after two hours of immersion, indicating that oxygen was consumed by the cathodic corrosion reaction on the metal surface [136, 137]. When the reaction time was extended to 24 hours, the corrosion activity further increased. For the Epoxy coating after 30 minutes of heating at 80 °C, the oxygen reduction current intensity at the damaged area after two hours of immersion was slightly higher than that of the scratched epoxy coating. This phenomenon can be explained by the shape memory of the scratches during heating. However, the corrosion activity intensified after 24 hours because the physical closure could not fully inhibit the corrosion process. In contrast, the healed Epoxy/PT-10 results exhibited nearly no corrosion activity within 24 hours, which demonstrates the synergistic effects of shape-memory recovery; hence, melted PCL can substantially enhance self-healing performance [138].

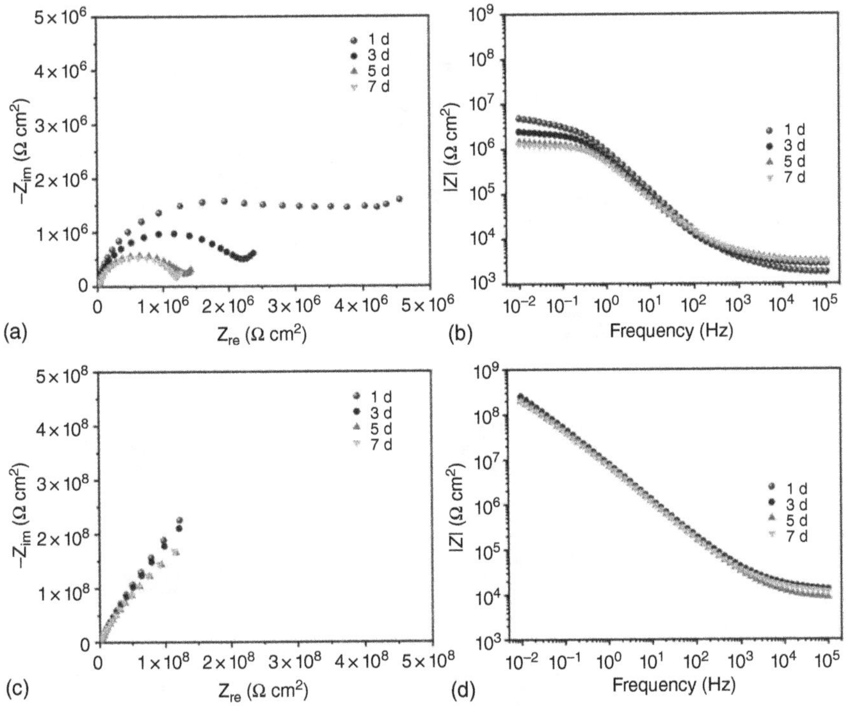

Figure 3.53 Nyquist diagrams of (a) irradiated epoxy coating and (c) healed Epoxy/PT-10 coating, and Bode impedance plots of (b) irradiated epoxy coating and (d) healed Epoxy/PT-10 coating during immersing in 3.5 wt% NaCl solutions [115].

3.4.3.4 Photothermal-induced Water Evaporation on Epoxy/PT Coatings

Water plays a necessary role in corrosion reactions, and the accumulation of water on a metal's surface leads to the formation of a thin liquid film that induces corrosion [139, 140]. Therefore, corrosion can be suppressed by accelerating water evaporation. Figure 3.55 presents the evaporation time of water droplets on the surface of the blank Epoxy and Epoxy/PT-10 coatings under laser irradiation. It can be seen that the evaporation time gradually increased with the increase in the volume of the water droplet. By comparison, the evaporation time of water droplets of different volumes on the blank Epoxy coating surfaces took much longer. The full evaporation time of 2 μl water droplets on the blank epoxy coating was ~135 seconds. When the water droplet volume increased to 10 μl, the evaporation time increased to ~260 seconds.

3.4.3.5 Outdoor Application Performance of Epoxy/PT Coatings

The self-healing ability of the Epoxy/PT-10 coating triggered by sunlight irradiation in an outdoor environment was evaluated by placing the coated sensors in a natural atmosphere [141, 142]. Outdoor exposure experiments were conducted in Beijing, China, from 15 : 00 on 23 June until 20 : 00 on 24 June 2021. As shown in Figure 3.56, the current output was very small for both sensors covered by the scratched Epoxy

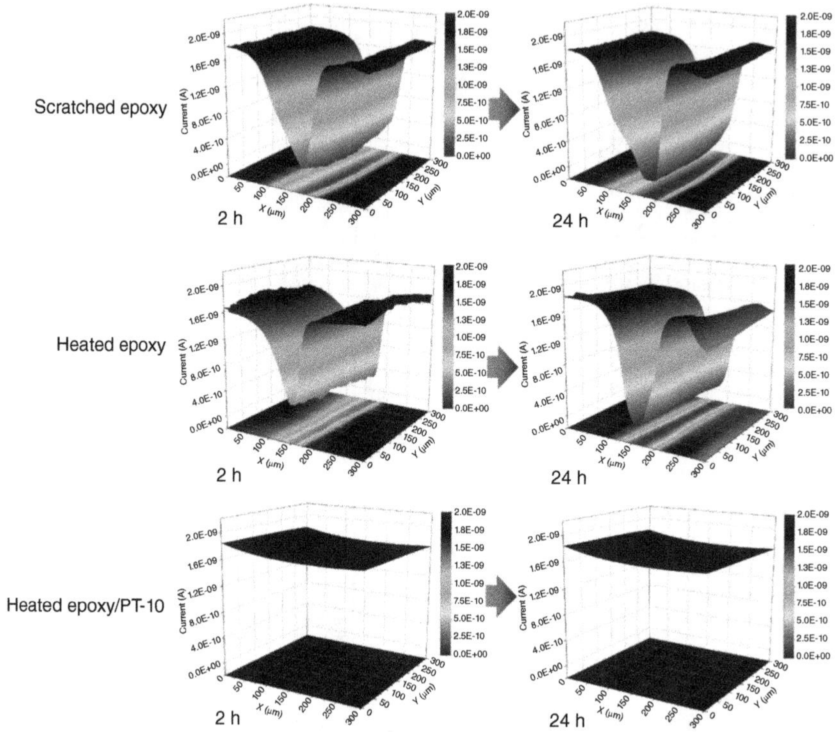

Figure 3.54 SECM current of different coatings [115].

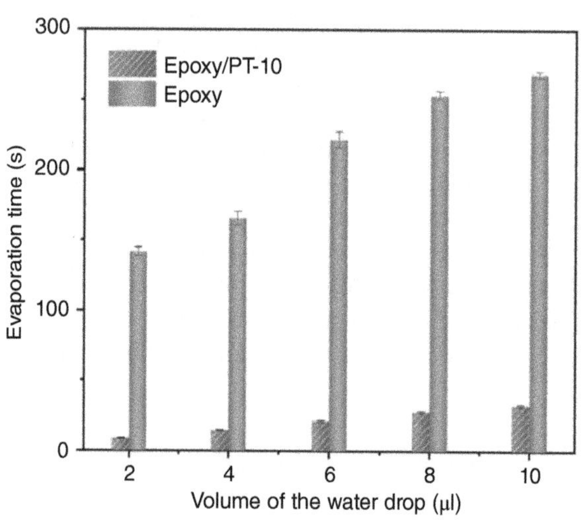

Figure 3.55 Evaporation time of water on coatings under NIR illumination.

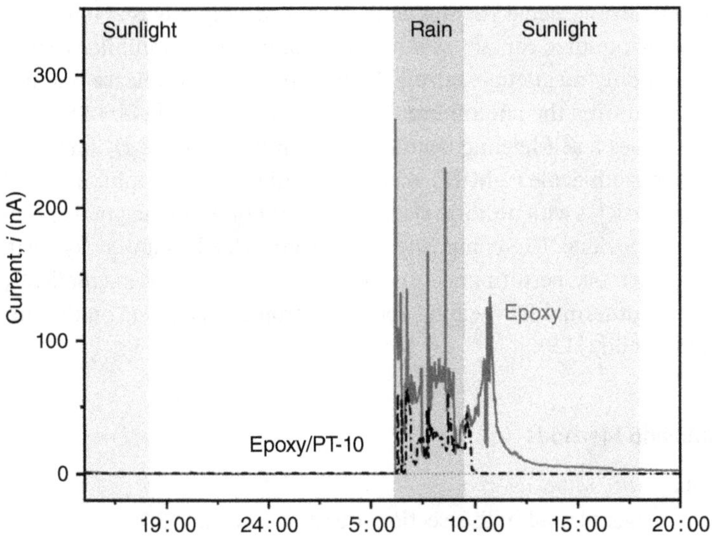

Figure 3.56 The current values of ACM sensors coated with scratched epoxy and Epoxy/PT-10 coatings during the outdoor exposure. Source: Wang et al. [115]/with permission of Elsevier.

and Epoxy/PT-10 coatings owing to the absence of corrosive media at the beginning of the test. Corrosion activity began at 18 : 40 on June 24, when the current increased dramatically as rain droplets created coating defects. The current outputs of the ACM sensor coated with the blank Epoxy coating fell in the range from 150 to 250 nA. In contrast, the ACM sensor coated with Epoxy/PT-10 showed much weaker corrosion activity during rainfall. The current output values were less than 50 nA, a reduction of more than three times compared with that of the blank epoxy-coated sensor. This phenomenon can be attributed to the photothermally induced self-healing effect activated by sunlight irradiation in the afternoon of 23 June. These results indicate that sunlight irradiation not only trigger the self-healing effects of the Epoxy/PT-10 coating but also accelerated water evaporation, endowing the coating with improved potential for outdoor applications.

3.5 Self-healing Coating Based on Fe_3O_4–MBT Photothermal Fillers

In recent years, magnetic Fe_3O_4 nanoparticles have attracted extensive attention in the fields of biomedicine, magnetic storage media, and magnetic resonance imaging owing to their superior chemical stability, strong superparamagnetism, good biocompatibility, and low toxicity [143–145]. Hollow Fe_3O_4 nanostructures are ideal carriers in drug delivery systems due to their high loading capacity and good magnetic responses, which reduce leakage during delivery. Various drugs, such as doxorubicin, paclitaxel, and acetylene, have been successfully loaded into hollow Fe_3O_4

nanostructures to realize targeted therapy and controlled drug release [146, 147]. As such, these nanostructures can act as a new type of corrosion inhibitor carrier that can achieve not only magnetic-controlled corrosion inhibitor release but also photothermal effects during the photothermal conversion of Fe_3O_4 [148, 149].

This section describes a self-healing coating based on magnetic Fe_3O_4 nanoparticles. 2-Mercaptobenzothiazole (2-MBT) was used as the corrosion inhibitor, and Fe_3O_4 hollow nanoparticles with uniform size, porous structure, and magnetic properties were used as carriers. The composite fillers were added to an epoxy resin matrix with shape-memory performance to construct a new type of magnetically controllable and photothermal self-healing corrosion protection coating on the surface of an aluminum alloy [150].

3.5.1 Materials and Methods

3.5.1.1 Materials

The materials and reagents used in this section are shown in Table 3.7.

3.5.1.2 Instruments

The instruments used in this section are shown in Table 3.8.

3.5.1.3 Preparation of Fe_3O_4 Hollow Nanoparticles

Monodisperse Fe_3O_4 hollow nanoparticles were synthesized using a hydrothermal method [151]. The specific experimental methods are as follows: In a 200 ml beaker, 1.00 g of $FeCl_3 \cdot 6H_2O$, 2.40 g of sodium citrate, and 0.80 g of urea were added and 80 ml of distilled water was added to dissolve the substances. Then, 1.2 g PAAS was added under continuous agitation until it completely dissolved. The solution was transferred into a 100 ml high-pressure reactor and sealed at 200 °C for four hours. The harvested products were then magnetically separated, washed four times with

Table 3.7 Experimental materials and reagents.

Materials	Purity	Company
Ferric chloride ($FeCl_3 \cdot 6H_2O$)	Analytically pure	Aladdin Corporation
Sodium citrate ($C_6H_5 \cdot Na_3O_7 \cdot 2H_2O$)	≥99%	Aladdin Corporation
Urea (CH_4N_2O)	99%	Aladdin Corporation
Sodium polyacrylate (PAAS)	Analytically pure	Aladdin Corporation
Anhydrous ethanol (C_2H_5OH)	Analytically pure	Sinopharm Group
Sodium chloride (NaCl)	Analytically pure	Sinopharm Group
2-Mercaptobenzothiazole (2-MBT)	98%	Aladdin Corporation
DGEBA	Analytically pure	Sigma Corporation
NGDE	Analytically pure	Aladdin Corporation
Jeffamine D230	Analytically pure	Aladdin Corporation

Table 3.8 Instruments and equipment.

Instrument	Production type	Manufacturer
Field emission transmission electron microscope	JEM-2100F	Japan JEOL
X-ray diffractometer	D/Max-2500	Japan Rigaku
Comprehensive physical property measurement system	PPMS Dyna Cool	United States Quantum Design
Fourier transform infrared spectrometer	Vector 22	Germany Bruker
Ultraviolet–visible spectrometer	U-3900/3900H	Japan Hitachi
Centrifuge	Allegra X-30	United States Beckman
Electrochemical workstation	PARSTAT-2273	United States Princeton
Laser scanning confocal microscope	VK-X2000 series	Japan KEYENCE
Thermocouple thermometer	TES-1310	China TES
Constant temperature blast drying oven	303-3S	China Baihui

distilled water and anhydrous ethanol, and dried in a vacuum oven at 50 °C until all liquid products were removed.

3.5.1.4 Preparation of Fe_3O_4 Nanoparticles Loaded with 2-MBT

Fe_3O_4/2-MBT nanocomposites were prepared by loading 2-MBT corrosion inhibitors into the Fe_3O_4 hollow nanoparticles via vacuum-negative pressure-assisted impregnation. Specific experimental methods were as follows: 0.50 g of Fe_3O_4 was dispersed into a 2-MBT ethanol solution with a concentration of 40 mg ml^{-1}, and ultrasonic treatment was conducted to form a uniform suspension. The suspension was then transferred to a flask and evacuated using a vacuum pump. The system was kept under a high vacuum for one hour, restored to normal pressure, and mechanically stirred for three hours. Subsequently, the sample was kept under vacuum for 1 hour and dried in air for 24 hours to obtain the Fe_3O_4/2-MBT nanocomposites.

3.5.1.5 Preparation of Fe_3O_4/2-MBT Containing Epoxy Coatings

An AA2024-T3 aluminum alloy was selected as the metal material for this experiment to investigate the effects of the Fe_3O_4/2-MBT filler content on the anticorrosion properties and self-healing ability of its coatings. The thickness of all coatings was controlled between 60 and 80 μm. The substrates were polished using 150 and 240 sandpaper until they exhibited a metallic luster. The polished metal substrates were placed in anhydrous ethanol for 20 minutes, covered with a plastic wrap, and sealed in a beaker. The shape-memory epoxy matrix was constructed in the ratio $n_{DGEBA}:n_{NGDE}:n_{D230} = 1:1:1$. 1.43 g of DGEBA, 0.91 g of NGDE (mixed evenly), and

0.0993 g (3 wt% relative to the weight of coating) Fe_3O_4/2-MBT nanocomposites were added. The mixture was fully stirred for 10–15 minutes to disperse it evenly, and a 0.97 g curing agent (D230) was added and stirred well. The coating was applied to the metal surface using a rotating machine for 30 seconds at 200 rpm. The coatings were prepared with and without magnetic fields. For the magnetic version, a magnet was placed under the metal substrate, and the Fe_3O_4/2-MBT nanocomposites were uniformly adsorbed onto the bottom of the coating as it was closer to the metal substrate [152]. The coatings were prepared by curing in an oven at 55 °C for 24 hours. This sample was labeled Fe_3O_4/2-MBT-epoxy. For the nonmagnetic version, a bare epoxy coating without nanofillers was prepared using the same method.

3.5.1.6 Characterization of Fe_3O_4 Hollow Nanoparticles

The microstructure and morphology of the Fe_3O_4 hollow nanoparticles were observed using TEM. Fe_3O_4 nanoparticles were dispersed in anhydrous ethanol and dripped onto an ultrathin carbon-supporting film for natural air drying. XRD was used to analyze the crystal face index of the Fe_3O_4 hollow nanoparticles. In addition, a comprehensive physical property measurement system (PPMS) was used to test the magnetic properties of the Fe_3O_4 nanoparticles, and the hysteresis loop of the sample was obtained to determine the appropriate magnetic field for operation.

3.5.1.7 Characterization of Fe_3O_4/2-MBT Nanoparticles

A Fourier transform infrared spectrometer was used to obtain the infrared spectra of the Fe_3O_4/2-MBT nanoparticles. By comparing the FTIR spectra of Fe_3O_4 and 2-MBT, the chemical bonds and functional groups of the different samples were analyzed to determine the loading behavior of 2-MBT into the Fe_3O_4 nanoparticles. A UV–Vis–NIR spectrometer was used to measure the absorption spectrum of the nanoparticles by analyzing and comparing the light absorption characteristics of samples with different concentrations in a 3.5 wt% NaCl aqueous solution at a specific pH value. The loading rate of 2-MBT molecules in the Fe_3O_4/2-MBT nanoparticles was then calculated.

3.5.1.8 Characterization of Corrosion Protection and Photothermal Properties of Coatings

Experiments with a 3.5 wt% NaCl immersion were conducted on intact composite coatings with different composite filler contents, and electrochemical studies were carried out using EIS to compare the protective effects and failure behaviors of different coatings. The key effect of the Fe_3O_4/2-MBT addition on the corrosion resistance of the self-healing coating was then analyzed. To investigate the influence of the distribution of Fe_3O_4/2-MBT nanoparticles in the coating on their self-healing properties, the EIS measurement data of the Fe_3O_4/2-MBT-epoxy coating prepared with and without a magnetic field were tested.

The photothermal response of the coating was measured using DSC and the temperature-raising curves of the coating samples [153]. To characterize the photothermal conversion performance of the photothermal self-healing coatings, the temperature probe of the thermocouple thermometer was fixed on the back of the metal substrate. An 808 nm laser light source was used to illuminate the corresponding coating surfaces at the position of the probe, and the real-time temperature data of the coating surface were recorded using a thermometer. A VK-X2000 series laser confocal microscope was used to observe the surface morphologies of the scratch coatings before and after photothermal healing. A scratch about 8 mm long and 80 μm wide was made on the coating surface using a No. 11 scalpel. The 808 nm NIR light with 2 W power was irradiated for 30 seconds to heal the coating.

3.5.2 Preparation and Characterization of Fe_3O_4/2-MBT Nanoparticles

3.5.2.1 Composition and Structural Characterization of Fe_3O_4 Hollow Nanoparticles

The morphology of the Fe_3O_4 nanoparticles was observed by TEM, as shown in Figure 3.57a. Fe_3O_4 particles were homogeneous monodisperse nanospheres with a diameter of 200 nm. Figure 3.57a shows the electron diffraction images obtained from the TEM image of Fe_3O_4. The electron diffraction rings of the Fe_3O_4 hollow nanoparticles corresponded to the (110), (220), (331), (400), (511), and (440) crystal faces of Fe_3O_4 from the inside to the outside, further indicating that the prepared hollow nanoparticles were Fe_3O_4 and had a polycrystalline structure [150].

The crystal structures of the Fe_3O_4 hollow nanoparticles were studied using XRD. As shown in Figure 3.58, the diffraction peaks appear to correspond to the (110), (220), (311), (400), (422), (511), and (440) crystal faces of fcc Fe_3O_4. Compared with the standard Fe_3O_4 spectrum, the obtained sample was confirmed to be Fe_3O_4 [154].

The magnetic properties of the Fe_3O_4 hollow nanoparticles were characterized using a Quantum Design PPMS at 300 K. According to the hysteresis loop shown in Figure 3.59, the Fe_3O_4 nanoparticles exhibited strong superparamagnetism. The saturation magnetization (Ms) was 71.11 emu g^{-1}, and the coercivity was 14.59 Oe.

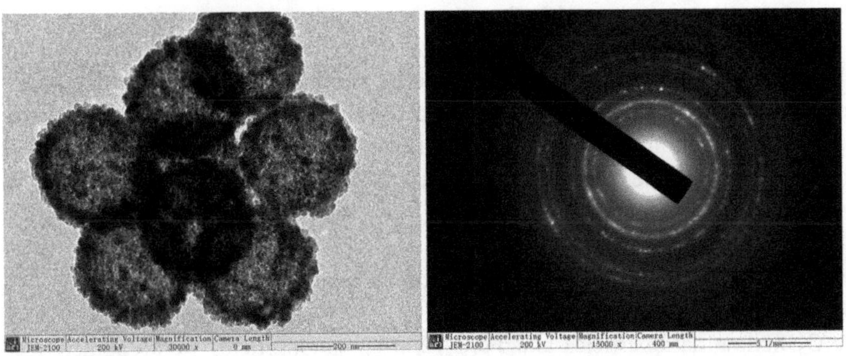

(a) (b)

Figure 3.57 (a) TEM observation image and (b) electron diffraction image of Fe_3O_4 nanoparticles.

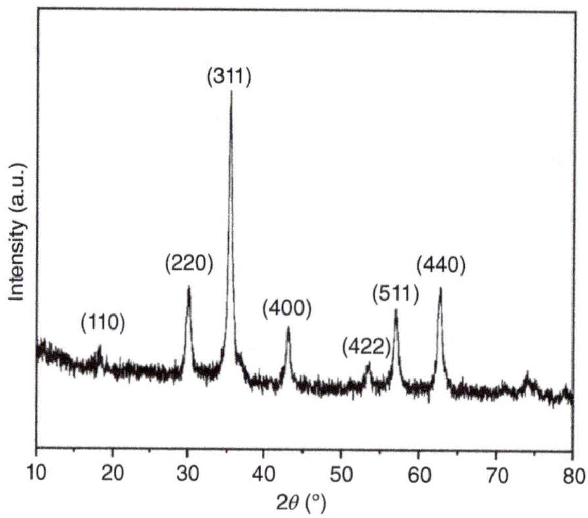

Figure 3.58 XRD pattern of Fe_3O_4 nanoparticles [150].

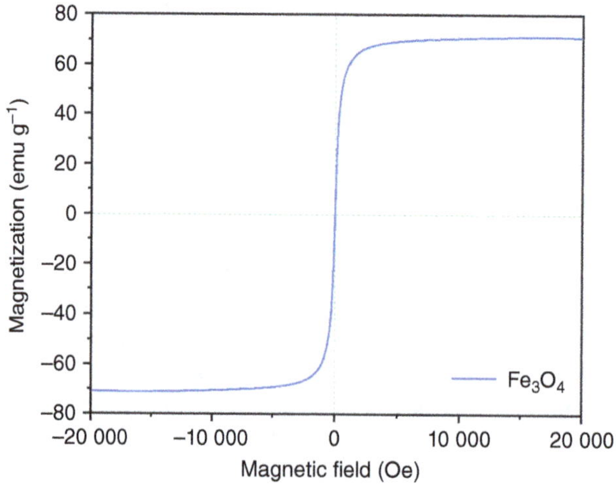

Figure 3.59 Hysteresis loop of Fe_3O_4 hollow nanoparticles.

The Fe_3O_4 hollow nanoparticles with superior paramagnetism exhibited outstanding advantages. For example, the position of the nanoparticles could be controlled to remain at the bottom of the coating so that the corrosion inhibitors would be closer to the metal substrate, which is conducive to adsorption on the metal surface after release.

3.5.2.2 Composition and Structural Characterization of Fe_3O_4/2-MBT Nanoparticles

Fe_3O_4 hollow nanoparticles and Fe_3O_4/2-MBT nanoparticles were characterized using an FTIR spectrometer. By comparing the three spectral lines in Figure 3.60, we determined that the sample was an ideal Fe_3O_4 magnetic particle containing the 2-MBT corrosion inhibitor. Its absorption peak at 580 cm^{-1} is the characteristic

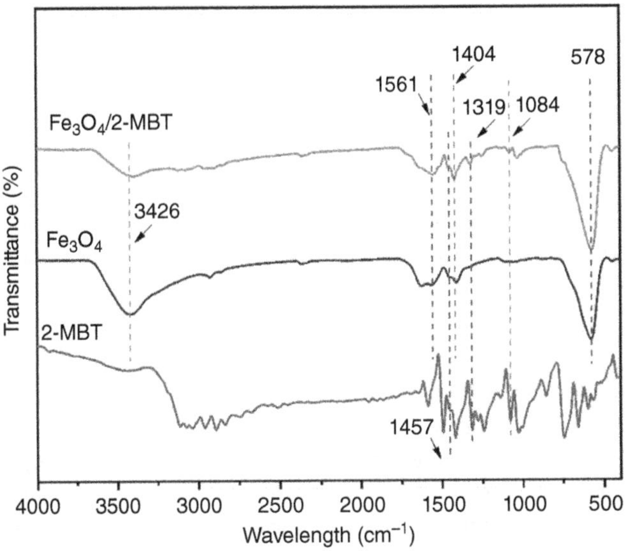

Figure 3.60 FTIR spectra of Fe_3O_4 nanoparticles, 2-MBT molecules, and Fe_3O_4/2-MBT nanoparticles [150].

vibration absorption peak of Fe–O in Fe_3O_4, whereas the 1084, 1319, and 1457 cm^{-1} peaks are the characteristic absorption peaks of 2-MBT, corresponding to the N=C–S tensile vibration of 2-MBT. The FTIR peaks at 1404 and 1561 cm^{-1} are the characteristic absorption peaks of COO^- from sodium polyacrylate. The 3426 cm^{-1} peak represents the tensile vibration of the O–H group [150].

The loading capacity of 2-MBT in the Fe_3O_4/2-MBT nanoparticles was characterized using UV–Vis absorption spectroscopy. Figure 3.61a shows the UV–Vis absorption spectra of 2-MBT molecules at different concentrations in a 3.5 wt% NaCl solution and the corresponding standard curve is shown in Figure 3.61b. The characteristic absorption peak value of 2-MBT corresponded to the different concentrations of 2-MBT, and a correlation scatter plot was drawn and fitted as a straight line [155].

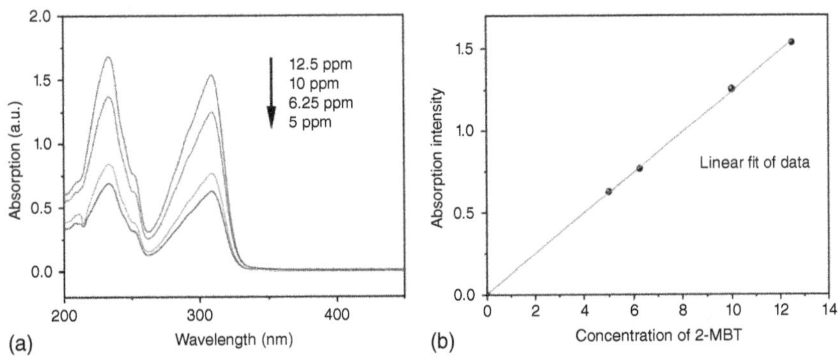

Figure 3.61 (a) UV–Vis absorption spectra and (b) standard curve of 2-MBT with different concentrations (5, 6.25, 10, and 12.5 ppm) in 3.5 wt% NaCl solution.

Standard curve fitting equation:

$$y = 0.123\,26x + 0.0034 \tag{3.2}$$

where y is the characteristic UV–Vis absorption peak value, and x is the concentration of 2-MBT.

The concentration of 2-MBT corresponds to the data point of the characteristic absorption peak at 309 nm. Therefore, by using the UV–Vis absorption intensity of the released 2-MBT from Fe_3O_4/2-MBT nanoparticles (Figure 3.62) and its corresponding standard curve (Figure 3.61b), the corrosion inhibitor loading rate in Fe_3O_4/2-MBT nanoparticles was calculated to be 7.4 wt%.

3.5.2.3 Corrosion Inhibition Properties of Fe_3O_4/2-MBT Nanoparticles

According to the Nyquist diagram, the arc radius of the AA2024-T3 sample in the presence of Fe_3O_4/2-MBT nanoparticles was much larger than that of the sample with 3.5 wt% NaCl solution without additives, showing better corrosion resistance when using the Fe_3O_4/2-MBT nanoparticles [156]. As shown in Figure 3.63, the aluminum alloy in the 3.5 wt% NaCl solution was corroded to some extent. Regarding the sample containing aluminum alloy, 2-MBT was successfully released from the Fe_3O_4/2-MBT nanoparticles to form an inhibitor layer on the metal surface. The arc radius of the AA2024-T3 sample in the presence of Fe_3O_4/2-MBT nanoparticles was larger, and it increased significantly with increasing immersion time.

A low-frequency impedance modulus ($|Z|_{0.01\text{Hz}}$) is an important parameter for evaluating the corrosion resistance of a material. Under the same voltage, the higher the $|Z|_{0.01\text{Hz}}$ value, the more difficult it is for the electrons to move in a directional manner and the more difficult it is to form an anode current, which results in better anticorrosion performance. From Figure 3.63a,b, it can be observed that the $|Z|_{0.01\text{Hz}}$ value of the sample in the 3.5 wt% NaCl solution was below $10^5\ \Omega\,\text{cm}^2$, whereas the $|Z|_{0.01\text{Hz}}$ of the sample in the presence of Fe_3O_4/2-MBT was always above $10^5\ \Omega\,\text{cm}^2$, and it increased gradually with time. This is because the 2-MBT corrosion inhibitor

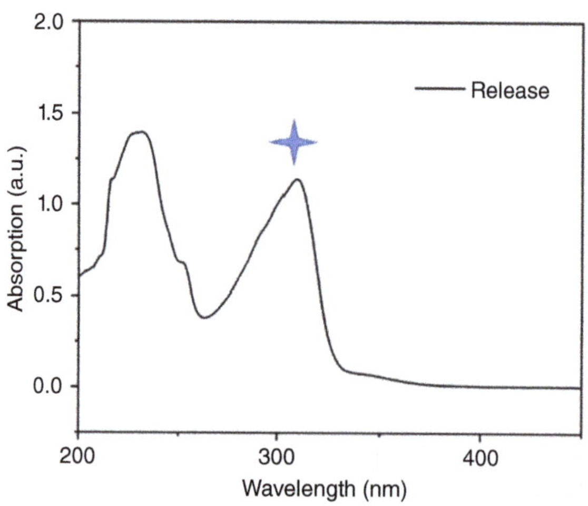

Figure 3.62 The UV–Vis absorption intensity of released 2-MBT from Fe_3O_4/2-MBT nanoparticles.

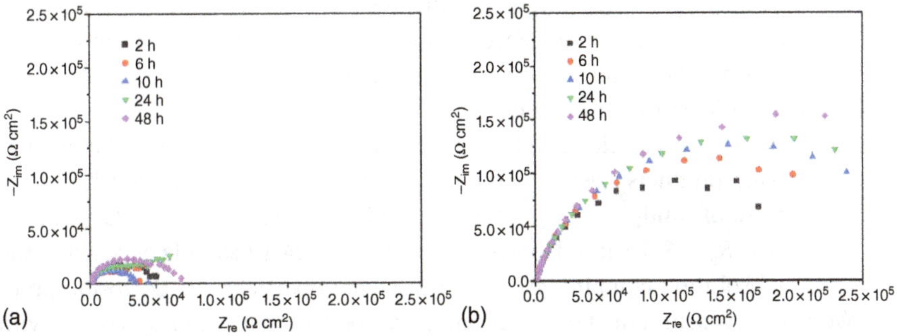

Figure 3.63 Nyquist diagram of AA2024-T3 aluminum alloys immersed in 3.5 wt% NaCl solution (a) and 3.5 wt% NaCl + 3 g l^{-1} Fe_3O_4/2-MBT solution (b) for 2, 6, 10, 24, and 48 hours, respectively [150].

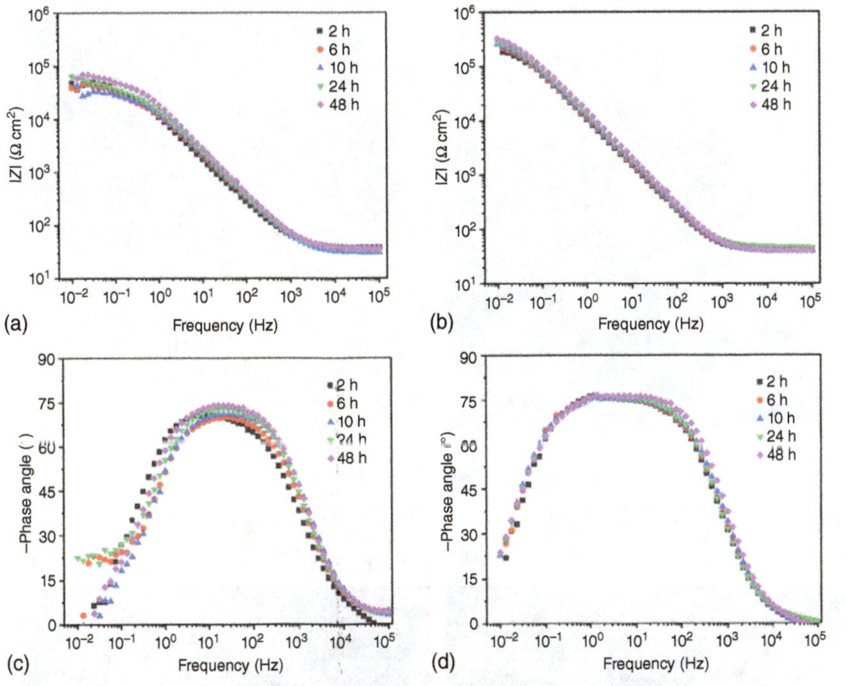

Figure 3.64 (a,c) Bode plots of AA2024-T3 aluminum alloys immersed in 3.5 wt% NaCl solution, and (b,d) Bode plots of AA2024-T3 aluminum alloys immersed in 3.5 wt% NaCl + 3 g l^{-1} Fe_3O_4/2-MBT solution for 2, 6, 10, 24, and 48 hours, respectively [150].

from Fe_3O_4/2-MBT was released over time and gradually adsorbed onto the metal substrate, effectively inhibiting the development of the corrosion process [157].

Regarding the AA2024 sample in the 3.5 wt% NaCl solution without additives, the $|Z|_{0.01Hz}$ value was low, and the frequency range of phase angles with high values was in the small range (Figure 3.64c). In contrast, for the sample containing Fe_3O_4/2-MBT, the $|Z|_{0.01Hz}$ value was higher, and the frequency range of the phase angle with high values was larger (Figure 3.64d). This is because the 2-MBT molecules were released from the Fe_3O_4/2-MBT nanoparticles into the salt solution and adsorbed onto the metal surface, slowing the reaction rate on the substrate and achieving an anticorrosion function.

From the CLSM observation results in Figure 3.65, it can be seen that a large number of corrosion products existed on the metal surface in the 3.5 wt% NaCl solution and the texture of sandpaper polishing was no longer visible. The surface roughness value was $R_a = 8.609\ \mu m$. However, for the AA2024-T3 sample exposed to the 3.5 wt% NaCl solution containing Fe_3O_4/2-MBT nanoparticles, no corrosion products were observed, and the texture of sandpaper polishing was clearly visible, with a surface roughness value $R_a = 0.679\ \mu m$. The significantly decreased R_a value can be

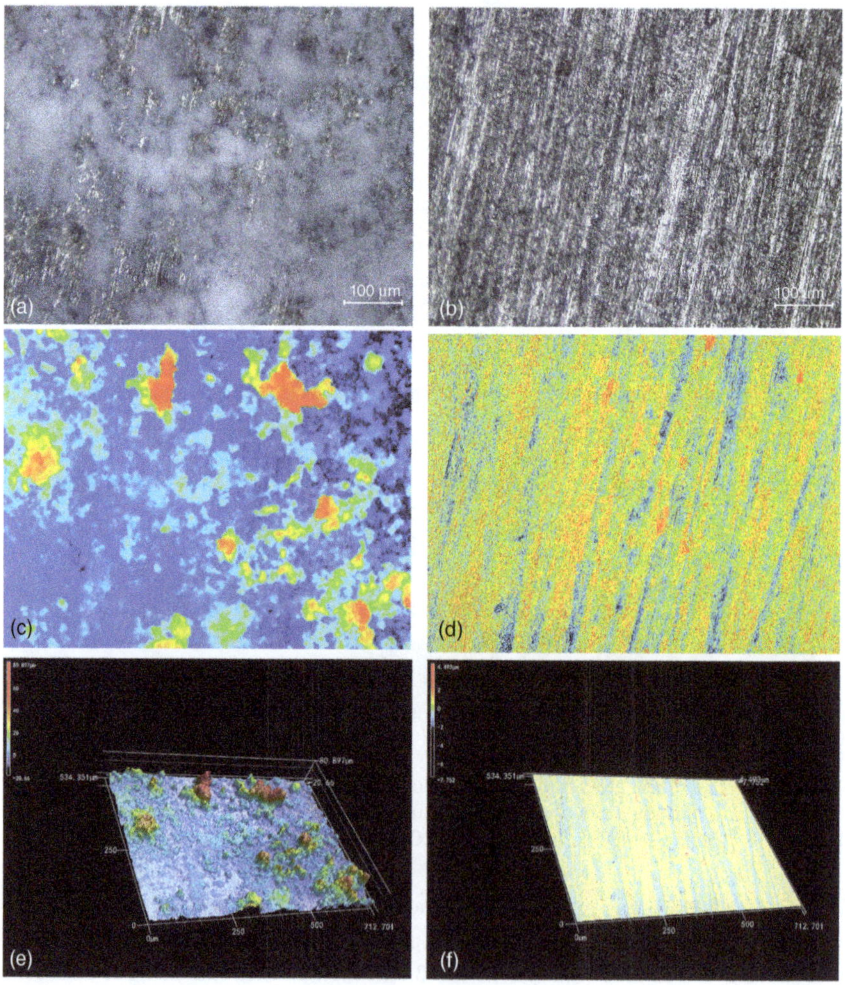

Figure 3.65 Confocal laser microscopy and surface roughness analysis of AA2024-T3 aluminum alloys in (a,c,e) 3.5 wt% NaCl solution, and (b,d,f) 3.5 wt% NaCl + 3 g l^{-1} Fe_3O_4/2-MBT solution after immersion for 72 hours.

ascribed to the suppressed corrosion activity on the metal surface. The results show that the Fe_3O_4/2-MBT nanoparticles as corrosion inhibitors can significantly reduce corrosion efficiency and improve the corrosion resistance of metallic materials [158]. Therefore, adding an appropriate amount of Fe_3O_4/2-MBT nanoparticles to the coating will facilitate the release and subsequent adsorption of corrosion inhibitors in the damaged areas of the coating and improve their active protection ability.

3.5.3 Preparation and Characterization of Fe_3O_4/2-MBT-epoxy Coating

3.5.3.1 Photothermal Response of Fe_3O_4/2-MBT Nanoparticle Containing Epoxy Coatings

First, the coating samples were tested using DSC measurements. The curves are shown in Figure 3.66. The DSC curve of the bare epoxy coating had an obvious heat absorption peak at 46 °C, which corresponds to its T_g, which refers to the relaxation phenomenon in which the amorphous part of the polymer changes from a glassy frozen state to a highly elastic thawed state. When the temperature was higher than T_g, the molecular segments of the polymer began to flow. Figure 3.66 shows that the DSC curve and T_g range of the Fe_3O_4/2-MBT-epoxy coatings containing the photothermal composite fillers remained almost unchanged, indicating that the presence of a small number of nanofillers did not influence the T_g value of the coating. Based on the DSC results, it was concluded that when the surface temperature of the coating was higher than the T_g value, the shape-memory effect of the epoxy matrix was triggered [155].

Thereafter, an NIR light source was applied to irradiate the coating surface, and the temperature change was recorded. Figure 3.67 shows that the surface temperature of the Fe_3O_4/2-MBT-epoxy coating samples increased rapidly within 60 seconds and exceeded the T_g of the coating. A stable state of relative equilibrium between heat generation and dissipation was reached at about 150 seconds, illustrating a surface temperature of 78.6 °C, which far exceeded the T_g value. Therefore, the coating successfully triggered the shape-memory effect of the epoxy matrix. For the bare

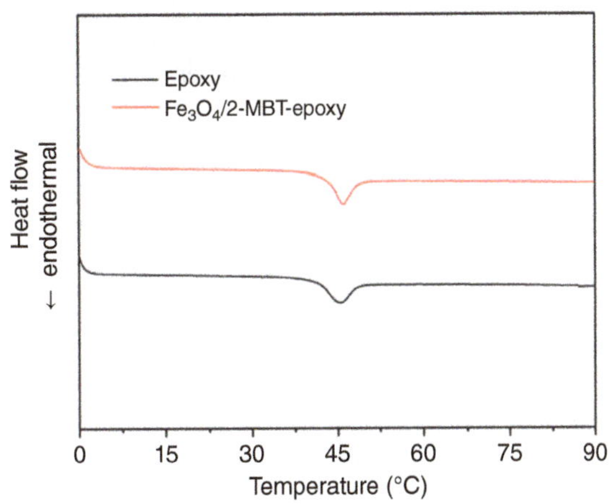

Figure 3.66 DSC curves of pure epoxy coating and the epoxy coating containing 3 wt% Fe_3O_4/2-MBT nanoparticles.

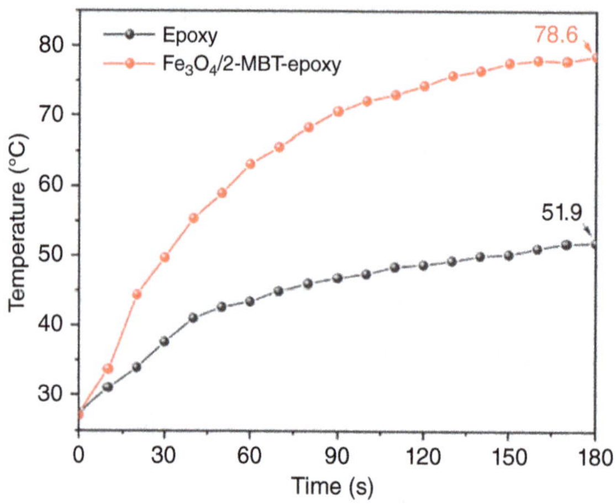

Figure 3.67 Surface temperature change of pure epoxy coating and the epoxy coating containing 3 wt% Fe_3O_4/2-MBT nanoparticles during NIR laser irradiation for 180 seconds.

epoxy coating, the temperature change was much lower, and the surface temperature reached the T_g value only after 120 seconds of laser irradiation. The analysis showed that the Fe_3O_4/2-MBT nanoparticles enhanced the photothermal absorption performance of the coating and rapidly increased the coating temperature to reach the T_g value. Under these conditions, the shape-memory effect was activated to achieve photothermal response self-healing performance [159].

Artificial scratches were then made to the coating surface and irradiated using an NIR laser. The experimental results are shown in Figure 3.68. For the bare epoxy coating, the scratch remained at about 80 μm before and after laser irradiation. For the epoxy coating containing Fe_3O_4/2-MBT nanoparticles, the scratch significantly narrowed from 80 to ~4 μm. Macroscale observations showed that the scratch was no longer visible to the naked eye, and only a thin scratch line was observed under a confocal laser microscope. The significantly narrowed coating scratch promoted the recovery of corrosion protection performance when the coating was damaged.

3.5.3.2 Characterization of Corrosion Resistance of Intact Coatings

To explore the corrosion resistance of the coatings, an immersion experiment in 3.5 wt% NaCl was carried out. After various immersion times, EIS measurements were performed on the coated samples. According to the Nyquist diagram in Figure 3.69, the arc radius of the capacitive reactance of the Fe_3O_4/2-MBT-epoxy samples was much greater than that of the epoxy sample, demonstrating its excellent anticorrosion properties. For the epoxy coating, the arc radius of the capacitive reactance decreased with immersion time, whereas that of the Fe_3O_4/2-MBT-epoxy sample remained unchanged [159].

Figure 3.70 shows that the $|Z|_{0.01Hz}$ values of the bare epoxy sample were about 10^8–10^9 Ω cm^2 and the $|Z|_{0.01Hz}$ values of the Fe_3O_4/2-MBT-epoxy samples were higher than 10^{10} Ω cm^2. The addition of the Fe_3O_4/2-MBT nanoparticles to the coating matrix prolonged the pathway of the corrosion solution toward the metal substrate. Simultaneously, the slow release and adsorption of corrosion inhibitors

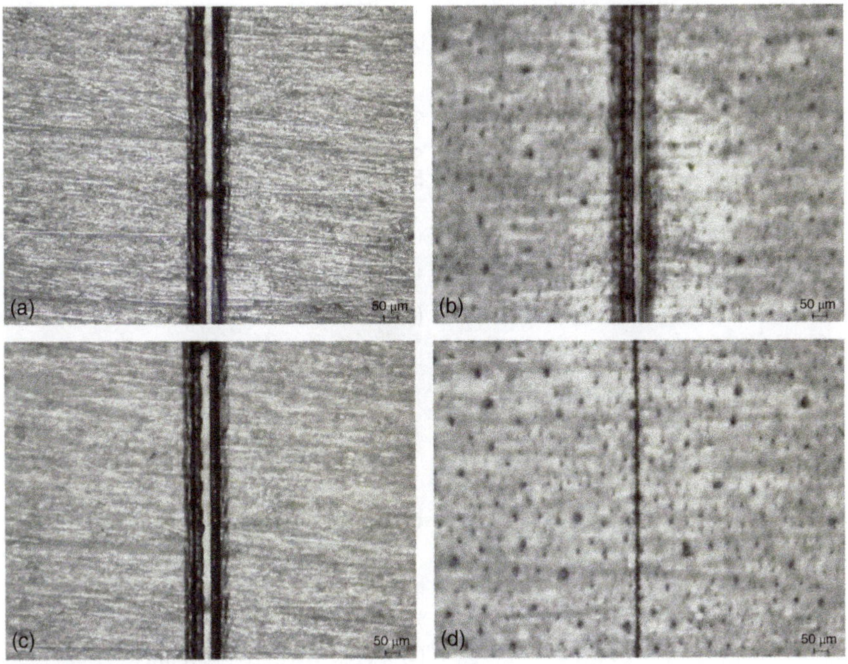

Figure 3.68 Coating scratches on pure epoxy coating (a) before and (c) after NIR laser irradiation, and coating scratches on Fe_3O_4/2-MBT-epoxy coating (b) before and (d) after NIR laser irradiation.

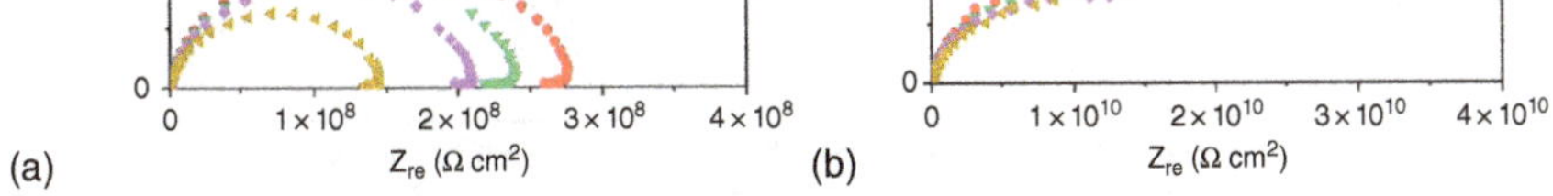

Figure 3.69 Nyquist diagrams of AA2024-T3 alloys coated with (a) pure epoxy coating and (b) epoxy coating containing 3 wt% Fe_3O_4/2-MBT nanoparticles during immersion in 3.5 wt% NaCl solution for 6, 24, 48, and 72 hours.

reduced the rate of corrosion reaction and inhibited further corrosion reactions at the coating–metal interface.

3.5.3.3 Characterization of the Self-healing Properties of Scratched Coatings

To investigate the self-healing properties of the bare epoxy and Fe_3O_4/2-MBT-epoxy coatings, they were subjected to manual scratching and NIR laser irradiation for 30 seconds to help the corrosion inhibitor spread to the metal substrate surface after

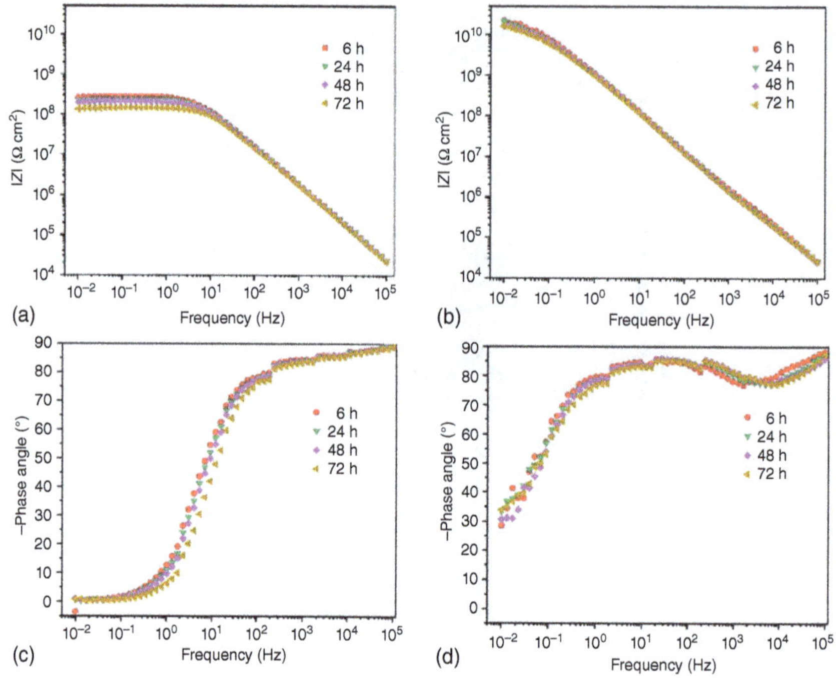

Figure 3.70 Bode plots of AA2024-T3 aluminum alloys coated with (a) pure epoxy coating and (b) epoxy coating containing 3 wt% Fe_3O_4/2-MBT nanoparticles during immersion in 3.5 wt% NaCl solution for 6, 24, 48, and 72 hours.

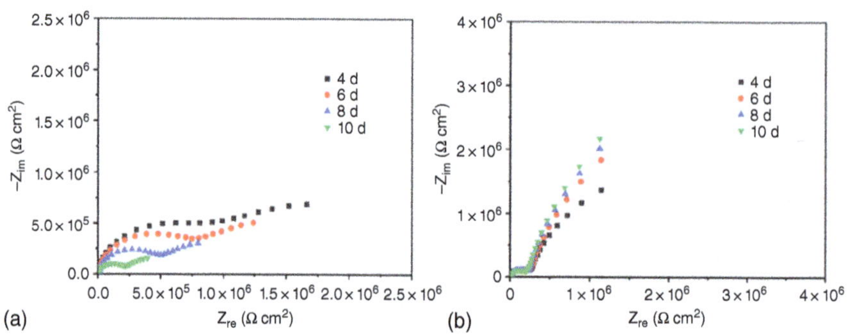

Figure 3.71 Nyquist diagrams of AA2024-T3 alloys with (a) scratched epoxy coating and (b) scratched epoxy coating containing 3 wt% Fe_3O_4/2-MBT nanoparticles during immersion in 3.5 wt% NaCl solution for 4, 6, 8, and 10 days.

release. The coating sample was immersed in a 3.5 wt% NaCl solution for 10 days. The Nyquist and Bode plots are illustrated in Figures 3.71 and 3.72 for the two coating samples after immersion for 4, 6, 8, and 10 days. According to the Nyquist diagram, the arc radius of the capacitive reactance of the Fe_3O_4/2-MBT-epoxy sample was much greater than that of the bare epoxy coating, demonstrating excellent self-healing performance [132]. The arc radius of the capacitive reactance of the

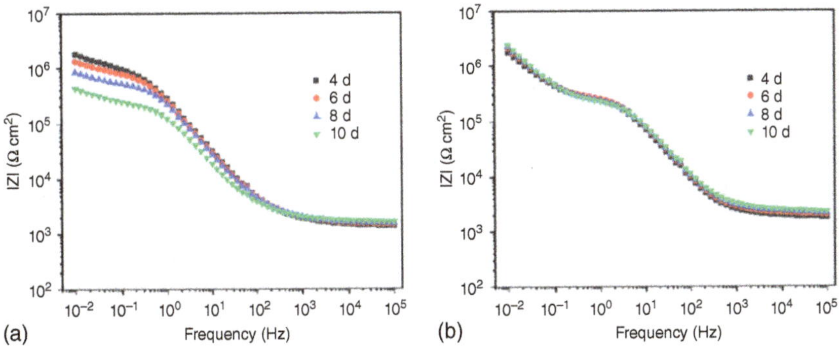

Figure 3.72 Bode plots of AA2024-T3 alloys coated with (a) scratched epoxy coating and (b) scratched epoxy coating containing 3 wt% Fe_3O_4/2-MBT nanoparticles during immersion in 3.5 wt% NaCl solution for 4, 6, 8, and 10 days.

scratched epoxy coating sample decreased significantly with prolonged immersion time, which can be ascribed to the severe corrosion activity at the coating scratch. For the Fe_3O_4/2-MBT-epoxy sample, the arc radius of the capacitive reactance increased slightly owing to the gradual release of corrosion inhibitors.

From the $|Z|_{0.01Hz}$ value comparison (see Figure 3.72), the $|Z|_{0.01Hz}$ values of scratched epoxy coating and Fe_3O_4/2-MBT-epoxy coating samples were both at $1.7 \times 10^6\ \Omega\,cm^2$ after four days of immersion. With the extension of the immersion time, the scratched epoxy coating exhibited a decreasing trend in the $|Z|_{0.01Hz}$ value, and the $|Z|_{0.01Hz}$ value was as low as $3 \times 10^5\ \Omega\,cm^2$ after 10 days. This is because, over time, Cl^-, H_2O, O_2, and other corrosive molecules and ions continued to invade and reach the surface of the metal matrix, resulting in an intensified corrosion phenomenon in the damaged coating area. In contrast, for the Fe_3O_4/2-MBT-epoxy samples, the $|Z|_{0.01Hz}$ values were maintained at $2.3 \times 10^6\ \Omega\,cm^2$ after 10 days of immersion. After photothermal healing, the scratch was physically closed due to the photothermal conversion of the Fe_3O_4/2-MBT-epoxy coating. In contrast, the 2-MBT corrosion inhibitors were released from the Fe_3O_4/2-MBT nanoparticles to form an inhibitor adsorption layer on the metal substrate surface in the damaged area of the coating. Hence, corrosion in the damaged area of the coating was effectively inhibited [110]. Based on an overall analysis of the EIS results, the Fe_3O_4/2-MBT-epoxy coating exhibited superior anticorrosion and self-healing properties.

After photothermal self-healing, the scratch in the Fe_3O_4/2-MBT-epoxy coating was significantly narrowed owing to the shape-memory effect, and the contact area between the metal substrate and corrosive media was reduced. However, the photothermal effect accelerated the release and diffusion of corrosion inhibitors from the Fe_3O_4/2-MBT nanoparticles into the coating defect, preventing further corrosion. The EIS test results confirmed the excellent self-healing properties of the Fe_3O_4/2-MBT-epoxy coatings, which were derived from the synergistic mechanism of shape-memory recovery at a scratch and the subsequent rapid release of corrosion inhibitors.

3.6 Chapter Summary

This chapter introduces four types of photothermally activated self-healing coatings (i.e. GO-CeO_2 filler, TiN-BTA@MSN, PCL@TiN, and Fe_3O_4/2-MBT) that were separately employed as coating fillers and added into a shape-memory epoxy matrix to construct photothermal response self-healing systems. The effects of the photothermal filler type, addition amount, and dispersion in the coating on the corrosion resistance, photothermal conversion performance, and self-healing performance of different coating systems were investigated and analyzed in detail. The addition of more photothermal fillers was found to improve the photothermal conversion of coatings, whereas it had the opposite effect on the barrier properties of the coating. This is because more fillers create a higher likelihood of poor dispersion in the coating. This can be explained by the agglomeration and uneven dispersion of excess PCL@TiN microspheres, resulting in more microdefects in the coating matrix. After balancing corrosion resistance and photothermal performance, the optimal coating sample was selected. Based on the release of corrosion inhibitors, melting of healing agents, and the shape-memory effect of the coating matrix, once the coating surface temperatures exceed the T_g of the epoxy matrix and the T_m of PCL under light irradiation, self-healing activity is activated effectively.

From the above mentioned analysis, it can be concluded that photothermally activated self-healing coatings hold great promise for practical applications as they can quickly repair coating defects under convenient light irradiation. Intrinsic self-healing approaches (i.e. physical or chemical reactions of polymer molecular chains or functional bonds) and extrinsic self-healing approaches (e.g. melting of healing agents and release of corrosion inhibitors) can be combined to further strengthen the self-healing performance and avoid the drawbacks of single-healing methods, resulting in outstanding self-healing effects and long-term corrosion protection.

Despite significant advances in photothermal self-healing coatings, developmental challenges still need to be addressed. First, the dispersion properties of photothermal fillers, their costs, and their environmental impacts should be considered. The development of low-cost, green, and facile surface-modified photothermal materials is urgently required for novel photothermal self-healing coatings. Second, sunlight should be considered the most practical light source when designing self-healing coatings. Therefore, the light absorption properties of photothermal fillers in the visible and NIR wavelength ranges should be considered when selecting optimal species. In addition, coatings may not experience only one-time damage; therefore, cyclic healing is necessary. To meet this need, the photothermal species must exhibit good stability and activity because intrinsic self-healing approaches can heal damage more than once without consuming the polymer components; therefore, they are suitable for cyclic self-healing. Finally, harsh conditions, such as underwater healing and healing in strongly acidic and alkaline solutions, should be considered to verify the feasibility of self-healing coatings under different application scenarios.

In summary, photothermal self-healing coatings have broad application prospects due to their precise and remote repair capabilities and instantaneous on–off characteristics. In the near future, preparation methods for photothermal self-healing coatings should be further optimized to improve photothermal conversion efficiency and reduce preparation costs. Self-healing coatings with multiple repair mechanisms should also be designed to improve their long-term protection ability and to enhance their physical, chemical, and mechanical surface properties, which are conducive to the hastened realization of industrial applications.

References

1 Habault, D., Zhang, H., and Zhao, Y. (2013). Light-triggered self-healing and shape-memory polymers. *Chemical Society Reviews* 42 (17): 7244–7256. https://doi.org/10.1039/C3CS35489J.

2 Xu, H., Han, P., Qin, A. et al. (2023). Recent advances and application prospects in photothermal materials. *Acta Chimica Sinica* 81 (10): 1420–1437. https://doi.org/10.6023/A23050232.

3 Wang, Y., Wang, J., Huang, L. et al. (2023). Photothermally activated self-healing coatings for corrosion protection: a review. *Progress in Organic Coatings* 185: 107886. https://doi.org/10.1016/j.porgcoat.2023.107886.

4 Dong, Y., Gong, M., Huang, D. et al. (2019). Shape memory, self-healing property, and NIR photothermal effect of epoxy resin coating with polydopamine@ polypyrrole nanoparticles. *Progress in Organic Coatings* 136: 1005232. https://doi.org/10.1016/j.porgcoat.2019.105232.

5 Wang, T., Wang, W., Feng, H. et al. (2022). Photothermal nanofiller-based polydimethylsiloxane anticorrosion coating with multiple cyclic self-healing and long-term self-healing performance. *Chemical Engineering Journal* 446: 137077. https://doi.org/10.1016/j.cej.2022.137077.

6 Lu, C., Ling, Z., Wang, C. et al. (2022). Multiple hydrogen bonding interactions toward rapidly self-healing, photothermal conversion elastomer composites. *Composites Part B: Engineering* 228: 109428. https://doi.org/10.1016/j.compositesb.2021.109428.

7 Edwards, R.S. and Coleman, K.S. (2013). Graphene synthesis: relationship to applications. *Nanoscale* 5 (1): 38–51. https://doi.org/10.1039/c2nr32629a.

8 Fiori, G., Bonaccorso, F., Iannaccone, G. et al. (2014). Electronics based on two-dimensional materials. *Nature Nanotechnology* 9 (10): 768–779. https://doi.org/10.1038/NNANO.2014.207.

9 Wang, J., Jin, X., Li, C. et al. (2019). Graphene and graphene derivatives toughening polymers: toward high toughness and strength. *Chemical Engineering Journal* 370: 831–854. https://doi.org/10.1016/j.cej.2019.03.229.

10 Cui, G., Bi, Z., Zhang, R. et al. (2019). A comprehensive review on graphene-based anti-corrosive coatings. *Chemical Engineering Journal* 373: 104–121. https://doi.org/10.1016/j.cej.2019.05.034.

11 George, J.S., Vijayan, P.P., Paduvilan, J.K. et al. (2022). Advances and future outlook in epoxy/graphene composites for anticorrosive applications. *Progress in Organic Coatings* 162: 106571. https://doi.org/10.1016/j.porgcoat.2021.106571.

12 Ma, L., Wang, X., Wang, J. et al. (2021). Graphene oxide–cerium oxide hybrids for enhancement of mechanical properties and corrosion resistance of epoxy coatings. *Journal of Materials Science* 56: 10108–10123. https://doi.org/10.1007/s10853-021-05932-z.

13 Zhang, H., Sun, W., Wang, L. et al. (2023). Photothermal cyclic self-healing coating with long-term corrosion protection based on bifunctional graphene. *Chemical Engineering Journal* 473: 145164. https://doi.org/10.1016/j.cej.2023.145164.

14 Han, X., Ren, L., Ma, Y. et al. (2022). A mussel-inspired self-repairing superhydrophobic coating with good anti-corrosion and photothermal properties. *Carbon* 197: 27–39. https://doi.org/10.1016/j.carbon.2022.05.056.

15 McAllister, M.J., Li, J.L., Adamson, D.H. et al. (2007). Single sheet functionalized graphene by oxidation and thermal expansion of graphite. *Chemistry of Materials* 19 (18): 4396–4404. https://doi.org/10.1021/cm0630800.

16 Daradmare, S., Raj, S., Bhattacharyya, A.R. et al. (2018). Factors affecting barrier performance of composite anti-corrosion coatings prepared by using electrochemically exfoliated few-layer graphene as filler. *Composites Part B: Engineering* 155: 1–10. https://doi.org/10.1016/j.compositesb.2018.08.023.

17 Abakah, R.R., Huang, F., Hu, Q. et al. (2021). Comparative study of corrosion properties of different graphene nanoplate/epoxy composite coatings for enhanced surface barrier protection. *Coatings* 11 (3): 285. https://doi.org/10.3390/coatings11030285.

18 Ma, J., Liu, J., Zhu, W. et al. (2018). Solubility study on the surfactants functionalized reduced graphene oxide. *Colloids and Surfaces A: Physicochemical and Engineering Aspects* 538: 79–85. https://doi.org/10.1016/j.colsurfa.2017.10.071.

19 Naficy, S., Jalili, R., Aboutalebi, S.H. et al. (2014). Graphene oxide dispersions: tuning rheology to enable fabrication. *Materials Horizons* 1 (3): 326–331. https://doi.org/10.1039/c3mh00144j.

20 Luo, J., Yang, L., Sun, D. et al. (2020). Graphene oxide "surfactant"-directed tunable concentration of graphene dispersion. *Small* 16 (45): 2003426. https://doi.org/10.1002/smll.202003426.

21 Chae, H.K., Siberio-Perez, D.Y., Kim, J. et al. (2004). A route to high surface area, porosity and inclusion of large molecules in crystals. *Nature* 427 (6974): 523–527. https://doi.org/10.1038/nature02311.

22 Yang, H., Shan, C., Li, F. et al. (2009). Convenient preparation of tunably loaded chemically converted graphene oxide/epoxy resin nanocomposites from graphene oxide sheets through two-phase extraction. *Journal of Materials Chemistry* 19 (46): 8856–8860. https://doi.org/10.1039/b915228h.

23 Ramezanzadeh, B., Bahlakeh, G., and Ramezanzadeh, M. (2018). Polyaniline-cerium oxide (PAni-CeO_2) coated graphene oxide for enhancement

of epoxy coating corrosion protection performance on mild steel. *Corrosion Science* 137: 111–126. https://doi.org/10.1016/j.corsci.2018.03.038.

24 Liu, Y., Jie, H., Liu, R. et al. (2021). Research on the preparation and anticorrosion properties of EP/CeO_2-GO nanocomposite coating. *Polymers* 13 (2): 183. https://doi.org/10.3390/polym13020183.

25 Huang, F., Wang, J., Chen, W. et al. (2018). Synergistic peroxidase-like activity of CeO_2-coated hollow Fe_3O_4 nanocomposites as an enzymatic mimic for low detection limit of glucose. *Journal of the Taiwan Institute of Chemical Engineers* 83: 40–49. https://doi.org/10.1016/j.jtice.2017.12.011.

26 Shen, Z., Xing, H., Wang, H. et al. (2018). Synthesis and enhanced electromagnetic absorption properties of Co-doped CeO_2/RGO nanocomposites. *Journal of Alloys and Compounds* 753: 28–34. https://doi.org/10.1016/j.jallcom.2018.04.195.

27 Xie, T. and Rousseau, I.A. (2009). Facile tailoring of thermal transition temperatures of epoxy shape memory polymers. *Polymer* 50 (8): 1852–1856. https://doi.org/10.1016/j.polymer.2009.02.035.

28 Luo, X. and Mather, P.T. (2013). Shape memory assisted self-healing coating. *ACS Macro Letters* 2 (2): 152–156. https://doi.org/10.1021/mz400017x.

29 Bu, Y., Li, K., Xiong, Z. et al. (2022). Evolution of layer distance and structural arrangement of graphene oxide with various oxygen content and functional types in low temperature: a ReaxFF molecular dynamics simulation. *Applied Surface Science* 572: 151390. https://doi.org/10.1016/j.apsusc.2021.151390.

30 Stankovich, S., Dikin, D.A., Piner, R.D. et al. (2007). Synthesis of graphene-based nanosheets via chemical reduction of exfoliated graphite oxide. *Carbon* 45 (7): 1558–1565. https://doi.org/10.1016/j.carbon.2007.02.034.

31 Dreyer, D.R., Park, S., Bielawski, C.W. et al. (2010). The chemistry of graphene oxide. *Chemical Society Reviews* 39 (1): 228–240. https://doi.org/10.1039/b917103g.

32 Verma, R. and Samdarshi, S.K. (2016). In situ decorated optimized CeO_2 on reduced graphene oxide with enhanced adsorptivity and visible light photocatalytic stability and reusability. *Journal of Physical Chemistry C* 120 (39): 22281–22290. https://doi.org/10.1021/acs.jpcc.6b04493.

33 Zhang, W., Li, B., and Hong, M. (2022). Effects of cerium oxide doping on microstructure and properties of Ni-GO-CeO_2 nanocomposite coatings. *Journal of Materials Research and Technology* 21: 3440–3450. https://doi.org/10.1016/j.jmrt.2022.10.149.

34 Kumar, S., Ojha, A.K., Patrice, D. et al. (2016). One-step in situ synthesis of CeO_2 nanoparticles grown on reduced graphene oxide as an excellent fluorescent and photocatalyst material under sunlight irradiation. *Physical Chemistry Chemical Physics* 18 (16): 11157–11167. https://doi.org/10.1039/c5cp04457j.

35 Chen, Y., Chen, Y., Qiu, C. et al. (2015). HAADF STEM observation of the Au/CeO_2 nanostructures. *Materials Letters* 141: 31–34. https://doi.org/10.1016/j.matlet.2014.11.048.

36 Bîru, E.I., Gârea, S.A., Nicolescu, A. et al. (2018). Advanced polybenzoxazine structures based on modified reduced graphene oxide. *Polymers* 10 (9): 941. https://doi.org/10.3390/polym10090941.

37 Dmonte, D.J., Pandiyarajan, A., Bhuvanesh, N. et al. (2018). Graphene oxide resorcinol hybrid material as fluorescent chemosensor for detection of cerium ion. *Materials Letters* 227: 154–157. https://doi.org/10.1016/j.matlet.2018.05.051.

38 da Silva, M.F.P., da Costa, H.C.D.F., Triboni, E.R. et al. (2012). Synthesis and characterization of CeO_2-graphene composite. *Journal of Thermal Analysis and Calorimetry* 107 (1): 257–263. https://doi.org/10.1007/s10973-011-1924-8.

39 Sun, H., Lin, D., and Gao, Z. (2017). Preparation and characterization of layered carbon hybrid framework materials pillared with titanocene. *Journal of Porous Materials* 24 (2): 291–296. https://doi.org/10.1007/s10934-016-0262-4.

40 Nosrati, R., Olad, A., and Shakoori, S. (2017). Preparation of an antibacterial, hydrophilic and photocatalytically active polyacrylic coating using TiO_2 nanoparticles sensitized by graphene oxide. *Materials Science & Engineering, C: Materials for Biological Applications* 80: 642–651. https://doi.org/10.1016/j.msec.2017.07.004.

41 Chuang, F. and Yang, S. (2008). Cerium dioxide/polyaniline core-shell nanocomposites. *Journal of Colloid and Interface Science* 320 (1): 194–201. https://doi.org/10.1016/j.jcis.2008.01.015.

42 Kumar, K.M., Mahendhiran, M., Diaz, M.C. et al. (2018). Green synthesis of Ce^{3+} rich CeO_2 nanoparticles and its antimicrobial studies. *Materials Letters* 214: 15–19. https://doi.org/10.1016/j.matlet.2017.11.097.

43 Chen, X., Hu, D., Zhang, Z. et al. (2019). In situ assembly of halloysite nanotubes@cerium oxide nanohybrid for highly UV-shielding and superhydrophobic coating. *Journal of Alloys and Compounds* 811: 151986. https://doi.org/10.1016/j.jallcom.2019.151986.

44 Wang, Y., Kang, Y., Ge, M. et al. (2018). Cerium and tin oxides anchored onto reduced graphene oxide for selective catalytic reduction of NO with NH_3 at low temperatures. *RSC Advances* 8: 36383–36391. https://doi.org/10.1039/c8ra05151h.

45 Xie, C., Jia, Y., Xue, A. et al. (2022). Anti-corrosion and self-healing behaviors of waterborne polyurethane composite coatings enhanced via chitosan-modified graphene oxide and phosphate intercalated hydrotalcite. *Progress in Organic Coatings* 168: 106881. https://doi.org/10.1016/j.porgcoat.2022.106881.

46 Ramezanzadeh, B., Niroumandrad, S., Ahmadi, A. et al. (2016). Enhancement of barrier and corrosion protection performance of an epoxy coating through wet transfer of amino functionalized graphene oxide. *Corrosion Science* 103: 283–304. https://doi.org/10.1016/j.corsci.2015.11.033.

47 Behzadnasab, M., Mirabedini, S.M., and Esfandeh, M. (2013). Corrosion protection of steel by epoxy nanocomposite coatings containing various combinations of clay and nanoparticulate zirconia. *Corrosion Science* 75: 134–141. https://doi.org/10.1016/j.corsci.2013.05.024.

48 Li, J., Cui, J., Yang, J. et al. (2016). Reinforcement of graphene and its derivatives on the anticorrosive properties of waterborne polyurethane coatings. *Composites Science and Technology* 129: 30–37. https://doi.org/10.1016/j.compscitech.2016.04.017.

49 Hu, G., Xiao, Y., and Ying, J. (2021). Nano-SiO_2 and silane coupling agent co-decorated graphene oxides with enhanced anti-corrosion performance of epoxy composite coatings. *International Journal of Molecular Sciences* 22 (20): 11087. https://doi.org/10.3390/ijms222011087.

50 Chai, Z., Chen, Y., Zhou, D. et al. (2022). Excellent corrosion resistance of FGO/Zn_2SiO_4 composite material in epoxy coatings. *Progress in Organic Coatings* 170: 106992. https://doi.org/10.1016/j.porgcoat.2022.106992.

51 Ozcan, U.E., Karabork, F., Yazman, S. et al. (2019). Effect of silica/graphene nanohybrid particles on the mechanical properties of epoxy coatings. *Arabian Journal for Science and Engineering* 44: 5723–5731. https://doi.org/10.1007/s13369-019-03724-x.

52 Becker, O., Varley, R., and Simon, G. (2002). Morphology, thermal relaxations and mechanical properties of layered silicate nanocomposites based upon high-functionality epoxy resins. *Polymer* 43 (16): 4365–4373. https://doi.org/10.1016/S0032-3861(02)00269-0.

53 Chen, J., Fang, L., Xu, Z. et al. (2016). Self-healing epoxy coatings curing with varied ratios of diamine and monoamine triggered via near-infrared light. *Progress in Organic Coatings* 101: 543–552. https://doi.org/10.1016/j.porgcoat.2016.09.020.

54 Fan, X., Ding, Y., Liu, Y. et al. (2019). Plasmonic $Ti_3C_2T_X$ MXene enables highly efficient photothermal conversion for healable and transparent wearable device. *ACS Nano* 13 (7): 8124–8134. https://doi.org/10.1021/acsnano.9b03161.

55 Bai, Y., Zhang, J., Wen, D. et al. (2020). A reconfigurable, self-healing and near infrared light responsive thermoset shape memory polymer. *Composites Science and Technology* 187: 107940. https://doi.org/10.1016/j.compscitech.2019.107940.

56 Montano, V., Vogel, W., Smits, A. et al. (2021). From scratch closure to electrolyte barrier restoration in self-healing polyurethane coatings. *ACS Applied Polymer Materials* 3 (5): 2802–2812. https://doi.org/10.1021/acsapm.1c00323.

57 Montano, V., Urban, M.W., van der Zwaag, S. et al. (2022). Local strain-induced energy storage as driving force for autogenous scratch closure. *Journal of Materials Chemistry A* 10 (13): 7073–7081. https://doi.org/10.1039/d1ta10441a.

58 Liu, S., Xie, H., Cao, Q. et al. (2022). Preparation of a novel IPDI/PUF@CeO_2 bi-functional microcapsules and its improvement for the self-healing and anti-corrosion performance in epoxy coatings. *Progress in Organic Coatings* 169: 106897. https://doi.org/10.1016/j.porgcoat.2022.106897.

59 Wang, H., Xu, J., Du, X. et al. (2022). Stretchable and self-healing polyurethane coating with synergistic anticorrosion effect for the corrosion protection of stainless steels. *Progress in Organic Coatings* 164: 106672. https://doi.org/10.1016/j.porgcoat.2021.106672.

60 Schramke, K.S., Qin, Y., Held, J.T. et al. (2018). Nonthermal plasma synthesis of titanium nitride nanocrystals with plasmon resonances at near-infrared wavelengths relevant to photothermal therapy. *ACS Applied Nano Materials* 1 (6): 2869–2876. https://doi.org/10.1021/acsanm.8b00505.

61 Sabzeghabae, A.N., Berrospe-Rodriguez, C., Wu, C. et al. (2022). Nonlinear absorption in plasmonic titanium nitride nanocrystals. *Advanced Optical Materials* 11 (1): 2201290. https://doi.org/10.1002/adom.202201290.

62 He, W., Ai, K., Jiang, C. et al. (2017). Plasmonic titanium nitride nanoparticles for *in vivo* photoacoustic tomography imaging and photothermal cancer therapy. *Biomaterials* 132: 37–47. https://doi.org/10.1016/j.biomaterials.2017.04.007.

63 Ifijen, I.H., Maliki, M. et al. (2022). A comprehensive review on the synthesis and photothermal cancer therapy of titanium nitride nanostructures. *Inorganic and Nano-Metal Chemistry* 53 (4): 366–387. https://doi.org/10.1080/24701556.2022.2068596.

64 Wang, X., Sha, C., Wang, W. et al. (2019). Functionalized biomass-derived composites for solar vapor generation. *Materials Research Express* 6 (12): 125613. https://doi.org/10.1088/2053-1591/ab586e.

65 Wang, B., Yu, P., Yang, Q. et al. (2022). Upcycling of biomass waste into photothermal superhydrophobic coating for efficient anti-icing and deicing. *Materials Today Physics* 24: 100683. https://doi.org/10.1016/j.mtphys.2022.100683.

66 Ma, L., Wang, J., Zhang, D. et al. (2021). Dual-action self-healing protective coatings with photothermal responsive corrosion inhibitor nanocontainers. *Chemical Engineering Journal* 404: 127118. https://doi.org/10.1016/j.cej.2020.127118.

67 Yeganeh, M., Asadi, N., Omidi, M. et al. (2019). An investigation on the corrosion behavior of the epoxy coating embedded with mesoporous silica nanocontainer loaded by sulfamethazine inhibitor. *Progress in Organic Coatings* 128: 75–81. https://doi.org/10.1016/j.porgcoat.2018.12.022.

68 Yeganeh, M., Saremi, M., Rezaeyan, H. et al. (2014). Corrosion inhibition of steel using mesoporous silica nanocontainers incorporated in the polypyrrole. *Progress in Organic Coatings* 77 (9): 1428–1435. https://doi.org/10.1016/j.porgcoat.2014.05.007.

69 Qu, F., Zhu, G., Lin, H. et al. (2006). Drug self-templated synthesis of ibuprofen/mesoporous silica for sustained release. *European Journal of Inorganic Chemistry* 19: 3943–3947. https://doi.org/10.1002/ejic.200501035.

70 Luo, L., Liang, Y., Erichsen, E.S. et al. (2018). Hierarchical mesoporous organosilica-silica core-shell nanoparticles capable of controlled fungicide release. *Chemistry – A European Journal* 24 (28): 7200–7209. https://doi.org/10.1002/chem.201800135.

71 Xu, J., Cao, Y., Fang, L. et al. (2018). A one-step preparation of inhibitor-loaded silica nanocontainers for self-healing coatings. *Corrosion Science* 140: 349–362. https://doi.org/10.1016/j.corsci.2018.05.030.

72 da Silva, B.P., Saji, V.S., Aoki, I.V. et al. (2022). Rapid and eco-friendly one-step synthesis of dodecylamine-encapsulated mesoporous silica nanocontainers. *Microporous and Mesoporous Materials* 341: 112109. https://doi.org/10.1016/j.micromeso.2022.112109.

73 Xiao, P., Li, G., Ma, S. et al. (2022). High energy conversion composites based on graphene material with excellent healing performances. *Journal of Applied Polymer Science* 139 (8): e51690. https://doi.org/10.1002/app.51690.

74 Wang, T., Feng, H., Wang, W. et al. (2022). Interfacial controllable heterojunctions nanosheets as photothermal catalyzer for cyclic photothermal self-healing of polydimethylsiloxane coating. *Composites Part B: Engineering* 240: 110002. https://doi.org/10.1016/j.compositesb.2022.110002.

75 Asadi, N., Naderi, R., Mandavian, M. et al. (2019). Halloysite nanotubes loaded with imidazole dicarboxylic acid to enhance protection properties of a polymer coating. *Progress in Organic Coatings* 127: 375–384. https://doi.org/10.1016/j.porgcoat.2018.11.035.

76 Tavandashti, N.P., Ghorbani, M., Shojaei, A. et al. (2016). Inhibitor-loaded conducting polymer capsules for active corrosion protection of coating defects. *Corrosion Science* 112: 138–149. https://doi.org/10.1016/j.corsci.2016.07.003.

77 Xu, D., Lou, C., Huang, J. et al. (2019). Effect of inhibitor-loaded halloysite nanotubes on active corrosion protection of polybenzoxazine coatings on mild steel. *Progress in Organic Coatings* 134: 126–133. https://doi.org/10.1016/j.porgcoat.2019.04.021.

78 Lin, B., Wang, J., Zhang, H. et al. (2022). Self-healing performance of ethyl-cellulose based supramolecular gel coating highly loaded with different carbon chain length imidazoline inhibitors in NaCl corrosion medium. *Corrosion Science* 197: 110084. https://doi.org/10.1016/j.corsci.2022.110084.

79 Yang, C., Xu, W., Meng, X. et al. (2021). A pH-responsive hydrophilic controlled release system based on ZIF-8 for self-healing anticorrosion application. *Chemical Engineering Journal* 415: 128985. https://doi.org/10.1016/j.cej.2021.128985.

80 Wang, L., Zhu, G., Wang, M. et al. (2019). Dual plasmonic Au/TiN nanofluids for efficient solar photothermal conversion. *Solar Energy* 184: 240–248. https://doi.org/10.1016/j.solener.2019.04.013.

81 Liu, J., Wang, A., Liu, S. et al. (2021). A titanium nitride nanozyme for pH-responsive and irradiation-enhanced cascade-catalytic tumor therapy. *Angewandte Chemie International Edition* 60 (48): 25328–25338. https://doi.org/10.1002/anie.202106750.

82 Zhu, Q., Xu, X., Huang, Y. et al. (2023). pH-responsive mesoporous silica nanocontainers based on Zn-BTA complexes as stoppers for controllable release of corrosion inhibitors and application in epoxy coatings. *Progress in Organic Coatings* 181: 107581. https://doi.org/10.1016/j.porgcoat.2023.107581.

83 Li, D., Peng, H., Lin, Z. et al. (2021). Corrosion protection coatings embedded with silane-functionalized rGO/SiO_2 nanocontainers: enhancing dispersive and corrosion-inhibitor loading capabilities. *Surface & Coatings Technology* 427: 127850. https://doi.org/10.1016/j.surfcoat.2021.127850.

84 Fu, J., Chen, T., Wang, M. et al. (2013). Acid and alkaline dual stimuli-responsive mechanized hollow mesoporous silica nanoparticles as smart nanocontainers for intelligent anticorrosion coatings. *ACS Nano* 7 (12): 11397–11408. https://doi.org/10.1021/nn4053233.

85 Ding, C., Liu, Y., Wang, M. et al. (2016). Self-healing, superhydrophobic coating based on mechanized silica nanoparticles for reliable protection of magnesium alloys. *Journal of Materials Chemistry A* 4 (21): 8041–8052. https://doi.org/10.1039/c6ta02575g.

86 Chen, T., Fu, J. et al. (2012). pH-responsive nanovalves based on hollow mesoporous silica spheres for controlled release of corrosion inhibitor. *Nanotechnology* 23 (23): 235605. https://doi.org/10.1088/0957-4484/23/23/235605.

87 Sedira, S., Achour, S., Avci, A. et al. (2014). Physical deposition of carbon doped titanium nitride film by DC magnetron sputtering for metallic implant coating use. *Applied Surface Science* 295: 81–85. https://doi.org/10.1016/j.apsusc.2014.01.010.

88 Cheng, G., Qian, J., Xia, R. et al. (2014). Comparative study on different coupling agents used as surfactants: effect on dispersion properties of TiN nanoparticles. *Micro & Nano Letters* 9 (7): 441–445. https://doi.org/10.1049/mnl.2013.0581.

89 Belachew, N. and Hinsene, H. (2020). Preparation of cationic surfactant-modified kaolin for enhanced adsorption of hexavalent chromium from aqueous solution. *Applied Water Science* 10 (1): 38. https://doi.org/10.1007/s13201-019-1121-7.

90 Karatas, D., Arslan, D.S., Unver, I.K. et al. (2019). Coating mechanism of AuNPs onto sepiolite by experimental research and MD simulation. *Coatings* 9 (12): 785. https://doi.org/10.3390/coatings9120785.

91 Wang, T., Du, J., Ye, S. et al. (2019). Triple-stimuli-responsive smart nanocontainers enhanced self-healing anticorrosion coatings for protection of aluminum alloy. *ACS Applied Materials & Interfaces* 11 (4): 4425–4438. https://doi.org/10.1021/acsami.8b19950.

92 Siva, T., Kumari, S.S.S., and Sathiyanarayanan, S. (2021). Dendrimer like mesoporous silica nano container (DMSN) based smart self healing coating for corrosion protection performance. *Progress in Organic Coatings* 154: 106201. https://doi.org/10.1016/j.porgcoat.2021.106201.

93 Hao, Y., Sun, W., Jiang, L. et al. (2021). Self-healing effect of epoxy coating containing mesoporous polyaniline hollow spheres loaded with benzotriazole. *Progress in Organic Coatings* 159: 106445. https://doi.org/10.1016/j.porgcoat.2021.106445.

94 Li, Q., Cao, L., Wang, W. et al. (2022). Photo-responsive rapid self-healing polyurethane elastomer with anticorrosion and antibacterial functions. *Composites Part A: Applied Science and Manufacturing* 163: 107213. https://doi.org/10.1016/j.compositesa.2022.107213.

95 Patra, R., Gautam, A., Gobi, K.V. et al. (2023). Hybrid silane coatings based on benzotriazole loaded aluminosilicate nanotubes for corrosion protection of mild steel. *Silicon* 15: 6981–6996. https://doi.org/10.1007/s12633-023-02556-7.

96 Zhang, M., Yu, X., Lin, Y. et al. (2021). Anti-corrosion coatings with active and passive protective performances based on v-COF/GO nanocontainers. *Progress in Organic Coatings* 159: 106415. https://doi.org/10.1016/j.porgcoat.2021.106415.

97 Wen, J., Lei, J., Chen, J. et al. (2020). Polyethylenimine wrapped mesoporous silica loaded benzotriazole with high pH-sensitivity for assembling self-healing anti-corrosive coatings. *Materials Chemistry and Physics* 253: 123425. https://doi.org/10.1016/j.matchemphys.2020.123425.

98 Zong, M., Yu, C., Li, J. et al. (2023). Redox and near-infrared light-responsive nanoplatform for enhanced pesticide delivery and pest control in rice: construction, efficacy, and potential mechanisms. *ACS Applied Materials & Interfaces* 15 (35): 41351–41361. https://doi.org/10.1021/acsami.3c08413.

99 Li, X., Gong, B., Zhang, J. et al. (2023). A photothermal and pH-responsive intelligent PSBG nanofiller for enhancing the barrier and self-healing performance of the SMP coatings. *Colloids and Surfaces A: Physicochemical and Engineering Aspects* 674: 131899. https://doi.org/10.1016/j.colsurfa.2023.131899.

100 Wang, P., Chen, S., Cao, Z. et al. (2017). NIR light-, temperature-, pH-, and redox-responsive polymer-modified reduced graphene oxide/mesoporous silica sandwich-like nanocomposites for controlled release. *ACS Applied Materials & Interfaces* 9 (34): 29005–29062. https://doi.org/10.1021/acsami.7b07468.

101 Skorb, E.V., Skirtach, A.G., Sviridov, D.V. et al. (2009). Laser-controllable coatings for corrosion protection. *ACS Nano* 3 (7): 1753–1760. https://doi.org/10.1021/nn900347x.

102 Nejad, H.B., Garrison, K.L., and Mather, P.T. (2016). Comparative analysis of shape memory-based self-healing coatings. *Journal of Polymer Science Part B: Polymer Physics* 54 (14): 1415–1426. https://doi.org/10.1002/polb.24061.

103 Wang, L., Deng, L., Zhang, D. et al. (2016). Shape memory composite (SMC) self-healing coatings for corrosion protection. *Progress in Organic Coatings* 97: 261–268. https://doi.org/10.1016/j.porgcoat.2016.04.041.

104 Wang, Y., Wang, Y., Wei, Q. et al. (2022). Light-responsive shape memory polymer composites. *European Polymer Journal* 173: 111314. https://doi.org/10.1016/j.eurpolymj.2022.111314.

105 Chen, J., Sun, D., Gu, T. et al. (2022). Photo-induced shape memory blend composites with remote selective self-healing performance enabled by polypyrrole nanoparticles. *Composites Science and Technology* 217: 109123. https://doi.org/10.1016/j.compscitech.2021.109123.

106 Huang, M. and Yang, J. (2014). Salt spray and EIS studies on HDI microcapsule-based self-healing anticorrosive coatings. *Progress in Organic Coatings* 77 (1): 168–175. https://doi.org/10.1016/j.porgcoat.2013.09.002.

107 Pourhashem, S., Rashidi, A., Vaezi, M.R. et al. (2017). Excellent corrosion protection performance of epoxy composite coatings filled with amino-silane functionalized graphene oxide. *Surface & Coatings Technology* 317: 1–9. https://doi.org/10.1016/j.surfcoat.2017.03.050.

108 Cheng, L., Liu, C., Wu, H. et al. (2021). A mussel-inspired delivery system for enhancing self-healing property of epoxy coatings. *Journal of Materials Science and Technology* 80: 36–49. https://doi.org/10.1016/j.jmst.2020.10.075.

109 Mahdavian, M. and Attar, M.M. (2006). Another approach in analysis of paint coatings with EIS measurement: phase angle at high frequencies. *Corrosion Science* 48 (12): 4152–4157. https://doi.org/10.1016/j.corsci.2006.03.012.

110 Liu, C., Qian, B., Hou, P. et al. (2021). Stimulus responsive zeolitic imidazolate framework to achieve corrosion sensing and active protecting in polymeric coatings. *ACS Applied Materials & Interfaces* 13 (3): 4429–4441. https://doi.org/10.1021/acsami.0c22642.

111 Honesty, N.R. and Gewirth, A.A. (2012). Shell-isolated nanoparticle enhanced Raman spectroscopy (SHINERS) investigation of benzotriazole film formation on Cu(100), Cu(111), and Cu (poly). *Journal of Raman Spectroscopy* 43 (1): 46–50. https://doi.org/10.1002/jrs.2989.

112 Ma, L., Wang, J., Ren, C. et al. (2020). Detection of corrosion inhibitor adsorption via a surface-enhanced Raman spectroscopy (SERS) silver nanorods tape sensor. *Sensors and Actuators B: Chemical* 321: 128617. https://doi.org/10.1016/j.snb.2020.128617.

113 Qian, H., Xu, D., Du, C. et al. (2017). Dual-action smart coatings with a self-healing superhydrophobic surface and anti-corrosion properties. *Journal of Materials Chemistry A* 5 (5): 2355–2364. https://doi.org/10.1039/c6ta10903a.

114 González-García, Y., Mol, J.M.C., Muselle, T. et al. (2011). SECM study of defect repair in self-healing polymer coatings on metals. *Electrochemistry Communications* 13 (2): 169–173. https://doi.org/10.1016/j.elecom.2010.12.005.

115 Wang, J., Ma, L., Huang, Y. et al. Photothermally activated self-healing protective coating based on the “close and seal” dual-action mechanisms. *Composites Part B: Engineering* 231: 109574. https://doi.org/10.1016/j.compositesb.2021.109574.

116 Huang, Y., Deng, L., Ju, P. et al. (2018). Triple-action self-healing protective coatings based on shape memory polymers containing dual-function microspheres. *ACS Applied Materials & Interfaces* 10 (27): 23369–23379. https://doi.org/10.1021/acsami.8b06985.

117 Qutachi, O., Vetsch, J.R., Gill, D. et al. (2014). Injectable and porous PLGA microspheres that form highly porous scaffolds at body temperature. *Acta Biomaterialia* 10 (12): 5090–5098. https://doi.org/10.1016/j.actbio.2014.08.015.

118 Fan, H. and Jin, Z. (2015). Hierarchical porous polycaprolactone microspheres generated via a simple pathway combining nanoprecipitation and hydrolysis. *Chemical Communications* 51 (82): 15114–15117. https://doi.org/10.1039/c5cc04586j.

119 Yang, Q., de Vries, M.H., Picchioni, F. et al. (2013). A novel method of preparing metallic Janus silica particles using supercritical carbon dioxide. *Nanoscale* 5 (21): 10420–10427. https://doi.org/10.1039/c3nr81280d.

120 Zeng, J. and Xuan, Y. (2018). Tunable full-spectrum photo-thermal conversion features of magnetic-plasmonic Fe_3O_4/TiN nanofluid. *Nano Energy* 51: 754–763. https://doi.org/10.1016/j.nanoen.2018.07.034.

121 Huang, L., Li, J., Yuan, W. et al. (2020). Near-infrared light controlled fast self-healing protective coating on magnesium alloy. *Corrosion Science* 163: 108257. https://doi.org/10.1016/j.corsci.2019.108257.

122 Yang, J., Wang, D., Wang, J. et al. (2022). Corrosion resistance and near-infrared light induced self-healing behavior of polycaprolactone coating with MIL-53@

TA on magnesium alloy. *Applied Surface Science* 585: 152729. https://doi.org/10.1016/j.apsusc.2022.152729.

123 Qian, H., Chang, W., Cui, T. et al. (2021). Multi-mode scanning electrochemical microscopic study of microbiologically influenced corrosion mechanism of 304 stainless steel by thermoacidophilic archaea. *Corrosion Science* 191: 109751. https://doi.org/10.1016/j.corsci.2021.109751.

124 Polcari, D., Dauphin-Ducharme, P., and Mauzeroll, J. (2016). Scanning electrochemical microscopy: a comprehensive review of experimental parameters from 1989 to 2015. *Chemical Reviews* 116 (22): 13234–13278. https://doi.org/10.1021/acs.chemrev.6b00067.

125 Pei, Z., Cheng, X., Yang, X. et al. (2021). Understanding environmental impacts on initial atmospheric corrosion based on corrosion monitoring sensors. *Journal of Materials Science and Technology* 64: 214–221. https://doi.org/10.1016/j.jmst.2020.01.023.

126 Pei, Z., Xiao, K., Chen, L. et al. (2020). Investigation of corrosion behaviors on an Fe/Cu-type ACM sensor under various environments. *Metals* 10 (7): 905. https://doi.org/10.3390/met10070905.

127 Li, Q., Xia, X., Pei, Z. et al. (2022). Long-term corrosion monitoring of carbon steels and environmental correlation analysis via the random forest method. *npj Materials Degradation* 6 (1): 1. https://doi.org/10.1038/s41529-021-00211-3.

128 Liu, T., Zhang, D., Ma, L. et al. (2022). Smart protective coatings with self-sensing and active corrosion protection dual functionality from pH-sensitive calcium carbonate microcontainers. *Corrosion Science* 200: 110254. https://doi.org/10.1016/j.corsci.2022.110254.

129 Hieda, J., Sakaguchi, A., Nakano, M. et al. (2019). Relationships between surface energy and charge of surface-modified titanium and HAp formation. *Applied Surface Science* 465: 509–516. https://doi.org/10.1016/j.apsusc.2018.09.162.

130 Milian, Y., Claros, M., Ushak, S. et al. (2023). Silica based Janus nanoparticles: synthesis methods, characterization, and applications. *Applied Materials Today* 34: 101901. https://doi.org/10.1016/j.apmt.2023.101901.

131 Farid, M.U., Kharraz, J.A., and An, A.K. (2021). Plasmonic titanium nitride nano-enabled membranes with high structural stability for efficient photothermal desalination. *ACS Applied Materials & Interfaces* 13 (3): 3805–3815. https://doi.org/10.1021/acsami.0c17154.

132 Shen, X., Dong, Z., Sim, C. et al. (2022). A comparative study on the self-healing characterizations and formulation optimization of polyurea coating. *Polymers* 14 (17): 3520. https://doi.org/10.3390/polym14173520.

133 Zou, Y., Fang, L., Chen, T. et al. (2018). Near-infrared light and solar light activated self-healing epoxy coating having enhanced properties using MXene flakes as multifunctional fillers. *Polymers* 10 (5): 474. https://doi.org/10.3390/polym10050474.

134 Cao, L., Wang, W., Li, Q. et al. (2022). Three-dimensional nanofibers network multifunctional material for photothermal self-healing protective coating.

Chemical Engineering Journal 440: 134943. https://doi.org/10.1016/j.cej.2022.134943.

135 Liu, C., Cheng, L., Hou, P. et al. (2022). Photothermal-responsive wormlike polydopamine-wrapped ethylene-vinyl acetate copolymer toward triple-action self-healing anticorrosion coating. *ACS Applied Polymer Materials* 4 (8): 6067–6079. https://doi.org/10.1021/acsapm.2c00875.

136 Wang, W., Xu, L., Sun, H. et al. (2015). Spatial resolution comparison of AC-SECM with SECM and their characterization of self-healing performance of hexamethylene diisocyanate trimer microcapsule coatings. *Journal of Materials Chemistry A* 3 (10): 5599–5607. https://doi.org/10.1039/c5ta00529a.

137 Jamali, S.S., Moulton, S.E., Tallman, D.E. et al. (2017). Self-healing characteristic of praseodymium conversion coating on AZNd Mg alloy studied by scanning electrochemical microscopy. *Electrochemistry Communications* 76: 6–9. https://doi.org/10.1016/j.elecom.2017.01.004.

138 Pilbáth, A., Szabó, T., Telegdi, J. et al. (2012). SECM study of steel corrosion under scratched microencapsulated epoxy resin. *Progress in Organic Coatings* 75 (4): 480–485. https://doi.org/10.1016/j.porgcoat.2012.06.006.

139 Katayama, H., Noda, K., Yamamoto, M. et al. (2001). Relationship between corrosion rate of carbon steel and water film thickness under thin layer of artificial sea water. *Journal of the Japan Institute of Metals* 65 (4): 298–302. https://doi.org/10.2320/jinstmet1952.65.4_298.

140 Hirohata, Y., Nishida, K., Haruna, T. et al. (2017). Development of measurement system for pH distribution on metal surface under atmospheric corrosion. *Journal of the Japan Institute of Metals and Materials* 81 (11): 495–501. https://doi.org/10.2320/jinstmet.JAW201704.

141 Mizuno, D., Suzuki, S., Fujita, S. et al. (2014). Corrosion monitoring and materials selection for automotive environments by using Atmospheric Corrosion Monitor (ACM) sensor. *Corrosion Science* 83: 217–225. https://doi.org/10.1016/j.corsci.2014.02.020.

142 Li, Z., Fu, D., Li, Y. et al. (2019). Application of an electrical resistance sensor-based automated corrosion monitor in the study of atmospheric corrosion. *Materials* 12 (7): 1065. https://doi.org/10.3390/ma12071065.

143 Long, Z., Li, C., Zhang, Q. et al. (2024). A novel synthesis of hollow Fe_3O_4 nanorods with a high saturation magnetization. *Materials Letters* 355: 135301. https://doi.org/10.1016/j.matlet.2023.135301.

144 Zhang, X., Qiu, C., Li, F. et al. (2023). Magnetic nanocellulose-based adsorbent for highly selective removal of malachite green from mixed dye solution. *International Journal of Biological Macromolecules* 253 (2): 126752. https://doi.org/10.1016/j.ijbiomac.2023.126752.

145 Nguyen, L.H., Nam, N.H., Tam, L.T. et al. (2023). Effect of Gd substitution on structure, optical and magnetic properties, and heating efficiency of Fe_3O_4 nanoparticles for magnetic hyperthermia applications. *Journal of Alloys and Compounds* 968: 172205. https://doi.org/10.1016/j.jallcom.2023.172205.

146 Eskandani, M., Derakhshankhah, H., Zare, S. et al. (2023). Enzymatically crosslinked magnetic starch-grafted poly(tannic acid) hydrogel for "smart" cancer treatment: an in vitro chemo/hyperthermia therapy study. *International Journal of Biological Macromolecules* 253 (6): 127214. https://doi.org/10.1016/j.ijbiomac.2023.127214.

147 Zhao, L., Liang, L., Guo, M. et al. (2021). Hepatocellular carcinoma targeting and pharmacodynamics of paclitaxel nanoliposomes modified by glycyrrhetinic acid and ferric tetroxide. *Current Topics in Medicinal Chemistry* 21 (14): 1268–1284. https://doi.org/10.2174/1568026621666210621090005.

148 Li, L., Yu, Z., Liu, J. et al. (2023). Swarming responsive photonic nanorobots for motile-targeting microenvironmental mapping and mapping-guided photothermal treatment. *Nano-Micro Letters* 15: 141. https://doi.org/10.1007/s40820-023-01095-5.

149 Shariati, B., Goodarzi, M.T., and Jalali, A. (2023). Gold nanorods incorporated into a MoS_2/Fe_3O_4 nanocomposite for photothermal therapy and drug delivery. *New Journal of Chemistry* 47: 20100–20108. https://doi.org/10.1039/D3NJ02894A.

150 Wang, J., Wu, S., and Ma, L. (2022). Corrosion resistant coating with passive protection and self-healing property based on Fe_3O_4-MBT nanoparticles. *Corrosion Communications* 7: 1–11. https://doi.org/10.1016/j.corcom.2021.12.005.

151 Dong, W., Liu, W., and Feng, Y. (2023). The chemical state of iron species influence on the performance of Fe–N–C bifunctional electrocatalyst for Zn–air batteries. *Nanotechnology* 35: 6. https://doi.org/10.1088/1361-6528/ad0909.

152 Bao, J., Zhang, T., and Wu, S. (2023). Hydrophilic magnetic $Ti_3C_2T_x$-based nanocomposite as an efficient boron adsorbent: synthesis, characterization, and application. *Journal of Hazardous Materials* 460: 132460. https://doi.org/10.1016/j.jhazmat.2023.132460.

153 Wang, X., Lei, W., and Zou, F. (2023). Ternary mixture thermochromic microcapsules for visible light absorption and photothermal conversion energy storage. *Applied Surface Science* 630: 157431. https://doi.org/10.1016/j.apsusc.2023.157431.

154 Huan, W., Dong, M., and Li, J. (2023). Magnetic properties of Al^{3+} doped Fe_3O_4 by solid-solid reaction and biological applications of GQDs grafted MNPs. *Journal of Magnetism and Magnetic Materials* 587: 171360. https://doi.org/10.1016/j.jmmm.2023.171360.

155 Ghorbani, M., Ebrahimnezhad-Khaljiri, H., Eslami-Farsani, R. et al. (2021). The synergic effect of microcapsules and titanium nanoparticles on the self-healing and self-lubricating epoxy coatings: a dual smart application. *Surfaces and Interfaces* 23: 100998. https://doi.org/10.1016/j.surfin.2021.100998.

156 Cho, Y.J., Kong, L.C., Islam, R. et al. (2020). Photothermal self-healing of gold nanoparticle-polystyrene hybrids. *Nanoscale* 12 (40): 20726–20736. https://doi.org/10.1039/d0nr05621a.

157 Li, K., Xiang, J., Zhou, J. et al. (2023). Self-healing and wear resistance stable superhydrophobic composite coating with electrothermal and photothermal effects for anti-icing. *Progress in Organic Coatings* 177: 107415. https://doi.org/10.1016/j.porgcoat.2023.107415.

158 Lutz, A., van den Berg, O., Wielant, J. et al. (2016). A multiple-action self-healing coating. *Frontiers in Materials* 2: 73. https://doi.org/10.3389/fmats.2015.00073.

159 Hao, Z., Chen, S., Lin, Z. et al. (2022). Anticorrosive composite self-healing coating enabled by solar irradiation. *Frontiers of Chemical Science and Engineering* 16 (9): 1355–1366. https://doi.org/10.1007/s11705-022-2147-1.

4

Development of pH- and Redox-activated Self-healing Coating

4.1 Introduction

Metal corrosion is a redox process, which results in pH changes and redox activity at the metal–coating interface. In this chapter, we introduce pH variation and redox activity–triggered intrinsic and extrinsic self-healing coatings. These coatings lead to the accelerated release of inhibitors from containers or polymer–inhibitor conjugates, thereby facilitating the formation of a protective layer on the metal surface. Methods for measuring the release behavior of corrosion inhibitors were investigated. We present several pH variations and redox activity–triggered self-healing coating systems, including coating preparation, filler preparation, surface characterization, macroscopic and microscopic electrochemical properties, and self-healing mechanisms. Finally, we discuss the advantages and disadvantages of pH variation and redox-activated self-healing coatings.

4.2 pH Responsive Self-healing Coating Based on the Synergistic Effect of Two Corrosion Inhibitors

Owing to the advantageous properties, such as high strength, stiffness, and ease of machining and casting, aluminum alloys present a broad range of applications in automobiles [1, 2], aerospace [3], household appliances, shipbuilding [4], and military hardware. The 2024-T3 aluminum alloy is widely used in aerospace applications [5–7]. Although an aluminum oxide film can protect the aluminum matrix, localized corrosion can easily occur in environments containing chloride ions and other corrosive media, leading to alloy failure [8, 9]. Active protection based on corrosion inhibitors has been widely employed due to its high efficiency and feasibility. The addition of an appropriate amount of effective corrosion inhibitors benefits corrosion inhibition [10–12]. Therefore, it is necessary to study the inhibitory effect of corrosion inhibitors on aluminum alloys and the synergistic effect of mixed corrosion inhibitors to analyze the corrosion mechanism and control of aluminum alloys.

Smart Protective Coatings for Corrosion Control, First Edition. Lingwei Ma and Dawei Zhang.

Organic coatings are widely utilized due to their exceptional anticorrosion properties and stability [13, 14]. These coatings effectively prolong the service life of metallic materials by physically isolating them from corrosive environments. In comparison with general environments [15], service conditions in aviation, aerospace, and marine fields are often more complex and harsh. The coatings in these settings are susceptible to the aging effects of the natural environment, including high temperature, high humidity, high salt, and strong ultraviolet irradiation [16]. In addition, external damage such as mechanical shock and scratching can occur [17–19]. Under these conditions, organic coatings are prone to both macroscopic damage and microscopic defects. If these issues are not discovered and effectively repaired in a timely manner, these defects can become rapid channels for the transmission of corrosive media, thereby leading to the loss of the coating's ability to protect the metal substrate [20, 21]. Therefore, it is crucial to develop self-healing anticorrosion coatings to promptly repair damaged coatings.

Incorporating corrosion inhibitors into the coating proves beneficial in realizing the self-healing effect and slowing down the corrosion reaction at coating defects. Corrosion inhibitors are typically introduced into coatings through microcarriers or nanocarriers [22]. However, there have been limited studies on the composition of corrosion inhibitors in micro- and nanocarriers. Moreover, the dosage and composition ratio of corrosion inhibitors are often suboptimally defined, preventing them from exerting their maximum corrosion inhibition effect and sometimes even causing antagonistic effects. This situation is not conducive to the advancement of inhibitor loading technology and the development of self-healing coatings. Therefore, determining the optimal proportion of mixed corrosion inhibitors is necessary to effectively guide the preparation of corrosion inhibitor carriers and self-healing coatings.

Among the various corrosion inhibitor containers, mesoporous SiO_2 ($mSiO_2$) is one of the most researched and widely used. Considerable attention has been directed toward $mSiO_2$ due to its unique advantages: (i) good compatibility with the coating and easy and uniform dispersion; (ii) large specific surface area and pore volume, enabling high corrosion inhibitor loading ability; (iii) easy modification for the stimulus-responsive release of corrosion inhibitors; and (iv) stable chemical and mechanical properties [23].

Commonly, methods used to prepare silica nanocapsules involve multiple steps. The initial step entails synthesizing nanosilica coated with templating agents, followed by the removal of the templating agent encapsulated in the silica capsule through either a sintering approach or ion exchange. This process results in the formation of $mSiO_2$ nanocapsules. Subsequently, corrosion inhibitor molecules are loaded into the nanocapsules through low-pressure loading [24]. However, the overall nanocapsule preparation process using these methods is complex. To simplify the preparation process and significantly reduce the experimental time, we adopted a one-step approach for the preparation of silica nanocapsules loaded with two types of corrosion inhibitors. By adjusting the experimental parameters, we obtained $mSiO_2$ capsules with high corrosion inhibitor–loading ratios, uniform

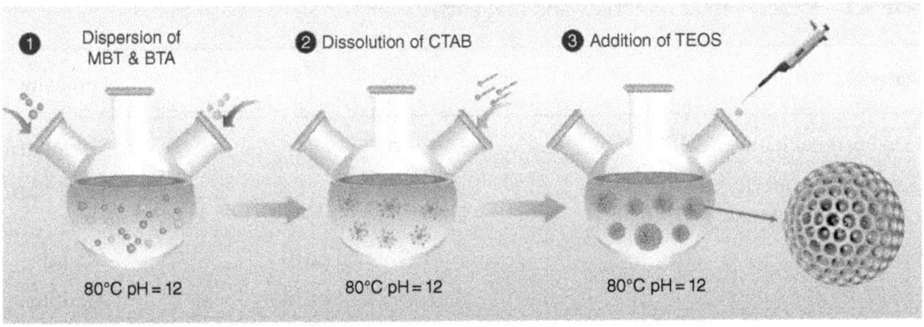

Figure 4.1 Schematic of MBT-BTA@mSiO_2 prepared by one-step synthesis method.

sizes, and good dispersion properties. The preparation process of silica nanocapsules is explained in Figure 4.1.

This section delves into the corrosion inhibition mechanism and self-healing of the coating based on the synergistic corrosion inhibition effect. To verify the corrosion inhibition performance, 2-mercaptobenzothiazole (MBT) and 1*H*-benzotriazole (BTA) corrosion inhibitors were selected, and their performances were assessed through electrochemical methods, SEM, and surface-enhanced Raman spectroscopy. Subsequently, two types of corrosion inhibitors were mixed to investigate the influence of their type, concentration, and ratio on the inhibition performance. The mechanism of synergistic corrosion inhibition was then elucidated [25].

The optimized corrosion inhibitor mixture was loaded into mesoporous silica nanocapsules using the Stober method. FTIR was employed to characterize the successful loading of MBT and BTA molecules. The morphologies of MBT-BTA@mSiO_2 were observed using SEM. The loading rate of the corrosion inhibitor was measured by ultraviolet spectroscopy, and the electrochemical properties of the MBT-BTA@mSiO_2 nanocapsules were assessed by electrochemical impedance spectroscopy (EIS).

Subsequently, MBT-BTA@mSiO_2, with optimal loading ratios of MBT and BTA, was incorporated into an epoxy coating, and its self-healing properties were characterized through coating scratch experiments. The morphological changes in the self-healing coatings were observed by SEM, and the adsorption of corrosion inhibitors on the aluminum alloy was characterized by XPS. Electrochemical experiments were conducted to verify the long-term protective ability of the coatings.

4.2.1 Materials and Methods

4.2.1.1 Materials

The materials and reagents used in this section are shown in Table 4.1.

4.2.1.2 Instruments

The instruments used in this section are shown in Table 4.2.

Table 4.1 Experimental materials and reagents.

Materials	Purity	Company
Ethyl orthosilicate (TEOS)	99.0%	Sinopharm
Hexadecyl trimethyl ammonium bromide (CTAB)	99.0%	Sinopharm
BTA	Analytical purity	Aladdin
MBT	Analytical purity	Aladdin
Ethanol (C_2H_5OH)	Analytical purity	Sinopharm

Table 4.2 The instruments and equipment.

Instrument	Production type	Manufacturer
SEM	FE-SEM G300	German Zeiss
TEM	JEM-2100F	Japan JEOL
UV/Vis/NIR spectrograph	Lambda 750	America PerkinElmer
Physical adsorption instrument	ASAP2460	America Micromeritics
Fourier transform infrared spectrometer	Vector 22	German, Bruker
Thermocouple thermometer	TES-1310	China, TES
Infrared thermal imaging camera	FLIRONEPRO	America FLIR
Electronic balances	ME204	Switzerland, Mettler Toledo
Cell crusher machine	S-450D	America SONICS
Acidimeter	PHS-3C	China Leici
Magnetic stirring oil bath	ZNCL-GS	China Jie-ang
Centrifuge	Allegra X-30	America Beckman
Electrochemical workstation	PARSTAT-2273	America Princeton
Constant temperature blast drying oven	303-3S	China, Baihui
Confocal laser scanning microscope	VK-X2000 series	Japan KEYENCE
UV/Vis/NIR spectroscopy	Lambda 750	America PerkinElmer
Simultaneous thermal analyzer	STA449F3	German Netzsch

4.2.1.3 Preparation of Corrosion Inhibitor-loaded SiO_2 Nanocapsules

In this study, silica nanocapsules were prepared using a modified Stober method [26]. TEOS and CTAB served as the source of silica and the porogen/templating agent for $mSiO_2$ nanocapsules, respectively. Subsequently, MBT-BTA@$mSiO_2$ was obtained through hydrolysis and polymerization of the silica material in an alkaline solution. Initially, a 250 ml aqueous solution at pH = 12 was stirred in a water bath at 80 °C. Then, 0.6 g of CTAB, 0.083 g of MBT, and 0.417 g of BTA were added, and the mixture was stirred for 30 minutes. Finally, 2.5 ml of TEOS was slowly added and reacted for two hours at 80 °C under stirring. After the reaction, the white suspension was centrifuged (8000 r min^{-1}, five minutes), washed twice, and then dried at 70 °C to obtain MBT-BTA@mSiO_2 with the original mass ratio of MBT/BTA (1 : 5). MBT-BTA@mSiO_2 with mass ratios of 2/4 and 3/3 were prepared by varying the amounts of MBT and BTA to 0.167 g MBT + 0.333 g BTA and 0.250 g MBT + 0.250 g BTA, respectively. These three coatings are denoted as MBT-BTA@m$SiO_2$1/5, MBT-BTA@m$SiO_2$2/4, and MBT-BTA@m$SiO_2$3/3.

In contrast, the BTA@mSiO_2 nanocapsules were prepared using a similar approach to the MBT-BTA@mSiO_2 composites, with the difference that only BTA was loaded. A solution containing hydrochloric acid and ethanol (1 : 50 v/v) was used to remove the templating agent from the prepared BTA@mSiO_2 and MBT-BTA@mSiO_2 nanocapsules to obtain pure mSiO_2 nanocapsules without corrosion inhibitor addition.

4.2.1.4 Corrosion Inhibitor Release Behavior from MBT-BTA@mSiO_2

To assess the loading of corrosion inhibitors in mSiO_2 and the release behavior of MBT and BTA, UV–Vis spectrometry measured the absorption peaks of the prepared nanocapsules. Distinct absorption peaks for MBT and BTA at 322 and 265 nm, respectively, were observed.

UV–Vis spectroscopy was used to evaluate the loading capacities of MBT-BTA@m$SiO_2$1/5 and BTA@mSiO_2. Essentially, 0.1 g of MBT-BTA@m$SiO_2$1/5 or BTA@mSiO_2 was immersed in the mixed solution of hydrochloric acid and ethanol (1 : 50 v/v) and stirred at 60 °C in the oil bath. Subsequently, the solution was centrifuged, and the supernatant was collected to measure the loading capacity through UV–Vis spectroscopy.

Standard UV–Vis spectrum curves of MBT and BTA were measured as follows: First, MBT and BTA solutions of different concentrations were prepared at varying pH conditions. Standard curves were plotted by matching UV–Vis peak intensities based on concentrations. The release behavior of the prepared materials was investigated using UV–Vis spectroscopy at specified time intervals. A value of 0.1 g of BTA@mSiO_2 or MBT-BTA@mSiO_2 was dispersed in a dialysis bag with 25 ml of 3.5 wt% NaCl solution at different pH levels, immersed in 200 ml of the solution. After specific intervals, the solution outside the dialysis bag was collected for measurement and then poured back.

4.2.1.5 Preparation of Self-healing Coatings

A 2024-T3 aluminum alloy was selected as the metal substrate. The coating system was formulated by combining DGEBA, NGDE, and D230 in a 1 : 1 : 1 ratio. Initially, DGEBA and NGDE were thoroughly mixed, and then MBT-BTA@mSiO_2 composite fillers were added at mass fractions of 1%, 2%, and 3%, respectively. The mixture was stirred for three hours to ensure proper dispersion of the fillers. Following this, the curing agent D230 was introduced and stirred. Using a dropper, a specific quantity of the mixed solution was applied to the surface of the substrate, spreading it completely and uniformly scraping it with a scraper. The coating was then cured at 60 °C for 30 hours, resulting in a thickness of approximately 80 μm for all prepared coating samples. The self-healing behavior was investigated using a medical scalpel to make a ~80-μm wide and 1-cm long scratch on the surface of coatings.

4.2.1.6 Electrochemical Measurements

EIS can be used to assess the macroscopic electrochemical activity of intact and healed coatings in NaCl solutions. The edges of the coatings were polished before testing to facilitate conductivity. A three-electrode system was used, and the exposed area of each sample was 1 cm^2.

4.2.2 Characterization of MBT-BTA@mSiO_2

4.2.2.1 Morphology and Structure Characterization of MBT-BTA@mSiO_2

Figure 4.2 shows SEM images of BTA@mSiO_2, MBT-BTA@m$SiO_2$1/5, and MBT-BTA@m$SiO_2$2/4. The BTA@mSiO_2 and MBT-BTA@m$SiO_2$1/5 nanocapsules exhibited a completely spherical morphology with a uniform size and well-dispersed nanoparticle morphology with size of approximately 80–130 nm. As shown in the figure, the nanocapsules tended to form nonspherical morphologies as the MBT content increased. The surface morphology of SiO_2 nanoparticles was significantly

(a) (b) (c)

Figure 4.2 SEM images of (a) BTA@mSiO_2, (b) MBT-BTA@m$SiO_2$1/5, (c) MBT-BTA@m$SiO_2$2/4.

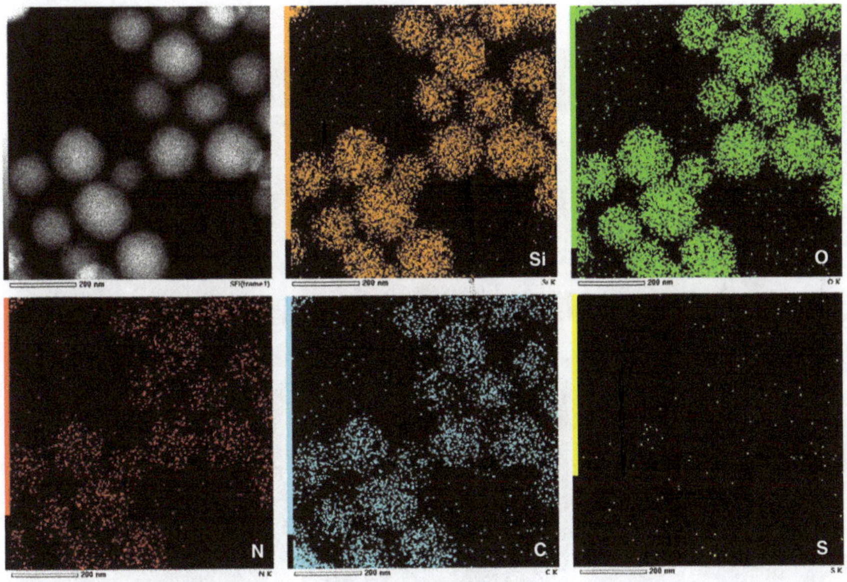

Figure 4.3 TEM image and EDS mapping of BTA@mSiO_2.

varied and partially agglomerated, which was ascribed to the addition of MBT that affected the formation of SiO_2.

The TEM image of BTA@mSiO_2, displayed in Figure 4.3, shows near-spherical nanocapsules. The results of the EDS mapping of BTA@mSiO_2 show that the distribution of Si and O corresponds to mSiO_2, and N and C correspond to the corrosion inhibitor molecules of BTA ($C_6H_5N_3$), confirming that the BTA molecules were successfully encapsulated to the SiO_2 nanocapsules [27, 28].

Figure 4.4 shows the TEM images and EDS mapping of MBT-BTA@m$SiO_2$1/5. The shapes of the silica spheres remained similar after loading with both MBT and BTA, exhibiting a nearly spherical structure. In the EDS energy spectra, the distribution of Si, O, C, N, and S corresponded to MBT-BTA@m$SiO_2$1/5, where the S element was derived from MBT, and the C and N elements corresponded to MBT and BTA, respectively, proving that MBT and BTA molecules were successfully loaded into the nanocapsules [29]. When the MBT content in the nanocapsules increased, the morphology of MBT-BTA@m$SiO_2$2/4 underwent severe agglomeration and changed from spherical to irregular (Figure 4.5). Similarly, S was derived from MBT, and C and N correspond to MBT and BTA, respectively. The increase in MBT led to more S than that in the MBT-BTA@m$SiO_2$1/5 nanocapsules. A typical sample of MBT-BTA@m$SiO_2$1/5 was selected for the characterization and labeled as MBT-BTA@mSiO_2.

To further evaluate the loading of MBT and BTA molecules in the silica nanocapsules, two samples, MBT-BTA@mSiO_2 and BTA@mSiO_2, were qualitatively characterized through FTIR. Figure 4.6 shows the FTIR spectra. The peaks at 1087 and 462 cm^{-1} were associated with the antisymmetric stretching vibration

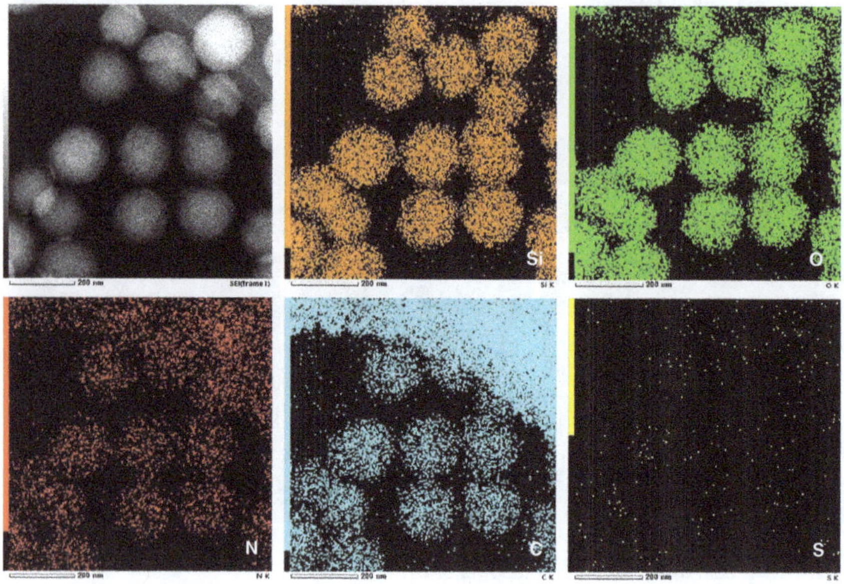

Figure 4.4 TEM image and EDS mapping of MBT-BTA@mSiO$_2$1/5.

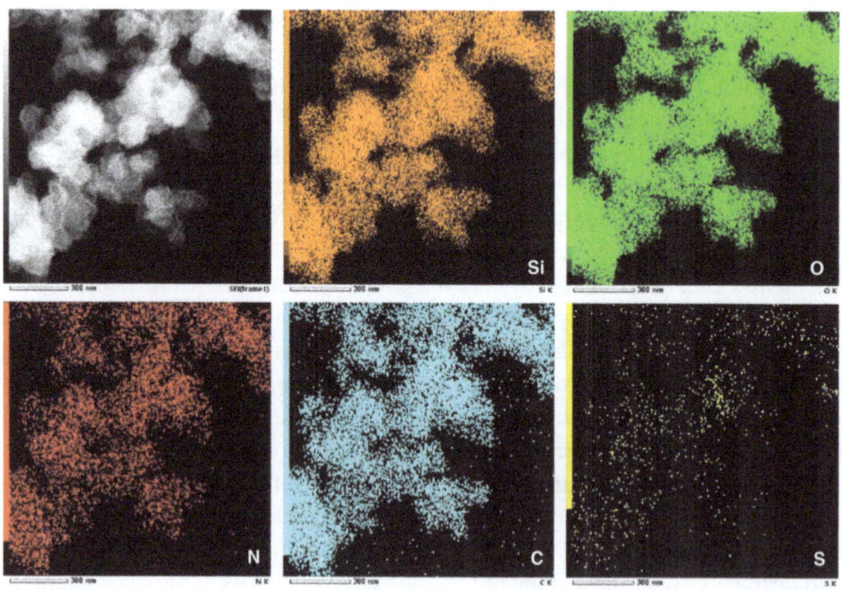

Figure 4.5 TEM image and EDS mapping of MBT-BTA@mSiO$_2$2/4.

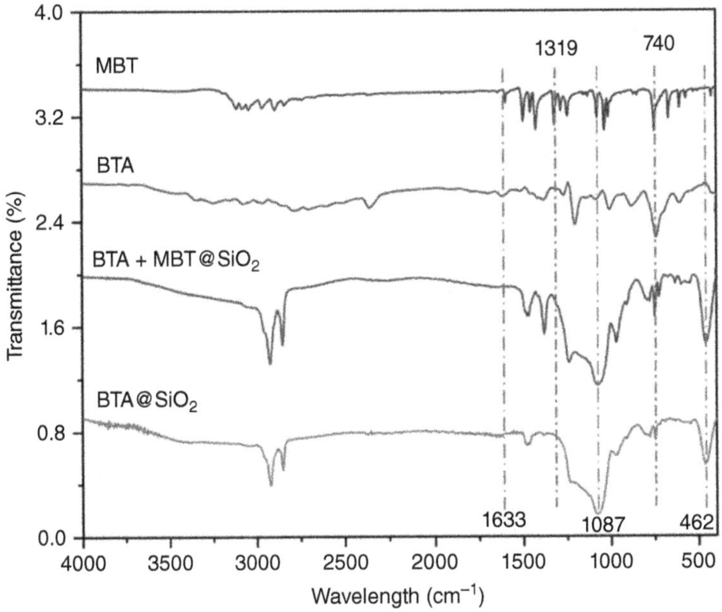

Figure 4.6 FTIR spectra of MBT-BTA@mSiO_2 and BTA@mSiO_2.

of Si—O—Si and bending vibration of Si—O bond, respectively, confirming the generation of mesoporous silica [30–33]. Stretching vibration of thiazole ring from MBT was observed at 1319 cm^{-1}, which confirmed the presence of MBT in MBT-BTA@mSiO_2. The bending vibrations within the C–H face of the benzene ring and N–H bending vibration of the BTA molecule were 740 and 1633 cm^{-1}, respectively [34, 35].

Nitrogen adsorption and desorption curves were obtained to determine the pore diameter distribution and structure of the silica nanostructures. Figure 4.7a,b shows the N_2 isothermal adsorption–desorption curves and the Barrett–Joyner–Halenda (BJH) pore diameter distribution of MBT-BTA@mSiO_2, respectively. Nitrogen adsorption by MBT-BTA@mSiO_2 was extremely low, indicating that there was no obvious porous structure. Similarly, the adsorption of nitrogen by BTA@mSiO_2 was low with a specific surface area of 68 $m^2\,g^{-1}$. The corrosion inhibitor and porogen were completely released after washing with the hydrogen chloride–ethanol solution, which resulted in a rapid increase in the BET-specific surface area of mSiO_2 to 635.3 $m^2\,g^{-1}$ and a narrow distribution centered at ~4.2 nm of a pore size. These results reveal that the prepared MBT-BTA@mSiO_2 possessed a rich porous structure [36–38], ensuring the efficient loading of corrosion inhibitor molecules into the mesoporous silica pores [39]. In summary, the MBT-BTA@mSiO_2 composite nanocapsules produced by the aforementioned one-step method had abundant pores, which enabled the successful loading of MBT and BTA corrosion inhibitor molecules into the pores of MBT-BTA@mSiO_2.

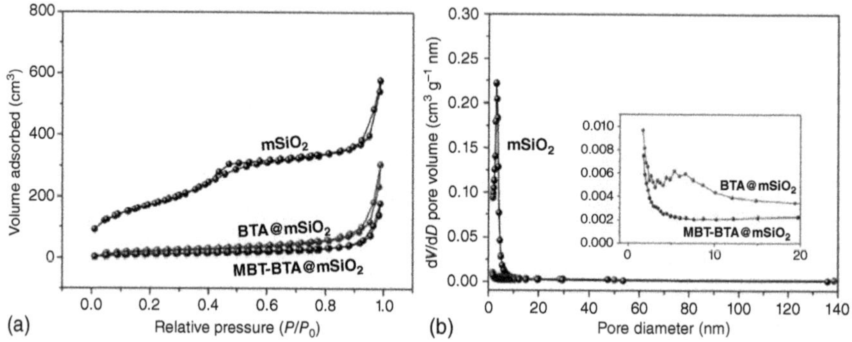

Figure 4.7 (a) N_2 isothermal adsorption desorption curves and (b) pore size distributions of MBT-BTA@mSiO_2, BTA@mSiO_2 and mSiO_2.

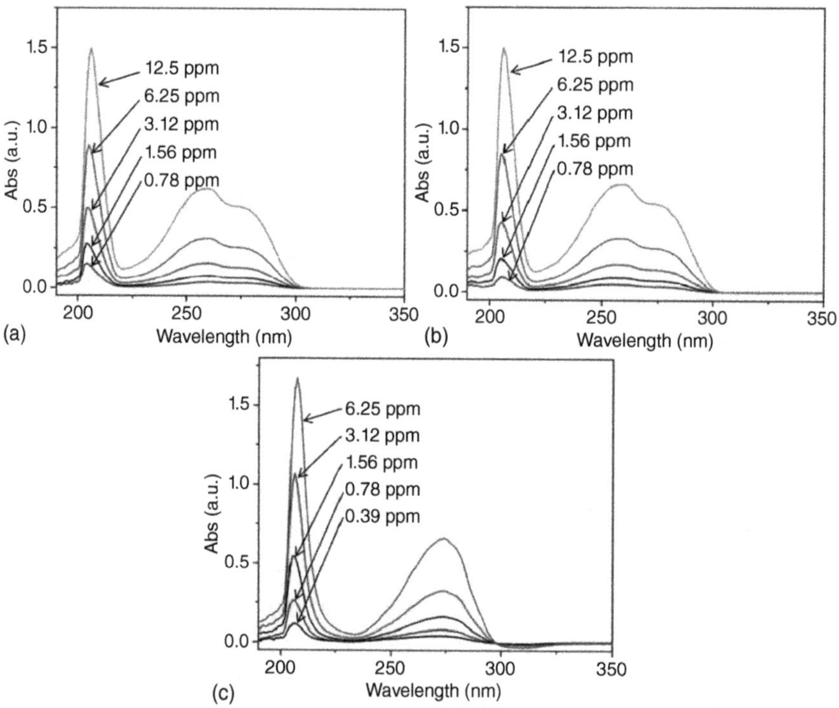

Figure 4.8 UV–Vis spectra of BTA in 3.5 wt% NaCl solution at (a) pH = 3, (b) pH = 7, (c) pH = 11.

4.2.2.2 Inhibitor Loading Capacity of MBT-BTA@mSiO_2

The loading ratio of the corrosion inhibitor in MBT-BTA@mSiO_2 was determined through UV–Vis absorption spectroscopy, and the inhibitor content in the nanocapsules was calculated by comparing with the standard curves. Figures 4.8 and 4.10 show the standard absorption spectra of BTA and MBT in NaCl solutions at different pH values, respectively. The relationship between the values of the

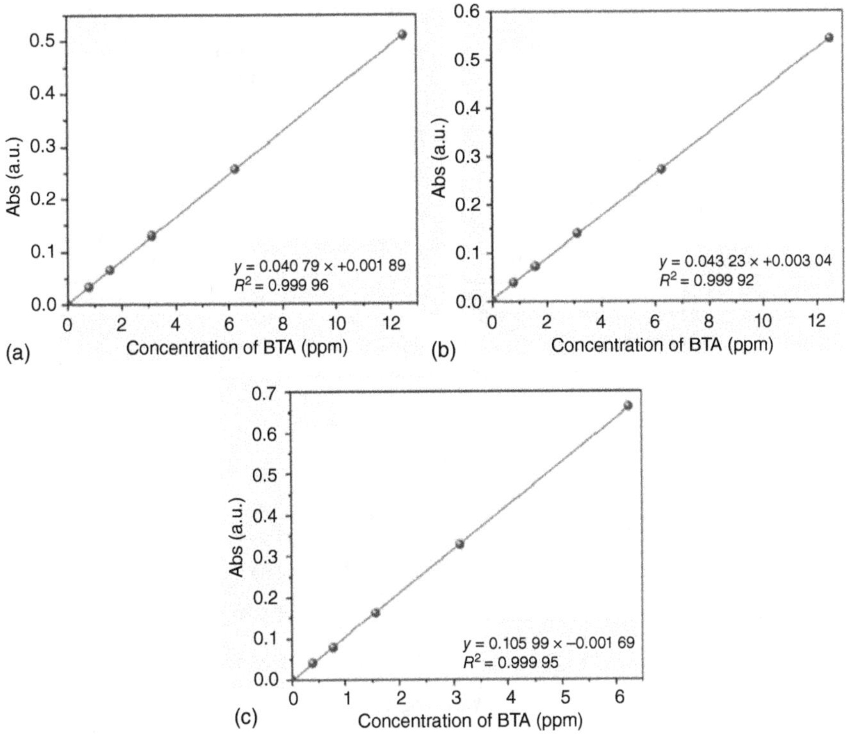

Figure 4.9 Standard UV–Vis absorption fitting curves of BTA in 3.5 wt% NaCl solution at (a) pH = 3, (b) pH = 7, (c) pH = 11.

characteristic absorption peaks and the concentrations of BTA and MBT was established by the linear fitting of the relevant scatter plots. Figures 4.9 and 4.11 show the standard UV–Vis absorption curves of BTA and MBT in different solutions.

BTA@mSiO_2 and MBT-BTA@mSiO_2 were immersed in hydrochloric acid and ethanol solutions (1 : 50 v/v), respectively, and stirred for 10 hours to ensure the full release of the inhibitors from the nanocapsules. The supernatant obtained from the mixed solution after centrifugation was analyzed through UV spectroscopy (see Figure 4.12). According to the standard fitting curves and results of the UV spectroscopy test, the loading capacities of MBT and BTA accounted for 3.29 and 11.42 wt% in the MBT-BTA@m$SiO_2$1/5, respectively. Similarly, the loading capacity of BTA accounted for 15.02 wt% in BTA@mSiO_2.

4.2.2.3 Release Behavior of MBT-BTA@mSiO_2

It is known that MBT-BTA@mSiO_2 has high loading capacity of corrosion inhibitor based on the above results. Subsequently, to evaluate the pH influence on the release behavior of corrosion inhibitor molecules from MBT-BTA@mSiO_2, UV–Vis absorption spectroscopy of solution with MBT-BTA@mSiO_2 was conducted to investigate the release behavior.

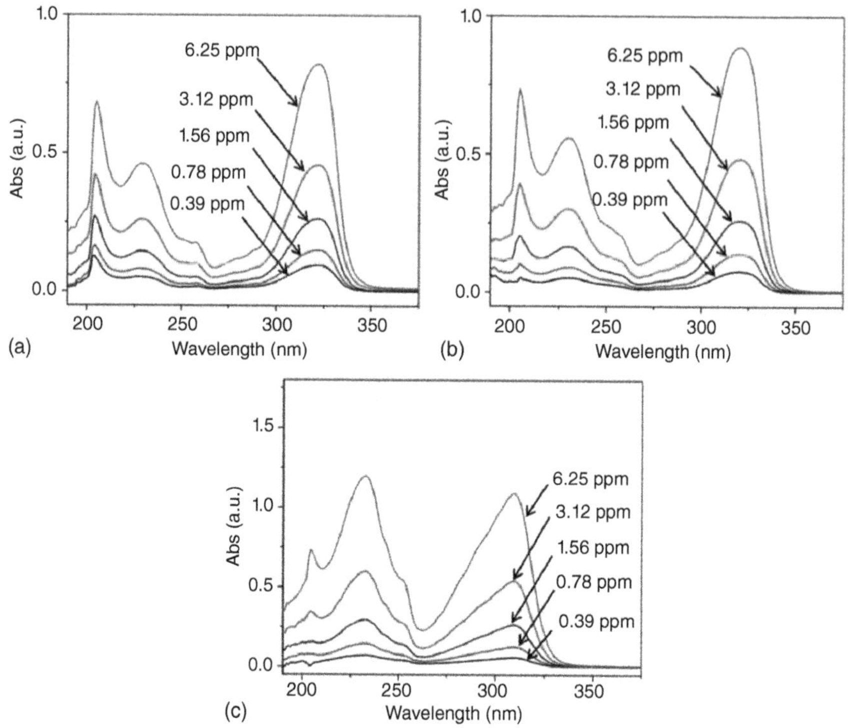

Figure 4.10 UV–Vis spectra of MBT in 3.5 wt% NaCl solution at (a) pH = 3, (b) pH = 7, (c) pH = 11.

Figure 4.13 shows that the UV absorption peaks of the MBT-BTA@mSiO_2 nanocapsules under different pH conditions increased continuously with time, implying that a growing amount of corrosion inhibitors was released [40, 41]. However, the release rates of BTA molecules at different pH values differed greatly [42–44]. Specifically, BTA was released rapidly and nearly completely under acidic or alkaline conditions within two hours, and the highest amount of BTA was released in an acidic environment. This is related to the fact that the OH^- adsorbed onto the silica surface destroyed the mSiO_2 structure [39, 45]. However, the overall MBT release was low due to the inherently low MBT content in the MBT-BTA@mSiO_2 [46, 47]. Similar to the BTA release, MBT was released faster and more rapidly under acidic conditions, but the UV absorption peaks changed slightly in neutral environments, indicating a small amount of release [48, 49]. Some researchers and scholars believe that the pH of epoxy coatings near a metal substrate changes when corrosion occurs in an NaCl solution, which leads to the formation of numerous tiny anodic and cathodic zones and different corrosion products on the metal substrate [50–52]. In the corroded anode of the noted tests, the metal ion concentration increased, and the pH decreased owing to hydrolysis [53]; inversely, the pH increased near the zone of the corroded cathode [54]. Therefore, MBT-BTA@mSiO_2 was still effectively released into the damaged coating area.

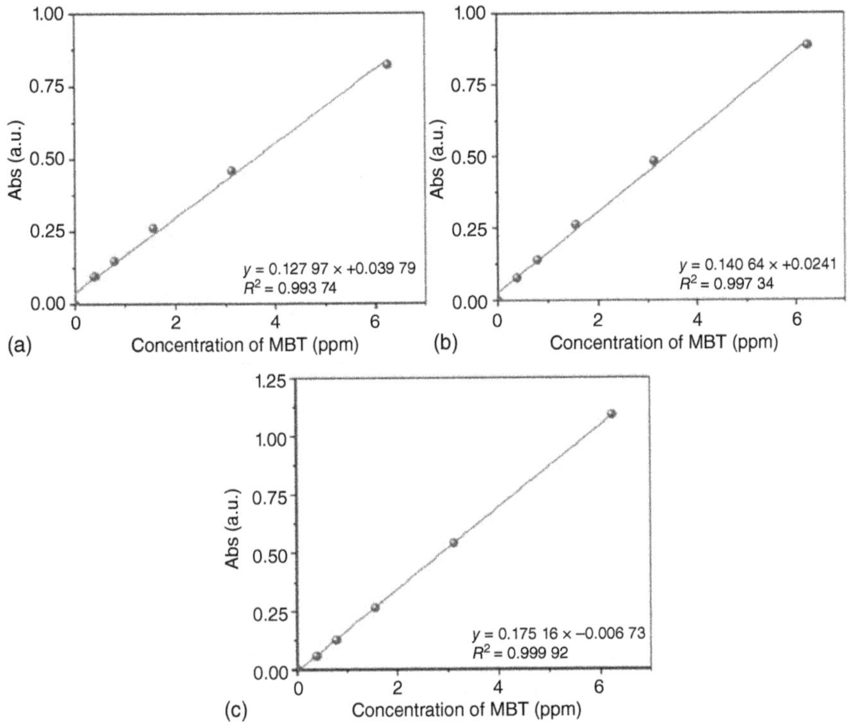

Figure 4.11 Standard UV–Vis absorption fitting curves of MBT in 3.5 wt% NaCl solution at (a) pH = 3, (b) pH = 7, (c) pH = 11.

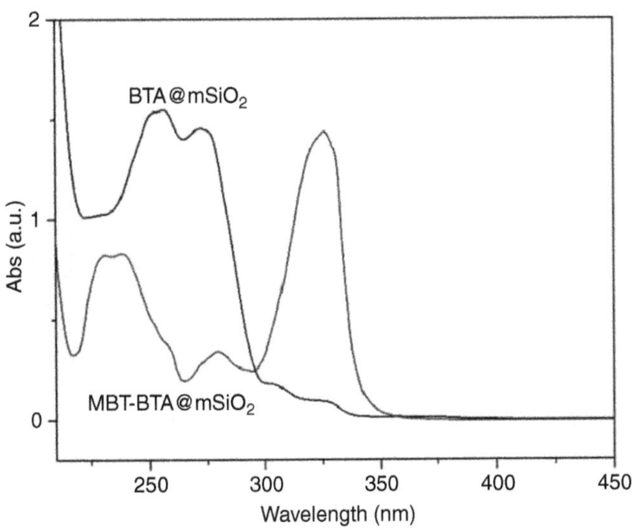

Figure 4.12 UV–Vis spectra of BTA@$mSiO_2$ and MBT-BTA@$mSiO_2$ after complete release in hydrochloric acid alcohol solution.

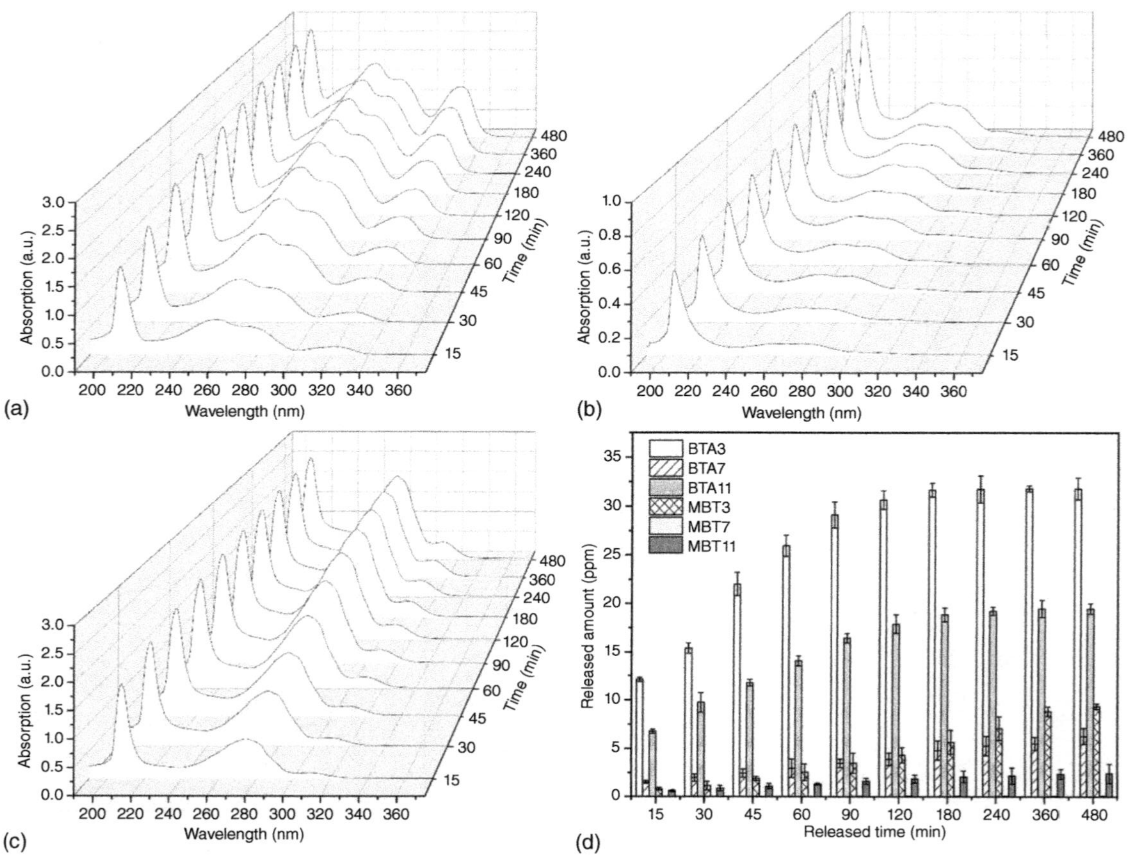

Figure 4.13 Corrosion inhibitor release from MBT-BTA@mSiO_2 in 3.5 wt% NaCl solutions at (a) pH = 3, (b) pH = 7, (c) pH = 11, and (d) release rates.

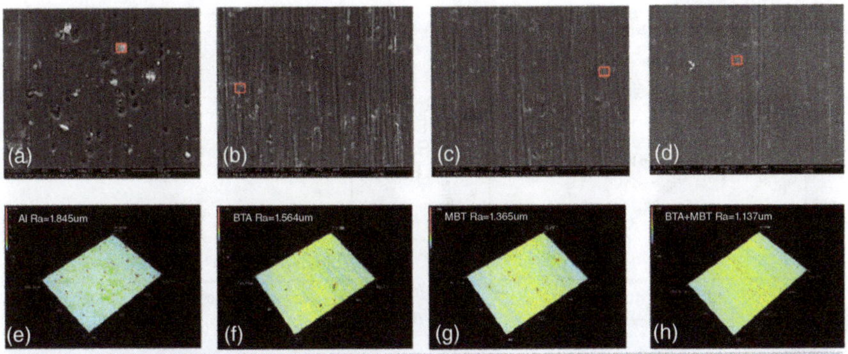

Atom (%)	Al	O	C	Cu	Mg	Mn
Al	51.33	21.83	19.95	4.47	1.29	1.13
BTA	70.76	2.51	7.38	9.92	9.14	0.28
MBT	78.51	1.07	9.27	5.31	4.33	0.51
MBT + BTA	92.22	0.77	2.05	1.72	2.79	0.45

Figure 4.14 SEM, CLSM and element content of 2024-T3 aluminum alloy immersed in 3.5 wt% NaCl solution for 72 hours without or with corrosion inhibitors: (a,e) blank sample, (b,f) 0.6 g l^{-1} BTA, (c,g) 0.4 g l^{-1} MBT, (d,h) 0.5 g l^{-1} BTA + 0.1 g l^{-1} MBT.

4.2.2.4 Corrosion Inhibition Performance of MBT-BTA@mSiO_2

Owing to the low solubility of MBT, its concentration when used alone was set to 0.4 g l^{-1}, when the corrosion inhibitor control variable was mixed. Due to the weak corrosion inhibition effect of BTA, a slightly higher concentration of 0.6 g l^{-1} was selected as the single concentration, and the total concentration of inhibitors was controlled to 0.6 g l^{-1} when MBT and BTA were combined. After immersing the aluminum alloy in 3.5 wt% NaCl containing different corrosion inhibitor combinations for 72 hours, the surface of the sample was removed using nitric acid and ultrasonically cleaned with ethanol. Then, the surface morphology was observed using SEM and CLSM. As shown in Figure 4.14, after corrosion in the 3.5 wt% NaCl solution, there were more pitting and corrosion products on the surface of the pure aluminum matrix without having added corrosion inhibitors, and the surface roughness reached 1.85 μm. However, the surface morphology of the aluminum alloy in the BTA and MBT solutions remained relatively intact, and the corrosion pits were nearly invisible. The surface roughness values were 1.56 and 1.37 μm, respectively, indicating that the corrosion inhibition effect improved in the presence of corrosion inhibitors [55, 56]. The combination of MBT and BTA at a mass ratio of 1 : 5 showed optimal performance, such that the surface morphology of the aluminum alloy was nearly free of corrosion, and the roughness was as low as 1.14 μm. EDS energy spectrum analysis showed different types of elements present on the alloy surface, and the higher oxygen content on the

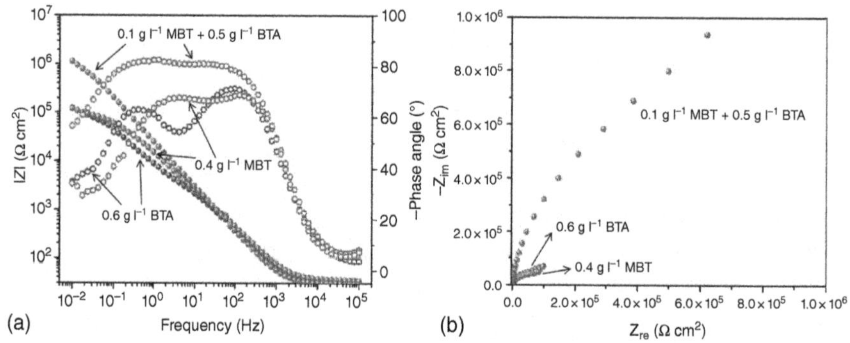

Figure 4.15 (a) Bode and (b) Nyquist plots of aluminum alloy after 60 days immersion in 3.5 wt% NaCl solution containing MBT, BTA, and their mixture.

aluminum alloy matrix without a corrosion inhibitor indicated the presence of more $Al(OH)_3$ or Al_2O_3 corrosion products [57, 58]. When corrosion inhibitors were added, the oxygen content was very low, and the aluminum alloy remained relatively intact. The Al alloy that came in contact with the MBT and BTA mixture showed the highest Al content and the lowest oxygen content, indicating that the mixture inhibitor had an excellent protective effect on the Al alloy matrix [59]. The AA2024-T3 surface contained Cu, which easily formed an S phase with Mg and Al, resulting in intergranular corrosion [60–62]. BTA prevents the corrosion of copper by forming complex, high-strength films with good densification and excellent corrosion inhibition performance [35, 63]. However, its formation speed is slow [64–67]. The MBT was directly adsorbed onto the copper surface at a high rate [68–70]. Hence, when these two copper corrosion inhibitors coexist in a system, they promote interfacial action and form a multicomponent protective film that produces excellent synergistic effects and better protection of the alloy matrix [71].

Electrochemical tests were conducted to compare the MBT:BTA = 1 : 5 mixture of corrosion inhibitors with MBT or BTA alone, as shown in Figure 4.15. After immersing in 3.5 wt% NaCl solution for two months, the impedance mode value of the compound inhibitor remained high; it remained greater than $10^6\ \Omega\,cm^2$ at a low-frequency impedance value of $|Z|_{0.01Hz}$. However, the $|Z|_{0.01Hz}$ value of MBT or BTA corrosion inhibitors decreased by an order of magnitude, and the Nyquist arc became significantly smaller. Two time constants appeared at the mid-frequency when only MBT or BTA were added, indicating that the inhibitor adsorption and the surface oxide layers were separated and could not provide excellent corrosion inhibition.

Electrochemical impedance spectra were measured on the surface of the AA2024-T3 in a 3.5 wt% NaCl solution containing MBT-BTA@mSiO_2 and BTA@mSiO_2 nanocapsules, which were labeled accordingly. Figure 4.16 shows that the $|Z|_{0.01Hz}$ value of the alloy sample immersed in the MBT-BTA@mSiO_2 solution was $2 \times 10^6\ \Omega\,cm^2$, which was significantly larger than that of the sample immersed in the BTA@mSiO_2 solution with a $|Z|_{0.01Hz}$ value of $9.1 \times 10^5\ \Omega\,cm^2$. This indicates that loading the corrosion inhibitor according to the concentration ratio with the

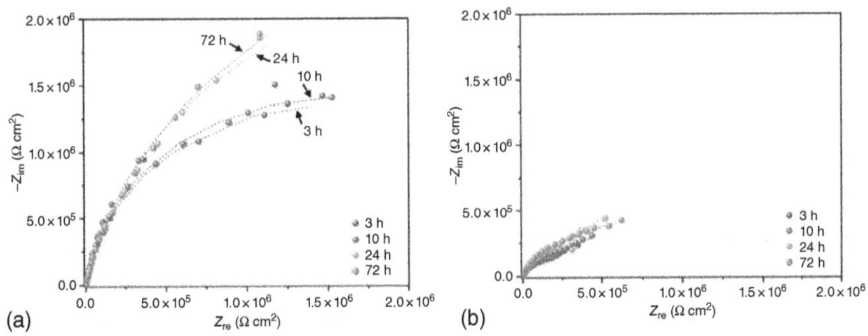

Figure 4.16 Nyquist plots of (a) MBT-BTA@mSiO$_2$ and (b) BTA@mSiO$_2$.

best inhibition performance improved the inhibition performance. The Nyquist arc radius of the MBT-BTA@mSiO$_2$ solution was significantly larger than that of the BTA@mSiO$_2$ solution, implying that the corrosion inhibition effect of the mixed inhibitors in the nanocapsules surpassed that of only the BTA corrosion inhibitor. Moreover, the Nyquist arc radii of the two samples gradually increased over time, suggesting a timely adsorption onto the AA2024 surface to impede corrosion [72–74].

When both MBT and BTA were loaded into the same mSiO$_2$ nanocapsules, the MBT-BTA@mSiO$_2$1/5 solution exhibited a homogeneous near-spherical structure with good dispersion at a mass ratio of 1 : 5 (MBT:BTA) and a uniform diameter in the range of 80–130 nm. However, a deviating trend in the spherical morphology was observed after increasing the MBT content. The loading capacities of MBT and BTA were approximately 3.29 and 11.42 wt% in the MBT-BTA@mSiO$_2$1/5, respectively, as determined by UV–Vis spectroscopy. Corrosion inhibitor release experiments were performed under different pH conditions, and the release rates of MBT and BTA were found to be the fastest, with the largest amounts occurring simultaneously under acidic conditions. The rates were slightly inferior at alkaline pH levels, and the release behavior was normal in a neutral environment. Nevertheless, the release of inhibitors can be ensured by varying the pH of the microenvironment around the corrosion area. The nanosilica loaded with the corrosion inhibitor was directly added to the saline solution, and the EIS results showed that the inhibition effect of MBT-BTA@mSiO$_2$ was better than that of BTA@mSiO$_2$, indicating that loading the corrosion inhibitor in accordance with the concentration ratio with the best corrosion inhibition performance strengthens the inhibition performance of the corrosion inhibitor carrier system.

4.2.3 Characterization and Property of MBT-BTA@mSiO$_2$/Epoxy Coating

To improve the corrosion resistance and mechanical properties of resin systems, nanofillers were added to the coating system to fill the microdefects. In this section, we propose the addition of prepared nanosilica fillers loaded with corrosion inhibitors in different ratios to the epoxy coatings, establishing an MBT-BTA@mSiO_2/epoxy composite coating system that is regulated by the amount of its addition to the epoxy coatings.

4.2.3.1 Electrochemical Performance of Scratched Coatings

EIS data with different contents of the MBT-BTA@m$SiO_2$1/5 filler were investigated. As shown in Figure 4.17, the corrosion resistance of the scratched coatings with MBT-BTA@mSiO_2 was significantly higher than that of the scratched blank coating. The impedance of the epoxy coating decreased after 10 hours, indicating that the blank epoxy coating was corroded by the NaCl solution [35, 75–78]. Then, the capacitive arc resistance decreased over time. Notably, the $|Z|_{0.01\text{Hz}}$ value decreased to $1\times10^4\,\Omega\,\text{cm}^2$ at the seven days, which was 2 orders of magnitude lower than the initial sample [79]. This demonstrates the effects of the largely exposed scratch on the blank epoxy coating, which occurred when the corrosive solution invaded the coating scratch, thereby causing a protection failure. The addition of 1 wt% MBT-BTA@m$SiO_2$1/5 nanofiller resulted in a $|Z|_{0.01\text{Hz}}$ value slightly lower than that of the blank epoxy coating at the beginning of immersion, but this decreasing trend significantly decelerated over time, and no significant decline of Nyquist arcs was observed. When 2 wt% MBT-BTA@m$SiO_2$1/5 nanofiller was added into the epoxy coating, the impedance value remained high, and the $|Z|_{0.01\text{Hz}}$ value reached $6.5\times10^5\,\Omega\,\text{cm}^2$ after seven days, with little change in the radius of the Nyquist capacitive arc resistance. This represents an overall good barrier performance because the loaded corrosion inhibitors MBT and BTA were slowly released to inhibit the corrosion behavior at the scratch area, resulting in coatings that maintained good protective effects [80–82]. However, a content of 3 wt% MBT-BTA@m$SiO_2$1/5 nanofiller in the coating led to a smaller Nyquist arc radius compared with the sample with 2% MBT-BTA@m$SiO_2$1/5 nanofiller addition, and the $|Z|_{0.01\text{Hz}}$ value reached $4.2\times10^5\,\Omega\,\text{cm}^2$ after seven days of immersion. This demonstrates that non-inevitable good performance results were achieved by the addition of MBT-BTA@mSiO_2 nanofiller. The nanoparticle dispersion in the coating deteriorated, and more corrosion channels were produced at the interface between the nanocapsules and the coating when superfluous MBT-BTA@mSiO_2 was added, leading to the degradation of the self-healing performance.

4.2.3.2 Surface Morphology Analysis of Scratched Coatings

To characterize the release and corrosion inhibition effects of MBT and BTA, surface analyses of the samples were performed after electrochemical testing. As shown in Figure 4.18, several corrosion pits were observed in the scratch area of the blank epoxy coating using a laser confocal microscope, suggesting that severe corrosion occurred in the 2024-T3 aluminum alloy. In contrast, the epoxy coating with the 2% MBT-BTA@mSiO_2 nanofiller was nearly free of corrosion pits around the scratch and showed a lower surface roughness than those without nanosilica.

4.2.3.3 Composition of Scratch Coatings

Figure 4.19 illustrates the SEM images and EDS patterns of the scratch area of the blank epoxy coating and the coating with 2% MBT-BTA@mSiO_2 nanofiller after

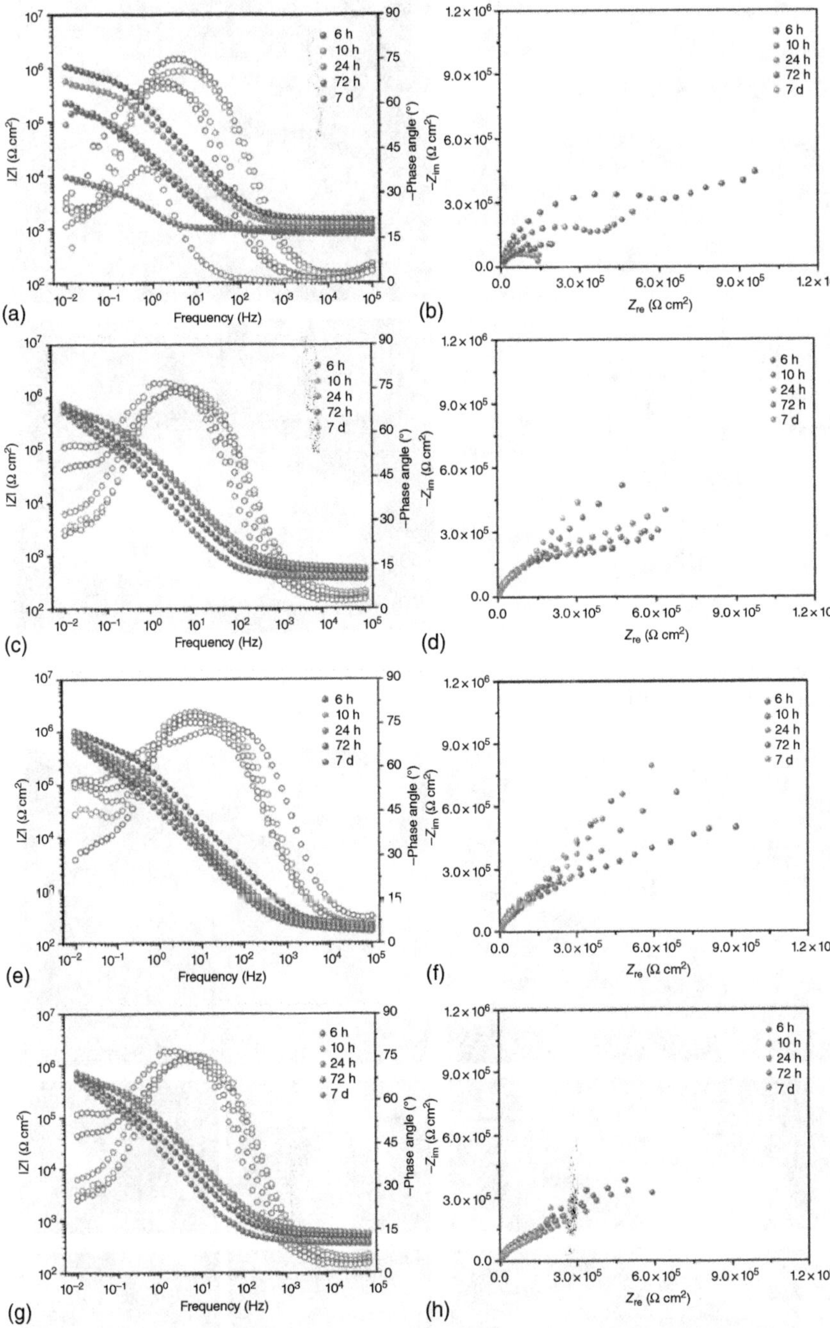

Figure 4.17 Bode and Nyquist plots of scratched coatings with different content of nanocapsules after 30 days of immersion in 3.5 wt% NaCl solution: (a,b) blank epoxy; (c,d) 1% MBT-BTA@mSiO_2; (e,f) 2% MBT-BTA@mSiO_2; (g,h) 3% MBT-BTA@mSiO_2. Source: Ma et al. [25]/with permission of Elsevier.

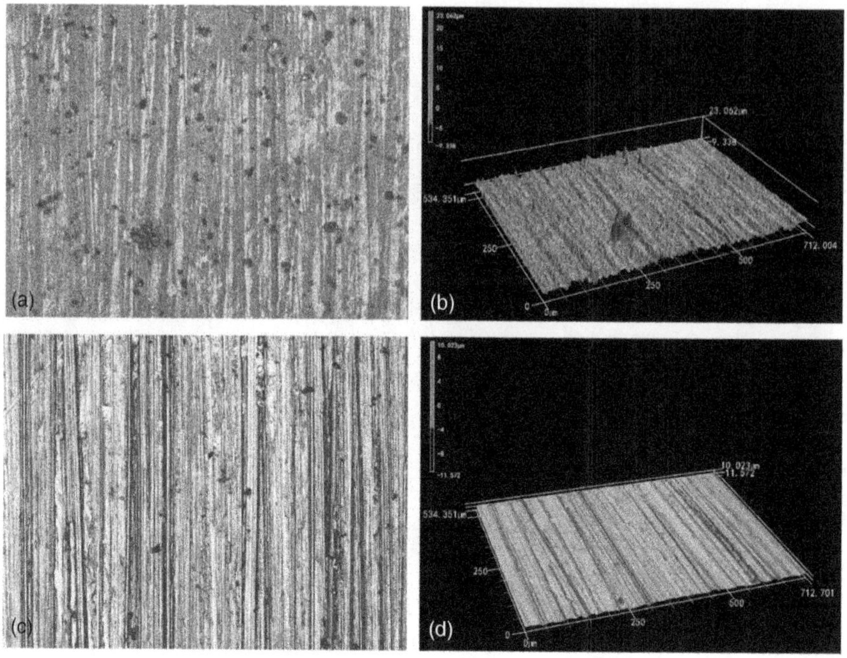

Figure 4.18 (a,b) CLSM at the scratch of the epoxy coating; (c,d) CLSM at the scratch of the coating with 2% MBT-BTA@mSiO_2.

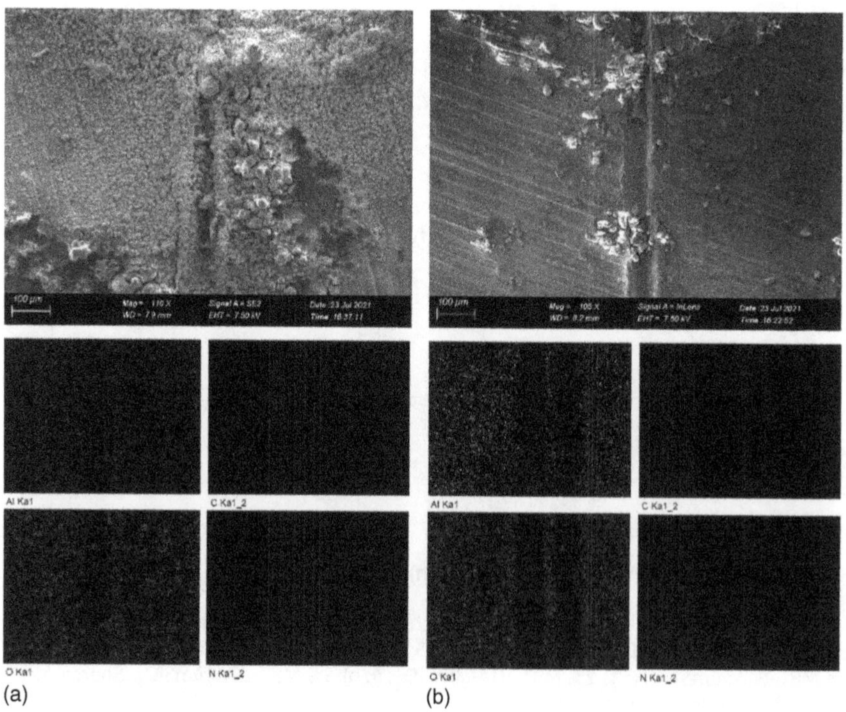

Figure 4.19 SEM and EDS images of scratched coatings after seven days of immersion in 3.5 wt% NaCl solution: (a) blank epoxy coating, (b) MBT-BTA@mSiO_2 coating.

seven days of immersion in a 3.5 wt% NaCl solution. In this figure, the scratch area of the epoxy coating without nanocapsules was covered with many corrosion products, whereas no obvious corrosion products were observed in the scratch area of the coating with nanofillers; this confirms a better protective performance on the metal substrate. EDS was utilized to further analyze the composition of the corrosion products in the scratches, revealing strong Al and oxygen signals in the scratches of the blank epoxy coating, reflecting the Al_2O_3 or $Al(OH)_3$ products. In contrast, no evident Al or oxygen signals were observed for the scratched coating with 2 wt% MBT-BTA@mSiO_2 nanofillers [83].

To further determine the adsorption of corrosion inhibitors on the metal surface, the scratched areas of the coating were analyzed by XPS after seven days of immersion in the NaCl solution. As shown in Figure 4.20a, the XPS peaks for Al, Cl, C, and O were detected in the scratched area of the blank epoxy coating. The addition of MBT-BTA@mSiO_2 resulted in the appearance of new elemental peaks corresponding to Si, S, and N, which can be attributed to the nanocapsules containing BTA and MBT corrosion inhibitors. Notably, the disappearance of the Cl 2p peak, compared with the XPS results of the blank epoxy coating, indicated that the interaction between the chloride ions and the metal substrate was inhibited [84]. Figure 4.20b illustrates the Al 2p peaks of the blank epoxy coating and MBT-BTA@mSiO_2 coating sample scratches. Two peaks at 73.8 and 74.7 eV corresponded with Al_2O_3 and $Al(OH)_3$, respectively [84–86]. The Al 2p signal intensity of the MBT-BTA@mSiO_2 samples was significantly lower than that of the blank epoxy coating due to the corrosion inhibitor coverage. The XPS spectrum of S 2p is shown in Figure 4.20c, where no S signal was observed with the blank coating sample. The S 2p spectrum could be clearly seen in the presence of MBT-BTA@mSiO_2, with peaks at 161.9 and 163.1 eV belonging to the heterocyclic ring of MBT [69, 87]. The N 1s spectrum (Figure 4.20d) corresponded with the MBT-BTA@mSiO_2 sample, which consisted of four peaks at 399.2, 399.6, 400.1, and 400.8 eV. The 399.2 and 399.6 eV peaks corresponded with the N=N and N—N bonding of the BTA corrosion inhibitor [88–90]. 400.1 eV was attributed to the C=N bonding of the MBT corrosion inhibitor. In addition, the 400.8 eV signal was attributed to the C—N bonds of the BTA and MBT molecules [91]. In summary, corrosion inhibitors were successfully released from the MBT-BTA@mSiO_2 solution and adsorbed into the alloy surface, thereby providing a synergistic effect of corrosion inhibition.

4.2.3.4 Long-term Protective Performance of Intact Coatings

Figure 4.21 shows the EIS results for the epoxy coatings with different contents of MBT-BTA@mSiO_2 added. Figure 4.21a,b shows the EIS spectra of the blank epoxy-coated sample. Initially, the impedance value of the epoxy coating was high, but the impedance arc radius decreased rapidly over time. In addition, the $|Z|_{0.01\text{Hz}}$ value decreased by 1 order of magnitude after one day. On the seven days, a second-time constant appeared in the Nyquist plot, indicating that the protection performance of the coating had deteriorated and that the corrosive solution contacted the metal substrate to induce further corrosion [92].

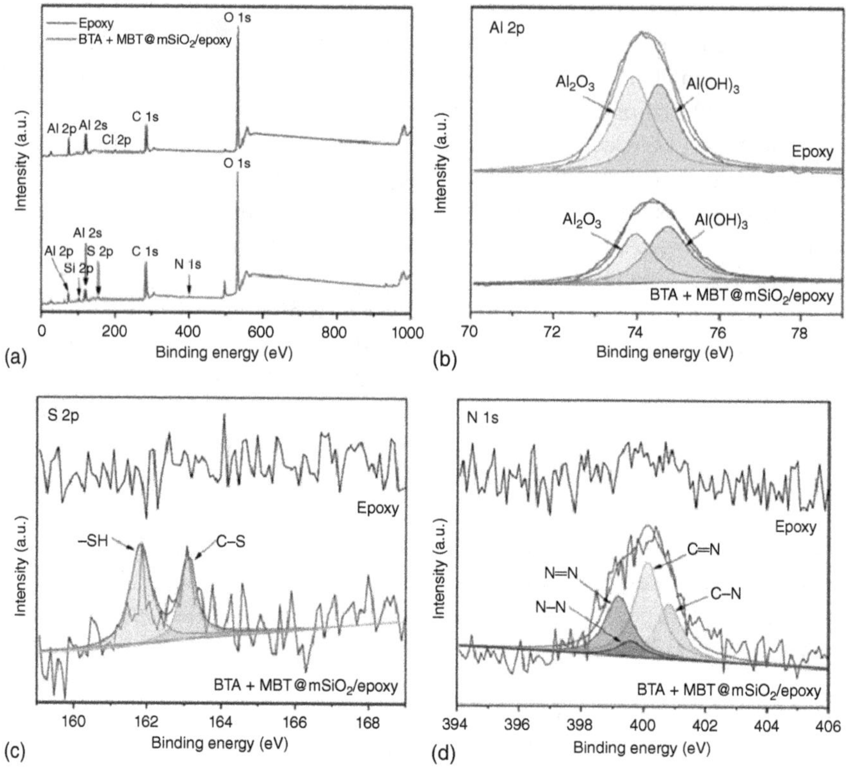

Figure 4.20 (a) Full XPS spectra of the scratch on the blank epoxy coating and epoxy coating with nanofillers; high-resolution XPS spectra of the scratch on the blank epoxy coating and epoxy coating with nanofillers: (b) Al2p peak, (c) S2p peak, and (d) N1s peak. Source: Ma et al. [25]/with permission of Elsevier.

Coatings with MBT-BTA@mSiO_2 nanofiller showed better protective effects, and the amount of nanofiller added affected the corrosion resistance of the coatings. When the addition of MBT-BTA@mSiO_2 was 1 wt%, the decreasing Nyquist impedance trend arc radius slowed, and the $|Z|_{0.01\text{Hz}}$ gradually lowered, remaining at $3.46 \times 10^8\ \Omega\,\text{cm}^2$ after 30 days. The best epoxy coating with the best protective effect was obtained by adding 2 wt% MBT-BTA@mSiO_2 nanofiller. As shown in Figure 4.21e,f, the electrochemical impedance of the coating decreased slightly over time. Interestingly, the $|Z|_{0.01\text{Hz}}$ value of the coating remained above $10^9\ \Omega\,\text{cm}^2$ after 30 days of immersion [93–96].

When the addition of MBT-BTA@mSiO_2 increased to 3 wt% in the coating, the long-term protection performance decreased. As shown in Figure 4.21g,h, the radius of the Nyquist impedance arc constantly decreased for 14 days, implying the diffusion of MBT and BTA from the nanofillers to the aluminum alloy surface [97]. The Nyquist impedance arc then decreased dramatically after 30 days of immersion, indicating the desorption of MBT and BTA from the surface of the substrate and significant degradation in protection performance.

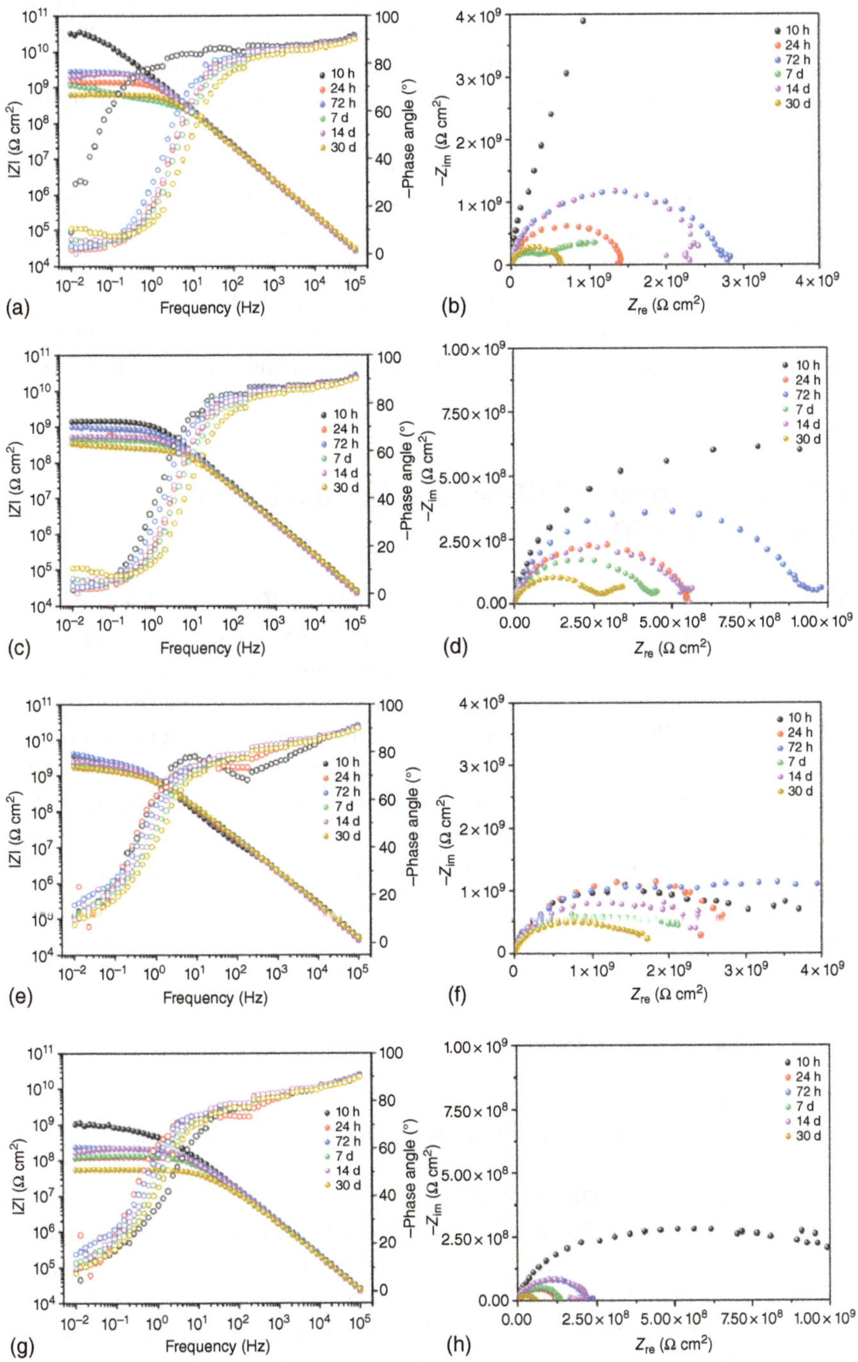

Figure 4.21 Bode and Nyquist plots of coating samples with different content of nanofillers after 30 days of immersion: (a,b) blank epoxy; (c,d) 1% MBT-BTA@mSiO_2; (e,f) 2% MBT-BTA@mSiO_2; (g,h) 3% MBT-BTA@mSiO_2.

Briefly, a coating of 2 wt% MBT-BTA@mSiO_2 nanofiller exhibited the best protective performance, with a $|Z|_{0.01Hz}$ value above $10^9\ \Omega\ cm^2$ after 30 days of immersion in a 3.5 wt% NaCl solution. The self-healing performance of the coating with 1 wt% MBT-BTA@mSiO_2 was limited due to the low corrosion inhibitor content, whereas the epoxy coating with 3 wt% nanofillers showed worse barrier properties, which can be ascribed to the agglomeration caused by the high content of nanocapsules in the coating. The epoxy coating with 2 wt% MBT-BTA@mSiO_2 demonstrated optimal long-term corrosion protection performance. The XPS results demonstrated the effective adsorption of corrosion inhibitors, highlighting how MBT and BTA were gradually released to effectively inhibit corrosion reactions. The SEM and EDS results showed that very few corrosion products were formed on the coating with 2 wt% MBT-BTA@mSiO_2 nanofillers and confirmed the superior protection performance over the blank epoxy coating.

4.3 pH Responsive Self-healing Coating Based on rGO@MS-P-BTA

A reduced graphene oxide/mesoporous silica (rGO@MS-P-BTA) structure assembled from a pH-responsive poly(*N*,*N*-dimethylaminoethyl methacrylate) (PDMAEMA) layer loaded with BTA inhibitors was designed. pH- and photothermal-responsive self-healing properties were achieved by the addition of rGO@MSP-BTA fillers to an epoxy coating. The structure of microcapsules was characterized by SEM; the successful assembly of PDMAEMA was characterized by TEM, zeta potential, and XPS; and the UV–Vis release profiles of rGO@MS-P-BTA microcapsules were measured at different pH solutions and with or without IR light irradiation. Morphological changes in the scratched and healed coatings were analyzed. In addition, EIS and corrosion monitoring tests were used to evaluate the self-healing performance of the coatings and analyze the corresponding mechanisms [98].

4.3.1 Materials and Methods

4.3.1.1 Materials

The materials and reagents used in this section are shown in Table 4.3.

4.3.1.2 Preparation of rGO@MS-P-BTA

First, 100 mg of GO was dispersed in 180 ml of deionized water at pH 12, and the solution was stirred at 80 °C for 12 hours to obtain rGO. CTAB (3.6 g) was added, and the pH was maintained at 12. Finally, 1.8 ml of TEOS was slowly added, and the mixture was heated at 80 °C for 24 hours. The solution was centrifuged and washed thrice. To obtain the mesoporous structure, the product was dispersed in a mixture of ethanol and HCl. After 12 hours of magnetic stirring, the residual CTAB was removed from the pore channels of the silica. The final rGO@MS product was washed and vacuum-dried.

Table 4.3 Experimental materials and reagents.

Materials	Purity	Company
Carbon steel	Q235	Hengde China
Graphene oxide	99.9%	Suzhou Tanfeng
BTA	99%	Aladdin
CuBr	99.99%	Sigma-Aldrich
Tris ((*N*,*N*-dimethylamino)ethyl)-amine	96%	Sigma-Aldrich
Ethyl 2-bromoisobutyrate	98%	Sigma-Aldrich
Dimethylaminoethyl methacrylate	99%	Sigma-Aldrich

To synthesize PDMAEMA, dimethylaminoethyl methacrylate (DMAEMA) was initiated using ethyl 2-bromoisobutyrate under the catalytic effects of CuBr and tris((*N*,*N*-dimethylamino)ethyl)amine [99]. At first, 5.81 g of DMAEMA, 0.08 g of cuprous bromide (CuBr) catalyst, 0.13 g of the ligand tris(2-dimethylaminoethyl)amine (Me6TREN), and 10 ml of tetrahydrofuran (THF) were sequentially added to a 100-ml, three-necked flask to form a mixture. After adding 0.11 g of ethyl 2-bromoisobutyrate (EBIB) initiator, the solution system was processed by eight cycles of "freezing–extraction–thawing," followed by a reaction in an oil bath at 70 °C for 13 hours. This solution was then diluted with a large amount of dichloromethane, and a neutral column was used to remove the copper species from the solution. The organic solvent was removed by heating and spinning, and the unreacted monomer was removed three times by precipitation in *n*-hexane. Finally, the product was dried at 40 °C for 48 hours.

To encapsulate BTA in the rGO@MS microcapsules, 100 mg of BTA and 200 mg of rGO@MS were dispersed in 40 ml of water to form a homogeneous solution. The solution was then stirred for four hours under negative pressure. The pH was then adjusted to 5, and 20 ml of the PDMAEMA solution was poured into the mixture, followed by stirring for 30 minutes. The mixture was washed to remove superfluous BTA from the microcapsules. Finally, the microcapsules were freeze-dried.

To study the effect of inhibitor release on the self-healing ability of the coatings, modified rGO@MS-P microcapsules without BTA inhibitors were prepared as follows: 100 mg of rGO@MS and 50 mg of PDMAEMA were dispersed in 30 ml of deionized water, and the pH was adjusted to 5 using a dilute HCl solution. The products were magnetically stirred at 200 rpm for 30 minutes, centrifuged, and freeze-dried.

4.3.1.3 Preparation of Self-healing Coating with rGO@MS-P-BTA Microcapsules

To prepare the corrosion-protection coating, rGO@MS-P-BTA was added to an epoxy matrix with the additional amount of 0.5, 1, and 2 wt%, respectively, relative to the quality of coating matrix. The coatings were coated onto Q235 carbon steel and cured at 55 °C for 24 hours. As a control, a blank epoxy coating was similarly prepared. The thickness of the dry coating was 60 ± 5 μm.

4.3.1.4 Characterization of rGO@MS-P-BTA

The morphologies of rGO@MS and rGO@MS-P microcapsules were characterized using TEM (JEM 2100F, JEOL). The surface charge was measured using a zeta potential analyzer (Zetasizer, Nano ZS90, Malvern). The specific surface areas and pore size distributions of the microencapsulated composites were studied using an ASAP 2460 physical adsorption instrument. UV–Vis absorption spectra were recorded using a UV-3900H spectrophotometer to investigate the release of BTA from the rGO@MS-P-BTA microcapsules. To investigate the effect of laser irradiation on the release of BTA, the rGO@MS-P-BTA solution was irradiated with a near-infrared laser at 808 nm (3 W cm^{-2}). The UV–Vis spectra of the solution before and after irradiation were measured.

4.3.1.5 Characterization of Self-healing Coatings

To investigate the self-healing performance of the coatings, the scratched and healed coatings were immersed in 3.5 wt% NaCl solution. The electrochemical performance of the coatings was evaluated using an electrochemical station (PARSTAT 2273, AMETEK) equipped with a three-electrode system.

The corrosion sensors were coated and placed in a salt spray chamber for corrosion monitoring tests according to the ASTM B117 standard. Pencil hardness and adhesion tests of the self-healing coatings were conducted according to standard GB/T 6739-2006 and application standard ISO 4624:2002, respectively [100].

4.3.2 Preparation and Characterization of rGO@MS-P-BTA

4.3.2.1 Morphology and Structure of rGO@MS-P-BTA Microcapsules

The preparation process of rGO@MS-P-BTA is illustrated in Figure 4.22. The silicon layer was grown on rGO by sol–gel condensation of TEOS as the silica source using CTAB as both the surfactant and template agent. After the removal of CTAB to obtain rGO@MS with a mesoporous structure, loading of BTA molecules was accomplished under negative vacuum pressure with vigorous stirring, and PDMAEMA was adsorbed onto the SiO_2 surface by electrostatic adsorption.

TEM images of the rGO@MS-P microcapsules are shown in Figure 4.23. The high-resolution TEM image of the rGO@MS microcapsules shows a highly ordered mesoporous structure. Successful adsorption of PDMAEMA was confirmed by the appearance of a polymer corona at the edge of rGO@MS-P after electrostatic adsorption. Moreover, the mesoporous structure was retained after the PDMAEMA modification. Figure 4.24 shows the SEM image of rGO@MS-P and the corresponding EDS results. The structure of rGO@MS-P was uniform, and there were abundant Si and O elements on the rGO@MS-P surface, demonstrating the successful combination of rGO and MS. N indicates successful adsorption of PDMAEMA.

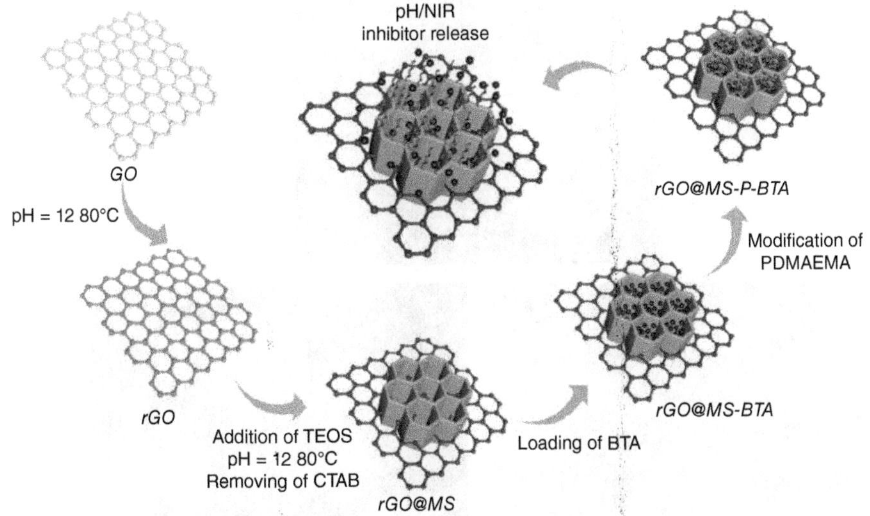

Figure 4.22 Preparation of rGO@MS-P-BTA.

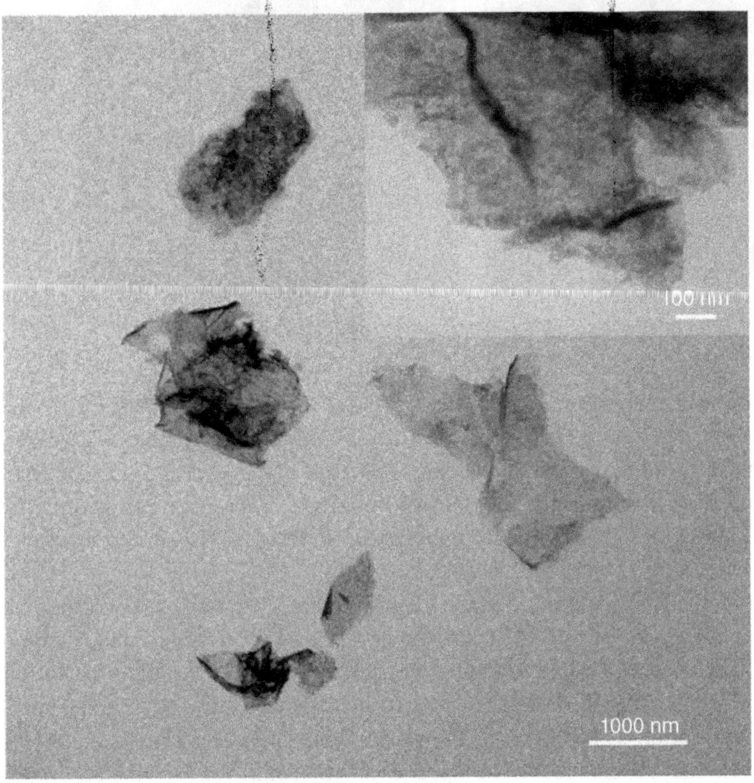

Figure 4.23 TEM image of rGO@MS-P.

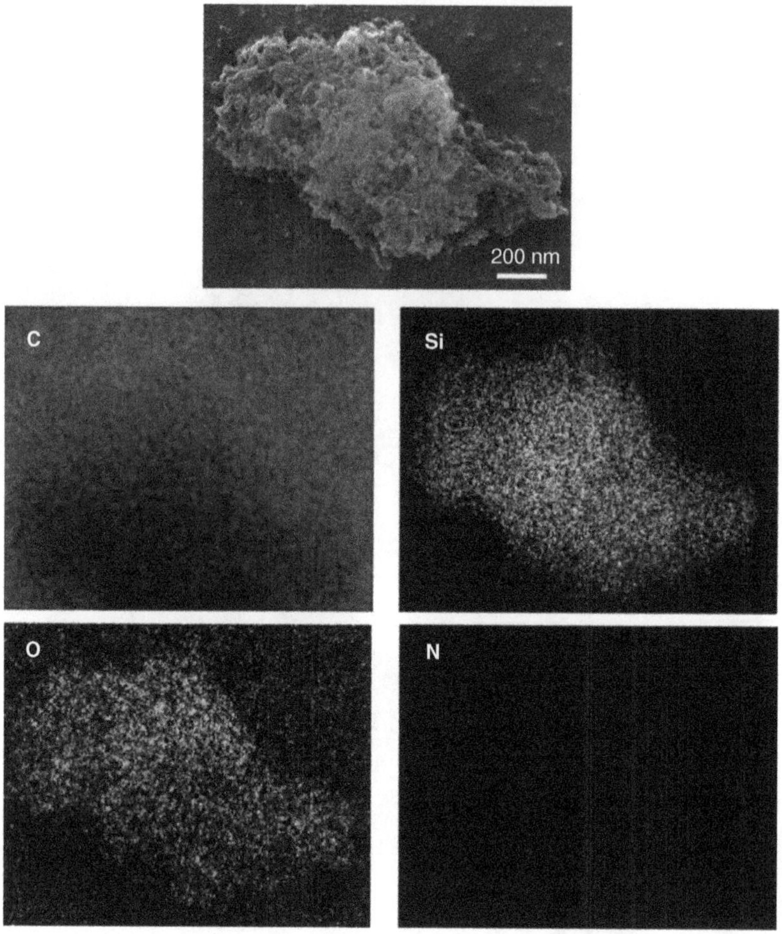

Figure 4.24 SEM image of rGO@MS-P and corresponding EDS results.

The adsorption between rGO@MS and PDMAEMA was characterized by the variation in surface charge based on electrostatic adsorption. Figure 4.25 shows that the surface potential of rGO@MS was −22.4 mV, then increased to −14.2 mV after loading BTA, and reversed to +31.3 mV for rGO@MS-P-BTA, confirming the successful loading of BTA and the immobilization of PDMAEMA on the surface of rGO@MS-BTA.

Meanwhile, the XPS spectra in Figure 4.26 show that, compared to rGO@MS, an N 1s peak appeared in the full XPS spectrum of rGO@MS-P-BTA and the C/O ratio increased in rGO@MS-P-BTA. A distinct N peak appeared in the high-resolution spectrum of N 1s of rGO@MS-P-BTA, which is consistent with the full XPS spectrum. In combination with the results of transmission electron microscopy pictures (Figure 4.23), the difference in the spectra of rGO@MS-P-BTA and rGO@MS was caused by the loading of BTA and the electrostatic adsorption of the polymer PDMAEMA [101].

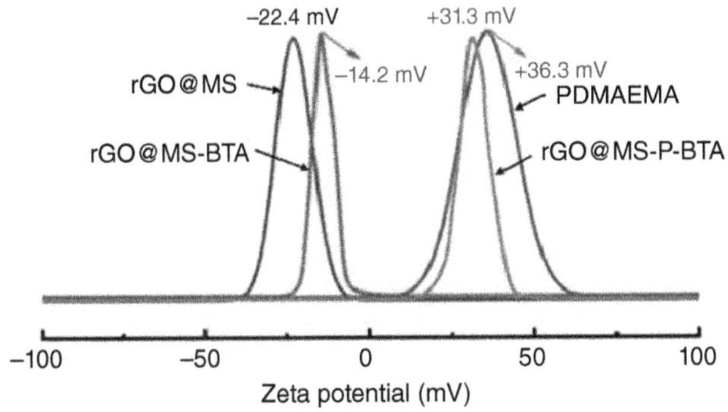

Figure 4.25 Zeta potentials of different samples.

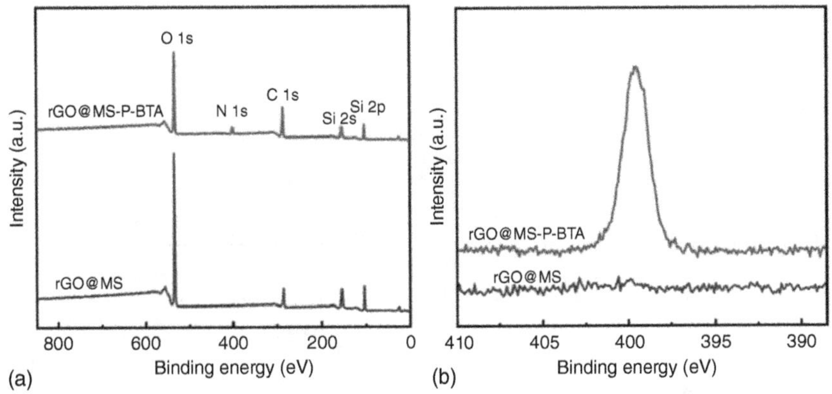

Figure 4.26 (a) XPS survey spectra and (b) N 1s high-resolution spectra of rGO@MS and rGO@MS-P-BTA.

Figure 4.27 depicts the N_2 adsorption–desorption and pore size distributions for different samples. rGO@MS exhibited the highest adsorption capacity for N_2 and the largest BET surface area. The typical *IV* curves verify the emergence of mesoporous structures [102, 103]. PDMAEMA adsorption resulted in the reduction of the pore volume of BJH from 1.4 to 0.5 $cm^3\,g^{-1}$, and the pore volume was further reduced to 0.43 $cm^3\,g^{-1}$ after loading BTA. In terms of pore size distribution, the average pore size of rGO@MS was 2.9 nm, which decreased to 2.2 and 2.1 nm after only adsorption of PDMAEMA and loaded BTA accompanied with the adsorption of PDMAEMA, respectively.

4.3.2.2 pH-Controlled Corrosion Inhibitor Release

During the synthesis of the microcapsules, the residual solution was collected and diluted to measure the BTA concentration using UV–Vis spectroscopy. The loading of BTA into rGO@MS-P-BTA was defined as the mass of BTA as a percentage of the total mass of the microcapsules, which was calculated to be ~13.2%. To investigate the release behavior of BTA, the microcapsules were dispersed in 3.5 wt% NaCl solutions at pH of 3, 7, and 11, respectively. The UV–Vis absorption spectrum of

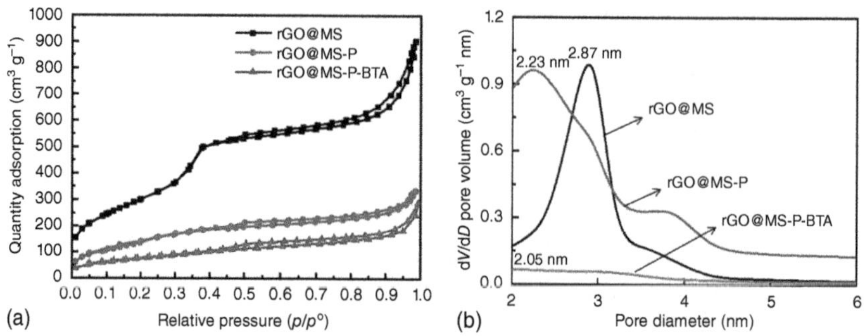

Figure 4.27 (a) N_2 adsorption–desorption isotherms, (b) pore size distribution curves [98].

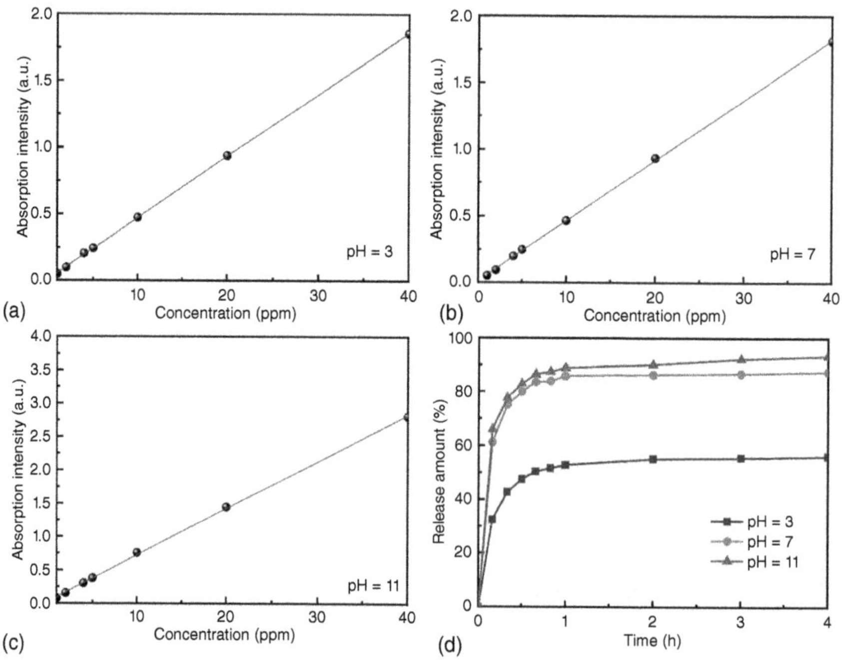

Figure 4.28 Standard curves of BTA at (a) pH = 3, (b) pH = 7, (c) pH = 11 solutions obtained from the corresponding UV–Vis spectra, (d) release of BTA from rGO@MS-P-BTA.

the solution was measured at certain intervals. The absorption intensity at 275 nm was used to establish a standard curve, as shown in Figure 4.28a–c. Figure 4.28d illustrates the release ratio of BTA from rGO@MS-P-BTA at different pH values. Notably, the release of BTA from the microcapsules was faster in acidic environments in neutral and alkaline environments, which was attributed to the protonation of PDMAEMA under acidic conditions, opening the nanopores and accelerating the release of internal BTA molecules [104, 105]. The pH often changed during the corrosion process, facilitating the release of internal molecules. The above results reveal that rGO@MS-P-BTA exhibits a pH-sensitive behavior that controls the release of BTA.

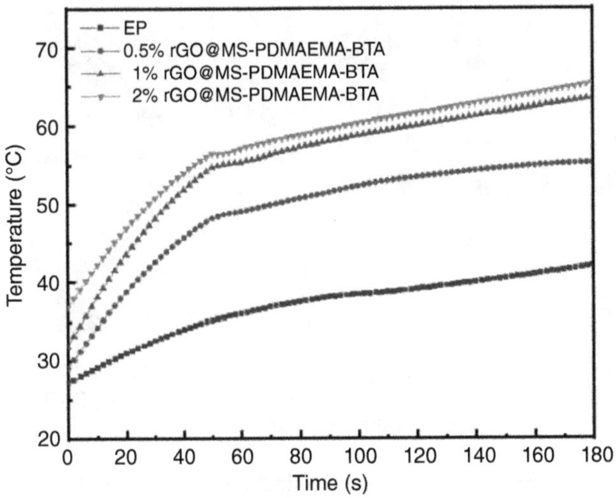

Figure 4.29 Surface temperature changes of coatings containing 0, 0.5, 1, and 2 wt% rGO@MS-P-BTA microcapsules during laser irradiation.

4.3.3 Self-healing Performance of rGO@MS-P-BTA Containing Coatings

4.3.3.1 Photothermal Effect of Coatings

When the temperature is higher than the glass transition temperature (T_g) of the shape memory polymer (SMP), it activates scratch closure of the coating, allowing the minification or even disappearance of scratches or cracks. To identify a suitable self-healing temperature, the T_g values of the coatings with rGO@MS-P-BTA microcapsules were investigated. T_g of blank epoxy coating was ~46 °C, and little effect was caused by the incorporation of small amount of rGO@MS-P-BTA fillers on the thermal behavior of coatings [106]. Because rGO has a good photothermal effect, irradiation of the coatings with rGO@MS-P-BTA was proposed to activate the shape memory effect. Figure 4.29 illustrates the surface temperature changes of the different coatings during laser irradiation. The surface temperature of the blank epoxy coating did not increase significantly because it had no photothermal conversion ability. When adding 0.5 wt% rGO@MS-P-BTA fillers into coatings, the photothermal property elevated the temperature of coating surface to be ~55 °C after laser irradiation. When 1 and 2 wt% rGO@MS-P-BTA fillers were added into coatings, the surface temperature further increased to be as high as ~62 °C. The temperature difference between the coatings containing 1 and 2 wt% rGO@MS-P-BTA fillers was not significant, which was ascribed to the balance of temperature rise and loss. When the surface temperature increased, the epoxy resin transformed to a rubbery state, started to relieve the strain energy stored in the deformed region, and exerted a contraction force to spatially close the defective regions, inducing the closure of scratches or cracks [107, 108].

Approximately 30-μm scratch was made on each coating and then irradiated by NIR (808 nm, 3 W cm^{-2}) for 15 seconds to investigate the self-healing behavior.

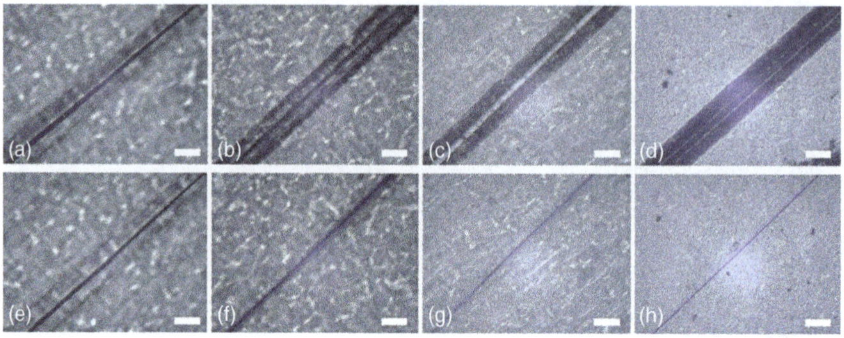

Figure 4.30 Optical images of scratched containing (a) 0, (b) 0.5, (c) 1, and (d) 2 wt% rGO@MS-P-BTA microcapsules, and SEM images of healed coatings containing (e) 0, (f) 0.5, (g) 1, (h) 2 wt% rGO@MS-P-BTA microcapsules; scale bar: 50 μm.

Figure 4.30 presents the optical images of the healed coatings with different amounts of rGO@MS-P-BTA. For the blank epoxy coating, the scratch width remained unchanged because the low surface temperature failed to activate the shape memory effect [109]. In contrast, coatings with rGO@MS-P-BTA exhibited a significant reduction in scratch width by photothermal healing [110]. In particular, because of the prominent photothermal effect of rGO@MS-P-BTA, scratches on the coatings containing 1–2 wt% rGO@MS-P-BTA narrowed to less than 10 μm [111].

Furthermore, different coatings were observed using confocal laser microscopy after photothermal healing (Figure 4.31). As seen in Figure 4.31a, the scratch width of the epoxy coating was maintained around 30 μm. The metallic luster at the bottom revealed the failure of the activation shape memory effect by NIR. When the rGO@MS-P-BTA ranged at 0.5 wt%, the width of the coating scratch decreased to some extent (Figure 4.31b). When the amount of rGO@MS-P-BTA microcapsules ranged between 1 and 2 wt%, the scratches on the coating were healed to a significant degree after irradiation with NIR for 15 seconds (Figure 4.31c,d).

4.3.3.2 Electrochemical Performance of Coatings with rGO@MS-P-BTA

Figure 4.32 summarizes the EIS results of the scratched and NIR-healed coatings during immersion in 3.5 wt% NaCl solution. The $|Z|_{0.01\text{Hz}}$ value of epoxy coating decreases from 1.5×10^5 (2 hours) to $1.4 \times 10^4\ \Omega\,\text{cm}^2$ (72 hours). However, with the addition of microcapsules, the coatings obtained higher $|Z|_{0.01\text{Hz}}$ values throughout the immersion period. After two hours, healed coatings containing 0.5, 1, 2 wt% of rGO@MS-P-BTA microcapsules had $|Z|_{0.01\text{Hz}}$ values of 8.5×10^5, 1.3×10^6, and $1.1 \times 10^6\ \Omega\,\text{cm}^2$, respectively. After 72 hours, they decreased to 3.5×10^4, 8.5×10^5, and $3.8 \times 10^4\ \Omega\,\text{cm}^2$, respectively. Higher $|Z|_{0.01\text{Hz}}$ values reflect superior corrosion protection of self-healing coatings with microcapsules. The self-healing coating exhibited the highest $|Z|_{0.01\text{Hz}}$ value and the best anticorrosion performance when the microcapsule content was 1 wt% in the coating. The $|Z|_{0.01\text{Hz}}$ values of the coatings with rGO@MS-P-BTA appeared to be significantly higher than those of the scratched epoxy coatings. Specifically, a blank epoxy coating with 0.5 wt% rGO@MS-P-BTA microcapsules and the coating with 2.0 wt% rGO@MS-P-BTA

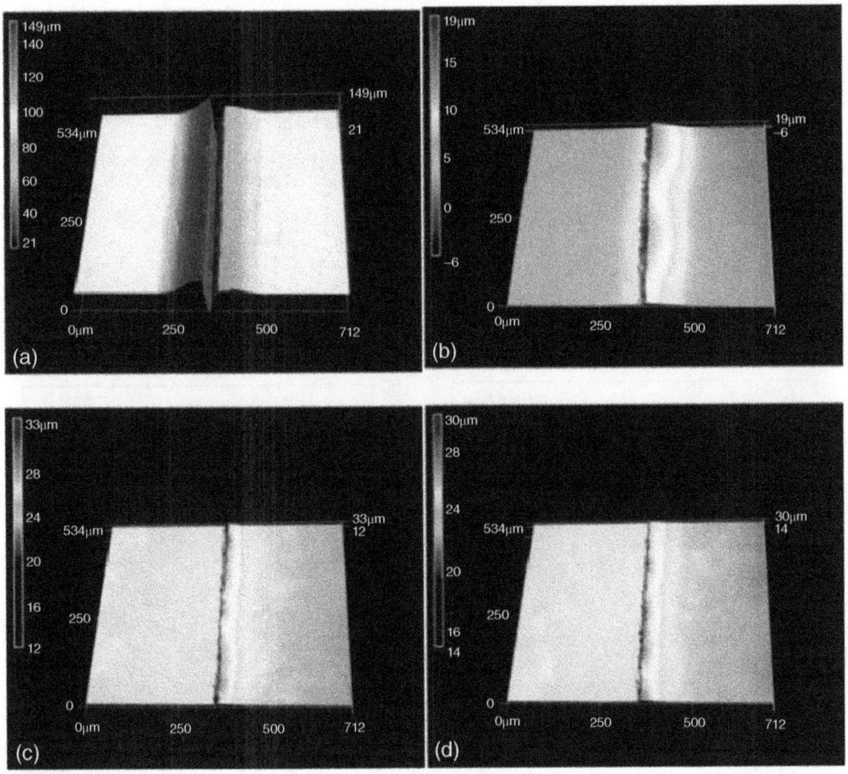

Figure 4.31 CLSM plots of scratches after photothermal healing: (a) blank epoxy coating, (b) 0.5 wt% rGO@MS-P-BTA, (c) 1 wt% rGO@MS-P-BTA, (d) 2 wt% rGO@MS-P-BTA.

microcapsules illustrated gradual decline of the corrosion protection properties [112, 113]. Remarkably, the $|Z|_{0.01\text{Hz}}$ values of the coating with 1.0 wt% rGO@MS-P-BTA microcapsules kept at high level of approximately $10^6\,\Omega\,\text{cm}^2$, reflecting the best self-healing performance. These results confirm the synergistic effect of the shape memory effect and corrosion inhibition of BTA for corrosion protection [114–116].

4.3.3.3 Corrosion Monitoring of Coatings with rGO@MS-P-BTA

The blank epoxy coating and coating with 1 wt% rGO@MS-P-BTA were covered onto the corrosion monitoring sensor and cured. After curing, scratches of the same length were made on the surface of the coating, and one of the sensors coated with 1 wt% of rGO@MS-P-BTA was photothermally healed, and all coatings were transferred into a salt spray tank for corrosion monitoring experiment.

Figure 4.33 exhibits the variation in corrosion current during the salt spray process. During salt spraying, the corrosion current of the scratched epoxy coating exhibited an upward trend, and the maximum value of the instantaneous corrosion current exceeded 1000 nA. Nevertheless, the corrosion currents of the 1 wt% rGO@MS-P-BTA scratched coating ranged from 100 to 300 nA, demonstrating

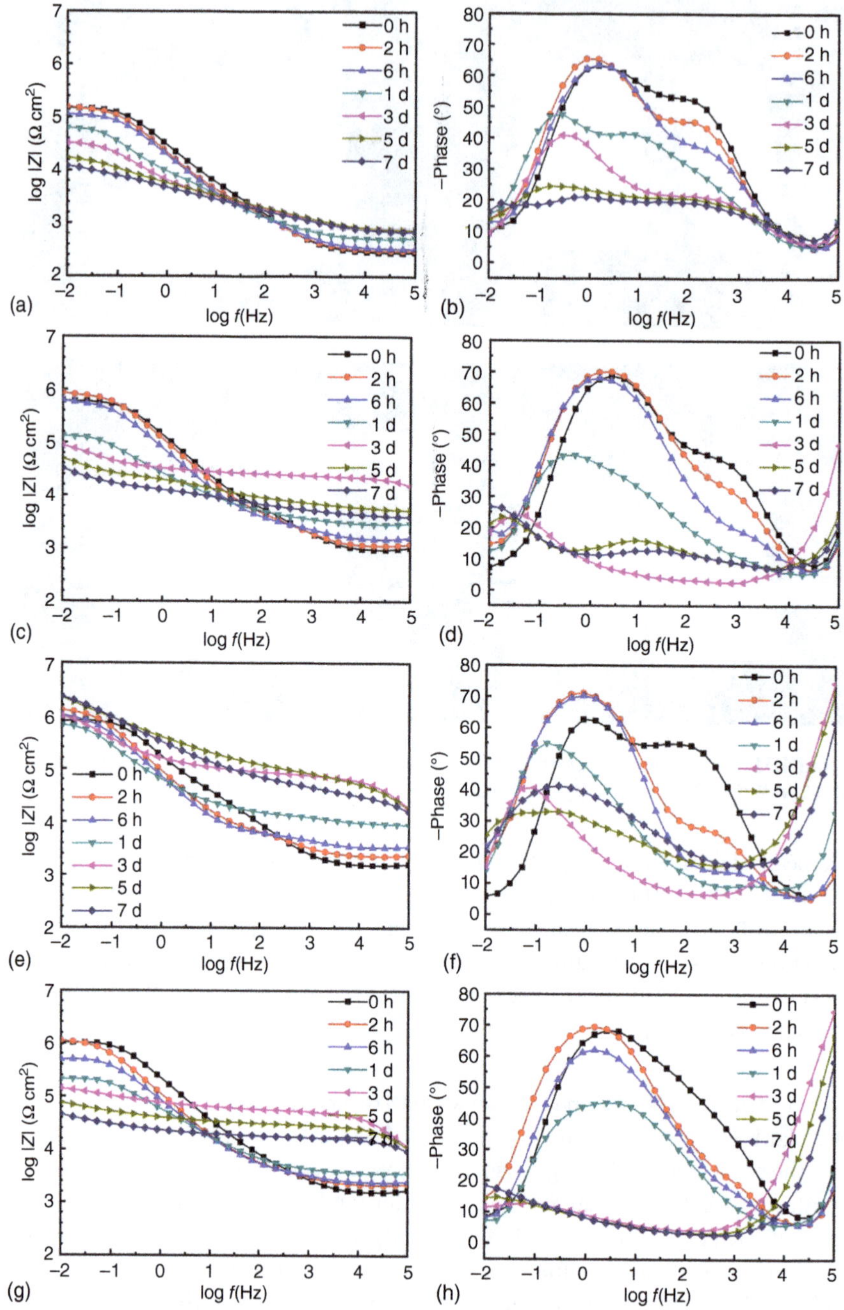

Figure 4.32 Impedance modulus and phase angle plots of (a,b) scratched epoxy coating, healed coating containing (c,d) 0.5 wt% rGO@MS-P-BTA, (e,f) 1 wt% rGO@MS-PBTA, and (g,h) 2 wt% rGO@MS-P-BTA.

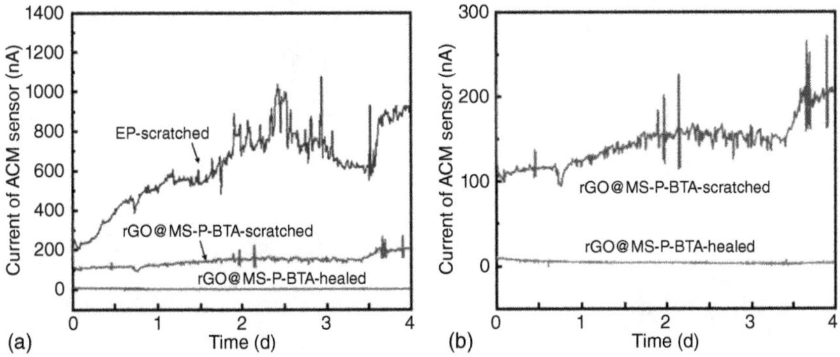

Figure 4.33 (a) Corrosion current curves of scratched and healed coatings with salt spray, (b) magnified corrosion current of scratched and healed 1 wt% rGO@MS-P-BTA coatings.

Table 4.4 Main experimental materials and reagents.

Contents of rGO@MS-P-BTA	0 wt%	0.5 wt%	1 wt%	2 wt%
Pencil hardness	H	H	2H	4H
Adhesion (MPa)	10.6 ± 0.8	10.3 ± 1.0	10.3 ± 1.1	9.5 ± 0.6

the contribution of BTA as the corrosion inhibitor. Notably, the corrosion current of the healed 1 wt% rGO@MS-P-BTA coating appeared to be as extremely low as 20 nA. The corrosion currents of the healed 1 wt% rGO@MS-P-BTA coating decreased by 2 orders of magnitude compared with the scratched epoxy coating and also decreased by 1 order of magnitude in comparison with the scratched rGO@MS-P-BTA coating, which was the consequence of the synergy of corrosion inhibition of the BTA and the shape memory effect to significantly enhance the anticorrosion of the coating.

4.3.3.4 Mechanical Properties of Coating with rGO@MS-P-BTA

As shown in Table 4.4, the hardness of the coating remained unchanged on addition of 0.5 wt% rGO@MS-P-BTA into the coating; and the hardness increased to 2H with the increase in addition percentage to 1 wt% and further increased to 4H when the addition amount of rGO@MS-P-BTA was 2 wt%. The adhesion of the blank epoxy coating was approximately 10.6 MPa, whereas that of the coating with 1 wt% rGO@MS-P-BTA was about 10.3 MPa, which implies a little impact of rGO@MS-P-BTA microcapsules on the adhesion of the coating.

4.3.3.5 Corrosion Protection Mechanism of Coatings with rGO@MS-P-BTA

The self-healing coatings containing 1 wt% rGO@MS-P-BTA microcapsules exhibited the optimal corrosion resistance to carbon steel; the schematic diagram is displayed in Figure 4.34. On coatings with physical scratches, irradiation with near-infrared light enables rapid photothermal conversion to elevate the surface

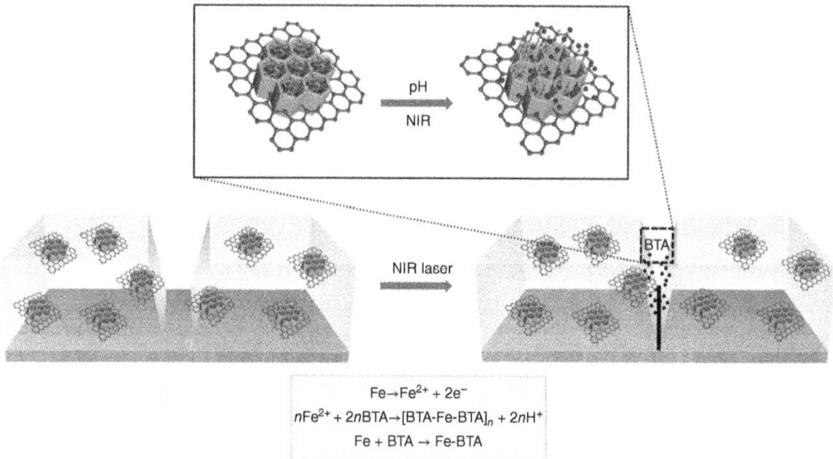

Figure 4.34 Schematic diagram of rGO@MS-P-BTA-based self-healing coating.

temperature of the coating, leading to the rapid triggering of the shape memory effect. Consequently, the coating matrix was restored, and the scratched area was considerably narrowed. Although this effect restricts the invasion of corrosive electrolytes and the maintenance of localized adhesion, it does not completely prevent the penetration of corrosive media. Notably, the corrosion reaction process induced a decrease in pH in the scoring region, which promoted the release of BTA in the microcapsules and suppressed the corrosion behavior via BTA adsorption onto the iron substrate or reaction with Fe^{2+} to form complexes. The combination of the shape memory effect stimulated by photothermal conversion, corrosion inhibitor release controlled by NIR irradiation, and pH response greatly enhanced the self-healing anticorrosion performance of the coating.

4.4 Redox and Saline Responsive Self-healing Coating Based on PANI/BTA Nanocapsules

Polyaniline (PANI) is a conductive polymer with broad development prospects for anticorrosion applications owing to its unique advantages, such as low cost, facile synthesis, and reversible redox properties. It has conjugated π-electron pairs and exists in three states: fully oxidized, intermediate oxidized, and fully reduced; thus, it can be used as an oxidizing agent to promote the formation of the oxide layer or passivate the metal surface [117]. PANI can also alter the oxygen–reduction reaction sites and hinder the degradation of interfacial bonds, improving the protective performance of the coatings. Novel PANI nanocapsules with NaCl response characteristics were synthesized [118]. This novel response mechanism is derived from sodium ion doping, which can lead to hydrogen bond breakage and PANI permeation expansion, promoting the release of internal inhibitors. Hao et al. [119] synthesized PANI nanofibers doped with phytic acid (PA) to encapsulate the corrosion inhibitor BTA. The PANI–PA and BTA functioned together to produce composite coatings with excellent shielding and self-healing properties. It is well known that the NaCl solution environment is a relatively common and harsh corrosion environment, and it is difficult to achieve good protective effects when relying solely on the action of corrosion inhibitors. Moreover, the drug release in nanocapsules is slow and untargeted, resulting in delayed corrosion inhibition. In addition, the coating may encounter problems such as an insufficient inhibitor content or excessive nanocapsule addition owing to large scratches on the coating, which may adversely affect its anticorrosion properties.

In this section, PANI nanocapsules containing BTA corrosion inhibitors were synthesized. A self-healing coating based on an SMP coating and PANI/BTA nanocapsules was designed. The PANI/BTA nanocapsules had three functions: (i) PANI functioned as an NaCl-responsive gatekeeper to control the release of BTA inhibitors, (ii) PANI prompted the formation of the Fe_2O_3 passive layer, and (iii) the thermoresponsive SMP coating narrowed the scratch after heating, which was beneficial for reducing the amount of corrosion inhibitors needed to fully cover the steel surface [120].

4.4.1 Materials and Methods

4.4.1.1 Materials

The materials and reagents used in this section are shown in Table 4.5.

4.4.1.2 Preparation and Characterization of PANI/BTA Nanocapsules

The specific synthesis process of PANI/BTA nanocapsules is as follows: 0.2 g of aniline and 2.14 g of toluene were weighed, dropped into the flask, and stirred evenly. Then, 25 ml of deionized water was dissolved in 0.125 g of sodium dodecyl sulfate,

Table 4.5 Experimental materials and reagents.

Experimental material	Specification/Purity	Manufacturer
Carbon steel	Q235	Dongguan Hengde Iron and Steel
Aniline	≥99.9%	Aladdin Chemical Co., Ltd.
Ammonium persulfate	≥98%	Aladdin Chemical Co., Ltd.
Benzotriazole	99%	Aladdin Chemical Co., Ltd.
Sodium dodecyl sulfate	≥98%	Aladdin Chemical Co., Ltd.
Bisphenol A diglycidyl ether	≥85%	Aladdin Chemical Co., Ltd.
Neopentyl glycol diglycidyl ether	≥40%	Aladdin Chemical Co., Ltd.
Polyether amine D-230	Mn ~ 230	Aladdin Chemical Co., Ltd.
Polyvinyl alcohol (PVA)	2 wt%	Shandong West Asia Chemical Co., Ltd.

0.1 g of BTA was poured into the flask, and ultrasonic treatment was conducted at 60 °C to obtain a uniform emulsion. Finally, 5 ml PVA solution was added, and the obtained emulsion was stirred under ice bath conditions. Deionized water solution of 10 ml with 5 g of ammonium persulfate was added by drop into the aforementioned emulsion. With the addition of ammonium persulfate solution, the emulsion gradually turned dark green. Nanocapsules were filtered and collected, washed with ethanol, and dried at 40 °C for 12 hours. For comparison, we also prepared PANI nanocapsules without BTA. Unlike aforementioned process, the corrosion inhibitor BTA was not added during the process.

The chemical structures of the PANI, PANI/BTA nanocapsules, and corrosion inhibitors were characterized via FTIR spectroscopy (VERTEX 70, Bruker). The microstructures of the PANI/BTA nanocapsules were examined using SEM (Quanta 250, FEI) and TEM (JEM 2100F, JEOL). The experiment involved heating the samples from 30 to 790 °C under nitrogen protection. The thermal stability of the nanocapsules and the drug loading rate of BTA were analyzed using TGA curves.

The release of BTA from the PANI/BTA nanocapsules was measured via UV–Vis spectroscopy (U-3900H, Hitachi). Before the test, different concentrations of BTA were added to deionized water, and a 3.5 wt% NaCl solution to establish the relationship between the concentration of BTA and the UV–Vis absorbance. For the BTA release test, 10 mg of PANI/BTA and 2 ml of water were added to a dialysis bag immersed in 50 ml of the 3.5 wt% NaCl solution under stirring. After a certain time interval, 250 μl of the NaCl solution was removed and placed in a trace colorimetric dish to measure its UV–Vis spectrum within the range of 210–500 nm. According to the changes in the absorbance of the measured solution at different times, amounts of BTA released at different times were calculated, and the release curve was drawn.

4.4.1.3 Preparation of Self-healing Coatings with PANI/BTA

Q235 steel was used as the coating substrate, and the sample was purchased from Dongguan Hengd steel. The epoxy coating was composed of bisphenol A diglycidyl ether, neopentyl glycol diglycidyl ether, and Jeffamine D230 at a molar ratio of 1 : 1 : 1. The self-healing coating solutions were prepared as follows: different amounts of PANI/BTA nanocapsules (0.25, 0.5, and 1 wt% relative to the coating matrix) were blended into the mixed solution of bisphenol A diglycidyl ether and neopentyl glycol diglycidyl ether. After two hours, Jeffamine D230 was added. The obtained suspension was spread over the Q235 surface and cured at 55 °C for 24 hours. Furthermore, a blank epoxy coating (EP) and a coating containing 0.5 wt% PANI were prepared for comparison. The average thickness of all the coatings was 50 ± 5 μm.

4.4.1.4 Self-healing Coating Performance Test

The water contact angles (WCAs) of PANI/BTA coatings were tested using a contact angle measuring instrument (OCA, OCA20, Dataphysics). To reduce errors, the WCA of each sample was measured at least five times and averaged. The surface morphology of coatings was analyzed by SEM and confocal laser scanning

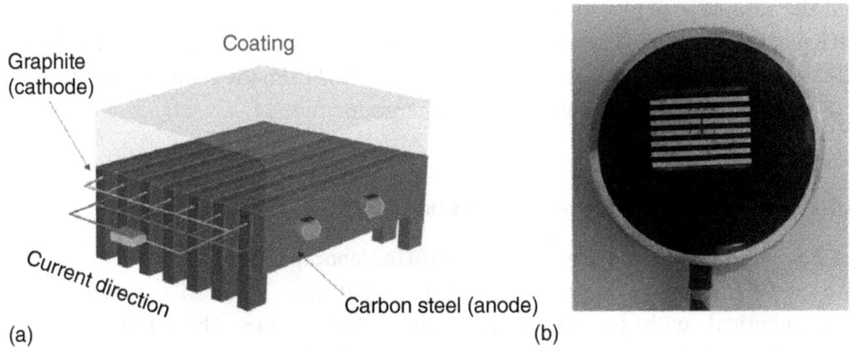

Figure 4.35 Corrosion sensor: (a) Structure diagram, (b) scratch diagram.

microscopy (CLSM, VK-X, Keyence). EIS (660E, CHI) was employed to assess the self-healing performance of PANI/BTA coatings.

Real-time changes in the corrosion current after coating scratches and repairs were assessed through corrosion monitoring experiments. A schematic of the sensor is shown in Figure 4.35, with a current identification of 0.1 nA and an acquisition frequency of 1 min^{-1}; the current collected by the corrosion monitoring sensor was the galvanic corrosion current. The sensor was composed of graphite and carbon steel electrodes arranged alternately with isolation plates inserted between them. The wires were removed from the back ends of the graphite and carbon steel and sealed with epoxy resin. The working principle was as follows: when a thin liquid film is generated on the sensor surface, the cathode and anode are conducted, and a galvanic corrosion current is generated owing to the different potentials of graphite and carbon steel, which are monitored and recorded by a micro-galvanometer. A self-healing coating was applied to the surface of the sensor, which was placed on a plastic shelf in a salt spray box (LRHS-108-RY, Linpin) at an inclination angle of approximately 20°. The salt spray test was performed in accordance with the ASTM B117 standard, and corrosion monitoring was performed on sensor samples with scratches and repairs. The scratch length on the coating was 4 mm. After the experiment, the corrosion products were analyzed using XPS (ESCALAB 250Xi, Thermo).

The pencil hardness was assessed according to the GB/T 6739-2006 standard for testing the pencil hardness of PANI/BTA nanocapsule coatings using a series of pencils with different hardness grades. First, the hardest 9H pencil was clamped onto the pencil clamp and tilted at an angle of 45° to the surface of the coated sample with the pen tip pressing tightly against the sample. Subsequently, a lateral force was applied to push the moving table forward, and the marks left by the pencil were visually inspected. If visible wear marks were observed, the next hardness-grade pencil was installed on the fixture, and the experiment was repeated. The test was terminated when the newly installed pencil failed to wear the coatings. The hardness of the coating was equal to the hardness of the pencil used last.

The adhesion experiment followed the ISO 4624:2002 standard, and the adhesion strength was measured using a pull-out instrument. First, a 10-mm diameter pull-out column was adhered to the coating surface with Ergo adhesive (1690, Ergo), and the sample was placed at room temperature for 24 hours to ensure complete curing of the adhesive. The column was then pulled out at a constant speed using an AT-M PosiTest tester until it was completely detached from the coating. Each sample was tested four times in parallel to ensure reproducibility.

4.4.2 Preparation and Characterization of PANI/BTA Nanocapsules

4.4.2.1 Morphology and Structure of PANI/BTA Nanocapsules

The SEM images in Figure 4.36a shows that the PANI/BTA nanocapsules were spherical, with an average diameter of 200–400 nm. The TEM images in Figure 4.36b,c shows hollow spheres with shell thicknesses of approximately 20–30 nm. The shell thickness was maintained within a reasonable range. Shells that are too thin cannot provide sufficient mechanical strength, whereas shells that are too thick cannot provide sufficient space for the storage of corrosion inhibitors [121].

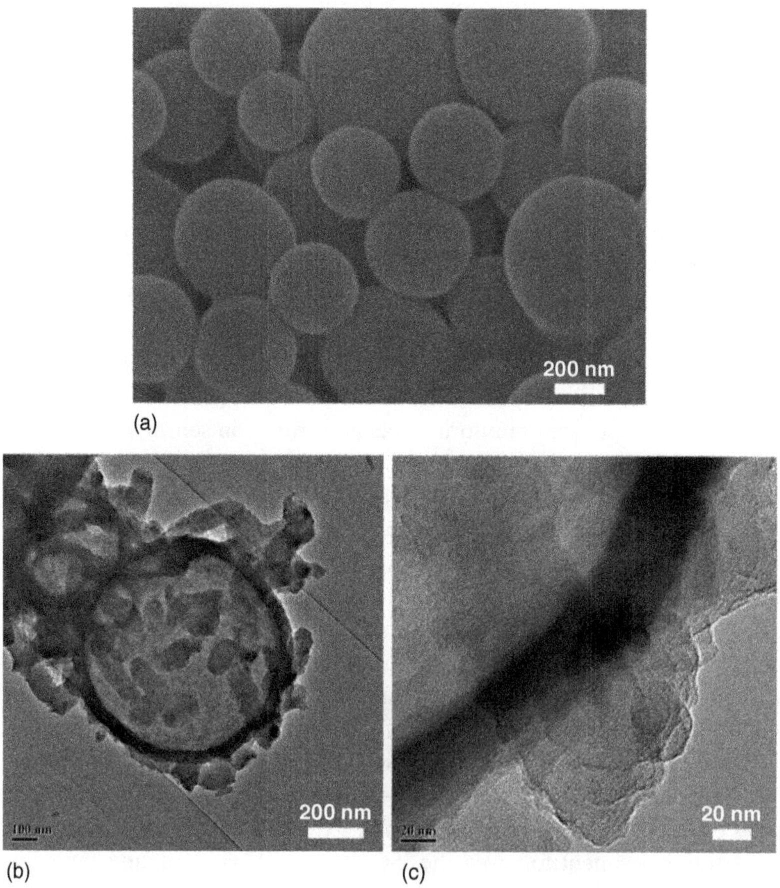

Figure 4.36 (a) SEM and (b,c) TEM images of PANI/BTA nanocapsules.

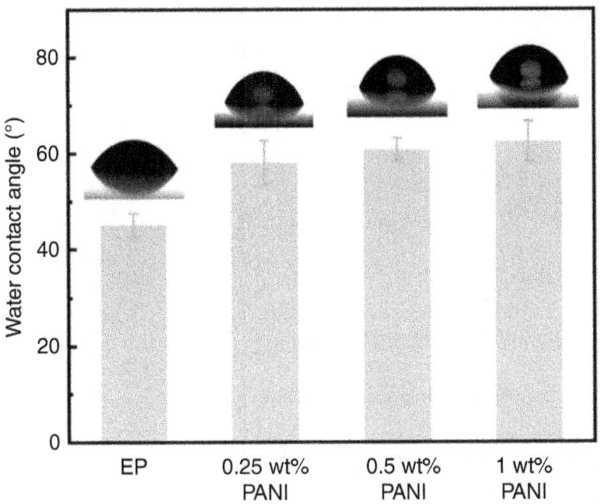

Figure 4.37 WCAs of PANI/BTA coatings.

4.4.3 Characterization of PANI/BTA-Based Epoxy Coating

4.4.3.1 Contact Angle Analysis of PANI/BTA Nanocapsule Coating

Wenzel's theory suggests that when θ is smaller than 90°, the surface has more tendency to become wet. Figure 4.37 shows the variation trend of WCA on the surface of PANI/BTA nanocapsule coatings. It can be observed that the WCA gradually increased with the PANI/BTA content. The WCA of epoxy coating was 45° indicating good wettability. With the addition of nanocapsules to 1 wt%, the WCA of the coating increased to 62.4°, which reduced the wettability of the epoxy resin surface to certain extent. This increase in the WCA has a certain helping effect on the improvement of the coating's protective ability.

4.4.3.2 Electrochemical Analysis of Scratched PANI/BTA Nanocapsule Coating

The open circuit potentials (OCPs) of the PANI/BTA nanocapsule coatings in a 3.5 wt% NaCl solution were recorded to evaluate the corrosion trend. In general, a higher OCP of the coatings corresponded to a weaker corrosion tendency. As shown in Figure 4.38, the OCP values of four types of coatings with different PANI/BTA contents ranged from −0.78 to −0.70 V during the initial immersion period. The OCP values exhibited an upward trend [122, 123]. The increase in the OCP value of the epoxy resin was due to the accumulation of corrosion products, whereas the increase in the OCP value of the PANI/BTA nanocapsule coating is ascribed to the passivation and corrosion inhibition effects of PANI/BTA [124]. Moreover, the OCP values of the PANI/BTA nanocapsule coatings were significantly higher than those of the blank EP coating, indicating that the PANI/BTA nanocapsules had a significant corrosion inhibition effect. The 0.5 wt% PANI/BTA coating exhibited the highest OCP value (−0.57 V) throughout the immersion process. With a further increase in the PANI/BTA nanocapsule content, the OCP value of the 1 wt% PANI/BTA coating was lower than that of 0.5 wt% PANI/BTA coating.

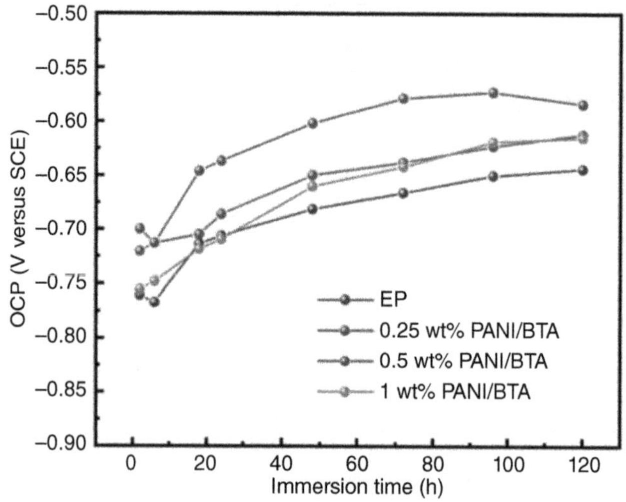

Figure 4.38 OCP variation trend of PANI/BTA nanocapsule scratch coating with different contents during the immersion process.

This phenomenon is attributed to local defects caused by excessive PANI/BTA nanocapsules inducing agglomeration, which made it easy for the corrosive media to reach the metal substrate through the coating defects, triggering corrosion reactions.

Figure 4.39 shows the EIS results of different scratched coatings during immersion in 3.5 wt% NaCl solution. The $|Z|_{0.01\text{Hz}}$ value of the blank EP coating was as low as $1.1 \times 10^5\ \Omega\,\text{cm}^2$ at two hours and decreased with time. For scratched coatings containing 0.25, 0.5, and 1 wt% PANI/BTA nanocapsules, the $|Z|_{0.01\text{Hz}}$ values were 3.9×10^5, 1.2×10^6, and $1.5 \times 10^6\ \Omega\,\text{cm}^2$, respectively, indicating improvement in the corrosion-resistant property. However, the $|Z|_{0.01\text{Hz}}$ values of all the coatings decreased with time, which is attributed to passivation and corrosion inhibition of the PANI/BTA nanocapsules. When the composite coatings were scratched, the inside nanocapsules broke, resulting in the rapid release of BTA molecules, which inhibited corrosion in a timely manner [116, 117]. The reaction equations are (4.1)–(4.3). Meanwhile, the PANI at the scratch obtained electrons lost by the metal substrate Fe, which transitioned from an oxidized state to a reduced state, further oxidizing Fe^0 into Fe^{2+}/Fe^{3+} and promoting the formation of the passive metal film Fe_2O_3, which inhibited metal corrosion. The reaction equations are shown as (4.4)–(4.8). This could be the reason for the higher $|Z|_{0.01\text{Hz}}$ values of the PANI/BTA nanocapsule coating during the initial EIS measurement stage. Subsequently, owing to the failure of the inhibitor layer and the passivation rate of Fe being lower than its corrosion rate, the $|Z|_{0.01\text{Hz}}$ values of the coating decreased. Notably, a 0.5 wt% PANI/BTA coating had the best corrosion protection performance. In the phase angle plots, the low-frequency peak in the range of 10^{-1}–10 Hz was assigned to the charge-transfer process on the exposed steel surface. The peak at a medium frequency (10–10^4 Hz) was associated with corrosion products. The high-frequency

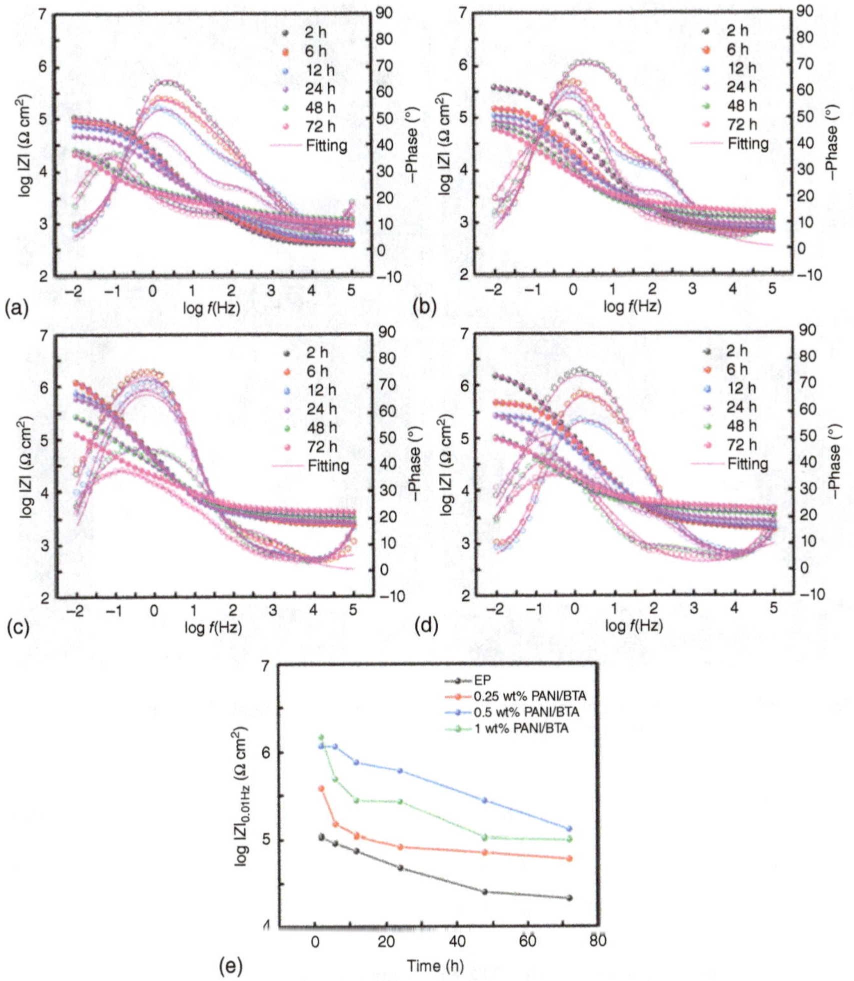

Figure 4.39 Impedance module values and phase angles of scratched coatings: (a) blank epoxy, (b) 0.25 wt% PANI/BTA coating, (c) 0.5 wt% PANI/BTA coating, (d) 1 wt% PANI/BTA, and (e) the variation trend of $|Z|_{0.01\,Hz}$ values. Source: Huang et al. [120]/with permission of Elsevier.

peak (10^4–10^5 Hz), which typically corresponds to pore resistance, was absent because of the wide-open scratches on all the coatings [125, 126].

$$Fe \rightarrow Fe^{2+} + 2e^- \tag{4.1}$$

$$nFe^{2+} + 2nBTA \rightarrow [BTA - Fe - BTA]_n + 2nH^+ \tag{4.2}$$

$$Fe + BTA \rightarrow Fe - BTA \tag{4.3}$$

$$Fe \rightarrow Fe^{2+} + 2e^- \tag{4.4}$$

$$Fe^{2+} \rightarrow Fe^{3+} + e^- \tag{4.5}$$

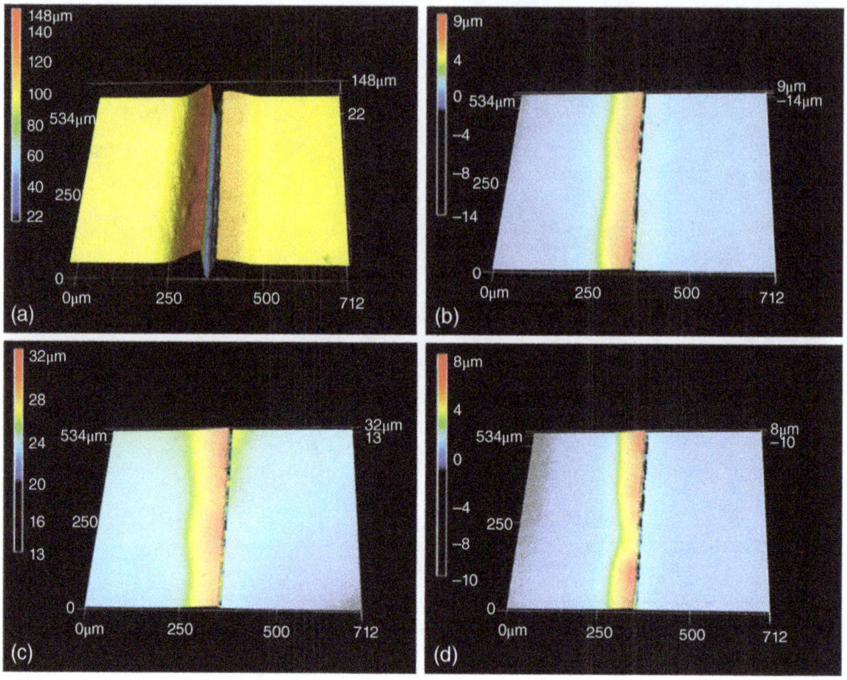

Figure 4.40 CLSM images of (a) scratched epoxy coating and healed coatings: (b) 0.25, (c) 0.5, and (d) 1 wt% PANI/BTA nanocapsule coating.

$$\mathrm{ES} + n\mathrm{H_2O} + ne^- \rightarrow \mathrm{LB} + n\mathrm{OH^-} \tag{4.6}$$

$$\mathrm{O_2} + 2\mathrm{H_2O} + \mathrm{LB} \rightarrow \mathrm{ES} + 4\mathrm{OH^-} \tag{4.7}$$

$$2\mathrm{Fe^{3+}} + 6\mathrm{OH^-} \rightarrow \mathrm{Fe_2O_3} + 3\mathrm{H_2O} \tag{4.8}$$

4.4.3.3 SMP Performance of PANI/BTA Nanocapsule Coating

To further increase the self-healing efficiency, the coatings were heated at 80 °C for 30 minutes to complete the scratch closure. The stereoscopic self-healing effect was observed using a confocal laser microscopy (Figure 4.40). A scalpel was used to create a V-shaped scratch on the surface of the coating, with protrusions appearing on both sides. The average width of the scratch was approximately 30 μm, and there was a clear metallic luster at the bottom, indicating that the metal substrate was in an exposed state. After 30 minutes of heating, the scratch was healed to a large extent, exhibiting a narrowed scratch width of <10 μm. This indicated the excellent shape memory effect of the SMP coating.

4.4.3.4 Electrochemical Analysis of Healed PANI/BTA Coatings

After heating the coatings at 80 °C for 30 minutes, the EIS measurement was conducted again. Figure 4.41 shows that the $|Z|_{0.01\mathrm{Hz}}$ values increased significantly after thermal healing. $|Z|_{0.01\mathrm{Hz}}$ value of the blank epoxy coating was $8.1 \times 10^5\ \Omega\,\mathrm{cm}^2$ after two hours of immersion, which was significantly higher than that of the scratched

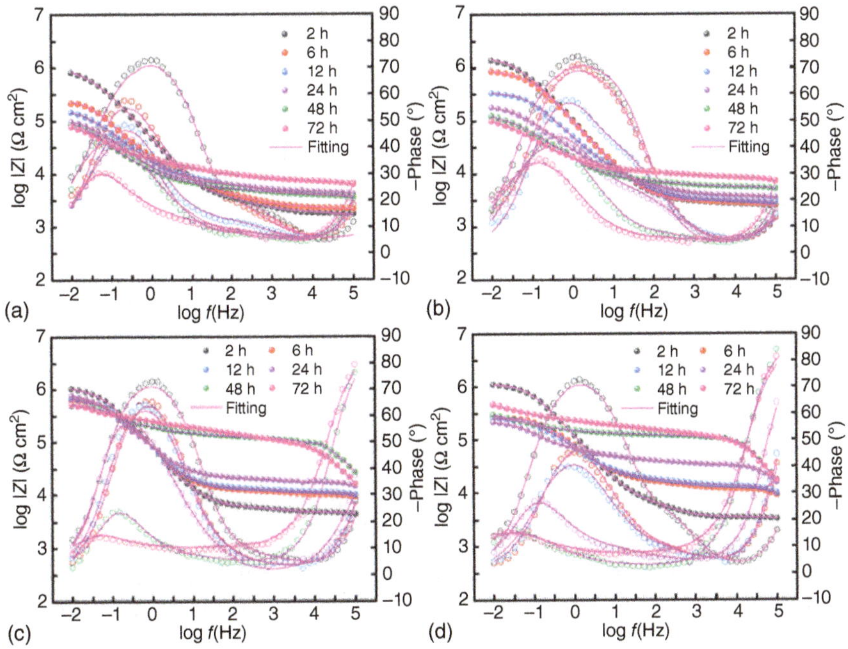

Figure 4.41 Bode plots of thermally healed coatings immersed in 3.5 wt% NaCl solution: (a) epoxy coating, (b) 0.25, (c) 0.5, and (d) 1 wt% PANI/BTA coating. Source: Huang et al. [120]/with permission of Elsevier.

one without heating (1.1×10^5 Ω cm^2). It indicated that the SMP effect could partially recover the corrosion resistance of the scratched coating. However, the $|Z|_{0.01\text{Hz}}$ value of the heated blank epoxy coating decreased to 7.7×10^4 Ω cm^2 after 72 hours because the narrowed SMP coating cannot completely recover the barrier effect against corrosive electrolyte. For the PANI/BTA nanocapsule coating, the $|Z|_{0.01\text{Hz}}$ values were higher, and the PANI/BTA coating showed the highest and most stabilized $|Z|_{0.01\text{Hz}}$ values, demonstrating an optimal self-healing performance. This was because under the premise of triggering shape memory effect to narrow scratches, PANI and BTA in nanocapsules were more conducive to play a role. The corrosion inhibitor BTA were adsorbed on the exposed metal surface or reacted with Fe^{2+} to form a complex, thus inhibiting metal corrosion [127–130], whereas PANI can play a passivation role inducing the formation of metal passive film Fe_2O_3, which changed the interfacial properties inside the resin [131–133]. When the corrosive electrolyte invades the metal substrate, the metal surface underwent a rapid oxidation reaction and releases electrons. At this time, the emeraldine salt of PANI (ES, oxidation state) obtained electrons that were reduced to the leucoemeraldine base (LB, reduction state) [134]. LB was oxidized to ES when it reacted with oxygen passing through the coating. During this process, the initial metal ions reacted with the hydroxide generated by the reaction to form a metal hydroxide, which was eventually converted into a stable metal oxide passive layer. Moreover, the reversibility of PANI redox allowed for an infinite cycle of this process, and the formation rate of passive film depended

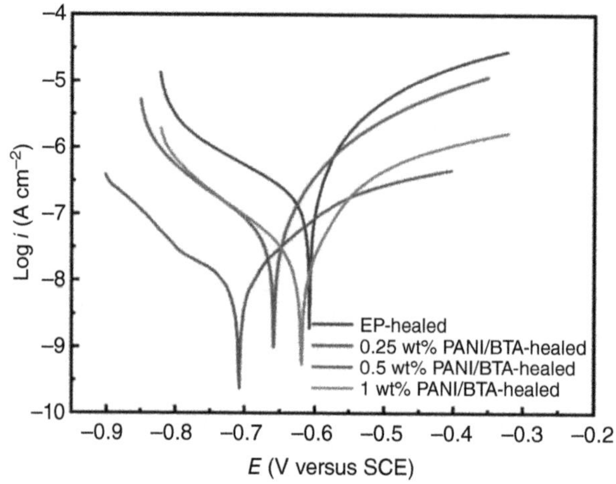

Figure 4.42 Polarization curve of the healed coating immersed in 3.5 wt% NaCl solution for 72 hours.

on the redox activity of the PANI-based material. Through the redox reaction, a dense passive layer was formed on the metal surface to slow down the corrosion development [135, 136].

The self-healing performance was confirmed by polarization curves (Figure 4.42). It is generally believed that a lower corrosion current (i_{corr}) corresponds to a stronger corrosion protection ability [137]. Table 4.6 presents the i_{corr} values of the different coatings. The i_{corr} value of the healed epoxy coating was 19.98 μA cm^{-2}; in comparison, those of the PANI/BTA nanocapsule coatings were significantly reduced. The i_{corr} of the healed 0.25 wt% PANI/BTA nanocapsule coating was 6.94 μA cm^{-2}, and that of the healed 1 wt% PANI/BTA nanocapsule coating was 2.93 μA cm^{-2}. Notably, the healed 0.5 wt% PANI/BTA nanocapsule coating had the lowest i_{corr} value of 1.26 μA cm^{-2}. These results indicate that the PANI/BTA nanocapsules can inhibit the corrosion reaction, which is attributed to the corrosion inhibition effect of BTA, the passivation effect of PANI, and the shape memory effect of the EP coating [138].

4.4.3.5 Corrosion Monitoring of PANI/BTA Nanocapsule Coatings

A blank EP coating and a 0.5 wt% PANI/BTA nanocapsule coating were coated onto the corrosion sensor. After curing, 4-mm scratches were made on the coating

Table 4.6 Corrosion potential and corrosion current of the healed coating with different contents of PANI/BTA nanocapsules immersed in 3.5 wt% NaCl solution for 72 hours.

Coating samples	E_{corr} (V_{SCE})	i_{corr} (10^{-2} μA cm^{-2})
Epoxy	−0.608	19.98
0.25 wt% PANI/BTA	−0.659	6.94
0.5 wt% PANI/BTA	−0.707	1.26
1 wt% PANI/BTA	−0.619	2.93

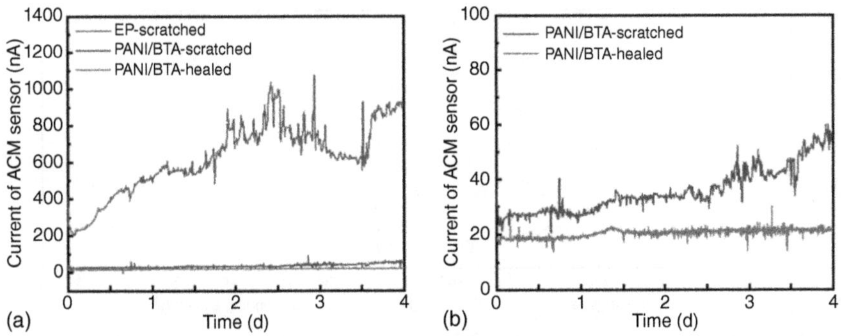

Figure 4.43 (a) Corrosion current curve of the scratch and healed coating under salt spray conditions, (b) enlarged view of scratched and healed 0.5 wt% PANI/BTA nanocapsule coating.

surfaces and one of the 0.5 wt% PANI/BTA nanocapsule scratch coatings were thermally healed. The scratched epoxy coating, scratched 0.5 wt% PANI/BTA nanocapsule coating, and healed 0.5 wt% PANI/BTA nanocapsule coating was placed in a salt spray tank for real-time corrosion monitoring.

Figure 4.43 shows the changes in the corrosion currents of the different coatings during the salt spray process. During salt spraying, the corrosion current of the scratched epoxy coating exhibited an increasing trend, with a maximum corrosion current of >1000 nA. However, the corrosion currents of the scratched 0.5 wt% PANI/BTA nanocapsule coating and healed 0.5 wt% PANI/BTA nanocapsule coating were <100 nA, exhibiting significant reductions. The corrosion current of the healed 0.5 wt% PANI/BTA nanocapsule coating was the lowest, remaining at approximately 20 nA, which indicated a good self-healing effect.

4.4.3.6 Passivation of PANI

To further explore the passivation function of PANI, a 2-day salt spray experiment was conducted on a scratched epoxy coating and 0.5 wt% PANI/BTA nanocapsule coating. Figure 4.44 shows the XPS results for the oxides that accumulated in the scratch. For the scratched epoxy coating, two peaks at 710.6 and 712.1 eV in the Fe $2p_{3/2}$ spectrum are ascribed to Fe_2O_3 and FeOOH, respectively. For the healed 0.5 wt% PANI/BTA coating, the peak at 712.1 eV disappeared, whereas a Fe $2p_{3/2}$ peak at 710.3 eV and a Fe^0 peak at 706.3 eV were detected [139, 140]. When the NaCl solution penetrated the coating scratch to trigger the corrosion reaction, the redox activity of PANI catalyzed the generation of ferric oxide and formed a dense rust layer, protecting the steel against further corrosion [134, 140–142]. Figure 4.44c,d shows the N 1s spectra of the two substrates. A weak peak at 399.4 eV with a peak half-width of 2.1 eV was observed for the scratched epoxy coating, which is attributed to the nitride phases in the carbon steel [143, 144]. The N 1s spectrum of 0.5 wt% PANI/BTA coating exhibited two peaks at 399.6 and 400.6 eV, which were related to the BTA sodium salt and BTA molecules adsorbed on the steel surface, respectively [145–147].

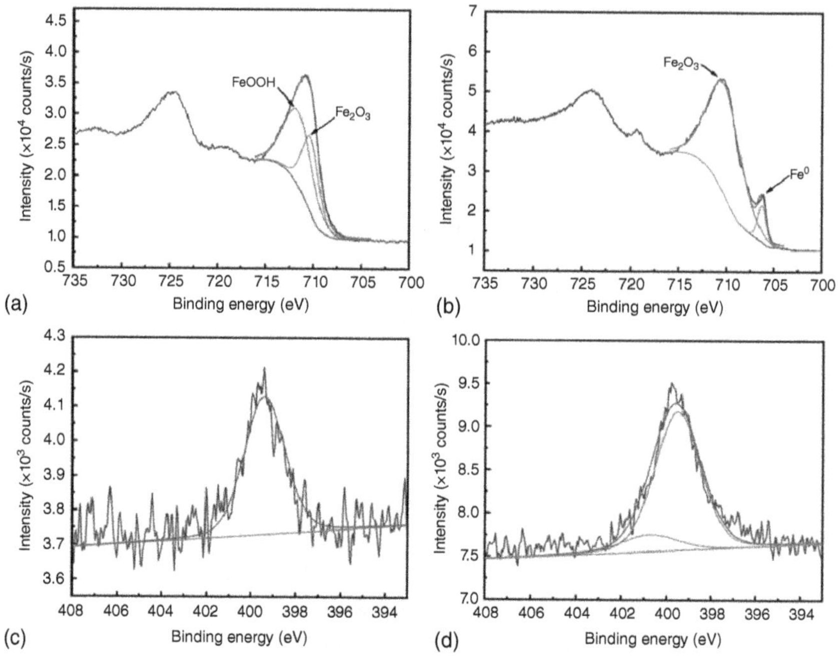

Figure 4.44 XPS spectra of Fe 2p of (a) blank epoxy coating, and (b) 0.5 wt% PANI/BTA coating; N 1s of (c) blank epoxy coating, and (d) 0.5 wt% PANI/BTA coating. Source: Huang et al. [120]/with permission of Elsevier.

Table 4.7 Mechanical properties of different PANI/BTA nanocapsule coatings.

PANI/BTA content	0 wt%	0.25 wt%	0.5 wt%	1 wt%
Pencil hardness	H	2H	2H	2H
Adhesive force	10.6 ± 0.8 MPa	10.0 ± 1.1 MPa	9.3 ± 1.5 MPa	8.3 ± 1.4 MPa

4.4.3.7 Mechanical Properties of PANI/BTA Nanocapsule Coatings

As shown in Table 4.7, with the addition of 0.25 wt% PANI/BTA nanocapsules to the coating, the hardness of the coating increased from H to 2H. When the weight percentage of PANI/BTA was increased, the hardness did not increase further but stabilized at 2H. The adhesion strength of the blank EP coating was approximately 10.6 MPa, and that of the 0.5 wt% PANI/BTA nanocapsule coating was approximately 9.3 MPa, indicating that the nanocapsules had little effect on the mechanical properties of the coating.

4.4.4 Self-healing Mechanism of PANI/BTA Nanocapsule Coatings

When corrosion occurs on the scratched coating, the hydrogen bonds of the PANI shells are broken, releasing the BTA molecules. BTA can be adsorbed onto a steel surface to form a complex with Fe^{2+}/Fe and inhibit corrosion. In addition, the

PANI nanocapsules near the scratched area can function as passivating agents on the metal surface owing to their redox reversibility. The ES oxidation state of PANI is reduced to the LB reduction state by receiving electrons and then reacts with oxygen to transform back into the ES state. Accompanying this redox process, Fe^{2+} transforms into Fe^{3+} and eventually converted into stable iron trioxide. Owing to the redox reversibility of PANI, the redox cycle continues, and a dense passive film is firmly formed on the metal surface to suppress corrosion. Furthermore, the coating matrix can exert a shape memory effect via heating to narrow the scratch. The synergistic effects between the PANI/BTA nanocapsules and SMP coating resulted in excellent self-healing performance for intelligent corrosion control.

4.5 Chapter Summary

In this chapter, two types of pH-activated self-healing coatings and one type of redox-activated self-healing coating are introduced. In view of the limited corrosion inhibition effect of a single inhibitor and the optimization of the corrosion inhibition performance, when different types of corrosion inhibitors are combined, a self-healing coating design based on the synergistic corrosion inhibition effect was developed. The optimal mixture inhibitors, BTA and MBT, were co-loaded into $mSiO_2$ capsules by Stober method, which achieved pH-activated quick release in both acid and alkaline solutions. EIS and surface analyses indicated that MBT and BTA had excellent corrosion inhibition effects; when the weight ratio of MBT to BTA was 1 : 5, the corrosion inhibition effect was optimal. MBT and BTA can promote the adsorption of corrosion inhibitors on the metal surface through intermolecular interactions to form a dense, less defective, and higher binding energy corrosion inhibition film, thereby preventing the corrosion of the aluminum alloy. Finally, by combining MBT-BTA@$mSiO_2$ with an epoxy coating, self-healing was realized through the efficient release of corrosion inhibitors and their synergistic effects. Once the scratched coating containing 2 wt% MBT-BTA@$mSiO_2$ was immersed in a 3.5 wt% NaCl solution for seven days, there were almost no corrosion products in the scratch area. The corrosion inhibitors in the filler were gradually released and adsorbed onto the metal surface, and the corrosion activity of the damaged coating was inhibited through a synergistic effect, significantly prolonging the service life of the coating.

For the SMP coating containing rGO@MS-P-BTA, pH-responsive PDMAEMA macromolecules were prepared via atomic radical transfer polymerization and anchored onto the rGO@MS via electrostatic assembly. The corrosion inhibitor in the rGO@MS-P-BTA microcapsules was released relatively rapidly in an acidic environment owing to the protonation of the PDMAEMA polymer chain segments in the outer layer into polycationic hydrophilic polymers under acidic conditions and the opening of the pores. Moreover, BTA was quickly released during NIR irradiation. These results indicate that the rGO@MS-P-BTA microcapsules possess pH and photothermal sensitivities that control the release of the BTA corrosion inhibitor. The rGO@MS-P-BTA containing epoxy coating exhibited a synergistic effect between the shape memory effect triggered by photothermal conversion and the release of corrosion inhibitors controlled by NIR irradiation and the pH response. The EIS results indicated that the coating achieved the optimal self-healing performance when 1 wt% rGO@MS-P-BTA microcapsules were incorporated. This smart coating with an ultrahigh healing efficiency provides a new strategy for satisfying the demand for corrosion protection in complex environments.

For the SMP coating containing PANI/BTA nanocapsules, the inside BTA molecules exhibited far faster release in the 3.5 wt% NaCl solution than in deionized water. This phenomenon is attributed to Na^+ doping, which leads to the fracture of hydrogen bonds in the PANI shell layer, permeation expansion of the polymer, and accelerated release of the corrosion inhibitor onto the PANI surface. The PANI/BTA coating doped with 0.5 wt% PANI/BTA nanocapsules exhibited the best self-healing performance, owing to the dual effects of the passivation and corrosion inhibition functions of the PANI/BTA nanocapsule and the shape memory effect of the epoxy coating. BTA was released from the nanocapsules and either adsorbed onto the metal surface or reacted with Fe^{2+} to form complexes. As a passivation promoter, PANI can rapidly transform hydroxyiron oxide into a stable passive iron oxide layer.

The pH-responsive, self-healing coatings represent a significant advancement in the field of smart materials, offering innovative solutions for prolonging the life and enhancing the performance of various substrates. These coatings are designed to respond to changes in pH levels, triggering self-healing mechanisms that repair physical damages automatically. This technology holds immense potential in corrosion protection field. The pH-triggered quick release behavior of corrosion inhibitors from micro/nanocapsules has been commonly reported in many studies because many materials have this specific characteristic. However, during the corrosion process, the pH value in the localized coating damage area may not change quickly, or the variation in the pH value may not satisfy the conditions for accelerating the inhibitor release from the micro/nanocapsules. Transitioning from theoretical models and laboratory-scale experiments to practical, real-world applications involves overcoming numerous challenges and harnessing cutting-edge research and technology. The selection of pH-responsive materials should consider the pH response time, pH response values, material cost, and compatibility with resins and capsules.

Redox-responsive self-healing coatings represent a groundbreaking development in smart material technology, offering dynamic responses to oxidative or reductive environmental changes. These coatings are designed to autonomously repair damage when exposed to specific redox conditions. PANI is the most commonly used material for redox-triggered self-healing coatings because of its ability to transform from an oxidation state (emeraldine salt of PANI) into a reduction state (LB of PANI). LB is oxidized into ES when it reacts with oxygen passing through the coating. During this process, the initial metal ions react with the hydroxide generated by the reaction to form a metal hydroxide, which is eventually converted into a stable metal oxide passive layer to slow corrosion development. Polypyrrole (PPy) has also been used in redox-activated self-healing coatings. However, redox-based materials have rarely been used. Because corrosion causes a redox reaction in the localized damaged coating, the design of novel redox-based materials for intelligent corrosion control applications is promising.

Following are the challenges of pH-responsive and redox-responsive self-healing coatings: (i) Sensitivity and selectivity – A primary challenge is developing coatings with high sensitivity and selectivity to pH changes and redox reactions. These coatings must be able to detect and respond accurately to minor pH fluctuations or intended redox conditions in their environment. Achieving this requires a deep understanding of the chemistry behind pH responsiveness, redox chemistry, and the integration of suitable chemical components into the coating matrix. (ii) Healing efficiency and speed – The efficiency and speed of the self-healing process are crucial for the practical utility of these coatings. The healing mechanism should be quick and robust to ensure minimal interruption in the functionality of the coated material. Developing highly responsive and efficient self-healing mechanisms while maintaining the coating's overall structural integrity is a complex task. (iii) Durability and longevity – Ensuring the long-term durability of pH-responsive and redox-responsive self-healing coatings under various environmental conditions, such as exposure to UV radiation, extreme temperatures, and mechanical wear and tear, is a significant hurdle. The coatings need to retain their self-healing properties over multiple cycles of damage and repair. (iv) Complexity and scalability of production – Many pH-responsive and redox-responsive self-healing coatings involve complex chemical compositions and manufacturing processes, making them expensive to produce. Streamlining these processes to make them more cost-effective and scalable is essential for their widespread application. (v) Environmental compatibility – The environmental impact of these coatings, especially in terms of the materials used and their disposal or degradation, is a growing concern. There is a need for eco-friendly materials and sustainable processes in the development of these coatings.

For the future development of pH-responsive and redox-responsive self healing coatings, a deeper understanding of the self-healing mechanism is crucial, particularly how the pH-triggered and redox-triggered processes function at the molecular level and interfacial level. Research should focus on the kinetics of the healing process, exploring ways to speed up the healing while ensuring the repair is durable and effective. Enhancing the long-term stability and performance of pH-triggered and redox-triggered self-healing coatings in real-world conditions is a key research area. This includes improving the corrosion resistance of the coatings to environmental factors and ensuring that their self-healing properties remain effective over time. Future developments in this area hinge on overcoming existing limitations through innovative material science, engineering advancements, and a commitment to sustainability and safety.

References

1 Pian, W., Zhou, Y., and Xiao, T. (2023). A review of the feasibility of aluminum alloys, carbon fiber composites and glass fiber composites for vehicle weight reduction in the automotive industry. *Journal of Physics: Conference Series* 2608 (1): 012005. https://doi.org/10.1088/1742-6596/2608/1/012005.

2 Djurdjevic, M., Manasijevic, S., Mihailović, M. et al. (2023). From bauxite as a critical material to the required properties of cast aluminum alloys for use in electro automotive parts. *Metals* 13 (11): 1796. https://doi.org/10.3390/met13111796.

3 Moreto, J.A., Marino, C.E.B., Bose Filho, W.W. et al. (2014). SVET, SKP and EIS study of the corrosion behaviour of high strength Al and Al–Li alloys used in aircraft fabrication. *Corrosion Science* 84: 30–41. https://doi.org/10.1016/j.corsci.2014.03.001.

4 Deyab, M.A., El-Rehim, S.S.A., Hassan, H.H. et al. (2020). Impact of rare earth compounds on corrosion of aluminum alloy (AA6061) in the marine water environment. *Journal of Alloys and Compounds* 820: 153428. https://doi.org/10.1016/j.jallcom.2019.153428.

5 Alba-Galvín, J.J., Bethencourt, M., Botana, F.J. et al. (2019). Modificaciones de la microestructura y la capa pasiva de la aleación 2024-T3 Al-Cu durante una limpieza química empleada en la industria aeroespacial. *Revista de Metalurgia* 55 (2): 144. https://doi.org/10.3989/revmetalm.144.

6 Li, C., Zhang, D., Gao, X. et al. (2021). Numerical simulation and experimental research on friction stir welding of 2024-T3 aeronautical aluminum alloy. *Journal of Adhesion Science and Technology* 35 (20): 2230–2248. https://doi.org/10.1080/01694243.2021.1882778.

7 Madhavi, Y., Rama Krishna, L., and Narasaiah, N. (2021). Corrosion-fatigue performance of hard anodized and MAO-coated 2024-T3 and 7075-T6 aerospace Al alloys. *Transactions of the Indian Institute of Metals* 74 (9): 2231–2243. https://doi.org/10.1007/s12666-021-02313-7.

8 Yadav, V.K., Gaur, V., and Singh, I.V. (2024). Effect of pre-corrosion on crack initiation behavior of friction stir welded aluminum alloy 2024-T3 in high and low cycle fatigue regimes. *International Journal of Fatigue* 179: 108037. https://doi.org/10.1016/j.ijfatigue.2023.108037.

9 Huang, I.-W., Hurley, B.L., Yang, F. et al. (2016). Dependence on temperature, pH, and Cl^- in the uniform corrosion of aluminum alloys 2024-T3, 6061-T6, and 7075-T6. *Electrochimica Acta* 199: 242–253. https://doi.org/10.1016/j.electacta.2016.03.125.

10 Jafari-Tarzanagh, Y., Seifzadeh, D., Khodayari, A. et al. (2022). Active corrosion protection of AA2024 aluminum alloy by sol-gel coating containing inhibitor-loaded mesoporous SBA-15. *Progress in Organic Coatings* 173: 107166. https://doi.org/10.1016/j.porgcoat.2022.107166.

11 Sahraoui, M., Boulkroune, M., Chibani, A. et al. (2022). Aqueous extract of *Punica granatum* fruit peel as an eco-friendly corrosion inhibitor for aluminium alloy in acidic medium. *Journal of Bio- and Tribo-Corrosion* 8 (2): 54. https://doi.org/10.1007/s40735-022-00658-0.

12 Ech-chihbi, E., Salim, R., Ouakki, M. et al. (2023). Corrosion resistance assessment of copper, mild steel, and aluminum alloy 2024-T3 in acidic solution by

a novel imidazothiazole derivative. *Materials Today Sustainability* 24: 100524. https://doi.org/10.1016/j.mtsust.2023.100524.

13 Jaldo Serrano, E., López-Sánchez, J., García-Galván, F. et al. (2023). Evaluation of low-toxic hybrid sol-gel coatings with organic pH-sensitive inhibitors for corrosion protection of AA2024 aluminium alloy. *Gels* 9 (4): 294. https://doi.org/10.3390/gels9040294.

14 Fernine, Y., Ech-chihbi, E., Arrousse, N. et al. (2021). *Ocimum basilicium* seeds extract as an environmentally friendly antioxidant and corrosion inhibitor for aluminium alloy 2024-T3 corrosion in 3 wt% NaCl medium. *Colloids and Surfaces A: Physicochemical and Engineering Aspects* 627: 127232. https://doi.org/10.1016/j.colsurfa.2021.127232.

15 Minhas, B., Dino, S., Huang, L. et al. (2022). Active corrosion protection by epoxy coating on Li_2CO_3-pretreated anodized aluminum alloy 2024-T3. *Frontiers in Materials* 8: 804328. https://doi.org/10.3389/fmats.2021.804328.

16 Al-Abed, S.R., Virkutyte, J., Ortenzio, J.N.R. et al. (2016). Environmental aging alters $Al(OH)_3$ coating of TiO_2 nanoparticles enhancing their photocatalytic and phototoxic activities. *Environmental Science: Nano* 3 (3): 593–601. https://doi.org/10.1039/C5EN00250H.

17 Xiao, X., Zhou, Z., Liu, C. et al. (2022). Microstructure and its effect on the intergranular corrosion properties of 2024-T3 aluminum alloy. *Crystals* 12 (3): 395. https://doi.org/10.3390/cryst12030395.

18 Yu, X.-X., Sun, J., Li, Z.-T. et al. (2020). Solidification behavior and elimination of undissolved Al_2CuMg phase during homogenization in Ce-modified Al–Zn–Mg–Cu alloy. *Rare Metals* 39 (11): 1279–1287. https://doi.org/10.1007/s12598-018-1172-1.

19 Staszczyk, A., Sawicki, J., and Adamczyk-Cieslak, B. (2019). A study of second-phase precipitates and dispersoid particles in 2024 aluminum alloy after different aging treatments. *Materials* 12 (24): 4168. https://doi.org/10.3390/ma12244168.

20 Li, Y., Feng, Z., Guo, H. et al. (2024). Waterline corrosion behaviors of carbon steel covered by organic coating with artificial defects in seawater. *Colloids and Surfaces A: Physicochemical and Engineering Aspects* 681: 132840. https://doi.org/10.1016/j.colsurfa.2023.132840.

21 Canton-Vitoria, R., Sayed-Ahmad-Baraza, Y., Humbert, B. et al. (2020). Pyrene coating transition metal disulfides as protection from photooxidation and environmental aging. *Nanomaterials* 10 (2): 363. https://doi.org/10.3390/nano10020363.

22 Liu, T., Ma, L., Wang, X. et al. (2021). Self-healing corrosion protective coatings based on micro/nanocarriers: a review. *Corrosion Communications* 1: 18–25. https://doi.org/10.1016/j.corcom.2021.05.004.

23 Wei, H., Wang, Y., Guo, J. et al. (2015). Advanced micro/nanocapsules for self-healing smart anticorrosion coatings. *Journal of Materials Chemistry A* 3 (2): 469–480. https://doi.org/10.1039/C4TA04791E.

24 Wang, H., Gan, M., Ma, L. et al. (2016). Synthesis of polyaniline-modified mesoporous-silica containers for anticorrosion coatings via in-situ polymerization and surface-protected etching. *Polymers for Advanced Technologies* 27 (7): 929–937. https://doi.org/10.1002/pat.3750.

25 Ma, L., Wang, J., Wang, Y. et al. (2022). Enhanced active corrosion protection coatings for aluminum alloys with two corrosion inhibitors co-incorporated in nanocontainers. *Corrosion Science* 208: 110663. https://doi.org/10.1016/j.corsci.2022.110663.

26 Ma, L., Wang, J., Zhang, D. et al. (2021). Dual-action self-healing protective coatings with photothermal responsive corrosion inhibitor nanocontainers. *Chemical Engineering Journal* 404: 127118. https://doi.org/10.1016/j.cej.2020.127118.

27 Recloux, I., Mouanga, M., Druart, M.-E. et al. (2015). Silica mesoporous thin films as containers for benzotriazole for corrosion protection of 2024 aluminium alloys. *Applied Surface Science* 346: 124–133. https://doi.org/10.1016/j.apsusc.2015.03.191.

28 Zhang, Y.-F., Hu, C., Zhao, Z.-B. et al. (2022). Fabrication of BTA-loaded mesoporous silica-encapsulated ZSM-5 molecular sieve for enhancing anti-corrosion performance of waterborne epoxy coatings. *Progress in Organic Coatings* 172: 107160. https://doi.org/10.1016/j.porgcoat.2022.107160.

29 Chenan, A., Ramya, S., George, R.P. et al. (2014). 2-Mercaptobenzothiazole-loaded hollow mesoporous silica-based hybrid coatings for corrosion protection of modified 9Cr–1Mo ferritic steel. *Corrosion* 70 (5): 496–511. https://doi.org/10.5006/1090.

30 Finšgar, M. (2020). Electrochemical, 3D topography, XPS, and ToF-SIMS analyses of 4-methyl-2-phenylimidazole as a corrosion inhibitor for brass. *Corrosion Science* 169: 108632. https://doi.org/10.1016/j.corsci.2020.108632.

31 Yeganeh, M., Asadi, N., Omidi, M. et al. (2019). An investigation on the corrosion behavior of the epoxy coating embedded with mesoporous silica nanocontainer loaded by sulfamethazine inhibitor. *Progress in Organic Coatings* 128: 75–81. https://doi.org/10.1016/j.porgcoat.2018.12.022.

32 Qian, B., Zheng, Z., Michailids, M. et al. (2019). Mussel-inspired self-healing coatings based on polydopamine-coated nanocontainers for corrosion protection. *ACS Applied Materials & Interfaces* 11 (10): 10283–10291. https://doi.org/10.1021/acsami.8b21197.

33 Wu, C., Liu, Q., Chen, R. et al. (2017). Fabrication of ZIF-8@SiO_2 micro/nano hierarchical superhydrophobic surface on AZ31 magnesium alloy with impressive corrosion resistance and abrasion resistance. *ACS Applied Materials & Interfaces* 9 (12): 11106–11115. https://doi.org/10.1021/acsami.6b16848.

34 Rajan, V. and Neelakantan, L. (2017). On the corrosion behavior of Al_2Cu by local electrochemical impedance spectroscopy using droplet cell microscopy. *Journal of Solid State Electrochemistry* 21 (2): 603–609. https://doi.org/10.1007/s10008-016-3388-z.

35 Zhang, X., Feng, Z., Xie, J. et al. (2023). Corrosion inspired inhibitor releasing for BTA@ZIF-8 modified epoxy coatings in seawater environment. *Colloids and Surfaces A: Physicochemical and Engineering Aspects* 673: 131866. https://doi.org/10.1016/j.colsurfa.2023.131866.

36 Almaghrabi, M., Alqurshi, A., Jadhav, S.A. et al. (2023). Evaluating thermogravimetric analysis for the measurement of drug loading in mesoporous silica nanoparticles (MSNs). *Thermochimica Acta* 730: 179616. https://doi.org/10.1016/j.tca.2023.179616.

37 Groen, J.C., Peffer, L.A.A., and Pérez-Ramı́ rez, J. (2003). Pore size determination in modified micro- and mesoporous materials. Pitfalls and limitations in gas adsorption data analysis. *Microporous and Mesoporous Materials* 60 (1): 1–17. https://doi.org/10.1016/S1387-1811(03)00339-1.

38 Zhang, T., Wei, H., Ling, R. et al. (2021). Efficient production of 5-hydroxymethylfurfural from glucose over silica-tin oxide composite catalysts. *Microporous and Mesoporous Materials* 311: 110717. https://doi.org/10.1016/j.micromeso.2020.110717.

39 Zhu, Q., Xu, X., Huang, Y. et al. (2023). pH-responsive mesoporous silica nanocontainers based on Zn-BTA complexes as stoppers for controllable release of corrosion inhibitors and application in epoxy coatings. *Progress in Organic Coatings* 181: 107581. https://doi.org/10.1016/j.porgcoat.2023.107581.

40 Jia, L., Li, Z., Shen, J. et al. (2015). Multifunctional mesoporous silica nanoparticles mediated co-delivery of paclitaxel and tetrandrine for overcoming multidrug resistance. *International Journal of Pharmaceutics* 489 (1–2): 318–330. https://doi.org/10.1016/j.ijpharm.2015.05.010.

41 Dement'eva, O.V., Senchikhin, I.N., Sedykh, E.M. et al. (2016). Mesostructured SiO_2-based nanocontainers synthesized on a functional template: capacity and rate of unloading. *Colloid Journal* 78 (1): 52–64. https://doi.org/10.1134/S1061933X16010051.

42 Huang, J., Zhu, Y., Ma, Y. et al. (2021). pH-triggered release performance of microcapsule-based inhibitor and its inhibition effect on the reinforcement embedded in mortar. *Materials* 14 (19): 5517. https://doi.org/10.3390/ma14195517.

43 Wen, J., Lei, J., Chen, J. et al. (2020). Polyethylenimine wrapped mesoporous silica loaded benzotriazole with high pH-sensitivity for assembling self-healing anti-corrosive coatings. *Materials Chemistry and Physics* 253: 123425. https://doi.org/10.1016/j.matchemphys.2020.123425.

44 Yang, C., Xu, W., Meng, X. et al. (2021). A pH-responsive hydrophilic controlled release system based on ZIF-8 for self-healing anticorrosion application. *Chemical Engineering Journal* 415: 128985. https://doi.org/10.1016/j.cej.2021.128985.

45 Liu, Z., Zhang, B., Yu, H. et al. (2022). A smart anticorrosive epoxy coating based on graphene oxide/functional mesoporous silica nanoparticles for controlled release of corrosion inhibitors. *Coatings* 12 (11): 1749. https://doi.org/10.3390/coatings12111749.

46 Cao, Y., Yuan, X., Wang, X. et al. (2021). Synthesis and controlled release kinetics of pH-sensitive hollow polyaniline microspheres encapsuled with the corrosion inhibitor. *Journal of Molecular Liquids* 342: 117497. https://doi.org/10.1016/j.molliq.2021.117497.

47 Cao, Y., Wu, H., Wang, X. et al. (2022). Novel long-acting smart anticorrosion coating based on pH-controlled release polyaniline hollow microspheres encapsulating inhibitor. *Journal of Molecular Liquids* 359: 119341. https://doi.org/10.1016/j.molliq.2022.119341.

48 Li, C., Zhao, X., Meng, C. et al. (2021). Application of hollow mesoporous organosilica nanoparticles as pH and redox double stimuli-responsive nanocontainer in the controlled release of corrosion inhibitor molecules. *Progress in Organic Coatings* 159: 106437. https://doi.org/10.1016/j.porgcoat.2021.106437.

49 Maia, F., Yasakau, K.A., Carneiro, J. et al. (2016). Corrosion protection of AA2024 by sol–gel coatings modified with MBT-loaded polyurea microcapsules. *Chemical Engineering Journal* 283: 1108–1117. https://doi.org/10.1016/j.cej.2015.07.087.

50 Liu, T., Li, W., Zhang, C. et al. (2021). Preparation of highly efficient self-healing anticorrosion epoxy coating by integration of benzotriazole corrosion inhibitor loaded 2D-COF. *Journal of Industrial and Engineering Chemistry* 97: 560–573. https://doi.org/10.1016/j.jiec.2021.03.012.

51 Angst, U.M., Elsener, B., Larsen, C.K. et al. (2011). Chloride induced reinforcement corrosion: electrochemical monitoring of initiation stage and chloride threshold values. *Corrosion Science* 53 (4): 1451–1464. https://doi.org/10.1016/j.corsci.2011.01.025.

52 Arjmand, F. and Adriaens, A. (2012). Influence of pH and chloride concentration on the corrosion behavior of unalloyed copper in NaCl solution: a comparative study between the micro and macro scales. *Materials* 5 (12): 2439–2464. https://doi.org/10.3390/ma5122439.

53 Yuan, B., Yan, C., Li, Z. et al. (2021). Dynamic pitting processes of 316 stainless steel in $NaCl + Na_2CO_3$ solution with digital holography. *Corrosion Communications* 4: 57–67. https://doi.org/10.1016/j.corcom.2022.02.001.

54 Rosliza, R., Senin, H.B., and Wan Nik, W.B. (2017). Electrochemical performances and corrosion inhibition of AA6061 in tropical seawater. *ASEAN Journal on Science and Technology for Development* 25 (2): 251–259. https://doi.org/10.29037/ajstd.248.

55 Zhang, K., Yang, W., Xu, B. et al. (2018). Inhibitory effect of konjac glucomanan on pitting corrosion of AA5052 aluminium alloy in NaCl solution. *Journal of Colloid and Interface Science* 517: 52–60. https://doi.org/10.1016/j.jcis.2018.01.092.

56 Singh, A., Lin, Y., Liu, W. et al. (2014). Plant derived cationic dye as an effective corrosion inhibitor for 7075 aluminum alloy in 3.5% NaCl solution. *Journal of Industrial and Engineering Chemistry* 20 (6): 4276–4285. https://doi.org/10.1016/j.jiec.2014.01.033.

57 Machkova, M., Matter, E.A., Kozhukharov, S. et al. (2013). Effect of the anionic part of various Ce(III) salts on the corrosion inhibition efficiency of AA2024 aluminium alloy. *Corrosion Science* 69: 396–405. https://doi.org/10.1016/j.corsci.2013.01.008.

58 Khlopyk, O., Korniy, S., Zin, I. et al. (2024). Corrosion inhibition of aluminium alloy in NaCl solution by glycerol from biodiesel production. *Chemical Engineering Communications* 211 (1): 124–132. https://doi.org/10.1080/00986445.2023.2223126.

59 Yasakau, K.A., Kuznetsova, A., Kallip, S. et al. (2018). A novel bilayer system comprising LDH conversion layer and sol-gel coating for active corrosion protection of AA2024. *Corrosion Science* 143: 299–313. https://doi.org/10.1016/j.corsci.2018.08.039.

60 Mondou, E., Proietti, A., Charvillat, C. et al. (2023). Understanding the mechanisms of intergranular corrosion in 2024 Al alloy at the polycrystal scale. *Corrosion Science* 221: 111338. https://doi.org/10.1016/j.corsci.2023.111338.

61 Zhu, Y. and Frankel, G.S. (2019). Effect of major intermetallic particles on localized corrosion of AA2060-T8. *Corrosion* 75 (1): 29–41. https://doi.org/10.5006/2867.

62 Wang, S.C. and Starink, M.J. (2005). Precipitates and intermetallic phases in precipitation hardening Al–Cu–Mg–(Li) based alloys. *International Materials Reviews* 50 (4): 193–215. https://doi.org/10.1179/174328005X14357.

63 Cao, F., He, Z., Wang, T. et al. (2023). The synergistic effect of sodium phytate and benzotriazole as corrosion inhibitors of 20SiMn steel in 3.5 wt-% NaCl solution. *Corrosion Engineering, Science and Technology* 58 (1): 36–48. https://doi.org/10.1080/1478422X.2022.2132616.

64 Li, J., Tao, Z., Cui, J. et al. (2022). Facile fabrication of dual functional graphene oxide microcapsules carrying corrosion inhibitor and encapsulating self-healing agent. *Polymers* 14 (19): 4067. https://doi.org/10.3390/polym14194067.

65 Prabhu, R., Roopashree, B., Jeevananda, T. et al. (2021). Synthesis and corrosion resistance properties of novel conjugated polymer-$Cu_2Cl_4L_3$ composites. *Materials Science for Energy Technologies* 4: 92–99. https://doi.org/10.1016/j.mset.2021.01.001.

66 Zhong, Z., Hong, R., and Wang, X. (2010). Construction of 3-aryl-1,2,4-benzotriazines via unprecedented rearrangement of bis(benzotriazol-1-yl)methylarenes. *Tetrahedron Letters* 51 (51): 6763–6766. https://doi.org/10.1016/j.tetlet.2010.10.093.

67 Goncharova, O.A., Luchkin, A.Y., Andreeva, N.P. et al. (2022). Mutual effect of components of protective films applied on copper and brass from octadecylamine and 1,2,3-benzotriazole vapors. *Materials* 15 (4): 1541. https://doi.org/10.3390/ma15041541.

68 Chiter, F., Costa, D., Maurice, V. et al. (2023). Corrosion inhibition at emergent grain boundaries studied by DFT for 2-mercaptobenzothiazole on bi-crystalline copper. *npj Materials Degradation* 7 (1): 5. https://doi.org/10.1038/s41529-022-00314-5.

69 Liu, Z., Fan, B., Zhao, J. et al. (2023). Benzothiazole derivatives-based supramolecular assemblies as efficient corrosion inhibitors for copper in artificial seawater: formation, interfacial release and protective mechanisms. *Corrosion Science* 212: 110957. https://doi.org/10.1016/j.corsci.2022.110957.

70 Zarebidaki, A., Mofidi, S.H.H., and Bahri, F.I. (2022). Effect of 2-mercaptobenzothiazole on the corrosion inhibition of Cu–10Ni alloy in 3 wt% NaCl solution. *Journal of Applied Electrochemistry* 52 (12): 1773–1788. https://doi.org/10.1007/s10800-022-01750-6.

71 Rosliza, R., Nora'aini, A., and Wan Nik, W.B. (2010). Study on the effect of vanillin on the corrosion inhibition of aluminum alloy. *Journal of Applied Electrochemistry* 40 (4): 833–840. https://doi.org/10.1007/s10800-009-0066-1.

72 Xie, P., He, Y., Zhong, F. et al. (2021). Cu-BTA complexes coated layered double hydroxide for controlled release of corrosion inhibitors in dual self-healing waterborne epoxy coatings. *Progress in Organic Coatings* 153: 106164. https://doi.org/10.1016/j.porgcoat.2021.106164.

73 Habib, S., Fayyad, E., Nawaz, M. et al. (2020). Cerium dioxide nanoparticles as smart carriers for self-healing coatings. *Nanomaterials* 10 (4): 791. https://doi.org/10.3390/nano10040791.

74 Nawaz, M., Habib, S., Khan, A. et al. (2020). Cellulose microfibers (CMFs) as a smart carrier for autonomous self-healing in epoxy coatings. *New Journal of Chemistry* 44 (15): 5702–5710. https://doi.org/10.1039/C9NJ06436B.

75 Liu, J., Lu, Z., Zhang, L. et al. (2020). Studies of corrosion behaviors of a carbon steel/copper-nickel alloy couple under epoxy coating with artificial defect in 3.5 wt.% NaCl solution using the WBE and EIS techniques. *Progress in Organic Coatings* 148: 105909. https://doi.org/10.1016/j.porgcoat.2020.105909.

76 Yang, S., Wang, J., Mao, W. et al. (2018). pH-responsive zeolitic imidazole framework nanoparticles with high active inhibitor content for self-healing anticorrosion coatings. *Colloids and Surfaces A: Physicochemical and Engineering Aspects* 555: 18–26. https://doi.org/10.1016/j.colsurfa.2018.06.035.

77 Wen, J., Lei, J., Chen, J. et al. (2020). An intelligent coating based on pH-sensitive hybrid hydrogel for corrosion protection of mild steel. *Chemical Engineering Journal* 392: 123742. https://doi.org/10.1016/j.cej.2019.123742.

78 Lei, Y., Jiang, Z.N., Zeng, X.Q. et al. (2023). Preparation of ZIF-67@DTMS NPs/Epoxy composite coating and its anti-corrosion performance for Q235 carbon steel in 3.5 wt% NaCl solution. *Colloids and Surfaces A: Physicochemical and Engineering Aspects* 656: 130370. https://doi.org/10.1016/j.colsurfa.2022.130370.

79 Obot, I.B., Macdonald, D.D., and Gasem, Z.M. (2015). Density functional theory (DFT) as a powerful tool for designing new organic corrosion inhibitors. Part 1: An overview. *Corrosion Science* 99: 1–30. https://doi.org/10.1016/j.corsci.2015.01.037.

80 El Basiony, N.M., Badr, E.E., Baker, S.A. et al. (2021). Experimental and theoretical (DFT&MC) studies for the adsorption of the synthesized Gemini cationic surfactant based on hydrazide moiety as X-65 steel acid corrosion inhibitor. *Applied Surface Science* 539: 148246. https://doi.org/10.1016/j.apsusc.2020.148246.

81 Abdrabo, W.S., Elgendy, B., Soliman, K.A. et al. (2020). Synthesis, assessment and corrosion protection investigations of some novel peptidomimetic cationic surfactants: empirical and theoretical insights. *Journal of Molecular Liquids* 315: 113672. https://doi.org/10.1016/j.molliq.2020.113672.

82 Lei, Y.H., Sheng, N., Hyono, A. et al. (2014). Effect of benzotriazole (BTA) addition on polypyrrole film formation on copper and its corrosion protection. *Progress in Organic Coatings* 77 (2): 339–346. https://doi.org/10.1016/j.porgcoat.2013.10.009.

83 Huang, Y., Zhao, C., Li, Y. et al. (2023). A smart sol-gel coating incorporating pH-responsive BTA-ZIF-8 MOF assembled hexagonal boron nitride for active/passive corrosion protection of 1060 aluminum alloy. *Surface and Coatings Technology* 474: 130072. https://doi.org/10.1016/j.surfcoat.2023.130072.

84 Uhart, A., Ledeuil, J.B., Gonbeau, D. et al. (2016). An Auger and XPS survey of cerium active corrosion protection for AA2024-T3 aluminum alloy. *Applied Surface Science* 390: 751–759. https://doi.org/10.1016/j.apsusc.2016.08.170.

85 Zhang, Q.H., Hou, B.S., Li, Y.Y. et al. (2021). Effective corrosion inhibition of mild steel by eco-friendly thiourea functionalized glucosamine derivatives in acidic solution. *Journal of Colloid and Interface Science* 585: 355–367. https://doi.org/10.1016/j.jcis.2020.11.073.

86 Ren, X., Xu, S., Gu, X. et al. (2021). Hyperbranched molecules having multiple functional groups as effective corrosion inhibitors for Al alloys in aqueous NaCl. *Journal of Colloid and Interface Science* 585: 614–626. https://doi.org/10.1016/j.jcis.2020.10.041.

87 Kopeć, M., Szczepanowicz, K., Mordarski, G. et al. (2015). Self-healing epoxy coatings loaded with inhibitor-containing polyelectrolyte nanocapsules. *Progress in Organic Coatings* 84: 97–106. https://doi.org/10.1016/j.porgcoat.2015.02.011.

88 Wang, T., Wang, Y., Sun, M. et al. (2020). Thermally treated zeolitic imidazolate framework-8 (ZIF-8) for visible light photocatalytic degradation of gaseous formaldehyde. *Chemical Science* 11 (26): 6670–6681. https://doi.org/10.1039/D0SC01397H.

89 Luo, Y., Fan, S., Yu, W. et al. (2018). Fabrication of $Au_{25}(SG)_{18}$–ZIF-8 nanocomposites: a facile strategy to position $Au_{25}(SG)_{18}$ nanoclusters inside and outside ZIF-8. *Advanced Materials* 30 (6): 1704576. https://doi.org/10.1002/adma.201704576.

90 Gladkikh, N., Makarychev, Y., Maleeva, M. et al. (2019). Synthesis of thin organic layers containing silane coupling agents and azole on the surface of mild steel. Synergism of inhibitors for corrosion protection of underground pipelines. *Progress in Organic Coatings* 132: 481–489. https://doi.org/10.1016/j.porgcoat.2019.04.004.

91 Shabanova, I.N., Reshetnikov, S.M., Naimushina, E.A. et al. (2021). X-ray photoelectron study of the mechanism of the anticorrosion effect of benzotriazole and mercaptobenzothiazole additives to mineral oil. *Journal of Surface Investigation: X-Ray, Synchrotron and Neutron Techniques* 15 (1): S186–S192. https://doi.org/10.1134/S1027451022020367.

92 Zhang, H., Sun, W., Wang, L. et al. (2023). Photothermal cyclic self-healing coating with long-term corrosion protection based on bifunctional graphene. *Chemical Engineering Journal* 473: 145164. https://doi.org/10.1016/j.cej.2023.145164.

93 Hosseini, M.G. and Aboutalebi, K. (2019). Epoxy coating with self-healing capability based on a 2-mercaptobenzothiazole-loaded CeO_2 nanocontainer. *Journal of Applied Polymer Science* 136 (15): 47297. https://doi.org/10.1002/app.47297.

94 Zhang, C.H., Li, R.M., Gao, L.L. et al. (2013). Corrosion protection of Mg–11Li–3Al–0.5RE alloy using hybrid epoxy/silica conversion coatings. *Corrosion Engineering, Science and Technology* 48 (4): 276–281. https://doi.org/10.1179/1743278212Y.0000000072.

95 Ganjaee Sari, M., Shamshiri, M., and Ramezanzadeh, B. (2017). Fabricating an epoxy composite coating with enhanced corrosion resistance through impregnation of functionalized graphene oxide-co-montmorillonite nanoplatelet. *Corrosion Science* 129: 38–53. https://doi.org/10.1016/j.corsci.2017.09.024.

96 Wang, Y., Zhu, Y., Li, C. et al. (2016). Smart epoxy coating containing Ce-MCM-22 zeolites for corrosion protection of Mg–Li alloy. *Applied Surface Science* 369: 384–389. https://doi.org/10.1016/j.apsusc.2016.02.102.

97 Guo, Y., Wang, J., Zhang, D. et al. (2019). pH-responsive self-healing anticorrosion coatings based on benzotriazole-containing zeolitic imidazole framework. *Colloids and Surfaces A: Physicochemical and Engineering Aspects* 561: 1–8. https://doi.org/10.1016/j.colsurfa.2018.10.044.

98 Huang, Y., Wang, P., Tan, W. et al. (2022). Photothermal and pH dual-responsive self-healing coating for smart corrosion protection. *Journal of Materials Science and Technology* 107: 34–42. https://doi.org/10.1016/j.jmst.2021.08.044.

99 Zhao, Z., Chen, H., Liu, X. et al. (2018). Development of high-efficient synthetic electric heating coating for anti-icing/de-icing. *Surface and Coatings Technology* 349: 340–346. https://doi.org/10.1016/j.surfcoat.2018.06.011.

100 Yan, J., Shi, J., Zhang, P. et al. (2018). Preparation and properties of epoxy/basalt flakes anticorrosive coatings. *Materials and Corrosion* 69 (11): 1669–1675. https://doi.org/10.1002/maco.201810039.

101 Zhou, X., Zhang, S., Song, Y. et al. (2024). A novel and green 3-amino-1,2,4-triazole modified graphene oxide nanomaterial for enhancing anti-corrosion performance of water-borne epoxy coatings on mild steel. *Progress in Organic Coatings* 187: 108106. https://doi.org/10.1016/j.porgcoat.2023.108106.

102 Xu, J.-B., Cao, Y.-Q., Fang, L. et al. (2018). A one-step preparation of inhibitor-loaded silica nanocontainers for self-healing coatings. *Corrosion Science* 140: 349–362. https://doi.org/10.1016/j.corsci.2018.05.030.

103 Wang, Z., Peng, W., Takenaka, Y. et al. (2017). Sandwich-type nanocomposite of reduced graphene oxide and periodic mesoporous silica with vertically aligned mesochannels of tunable pore depth and size. *Advanced Functional Materials* 27 (47): 1704066. https://doi.org/10.1002/adfm.201704066.

104 Lamprakou, Z., Bi, H., Weinell, C.E. et al. (2022). Smart epoxy coating with mesoporous silica nanoparticles loaded with calcium phosphate for corrosion protection. *Progress in Organic Coatings* 165: 106740. https://doi.org/10.1016/j.porgcoat.2022.106740.

105 Qian, B., Michailidis, M., Bilton, M. et al. (2019). Tannic complexes coated nanocontainers for controlled release of corrosion inhibitors in self-healing coatings. *Electrochimica Acta* 297: 1035–1041. https://doi.org/10.1016/j.electacta.2018.12.062.

106 Lu, H., Yao, Y., Huang, W.M. et al. (2014). Significantly improving infrared light-induced shape recovery behavior of shape memory polymeric nanocomposite via a synergistic effect of carbon nanotube and boron nitride. *Composites Part B: Engineering* 62: 256–261. https://doi.org/10.1016/j.compositesb.2014.03.007.

107 Wang, B., Liu, J., Liu, M. et al. (2023). Preparation and corrosion resistance of shape memory self-healing coatings responsive to near-infrared light. *Polymer Testing* 126: 108146. https://doi.org/10.1016/j.polymertesting.2023.108146.

108 Liu, J., Zhang, X., Liu, S. et al. (2023). Allyl-modified epoxy networks that break and reorganize at high temperatures to avoid T reduction. *Chemical Engineering Journal* 455: 140674. https://doi.org/10.1016/j.cej.2022.140674.

109 Cai, Z., Zeng, J., Guo, T. et al. (2023). Dual responsive self-healing hydrogels with wide stability and excellent mechanical strength based on aliphatic polycarbonate. *Heliyon* 9 (4): e15070. https://doi.org/10.1016/j.heliyon.2023.e15070.

110 Wu, D., Ma, L., Zhang, F. et al. (2020). Durable deicing lubricant-infused surface with photothermally switchable hydrophobic/slippery property. *Materials & Design* 185: 108236. https://doi.org/10.1016/j.matdes.2019.108236.

111 Lorero, I., Rodríguez, A., Campo, M. et al. (2022). Thermally remendable, weldable, and recyclable epoxy network crosslinked with reversible Diels-alder bonds. *Polymer* 259: 125334. https://doi.org/10.1016/j.polymer.2022.125334.

112 Huang, H., Huang, X., Xie, Y. et al. (2019). Fabrication of h-BN-rGO@PDA nanohybrids for composite coatings with enhanced anticorrosion performance. *Progress in Organic Coatings* 130: 124–131. https://doi.org/10.1016/j.porgcoat.2019.01.059.

113 Cui, M., Ren, S., Chen, J. et al. (2017). Anticorrosive performance of waterborne epoxy coatings containing water-dispersible hexagonal boron nitride (h-BN) nanosheets. *Applied Surface Science* 397: 77–86. https://doi.org/10.1016/j.apsusc.2016.11.141.

114 Chen, X., Wang, W., Chen, S. et al. (2022). Effect of microcrystalline cellulose on mechanical property and shape memory property of water-borne epoxy. *Pigment & Resin Technology* 51 (6): 641–648. https://doi.org/10.1108/PRT-06-2021-0059.

115 Li, D., Peng, H., Lin, Z. et al. (2021). Corrosion protection coatings embedded with silane-functionalized rGO/SiO_2 nanocontainers: enhancing dispersive and corrosion-inhibitor loading capabilities. *Surface and Coatings Technology* 427: 127850. https://doi.org/10.1016/j.surfcoat.2021.127850.

116 Zhou, X., Zhu, M., Ma, B. et al. (2023). Shape optimization of thermal shape memory epoxy resin and its mechanism for improving the self-healing of asphalt mixtures. *Construction and Building Materials* 401: 132863. https://doi.org/10.1016/j.conbuildmat.2023.132863.

117 Meng, X., Hou, L., Jin, H. et al. (2023). Study on corrosion protection properties of PANI/ZnO/Zn/graphene coating on aluminum alloy. *Diamond and Related Materials* 136: 110067. https://doi.org/10.1016/j.diamond.2023.110067.

118 Dolatkhah, A. and Wilson, L.D. (2018). Salt-responsive Fe_3O_4 nanocomposites and phase behavior in water. *Langmuir* 34 (1): 341–350. https://doi.org/10.1021/acs.langmuir.7b03613.

119 Hao, Y., Zhao, Y., Yang, X. et al. (2019). Self-healing epoxy coating loaded with phytic acid doped polyaniline nanofibers impregnated with benzotriazole for Q235 carbon steel. *Corrosion Science* 151: 175–189. https://doi.org/10.1016/j.corsci.2019.02.023.

120 Huang, Y., Liu, T., Ma, L. et al. (2022). Saline-responsive triple-action self-healing coating for intelligent corrosion control. *Materials & Design* 214: 110381. https://doi.org/10.1016/j.matdes.2022.110381.

121 Feng, Y., Li, H., Zhang, M. et al. (2022). Preparation of BTA@PDA/PANI microcapsules and anti-corrosion performance of self-healing epoxy coatings on low carbon steel. *Colloids and Surfaces A: Physicochemical and Engineering Aspects* 649: 129481. https://doi.org/10.1016/j.colsurfa.2022.129481.

122 Pour-Ali, S., Dehghanian, C., and Kosari, A. (2014). In situ synthesis of polyaniline–camphorsulfonate particles in an epoxy matrix for corrosion protection of mild steel in NaCl solution. *Corrosion Science* 85: 204–214. https://doi.org/10.1016/j.corsci.2014.04.018.

123 Sababi, M., Pan, J., Augustsson, P.-E. et al. (2014). Influence of polyaniline and ceria nanoparticle additives on corrosion protection of a UV-cure coating on carbon steel. *Corrosion Science* 84: 189–197. https://doi.org/10.1016/j.corsci.2014.03.031.

124 Kamaraj, K., Karpakam, V., Sathiyanarayanan, S. et al. (2011). Synthesis of tungstate doped polyaniline and its usefulness in corrosion protective coatings. *Electrochimica Acta* 56 (25): 9262–9268. https://doi.org/10.1016/j.electacta.2011.08.005.

125 Yıldız, R. (2015). An electrochemical and theoretical evaluation of 4,6-diamino-2-pyrimidinethiol as a corrosion inhibitor for mild steel in HCl solutions. *Corrosion Science* 90: 544–553. https://doi.org/10.1016/j.corsci.2014.10.047.

126 Yu, D., Wang, J., Hu, W. et al. (2017). Preparation and controlled release behavior of halloysite/2-mercaptobenzothiazole nanocomposite with calcined halloysite as nanocontainer. *Materials & Design* 129: 103–110. https://doi.org/10.1016/j.matdes.2017.05.033.

127 Yang, J., Zou, B., Fang, Q. et al. (2022). BTA loaded in various CeO_2 nanocontainers in epoxy coating for long-term anticorrosion of low carbon steel. *Progress in Organic Coatings* 172: 107156. https://doi.org/10.1016/j.porgcoat.2022.107156.

128 Nwokolo, I.K., Shi, H., Ikeuba, A.I. et al. (2022). Synthesis, characterization and investigation of anticorrosion properties of an innovative metal–organic framework, ZnMOF-BTA, on carbon steel in HCl solution. *Coatings* 12 (9): 1288. https://doi.org/10.3390/coatings12091288.

129 Chen, G., Jin, B., Li, Y. et al. (2022). A smart healable anticorrosion coating with enhanced loading of benzotriazole enabled by ultra-highly exfoliated graphene and mussel-inspired chemistry. *Carbon* 187: 439–450. https://doi.org/10.1016/j.carbon.2021.11.048.

130 Turano, M., Walker, M., Grillo, F. et al. (2022). Adsorption of the prototypical organic corrosion inhibitor benzotriazole on the Cu(100) surface. *Corrosion Science* 207: 110589. https://doi.org/10.1016/j.corsci.2022.110589.

131 Liu, T., Li, J., Li, X. et al. (2019). Effect of self-assembled tetraaniline nanofiber on the anticorrosion performance of waterborne epoxy coating. *Progress in Organic Coatings* 128: 137–147. https://doi.org/10.1016/j.porgcoat.2018.11.033.

132 Sathiyanarayanan, S., Karpakam, V., Kamaraj, K. et al. (2010). Sulphonate doped polyaniline containing coatings for corrosion protection of iron. *Surface and Coatings Technology* 204 (9): 1426–1431. https://doi.org/10.1016/j.surfcoat.2009.09.037.

133 Kamaraj, K., Sathiyanarayanan, S., Muthukrishnan, S. et al. (2009). Corrosion protection of iron by benzoate doped polyaniline containing coatings. *Progress in Organic Coatings* 64 (4): 460–465. https://doi.org/10.1016/j.porgcoat.2008.08.008.

134 Gupta, G., Birbilis, N., Cook, A.B. et al. (2013). Polyaniline-lignosulfonate/epoxy coating for corrosion protection of AA2024-T3. *Corrosion Science* 67: 256–267. https://doi.org/10.1016/j.corsci.2012.10.022.

135 Qiu, S., Chen, C., Cui, M. et al. (2017). Corrosion protection performance of waterborne epoxy coatings containing self-doped polyaniline nanofiber. *Applied Surface Science* 407: 213–222. https://doi.org/10.1016/j.apsusc.2017.02.142.

136 Liu, T., Zhao, H., Li, J. et al. (2020). POSS-tetraaniline based giant molecule: synthesis, self-assembly, and active corrosion protection of epoxy-based organic coatings. *Corrosion Science* 168: 108555. https://doi.org/10.1016/j.corsci.2020.108555.

137 Zhang, G., Wu, L., Tang, A. et al. (2018). Active corrosion protection by a smart coating based on a MgAl-layered double hydroxide on a cerium-modified plasma electrolytic oxidation coating on Mg alloy AZ31. *Corrosion Science* 139: 370–382. https://doi.org/10.1016/j.corsci.2018.05.010.

138 Zhang, B., He, C., Wang, C. et al. (2015). Synergistic corrosion inhibition of environment-friendly inhibitors on the corrosion of carbon steel in soft water. *Corrosion Science* 94: 6–20. https://doi.org/10.1016/j.corsci.2014.11.035.

139 Descostes, M., Mercier, F., Thromat, N. et al. (2000). Use of XPS in the determination of chemical environment and oxidation state of iron and sulfur samples: constitution of a data basis in binding energies for Fe and S reference compounds and applications to the evidence of surface species of an oxidized pyrite in a carbonate medium. *Applied Surface Science* 165 (4): 288–302. https://doi.org/10.1016/S0169-4332(00)00443-8.

140 Biesinger, M.C., Payne, B.P., Grosvenor, A.P. et al. (2011). Resolving surface chemical states in XPS analysis of first row transition metals, oxides and hydroxides: Cr, Mn, Fe, Co and Ni. *Applied Surface Science* 257 (7): 2717–2730. https://doi.org/10.1016/j.apsusc.2010.10.051.

141 Karpakam, V., Kamaraj, K., Sathiyanarayanan, S. et al. (2011). Electrosynthesis of polyaniline–molybdate coating on steel and its corrosion protection performance. *Electrochimica Acta* 56 (5): 2165–2173. https://doi.org/10.1016/j.electacta.2010.11.099.

142 Chen, Y., Wang, X.H., Li, J. et al. (2007). Long-term anticorrosion behaviour of polyaniline on mild Steel. *Corrosion Science* 49 (7): 3052–3063. https://doi.org/10.1016/j.corsci.2006.11.007.

143 Ramezanzadeh, B., Ghasemi, E., Askari, F. et al. (2015). Synthesis and characterization of a new generation of inhibitive pigment based on zinc acetate/benzotriazole: solution phase and coating phase studies. *Dyes and Pigments* 122: 331–345. https://doi.org/10.1016/j.dyepig.2015.07.013.

144 Hao, Y., Zhao, Y., Li, B. et al. (2020). Self-healing effect of graphene@PANI loaded with benzotriazole for carbon steel. *Corrosion Science* 163: 108246. https://doi.org/10.1016/j.corsci.2019.108246.

145 Chen, Z., Huang, L., Zhang, G. et al. (2012). Benzotriazole as a volatile corrosion inhibitor during the early stage of copper corrosion under adsorbed thin electrolyte layers. *Corrosion Science* 65: 214–222. https://doi.org/10.1016/j.corsci.2012.08.019.

146 Wang, J.-L., Ke, C., Pohl, K. et al. (2015). The unexpected role of benzotriazole in mitigating magnesium alloy corrosion: a nucleating agent for crystalline nanostructured magnesium hydroxide film. *Journal of the Electrochemical Society* 162 (8): 403–411. https://doi.org/10.1149/2.0781508jes.

147 Mezzi, A., Angelini, E., De Caro, T. et al. (2012). Investigation of the benzotriazole inhibition mechanism of bronze disease. *Surface and Interface Analysis* 44 (8): 968–971. https://doi.org/10.1002/sia.4841.

5

Development of Ion Exchange Activated Self-healing Coating

5.1 Introduction

Chloride ion solution is the most common corrosive substance in the environment. If corrosion inhibitors demonstrate accelerated release in chloride ion-containing solutions, this will improve the self-healing property of coatings based on their inhibition effect [1–8]. In this chapter, the ion exchange effect based on layered double hydroxide (LDH) is studied, and the corresponding self-healing coating is designed. Further, the coating system is introduced, including the coating preparation, filler preparation, surface characterization, macroscopic and microscopic electrochemical properties, and self-healing performance. Finally, the advantages and disadvantages of ion exchange activated self-healing coatings are discussed.

5.2 Ion Exchange Activated Self-healing Coating Using GO-LDH Fillers

The modification of graphene oxide (GO) based on its original structure has attracted considerable attention and further expanded the application of graphene composite coatings [9–11]. Currently, materials that can be compounded with GO include small organic molecules [12–14], polymers [15–17], and nanoparticles [18-20]. Combining nanomaterials and GO can increase the interlayer spacing of the GO sheets and promote their dispersion [21–23]. LDH nanoparticles can combine with the negatively charged GO by electrostatic action owing to the positive charge of the surface layer [24–26]. GO, loaded with LDH nanoparticles (GO-LDH), is an excellent carrier that can improve the utilization of corrosion inhibitors and extend their release time.

In this chapter, we study a series of GO-LDH composites with different GO addition ratios, which were prepared using coprecipitation and ion exchange methods [27–29]. The prepared GO-LDH composite filler was added to waterborne epoxy resin using the solvent evaporation method [30–32]. To determine the optimal filler amount, different proportions of the prepared fillers were added to the waterborne epoxy coating, and the coatings were scratched to create artificial defects. A salt spray test was later conducted to select the optimal filler ratio and

Smart Protective Coatings for Corrosion Control, First Edition. Lingwei Ma and Dawei Zhang.

Table 5.1 Experimental materials and reagents.

Materials	Purity	Company
Magnesium nitrate hexahydrate	Analytically pure	Sinopharm Chemical Reagent Co., Ltd.
Aluminum nitrate hexahydrate	Analytically pure	Sinopharm Chemical Reagent Co., Ltd.
Sodium nitrate	Analytically pure	Sinopharm Chemical Reagent Co., Ltd.
Sodium metavanadate	Analytically pure	Sinopharm Chemical Reagent Co., Ltd.
Graphene oxide	Analytically pure	Suzhou Carbon Feng Graphene Technology Co., Ltd.
Sodium hydroxide	Analytically pure	Shandong West Asia Chemical Co., Ltd.
Anhydrous ethanol	Analytically pure	Sinopharm Chemical Reagent Co., Ltd.
Acetone	Analytically pure	
Waterborne epoxy resin curing agent	Industrial grade	Guangzhou Haoyun Chemical Co., Ltd.

an additional amount for the coating. The self-healing properties of the composite coatings were characterized using electrochemical impedance spectroscopy (EIS) measurements.

5.2.1 Materials and Methods

5.2.1.1 Materials

The materials and reagents used in this section are shown in Table 5.1.

5.2.1.2 Preparation of GO-LDH Composites

The detailed preparation process of the modified GO, loaded with LDH is as follows: 100, 200, and 300 mg of GO powders were separately added to 100 ml deionized water, and ultrasonic dispersion was performed at 100 W power to obtain a uniform suspension, and then 2.56 g of $Mg(NO_3)_2{\cdot}6H_2O$ and 1.875 g of $Al(NO_3)_3{\cdot}9H_2O$ were added in each suspension. Additionally, 0.4 M $NaNO_3$ was gradually added to the solution, while the pH of the suspension was maintained at 10 ± 0.3 using NaOH. After stirring, using a magnetic agitator, for 30 minutes, the suspension was reacted in a condensing reflux unit at 80 °C for 24 hours. During the reaction, Ar gas was introduced to obtain GO-LDH precipitation crystals without corrosion inhibitor loading. The GO-LDH obtained without the corrosion inhibitor was centrifuged and washed four times. The obtained precipitate was dried in a vacuum freeze-drying system for 24 hours, and GO-LDH intercalated with the metavanadate inhibitor was obtained via an anion exchange reaction. The GO-LDH precipitate was poured

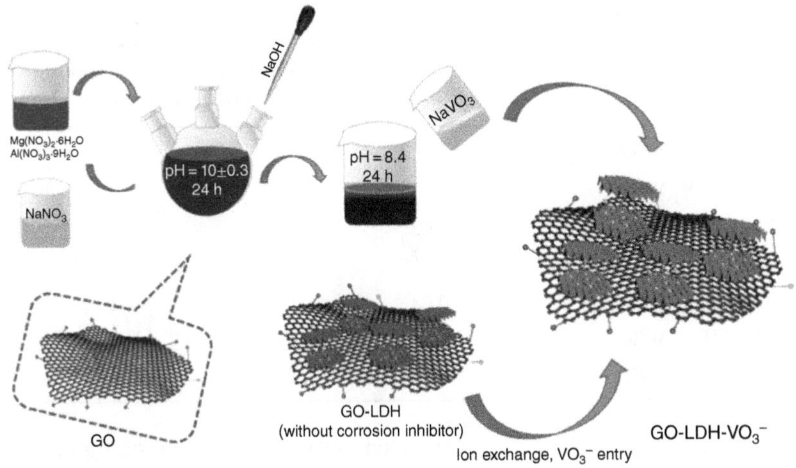

Figure 5.1 GO-LDH synthesis process.

Table 5.2 Samples with different GO additive amounts.

Sample names	GO (mg)	$Mg(NO_3)_2 \cdot 6H_2O$ (g)	$Al(NO_3)_3 \cdot 9H_2O$ (g)
GO-LDH-V_1	100	2.56	1.875
GO-LDH-V_2	200	2.56	1.875
GO-LDH-V_3	300	2.56	1.875

into a 0.1 M $NaVO_3$ solution and stirred at 40 °C for 24 hours to complete the ion displacement reaction. In the preparation process, the pH value of the solution was maintained at 8.4 ± 0.3 by adding NaOH. The unreacted $NaVO_3$ solution was removed by centrifugation and rinsed with water until the pH became 7.0. Finally, the GO-LDH was dried for 1 day, and pure vanadate (VO_3^-)-intercalated double hydroxide (LDH) without GO was prepared using the aforementioned method. The synthesis of GO-LDH is shown in Figure 5.1. The names of the samples prepared using different GO contents are presented in Table 5.2.

5.2.1.3 Coating Preparation

AA2024-T3 aluminum alloy was used as the coating substance. Before applying the coating, the aluminum plate was first polished using 150-mesh sandpaper to remove the surface oil and relatively loose oxide film on the surface. Hence, the polished aluminum plate had a bright metallic luster. The aluminum alloy was further polished using 240-mesh sandpaper to smoothen and brighten its surface. The polished metal substrate was later wiped with acetone, completely immersed in anhydrous ethanol, and placed in an ultrasonic instrument for ultrasonic vibration for 15 minutes to remove surface impurities. After several rounds of cleaning and wiping using anhydrous ethanol, the aluminum sheet was immersed in it again for later use to prevent the oxidation of its surface, while waiting for the coating application to proceed.

Table 5.3 Names of prepared coating samples.

Sample names	GO-LDH (g)	Waterborne epoxy resin (g)	Curing agent (g)
Pure waterborne epoxy coating (WEC)	0	3	1
0.5 wt%GO-LDH-V_1 WEC	0.015	3	1
1.0 wt%GO-LDH-V_1 WEC	0.03	3	1
0.5 wt%GO-LDH-V_2 WEC	0.015	3	1
1.0 wt%GO-LDH-V_2 WEC	0.03	3	1
0.5 wt%GO-LDH-V_3 WEC	0.015	3	1
1.0 wt%GO-LDH-V_3 WEC	0.03	3	1

Different proportions of GO-LDH were added to a certain amount of anhydrous ethanol, and an ultrasonic instrument was used to obtain a uniformly dispersed GO-LDH mixed solution. A certain amount of waterborne epoxy resin was weighed, and the mixture was placed in an ultrasonic cleaning device for 15 minutes to obtain a uniformly mixed solution. Anhydrous ethanol was evaporated from the mixture using a rotary evaporator to obtain a uniformly dispersed suspension. A curing agent was later added and mixed evenly. The evenly mixed mixture was coated onto the base of the aluminum sheet using a scraping rod, while the curing process was carried out using an oven with the temperature set at 55 °C for 24 hours. The amount of the GO-LDH filler was 0.5 and 1.0 wt%, respectively. Further, a pure waterborne epoxy coating (WEC) without filler was prepared using a similar process. The different proportions of GO-LDH are presented in Table 5.3. The coating samples were successively named WEC, 0.5 wt%GO-LDH-V_1, 1.0 wt%GO-LDH-V_1, 0.5 wt%GO-LDH-V_2, 1.0 wt%GO-LDH-V_2, 0.5 wt%GO-LDH-V_3, and 1.0 wt%GO-LDH-V_3.

5.2.1.4 Morphology Characterization

An FE-SEM Zeiss G300 scanning electron microscope was used to observe the microstructures of the GO, LDH, and GO-LDH composites and composite coatings. Before observation, the samples were ultrasonically dispersed in anhydrous ethanol and placed on a clean silicon wafer to air-dry. Before the experiment, the surfaces of the samples were sprayed with gold to obtain the observation results. An electron excitation power of 10 keV was used to observe the GO, LDH, and GO-LDH composites, and an electron excitation power of 15 keV was used to observe the microscopic morphology of the coating surface. The microstructures of the GO, LDH, and GO-LDH composites were observed using a JEOL JEM-2100 transmission electron microscope. A KEYENCE VK-X2000 series optical microscope was used to observe the three-dimensional morphology of the coatings.

5.2.1.5 Chemical Structure Characterization

The presence of the LDH nanoparticles in GO was investigated by X-ray photoelectron spectroscopy (XPS) using a Thermo SCIENTIFIC ESCALAB 250Xi instrument.

Other peaks were corrected based on the position of the C1s signal (285 eV). The chemical structure of the fillers was analyzed using a Fourier-transform infrared spectrometer (FT-IR, WQF520).

5.2.1.6 Electrochemical Measurement

EIS was used to evaluate the corrosion characteristics of the coating using an electrochemical workstation PARSTAT-2273 produced by the Princeton Company. A standard three-electrode system was used for this experiment. An aqueous solution with a concentration of 3.5 wt% NaCl was used as the electrolyte solution. A PowerSuite electrochemical test system was used as the test software. In the scanning range of ±200 mV relative to the open circuit potential (E_{corr} versus SCE), the potential (from the cathode to the anode side) was subjected to the action potential polarization test, and the scanning rate of the experiment was 0.66 mV s^{-1}. The polarization curve was tested thrice to ensure the repeatability and reproducibility of the experimental data.

5.2.1.7 Neutral Salt Spray Test

The salt spray test can be used to evaluate the degree of protection provided by the coating on the substrate under accelerated conditions. This method has been widely used as a test method for evaluating batch products or screening coatings to determine the thickness uniformity and porosity of different protective coatings. The prepared coating sample was placed in a salt spray tank at an angle of 30 ± 5° from the vertical surface. The concentration of the NaCl solution added was 5 ± 0.5%, the pH value was 6.8–7.2, the temperature of the test chamber was 35 °C, the temperature of the pressure barrel was 47 °C, and the spray pressure was 1.2 MPa. A CK/YWX-90C salt spray corrosion test chamber with an inlet pressure of 2 MPa was used for 200 hours to conduct the corrosion test. Here, samples of different coatings were removed at certain experimental intervals, and the electrochemical workstation PARSTAT-2273 was used to test the EIS data of the coatings.

5.2.1.8 Analysis of the Corrosion Inhibitor Film

A focused ion beam (FIB) was used to cut the sample cross section to obtain a smooth and flat one. The instrument model was TESCAN LYRA 3, the acceleration voltage was 15 keV, and the tilt angle was 50°. Composition analysis of the coating and substrate bonding interface was performed using electron probe microanalysis (EPMA, instrument model JXA-8530F PLUS).

5.2.2 Characterization of GO-LDH

5.2.2.1 Structure and Morphology Analysis

Figure 5.2 shows the infrared spectra of GO, GO-LDH (without corrosion inhibitor), and GO-LDH (with corrosion inhibitor). Regarding the pure GO samples, absorption peaks were observed at 3450, 1724, 1634, and 1058 cm^{-1}. The existence of these four peaks was related to oxygen-containing functional groups. The adsorption band at 3450 cm^{-1} indicated the stretching vibration of the O—H hydroxyl group [33–35],

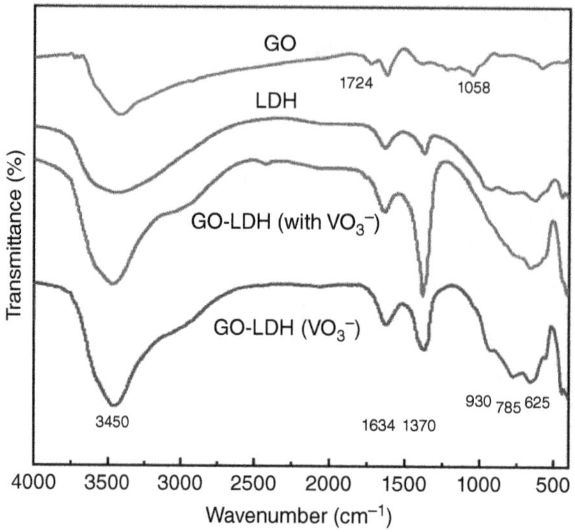

Figure 5.2 FTIR of GO, GO-LDH (without corrosion inhibitor VO_3^-), GO-LDH (with corrosion inhibitor VO_3^-).

while that at 1724 cm^{-1} belonged to the C=O stretching vibration. The characteristic peak at 1634 cm^{-1} was attributed to the C—H bond, while the stretching vibration peak of C—O was observed at 1058 cm^{-1} [36–39]. Regarding the GO-LDH sample, the adsorption band at 3450 cm^{-1} showed the contraction vibration of the LDH layer, interlayer water molecules, and -OH in GO [40, 41]. The typical adsorption peak at 1373 cm^{-1} is for the triple stretching pattern of NO_3^-. The peak at 625 cm^{-1} corresponds with the stretching and deformation pattern of Mg–O and Al–O [42–44]. The peak at 1724 cm^{-1} for GO disappeared, thereby indicating a decrease in the active oxide groups of GO and the successful synthesis of LDH. The FTIR spectra of the GO-LDH (containing the corrosion inhibitor VO_3^-) and GO-LDH samples were similar, although the corresponding NO_3^- anion absorption peak intensity of the former was reduced considerably. Further, the characteristic bond of metavanadate was at 930 and 785 cm^{-1}, which was related to the addition of metavanadate ions owing to $NaVO_3$ [45, 46]. VO_3^- exchanged NO_3^- anions and was successfully intercalated into the LDH. These results show that the GO-LDH was successfully prepared.

The surface morphologies and chemical states before and after synthesizing GO and LDH were analyzed using XPS. Compared to the pure GO spectrum, the presence of C, O, Mg, Al, and V in GO-LDH confirms the successful combination of LDH nanocontainers and GO layers, as shown in Figure 5.3a. The high-resolution elemental XPS peaks of C, O, Mg, Al, and V in GO-LDH are presented in Figure 5.3 b-f. C 1s spectrum can be divided into three peaks: 284.8, 286.7, and 289.5 eV, which are attributed to the C—C, C—O, and O—C=O bonds, respectively, in GO nanosheets [47–49]. The three peaks in the O 1s spectrum at 531.2, 532.6, and 533.1 eV can be assigned to O—C=O, C=O, and C—OH bonds [50–52], respectively. In the Mg 1s, Al 2s, and V 2s spectra, the peaks centered at 1304.4, 74.2, and 517 eV, which correspond to the Mg—O, Al—O, and V—O bonds, respectively. This proves the successful combination of the LDH and GO layer [53–55].

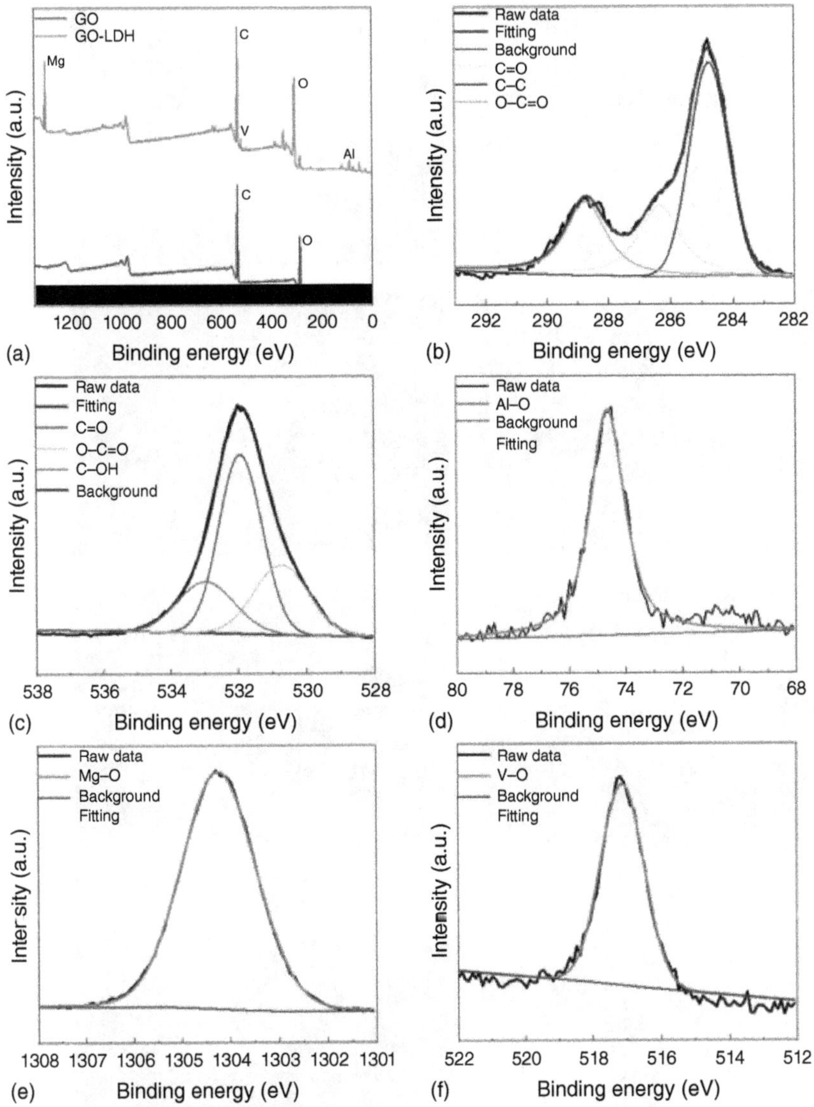

Figure 5.3 XPS spectra of the surface morphology and chemical states before and after the synthesis of GO and LDH: (a) the overall spectra of GO and GO-LDH; elemental spectra of (b) C, (c) O, (d) Al, (e) Mg, (f) V.

The microstructures of GO, LDH, and GO-LDH were observed using scanning electron microscopy (SEM) and transmission electron microscopy (TEM), as shown in Figures 5.4 and 5.5. The pure LDH formed an irregular hexagonal shape, while the gaps within the LDH were small and stacked together (Figure 5.4a,b). This shows that the strong interaction force between the LDHs is prone to agglomeration. As shown in Figure 5.4c,d, the GO has a layered structure with folded edges. In Figure 5.4e,f, GO-LDH-V_1 exhibits a layered wave shape with a rough surface owing

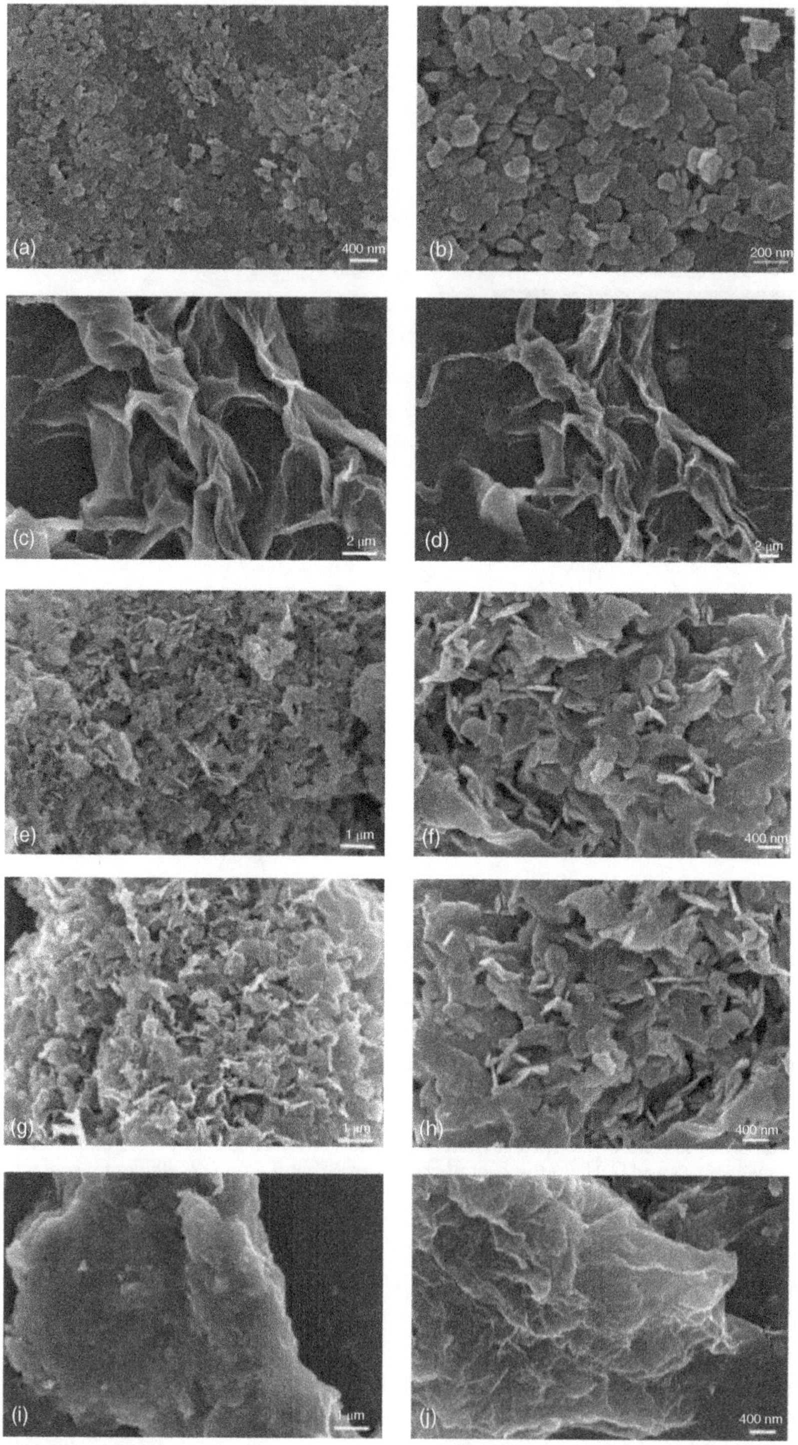

Figure 5.4 SEM images of (a,b) LDH; (c,d) GO; (e,f) GO-LDH-V_1; (g,h) GO-LDH-V_2; (i,j) 6 : 1 GO-LDH-V_3.

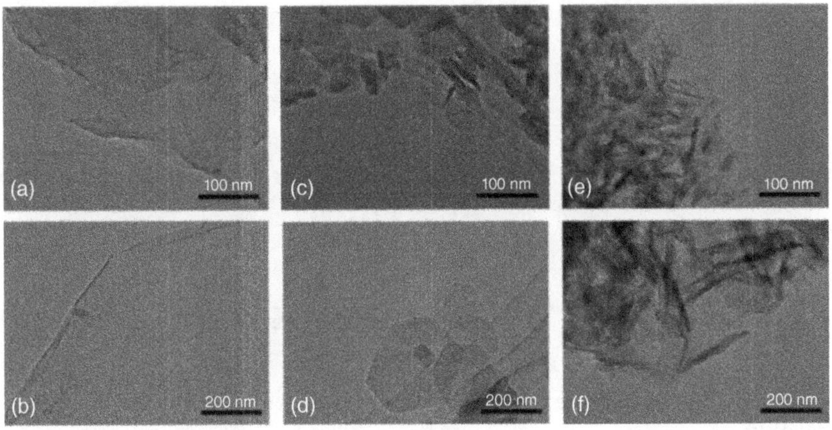

Figure 5.5 TEM images of (a,b) GO; (c,d) LDH; (e,f) GO-LDH-V_2.

to the distribution of LDH on GO. Further, the distance between the layers increases, thereby indicating the formation of a relatively loose structure. However, LDH accumulation remained considerable. As shown in Figure 5.4g,h, the layer spacing of the GO-LDH-V_2 composite filler was further increased, thereby improving the accumulation of LDH to a certain extent. Compared to the GO-LDH-V_1 sample, its structure was more porous, and its dispersion performance in the solution was better. As illustrated in Figure 5.4i,j, the GO-LDH-V_3 composite filler is stacked together without forming a relatively loose structure owing to the poor dispersion performance caused by adding a large amount of GO. These fillers were also prone to agglomeration and precipitation [56–58].

As shown in Figure 5.5, the TEM image of the composite reveals that GO has a ductile layered structure with a large specific surface area, which folds on its surface, while LDH has a hexagonal structure. In the GO-LDH structure, the crystal structure of LDH was evenly distributed on GO, thereby increasing the layer spacing of the GO owing to the electrostatic attraction between GO and LDH.

5.2.2.2 Analysis of Corrosion Inhibitor Release

The determined amount of $NaVO_3$ was dissolved in 0.05 M NaCl to obtain the $NaVO_3$ solution with the determined concentration, and its calibration curve was measured, as shown in Figure 5.6a. 50 mg of GO-LDH-V_2 was placed in a dialysis bag, and 50 ml of 0.05 M NaCl solution was added. The solution samples were then periodically removed and placed in a colorimetric dish, and the concentration of vanadate in the extracted solution was determined using UV–Vis spectrophotometry (see Figure 5.6b). It can be seen that the VO_3^- release behavior of the LDH within 48 hours, indicating that the ion exchange between VO_3^- and Cl^- rapidly increased the concentration of VO_3^- ions to 77 ppm after five hours of immersion. When the immersion time was extended to 10 hours, the VO_3^- concentration further increased to 117.86 ppm. After 30 hours, the release of VO_3^- gradually increased to 172.98 ppm. Subsequently, the release of VO_3^- gradually reached equilibrium at 173.35 ppm without significant changes, indicating that the inhibitor

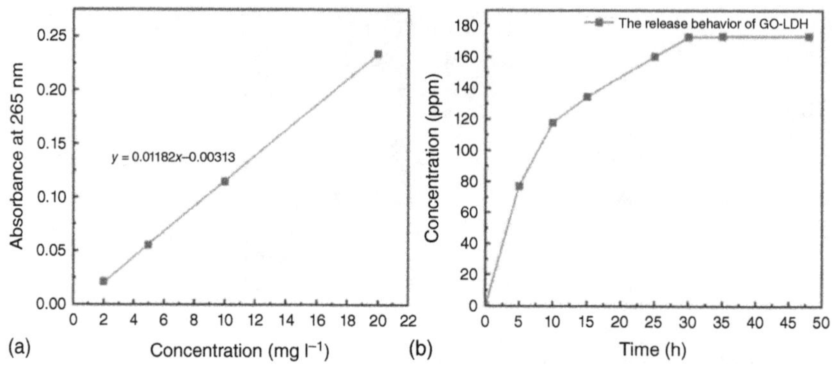

Figure 5.6 (a) Standard release curve of metavanadate; (b) release curve of metavanadate in GO-LDH-V_2.

molecules were released from the nanocontainer when the corrosive medium invaded. Preliminary calculations indicated that the corrosion inhibitor loading rate was 17.3%.

5.2.2.3 Evaluation of the Corrosion Inhibition of GO-LDH

EIS was used to evaluate the anti-corrosion effect of the synthesized GO-LDH on the cleaned AA2024-T3 sample, which was immersed in a 0.05 M NaCl solution with and without 50 mg GO-LDH. Then, the EIS data were recorded and compared. Before the experiment, AA2024-T3 was placed in a 0.1 M NaOH solution for 120 s. Then, it was cleaned with deionized water and placed in a 0.1 M HNO_3 solution for 480 s. After rinsing with deionized water, the AA2024-T3 aluminum plate was sonicated in anhydrous ethanol for five minutes and dried using a hair dryer. Figure 5.7 shows that during the initial immersion time, the EIS impedance values of the samples in the solution containing GO-LDH were higher than those in the pure NaCl solution. As the immersion time increased, the impedance values of the sample in the NaCl solution decreased. On the 14th day, the low-frequency impedance modulus ($|Z|_{0.01\ \mathrm{Hz}}$) of the sample in the NaCl solution was $1.6 \times 10^4\ \Omega\,\mathrm{cm}^2$, whereas that of the sample in a solution containing GO-LDH was $4.7 \times 10^4\ \Omega\,\mathrm{cm}^2$, which was about three times that of the NaCl solution. This indicates that GO-LDH had a good corrosion inhibition effect on the aluminum alloy.

To further determine whether VO_3^- had a protective effect on the surface of AA2024-T3, polarization curve tests were conducted on the experimental samples immersed in NaCl solution for 14 days, as shown in Figure 5.8. The results indicated that, compared with the samples immersed in the pure NaCl solution, the corrosion potential of the samples immersed in the GO-LDH-V_2 powder shifted in the negative direction, and the current densities of both the anode and cathode branches decreased significantly. For the samples immersed in the NaCl solution with the addition of GO-LDH-V_2, the passivation zone of the anode branch was significantly enhanced. The addition of vanadate to the corrosive medium played a protective role in AA2024-T3, resulting in good corrosion resistance. In addition, the change in the shape of the anode branch was greater than that of the cathode

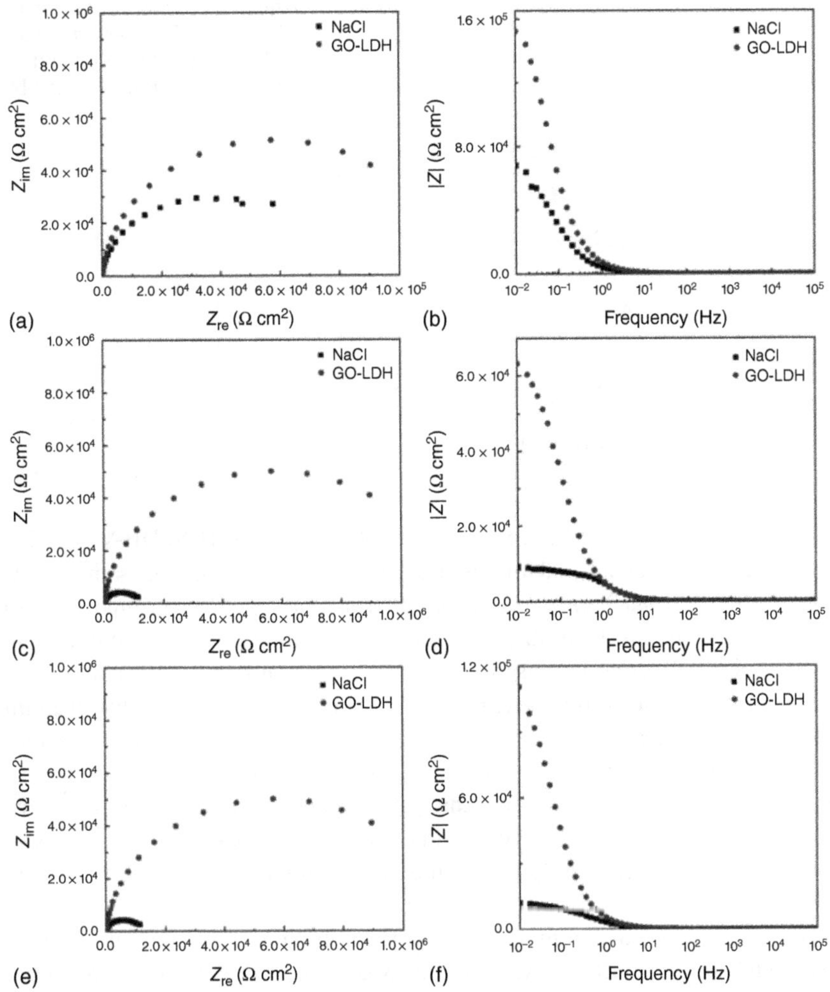

Figure 5.7 Typical EIS curves of AA2024-T3 in NaCl solution and in NaCl solution with GO-LDH-V_2 added: Nyquist diagram (a) 1; (c) 7; (e) 14 days; Bode diagram (b) 1; (d) 7; (f) 14 days.

branch, indicating that the main inhibition mechanism was anodic. This indicates that GO-LDH-V_2 reduced the anode reaction rate by forming a protective film at the anode site.

5.2.3 Barrier and Self-healing Performance of Waterborne Epoxy Coatings Containing GO-LDH-V

Figure 5.9 shows the EIS data of the WEC and WEC coatings with different GO-LDH contents after 0, 10, 50, and 100 hours of a salt spray test for 200 hours. The $|Z|_{0.01\,Hz}$ values were used to determine and compare the corrosion resistance of the coatings [59–62]. Nyquist plots showed that at the beginning of the salt

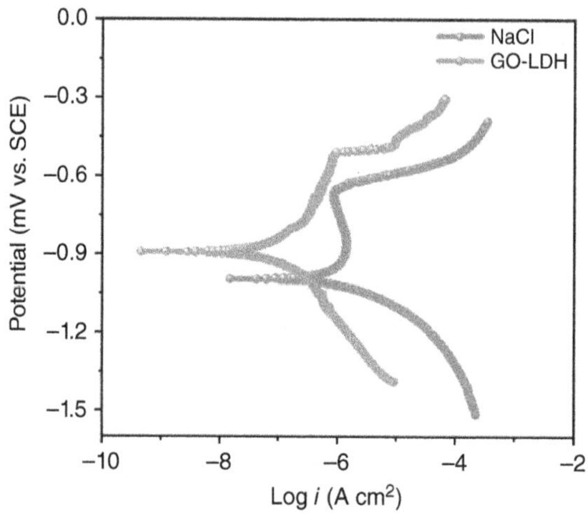

Figure 5.8 Polarization curves of AA2024-T3 immersed in pure 3.5% NaCl solution with and without GO-LDH powder for 14 days.

spray test, the initial capacitance arc radius of the 1.0 wt% GO-LDH-V_1 coating was the largest, whereas that of the 1.0 wt% GO-LDH-V_2 coating sample was the smallest. The capacitance arcs of the other samples were superimposed with little difference. In the Bode diagram, the $|Z|_{0.01\,Hz}$ values of each sample were not significantly different because the corrosion medium had not yet entered the coating/metal interface through scratch [63]. After 10 hours, the Nyquist diagram showed that the capacitance arc radius of pure WEC was the smallest, whereas that of the WEC coating with 0.5 wt% GO-LDH-V_2 powder exhibited the largest capacitance arc, where a second-time constant appeared at this moment. Generally, for a pure coating system, the second-time constant indicates that the corrosive medium penetrated the coating/metal interface to initiate the corrosion reaction [64–66]. For coatings containing corrosion inhibitors, the second-time constant represents the protective film formed by the inhibitors on the coating/metal substrate [67–69]. The $|Z|_{0.01\,Hz}$ value of the WEC rapidly dropped from the initial 6.8×10^6 to $5.2\times10^5\ \Omega\,cm^2$, whereas the $|Z|_{0.01\,Hz}$ value of the 0.5 wt% GO-LDH-V_2 coating sample changed from 6.6×10^6 to $6.5\times10^6\ \Omega\,cm^2$ with little variation, which was related to the release of corrosion inhibitors in the coating to form a protective film on the metal surface [70, 71]. As the salt spray duration increased to 50 hours, the Nyquist diagram shows that the capacitance arc radius of the pure WEC coating further decreased, and the $|Z|_{0.01\,Hz}$ value was further reduced to $4.9\times10^4\ \Omega\,cm^2$, which was related to the passive film on the aluminum surface. The capacitance arc size of the coating sample with 0.5 wt% GO-LDH-V_2 and the coating sample with 1 wt% GO-LDH-V_2 were similar, whose $|Z|_{0.01\,Hz}$ values also exhibited the same characteristics at near $6.0\times10^6\ \Omega\,cm^2$. Figure 5.9 shows that, with the further increase of salt spray time to 100 hours, the capacitance arc and $|Z|_{0.01\,Hz}$ of the coating with 0.5 wt% GO-LDH-V_2 were much higher than those of the other coating samples. Consequently, the coating sample with the addition of 0.5 wt% GO-LDH-V_2 powder had the best protective performance among the coatings with GO-LDH-V powders.

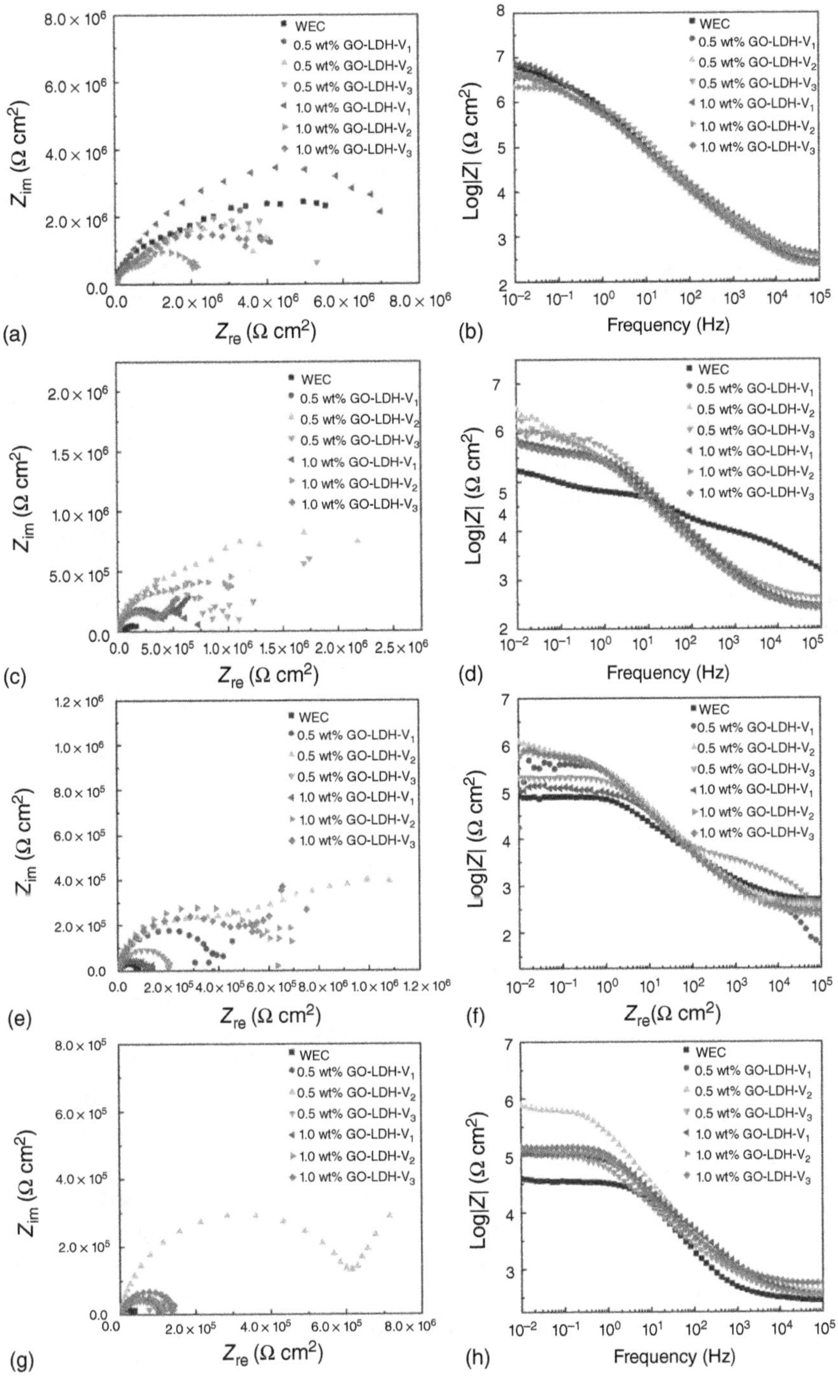

Figure 5.9 EIS results of different coatings after salt spray test: Nyquist diagram (a) 0; (c) 10; (e) 50; (g) 100 hours; Bode diagram (b) 0; (d) 10; (f) 50; (h) 100 hours.

This result can be explained as follows. When the amount of GO added in the process of GO-LDH powder preparation was 100 mg (V_1), it did not act as a good barrier to the coating due to its low GO content. However, when the GO content reached 300 mg (V_3), the barrier performance decreased due to the poor dispersion of GO in the powder. The GO-LDH with 200 mg of GO (V_2) ensured good dispersion in the powder. Moreover, as the overall GO-LDH content increased from 0.5 to 1.0 wt%, the corrosion resistance of the coating showed a downward trend, which may have been caused by the excessive addition of GO-LDH filler that led to poor dispersion performance and severe agglomeration within the coating. This resulted in the formation of large pores inside the coating, increasing the opportunity for the solution to reach the metal substrate.

As shown in Figure 5.10, the color of the coating scratch became darker with the increase in time of salt spray, indicating that the degree of corrosion and corrosion products on the metal substrate gradually increased. When the salt spray time increased to 100 hours, the WEC coating exhibited its deepest color and numerous corrosion products at the scratch, whereas the coating with 0.5 wt% GO-LDH-V_2 had the best protection effect and the least corrosion products at the scratch. The analysis of the corrosion products on the coating substrate shown in Figure 5.11 confirms this conclusion. The adhesion around the scratch of the WEC coating was the lowest, making it easy to peel from the substrate, whereas the coating with 0.5 wt% GO-LDH-V_2 retained a high adhesion strength. Hence, it could not be easily removed. These results are consistent with the EIS experimental results for the coating.

Figure 5.12 shows the Nyquist and Bode curves of the WEC, GO/WEC, LDH/WEC, 0.5 wt% GO-LDH-V_2, and 1.0 wt% GO-LDH-V_2 samples. Figure 5.12a,b show the EIS changes in the WEC samples after immersion in a 3.5% NaCl solution for different durations. As shown in Figure 5.12a, for the pure WEC samples, the impedance arc radius rapidly decreased with increasing immersion time. The second-time constant appeared in the Nyquist diagram from day 14 onwards, indicating that the corrosive medium penetrated the coating substrate and that further corrosion developed. Figure 5.12b shows that after 7 days of immersion, the $|Z|_{0.01\,\text{Hz}}$ value of the WEC sample was $5.2 \times 10^8\ \Omega\,\text{cm}^2$, and the $|Z|_{0.01\,\text{Hz}}$ value decreased rapidly. On the 60th day, the $|Z|_{0.01\,\text{Hz}}$ value decreased to $5.4 \times 10^5\ \Omega\,\text{cm}^2$, at which time it could be considered that the coating had lost its protective effect. In the Nyquist and Bode plots of the 0.5 wt% GO/WEC (Figure 5.12c,d), when GO was added to the pure waterborne epoxy coating, the impedance arc radius in the Nyquist plot decreased slowly for 60 days (Figure 5.12c). The second-time constant did not appear until day 28, indicating that the GO layer can form a labyrinth, effectively delaying the penetration of corrosive media. When the immersion time was extended to the 60th day, the $|Z|_{0.01\,\text{Hz}}$ value decreased to $2.8 \times 10^7\ \Omega\,\text{cm}^2$ (Figure 5.12d). The barrier effect of the waterborne epoxy coating was improved by the addition of GO to the coating substrate. In the latter coating fitting parameters, the coating resistance value, R_c, decreased from 3.4×10^8 to $1.3 \times 10^6\ \Omega\,\text{cm}^2$, which was much higher than the coating resistance, R_c ($4.14 \times 10^4\ \Omega\,\text{cm}^2$) of the pure WEC, indicating that the addition of GO improved the barrier effect of waterborne epoxy coatings to some extent,

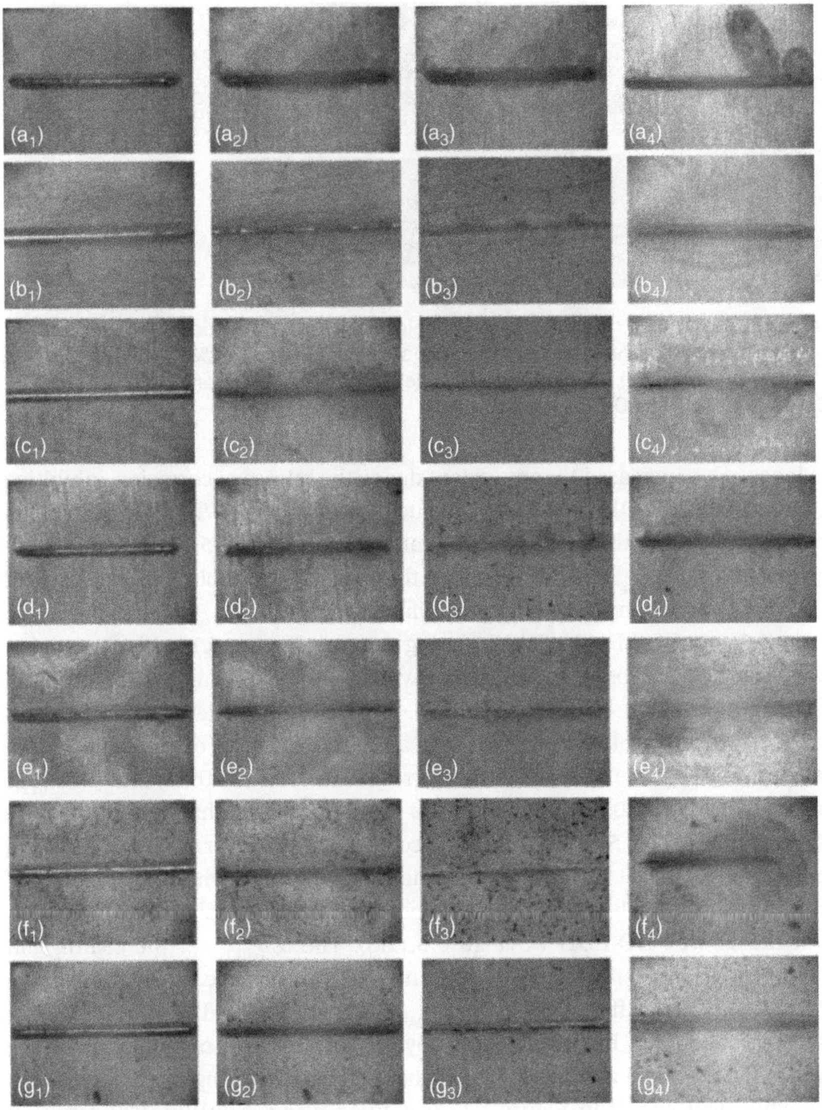

Figure 5.10 Scratched coating surface morphology after salt spray test for 0, 10, 50, 100: (a_1–a_4) pure waterborne epoxy coating; (b_1–b_4) 0.5 wt%GO-LDH-V_1; (c_1–c_4) 0.5 wt%GO-LDH-V_2; (d_1–d_4) 0.5 wt%GO-LDH-V_3; (e_1–e_4) 1.0 wt%GO-LDH-V_1; (f_1–f_4) 1.0 wt%GO-LDH-V_2; (g_1–g_4) 1.0 wt%GO-LDH-V_3.

making it difficult for corrosive media to enter. This resulted in excellent barrier performance and good corrosion resistance [72, 73]. The EIS data for the LDH/WEC coating samples are shown in Figure 5.12e,f. Owing to the absence of GO, the coating sample with LDH appeared as a third constant after 14 days, indicating that the LDH could not prevent the penetration of corrosive substances. However, the addition of LDH slowed the decreasing $|Z|_{0.01\ \mathrm{Hz}}$. When the immersion time reached 40 days, the

Figure 5.11 Metal substrate corrosion morphology at scratched coating area after salt spray test for 100 hours: (a) pure waterborne epoxy coating; (b) 0.5 wt%GO-LDH-V_1; (c) 0.5 wt%GO-LDH-V_2; (d) 0.5 wt%GO-LDH-V_3; (e) 1.0 wt%GO-LDH-V_1; (f) 1.0 wt% GO-LDH-V_2; (g) 1.0 wt%GO-LDH-V_3.

impedance value increased to some extent due to the release of corrosion-inhibiting VO_3^- ions from the LDH to form a vanadate protective film. When the immersion time reached 60 days, the $|Z|_{0.01\,Hz}$ of the sample dropped to $8.5\times10^6\,\Omega\,cm^2$. Compared with the $|Z|_{0.01\,Hz}$ $5.4\times10^5\,\Omega\,cm^2$ of the pure WEC sample, the coating with LDH showed certain anti-corrosion performance.

In the 0.5 wt% GO-LDH-V_2/WEC coating samples (Figure 5.12g,h), the coating showed the best corrosion resistance. Owing to the physical barrier effect of GO and the formation of a vanadate film, a third-time constant appeared at 28 days. Afterward, the release of VO_3^- ions and the formation of vanadate on the metal surface significantly reduced the corrosion rate. After 60 days, the $|Z|_{0.01\,Hz}$ remained above $3.2\times10^7\,\Omega\,cm^2$, which was 100 times higher than that of the pure WEC coating. For the 0.5 wt% GO-LDH-V_2 coating, the GO layer and LDH provided a dual protective effect on corrosive substances. However, when the amount of GO-LDH-V_2 reached 1.0 wt%, its corrosion resistance was slightly lower than that of the 0.5 wt% GO-LDH-V_2 (Figure 5.12i,j). The $|Z|_{0.01\,Hz}$ of the coating did not show the same obvious increasing trend that was observed with the 0.5 wt% GO-LDH-V_2 coating. After 60 days, the $|Z|_{0.01\,Hz}$ value decreased to $1.7\times10^6\,\Omega\,cm^2$. This phenomenon may have been caused by the large number of nanofillers in the coating, which resulted in more diffusion channels at the coating-metal interface.

The equivalent circuit in Figure 5.13 was fitted using ZsimWin 3.30 software. The values corresponding to each fitting element in the equivalent circuit are listed in Table 5.4, where R_s, R_c, R_{inh}, R_{ct}, CPE_{ct}, CPE_{inh}, and CPE_{dl} represent solution resistance, coating resistance, corrosion inhibitor protective film, passive film resistance, charge transfer resistance, coating constant phase element, corrosion inhibitor protective film, passive film constant phase element, and double-layer constant phase element, respectively. R_c was used to measure the physical barrier performance of the coatings. Through fitting, the R_c values of all samples showed a decreasing trend with increasing immersion time. Compared with the pure WEC coating samples, the R_c values of the 0.5 wt% GO-LDH coating samples significantly increased, indicating the best physical barrier performance. Compared with the other samples, the R_{inh} value of the 0.5 wt% GO-LDH coating sample

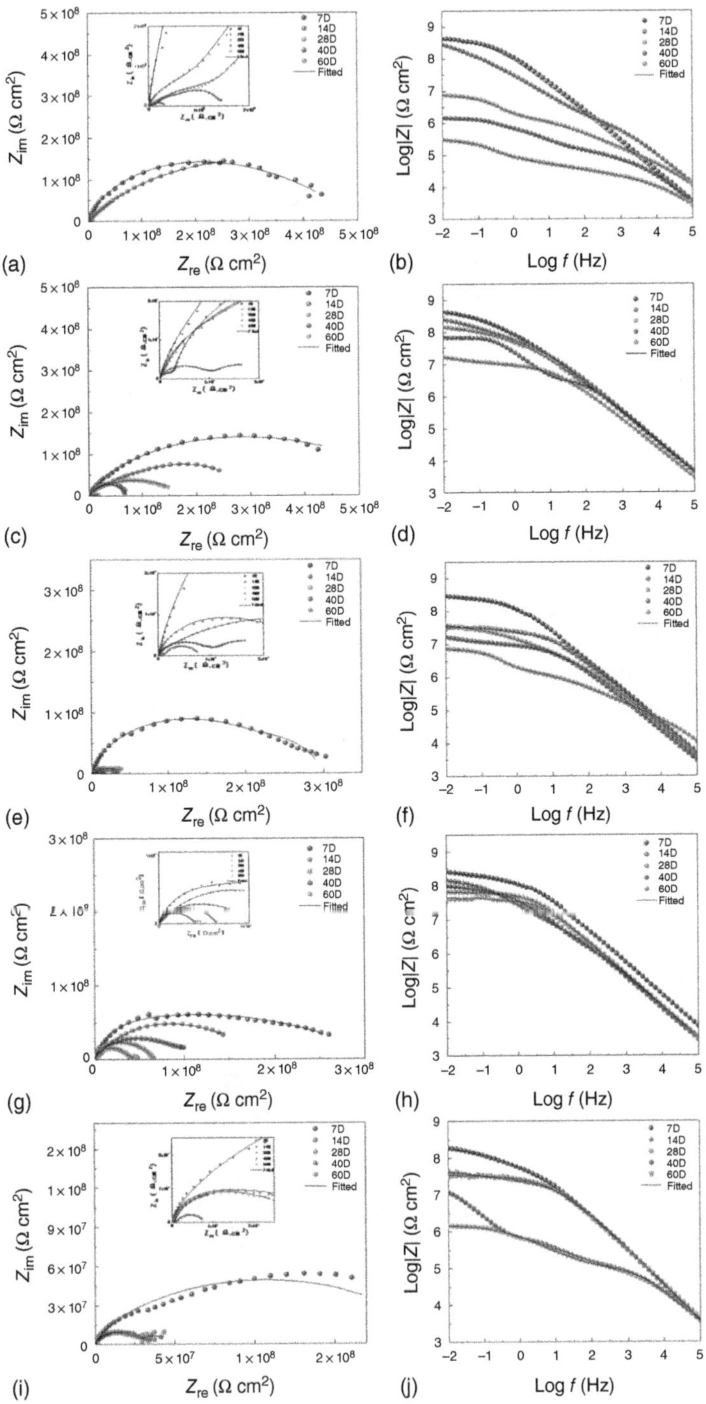

Figure 5.12 Nyquist and Bode curves of coatings under different immersion times: (a,b) WEC; (c,d) GO/WEC; (e,f) LDH/WEC; (g,h) 0.5 wt%GO-LDH-V_2; (i,j) 1.0 wt%GO-LDH-V_2.

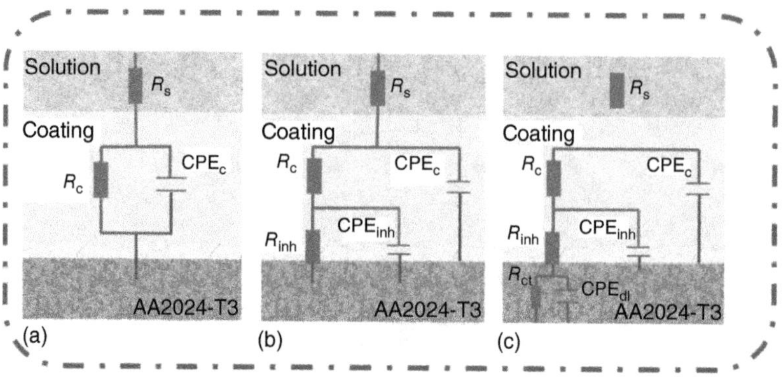

Figure 5.13 Equivalent circuit fitting.

was more stable and higher. This is related to the excellent protective effect of vanadate inhibitors, which can repair defects in the oxide layer on the surfaces of Al alloys. The R_{ct} value was used to characterize the corrosion resistance of the metal matrix [74, 75]. Over time, the R_{ct} values of all samples showed a downward trend. The R_{ct} value of the WEC coating reduced from 2.3×10^8 to $2.1 \times 10^5\ \Omega\,cm^2$, indicating an increase in the corrosion rate. Compared with the other coating samples, the 0.5 wt% GO-LDH coating sample exhibited a higher R_{ct} during the full immersion process, indicating that the resistance to electron transfer was increased due to the enhanced barrier properties of the GO layer and the formation of a corrosion-inhibiting adsorption layer in the filler. In contrast to the changing trends in R_c and R_{ct}, CPE_{inh} and CPE_{ct} showed upward trends over time.

To investigate the self-healing effects of vanadate ions on the coating, the coating was scratched with a sharp blade (120 μm × 5 mm) and immersed in a 3.5 wt% NaCl solution for EIS testing. Figure 5.14 shows the EIS images of WEC, 0.5 wt% GO-LDH-V_2/WEC, and 1.0 wt% GO-LDH-V_2/WEC under different immersion times. In Figure 5.14, the Bode plots show two time constants that indicate the corrosion process occurring on the aluminum alloy substrate in the scratch area. The first time constant represents the penetration of the electrolyte, and the second represents the penetration of the solution into the coating–substrate interface, where a mixed film composed of corrosion inhibitors and corrosion products began to develop. For the pure epoxy coating, the $|Z|_{0.01\,Hz}$ decreased rapidly over time, reaching $7.6 \times 10^4\ \Omega\,cm^2$ after 120 hours. For the 0.5 wt% GO-LDH-V_2 coating, after 36 hours of immersion, the $|Z|_{0.01\,Hz}$ value showed an obvious upward trend and slowly declined afterward. This is related to the fact that vanadate can inhibit the local corrosion of AA2024 by forming insoluble complexes at the active site, such as S-phase (Al_2CuMg) intermetallic compounds. The 1.0 wt% GO-LDH-V_2 coating also showed an increasing trend after 36 hours of immersion, but the increase was not significant. After 120 hours, $|Z|_{0.01\,Hz}$ of the WEC, 0.5 wt% GO-LDH-V_2/WEC, and 1.0 wt% GO-LDH-V_2/WEC decreased to 7.6×10^4, 5.4×10^5, and $1.8 \times 10^5\ \Omega\,cm^2$, respectively. In summary, the coating with the addition of 1.0 wt% GO-LDH-V_2 did not show better self-healing effects than the 0.5 wt% GO-LDH-V_2 coating, which is consistent with our previous long-term corrosion resistance test results.

Table 5.4 EIS fitting data of long-term immersion experiments.

Samples	Time (day)	R_c ($\Omega \cdot cm^2$)	CPE_c ($S\ s^n\ cm^{-2}$)	R_{inh} ($\Omega\ cm^2$)	CPE_{inh} ($S\ s^n\ cm^{-2}$)	R_{ct} ($\Omega\ cm^2$)	CPE_{ct} ($S\ s^n\ cm^{-2}$)
WEC	7	3.57×10^8	7.31×10^{-10}	—	—	—	—
	14	2.26×10^8	8.63×10^{-10}	—	—	2.25×10^8	2.75×10^{-9}
	28	9.39×10^5	9.85×10^{-10}	—	—	5.51×10^6	9.93×10^{-8}
	40	1.33×10^5	1.01×10^{-9}	—	—	1.07×10^6	6.96×10^{-8}
	60	4.14×10^4	3.25×10^{-9}	—	—	2.12×10^5	2.12×10^{-6}
GO/WEC	7	3.43×10^8	3.12×10^{-10}	—	—	—	—
	14	1.78×10^8	4.41×10^{-10}	—	—	—	—
	28	1.43×10^7	4.45×10^{-10}	—	—	5.28×10^7	1.84×10^{-9}
	40	3.23×10^5	5.01×10^{-10}	—	—	2.06×10^7	5.60×10^{-9}
	60	1.34×10^5	6.92×10^{-10}	—	—	4.70×10^6	2.29×10^{-8}
LDH/WEC	7	2.44×10^8	4.15×10^{-10}	—	—	—	—
	14	4.88×10^6	4.36×10^{-10}	1.65×10^7	7.62×10^{-10}	6.21×10^7	5.27×10^{-8}
	28	2.07×10^6	5.34×10^{-10}	8.93×10^6	1.82×10^{-9}	2.22×10^7	2.53×10^{-8}
	40	1.34×10^6	6.69×10^{-10}	7.38×10^6	1.20×10^{-9}	6.70×10^6	1.17×10^{-7}
	60	1.37×10^5	1.97×10^{-9}	9.92×10^5	2.31×10^{-9}	3.51×10^6	2.23×10^{-7}
0.5 wt%GO-LDH-V_2/WEC	7	1.98×10^8	2.85×10^{-10}	—	—	—	—
	14	1.14×10^8	5.28×10^{-10}	—	—	—	—
	28	3.11×10^7	5.57×10^{-10}	5.51×10^7	3.27×10^{-10}	2.46×10^7	1.36×10^{-9}
	40	7.67×10^6	5.52×10^{-10}	7.79×10^7	1.26×10^{-9}	1.06×10^7	3.25×10^{-9}
	60	3.62×10^6	7.57×10^{-10}	3.55×10^7	1.36×10^{-9}	6.52×10^6	5.31×10^{-9}
1.0 wt%GO-LDH-V_2/WEC	7	1.39×10^8	4.42×10^{-10}	—	—	—	—
	14	2.41×10^6	4.53×10^{-10}	1.75×10^7	3.59×10^{-10}	1.49×10^7	1.24×10^{-8}
	28	1.60×10^6	4.95×10^{-10}	1.62×10^7	3.44×10^{-9}	1.36×10^7	1.26×10^{-8}
	40	1.10×10^5	5.66×10^{-10}	4.80×10^5	1.33×10^{-8}	8.65×10^6	1.66×10^{-7}
	60	8.03×10^4	6.73×10^{-10}	2.00×10^5	2.93×10^{-8}	1.00×10^6	6.34×10^{-7}

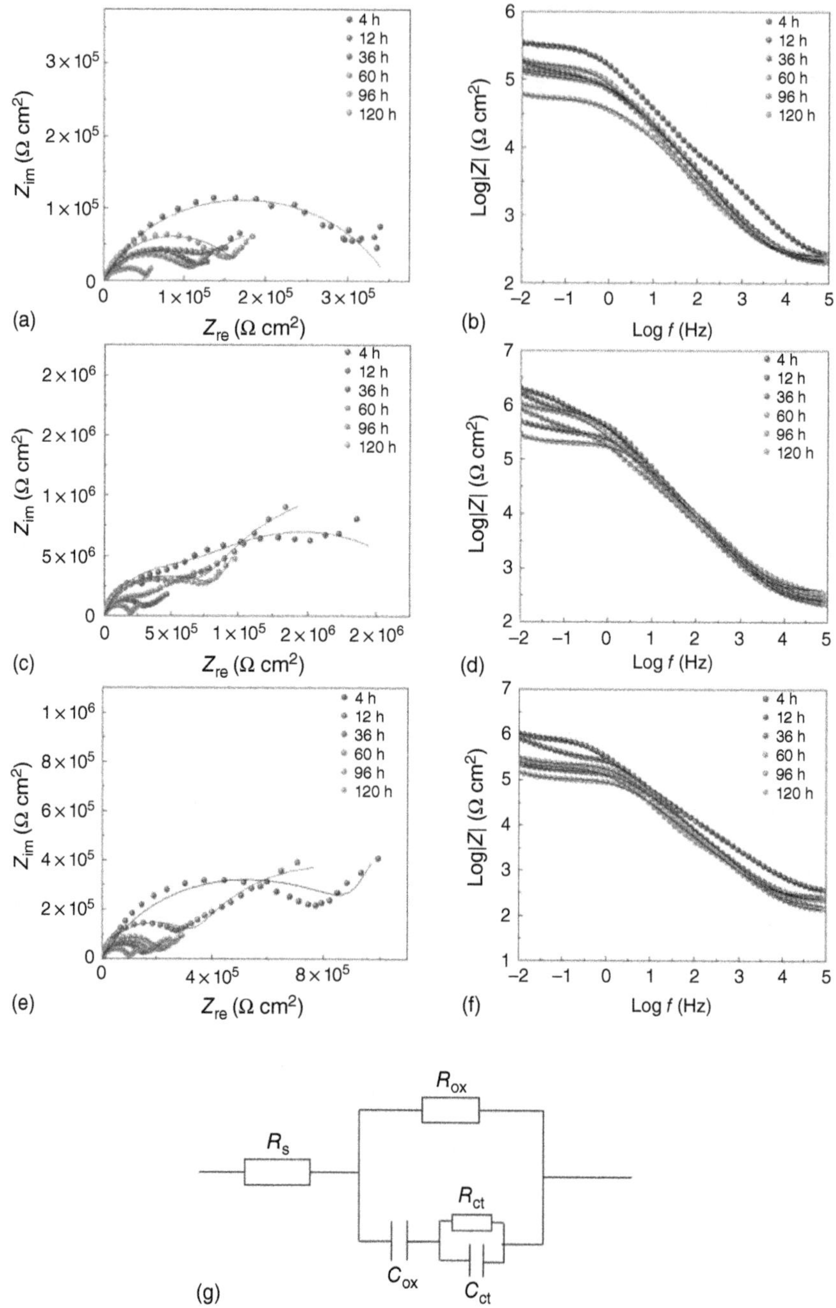

Figure 5.14 Nyquist and Bode diagrams of coatings during immersion: (a,b) WEC coating; (c,d) 0.5 wt%GO-LDH-V_2/WEC coating; (e,f) 1.0 wt%GO-LDH-V_2/WEC coating; (g) coating fitting circuit.

Table 5.5 EIS fitting data of scratch immersion experiments.

Samples	Time (h)	R_{ox}(Ω cm²)	C_{ox}(S sⁿ cm⁻²)	R_{ct}(Ω cm²)	C_{ct}(S sⁿ cm⁻²)
WEC	4	4.27×10^4	1.93×10^{-6}	2.58×10^5	6.15×10^{-7}
	12	2.63×10^4	4.50×10^{-6}	1.15×10^5	1.88×10^{-6}
	36	3.14×10^4	7.22×10^{-6}	8.82×10^5	4.47×10^{-6}
	60	4.15×10^4	6.91×10^{-6}	6.87×10^5	5.20×10^{-6}
	96	2.38×10^4	9.92×10^{-6}	3.57×10^5	7.42×10^{-6}
	120	1.95×10^4	2.26×10^{-5}	3.34×10^5	1.06×10^{-5}
0.5 wt%GO-LDH-V_2	4	7.26×10^5	1.70×10^{-7}	1.32×10^6	2.81×10^{-6}
	12	6.30×10^5	1.35×10^{-7}	1.32×10^6	6.07×10^{-6}
	36	2.84×10^5	1.66×10^{-7}	5.92×10^5	8.21×10^{-7}
	60	5.17×10^5	2.71×10^{-7}	5.58×10^5	9.85×10^{-6}
	96	2.89×10^5	4.09×10^{-7}	2.54×10^5	1.76×10^{-6}
	120	8.81×10^5	6.99×10^{-7}	1.96×10^5	2.39×10^{-6}
1.0 wt%GO-LDH-V_2	4	2.31×10^5	8.81×10^{-8}	7.96×10^5	3.02×10^{-7}
	12	1.04×10^5	1.43×10^{-7}	5.87×10^5	3.60×10^{-7}
	36	8.60×10^4	5.07×10^{-7}	1.40×10^5	2.70×10^{-6}
	60	9.43×10^4	8.81×10^{-7}	1.38×10^5	5.10×10^{-6}
	96	6.42×10^4	1.62×10^{-6}	1.35×10^5	6.41×10^{-6}
	120	4.40×10^4	3.25×10^{-6}	9.57×10^4	5.25×10^{-6}

Throughout the entire immersion stage, both Bode and Nyquist diagrams can be constructed using the circuit shown in Figure 5.14g to establish an equivalent circuit. The numerical parameters of the fitting elements in the equivalent circuits are listed in Table 5.5. For pure waterborne epoxy coatings, R_{ox} and C_{ox} represent the resistance and double-layer capacitance of corrosion products and oxide layers, whereas for the 0.5 wt% GO-LDH-V_2 and 1.0 wt% GO-LDH-V_2 coating samples, they represent the resistance and double-layer capacitance of the layer of the mixed corrosion products and inhibitors. For the pure waterborne epoxy resins, the R_{ox} and R_{ct} values decreased over time due to the corrosive medium reaching the metal substrate via scratching and coating. After adding 0.5 wt% GO-LDH-V_2 nanosheets to the waterborne epoxy resin, R_{ox} increased significantly throughout the immersion process, mainly due to the formation of the corrosion inhibitor film, which enhanced the diffusion barrier effect of the mixed layer on the corrosive medium. The R_{ct} values showed the same trend, indicating that the physical barrier effect of the vanadate film limited the electrochemical reaction on the metal surface. Vanadate ions have a good corrosion inhibition effect, forming a protective film on the metal substrate that can effectively control the peeling phenomenon caused by the reduced adhesion of the coating. When the immersion time was further extended to 96 hours, the R_{ox} value of the sample suddenly decreased, which may be attributed to the destruction of the protective film. When 1.0 wt% GO-LDH-V_2 nanosheets were added to the waterborne epoxy resin, the R_c and R_{ct} values increased slightly, and subsequently decreased rapidly after 36 hours, indicating that the formed protective film was unstable and that the GO layer did not provide good protection. In addition, the R_{ox} and R_{ct} values of the 0.5 wt% GO-LDH-V_2 coating sample were higher than those of the other coating samples, demonstrating the good barrier and self-healing properties of the filler.

5.2.4 Protection Mechanism of GO-LDH-V Waterborne Epoxy Coating

5.2.4.1 Electron Probe Microanalysis

Research on the protection mechanisms of coatings can be divided into the ion exchange behaviors of vanadate and chloride ions in the coating. EPMA was used to provide results for the 0.5 wt% GO-LDH-V_2 coating, which was immersed for different periods, as shown in Figure 5.15. After 1 day, chloride ions were found in the coating, indicating that the corrosive medium had infiltrated it. At this point, the color of vanadium in the coating was darker, indicating its increased content. After immersion for 5 days, the Cl^- content in the coating significantly increased, whereas the vanadium content significantly decreased, indicating that a portion of the corrosion inhibitor was released from the coating into the solution, resulting in a decrease in inhibitor contents. This result further demonstrates that the composite coating can store chloride ions and release metavanadate into the LDH during corrosion.

5.2.4.2 Scratch Morphology and EDS Spectrum Analysis

Figure 5.16 shows the corrosion morphology of the scratched coating after immersion in 3.5 wt% NaCl solution for 120 hours. The film-forming behavior of vanadate on metal surfaces was also studied. It can be seen that the pure waterborne epoxy

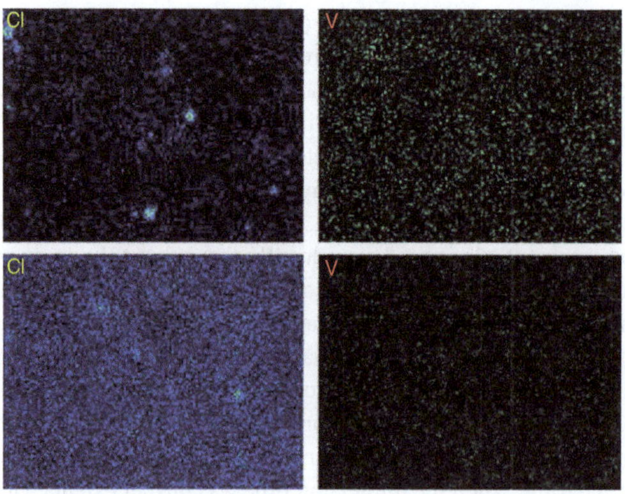

Figure 5.15 Electron probe microanalysis of Cl and V elements in the coating.

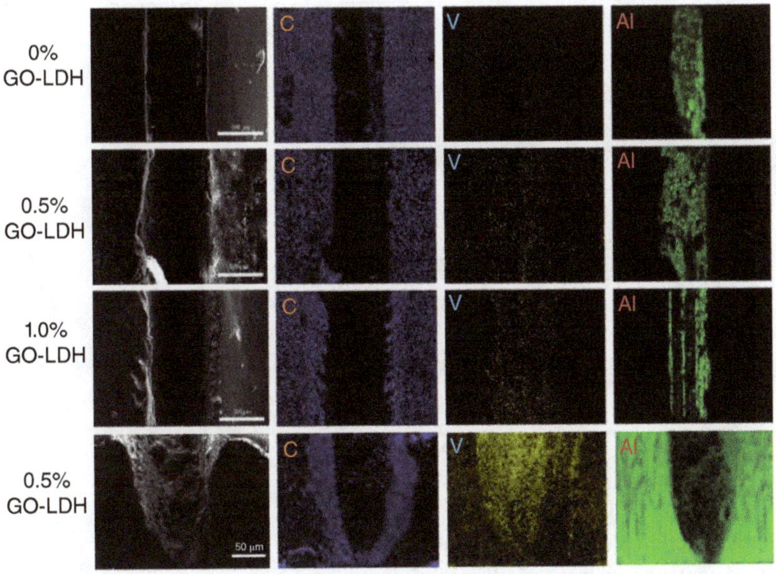

Figure 5.16 Morphology and energy spectrum analysis at the coating scratch.

coating had no V distribution in the coating defects. However, it was observed in the scratched 0.5 wt% GO-LDH-V_2 coating, indicating that the vanadium ions released from the LDH successfully adhered to the metal surface. In the 1.0 wt% GO-LDH-V_2 coating, the vanadium content at the scratch increased. This phenomenon indicates that the V content on the metal surface increased with an increase in the amount of filler. Elemental surface scanning of the scratched section showed that a large amount of V was adsorbed on the surface of the metal substrate at the scratch.

5.2.4.3 Raman Analysis of the Scratched Coating Surface

Additionally, surface-enhanced Raman scattering (SERS) spectroscopy demonstrated that the vanadate-adsorbed layer was located at the bottom of the scratch (Figure 5.17). The coating containing vanadium element has a typical vanadate peak. The peaks at 944 and 695 cm^{-1}, reflect V^{5+} compounds with V^{4+} band species at 835 cm^{-1} [76]. The peak strength increased with an increase in GO-LDH-V_2, whereas there was no vanadium peak in the pure waterborne epoxy coating. Previous studies have shown that the degree of corrosion inhibition of vanadate strongly depends on its morphology, among which tetrahedral coordination compounds have the best corrosion inhibition effects [77–79]. The V^{5+} and V^{4+} compounds were tetrahedral coordination species, indicating that the corrosion inhibitor loaded in GO-LDH-V_2 provided good protection to the aluminum alloy surface. The peak value near 588 cm^{-1} was consistent with Al oxide [80]. These results show that the vanadate corrosion inhibitor released by the LDH nanocontainers in the coating effectively binds to the metal substrate to form a corrosion inhibition film. Previous electrochemical experimental results have also shown that vanadate film has a certain corrosion inhibition effect.

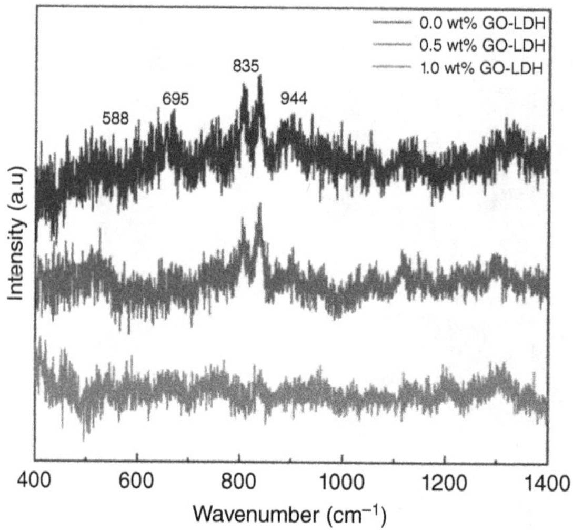

Figure 5.17 SERS analysis of coating scratch.

5.2.4.4 Cross Section Morphology of Coating

The prepared coating was frozen in liquid nitrogen, and its microscopic morphology was observed using SEM, as shown in Figure 5.18. In Figure 5.18a,b, a large number of voids were observed on the cross-section of the pure waterborne epoxy coating, which was responsible for the poor barrier performance. The GO/WEC cross-section in Figure 5.18c,d shows the presence of GO, which, owing to its dispersion stability, was dispersed in a layered form in the waterborne epoxy coating. This hindered the infiltration of corrosive media and gave the coating good barrier performance. In Figure 5.18e,f, the cross-section of the waterborne epoxy coating with the addition of the 0.5 wt% GO-LDH filler showed the layered morphology of GO sheets, and the dense distribution of LDH nanocontainers can be seen on the cross section. Compared with the coatings with GO alone, the phenomenon of GO layer stacking in the coatings with the GO-LDH composite improved.

5.2.5 Coating Protection Mechanism

The protection mechanisms of WEC, GO/WEC, LDH/WEC, and GO-LDH-V_2/WEC coatings are shown in Figure 5.19. For the pure waterborne epoxy coating, the defects such as pores and voids generated during coating the curing process make it easy for corrosive medium such as chloride ions, oxygen, and water to reach the metal substrate and undergo oxidation–reduction corrosion reactions on the metal surface. In GO/WEC coatings, the barrier effect of the GO layer to some extent hinders the penetration of corrosive media in the coating, thereby improving the corrosion resistance of GO/WEC. However, the limited dispersion capacity of GO layers limits the full utilization of its excellent barrier performance. Therefore, the corrosion resistance of GO/WEC is slightly improved compared to WEC. In the 0.5 wt%LDH/WEC sample, the coating achieved the capture of chloride ions in the corrosive medium through the ion exchange capability of LDH. Due to the existence of LDH nanocontainer, the corrosion resistance of the coating is

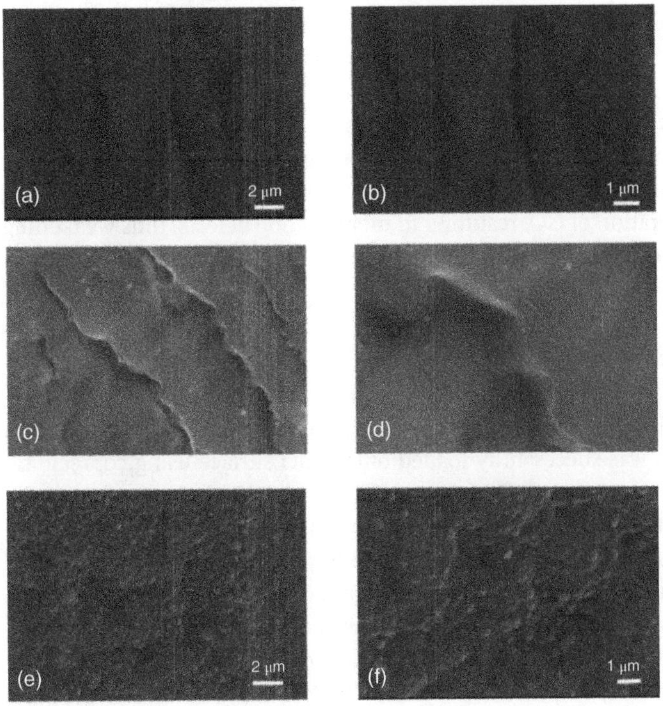

Figure 5.18 Morphology analysis of coating cross section: (a,b) WEC; (c,d) GO-LDH-V_2/WEC; (e,f) GO/WEC.

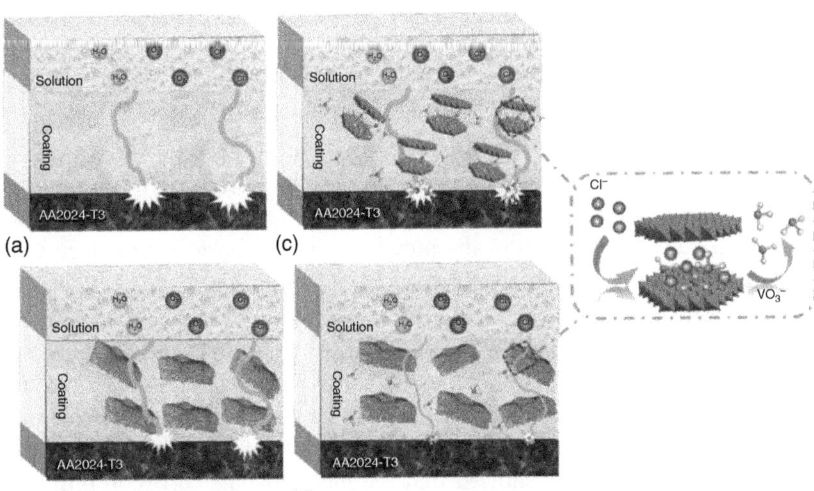

Figure 5.19 Coating protection mechanism: (a) pure waterborne epoxy coating; (b) GO/WEC coating; (c) LDH/WEC coating; (d) GO-LDH-V_2/WEC coating.

better than that of the pure waterborne epoxy coating. In GO-LDH-V_2/WEC, the protective function of the coating can be divided into two aspects. On the one hand, the GO layer in the GO-LDH fillers can provide a good shielding effect. On the other hand, when chloride ions penetrate into the coating, LDH can capture chloride ions and allow the metavanadate corrosion inhibitors to precipitate from the interlayer structure and form a protective film on the surface of AA2024-T3 to reduce the corrosion rate. However, excessive addition of GO-LDH composites can lead to the agglomeration of GO, resulting in more coating defects, thus weakening their excellent anti-corrosion properties.

5.3 Chapter Summary

In this chapter, an LDH nanocontainer containing sustained-release metavanadate corrosion inhibitors was successfully loaded onto a GO surface using coprecipitation and ion exchange methods. SEM, TEM, FTIR, and XPS results confirmed the successful combination of LDH and GO and the loading of corrosion inhibitors in the nanocontainer. Waterborne epoxy coatings with different proportions and GO-LDH contents were prepared, and the effects of these proportions on the corrosion resistance of the coatings were studied. For the coatings with physical scratch defects, the EIS experiments showed that, with the extension of salt spray time, the $|Z|_{0.01\,\mathrm{Hz}}$ value of the coating samples presented a gradually decreasing trend; however, the declining trend of the pure waterborne epoxy coatings was particularly obvious. For the samples with different filler contents, the initial protective performance of the coating was similar. With the extension of the salt spray test time, it was found that the 0.5 wt% GO-LDH-V_2 waterborne epoxy coating exhibited the best protective performance after the 200 hours of salt spray test. However, the 0.5 wt% GO-LDH-V_3 coating sample did not play a protective role owing to the agglomeration caused by the high amount of GO added. Thus, the experiment finally determined that the optimal amount of added GO-LDH was 0.5 wt% of the GO to prepare a 200-mg l^{-1} GO-LDH concentration, which offers good corrosion protection.

The long-term immersion experiments showed that the $|Z|_{0.01\,\mathrm{Hz}}$ of the waterborne epoxy coating with the addition of GO-LDH decreased to $5.8 \times 10^7\ \Omega\,\mathrm{cm}^2$ after 60 days of immersion, whereas that of the pure waterborne epoxy coating decreased to $5.4 \times 10^5\ \Omega\,\mathrm{cm}^2$. This was related to the uniform distribution of the large specific surface area of GO within the coating, which increased its density and formed a labyrinth effect, thus prolonging the penetration pathways of corrosive media (e.g. water, chloride ions, and oxygen). Hence, the barrier performance of the coating was greatly enhanced. In addition, the scratch self-healing experiment showed that the $|Z|_{0.01\,\mathrm{Hz}}$ value of the coating with 0.5 wt% GO-LDH-V_2 was the highest, and EDS and Raman spectroscopy showed that the metavanadate corrosion inhibitor loaded in the LDH effectively released to the metal substrate surface, providing good protection for the metal substrate.

LDH has been a research hotspot for many years because it can be used as the carrier of catalysts, drugs, and corrosion inhibitors owing to its adjustable chemical composition and easy exchange of anions between layers. In the field of metal corrosion and protection, LDH is suitable for the preparation of physically isolated and self-healing coatings due to its excellent barrier and ion exchange properties. Currently, the formation and protection mechanisms of LDH have not been studied thoroughly, changes in the coating structure during coating failures are not well understood, and the influence of different factors is unclear. Therefore, more research is needed on these basic aspects. Different preparation processes and agents may affect the amount of corrosion inhibitor intercalated in the LDH layers and their release rate; however, there are few reports on this. The durability of LDH in coating systems in corrosive environments has always been a limiting factor. After long-term immersion in a corrosive solution, the LDH may destroy its original layered structure, or when the corrosive ion adsorption reaches saturation, it will no longer adsorb and lose its ion exchange and protection performance. Therefore, improving durability should be the focus of future research. For the corrosion inhibitors in the LDH structure, the synergistic effect of corrosion inhibitors in solution has been studied extensively. Studies have shown that the effect of a single corrosion inhibitor is generally not ideal because the dosage and cost are high. In most cases, when two or more types of corrosion inhibitors are added to a corrosion solution simultaneously, the corrosion inhibition effect is significantly improved over the use of a single inhibitor, and the concentration is also reduced. Therefore, two or more corrosion inhibitors should be added to the LDH structure simultaneously, and their synergistic effect is expected to further improve the corrosion protection effect of the coating. Depending on the number of multiple corrosion inhibitors embedded, it is possible to stabilize the layered structure or intercalate LDH with different corrosion inhibitors to build a multiple corrosion protection system based on synergistic effects. Doing so may alleviate the problem of LDH film durability. Furthermore, the metal salt used in the synthesis of LDH is mostly nitrate. However, due to the hazardous nature of magnesium nitrate, its use is limited, and magnesium chloride salt is very cheap. China's Salt Lake contains rich magnesium chloride resources; hence, the process and effect of using chlorine salt to synthesize LDH should be studied in the future for large-scale LDH applications.

In conclusion, LDH with intercalated corrosion inhibitors has broad application prospects for metal protection. A systematic study of environmentally friendly corrosion inhibitors is important, and a relevant database of LDH nanocontainer information should be established to help researchers obtain the synthesis parameters and anion exchange capacities needed to produce the best experimental results in terms of self-healing coating designs. Another consideration is the long-term service performance of the coatings. As such, the addition of LDH to these coatings requires consideration of the impacts of adhesion and/or bubbles. Further research will help improve corrosion protection levels and develop self-healing coatings with ion exchange-activated protection mechanisms.

References

1 Pulikkalparambil, H., Siengchin, S., and Parameswaranpillai, J. (2018). Corrosion protective self-healing epoxy resin coatings based on inhibitor and polymeric healing agents encapsulated in organic and inorganic micro and nanocontainers. *Nano-Structures & Nano-Objects* 16: 381–395. https://doi.org/10.1016/j.nanoso.2018.09.010.

2 Farh, H., Ben Seghier, M., and Zayed, T. (2023). A comprehensive review of corrosion protection and control techniques for metallic pipelines. *Engineering Failure Analysis* 143: 106885. https://doi.org/10.1016/j.engfailanal.2022.106885.

3 Natishan, P. and O'Grady, W. (2014). Chloride ion interactions with oxide-covered aluminum leading to pitting corrosion: a review. *Journal of the Electrochemical Society* 161: 421–432. https://doi.org/10.1149/2.1011409jes.

4 Umoren, S. and Solomon, M. (2015). Effect of halide ions on the corrosion inhibition efficiency of different organic species – a review. *Journal of Industrial and Engineering Chemistry* 21: 81–100. https://doi.org/10.1016/j.jiec.2014.09.033.

5 Zhu, Y., Ma, Y., Yu, Q. et al. (2017). Preparation of pH-sensitive core-shell organic corrosion inhibitor and its release behavior in simulated concrete pore solutions. *Materials & Design* 119: 254–262. https://doi.org/10.1016/j.matdes.2017.01.063.

6 Mohammadi, I., Shahrabi, T., Mahdavian, M. et al. (2022). Construction of an epoxy coating with excellent protection performance on the AA 2024-T3 using ion-exchange materials loaded with eco-friendly corrosion inhibitors. *Progress in Organic Coatings* 166: 106786. https://doi.org/10.1016/j.porgcoat.2022.106786.

7 Yan, D., Wang, Y., Liu, J.L. et al. (2020). Self-healing system adapted to different pH environments for active corrosion protection of magnesium alloy. *Journal of Alloys and Compounds* 824: 153918. https://doi.org/10.1016/j.jallcom.2020.153918.

8 Liu, X., Wang, J., and Hu, W.B. (2020). Preparation and controlled inhibition behavior of Fe_3O_4/CS/inhibitors nanocomposite for carbon steel in 3.5% NaCl solution. *Colloids and Surfaces A: Physicochemical and Engineering Aspects* 601: 124985. https://doi.org/10.1016/j.colsurfa.2020.124985.

9 Ding, R., Li, W., Wang, X. et al. (2018). A brief review of corrosion protective films and coatings based on graphene and graphene oxide. *Journal of Alloys and Compounds* 764: 1039–1055. https://doi.org/10.1016/j.jallcom.2018.06.133.

10 Kumar, S.S.A., Bashir, S., Ramesh, K. et al. (2021). New perspectives on graphene/graphene oxide based polymer nanocomposites for corrosion applications: the relevance of the graphene/polymer barrier coatings. *Progress in Organic Coatings* 154: 106215. https://doi.org/10.1016/j.porgcoat.2021.106215.

11 Chauhan, D., Quraishi, M., Ansari, K. et al. (2020). Graphene and graphene oxide as new class of materials for corrosion control and protection: present status and future scenario. *Progress in Organic Coatings* 147: 105741. https://doi.org/10.1016/j.porgcoat.2020.105741.

12 Gupta, R., Malviya, M., Verma, C. et al. (2017). Aminoazobenzene and diaminoazobenzene functionalized graphene oxides as novel class of corrosion inhibitors

for mild steel: experimental and DFT studies. *Materials Chemistry and Physics* 198: 360–373. https://doi.org/10.1016/j.matchemphys.2017.06.030.

13 Nikpour, B., Ramezanzadeh, B., Bahlakeh, G. et al. (2017). Synthesis of graphene oxide nanosheets functionalized by green corrosion inhibitive compounds to fabricate a protective system. *Corrosion Science* 127: 240–259. https://doi.org/10.1016/j.corsci.2017.08.029.

14 Meng, W., Gall, E., Ke, F. et al. (2015). Structure and interaction of graphene oxide-cetyltrimethylammonium bromide complexation. *Journal of Physical Chemistry C* 119: 21135–21140. https://doi.org/10.1021/acs.jpcc.5b04102.

15 Xie, C., Jia, Y., Xue, M. et al. (2022). Anti-corrosion and self-healing behaviors of waterborne polyurethane composite coatings enhanced via chitosan-modified graphene oxide and phosphate intercalated hydrotalcite. *Progress in Organic Coatings* 168: 106881. https://doi.org/10.1016/j.porgcoat.2022.106881.

16 Li, J., Jiang, Z., Gan, L. et al. (2018). Functionalized graphene/polymer composite coatings for autonomous early-warning of steel corrosion. *Composites Communications* 9: 6–10. https://doi.org/10.1016/j.coco.2018.04.002.

17 Liu, P., Gong, K., Xiao, P. et al. (2000). Preparation and characterization of poly(vinyl acetate)-intercalated graphite oxide nanocomposite. *Journal of Materials Chemistry* 10: 933–935. https://doi.org/10.1039/a908179h.

18 Amrollahi, S., Ramezanzadeh, B., Yari, H. et al. (2019). In-situ growth of ceria nanoparticles on graphene oxide nanoplatelets to be used as a multifunctional (UV shield/radical scavenger/anticorrosive) hybrid compound for exterior coatings. *Progress in Organic Coatings* 136: 105241. https://doi.org/10.1016/j.porgcoat.2019.105241.

19 Cui, G., Bi, Z.X., Zhang, R. et al. (2019). A comprehensive review on graphene-based anti-corrosive coatings. *Chemical Engineering Journal* 373: 104–121. https://doi.org/10.1016/j.cej.2019.05.034.

20 Gupta, V., Atar, N., Yola, M. et al. (2014). A novel magnetic Fe@Au core-shell nanoparticles anchored graphene oxide recyclable nanocatalyst for the reduction of nitrophenol compounds. *Water Research* 48: 210–217. https://doi.org/10.1016/j.watres.2013.09.027.

21 Xiong, L., Yu, M., Li, Y. et al. (2020). Modified salicylaldehyde@ZIF-8/graphene oxide for enhancing epoxy coating corrosion protection property on AA2024-T3. *Progress in Organic Coatings* 142: 105562. https://doi.org/10.1016/j.porgcoat.2020.105562.

22 Ramezanzadeh, B., Haeri, Z., Ramezanzadeh, M. et al. (2016). A facile route of making silica nanoparticles-covered graphene oxide nanohybrids (SiO_2-GO); fabrication of SiO_2-GO/epoxy composite coating with superior barrier and corrosion protection performance. *Chemical Engineering Journal* 303: 511–528. https://doi.org/10.1016/j.cej.2016.06.028.

23 Chen, L., Li, N., Wen, Z. et al. (2018). Graphene oxide based membrane intercalated by nanoparticles for high performance nanofiltration application. *Chemical Engineering Journal* 347: 12–18. https://doi.org/10.1016/j.cej.2018.04.069.

24 Shen, G., Zhang, L., Gu, Z. et al. (2022). Zinc aluminum-layered double hydroxide (LDH)-graphene oxide (GO) lubricating and corrosion-resistant composite coating on the surface of magnesium alloy. *Surface and Coatings Technology* 437: 128354. https://doi.org/10.1016/j.surfcoat.2022.128354.

25 Wang, Q., Yu, Z., Liu, Y. et al. (2021). Electrostatic self-assembly method to prepare intercalated graphene oxide composite membrane to improve hydrophilicity and flux. *Diamond and Related Materials* 117: 180492. https://doi.org/10.1016/j.diamond.2021.108492.

26 Zeng, H., Yu, Z., Peng, Y. et al. (2020). Environmentally friendly electrostatically driven self-assembled LDH/GO/PVDF composite membrane for water treatment. *Applied Clay Science* 183: 105322. https://doi.org/10.1016/j.clay.2019.105322.

27 Li, L., Chen, S., Guo, X. et al. (2022). Design and fabrication of a novel LDH@GO nanohybrid material for its application potentials in polypropylene. *Journal of Polymer Research* 29: 85. https://doi.org/10.1007/s10965-022-02936-4.

28 Su, Y., Qiu, S., Yang, D. et al. (2020). Active anti-corrosion of epoxy coating by nitrite ions intercalated MgAl LDH. *Journal of Hazardous Materials* 391: 122215. https://doi.org/10.1016/j.jhazmat.2020.122215.

29 Huang, Q., Chen, Y., Yu, H. et al. (2018). Magnetic graphene oxide/MgAl-layered double hydroxide nanocomposite: one-pot solvothermal synthesis, adsorption performance and mechanisms for Pb^{2+}, Cd^{2+}, and Cu^{2+}. *Chemical Engineering Journal* 341: 1–9. https://doi.org/10.1016/j.cej.2018.01.156.

30 Wu, Y., Jiang, F., Qiang, Y. et al. (2021). Synthesizing a novel fluorinated reduced graphene oxide-CeO_2 hybrid nanofiller to achieve highly corrosion protection for waterborne epoxy coatings. *Carbon* 176: 39–51. https://doi.org/10.1016/j.carbon.2021.01.135.

31 Zhou, S., Wu, Y., Zhao, W. et al. Comparative corrosion resistance of graphene sheets with different structures in waterborne epoxy coatings. *Colloids and Surfaces A: Physicochemical and Engineering Aspects* 556: 273–283. https://doi.org/10.1016/j.colsurfa.2018.08.045.

32 Cui, M., Ren, S., Zhao, H. et al. (2018). Polydopamine coated graphene oxide for anticorrosive reinforcement of water-borne epoxy coating. *Chemical Engineering Journal* 335: 255–266. https://doi.org/10.1016/j.cej.2017.10.172.

33 Qi, Y., Yang, M., Xu, W. et al. (2017). Natural polysaccharides-modified graphene oxide for adsorption of organic dyes from aqueous solutions. *Journal of Colloid and Interface Science* 486: 84–96. https://doi.org/10.1016/j.jcis.2016.09.058.

34 Guerrero-Contreras, J. and Caballero-Briones, F. (2015). Graphene oxide powders with different oxidation degree, prepared by synthesis variations of the Hummers method. *Materials Chemistry and Physics* 153: 209–220. https://doi.org/10.1016/j.matchemphys.2015.01.005.

35 Lu, N., Yin, D., Li, Z. et al. (2011). Structure of graphene oxide: thermodynamics versus kinetics. *Journal of Physical Chemistry C* 115: 11991–11995. https://doi.org/10.1021/jp204476q.

36 Krishnamoorthy, K., Veerapandian, M., Yun, K. et al. (2013). The chemical and structural analysis of graphene oxide with different degrees of oxidation. *Carbon* 53: 38–49. https://doi.org/10.1016/j.carbon.2012.10.013.

37 Wang, H., Yuan, X., Wu, Y. et al. (2013). Adsorption characteristics and behaviors of graphene oxide for Zn(II) removal from aqueous solution. *Applied Surface Science* 279: 432–440. https://doi.org/10.1016/j.apsusc.2013.04.133.

38 Pei, Z., Li, L., Sun, L. et al. (2013). Adsorption characteristics of 1,2,4-trichlorobenzene, 2,4,6-trichlorophenol, 2-naphthol and naphthalene on graphene and graphene oxide. *Carbon* 51: 156–163. https://doi.org/10.1016/j.carbon.2012.08.024.

39 Behbahani, E., Dashtian, K., Ghaedi, M. et al. (2021). Fe_3O_4-$FeMoS_4$: promise magnetite LDH-based adsorbent for simultaneous removal of Pb (II), Cd (II), and Cu (II) heavy metal ions. *Journal of Hazardous Materials* 410: 124560. https://doi.org/10.1016/j.jhazmat.2020.124560.

40 Shan, R., Yan, L., Yang, K. et al. (2014). Magnetic Fe_3O_4/MgAl-LDH composite for effective removal of three red dyes from aqueous solution. *Chemical Engineering Journal* 252: 38–46. https://doi.org/10.1016/j.cej.2014.04.105.

41 Wang, L., Xu, X., Evans, D. et al. (2010). Synthesis and selective IR absorption properties of iminodiacetic-acid intercalated MgAl-layered double hydroxide. *Journal of Solid State Chemistry* 183: 1114–1119. https://doi.org/10.1016/j.jssc.2010.03.022.

42 Guo, S., Li, D., Zhang, W. et al. (2004). Preparation of an anionic azo pigment-pillared layered double hydroxide and the thermo- and photostability of the resulting intercalated material. *Journal of Solid State Chemistry* 117: 4597–4604. https://doi.org/10.1016/j.jssc.2004.09.028.

43 Du, L., Qu, B., Meng, Y. et al. (2006). Structural characterization and thermal and mechanical properties of poly(propylene carbonate)/MgAl-LDH exfoliation nanocomposite via solution intercalation. *Composites Science and Technology* 66: 913–918. https://doi.org/10.1016/j.compscitech.2005.08.012.

44 Voll, D. and Beran, A. (2002). Dehydration process and structural development of cordierite ceramic precursors derived from FTIR spectroscopic investigations. *Physics and Chemistry of Minerals* 29: 545–551. https://doi.org/10.1007/s00269-002-0266-2.

45 Blonska-Tabero, A., Filipek, E., Bosacka, M. et al. (2022). Reinvestigation of thermal and other properties of compounds from the V_2O_5-SrO system. *Journal of Thermal Analysis and Calorimetry* 147: 9903–9911. https://doi.org/10.1007/s10973-022-11363-y.

46 Reddy, C., Yeo, I., and Mho, S. (2008). Synthesis of sodium vanadate nanosized materials for electrochemical applications. *Journal of Physics and Chemistry of Solids* 69: 1261–1264. https://doi.org/10.1016/j.jpcs.2007.10.072.

47 Al-Gaashani, R., Najjar, A., Zakaria, Y. et al. (2019). XPS and structural studies of high quality graphene oxide and reduced graphene oxide prepared by different chemical oxidation methods. *Ceramics International* 45: 14439–14448. https://doi.org/10.1016/j.ceramint.2019.04.165.

48 Stobinski, L., Lesiak, B., Malolepszy, A. et al. (2014). Graphene oxide and reduced graphene oxide studied by the XRD, TEM and electron spectroscopy methods. *Journal of Electron Spectroscopy and Related Phenomena* 195: 145–154. https://doi.org/10.1016/j.elspec.2014.07.003.

49 Chen, J., Zhang, X., Zheng, X. et al. (2013). Size distribution-controlled preparation of graphene oxide nanosheets with different C/O ratios. *Materials Chemistry and Physics* 139: 8–11. https://doi.org/10.1016/j.matchemphys.2012.12.025.

50 Ederer, J., Janos, P., Ecorchard, P. et al. (2017). Determination of amino groups on functionalized graphene oxide for polyurethane nanomaterials: XPS quantitation vs. functional speciation. *RSC Advances* 7: 12464–12473. https://doi.org/10.1039/C6RA28745J.

51 Mohammadkhani, R., Ramezanzadeh, M., Saadatmandi, S. et al. (2020). Designing a dual-functional epoxy composite system with self-healing/barrier anti-corrosion performance using graphene oxide nano-scale platforms decorated with zinc doped-conductive polypyrrole nanoparticles with great environmental stability and non-toxicity. *Chemical Engineering Journal* 382: 122819. https://doi.org/10.1016/j.cej.2019.122819.

52 Sitko, R., Turek, E., Zawisza, B. et al. (2013). Adsorption of divalent metal ions from aqueous solutions using graphene oxide. *Dalton Transactions* 42: 5682–5689. https://doi.org/10.1039/c3dt33097d.

53 Mittal, V., Bera, S., Nithya, R. et al. (2004). Solid state synthesis of Mg-Ni ferrite and characterization by XRD and XPS. *Journal of Nuclear Materials* 335: 302–310. https://doi.org/10.1016/j.jnucmat.2004.05.010.

54 Figueiredo, N., Carvalho, N., and Cavaleiro, A. (2011). An XPS study of Au alloyed Al-O sputtered coatings. *Applied Surface Science* 257: 5793–5798. https://doi.org/10.1016/j.apsusc.2011.01.104.

55 Silversmit, G., Depla, D., Poelman, H. et al. (2004). Determination of the V2p XPS binding energies for different vanadium oxidation states (V^{5+} to V^{0+}). *Journal of Electron Spectroscopy and Related Phenomena* 135: 167–175. https://doi.org/10.1016/j.elspec.2004.03.004.

56 Chen, Q., Guo, Z., Cheng, F. et al. (2022). Effect of GO agglomeration on the mechanical properties of graphene oxide and nylon 66 composites and micromechanical analysis. *Polymer Composites* 43: 8356–8367. https://doi.org/10.1002/pc.27006.

57 Ma, Y., Ye, Y., Wan, H. et al. (2020). Chemical modification of graphene oxide to reinforce the corrosion protection performance of UV-curable polyurethane acrylate coating. *Progress in Organic Coatings* 141: 105547. https://doi.org/10.1016/j.porgcoat.2020.105547.

58 Naseem, Z., Shamsaei, E., Sagoe-Crentsil, K. et al. (2022). Antifoaming effect of graphene oxide nanosheets in polymer-modified cement composites for enhanced microstructure and mechanical performance. *Cement and Concrete Research* 158: 106843. https://doi.org/10.1016/j.cemconres.2022.106843.

59 Toorani, M., Aliofkhazraei, M., Mahdavian, M. et al. (2021). Superior corrosion protection and adhesion strength of epoxy coating applied on AZ31 magnesium alloy pre-treated by PEO/Silane with inorganic and organic corrosion inhibitors. *Corrosion Science* 178: 109065. https://doi.org/10.1016/j.corsci.2020.109065.

60 Zhang, F., Ju, P., Li, X. et al. (2018). Self-healing mechanisms in smart protective coatings: a review. *Corrosion Science* 144: 74–88. https://doi.org/10.1016/j.corsci.2018.08.005.

61 Montemor, M. (2014). Functional and smart coatings for corrosion protection: a review of recent advances. *Surface and Coatings Technology* 258: 17–37. https://doi.org/10.1016/j.surfcoat.2014.06.031.

62 Li, H., Qiang, Y., Zhao, W. et al. (2021). 2-Mercaptobenzimidazole-inbuilt metal-organic-frameworks modified graphene oxide towards intelligent and excellent anti-corrosion coating. *Corrosion Science* 191: 109715. https://doi.org/10.1016/j.corsci.2021.109715.

63 Zhu, Q., Zhao, Y., Miao, B. et al. (2022). Hydrothermally synthesized ZnO-RGO-PPy for water-borne epoxy nanocomposite coating with anticorrosive reinforcement. *Progress in Organic Coatings* 172: 107153. https://doi.org/10.1016/j.porgcoat.2022.107153.

64 Liu, X., Xiong, J., Lv, Y. et al. (2009). Study on corrosion electrochemical behavior of several different coating systems by EIS. *Progress in Organic Coatings* 64: 497–503. https://doi.org/10.1016/j.porgcoat.2008.08.012.

65 Ding, R., Zheng, Y., Yu, H. et al. (2018). Study of water permeation dynamics and anti-corrosion mechanism of graphene/zinc coatings. *Journal of Alloys and Compounds* 748: 481–495. https://doi.org/10.1016/j.jallcom.2018.03.160.

66 Hamlaoui, Y., Pedraza, F., and Tifouti, L. (2008). Corrosion monitoring of galvanised coatings through electrochemical impedance spectroscopy. *Corrosion Science* 50: 1558–1566. https://doi.org/10.1016/j.corsci.2008.02.010.

67 Zheludkevich, M., Shchukin, D., Yasakau, K. et al. (2007). Anticorrosion coatings with self-healing effect based on nanocontainers impregnated with corrosion inhibitor. *Chemistry of Materials* 19: 402–411. https://doi.org/10.1021/cm062066k.

68 Feng, Y. and Cheng, Y. (2017). An intelligent coating doped with inhibitor-encapsulated nanocontainers for corrosion protection of pipeline steel. *Chemical Engineering Journal* 315: 537–551. https://doi.org/10.1016/j.cej.2017.01.064.

69 Bastos, A., Ferreira, M., and Simoes, A. (2005). Comparative electrochemical studies of zinc chromate and zinc phosphate as corrosion inhibitors for zinc. *Progress in Organic Coatings* 52: 339–350. https://doi.org/10.1016/j.porgcoat.2004.09.009.

70 Li, Y., Lu, X., Serdechnova, M. et al. (2023). Incorporation of LDH nanocontainers into plasma electrolytic oxidation coatings on Mg alloy. *Journal of Magnesium and Alloys* 11: 1236–1246. https://doi.org/10.1016/j.jma.2021.07.015.

71 Chang, K., Hsu, M., Lu, H. et al. (2014). Room-temperature cured hydrophobic epoxy/graphene composites as corrosion inhibitor for cold-rolled steel. *Carbon* 66: 144–153. https://doi.org/10.1016/j.carbon.2013.08.052.

72 Zhou, S., Yan, J., Yan, H. et al. (2022). ZrO_2-anchored rGO nanohybrid for simultaneously enhancing the wear resistance and anticorrosion performance of multifunctional epoxy coatings. *Progress in Organic Coatings* 166: 106795. https://doi.org/10.1016/j.porgcoat.2022.106795.

73 Wang, L., Chen, Y., Luo, J. et al. (2022). Synthesis of graphene oxide functionalized by phytic acid for anticorrosive reinforcement of waterborne epoxy coating. *Journal of Applied Polymer Science* 139: 51910. https://doi.org/10.1002/app.51910.

74 Yan, Y., Chu, X., Luo, X. et al. (2021). A homogenous microstructural Mg-based matrix model for orthopedic application with generating uniform and smooth corrosion product layer in Ringer's solution: study on biodegradable behavior of Mg–Zn alloys prepared by powder metallurgy as a case. *Journal of Magnesium and Alloys* 9: 225–240. https://doi.org/10.1016/j.jma.2020.03.010.

75 He, Z., Chen, B., Zhou, B. et al. (2023). Effect of TiC precipitation on the corrosion behavior of Monel K500 alloy in 3.5 wt.% NaCl solution. *Corrosion Science* 211: 110886. https://doi.org/10.1016/j.corsci.2022.110886.

76 Shvets, P., Dikaya, O., Maksimova, K. et al. (2019). A review of Raman spectroscopy of vanadium oxides. *Journal of Raman Spectroscopy* 50: 1226–1244. https://doi.org/10.1002/jrs.5616.

77 Ralston, K., Chrisanti, S., Young, T. et al. (2008). Corrosion inhibition of aluminum alloy 2024-T3 by aqueous vanadium species. *Journal of the Electrochemical Society* 155: C350–C359. https://doi.org/10.1149/1.2907772.

78 Hurley, B., Ralston, K., and Buchheit, R. (2014). Corrosion inhibition of zinc by aqueous vanadate species. *Journal of the Electrochemical Society* 161: C471–C475. https://doi.org/10.1149/2.0381410jes.

79 Ralston, K., Young, T., and Buchheit, R. (2009). Electrochemical evaluation of constituent intermetallics in aluminum alloy 2024-T3 exposed to aqueous vanadate inhibitors. *Journal of the Electrochemical Society* 156: C135–C146. https://doi.org/10.1149/1.3076147.

80 Misra, A., Bist, H., Navati, M. et al. (2001). Thin film of aluminum oxide through pulsed laser deposition: a micro-Raman study. *Materials Science and Engineering B* 79: 49–54. https://doi.org/10.1016/S0921-5107(00)00554-7.

6

Development of pH-responsive and Metal Ion-responsive Self-reporting Coating

6.1 Introduction

Self-healing coatings mentioned in Chapters 2–5 can repair various types of macro/microdamage on the coating surface to extend the service life of the metal substrate [1, 2]. However, the self-healing ability of coatings is limited by the number of healing cycles and applicable damage sizes, and many self-healing processes of coatings are irreversible [3–5]. Once the healing agent is released, the original coating site often cannot be used to repair the subsequent damage the next time [3, 6, 7]. Therefore, when the coating is damaged, or corrosion occurs on the metal substrate, timely warning of the damage and determination of the exact damage location are crucial. At the initial stage, micro–nano damage to the coating is difficult to detect without specialized instruments. Nondestructive testing of coatings can be performed regularly by means of infrared thermal imaging, magnetic detection, electrochemical measurement, ultrasonic detection, and X-ray scanning so the coating damage and the development of corrosion can be effectively monitored. These methods can be used to detect microscopic damage before it expands and causes a major disaster; however, they typically involve large equipment and complex labor costs.

Researchers often turn to the natural world when faced with problems, and feedback mechanisms that indicate the structural integrity of tissues are crucial for the long-term survival of living organisms. Thus far, numerous perceptual mechanisms have evolved for assessing their physical state. These signals can help individuals protect their bodies, seek medical assistance, and respond to injuries and illnesses. Inspired by the biological early sensing mechanism, the concept of a self-reporting coating has been proposed for the early detection of corrosion [8–11]. Such smart coatings often incorporate fluorescent molecules or color-rendering dyes to provide reliable early warnings of environmental changes or mechanical damage associated with corrosion [12, 13]. According to the different early warning triggering mechanisms, self-reporting coatings can be divided into corrosion-triggering and mechanical-triggering types.

In this chapter, pH- and metal ion-responsive self-reporting coatings are introduced. The indicator molecules were sensitive to pH variations or metallic cations within local defects. Common pH-indicating molecules used for corrosion sensing

Smart Protective Coatings for Corrosion Control, First Edition. Lingwei Ma and Dawei Zhang.

include phenolphthalein (PhPh), bromocresol green, and cresol red. PhPh has a high color contrast during pH variation, being colorless when the pH is <8.2 and bright pink in the pH range of 8.2–12.0. Therefore, it can detect the generation of OH^- during the corrosion process. Several self-reporting coatings containing pH indicators have been introduced. For the metal ion-responsive self-reporting coating, phenanthroline (Phen) and its derivatives exhibited a red color owing to a chromogenic reaction with Fe^{2+}. Consequently, steel corrosion can be easily detected via interaction with the Phen molecules. In addition, Phen has an inhibitory effect against steel corrosion. Phen and its derivatives were loaded into different macro/nanocontainers, and their self-reporting and self-healing properties were investigated in detail. The advantages and disadvantages of pH-responsive and metal ion-responsive self-reporting coatings are discussed.

6.2 pH-responsive Self-reporting Coatings Based on PhPh Indicator

Corrosion-triggered self-reporting coatings primarily consider the corrosion reaction when the coating is damaged; that is, metal corrosion occurs when the metal matrix is exposed to an external corrosive environment. Metal corrosion typically involves the following reactions: Anodic reactions occur during metal electron loss oxidation, generating related metal ions, which may be accompanied by the hydrolysis and acidification of metal ions, depending on the environmental conditions. In the cathode reaction, oxygen or water in the etched area is reduced by electrons, forming hydroxide ions. Therefore, metal ions are inevitably generated once corrosion occurs, and the pH changes in the corroded area.

PhPh is the most widely used pH indicator for corrosion monitoring. When the pH value is 8.2–12.0, the indicator appears bright pink, and at other pH values, it remains colorless. PhPh can detect the formation of hydroxyl ions associated with cathodic corrosion reactions. The direct addition of a pH indicator to the coating matrix is often unsuitable, as it is likely to cause premature leakage of the pH indicator and even degrade the overall performance of the resin matrix. This was verified in a previous study in which two water-based coating systems were prepared. In one of the systems, SiO_2 nanocapsules were used to encapsulate PhPh as a filler and added to the coating on an AA2024 aluminum plate, and in the other, PhPh was directly added to the coating as a contrast sample. Electrochemical measurements indicated that the properties of the coatings with PhPh added directly were inferior to those of the pure coatings. Fourier transform infrared (FTIR) spectroscopy indicated that PhPh loaded directly into the coating would reduce the degree of cross-linking of the resin during the curing process. Therefore, the use of appropriate containers for PhPh encapsulation is crucial. Finally, salt spray tests indicated the early self-reporting ability of the coating; that is, the scratched area gradually turned pink owing to the corrosion process. Similarly, Maia et al. [13] reported an epoxy coating developed on mesoporous nano-SiO_2 containers containing PhPh and compared the corrosion warning effect of this coating applied on aluminum alloy and magnesium alloy substrates. The results indicated that the damage to both alloys can be successfully predicted; however, the color change of the magnesium alloy damaged coating was more obvious than that of the aluminum alloy coating. This was due to the higher degree of corrosion of the magnesium alloy, which resulted in more free hydroxide ions reacting with PhPh. However, PhPh indicators are not suitable for all metals, such as carbon steel, and when corrosion occurs around the scratch, it is usually acidic. Therefore, the application of PhPh should consider the corrosion activity and corrosion behavior of the metal substrate that needs to be protected; this design principle is also applicable to other pH indicators [14].

In short, PhPh has considerable potential for identifying the early corrosion of various metal substrates. A key issue is the effect of pH on discoloration in a strongly alkaline environment. As corrosion progresses, the local pH of the damaged area of the coating may change, which affects the color of the indicator. Therefore, the distribution of pH indicator particles must be optimized to offer better coloring dynamics and long-term durability. In addition, the selection of a pH indicator with a color change at an appropriate pH is crucial. Some pH indicators vary in color at pH values of 4–5 and 6–7, which are unsuitable for detecting corrosion. Currently, there are few options for pH indicators; therefore, the development and synthesis of new pH indicators are essential [15–18].

6.3 Metal Ion-responsive Self-reporting Coatings Based on CDs

Carbon dots (CDs) are carbon-based nanomaterials with zero-dimensional structures (diameters of <10 nm) [19]. They are usually spherical and consist of sp^2/sp^3 hybrid carbon atoms and oxygen, hydrogen, and nitrogen atoms, which are mainly attached to the surface [20]. In 2004, Xu et al. accidentally discovered CDs when they were electrophoretically purified, prepared single-wall carbon nanotubes for the first time, and determined their exact composition [21, 22]. CDs can be classified as graphene quantum dots (GQDs), carbon quantum dots (CQDs), carbon nanodots (CNDs), and carbonized polymer dots (CPDs) [23].

Owing to the abundance and controllability of CDs surface groups, the properties of CDs can be enhanced, or new characteristics can be introduced by modifying the surface groups to enrich the versatility of CDs [24, 25]. In this section, the CDs surface hydrate was modified by sulfosalicylic acid (SSA), which provided CDs with excellent corrosion inhibition performance and sensitivity to corrosion induction and played the dual role of corrosion warning and slowing the corrosion of the metal matrix. Sulfonate groups have been used for corrosion protection because of their passivation and corrosion inhibition abilities. Liu et al. [26] used polyaniline with sulfosalicylic acid (PANI–SSA) as a coating filler in an epoxy matrix, which was applied to the protection of a 5083 aluminum alloy. The high electrical conductivity and redox activity of PANI–SSA can promote the formation of a protective oxide layer. When PANI–SSA was mixed with epoxy resin, the coating exhibited good barrier and bonding properties. With regard to corrosion-sensing sensitivity, SSA and

Fe^{3+} can form purplish-red ferrosulfonyl salicylate (III) in solution with a pH of 2–3, and red complexes can be formed in solution with a pH of 4–9; thus, iron ions can be monitored during corrosion through color changes.

In this study, novel sulfosalicylic acid–modified carbon dots (CDs–SSA) were synthesized using ammonium citrate and sulfosalicylic acid [27]. The CDs–SSA complex exhibited both corrosion inhibition and corrosion sensing capabilities. The structure and composition of the modified CDs were characterized. The corrosion inhibition effect was evaluated via electrochemical measurements and surface analyses. Next, the fluorescence response between the CDs–SSA complex and Fe^{2+}/Fe^{3+} ions was examined under irradiation from a 365-nm ultraviolet (UV) lamp. To evaluate the corrosion-sensing ability of the complexes toward Fe^{2+}/Fe^{3+} ions, the fluorescence emission spectra of the CDs–SSA were measured. The experimental results indicated that CDs–SSA had anodic passivation behavior, and the electrochemical impedance spectroscopy (EIS) low-frequency impedance modulus was significantly increased, leading to corrosion inhibition efficiencies of >99% in a hydrochloric acid (HCl) solution. The fluorescence emission spectrum data indicated that the limit of CDs–SSA for Fe^{2+} and Fe^{3+} ion detection was 5×10^{-4} and 1×10^{-5} M, respectively, with a significant fluorescence quenching effect. Therefore, the CDs–SSA complex could effectively detect the early stages of corrosion with high sensitivity. The preferred CDs–Phen complex was then loaded into the epoxy coating to evaluate its self-healing performance. The coatings scratched with 0.5 wt% filler were immersed in a saline solution, and they exhibited significant fluorescence quenching in the damaged areas. Therefore, the coatings can warn of corrosion initiation and indicate the corrosion sites, achieving a self-reporting effect.

6.3.1 Materials and Methods

6.3.1.1 Materials

The materials and reagents used in this section are shown in Table 6.1.

6.3.1.2 Instruments

The instruments used in this section are shown in Table 6.2.

6.3.1.3 Preparation and Characterization of Carbon Dots

The procedure for CDs preparation is shown in Figure 6.1. First, 1.5 g of ammonium citrate was added to a beaker and heated in an oven at 180 °C for different durations (1, 1.5, and 2 hours). The samples were labeled CDs1, CDs1.5, and CDs2. The reaction products were dissolved in deionized water and centrifuged. After the supernatant was removed, deionized water was added to the pellet, and centrifugation was performed again. Subsequently, the mixture was refined via dialysis in water. Thereafter, the product was transferred to an oven and dried, yielding a black CDs powder.

Table 6.1 Experimental materials and reagents.

Materials	Purity	Company
Ammonium citrate	Analytically pure, 98.5%	Aladdin Chemical Co., Ltd.
N-hydroxysuccinimide (NHS)	Analytically pure	Aladdin Chemical Co., Ltd.
1-Ethyl-(3-dimethyl-aminopropyl) carbodiimide hydrochloride (EDC)	Analytically pure	Aladdin Chemical Co., Ltd.
Ferrous chloride, anhydrous	99.5%	Aladdin Chemical Co., Ltd.
$FeCl_3 \cdot 6H_2O$	99%	Aladdin Chemical Co., Ltd.
Hydrochloric acid	Guaranteed reagent	Sinopharm Chemical Reagent Co., Ltd.
Sulfosalicylic acid	Analytically pure	Sinopharm Chemical Reagent Co., Ltd.
Hydrochloric acid	Guaranteed reagent	Sinopharm Chemical Reagent Co., Ltd.
Tetrahydrofuran	Analytically pure	Sinopharm Chemical Reagent Co., Ltd.
Absolute ethanol	Analytically pure	Modern Oriental Technology Co., Ltd.

Table 6.2 The instruments and equipment.

Instrument	Production type	Manufacturer
Constant temperature blast drying oven	303-3S	China Baihui
Electronic balance	ME204	Switzerland Mettler Toledo
Centrifuge	Allegra X-30	USA Beckman
Electrochemical Workstation	PARSTAT-2273	USA Princeton
Laser scanning confocal microscope (CLSM)	VK-X2000 series	Japan KEYENCE
Atomic force microscope (AFM)	Dimension3100	USA Veeco
Transmission electron microscopy (TEM)	JEM-2100F	Japan JEOL
UV/Vis/NIR spectrometer	Lambda 750	USA PerkinElmer
Fourier transform infrared spectrometer (FTIR)	Vector 22	Germany Bruker

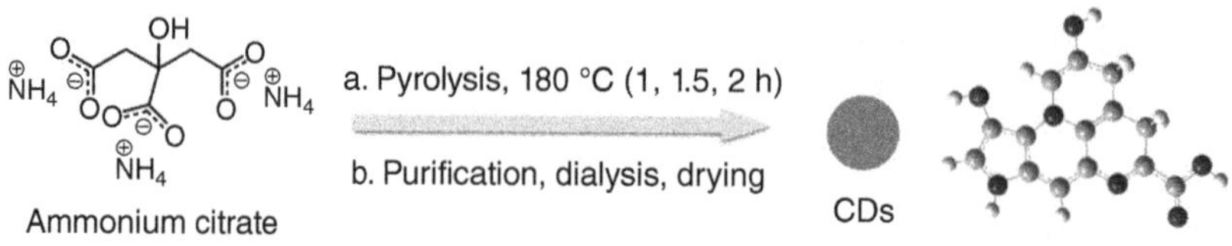

Figure 6.1 Preparation process of carbon dot.

6.3.1.4 Corrosion Inhibition Performance of CDs

The corrosion inhibition effects of different CDs in a 1 M HCl solution were evaluated via potentiodynamic polarization (PDP) and EIS measurements. The corroded surface characteristics of the Q235 steel were recorded after immersion in a 1 M HCl solution with and without CDs. After the metal surface was cleaned with ethanol, the surface roughness was measured via confocal laser scanning microscopy (CLSM).

6.3.1.5 Preparation and Characterization of CDs–SSA

First, 50 mg of CDs1.5 was placed in a conical flask, 50 ml of tetrahydrofuran was added, and the mixture was continuously stirred using a magnetic stirrer. Subsequently, 0.060 g of NHS, 0.133 g of SSA, and 0.100 g of carbodiimide hydrochloride (EDC) were added successively. After 48 hours of stirring, the mixed solution was heated at 65 °C for 12 hours, and the tetrahydrofuran solvent and other substances were removed to obtain the final CDs–SSA complex. The preparation process is shown in Figure 6.2. Transmission electron microscopy (TEM), atomic force microscopy (AFM), FTIR spectroscopy, and UV–visible (UV–Vis) spectroscopy were used to characterize the morphology and structure of the CDs–SSA.

6.3.1.6 Corrosion Inhibition and Corrosion Sensitivity of CDs–SSA

The electrochemical PDP and EIS measurement methods are described in Section 6.3.1.4. The surface corrosion morphology and the composition of the corrosion products were examined via SEM and EDS, and the surface roughness was measured via CLSM. The compositions of the surface-adsorbed films were analyzed via X-ray photoelectron spectroscopy (XPS).

The water contact angle (WCA) of the Q235 steel surface was determined using a contact-angle measuring instrument (DataPhysics OCA25). It was measured by dropping 2 μl of deionized water onto the surface of the sample. Images of static water were captured using a high-resolution camera. For each sample, the final WCA was the average of measurements from three different regions.

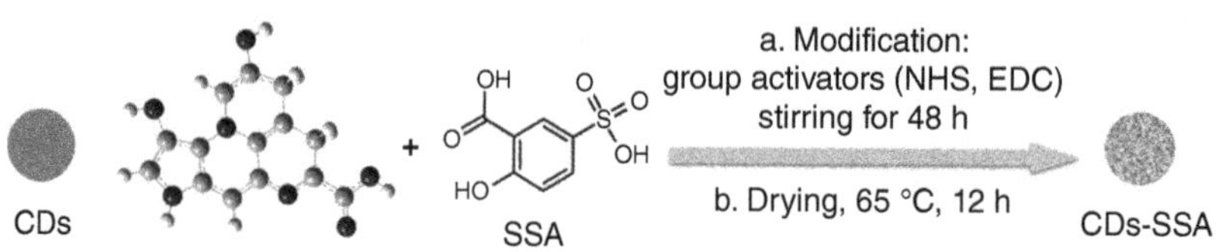

Figure 6.2 Diagram of the preparation process of CDs–SSA.

CD (100 ppm) and CDs–SSA (100 ppm) solutions were separated in glass bottles. Meanwhile, $FeCl_2$ and $FeCl_3 \cdot 6H_2O$ were dissolved in deionized water to obtain Fe^{2+} and Fe^{3+} ionic solutions with different concentrations. After 30 μl of a metal-ion solution was added to 15 ml of the CDs or CDs–SSA solution, photographs were taken under 365-nm UV excitation to investigate the fluorescence-quenching characteristics of CDs and CDs–SSA for Fe^{2+} and Fe^{3+} ions.

6.3.1.7 Self-reporting Performance of Coatings

In a coating scratch experiment, a medical scalpel was used to scrape the coating surface to form defects having a width of approximately 80 μm and a length of 1 cm. The coatings were immersed in a 3.5 wt% NaCl solution and their self-reporting performance was evaluated under irradiation from a 365-nm UV lamp.

6.3.2 Characterization of CDs

6.3.2.1 Structure and Morphology Analysis

TEM images and the particle sizes of the CDs1, CDs1.5, and CDs2 samples are shown in Figure 6.3. In general, when the reaction time was increased from 1 to 1.5 hours and then to 2 hours, the average particle size of the sample first decreased and then increased. At 1 hour, the CDs exhibited a uniform spherical morphology with a size of 2.8–4.2 nm. The average size of the CDs1.5 sample was reduced to 1.8–2.8 nm, with a uniform distribution. However, the average size of CDs2 was increased to 5–8 nm. The AFM images confirmed the distributions of CDs1, CDs1.5, and CDs2. As shown in Figure 6.4, the shapes of all the samples were granular, and the dispersion was uniform. The height distribution of the CDs1 samples was 3–4 nm. With an increase in the reaction time to 1.5 hours, the CDs1.5 particle height decreased to 1–2 nm.

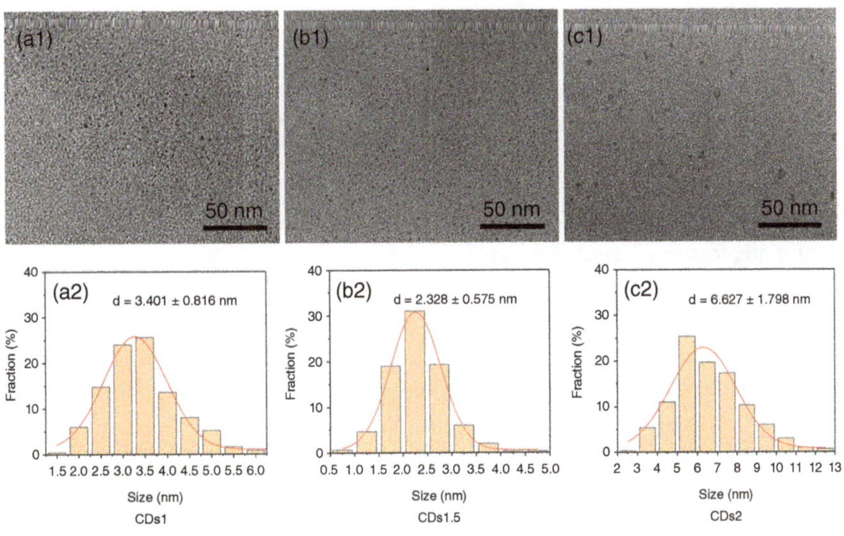

Figure 6.3 TEM images and grain sizes of CDs samples: (a_1,a_2) CDs1; (b_1,b_2) CDs1.5; (c_1,c_2) CDs2.

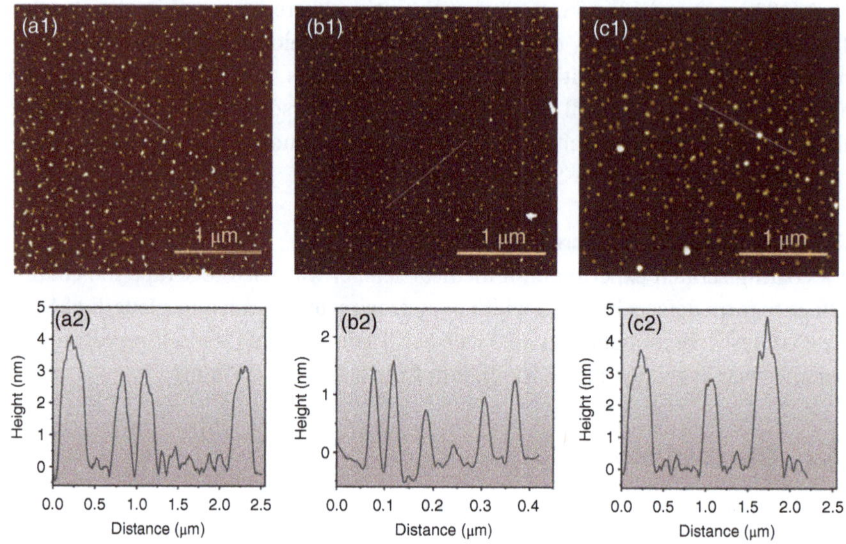

Figure 6.4 AFM images and height distribution of CDs samples: (a_1,a_2) CDs1; (b_1,b_2) CDs1.5; (c_1,c_2) CDs2.

The CDs2 sample had a height distribution of approximately 4–5 nm, which was the highest among all the samples.

For the formation of CDs, in the early stages of the reaction, the reactant precursors underwent dehydration, resulting in aggregation and slight condensation of the decomposition products to form large polymer nanoparticles. As heating proceeded, the polymer nanoparticles shrank owing to continuous intramolecular dehydration [28]. At this stage, C—C bonds were formed, and aromatic clusters were formed inside the polymer. Nucleation of the CDs occurred once the cluster concentration reached a critical oversaturation point. At this stage, the aromatic clusters spread to the particle surface, and simultaneously, various functional groups were passivated to form atomic nuclei. Over time, the polymer nanoparticles tended to die out and transform, and a decrease in the proportion of polymers led to a decrease in the particle size, polycrystallinity, and a more uniform distribution of CDs. As the reaction time increased, an increasing number of CDs molecules tended to aggregate and accumulate, forming larger fragments.

6.3.2.2 Electrochemical Analysis

Figure 6.5 shows the PDP curves of Q235 steel in a 1 M HCl solution with and without different CDs inhibitors. When the inhibitors were added, the polarization curve shifted in the negative potential direction, but its main characteristics remained unchanged. For the samples containing CDs, the corrosion inhibitor mainly affected the cathode reaction, and the anode and cathode current densities decreased significantly. The corrosion inhibitor adsorbed on the surface of Q235 steel can block the active corrosion center, inhibiting both the reduction reaction of hydrogen ions at the cathode and the dissolution reaction of the anode steel.

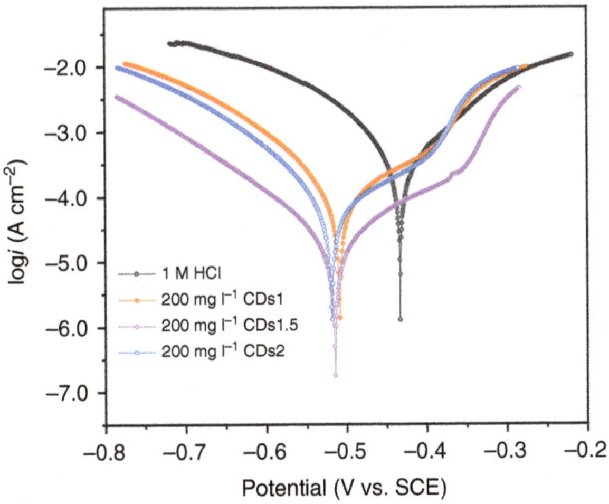

Figure 6.5 Potentiodynamic polarization curve of Q235 steel immersed in 1 M HCl solution containing different CDs samples for 24 hours.

The corrosion potential (E_{corr}), corrosion current density (i_{corr}), anode polarization curve slope (β_a), cathode polarization curve slope (β_c), and corrosion inhibition efficiency (IE) were calculated using the PDP curve, and the calculation results are presented in Table 6.3. Among them, IE can calculate the formula:

$$\mathrm{IE} = \frac{i^0_{corr} - i_{corr}}{i^0_{corr}} \times 100\% \tag{6.1}$$

where i^0_{corr} and i_{corr} represent the corrosion current density of solution without and with CDs, respectively. As indicated by Figure 6.5 and Table 6.3, when the CDs were introduced, the anode and cathode reactions shifted to a lower i_{corr}, suggesting that the CDs were mixed-type corrosion inhibitors. In addition, the degree of inhibition of the anode reaction was lower than that of the cathode reaction, indicating that the cathode reaction was primarily inhibited.

To explore the corrosion inhibition performance and mechanism of CDs, the EIS spectra of all the samples at different times were examined, as shown in Figure 6.6. In the Nyquist diagram, with the addition of CDs, the radius of the capacitor was increased compared with that of the pure HCl solution, indicating that CDs can

Table 6.3 PDP parameters of Q235 steel in different test solutions.

Sample	Conc. (mg l^{-1})	E_{corr} (V)	i_{corr} (μA cm^{-2})	β_a (mV dec^{-1})	$-\beta_c$ (mV dec^{-1})	IE %
Blank	—	−0.4331	254.21	86.8	−65.9	—
CDs1	200	−0.5098	92.63	159.9	−84.3	63.56
CDs2	200	−0.5189	59.11	150.9	−79.3	76.75
CDs1.5	200	−0.5151	21.26	134.5	−102.5	91.64

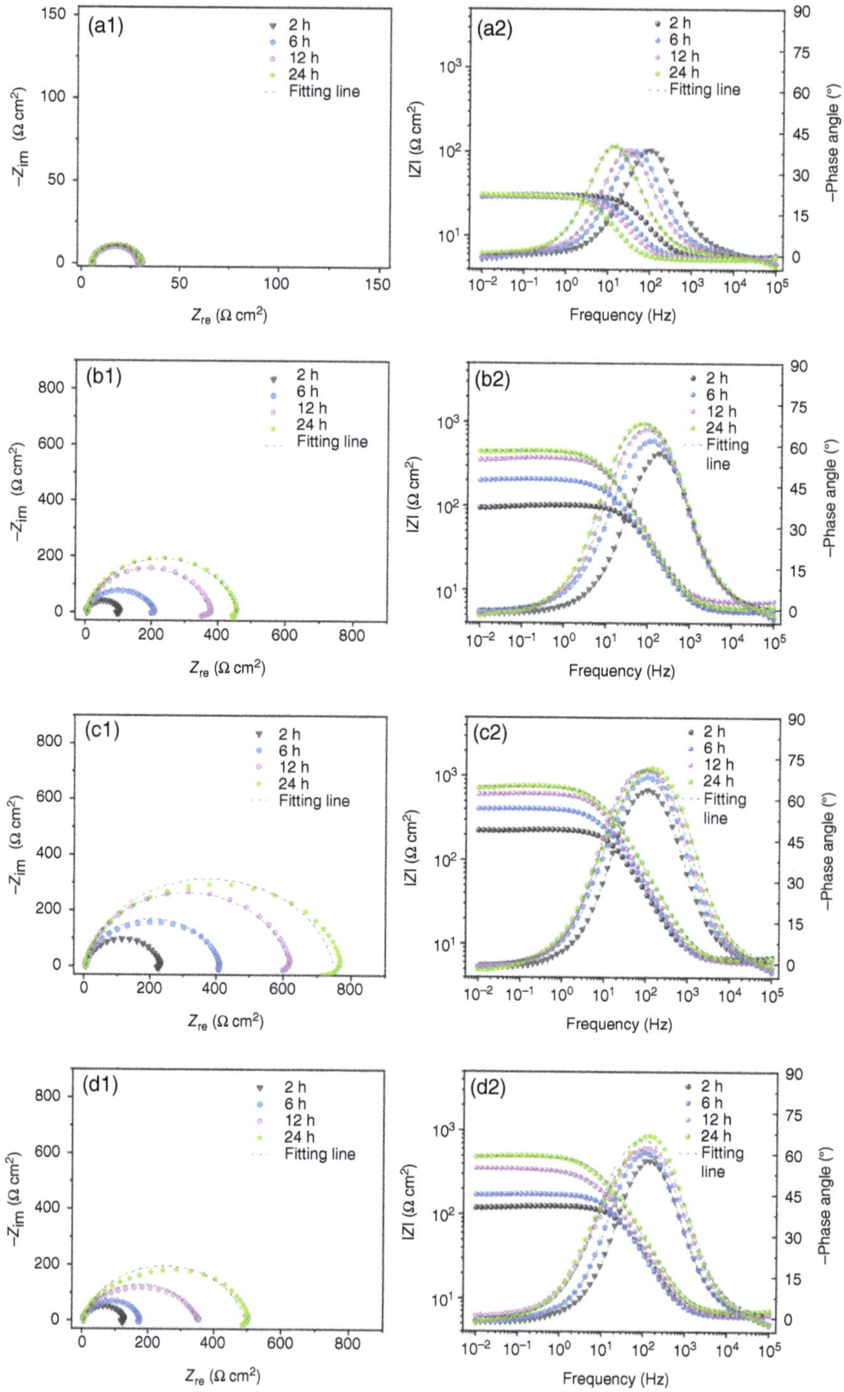

Figure 6.6 Nyquist and Bode diagrams of Q235 steel immersed in 1 M HCl solution for 24 hours: (a_1,a_2) no corrosion inhibitor; (b_1,b_2) 200 mg l^{-1} CDs1; (c_1,c_2) 200 mg l^{-1} CDs1.5; (d_1,d_2) 200 mg l^{-1} CDs2.

form a film on the electrode surface and play a role in corrosion inhibition. In addition, the arc radius of the capacitor increased over time because the steel surface was covered with more inhibitors. Thus, with an increase in the inhibitor concentration, the density of the protective film increased. In the Bode diagram, the impedance modulus at 0.01 Hz was inversely related to the corrosion rate of the electrode. After the addition of CDs, the $|Z|_{0.01\,Hz}$ was significantly higher than that when the substrate was immersed in pure HCl solution. Within 24 h, the $|Z|_{0.01\,Hz}$ for the sample immersed in pure HCl solution changed a little, stabilizing at approximately 30 Ω cm^2; the $|Z|_{0.01\,Hz}$ values of CDs1, CDs1.5, and CDs2 were 450, 714, and 486 Ω cm^2, respectively, at 24 hours. The peak height of the corresponding phase angle curve increased with the adsorption time, indicating the gradual adsorption of CDs at the steel–solution interface. The electrochemical results indicate that CDs can provide effective protection for the steel in HCl solution and that CDs1.5 has the best corrosion inhibition effect.

6.3.2.3 Surface Corrosion Morphology Characterization

The corrosion morphologies of the Q235 steel in different CDs solutions are shown in Figure 6.7. Carbon steel exhibited different characteristics under different corrosion inhibitors. After the steel was immersed in a 1 M HCl solution for 24 hours, owing to the continuous erosion effect of corrosive ions, severe corrosion, and spalling were observed on the steel surface, and the surface roughness reached 2.372 μm. The prepared CDs promoted the formation of an adsorption film, inhibited the corrosion of the steel surface, and significantly reduced the surface roughness. The surface roughness of CDs1, CDs1.5, and CDs2 was 1.875, 1.370, and 1.478 μm, which was 20.95%, 42.24%, and 37.69% lower, respectively, than that in the HCl solution without inhibitors. Among the samples tested, CDs1.5 exhibited the best corrosion inhibition, which was consistent with the EIS results.

6.3.3 Preparation and Properties of CDs–SSA

In the previous section, the morphology and structure of CDs prepared with different reaction times were explored, and it was determined that CDs prepared in 1.5 hours had the smallest particle size, uniform distribution and best corrosion inhibition performance. Hence, in subsequent experiments, CDs with a reaction time of 1.5 hours were used as the reaction substrates to synthesize the modified CDs. Sulfonate groups have been shown to be useful for corrosion protection owing to their passivation ability and corrosion inhibition. Sulfonyl salicylic acid (SSA) can be chelated using Fe^{2+} and Fe^{3+}, and the complex ions formed can resist the further corrosion of the steel matrix in corrosive solutions. The introduction of sulfonate groups onto the surface of the CDs can simultaneously improve their optical properties and enable sensitive corrosion monitoring. In this section, a new complex (CDs–SSA) was obtained using a chemical bonding reaction between SSA and CDs, thereby endowing CDs–SSA with corrosion inhibition properties and sensitivity to corrosion induction.

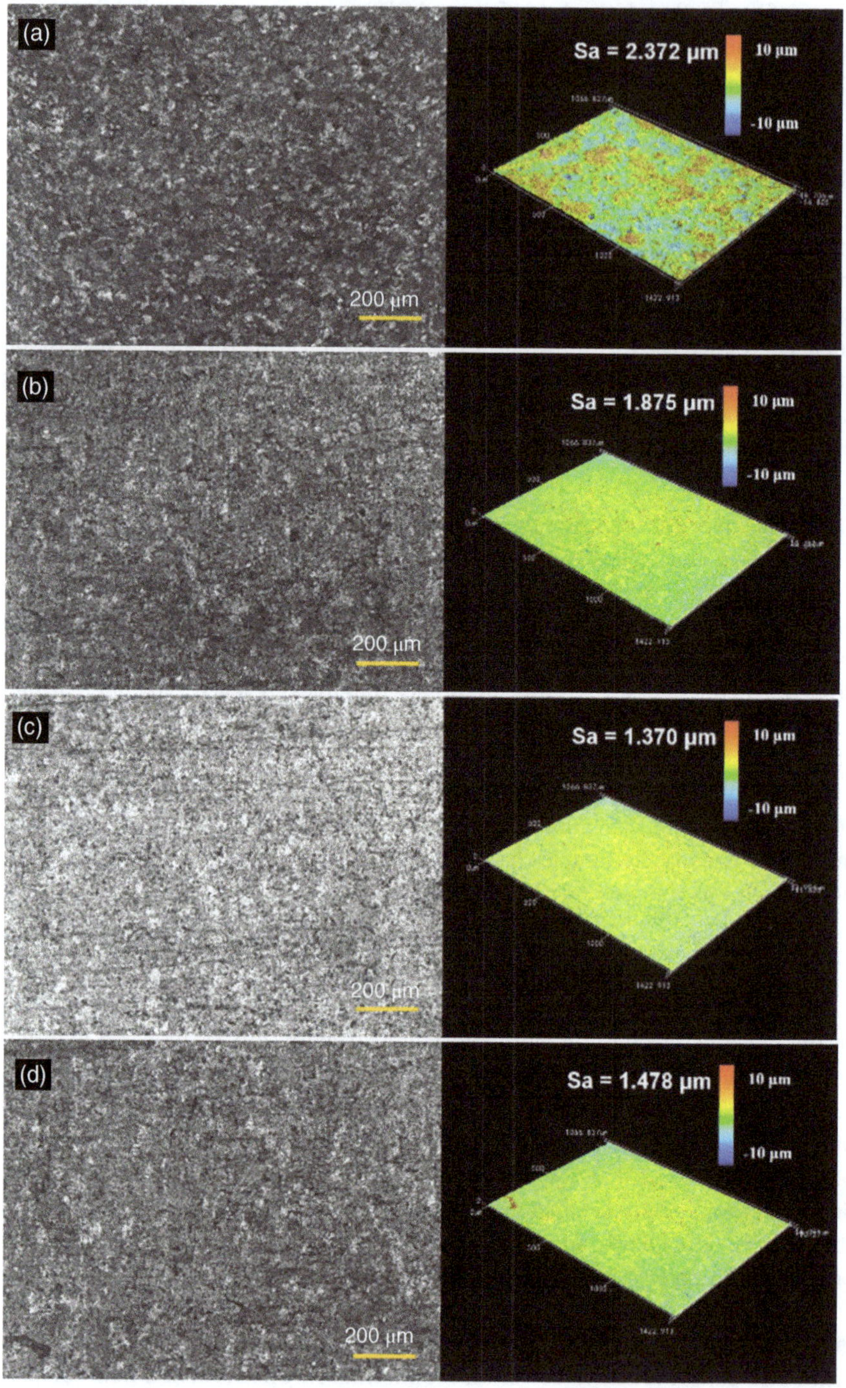

Figure 6.7 Laser confocal morphology and surface roughness of Q235 steel immersed in 1 M HCl solution for 24 hours: (a) no corrosion inhibitor; (b) 200 mg l^{-1} CDs1; (c) 200 mg l^{-1} CDs1.5; (d) 200 mg l^{-1} CDs2.

6.3.3.1 Morphology and Structural Characterization of CDs–SSA

The fabricated CDs–SSA particles were spherical and uniformly distributed, with an average particle size of 20.4 nm (Figure 6.8). The AFM images in Figure 6.8 further confirm that the morphology of CDs–SSA was similar to that of the CDs. However, the average particle size increased considerably, while the particle height distribution was in the range 3–4 nm.

The chemical structures of CDs and CDs–SSA are illustrated in Figure 6.9. The CDs showed some characteristic peaks at 1184, 1349, 1400, 1598, 1707, and 3420 cm^{-1}. New FTIR peaks were observed at 600, 840, and 1228 cm^{-1} for CDs–SSA complexes [29]. The peak at 1712 cm^{-1} was attributed to the ester carbonyl group produced by the CDs–SSA complex. Compared to that of the carboxyl group on the CDs (1707 cm^{-1}), the C=O stretching vibration of CDs–SSA moved slightly toward higher frequencies, wherein the spectral peak was stronger. These new peaks indicate that SSA was successfully grafted onto the CDs. For the UV–Vis spectrum in Figure 6.9c, after grafting SSA onto the CDs, the UV–Vis peak of the CDs was retained, whereas the characteristic peak of SSA at 297 nm was blue-shifted to 282 nm. This was attributed to the formation of an acetate carbonyl group on the CDs–SSA complex, thereby resulting in the redistribution of the electron cloud [30].

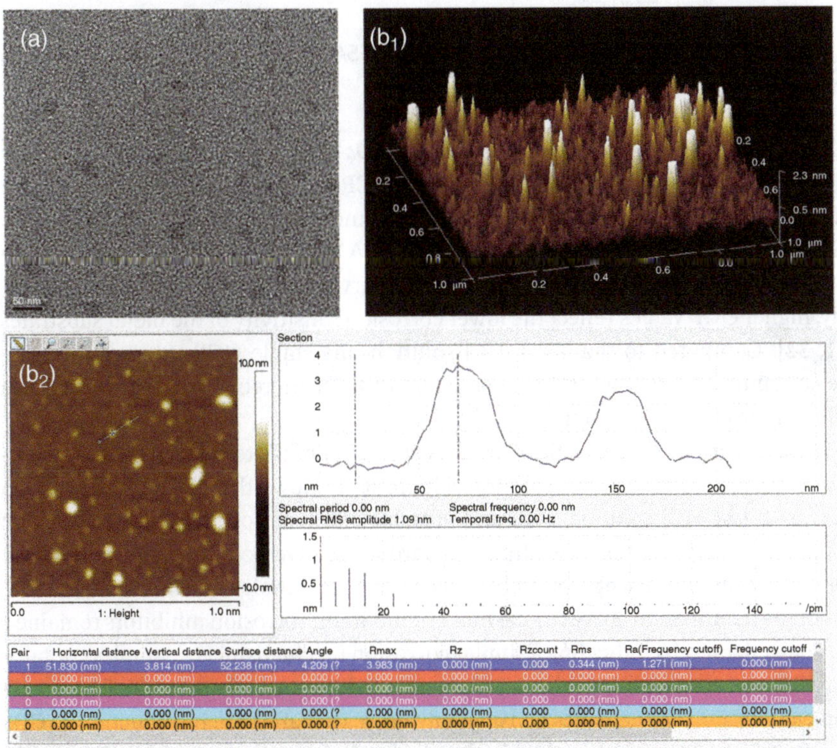

Figure 6.8 (a) TEM image of CDs; (b_1) AFM height profile and (b_2) particle size distribution of CDs–SSA.

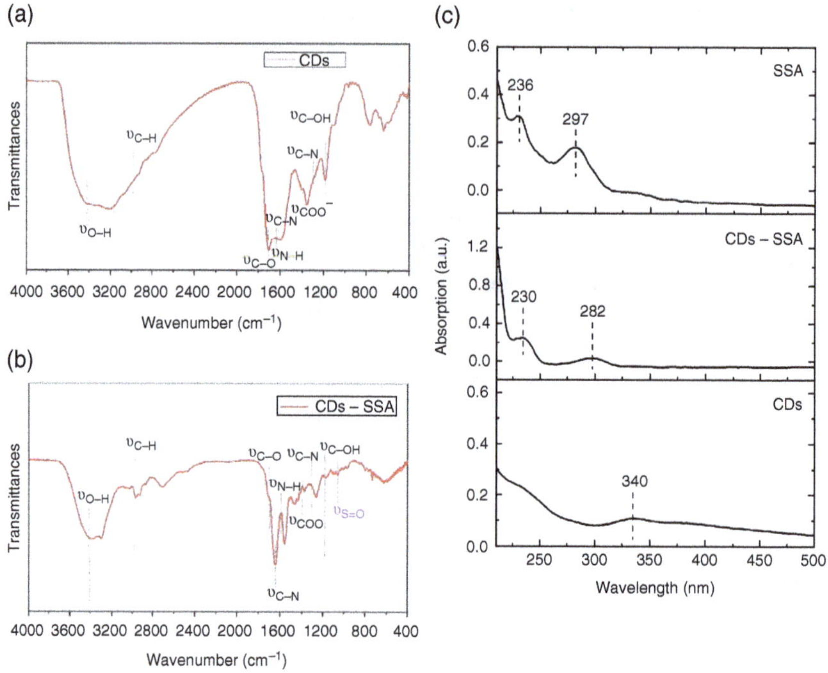

Figure 6.9 (a) FTIR spectra of (a) CDs and (b) CDs–SSA; (c) UV–Vis spectra of CDs and CDs–SSA.

6.3.3.2 Analysis of Electrochemical Results of CDs–SSA

Figure 6.10 shows the open circuit potential (OCP) results of Q235 carbon steel in a 1 M HCl solution without a corrosion inhibitor and with 200 mg l^{-1} CDs, 200 mg l^{-1} SSA, 100 mg l^{-1} CDs, 100 mg l^{-1} SSA (CDs + SSA), and 200 mg l^{-1} CDs–SSA. With the extension of immersion time, all the OCP curves tended to be stabilized. Generally, higher OCP values reflect the lower corrosion sensitivity of the metal substrates [31, 32]. Compared to the blank HCl solution, the stable OCP value of the sample containing corrosion inhibitors shifted positively, thereby reaching a maximum value in the CDs–SSA sample.

These results showed that the CDs–SSA corrosion inhibitor possessed a good corrosion resistance in the HCl solution. Figure 6.11 shows the PDP curves of Q235 steel in a 1 M HCl solution with and without different corrosion inhibitors. The polarization curve of the steel shifted toward a negative potential after adding the corrosion inhibitor. Except for those of the CDs–SSA samples, the main characteristics of the polarization curves in the media containing corrosion inhibitors remained unchanged, thereby indicating a similar corrosion mechanism. For samples containing SSA, the corrosion inhibitors mainly affected the cathode reaction. The anode and cathode current densities considerably decreased after the introduction of CDs or CDs + SSA. Compared to the blank sample, the corrosion potential (E_{corr}) shift was less than 85 mV, thereby indicating that the mixture of CDs and CDs + SSA can be used as a mixed-type corrosion inhibitor [33]. The corrosion inhibitor adsorbed

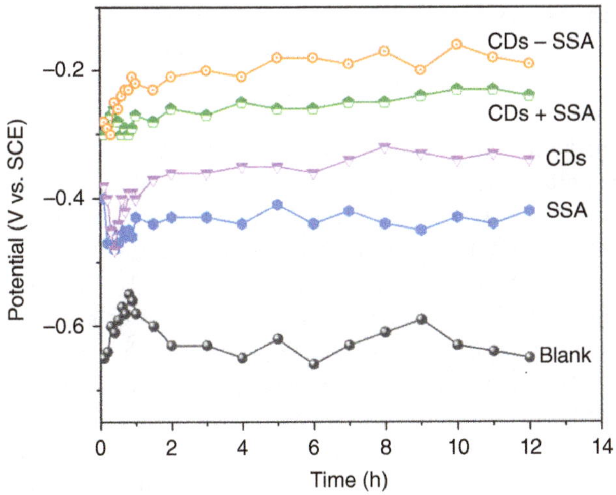

Figure 6.10 OCP curve of Q235 steel in 1 M HCl solution without corrosion inhibitor and containing 200 mg l^{-1} SSA, 200 mg l^{-1} CDs, 100 mg l^{-1} CDs + 100 mg l^{-1} SSA and 200 mg l^{-1} CDs-SSA.

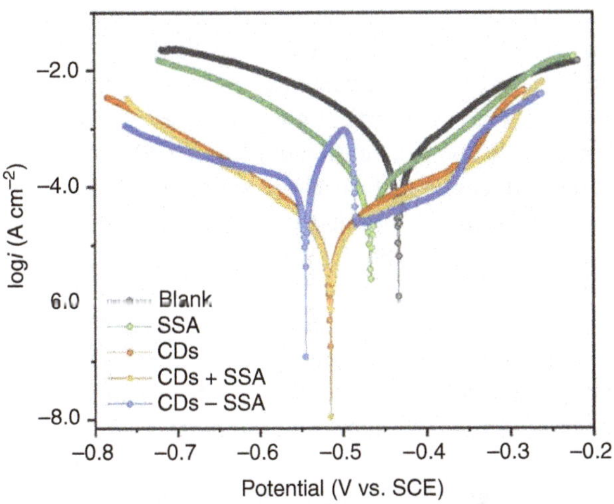

Figure 6.11 Potentiodynamic polarization of Q235 steel in 1 M HCl solution without corrosion inhibitor and containing 200 mg l^{-1} SSA, 200 mg l^{-1} CDs, 100 mg l^{-1} CDs + 100 mg l^{-1} SSA and 200 mg l^{-1} CDs–SSA. Source: Wu et al. [27]/with permission of Elsevier.

on Q235 steel can block the active corrosion center, thereby inhibiting not only the reduction reaction of hydrogen ions in the cathode but also the dissolution reaction of the anode steel. Anodic passivation was observed in the CDs–SSA sample. With a positive increase in E_{corr} potential, the polarization curve of the CDs–SSA passed through the active region, the active-passivation transition region, the passivation region, and the superpassivation region [34, 35]. After reaching the main passivation potential (E_{pp}) of −0.497 V, the corrosion current density decreased rapidly, thereby resulting in a complete passivation film at the passivation potential (E_p) of −0.481 V. Further, the current density decreased from a critical passivation current density (i_{cp}) of 954.98 μA cm^{-2} to a passivation current (i_p) of 23.44 μA cm^{-2}, thereby resulting in no corrosion of the metal substrate.

In the passivation region, the current density increased slightly with an increase in the potential. When the potential was increased to a passivation breakdown potential higher than −0.367 V, a sudden increase in the current density was observed, thereby indicating that the passivation film was destroyed. Table 6.4 presents the E_{corr}, i_{corr}, β_a, and β_c for different samples. The i_{corr} was the highest in the blank HCl solution and exhibited a considerable downward trend. It also decreased by a factor of 10 in the presence of the corrosion inhibitor. The considerable decrease in the i_{corr} value and passivation behavior indicates that the accumulation of corrosion inhibitors on the metal surface can effectively inhibit the corrosion process.

To further study the corrosion inhibition effect and mechanism of different corrosion inhibitors on Q235 carbon steel in the HCl solution, EIS measurements were performed. Figure 6.12 shows the Nyquist and Bode diagrams of Q235 steel after immersion in 1 M HCl in the absence and presence of corrosion inhibitors. As shown in the Nyquist diagram, the diameter of the capacitive reactance arc increased after adding the corrosion inhibitors compared to that of the blank sample, as shown in Figure 6.6. When only CDs or SSA were added, the diameter of the capacitive reactance arc was slightly greater than that of the steel in the blank HCl solution. When the CDs + SSA mixture or the CDs–SSA complex was present, the diameter of the capacitive reactance arc increased further, and the largest diameter was observed for the CDs–SSA sample. With the extension of the immersion time from 6 to 12 hours and 24 hours, the diameter of the capacitive reactance arc showed an increasing trend under the action of corrosion inhibitors. This may be owing to the gradual adsorption of corrosion inhibitors onto the steel surface, thereby hindering the electron transfer process and inhibiting metal corrosion [16, 36]. Conversely, the diameter of the capacitive reactance arc of the steel sample in the blank HCl solution was minimal and remained almost constant. The impedance mode values increased after adding the corrosion inhibitors. By combining the results shown in Figures 6.6 and 6.12 (after immersion for 12 and 24 h), the order of $|Z|_{0.01\,Hz}$ values decreased as follows: CDs–SSA > CDs + SSA > CDs > SSA > control sample. The $|Z|_{0.01\,Hz}$ of CDs–SSA was approximately 65 times higher than that of the $|Z|_{0.01\,Hz}$ of

Table 6.4 PDP parameters of Q235 steel in different test solutions.

	Concentration (mg l^{-1})	E_{corr} (V vs. SCE)	i_{corr} (μA cm^{-2})	β_a (mV dec^{-1})	β_c (mV dec^{-1})
Blank HCl	0	−0.432	256.98	91.5	−70.6
SSA	200	−0.467	90.99	116.8	−68.7
CDs	200	−0.515	19.26	132.1	−105.0
CDs + SSA	100 + 100	−0.515	12.98	117.8	−91.2
CDs–SSA	200	−0.544	—	—	—

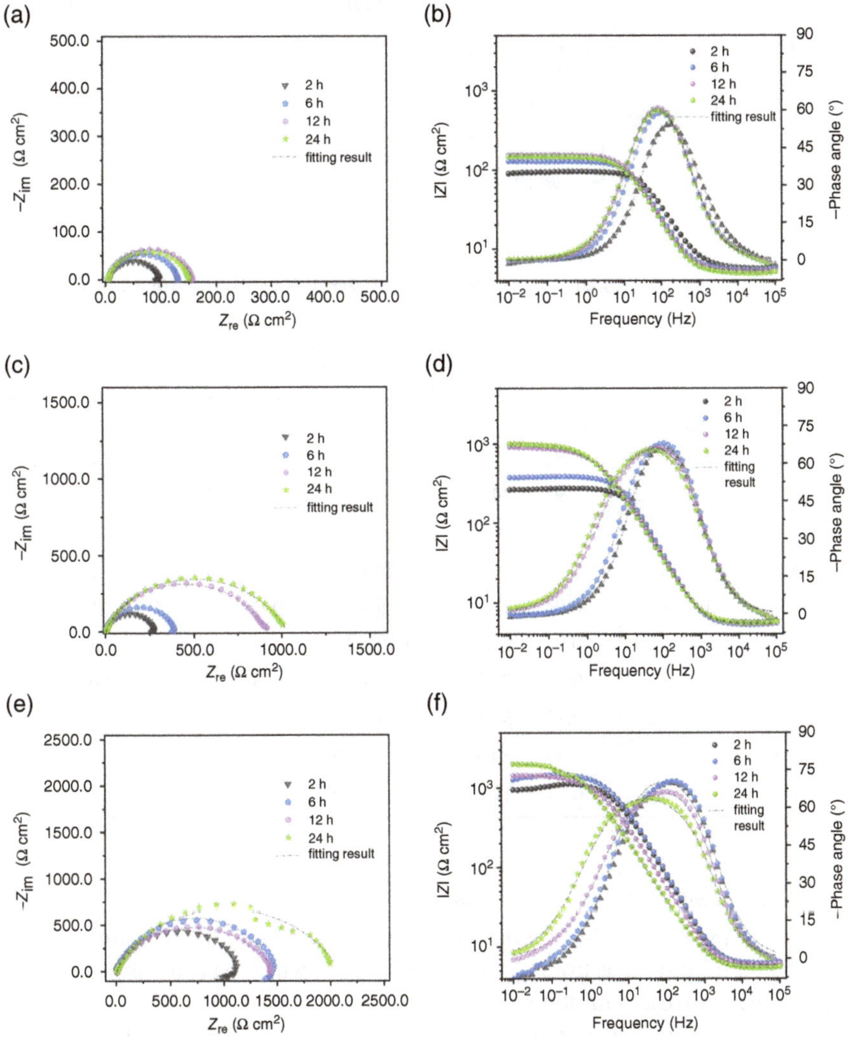

Figure 6.12 Nyquist and Bode diagrams of Q235 steel in 1 M HCl solution for different time: (a,b) 200 mg l^{-1} SSA, (c,d) 100 mg l^{-1} CDs + 100 mg l^{-1} SSA, (e,f) 200 mg l^{-1} CDs-SSA.

steel in blank HCl solution after 24 hours of immersion, thereby indicating the best corrosion inhibition effect of the CDs–SSA composite sample. In the phase angle diagram, the addition of a corrosion inhibitor resulted in an increase in the phase angle; the maximum frequency range was broadened, and strong adsorption of corrosion inhibitors at the steel–solution interface was demonstrated [37].

The equivalent circuit (EEC) as shown in Figure 6.13 was used to fit the EIS data, and the corresponding fitting parameters are presented in Table 6.5. In the EEC, R_s is the solution resistance, R_f is the film resistance, and R_{ct} is the charge-transfer resistance. CPE_f and CPE_{ct} are constant-phase elements used to replace the ideal capacitance of the surface film and electrode interface double layer, respectively. R_{ct}

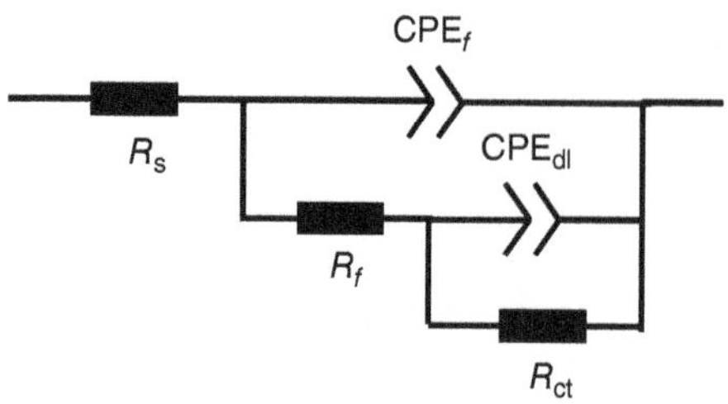

Figure 6.13 EEC used to fit the EIS data.

Table 6.5 EIS parameters of Q235 steel in 1 M HCl solution without and with different corrosion inhibitors.

Time (h)	Sample	CPE$_f$ (Ω^{-1} cm^{-2} S^n)	n_f	R_f (Ω cm^2)	CPE$_{dl}$ (Ω^{-1} cm^{-2} S^n)	n_{ct}	R_{ct} (Ω cm^2)	$\chi^2 \times 10^{-3}$	IE %
12	Blank HCl	7.83×10^{-4}	0.92	10.7	2.46×10^{-3}	0.98	18.6	3.1	—
	SSA	1.26×10^{-4}	0.91	44.3	4.96×10^{-4}	0.66	107.9	2.6	82.8
	CDs	7.60×10^{-5}	0.97	70.1	1.01×10^{-4}	0.64	329.4	1.8	94.3
	CDs + SSA	6.28×10^{-5}	0.92	78.4	1.48×10^{-4}	0.62	849.6	2.9	97.8
	CDs-SSA	3.69×10^{-5}	0.93	111.5	1.18×10^{-4}	0.65	1351.2	2.1	98.6
24	Blank HCl	1.47×10^{-3}	0.93	9.4	1.18×10^{-3}	0.94	16.4	1.6	—
	SSA	1.25×10^{-4}	0.92	70.5	3.52×10^{-4}	0.60	74.9	2.0	78.1
	CDs	9.73×10^{-5}	0.97	85.4	1.74×10^{-4}	0.56	432.5	1.4	96.2
	CDs + SSA	8.71×10^{-5}	0.89	90.4	1.47×10^{-4}	0.64	934.7	3.1	98.2
	CDs-SSA	4.30×10^{-5}	0.87	122.3	1.05×10^{-4}	0.68	1920.0	2.5	99.1

can be used to reflect the corrosion rate, with a higher value indicating a lower rate of corrosion [37, 38]. In the absence of a corrosion inhibitor, the R_{ct} value of steel in the blank solution is as low as 18.60 Ω cm^2 at 12 h, which considerably increases after adding a corrosion inhibitor (Table 6.5).

The results showed that the R_{ct} value reached the maximum (1351.2 Ω cm^2) after adding the CDs–SSA corrosion inhibitor for 12 hours. It further increased to 1920.0 Ω cm^2 with the extension of the immersion time to 24 hours. Comparing the two corrosion inhibitors, the R_{ct} value of the CDs–SSA composite corrosion inhibitor was doubled, thereby indicating that the composite corrosion inhibitor can effectively improve the corrosion inhibition effect. As regards the CPE$_{dl}$ value, the trend was contrary to the R_{ct} value. After immersion in the blank HCl solution for 24 h, the CPE$_{dl}$ value of steel was approximately 1.18×10^{-3} Ω^{-1} cm^{-2} S^n value, whereas it decreased considerably in the presence of SSA, CDs, CDs + SSA, and CDs–SSA. The considerable increase in the R_{ct} value and the decrease in the CPE$_{dl}$ value is owing to the adsorption of corrosion inhibitors at the steel–solution interface and the increase in the thickness of the double electrode layer [36, 39].

Table 6.5 also presents the *IE* values of the different corrosion inhibitors. The IE value of the sample containing SSA, CDs and CDs + SSA at 12 hours were 82.8%,

94.3%, and 97.8%, respectively. The IE values of the samples containing CDs–SSA were the highest (98.6% at 12 hours and 99.1% at 24 hours), thereby indicating that CDs–SSA had good corrosion inhibition properties and could reduce the corrosion rate of Q235 carbon steel in the HCl solution. Notably, the synthetic effect of the CDs–SSA inhibitor has a higher IE value than most similar CDs-based inhibitors reported in the literature.

6.3.3.3 Surface Corrosion Morphology Characterization

Figure 6.14a–d show the 3D surface morphology and surface roughness (S_a) of Q235 steel after immersing in 1 M HCl solution without and with corrosion inhibitors for 24 hours. Owing to the continuous corrosion effect caused by the corrosive medium, the surface of the blank steel was extremely corroded, forming numerous cracks, pits, and bumps. Here, the surface roughness was as high as 8.302 μm. When the steel was immersed in an HCl solution containing SSA, the corrosion was slightly suppressed, and the number of pits and cracks was reduced. When the steel was immersed in the CDs + SSA mixed solution, or the CDs–SSA solution, the surface of the steel was smoother, and there was no obvious corrosion morphology. The surface roughness of the steel in 1 M HCl solution containing SSA, CDs + SSA, and CDs–SSA was 3.303, 1.403, and 0.267 μm, which was 60.2%, 83.1%, and 96.8% lower than that of the blank sample, respectively. Here, polished scratches can be observed on the CDs–SSA sample, thereby indicating that CDs–SSA had the best corrosion inhibition performance on carbon steel.

WCA measurements reflect the hydrophobic properties of metal surfaces adsorbed by corrosion inhibitors. Figure 6.15 shows the WCA of Q235 carbon steel before and after immersion in a 1 M HCl solution containing different corrosion inhibitors. The WCA of the polished steel was 77.5°, which was considerably reduced to 31.7° after immersion in a 1 M HCl solution for 24 hours. This indicates that during severe corrosion, the hydrophilicity of the steel was enhanced, water droplets easily diffused, and more free iron ions on the corroded surface had synergistic effects with corrosion molecules in the solution [25]. With the addition of the corrosion inhibitors, SSA, CDs, CDs + SSA, and CDs–SSA to the solution, the WCA increased to 43.2°, 54.1°, 59.2°, and 84.7°, respectively. Here, the maximum WCA is obtained for the CDs–SSA sample, thereby indicating that the adsorption can improve the hydrophobic properties of the steel surface and isolate the metal from corrosion in aggressive media [23].

SEM was used to observe the two-dimensional morphology of Q235 steel after immersion in a 1 M HCl solution for 24 h in the absence and presence of corrosion inhibitors (Figure 6.16). In the blank solution, the steel surface exhibited severe corrosion with pits and irregular grooves. In a 200 mg l^{-1} SSA solution, the surface morphology of the steel was similar to that of the blank sample, owing to the poor corrosion inhibition performance of SSA. Conversely, after adding 200 CDs, 100 mg l^{-1} CDs + 100 mg l^{-1} SSA, and 200 mg l^{-1} CDs–SSA, the corrosion severity and number of defects considerably reduced, thereby weakening the corrosion effect.

The results of the energy spectrum analysis (Table 6.6) confirm the elemental composition of the steel surface. In the blank samples, the presence of O and Cl in

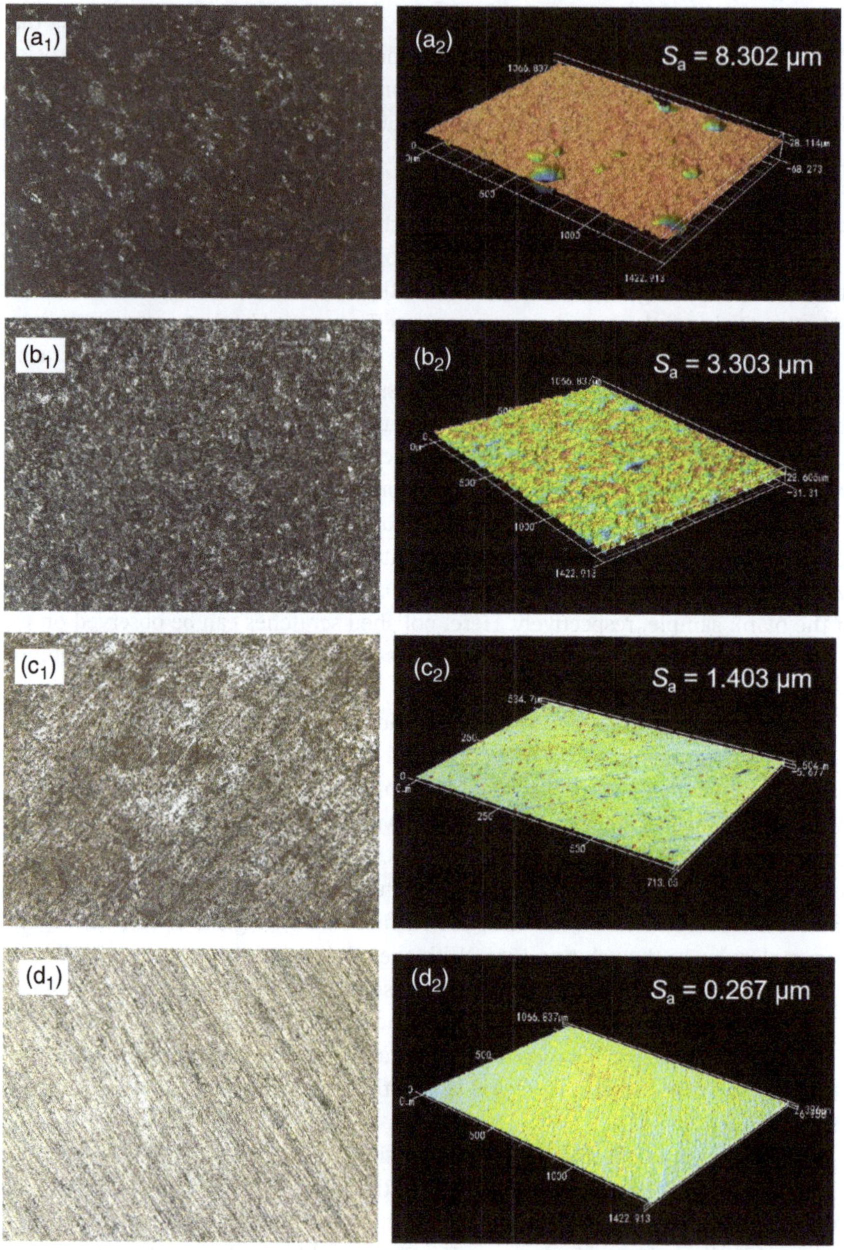

Figure 6.14 3D confocal microscope image and surface roughness of Q235 steel after immersing in 1 M HCl solution for 24 hours: (a_1,a_2) without inhibitor, (b_1,b_2) 200 mg l^{-1} SSA, (c_1,c_2) 100 mg l^{-1} CDs + 100 mg l^{-1} SSA, (d_1,d_2) 200 mg l^{-1} CDs-SSA.

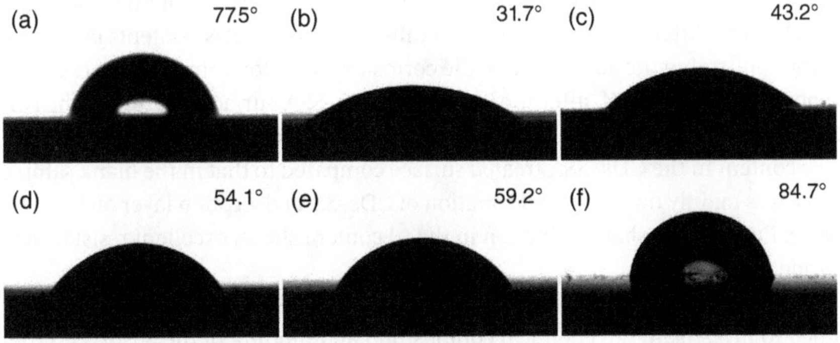

Figure 6.15 WCAs of Q235 carbon steel (a) before and (b) after immersion in 1 M HCl solution; WCAs of Q235 carbon steel after immersion in 1 M HCl solution containing (c) SSA, (d) CDs, (e) CDs + SSA, (f) CDs–SSA.

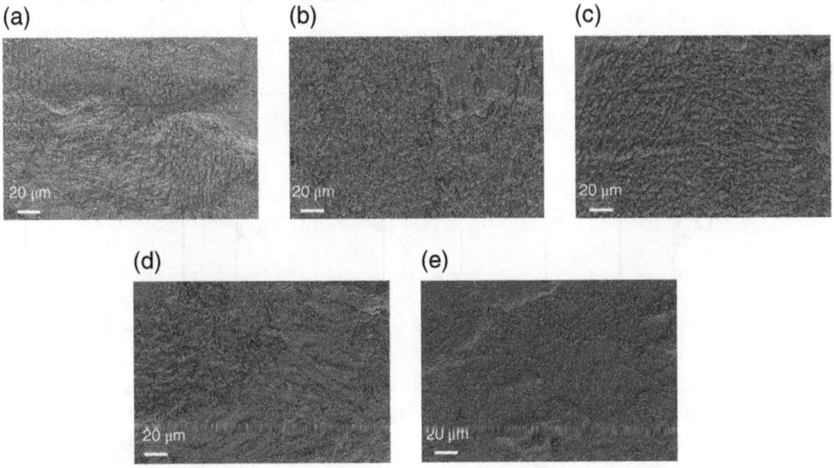

Figure 6.16 SEM image of Q235 steel immersed in 1 M HCl solution, in which (a) no corrosion inhibitor, (b) 200 mg l^{-1} SSA, (c) 200 mg l^{-1} CDs, (d) 100 mg l^{-1} CDs + 100 mg l^{-1} SSA, And (e) 200 mg/L CDs-SSA.

Table 6.6 Surface element content of Q235 steel after immersing in different solutions for 24 hours.

Elemental composition (wt%)	Fe	C	O	N	S	Cl
Blank HCl solution	60.40	26.53	12.20	0.50	—	0.36
SSA	57.98	28.55	12.52	0.68	0.11	0.36
CDs	46.92	39.76	9.29	3.71	—	0.32
CDs + SSA	53.58	32.09	12.06	1.98	0.11	0.18
CDs–SSA	51.68	37.20	7.29	3.64	0.12	0.06

addition to Fe and C from the steel matrix indicates the formation of iron oxides and ferric chloride. After adding the corrosion inhibitor, the C and N contents increased, thereby confirming the adsorption of the corrosion inhibitor onto the metal surface. As shown in Figure 6.16f, after the 200 mg l^{-1} CDs–SSA surface treatment, the surface of Q235 steel comprised Fe, C, O, N, S, Cl, and other elements. The decrease in the Fe content in the CDs–SSA treated surface compared to that in the blank sample surface was mainly owing to the formation of CDs–SSA adsorption layer on the steel surface. Further, the sharp reduction in the Cl content shows excellent resistance to chloride ion attack.

The XPS spectra of Q235 steel immersed in a 1 M HCl solution for 24 h were studied to investigate the chemical composition and binding of the adsorbed layer. Figure 6.17 illustrates the full XPS spectra of the steel surfaces with and without different corrosion inhibitors. In the full XPS spectrum of the steel sample without corrosion inhibitors, the surface of Q235 steel in 1 M HCl solution contained Fe, O, and C, the XPS spectrum of the sample containing CDs exhibited Fe, O, C, and

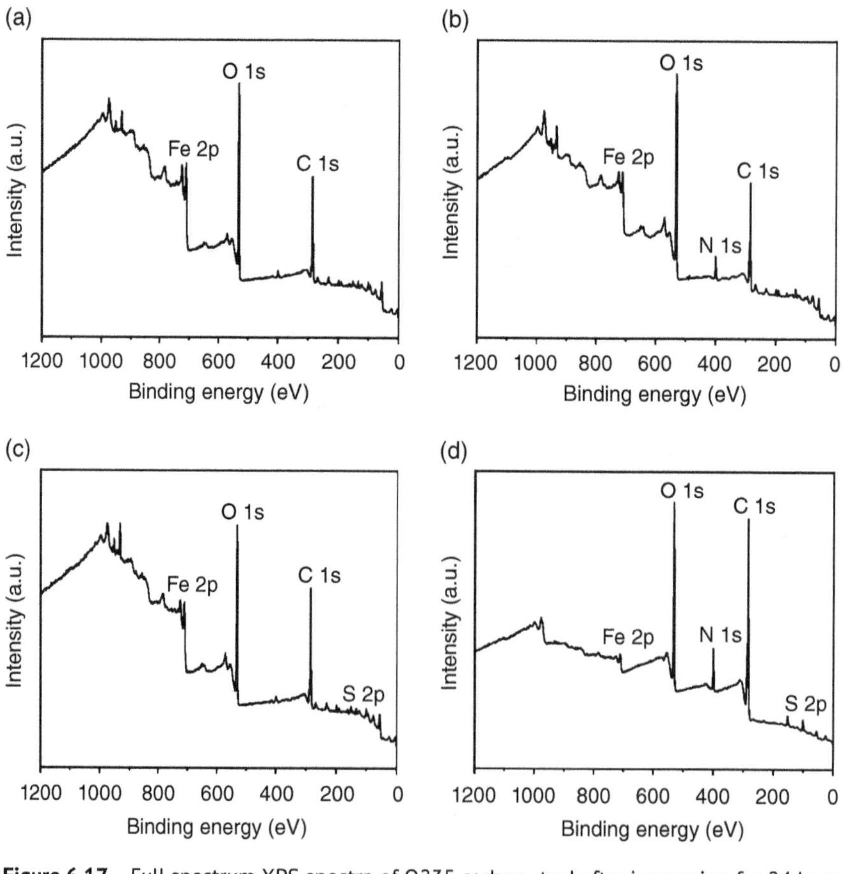

Figure 6.17 Full spectrum XPS spectra of Q235 carbon steel after immersing for 24 hours in 1 M HCl solution (a) without corrosion inhibitors, and with (b) 200 mg l^{-1} CDs, (c) 200 mg l^{-1} SSA, and (d) 200 mg l^{-1} CDs–SSA.

N, the XPS spectrum of the sample containing SSA exhibited elemental Fe, O, C, and S, and the XPS spectrum of the sample containing CDs–SSA exhibited Fe, O, C, N, and S. These results confirm that the corrosion inhibitors were successfully adsorbed onto the steel surface during the immersion process.

Figures 6.18, 6.19, and 6.20 show the high-resolution XPS spectra of Q235 carbon steel after immersion for 24 h in a 1 M HCl solution containing 200 CDs, 200 SSA, and 200 mg l^{-1} CDs–SSA, respectively. In the CDs-containing sample, the characteristic XPS peaks were observed for N, O, Fe, and C. In the SSA-containing sample, characteristic XPS peaks were observed for S, O, Fe, and C. For the CDs–SSA sample, the N 1s spectrum can be divided into three peaks at 399.4, 400.4, and 401.9 eV, corresponding to the N—C, N—Fe, and N—H bonds, respectively [38, 39]. The S 2p spectrum is divided into two peaks at 161.5 and 162.6 eV, corresponding to the S–Fe and C–S bonds, respectively [40]. The formation of N—Fe and S—Fe bonds confirmed the interaction between CDs–SSA and the steel surface, which formed a corrosion-inhibition film that prevented the corrosion of carbon steel [22, 41]. The C 1s spectrum comprises four peaks located at 284.5, 285.6, 286.7, and 288.2 eV, respectively [15, 42]. The O 1s spectrum has three deconvolution peaks at 530.1, 531.7, and

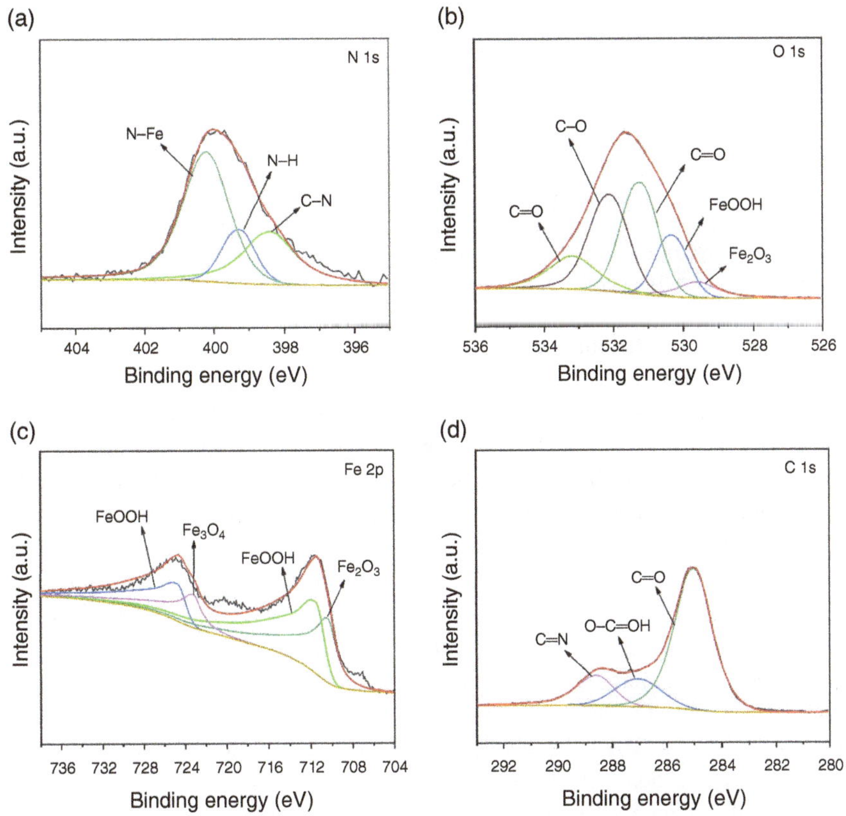

Figure 6.18 High-resolution XPS spectra of Q235 carbon steel after immersing for 24 hours in 1 M HCl solution containing 200 mg l^{-1} CDs: (a) S 2p, (b) O 1s, (c) Fe 2p, (d) C 1s.

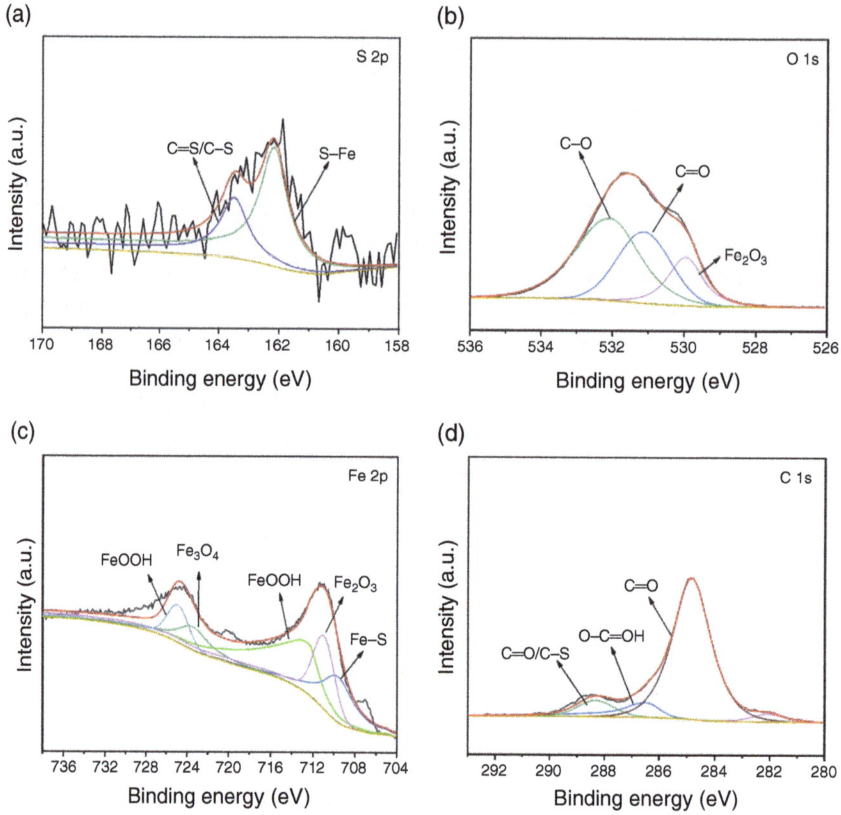

Figure 6.19 High-resolution XPS spectra of Q235 carbon steel after immersing for 24 hours in 1 M HCl solution containing 200 mg l^{-1} SSA: (a) S 2p, (b) O 1s, (c) Fe 2p, (d) C 1s.

533.0 eV, which are attributed to the FeO/O···Fe, C—O, and C—O bonds, respectively. Simultaneously, the high-resolution spectrum of Fe $2p_{3/2}$ can simulate the synthesis of four peaks of Fe—S, Fe···N_x, Fe_2O_3, and FeOOH with binding energies of 709.2, 710.1, 711.7, and 713.9 eV, respectively [43]. The abovementioned XPS results confirm that the chemical reaction of CDs–SSA with the steel surface forms coordination bonds, corresponding with electrochemical testing and surface characterization results.

6.3.3.4 Corrosion Sensitivity of CDs–SSA

After verifying the effectiveness of the CDs–SSA complex as a corrosion inhibitor, its sensitivity to pH changes, and its ability to detect Fe^{2+} and Fe^{3+} were evaluated. Figure 6.21 shows fluorescence images of the CDs and CDs–SSA solutions at different pH values. The color change in an alkaline environment facilitates the sensing of the cathodic corrosion process in a localized area. Figure 6.22a shows a mixture of 100 ppm CDs and different concentrations of Fe^{2+} ions (1×10^{-6}, 1×10^{-5}, 1×10^{-4}, 5×10^{-4}, 1×10^{-3}, and 1×10^{-2} M) excited using a 365 nm UV light. The CDs exhibited clear fluorescence emission, although the addition of

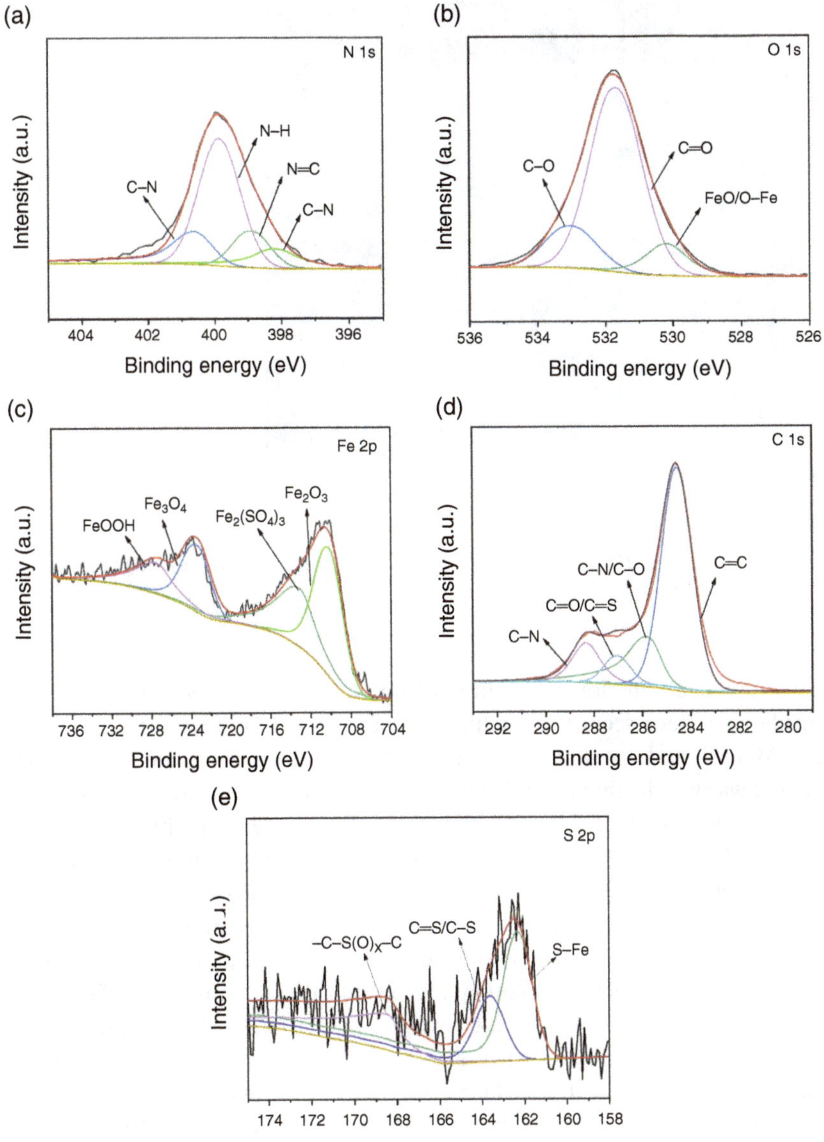

Figure 6.20 High-resolution XPS spectra of Q235 carbon steel after immersing for 24 hours in 1 M HCl solution containing 200 mg l^{-1} CDs–SSA: (a) S 2p, (b) O 1s, (c) Fe 2p, (d) C 1s.

Fe^{2+} ions had no considerable effect on the emission intensity, thereby indicating that the sensitivity of the CDs to Fe^{2+} ions was not high. Regarding Fe^{3+} ions, Figure 6.22b shows that when the concentration of Fe^{3+} ions was higher than 1×10^{-4} M, the fluorescence emission of CDs was quenched, although the decrease in fluorescence intensity did not enable the sensitive detection of Fe^{3+}. Conversely, Figure 6.22c,d show that CDs–SSA exhibited a brighter

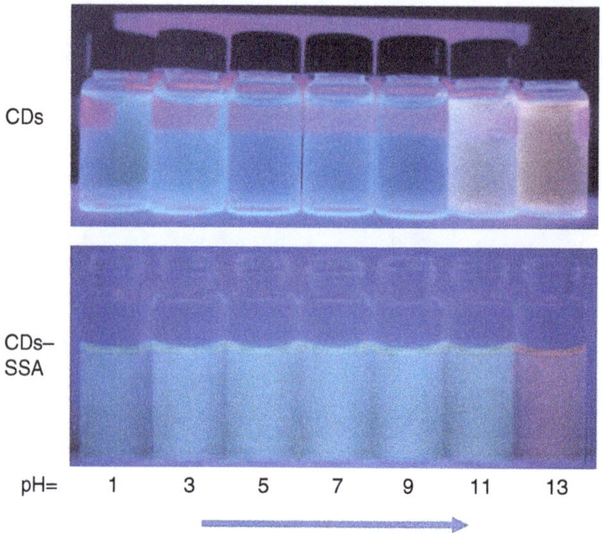

Figure 6.21 Images of 100 ppm CDs and CDs–SSA at different pH values, adjusted by adding HCl or NaOH.

blue fluorescence emission when interacting with Fe^{2+} and Fe^{3+} ions and had a considerable fluorescence quenching effect. When CDs–SSA interacted with 5×10^{-4} M Fe^{2+} and 1×10^{-5} M Fe^{3+} or higher concentrations, the fluorescence quenching was visible, thereby indicating that the detection sensitivity of CDs–SSA was higher than that of CDs. As shown in Figure 6.22g–h, the fluorescence intensity of CDs–SSA decreases sharply with an increase in metal ion concentration, thereby achieving a considerable fluorescence quenching effect. After introducing CDs–SSA complexes into SSA, sulfur atoms in the sulfonic acid groups had low electronegativity and large radii. Further, they were prone to lose their outermost electrons, thereby resulting in the coordination of Fe^{2+}/Fe^{3+} ions with sulfonic acid groups [44]. Compared to CDs, CDs–SSA complexes have more active surface groups. As such, Fe^{2+}/Fe^{3+} ions are easily adsorbed onto the surface of CDs–SSA to form non-fluorescent chelates, which have a considerable fluorescence quenching effect. Hence, CDs–SSA can detect Fe^{2+} and Fe^{3+} ions at the initial stage of corrosion of the steel matrix, thereby predicting the corrosion phenomena.

6.3.4 Self-reporting and Self-healing Properties of CDs–SSA-Based Epoxy Coating

Figure 6.23 shows the self-reporting capabilities of the coatings. When CDs–SSA was dropped onto the steel surface, and immersed the steel in the corrosion solution, the metal substrate corroded, thereby producing Fe^{2+} ions. Here, the H_2O_2 solution was added to the surface of the coating, and the site in contact with the H_2O_2 solution

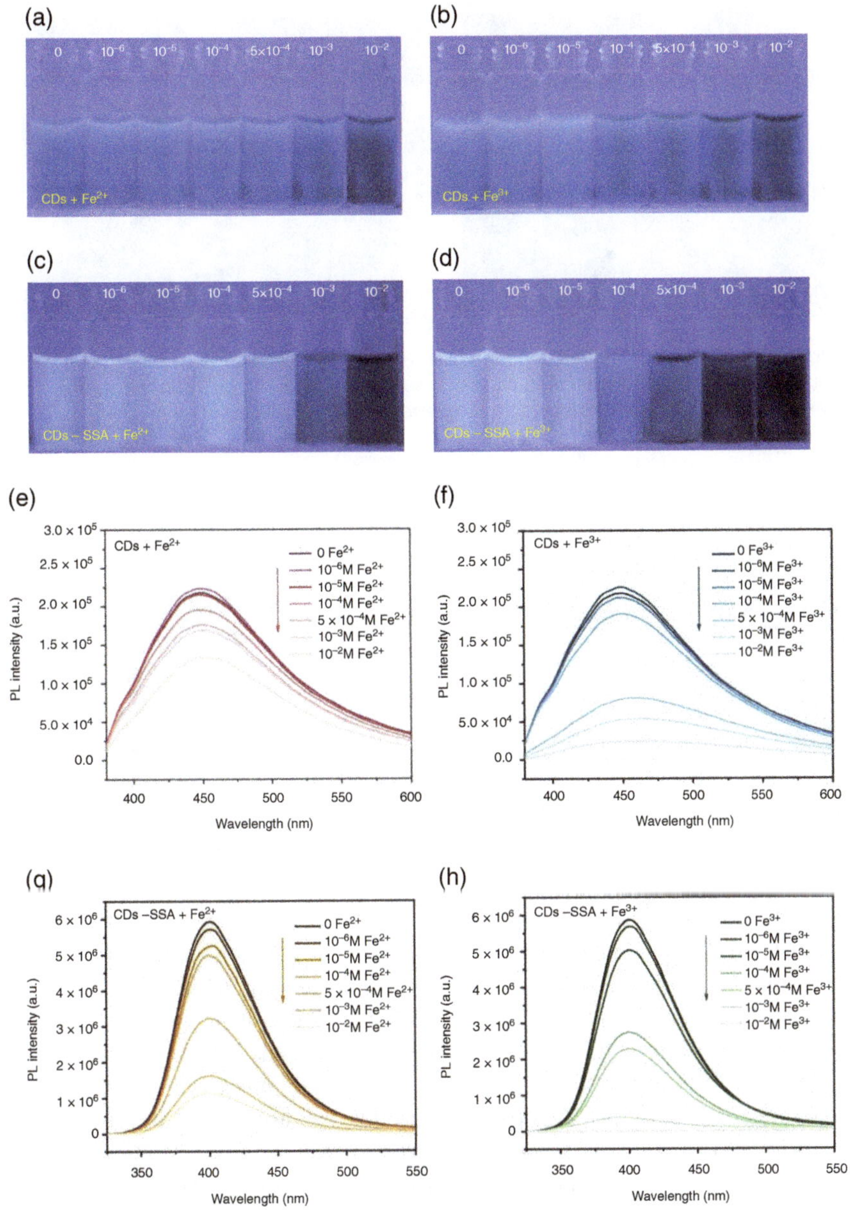

Figure 6.22 Images of 100 ppm CDs after reaction with different concentrations (unit: M) of (a) Fe^{2+} and (b) Fe^{2+} ions under excitation of 365 nm ultraviolet light; image after reaction of 100 ppm CDs–SSA and (c) Fe^{2+} (d) Fe^{2+} ions; Corresponding fluorescence spectra: (e,f) CDs; (g,h) CDs–SSA. Source: Wu et al. [27]/with permission from Elsevier.

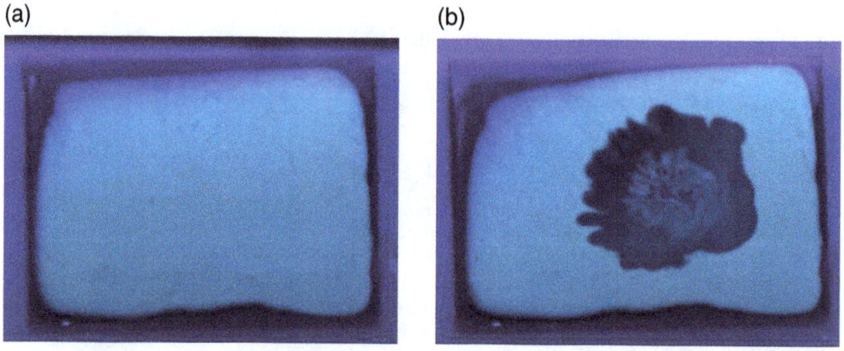

Figure 6.23 Fluorescence quenching effect of the CDs–SSA on steel surface: (a) CDs-SSA dropping, and immersed the steel in corrosion solution; (b) adding H_2O_2 on the steel surface, and the area became fluorescence quenched.

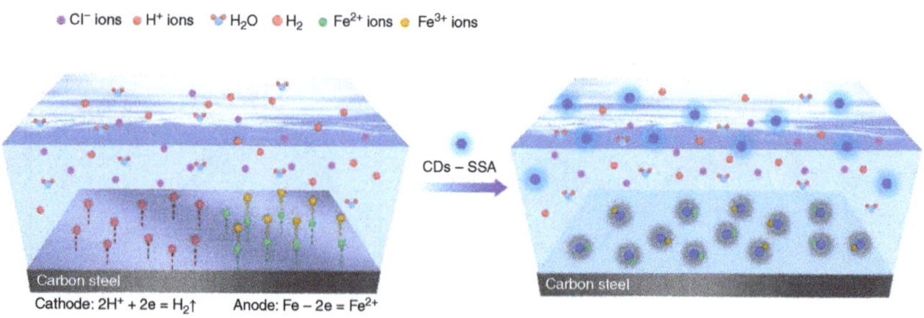

Figure 6.24 Self-healing effect of the CDs–SSA based epoxy coating.

quickly turned black because the Fe^{2+} ions were oxidized to Fe^{3+} ions. Further, fluorescence quenching occurred when the Fe^{3+} ions reacted with the CDs–SSA on the steel surface. This phenomenon indicates the corroded areas on the metal substrate. This phenomenon will also occur even without the addition of the H_2O_2 solution; however, the conversion of Fe^{2+} ions into Fe^{3+} ions would be longer. Owing to the outstanding corrosion inhibition effect of CDs–SSA, the coating with CDs–SSA can provide good self-healing effects. Once the coating is damaged with microscale or larger scratches, the CDs–SSA corrosion inhibitors can be released at the damaged part of the coating [45]. As shown in Figure 6.24, the corrosion reaction stops at the exposed metal surface, thereby demonstrating the good self-healing and active protection performance of the coating [46].

The effect of CDs–SSA addition on the corrosion protection effect of the coating was further studied using EIS. Figure 6.25 shows the EIS results of the intact blank epoxy coating containing 0.5 wt% CDs–SSA after immersion in 3.5 wt% NaCl solution for 12–120 hours. The $|Z|_{0.01\,Hz}$ values of the epoxy coating were maintained at a $\sim 2 \times 10^8\ \Omega\,cm^2$ level within the immersion period, and those of the epoxy coating containing 0.5 wt% CDs–SSA were maintained at a $\sim 2 \times 10^9\ \Omega\,cm^2$ level. This indicates that the addition of CDs–SSA hampered the penetration of the corrosive media,

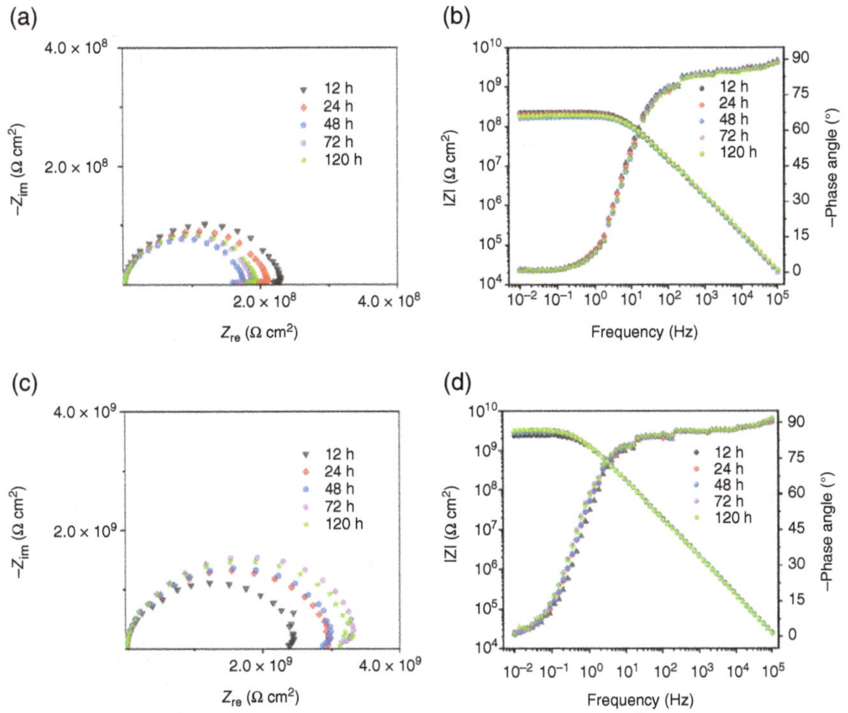

Figure 6.25 Nyquist plots and Bode diagrams of (a,b) intact blank epoxy coating and (c,d) intact epoxy coating containing 0.5 wt% CDs-SSA during immersion in 3.5 wt% NaCl solution.

thereby optimizing the barrier properties of the coating. Figure 6.26 shows the EIS results of the intact blank epoxy coating and that containing 0.5 wt% CDs–SSA after immersion in 3.5 wt% NaCl solution for 2–48 hours. The arc radius of the capacitive reactance of the blank epoxy coating decreased with time, and the $|Z|_{0.01\,\text{Hz}}$ values decreased to some extent. This is owing to the penetration of the corrosive media through the coating scratch. After adding the CDs–SSA corrosion inhibitors to the coating, the $|Z|_{0.01\,\text{Hz}}$ values remained high. With the release of the CDs–SSA, the decreased rate of the $|Z|_{0.01\,\text{Hz}}$ values reduced considerably and were 8.2×10^5 and $8.1 \times 10^5\ \Omega\,\text{cm}^2$ after 24 and 48 hours. The $|Z|_{0.01\,\text{Hz}}$ values of the composite coating were 10 times higher than that of the blank coating. Hence, with an appropriate amount of CDs–SSA loaded in the coating, the corrosion resistance and self-healing properties can be ensured.

6.4 Metal Ion-responsive Self-reporting Coatings Based on UF/Phen Microcapsules

In the actual application process, it is important to realize the corrosion-sensing property of coating damage. Even small scratches (nanoscale or microscale) can

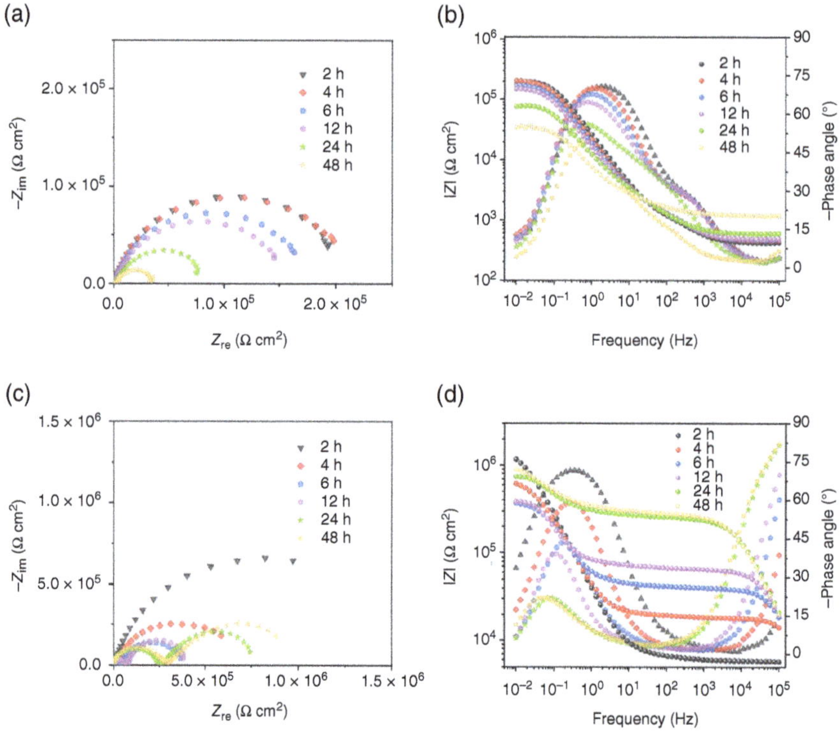

Figure 6.26 Nyquist plots and Bode diagrams of (a,b) scratched blank epoxy coating and (c,d) scratched epoxy coating containing 0.5 wt% CDs–SSA during immersion in 3.5 wt% NaCl solution.

significantly degrade the overall performance of a material and lead to catastrophic failure [47–50]. In view of the abovementioned situation, it is necessary for inspectors to detect, identify, and manually repair damage promptly [51]. However, under normal circumstances, it is difficult to observe coating damage with the naked eye without the assistance of professional instruments. It must be detected via electrochemical measurement, thermal imaging, ultrasonic testing, etc.; however, these methods not only require a large amount of equipment but also have difficulty with damage detection over large areas or with manual access. In addition, some of these techniques are complex and time-consuming. Therefore, the development of self-reporting coatings has important practical applications [52, 53]. Self-reporting coatings can conveniently and quickly warn of corrosion under the coating film and provide efficient and visual signals for coating damage or early corrosion monitoring in a nondestructive manner, which has considerable practical significance.

Few studies have been performed on smart coatings that can realize damage warnings visible to the naked eye and have self-healing functions. Phen, an indicator of Fe^{2+}, exhibits an orange-red color through its coordination with Fe^{2+} and has a corrosion-inhibitory effect on carbon steel [52, 54]. In this study, Phen–urea–

formaldehyde resin (UF/Phen) microcapsules with corrosion-sensing and self-healing functions were prepared and incorporated into an epoxy coating [55]. When corrosion occurs at the scratch, Fe^{2+} reacts with Phen to form an orange–red complex, which simultaneously realizes the dual functions of early corrosion sensing and corrosion inhibition. In addition, by adjusting the microcapsule content and curing method, the T_g of the shape-memory coating can be reduced, which can be triggered under sunlight to promote the self-healing effect of the coating. The effects of different contents of UF/Phen microcapsules on the self-healing properties of the coating were studied using EIS, and a detailed corrosion inhibition process was revealed. The ratio of Fe^{2+} to Phen was adjusted to clarify the law of color development, and the early self-reporting effect of the coating was examined through salt spray experiments.

6.4.1 Materials and Methods

6.4.1.1 Materials

The materials and reagents used in this section are shown in Table 6.7.

6.4.1.2 Preparation of UF/Phen Microcapsules

The preparation process for urea–formaldehyde microcapsules coated with Phen was as follows: 0.15 g of polyvinyl alcohol, 0.1 g of sodium lauryl sulfate, 0.15 g of NH_4Cl, and 0.15 g of resorcinol were poured into a beaker filled with 75 ml of water and stirred for 10 minutes. Subsequently, 5 ml of a benzyl alcohol (BnOH) solution containing 0.4 g of phen was poured into the abovementioned solution

Table 6.7 Experimental materials and reagents.

Materials	Purity	Company
Carbon steel	Q235	Dongguan Hengde Steel
Urea	99%	McLean Reagent Co., Ltd.
NH_4Cl	99.8%	McLean Reagent Co., Ltd.
Resorcinol	AR	McLean Reagent Co., Ltd.
Formaldehyde solution	37%	Sinopharm Chemical Reagent Co., Ltd.
Polyvinyl alcohol	1788	West Asia reagent
Sodium dodecyl sulfate	≥98%	Aladdin Reagent Co., Ltd.
Benzyl alcohol	≥99.5%	Aladdin Reagent Co., Ltd.
Bisphenol A diglycidyl ether	≥85%	Aladdin Reagent Co., Ltd.
Neopentyl glycol diglycidyl ether	≥40%	Aladdin Reagent Co., Ltd.
Polyetheramine D-230	Mn∼230	Aladdin Reagent Co., Ltd.

and homogenized at 5000 rpm for five minutes to form a homogeneous emulsion. The pH of the emulsion was adjusted to 3.4–3.5, and the emulsion was poured into a 250 ml single-necked flask. Then, magnetic stirring was performed at 400 rpm. During the stirring process, 3.6 ml of formaldehyde solution was added dropwise, and the temperature was slowly increased to 55 °C and maintained for four hours. After the reaction, the solution was cooled to room temperature, centrifuged, and washed with deionized water. It was frozen in a refrigerator for two hours and then in a freeze dryer for 24 hours. Subsequently, it was ground to obtain the UF/Phen microcapsules.

6.4.1.3 Characterization of UF/Phen Microcapsules

The morphology and particle size of the UF/Phen microcapsules were examined using scanning electron microscopy (SEM, Quanta 250, FEI) and Nano-measure software, respectively. The chemical structures and compositions of Phen, UF, and UF/Phen microcapsules were characterized using FTIR spectroscopy (VERTEX 70, Bruker).

UF, Phen, and UF/Phen were subjected to thermogravimetric analysis (TGA, STA 449 F5, NETZSCH). The experiment involved heating the samples from 30 to 790 °C under the protection of nitrogen. The thermal stability of the microcapsules and the inhibitor loading rate of Phen were analyzed using thermogravimetric curves.

6.4.1.4 Preparation of Self-reporting and Self-healing Coating Based on UF/Phen Microcapsules

Q235 carbon steel ($40 \times 30 \times 1\ mm^3$) was treated before use to remove oxides and grease attached to the metal surface. The specific preparation process was as follows: 5, 10, and 15 wt% UF/Phen microcapsules (accounting for the coating weight) were poured into the mixed solution of neopentyl glycol, and bisphenol A and sonicated for 30 minutes and then magnetically stirred for four hours until the dispersion was uniform. Next, the curing agent D230 was added dropwise, followed by stirring for 10 minutes. The coating solution was uniformly coated on the metal substrate and cured at room temperature for three days. The cured film thickness of all coatings was $100 \pm 10\ \mu m$.

6.4.1.5 Self-reporting and Self-healing Performance Analysis

The thermal properties of the coatings were analyzed using differential scanning calorimetry (DSC, DSC25, and TA). The surface morphologies of the coatings before and after healing were characterized using SEM (Quanta 250, FEI). EIS (660E, CHI) was used to evaluate the anticorrosion performance of the self-healing coatings containing different amounts of UF/Phen microcapsules.

6.4.1.6 Corrosion Sensing Behavior of Coatings Based on UF/Phen Microcapsules

Phen (200 ppm) and $FeCl_2$ solutions with different concentrations were prepared. 100 μL of the $FeCl_2$ solution with the abovementioned concentrations was added dropwise to 3 ml of a Phen aqueous solution to observe the color change. The images of the X-shaped scratches on the epoxy coating and epoxy coatings containing different UF/Phen microcapsules contents were photographed and placed in a salt spray box (LRHS-108-RY, Linpin) to conduct salt spray tests. During the experiment, the samples were removed at different durations to obtain images.

6.4.2 Characterization of UF/Phen Microcapsules

6.4.2.1 Morphology of UF/Phen Microcapsules and Release of Corrosion Inhibitors

As shown in Figure 6.27a,b, the UF/Phen microcapsules had a spherical shape. Their particle size distribution is shown in Figure 6.27c; it was narrow, with a diameter range of 0.75–5.93 μm and an average particle size of 2.63 μm. The TEM images in Figure 6.28 show that the microcapsules were composed of three elements, C, O, and N, with uniform elemental distributions.

Figure 6.29 shows the FTIR spectra of UF microcapsules, Phen, and UF/Phen microcapsules. The characteristic peak of —NH was at 3330 cm^{-1}. The —CO—NH—

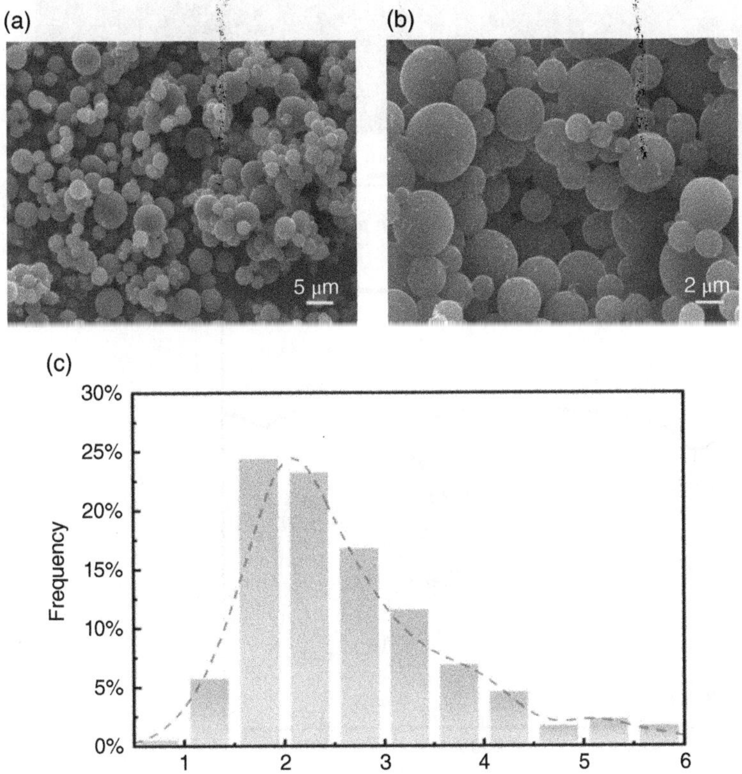

Figure 6.27 (a,b) SEM images of UF/Phen microcapsules; (c) particle size distribution of UF/Phen microcapsules. Source: Wang et al. [55]/with permission from Elsevier.

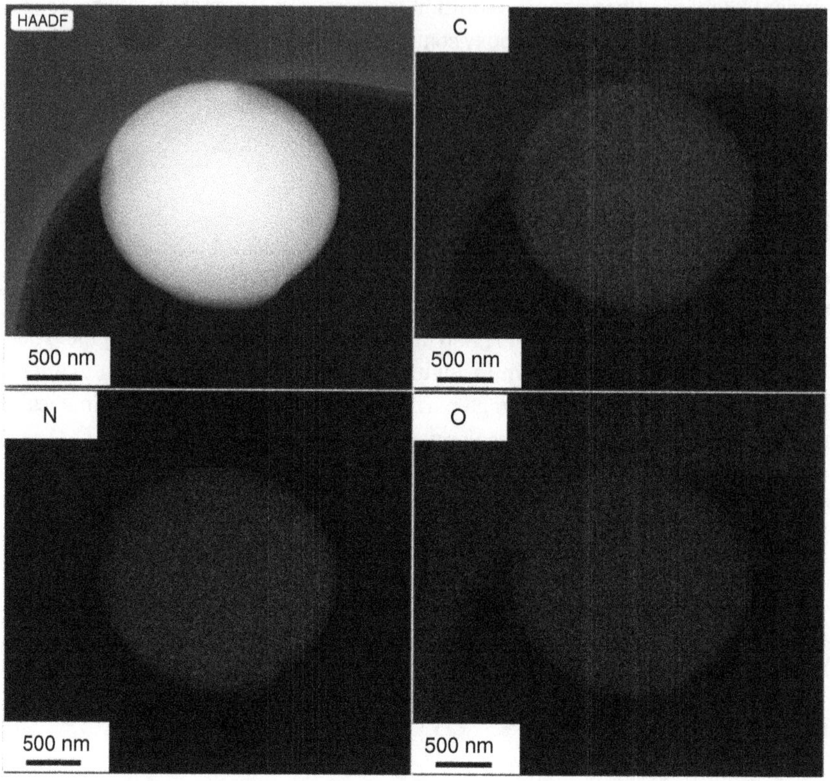

Figure 6.28 TEM and energy spectrum of UF/Phen microcapsules.

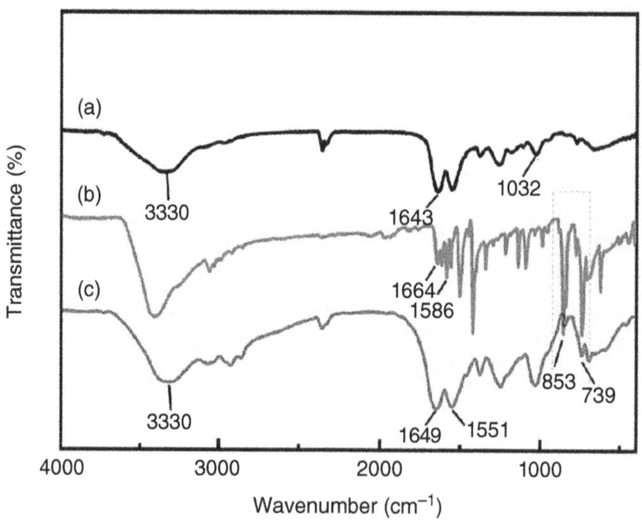

Figure 6.29 FTIR spectra of (a) UF microcapsules, (b) Phen, and (c) UF/Phen microcapsules [55].

absorption band and the stretching vibration peak of —C—O—C— were observed at 1643 and 1027 cm^{-1}, respectively. For Phen, stretching vibration peaks of the C=N bonds and C=C on the benzene ring were observed at 1664 and 1586 cm^{-1}, respectively. The absorption band of the C—H bond appeared at 739 cm^{-1}, corresponding to the out-of-plane bending vibration of the C—H bond on the benzene ring. This proved that Phen was successfully loaded into the microcapsules [56].

Figure 6.30 shows the TGA and DTG curves of the UF/Phen microcapsules, UF microcapsules, BnOH, and Phen. The thermal decomposition temperature of BnOH was 188 °C, and the maximum decomposition temperature was 176 °C. The thermal decomposition temperature of Phen was 348 °C, and the maximum decomposition temperature was 337 °C. The mass loss of UF microcapsules was 74% in the temperature range of 30–790 °C and the maximum decomposition occurred at 250 °C. The mass loss of UF/Phen microcapsules in the temperature range of 30–790 °C was 82.3%. Because of the presence of UF and Phen, the maximum decomposition temperature of the UF/Phen microcapsules shifted to 276 °C. When the temperature increased to 188 °C, BnOH was completely decomposed, the mass loss of UF was 6%, the mass loss of Phen was 10%, and the mass loss of UF/Phen microcapsules was 18.9%. The UF mass fraction in UF/Phen is defined as a, the BnOH mass fraction is defined as b, and the Phen mass fraction is defined as c; accordingly, we obtained the equation, $0.06a + b + 0.1c = 18.9\%$. When the temperature increased to 348 °C, Phen was completely decomposed, and there was 60% mass loss for UF, the mass loss of UF/Phen microcapsules was 72%. We can get the following equation: $0.6a + b + c = 72\%$, $a + b + c = 100\%$. Therefore, the loading rate of Phen was calculated to be 17%.

Figure 6.31 shows the UV spectra of Phen at different concentrations in 3.5 wt% NaCl solutions at pH = 3, 7, and 11. The characteristic peak of Phen was clearly observed under the different solution conditions. The peak position of the UV–Vis spectrum of Phen was the same under neutral and alkaline conditions, and there was a shift under acidic conditions. With an increase in the Phen concentration, the UV peak intensity gradually increased; however, the position of the UV peak did

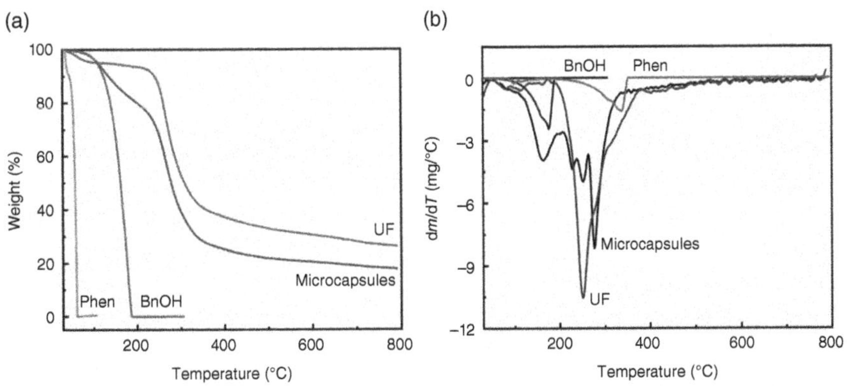

Figure 6.30 (a) TGA curves of UF/Phen, UF microcapsules, BnOH and Phen; (b) DTG curves of UF/Phen, UF microcapsules, BnOH and Phen.

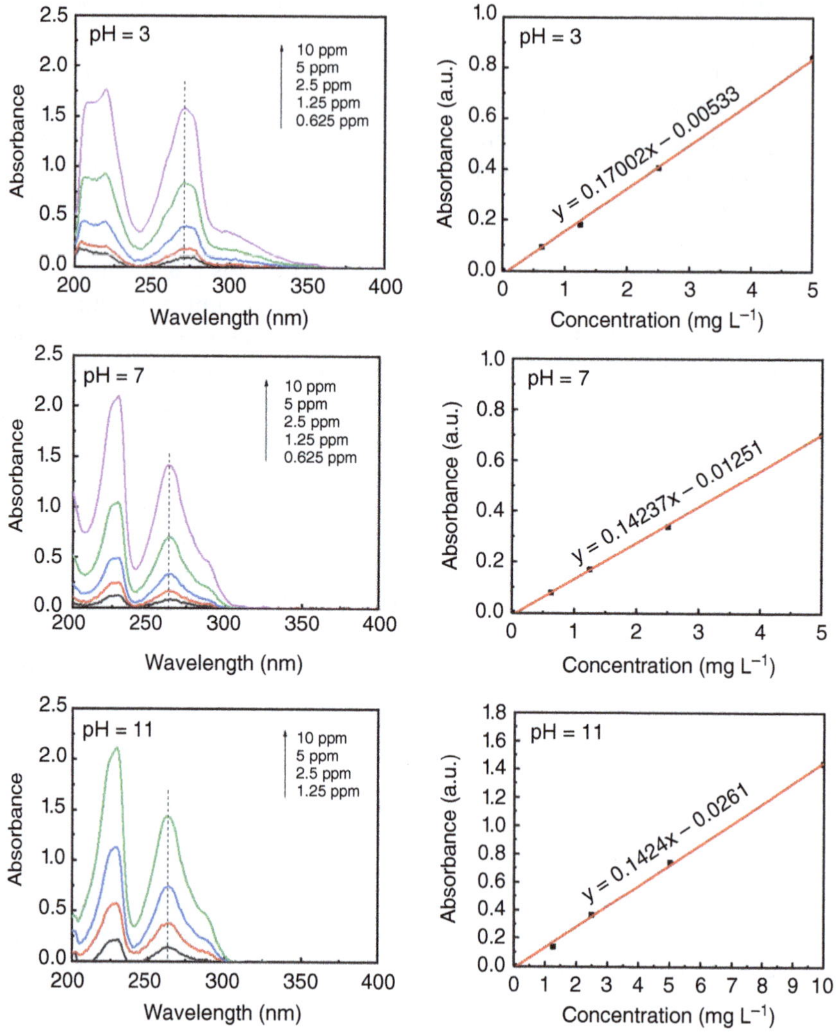

Figure 6.31 UV–Vis diagram of Phen and its corresponding standard curve at pH = 3, 7, 11.

not change. By correlating the UV peak intensity of Phen with the corresponding concentration, the data shown on the right-hand side of the figure were obtained. A strong correlation was observed under different pH conditions. Therefore, the correlation data can obtain the Phen concentration corresponding to the absorption peak at a certain concentration.

Figure 6.32 presents the release curves of the UF/Phen microcapsules in 3.5 wt% NaCl solutions at pH = 3, 7, and 11. As shown, at the initial stage of release, the release rates of Phen in the acidic and neutral solutions were higher than that in the alkaline solution. After 30 hours of release, the release amount was the largest under neutral conditions, and the release amount was smaller under acidic conditions.

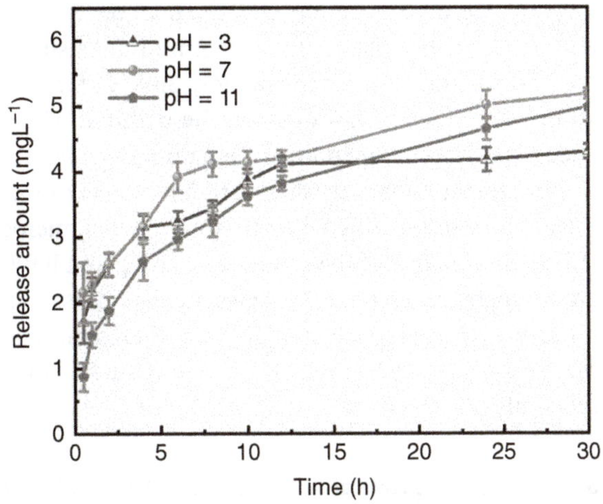

Figure 6.32 Release curves of Phen at pH = 3, 7, 11.

6.4.2.2 Thermal Performance and Shape-Memory Effect of Coating Based on UF/Phen Microcapsules

The self-healing mechanism is based on the scratch closure of the epoxy matrix and the corrosion inhibition effect of Phen. To determine the healing temperature, the DSC curves of the epoxy coating and coatings with 5, 10, and 15 wt% UF/Phen microcapsules were measured. As shown in Figure 6.33, the T_g of the epoxy coating cured at room temperature was 35.3 °C, and those of the coatings with 5, 10, and 15 wt% of UF/Phen microcapsules were 32.5, 29.0, and 28.3 °C, respectively. A lower coating T_g is beneficial for triggering the shape-memory effect under sunlight irradiation.

The epoxy coating and the coating with 10 wt% UF/Phen microcapsules were placed outdoors to receive sunlight. Figure 6.33b presents the surface temperature

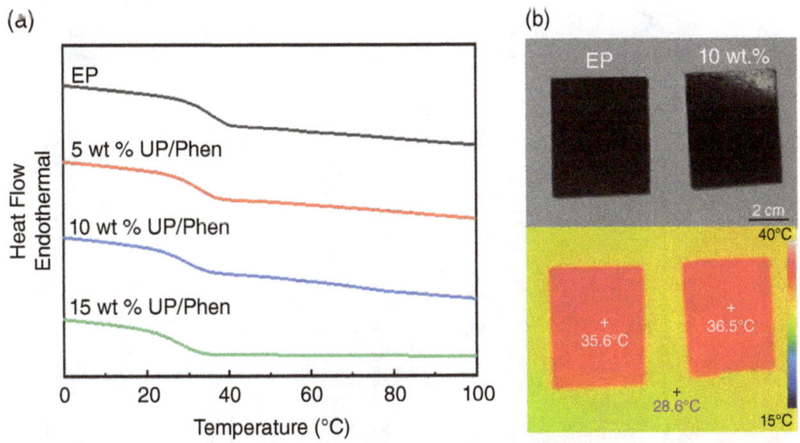

Figure 6.33 (a) DSC curves of epoxy coating, 5 wt% UF/Phen microcapsule coating, 10 wt% UF/Phen microcapsule coating and 15 wt% UF/Phen microcapsule coating; (b) surface temperature of epoxy coatings and coating with 10 wt% UF/Phen microcapsules after outdoor light irradiation for five minutes.

maps of the epoxy coating and the coating with 10 wt% UF/Phen microcapsule after sunlight irradiation for five minutes. As shown, the surface temperature of the two coatings was approximately 36.5 °C, which was higher than the ambient temperature (28.6 °C). Moreover, the addition of UF/Phen microcapsules did not affect its surface temperature after sunlight irradiation, which was higher than the trigger temperature of its shape memory effect (~35.3 °C for the epoxy coating and ~29.0 °C for the coating with 10 wt% UF/Phen microcapsules).

The SEM images of the scratched coating and light-triggered self-healing coating are compared in Figure 6.34. The scratch widths of all the healed coatings were significantly reduced to <10 μm. This was due to the shape-memory effect of the epoxy matrix triggered by heating of the coating surface under light irradiation, which significantly narrowed the scratches [57–60].

6.4.2.3 Electrochemical Analysis of Self-healing Coating Based on UF/Phen Microcapsules

The effects on different contents of UF/Phen microcapsules on the self-healing of the coating were studied via EIS. Figure 6.35 shows the EIS results of the light-

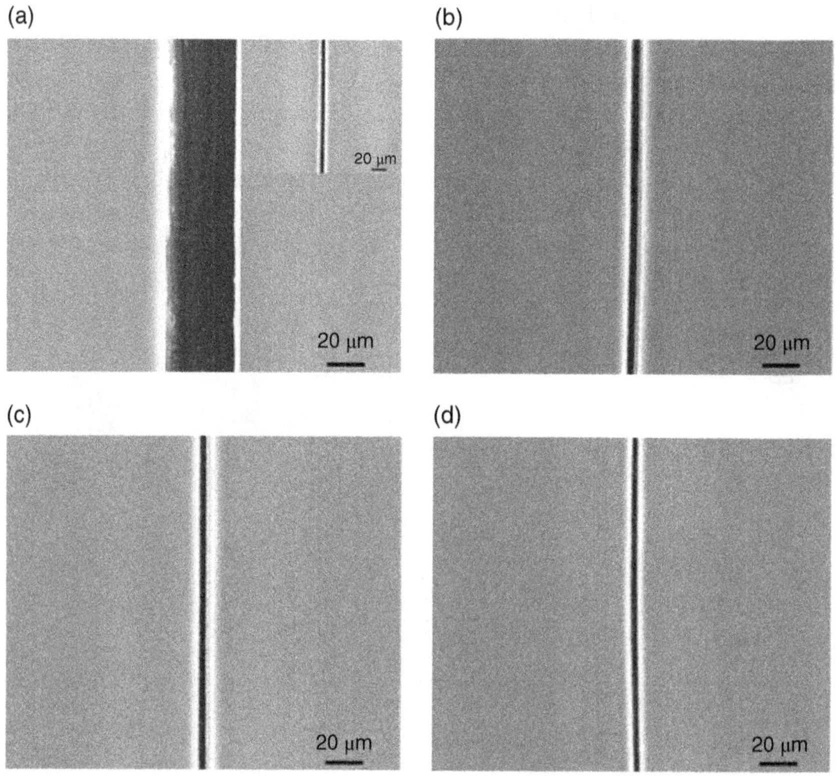

Figure 6.34 SEM images of different coatings: (a) scratched epoxy coating; (b) healed 5 wt% UF/Phen microcapsule coating; (c) healed 10 wt% UF/Phen microcapsule coating; (d) healed 15 wt% UF/Phen microcapsule coating.

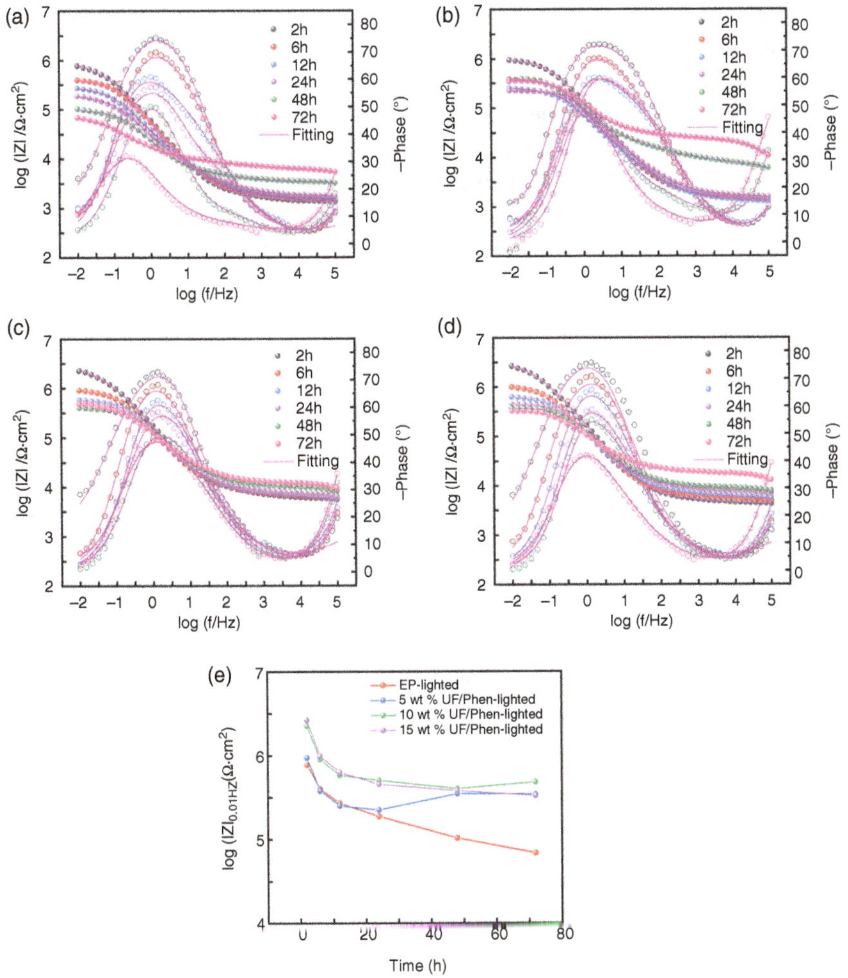

Figure 6.35 Bode diagrams of coatings with different contents of UF/Phen microcapsules during immersion: (a) light irradiated epoxy coating; (b) light healed 5 wt% UF/Phen microcapsule coating; (c) light healed 10 wt% UF/Phen microcapsule coating; (d) light healed 15 wt% UF/Phen microcapsule coating; (e) low-frequency impedance modulus variation trend. Source: Wang et al. [55]/with permission of Elsevier.

irradiated epoxy coating, healed 5 wt% UF/Phen microcapsule coating, healed 10 wt% UF/Phen microcapsule coating, and healed 15 wt% UF/Phen microcapsule coating after immersion in a 3.5 wt% NaCl solution for 2–72 hours [61–63]. The $|Z|_{0.01\,Hz}$ value of the epoxy coating after light-triggered healing was at a relatively high level at the initial stage of immersion (Figure 6.35a), i.e., $7.8 \times 10^5\,\Omega\,cm^2$ (two hours), but it decreased significantly with time, reaching $6.9 \times 10^4\,\Omega\,cm^2$ (72 hours). With the addition of 5 wt% microcapsules, the impedance values remained similar to that of the blank epoxy coating. With the release of the Phen, the reduction in $|Z|_{0.01\,Hz}$ value slowed significantly, and $|Z|_{0.01\,Hz}$ increased to

$3.4 \times 10^5\ \Omega\,cm^2$ in the later stage. When the microcapsule content was further increased to 10 wt%, the $|Z|_{0.01\,Hz}$ was relatively high throughout the immersion process, and the $|Z|_{0.01\,Hz}$ value ranged from $2.3 \times 10^6\ \Omega\,cm^2$ (immersed for two hours) to $4.8 \times 10^5\ \Omega\,cm^2$ (immersed for 72 hours). When the microcapsule content was increased to 15 wt%, the $|Z|_{0.01\,Hz}$ value was $2.7 \times 10^6\ \Omega\,cm^2$ (two hours) at the beginning, but it exhibited a downward trend during the subsequent immersion time, decreasing to $3.3 \times 10^5\ \Omega\,cm^2$. This was due to poor dispersibility caused by the excessive microcapsule content, which resulted in more corrosion channels at the interface between the microcapsules and the coating. Figure 6.35e shows the changing trend of the $|Z|_{0.01\,Hz}$ values of coatings with different microcapsule contents. The light-irradiated epoxy coating exhibited a continuous reduction in the impedance modulus, which is ascribed to the gradual penetration of the corrosive media. With the incorporation of the UF/Phen microcapsules, the impedance modulus values increased for all the composite coatings. With an increase in the immersion time, the $|Z|_{0.01\,Hz}$ values stabilized or even increased, which can be explained by the gradual release of Phen from the microcapsules [64–66]. With an appropriate amount of UF/Phen microcapsules loaded into the coating, both corrosion resistance and self-healing properties can be ensured [67–69].

The EIS results were fitted using the equivalent circuit shown in Figure 6.36a. The R_{ct} value can reflect the electron-transfer ability of the electrochemical reaction. A higher R_{ct} value indicates more difficult electron transfer and better corrosion resistance [68, 70, 71]. As shown in Figure 6.36b, during 72 hours of immersion, the R_{ct} values of all the healed coatings exhibited a downward trend, whereas the R_{ct} values of the UF/Phen microcapsule composite coatings were higher. The order of R_{ct} values is as follows: EP coating < 5 wt% UF/Phen coating < 15 wt% UF/Phen coating < 10 wt% UF/Phen coating. Thus, the coating with a UF/Phen microcapsule content of 10 wt% had the best corrosion resistance. During immersion of the healed coating, Phen was released from the UF/Phen microcapsules and adsorbed onto the steel surface, forming a complex. In addition, the scratched UF/Phen microcapsule coating quickly activated the

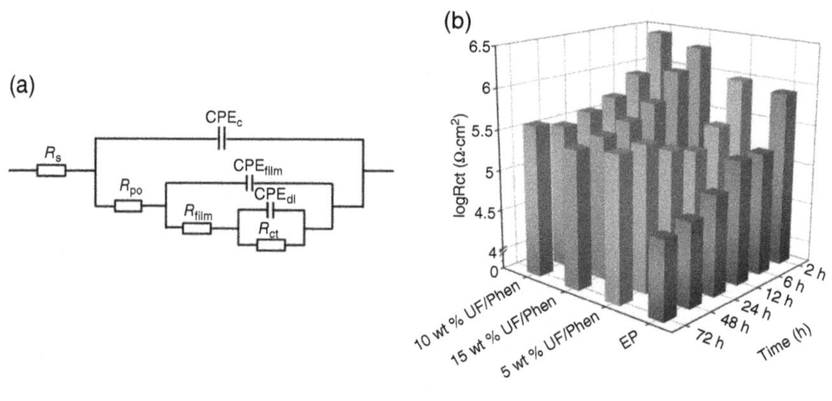

Figure 6.36 (a) Equivalent circuit of the coating immersed in 3.5 wt% NaCl solution; (b) change trend diagram of R_{ct} values.

shape-memory effect under sunlight irradiation [72–74]. These two methods worked together to improve the anticorrosion performance of the scratched coatings. Compared with the 5 wt% UF/Phen coating, the 10 wt% UF/Phen coating had more microcapsules, which provided more corrosion inhibitors and enhanced the protection performance of the damaged coating. Further increasing the microcapsule content to 15 wt% would easily cause the agglomeration of microcapsules, increase the corrosion channels, and adversely affect the barrier performance of the coating. Therefore, the addition of 10 wt% UF/Phen microcapsules optimized the protective performance of the coating.

6.4.2.4 Corrosion Sensing of Coating Based on UF/Phen Microcapsules

To illustrate the high sensitivity of Phen to localized corrosion (Fe^{2+} produced in the early stage of corrosion), the Phen solution (200 ppm aqueous solution) was mixed with different concentrations of a $FeCl_2$ solution (0, 50, 100, 200, 300, 400, 500, 1000 ppm, 2000, 4000, and 5000 ppm), and the mixture was analyzed via macroscopic photographs (Figure 6.37a) and UV–Vis spectra (Figure 6.37b). The degree of color development and the interaction between Phen and the Fe^{2+} ions were studied [75–77]. After 100 μl of the $FeCl_2$ solution was added to 3 ml of the Phen solution, an obvious absorption peak appeared at 512 nm, which was caused by the formation of a

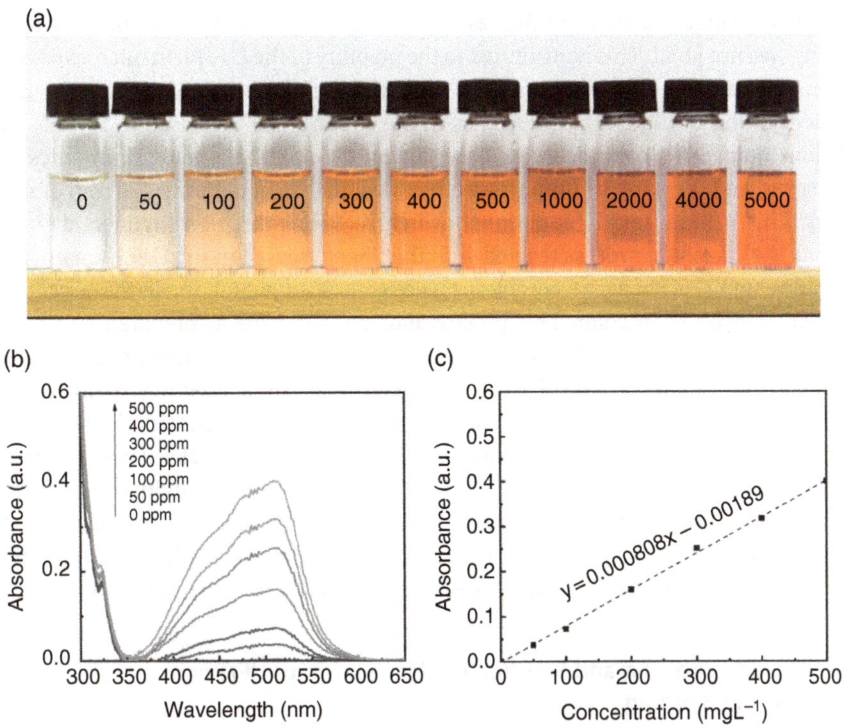

Figure 6.37 (a) Color images of Phen solution at different Fe^{2+} concentrations; (b) UV–Vis spectrum of complexation of Fe^{2+} and Phen in aqueous solution; (c) complexation of Fe^{2+} and Phen in aqueous solution standard curve line.

$[Fe(Phen)_2]^{2+}$ complex. As the $FeCl_2$ concentration increased, the absorbance of the $[Fe(Phen)_2]^{2+}$ complex increased, and the color of the Phen solution changed significantly after the addition of Fe^{2+} ions. The standard curve of the complex is shown in Figure 6.37c. When the concentration of Fe^{2+} reached 2000 ppm, the degree of color development in the Phen solution reached its highest level, and the maximum absorption intensity was observed in the UV–Vis spectrum [78–80]. The molar ratio of Fe^{2+} to Phen was approximately 0.5. When the concentration of Fe^{2+} was further increased, the degree of color development did not increase but decreased, possibly owing to a change in the coordination mode of Fe^{2+} and Phen. The color gradient diagram and UV–Vis experimental results indicate that Phen can effectively sense Fe^{2+} ions produced by localized corrosion reactions [81–83].

To verify the corrosion self-reporting behavior of the UF/Phen microcapsule coating after damage, a coating solution containing UF/Phen microcapsules was first coated onto a steel sheet. Prior to testing, scratches were created on the coated surfaces to expose the steel substrates. The coating sample was placed in a salt spray box for a salt spray test, and a color change in the scratched area of the coating was observed. As shown in Figure 6.38a, no obvious corrosion was observed in the blank epoxy coating after 15 minutes of salt spraying, and visible corrosion appeared in some scratched areas after 30 minutes, which rapidly deepened with an increase in the salt spray time. The scratched area of the coating with 5 wt% UF/Phen was obviously orange-red after 15 minutes of salt spray (Figure 6.38b), but the color rendering was not ideal. This is attributed to the inability of the UF/Phen microcapsules to provide sufficient Phen molecules to react with Fe^{2+} ions per unit area. When the content of the UF/Phen microcapsules in the coating was increased to 10 wt%, an obvious orange-red color could be observed by the naked eye after 15 minutes of salt spray, and the early self-reporting effect was maintained during 120 min of salt spray. This is because a large number of free Fe^{2+} ions produced by metallic iron during the corrosion process reacted with the Phen released in the UF/Phen microcapsules, and $[Fe(Phen)_2]^{2+}$ complexes were generated, which made the scratches appear orange-red in color. This process indicates that the as-prepared UF/Phen microcapsules can respond to the Fe^{2+} generated by localized corrosion, exhibiting self-reporting behavior [56, 84]. These characteristics can be used to visually monitor coating surfaces for damage or microcracks so that coating defects can be found and repaired in time. When a coating is damaged or has microcracks, it can automatically restore its barrier performance to a certain extent under sunlight, significantly reducing the labor required for repair. When corrosion reactions occur, compared with pure epoxy coatings, UF/Phen microcapsule coatings can achieve early sensing of corrosion and improve the anticorrosion performance of the coating [85–87].

6.4.3 Self-reporting and Self-healing Mechanism of UF/Phen Microcapsule Coating

The self-reporting and self-healing mechanisms of the UF/Phen microcapsule coatings are shown in Figure 6.39. When the coating has cracks or scratches, the carbon steel is corroded owing to corrosion of the corrosive medium. At this time, the Phen

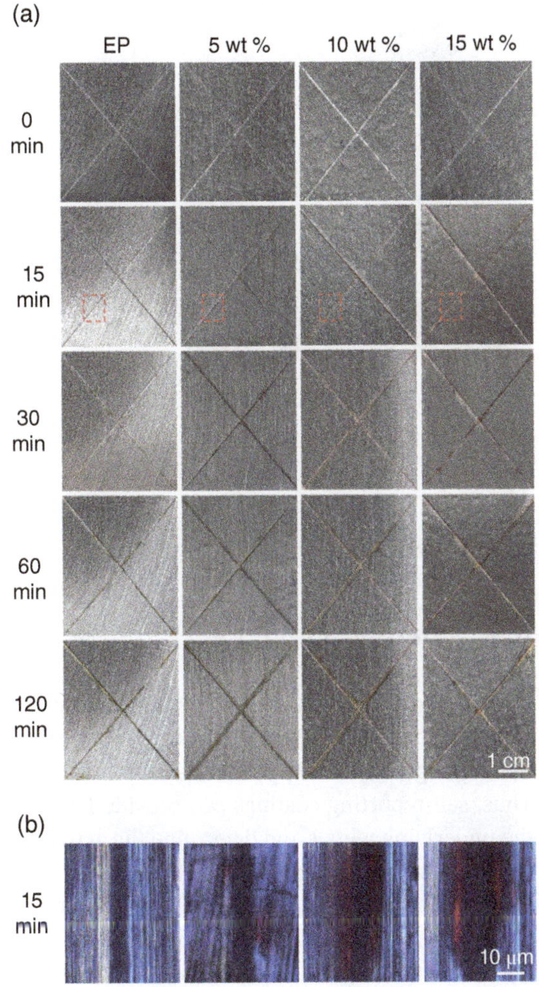

Figure 6.38 Corrosion self-reporting photos of X-type scratch coating during 120 minutes salt spray: (a) epoxy coating containing 5, 10, 15 wt% UF/Phen microcapsules; (b) enlarged view of the selected area at 15 minutes. Source: Wang et al. [55]/with permission from Elsevier.

in the microcapsules at the scratches seeps out from the damaged area, and the Fe^{2+} ions produced by the corrosion of the damaged area of the coating can react with Phen to form a $[Fe(Phen)_2]^{2+}$ complex with an obvious orange-red color and a corrosion inhibition effect [88–90]. This improves the protective performance of the coating while providing an early warning of the corrosion behavior. It is worth noting that the shape-memory epoxy coating can trigger the scratch closure to narrow the scratches under sunlight, and together with the corrosion inhibitor Phen, it can inhibit the corrosion reaction and improve the anticorrosion performance [90–92]. The color development and corrosion inhibition of Phen in the microcapsules and the scratch closure of the epoxy matrix are the keys to realizing the dual functions of self-reporting and self-healing of the coating and further improving its long-term protection ability [82, 93, 94].

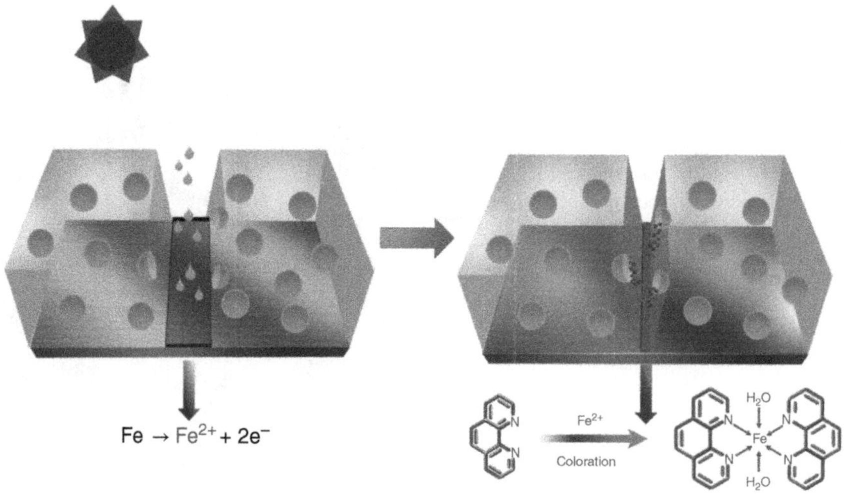

Figure 6.39 Schematic diagram of the self-reporting and self-healing mechanism of UF/Phen microcapsule coating.

6.5 Chapter Summary

In this chapter, various self-reporting coatings based on different types of corrosion indicators are introduced. These indicators can exhibit fluorescence signals or color changes activated by pH variations or metallic cations within local defects [16]. Thus, self-reporting coatings can provide fast feedback on the corrosion status of the underlying metals and determine the active corrosion sites before the corrosion products are observable [26, 27]. PhPh is the most commonly used pH-indicating molecule for corrosion sensing and can be used to sense the generation of hydroxide anions associated with the cathodic corrosion reaction. It was loaded into different types of containers, and the amount of optical addition to the coating was investigated to ensure both corrosion sensing and corrosion resistance [95, 96].

For the metallic cations-activated self-reporting coatings, CDs–SSA complex was designed to react with Fe^{2+} ions. First, new CDs–SSA complex was prepared, which can be used as both a corrosion inhibitor and a corrosion indicator to protect carbon steel from corrosion damage and warn of early corrosion reactions. TEM and AFM indicated that the CDs–SSA had a spherical morphology and was evenly distributed, with a particle size of approximately 20.37 nm and a height distribution of 3–4 nm. FTIR and UV–Vis confirmed that SSA was successfully grafted onto the surface of the CDs, and the results of PDP and EIS indicated that the presence of CDs–SSA led to the anode passivation of the carbon steel surface during the PDP test. The i_{corr} value was significantly reduced compared with that of the blank HCl solution. In the EIS test, the $|Z|_{0.01\,Hz}$ and R_{ct} values of the CDs–SSA samples were the highest, and the CPE_{dl} values were the lowest, indicating that the CDs–SSA complexes had good corrosion inhibition properties for the steel substrate. After immersion in 1 M HCl for 24 hours, the corrosion inhibition rate of 200 mg l^{-1} CDs–SSA reached

99.1%. CDs–SSA can form hydrophobic physicochemical adsorption films on steel surfaces, which are effective for corrosion inhibition. Additionally, CDs–SSA had a significant fluorescence quenching effect when reacting with Fe^{2+}/Fe^{3+} ions, and the fluorescence quenching was obvious when interacting with 5×10^{-4} M Fe^{2+} and 1×10^{-5} M Fe^{3+} or higher concentrations, which can intuitively sense the early corrosion phenomenon. Therefore, CDs–SSA has a high corrosion inhibition efficiency and sensitivity to corrosion induction.

The epoxy coating containing UF/Phen microcapsules were spherical, with an average particle size of 2.63 μm, and the loading rate of Phen was 17% according to the TGA. Salt spray experiments were carried out on the UF/Phen microcapsule coating, and it was found that the 10 wt% UF/Phen microcapsule coating had the best self-reporting effect. The coating displayed a visible corrosion warning signal within 15 minutes and maintained a self-reporting effect for 120 minutes. When the coating was severely damaged, and it was difficult for the self-healing mechanism to exert a long-term effect, the damaged area facilitated timely manual repair. Regarding the self-healing property, after five minutes of sunlight irradiation, the temperature of the coating surface increased to >36 °C, which was sufficient to trigger the shape-memory effect of the room temperature-cured epoxy coating. An SEM image of the layer indicated that the scratches on the coating had narrowed. Electrochemical analysis of the UF/Phen microcapsule coating indicated that the best self-healing effect was achieved when the microcapsule content was 10 wt%.

The color indicator added to the self-reporting coating allows corrosion to be observed with the naked eye, making it suitable for practical industrial applications. However, the choice of color indicators is limited. For example, some color indicators change color under strongly acidic or alkaline conditions, making them difficult to distinguish during the corrosion process because the pH value change is a gradual and less intense process. Molecular modifications can be performed on PhPh molecules to endow them with corrosion inhibition or better distribution properties in the coating. With regard to metal-ion indicators, only Phen and its derivatives have been confirmed to react with ferrous ions, whereas color indicators that react with Al^{3+} and Mg^{2+} are not commonly reported. Therefore, the development of new color-developing agents is crucial. In this chapter, modified CDs are reported for the detection of Fe^{2+} and Fe^{3+} ions generated during the corrosion process. Further modifications can be designed to make them highly sensitive to Al^{3+} and Mg^{2+}, as well as pH variations. In addition, to increase the color indication efficiency, it is helpful to study the corrosion-related microenvironment changes at the coating defects, such as the spatial distributions of the pH and metal ions and their changes over time [97, 98]. These factors are related to not only the type of metal substrate but also the composition of the coating and the size and depth of the defect. The effects of coating transparency, coating thickness, concentration of micro- and nanocapsules, and stability of the color signal on the coloring effect and intensity were studied. In the case where the microcracks in the coating have not yet reached the metal surface, even if the protective properties of the coating are partially destroyed, a damage indication signal cannot be observed. Therefore, the self-reporting mechanism directly related to damage to the coating surface needs to be further explored [99–101]. According to current research, the application of color indicators in the field of anticorrosion coatings still faces the following problems. Single-function self-reporting or self-healing coatings no longer satisfy the application requirements. Healing the coating during reporting can significantly increase the efficiency of the coating [55, 65]. The color of the coating surface should also be considered because the self-reporting signals should not be the same as those of the coating surface, which influences the self-reporting effect. Furthermore, the selection of the color indicator should consider the contrast with the original color of the coating; using a proper fluorescence indicator is another way to solve this problem [36, 39].

References

1 Li, X., Zhang, D., Liu, Z. et al. (2015). Materials science: share corrosion data. *Nature* 527 (7579): 441–442. https://doi.org/10.1038/527441a.

2 Wang, P., Ma, L., Cheng, X. et al. (2021). Influence of grain refinement on the corrosion behavior of metallic materials: a review. *International Journal of Minerals, Metallurgy, and Materials* 28 (7): 1112–1126. https://doi.org/10.1007/s12613-021-2308-0.

3 Zhang, F., Ju, P., Pan, M. et al. (2018). Self-healing mechanisms in smart protective coatings: a review. *Corrosion Science* 144: 74–88. https://doi.org/10.1016/j.corsci.2018.08.005.

4 Chen, C., He, Y., Xiao, G. et al. (2021). Co-modification of epoxy based polyhedral oligomeric silsesquioxanes and polyaniline on graphene for enhancing corrosion resistance of waterborne epoxy coating. *Colloids and Surfaces A: Physicochemical and Engineering Aspects* 614: 126190. https://doi.org/10.1016/j.colsurfa.2021.126190.

5 Wang, J., Ma, L., Huang, Y. et al. (2022). Photothermally activated self-healing protective coating based on the "close and seal" dual-action mechanisms. *Composites Part B: Engineering* 231: 109574. https://doi.org/10.1016/j.compositesb.2021.109574.

6 Wilson, N.G., Raveendran, J., and Docoslis, A. (2021). Portable identification of fentanyl analogues in drugs using surface-enhanced Raman scattering. *Sensors and Actuators B: Chemical* 330: 129303. https://doi.org/10.1016/j.snb.2020.129303.

7 Song, P., Shen, S., Li, C.-C. et al. (2015). Insight in layer-by-layer assembly of cysteamine and l-cysteine on the copper surface by electrochemistry and Raman spectroscopy. *Applied Surface Science* 328: 86–94. https://doi.org/10.1016/j.apsusc.2014.12.015.

8 Shchukin, D.G. (2013). Container-based multifunctional self-healing polymer coatings. *Polymer Chemistry* 4 (18): 4871. https://doi.org/10.1039/c3py00082f.

9 White, S.R., Sottos, N.R., Geubelle, P.H. et al. (2001). Autonomic healing of polymer composites. *Nature* 409 (6822): 794–797. https://doi.org/10.1038/35057232.

10 Lu, P., Saal, J.E., Olson, G.B. et al. (2018). Computational materials design of a corrosion resistant high entropy alloy for harsh environments. *Scripta Materialia* 153: 19–22. https://doi.org/10.1016/j.scriptamat.2018.04.040.

11 Wang, M., Liu, M., and Fu, J. (2015). An intelligent anticorrosion coating based on pH-responsive smart nanocontainers fabricated via a facile method for protection of carbon steel. *Journal of Materials Chemistry A* 3 (12): 6423–6431. https://doi.org/10.1039/c5ta00417a.

12 Song, Y.K., Lee, T.H., Lee, K.C. et al. (2020). Coating that self-reports cracking and healing using microcapsules loaded with a single AIE fluorophore. *Applied Surface Science* 511: 145556. https://doi.org/10.1016/j.apsusc.2020.145556.

13 Maia, F., Tedim, J., Bastos, A.C. et al. (2013). Nanocontainer-based corrosion sensing coating. *Nanotechnology* 24 (41): 415502. https://doi.org/10.1088/0957-4484/24/41/415502.

14 Banerjee, S.N. and Misra, S. (1989). 1,10-Phenanthroline as corrosion inhibitor for mild steel in sulfuric acid solution. *Corrosion* 45 (9): 780–783. https://doi.org/10.1021/ie102034c.

15 Cui, M., Ren, S., Xue, Q. et al. (2017). Carbon dots as new eco-friendly and effective corrosion inhibitor. *Journal of Alloys and Compounds* 726: 680–692. https://doi.org/10.1016/j.jallcom.2017.08.027.

16 Ye, Y., Yang, D., and Chen, H. (2019). A green and effective corrosion inhibitor of functionalized carbon dots. *Journal of Materials Science and Technology* 35 (10): 2243–2253. https://doi.org/10.1016/j.jmst.2019.05.045.

17 Yang, D., Ye, Y., Su, Y. et al. (2019). Functionalization of citric acid-based carbon dots by imidazole toward novel green corrosion inhibitor for carbon steel. *Journal of Cleaner Production* 229: 180–192. https://doi.org/10.1016/j.jclepro.2019.05.030.

18 Wu, X., Li, J., Deng, C. et al. (2022). Novel carbon dots as effective corrosion inhibitor for N80 steel in 1 M HCl and CO_2-saturated 3.5 wt% NaCl solutions. *Journal of Molecular Structure* 1250: 131897. https://doi.org/10.1016/j.molstruc.2021.131897.

19 Jia, X., Li, J., and Wang, E. (2012). One-pot green synthesis of optically pH-sensitive carbon dots with upconversion luminescence. *Nanoscale* 4 (18): 5572. https://doi.org/10.1039/c2nr31319g.

20 Yu, Y., Zeng, Q., Tao, S. et al. (2023). Carbon dots based photoinduced reactions: advances and perspective. *Advanced Science* 10: 2207621. https://doi.org/10.1002/advs.202207621.

21 Himaja, A.L., Karthik, P.S., and Singh, S.P. (2015). Carbon dots: the newest member of the carbon nanomaterials family. *The Chemical Record* 15 (3): 595–615. https://doi.org/10.1002/tcr.201402090.

22 Zhao, P., Zhu, M., He, Z. et al. (2021). Corrosion inhibition of eco-friendly nitrogen-doped carbon dots for carbon steel in acidic media: Performance and mechanism investigation. *Journal of Molecular Liquids* 342: 117583. https://doi.org/10.1016/j.molliq.2021.117583.

23 Zhao, P. and Zhu, L. (2018). Dispersibility of carbon dots in aqueous and/or organic solvents. *Chemical Communications* 54 (43): 5401–5406. https://doi.org/10.1039/c8cc02279h.

24 Ding, H., Zhou, X., Wei, J. et al. (2020). Carbon dots with red/near-infrared emissions and their intrinsic merits for biomedical applications. *Carbon* 167: 322–344. https://doi.org/10.1016/j.carbon.2020.06.024.

25 Liu, R., Wu, D., Liu, S. et al. (2009). An aqueous route to multicolor photoluminescent carbon dots using silica spheres as carriers. *Angewandte Chemie International Edition* 48 (25): 4598–4601. https://doi.org/10.1002/anie.200900652.

26 Liu, S., Liu, L., Wei, X. et al. (2020). Oxide film formed on Al alloy beneath sulfosalicylic acid doped polyaniline incorporated into epoxy organic coating. *Applied Surface Science* 512: 144840. https://doi.org/10.1016/j.apsusc.2019.144840.

27 Wu, S., Wang, J., Liu, T. et al. (2023). Sulfosalicylic acid modified carbon dots as effective corrosion inhibitor and fluorescent corrosion indicator for carbon steel in HCl solution. *Colloids and Surfaces A: Physicochemical and Engineering Aspects* 661: 130951. https://doi.org/10.1016/j.colsurfa.2023.130951.

28 Papaioannou, N., Titirici, M.-M., and Sapelkin, A. (2019). Investigating the effect of reaction time on carbon dot formation, structure, and optical properties. *ACS Omega* 4 (26): 21658–21665. https://doi.org/10.1021/acsomega.9b01798.

29 Gomaa, M.M., Hugenschmidt, C., Dickmann, M. et al. (2018). Crosslinked PVA/SSA proton exchange membranes: correlation between physiochemical properties and free volume determined by positron annihilation spectroscopy. *Physical Chemistry Chemical Physics* 20 (44): 28287–28299. https://doi.org/10.1039/c8cp05301d.

30 Cheng, S., Zhang, J., Liu, Y. et al. (2021). One-step synthesis of N, S-doped carbon dots with orange emission and their application in tetracycline antibiotics, quercetin sensing, and cell imaging. *Microchimica Acta* 188 (10): 325. https://doi.org/10.1007/s00604-021-04969-w.

31 Cordier, P., Tournilhac, F., Soulié-Ziakovic, C. et al. (2008). Self-healing and thermoreversible rubber from supramolecular assembly. *Nature* 451 (7181): 977–980. https://doi.org/10.1038/nature06669.

32 Liu, J., Zhou, Z., Su, X. et al. (2020). Stiff UV-Curable self-healing coating based on double reversible networks containing Diels-Alder cross-linking and hydrogen bonds. *Progress in Organic Coatings* 146: 105699. https://doi.org/10.1016/j.porgcoat.2020.105699.

33 Cen, H., Chen, Z., and Guo, X. (2019). N, S co-doped carbon dots as effective corrosion inhibitor for carbon steel in CO_2-saturated 3.5% NaCl solution. *Journal of the Taiwan Institute of Chemical Engineers* 99: 224–238. https://doi.org/10.1016/j.jtice.2019.02.036.

34 Xing, B., Ding, Q., Jin, B. et al. (2022). Corrosion resistance and passivation behavior of CoCrFeNi-TiAl high-entropy alloy coatings in acidic solutions.

Journal of Thermal Spray Technology 31 (5): 1673–1682. https://doi.org/10.1007/s11666-022-01380-6.

35 Bellezze, T., Giuliani, G., Viceré, A. et al. (2018). Study of stainless steels corrosion in a strong acid mixture. Part 2: anodic selective dissolution, weight loss and electrochemical impedance spectroscopy tests. *Corrosion Science* 130: 12–21. https://doi.org/10.1016/j.corsci.2017.10.010.

36 Ma, L., Wang, J., Ren, C. et al. (2020). Detection of corrosion inhibitor adsorption via a surface-enhanced Raman spectroscopy (SERS) silver nanorods tape sensor. *Sensors and Actuators B: Chemical* 321: 128617. https://doi.org/10.1016/j.snb.2020.128617.

37 Wang, Y., Wang, J., Ma, L. et al. (2021). Qualitative and quantitative detection of corrosion inhibitors using surface-enhanced Raman scattering coupled with multivariate analysis. *Applied Surface Science* 568: 150967. https://doi.org/10.1016/j.apsusc.2021.150967.

38 Zeng, Y., Kang, L., Wu, Y. et al. (2022). Melamine modified carbon dots as high effective corrosion inhibitor for Q235 carbon steel in neutral 3.5 wt% NaCl solution. *Journal of Molecular Liquids* 349: 118108. https://doi.org/10.1016/j.molliq.2021.118108.

39 Liu, Z., Jia, R., Jian, Y. et al. (2021). N-doped carbon dots as a multifunctional platform for real-time corrosion monitoring and inhibition. *SSRN Electronic Journal* https://doi.org/10.1016/j.colsurfa.2022.129499.

40 Cui, M. and Li, X. (2021). Nitrogen and sulfur Co-doped carbon dots as ecofriendly and effective corrosion inhibitors for Q235 carbon steel in 1 M HCl solution. *RSC Advances* 11 (35): 21607–21621. https://doi.org/10.1039/d1ra02775a.

41 Cao, S., Liu, D., Wang, T. et al. (2021). Nitrogen-doped carbon dots as high-effective inhibitors for carbon steel in acidic medium. *Colloids and Surfaces A: Physicochemical and Engineering Aspects* 616: 126280. https://doi.org/10.1016/j.colsurfa.2021.126280.

42 Saraswat, V. and Yadav, M. (2021). Improved corrosion resistant performance of mild steel under acid environment by novel carbon dots as green corrosion inhibitor. *Colloids and Surfaces A: Physicochemical and Engineering Aspects* 627: 127172. https://doi.org/10.1016/j.colsurfa.2021.127172.

43 Cen, R., Liu, M., Lu, J. et al. (2022). Synthesis and characterization of a sensitive and selective Fe^{3+} fluorescent sensor based on novel sulfonated calix [4] arene-based host-guest complex. *Chinese Chemical Letters* 33 (5): 2469–2472. https://doi.org/10.1016/j.cclet.2021.12.005.

44 Padhan, S., Rout, T.K., and Nair, U.G. (2022). N-doped and Cu, N-doped carbon dots as corrosion inhibitor for mild steel corrosion in acid medium. *Colloids and Surfaces A: Physicochemical and Engineering Aspects* 653: 129905. https://doi.org/10.1016/j.colsurfa.2022.129905.

45 Wangm, F.T., Wangm, L.N., Xu, J. et al. (2021). Synthesis and modification of carbon dots for advanced biosensing application. *Analyst* 146 (14): 4418–4435. https://doi.org/10.1039/d1an00466b.

46 Haddadi, S.A., Hu, S., Ghaderi, S. et al. (2021). Amino-functionalized MXene nanosheets doped with Ce (III) as potent nanocontainers toward self-healing epoxy nanocomposite coating for corrosion protection of mild steel. *ACS Applied Materials & Interfaces* 13 (35): 42074–42093. https://doi.org/10.1021/acsami.1c13055.

47 Liu, X., Li, W., Wang, W. et al. (2018). Synthesis and characterization of pH-responsive mesoporous chitosan microspheres loaded with sodium phytate for smart water-based coatings. *Materials and Corrosion* 69 (6): 736–748. https://doi.org/10.1002/maco.201709840.

48 Sun, Y., Zhou, B., Lin, Y. et al. (2006). Quantum-sized carbon dots for bright and colorful photoluminescence. *Journal of the American Chemical Society* 128 (24): 7756–7757. https://doi.org/10.1021/ja062677d.

49 Hou, B., Li, X., Ma, X. et al. (2017). The cost of corrosion in China. *npj Materials Degradation* 1 (1): 1–10. https://doi.org/10.1038/s41529-017-0005-2.

50 Ye, K., Bi, Z., Cui, G. et al. (2020). External self-healing coatings in anticorrosion applications: a review. *Corrosion* 76 (3): 279–298. https://doi.org/10.5006/3430.

51 Liu, C., Jin, Z., Cheng, L. et al. (2020). Synthesis of nanosensors for autonomous warning of damage and self-repairing in polymeric coatings. *Nanoscale* 12 (5): 3194–3204. https://doi.org/10.1039/c9nr09221h.

52 Ma, L., Ren, C., Wang, J. et al. (2020). Self-reporting coatings for autonomous detection of coating damage and metal corrosion: a Review. *Chemical Engineering Journal* 421: 127854. https://doi.org/10.1016/j.cej.2020.127854.

53 Cheng, L., Liu, C., Zhao, H. et al. (2023). Hierarchically self-reporting and self-healing photothermal responsive coatings towards smart corrosion protection. *Chemical Engineering Journal* 467: 143463. https://doi.org/10.1016/j.cej.2023.143463.

54 Dir, B. and Kellou-Kerkouche, F. (2018). Experimental and theoretical studies on corrosion inhibition performance of phenanthroline for cast iron in acid solution. *Journal of Electrochemical Science and Technology* 9 (4): 260–275. https://doi.org/10.1007/s00894-019-4084-6.

55 Wang, J., Huang, Y., Ma, L. et al. (2022). Corrosion-sensing and self-healing dual-function coating based on 1,10-phenanthroline loaded urea formaldehyde microcapsules for carbon steel protection. *Colloids and Surfaces A: Physicochemical and Engineering Aspects* 652: 129855. https://doi.org/10.1016/j.colsurfa.2022.129855.

56 Zheng, C., Ren, H., Cui, Z. et al. (2009). Synthesis and characterization of nanoscale Terbium (III)-trimesic acid (TMA)-1, 10-phenanthroline (phen) luminescent complex. *Journal of Alloys and Compounds* 477 (1-2): 333–336. https://doi.org/10.1016/j.jallcom.2008.09.201.

57 Nji, J. and Li, G. (2012). Damage healing ability of a shape-memory-polymer-based particulate composite with small thermoplastic contents. *Smart Materials and Structures* 21 (2): 025011. https://doi.org/10.1088/0964-1726/21/2/025011.

58 Luo, X., Mather, P., and T. (2013). Shape memory assisted self-healing coating. *ACS Macro Letters* 2 (2): 152–156. https://doi.org/10.1021/mz400017x.

59 Birjandi, N.H., Garrison, K.L., and Mather, P.T. (2016). Comparative analysis of shape memory-based self-healing coatings. *Journal of Polymer Science Part B: Polymer Physics* 54 (14): 1415–1426. https://doi.org/10.1002/polb.24061.

60 Wang, L., Deng, L., Zhang, D. et al. (2016). Shape memory composite (SMC) self-healing coatings for corrosion protection. *Progress in Organic Coatings* 97: 261–268. https://doi.org/10.1016/j.porgcoat.2016.04.041.

61 Wu, D., Ma, L., Zhang, F. et al. (2020). Durable deicing lubricant-infused surface with photothermally switchable hydrophobic/slippery property. *Materials & Design* 185: 108236. https://doi.org/10.1016/j.matdes.2019.108236.

62 Jiang, D., Xia, X., Hou, J. et al. (2019). A novel coating system with self-reparable slippery surface and active corrosion inhibition for reliable protection of Mg alloy. *Chemical Engineering Journal* 373: 285–297. https://doi.org/10.1016/j.cej.2019.05.046.

63 Raj, R., Morozov, Y., Calado, L.M. et al. (2019). Inhibitor loaded calcium carbonate microparticles for corrosion protection of epoxy-coated carbon steel. *Electrochimica Acta* 319: 801–812. https://doi.org/10.1016/j.electacta.2019.07.059.

64 Wang, M., Wang, J., and Hu, W. (2020). Preparation and corrosion behavior of Cu-8-HQ@HNTs/epoxy coating. *Progress in Organic Coatings* 139: 105434. https://doi.org/10.1016/j.porgcoat.2019.105434.

65 Ma, L., Wang, X., Wang, J. et al. (2021). Graphene oxide-cerium oxide hybrids for enhancement of mechanical properties and corrosion resistance of epoxy coatings. *Journal of Materials Science* 56: 10108–10123. https://doi.org/10.1007/s10853-021-05932-z.

66 Lin, Y., Chen, R., Zhang, Y. et al. (2020). Sandwich-like polyvinyl alcohol (PVA) grafted graphene: a solid-inhibitors container for long term self-healing coatings. *Chemical Engineering Journal* 383: 123203. https://doi.org/10.1016/j.cej.2019.123203.

67 Cheng, L., Wu, H., Li, J. et al. (2021). Polydopamine modified ultrathin hydroxyapatite nanosheets for anti-corrosion reinforcement in polymeric coatings. *Corrosion Science* 178: 109064. https://doi.org/10.1016/j.corsci.2020.109064.

68 Wen, J., Lei, J., Chen, J. et al. (2020). An intelligent coating based on pH-sensitive hybrid hydrogel for corrosion protection of mild steel. *Chemical Engineering Journal* 392: 123742. https://doi.org/10.1016/j.cej.2019.123742.

69 Wang, N., Zhang, Y., Chen, J. et al. (2017). Dopamine modified metal-organic frameworks on anti-corrosion properties of waterborne epoxy coatings. *Progress in Organic Coatings* 109: 126–134. https://doi.org/10.1016/j.porgcoat.2017.04.024.

70 Ma, Y., Zhang, Y., Liu, J. et al. (2020). GO-modified double-walled polyurea microcapsules/epoxy composites for marine anticorrosive self-healing coating. *Materials & Design* 189: 108547. https://doi.org/10.1016/j.matdes.2020.108547.

71 Liu, T., Liu, Y., Ye, Y. et al. (2019). Corrosion protective properties of epoxy coating containing tetraaniline modified nano-α-Fe_2O_3. *Progress in Organic Coatings* 132: 455–467. https://doi.org/10.1016/j.porgcoat.2019.04.010.

72 Huang, Y., Deng, L., Ju, P. et al. (2018). Triple-action self-healing protective coatings based on shape memory polymers containing dual-function microspheres. *ACS Applied Materials & Interfaces* 10 (27): 23369–23379. https://doi.org/10.1021/acsami.8b06985.

73 Qian, H., Xu, D., Du, C. et al. (2017). Dual-action smart coatings with a self-healing superhydrophobic surface and anti-corrosion properties. *Journal of Materials Chemistry A* 5 (5): 2355–2364. https://doi.org/10.1039/c6ta10903a.

74 Chen, J., Fang, L., Xu, Z. et al. (2016). Self-healing epoxy coatings curing with varied ratios of diamine and monoamine triggered via near-infrared light. *Progress in Organic Coatings* 101: 543–552. https://doi.org/10.1016/j.porgcoat.2016.09.020.

75 Liu, G. and Wheat, H.G. (2010). Coatings for early corrosion detection. *ECS Transactions* 28 (24): 239. https://doi.org/10.1149/1.3496435.

76 Liu, G. and Wheat, H.G. (2009). Use of a fluorescent indicator in monitoring underlying corrosion on coated aluminum 2024-T4. *Journal of the Electrochemical Society* 156 (4): C160. https://doi.org/10.1149/1.3078395.

77 Robb, M.J., Li, W., Gergely, R.C. et al. (2016). A robust damage-reporting strategy for polymeric materials enabled by aggregation-induced emission. *ACS Central Science* 2 (9): 598–603. https://doi.org/10.1021/acscentsci.6b00198.

78 Chen, S., Han, T., Zhao, Y. et al. (2019). A facile strategy to prepare smart coatings with autonomous self-healing and self-reporting functions. *ACS Applied Materials & Interfaces* 12 (4): 4870–4877. https://doi.org/10.1021/acsami.9b18919.

79 Rohwerder, M. and Michalik, A. (2007). Conducting polymers for corrosion protection: What makes the difference between failure and success? *Electrochimica Acta* 53 (3): 1300–1313. https://doi.org/0.1016/j.electacta.2007.05.026.

80 Rohwerder, M. (2009). Intelligent corrosion protection by conducting polymers. In: *SMART COATINGS II*, 274–287. Orlando, FL: ACS Publications.

81 Situ, Y., Ji, W., Liu, C. et al. (2019). Synergistic effect of homogeneously dispersed PANI-TiN nanocomposites towards long-term anticorrosive performance of epoxy coatings. *Progress in Organic Coatings* 130: 158–167. https://doi.org/10.1016/j.porgcoat.2019.01.034.

82 Liu, T., Zhao, H., Li, J. et al. (2020). POSS-tetraaniline based giant molecule: Synthesis, self-assembly, and active corrosion protection of epoxy-based organic coatings. *Corrosion Science* 168: 108555. https://doi.org/10.1016/j.corsci.2020.108555.

83 Dolatkhah, A. and Wilson, L.D. (2018). Salt-responsive Fe_3O_4 nanocomposites and phase behavior in water. *Langmuir* 34 (1): 341–350. https://doi.org/10.1021/acs.langmuir.7b03613.

84 de Morais, C.D.M., Silva, S.R.B., Vieira, D.S. et al. (2016). Integrating a smartphone and molecular modeling for determining the binding constant and stoichiometry ratio of the iron (II)-phenanthroline complex: an activity for analytical and physical chemistry laboratories. *Journal of Chemical Education* 93 (10): 1760–1765. https://doi.org/10.1021/acs.jchemed.6b00112.

85 Calado, L.M., Taryba, M.G., Morozov, Y. et al. (2021). Cerium phosphate-based inhibitor for smart corrosion protection of WE43 magnesium alloy. *Electrochimica Acta* 365: 137368. https://doi.org/10.1016/j.electacta.2020.137368.

86 Dastgheib, A., Zarebidaki, A., and Attar, M.M. (2020). Enhanced barrier properties of epoxy coating containing $CaCO_3$ microparticles modified with cerium nitrate. *Progress in Organic Coatings* 144: 105660. https://doi.org/10.1016/j.porgcoat.2020.105660.

87 Cai, G., Xiao, S., Deng, C. et al. (2021). CeO_2 grafted carbon nanotube via polydopamine wrapping to enhance corrosion barrier of polyurethane coating. *Corrosion Science* 178: 109014. https://doi.org/10.1016/j.corsci.2020.109014.

88 Morozov, Y., Calado, L., Shakoor, R. et al. (2019). Epoxy coatings modified with a new cerium phosphate inhibitor for smart corrosion protection of steel. *Corrosion Science* 159: 108128. https://doi.org/10.1016/j.corsci.2019.108128.

89 Bahlakeh, G., Ramezanzadeh, B., and Ramezanzadeh, M. (2017). Cerium oxide nanoparticles influences on the binding and corrosion protection characteristics of a melamine-cured polyester resin on mild steel: an experimental, density functional theory and molecular dynamics simulation study. *Corrosion Science* 118: 69–83. https://doi.org/10.1016/j.corsci.2017.01.021.

90 Wang, W., Li, W., Fan, W. et al. (2018). Accelerated self-healing performance of magnetic gradient coating. *Chemical Engineering Journal* 332: 658–670. https://doi.org/10.1016/j.cej.2017.09.112.

91 Wang, P., Chen, S., Cao, Z. et al. (2017). NIR light-, Temperature-, pH-, and Redox-responsive polymer-modified reduced graphene oxide/mesoporous silica sandwich-like nanocomposites for controlled release. *ACS Applied Materials & Interfaces* 9 (34): 29055–29062. https://doi.org/10.1021/acsami.7b07468.

92 Guo, W., Umar, A., Du, Y. et al. (2020). Surface modification of bentonite with polymer brushes and its application as an efficient adsorbent for the removal of hazardous dye orange I. *Nanomaterials* 10 (6): 1112. https://doi.org/10.3390/nano10061112.

93 Chen, Y., Wang, X.H., Li, J. et al. (2007). Long-term anticorrosion behaviour of polyaniline on mild steel. *Corrosion Science* 49 (7): 3052–3063. https://doi.org/10.1016/j.corsci.2006.11.007.

94 Lendlein, A. and Kelch, S. (2002). Shape-memory polymers. *Angewandte Chemie International Edition* 41 (12): 2034–2057. https://doi.org/10.1002/1521-3773(20020617)41:12<2034::aid-anie2034>3.0.co;2-m.

95 Chen, T., Chen, R., Jin, Z. et al. (2015). Engineering hollow mesoporous silica nanocontainers with molecular switches for continuous self-healing anticorrosion coating. *Journal of Materials Chemistry A* 3 (18): 9510–9516. https://doi.org/10.1039/c5ta01188d.

96 Gao, L., He, J., Hu, J. et al. (2015). Photoresponsive self-healing polymer composite with photoabsorbing hybrid microcapsules. *ACS Applied Materials & Interfaces* 7 (45): 25546–25552. https://doi.org/10.1021/acsami.5b09121.

97 Turkenburg, D. and Fischer, H. (2015). Diels-Alder based, thermo-reversible cross-linked epoxies for use in self-healing composites. *Polymer* 79: 187–194. https://doi.org/10.1016/j.polymer.2015.10.031.

98 Bastos, A., Simões, A., González, S. et al. (2004). Imaging concentration profiles of redox-active species in open-circuit corrosion processes with the scanning electrochemical microscope. *Electrochemistry Communications* 6 (11): 1212–1215. https://doi.org/10.1016/j.elecom.2004.09.022.

99 Hillewaere, X.K.D. and Du Prez, F.E. (2015). Fifteen chemistries for autonomous external self-healing polymers and composites. *Progress in Polymer Science* 49: 121–153. https://doi.org/10.1016/j.progpolymsci.2015.04.004.

100 Roy, N., Bruchmann, B., and Lehn, J.M. (2015). DYNAMERS: dynamic polymers as self-healing materials. *Chemical Society Reviews* 44 (11): 3786–3807. https://doi.org/10.1039/c5cs00194c.

101 García, S.J., Fischer, H.R., and van der Zwaag, S. (2011). A critical appraisal of the potential of self healing polymeric coatings. *Progress in Organic Coatings* 72 (3): 211–221. https://doi.org/10.1016/j.porgcoat.2011.06.016.

7

Development of Mechanically Responsive Self-reporting Coating

7.1 Introduction

In addition to the fluorescence/coloration effects activated by pH variation or metal ions, some chemicals can be released from the ruptured capsules that directly react with the coating components, indicating coating damage [1–3]. Two types of polymers are used to fabricate mechanically responsive self-reporting coatings: intrinsic and extrinsic [4–6]. Intrinsic self-reporting polymers directly embed mechanically responsive molecules into a polymer backbone or cross-linker units. For example, spiropyran, one of the most commonly used mechanical color-changing molecules, is immobilized on the backbone of a polyacrylate and undergoes a ring-opening reaction in response to an external force, resulting in a color change from yellow to purple [7].

Extrinsic self-reporting polymers are usually incorporated into microcapsules containing an indicator and can be activated by capsule rupture [8–10]. For example, polyaniline nanocapsules containing fluorescein isothiocyanate have been incorporated into epoxy coatings to achieve salt-responsive self-reporting [11]. NaCl allows the phase transformation of polyaniline to open its channels for the release of isothiocyanate, resulting in green fluorescence signals on the damaged coating surface upon exposure to ultraviolet (UV) light. As such microcontainers can be easily integrated into any coating of interest without chemical modification, extrinsic self-reporting polymers are more suitable for detecting polymer damage. Relevant examples used as mechanically responsive self-reporting coatings are presented in this chapter, with a focus on the application of aggregation-induced emission (AIE) chemicals [12–14], crystal violet lactone (CVL) [9, 15, 16], and 2′,7′-dichlorofluorescein (DCF) [17–19] as damage indicators.

For the application of AIE, mechanical damage rapidly generates localized fluorescent signals that are easily observed under UV light, providing excellent contrast between the intact and damaged areas of the material. For example, to demonstrate this concept, Robb et al. [20] investigated the AIE luminogen 1,1,2,2-tetraphenylethylene (TPE). When TPE was dissolved in hexyl acetate, the solution became colorless and did not fluoresce under UV illumination. However, bright blue fluorescence with an emission maximum of 450 nm was observed because a solid residue formed when the solvent evaporated. Transparent

Smart Protective Coatings for Corrosion Control, First Edition. Lingwei Ma and Dawei Zhang.

epoxy coatings (10 wt% TPE microcapsules) were prepared to investigate their mechanically responsive self-reporting capabilities. Photographs of the scratched coating areas exhibited a distinct blue color in contrast to the intact coating region. In addition, the indication response of epoxy coatings with different depths (94–376 μm) was characterized using fluorescence microscopy. As the depth increased, the area of the fluorescent region and average intensity increased proportionally. TPE microcapsules provide excellent detection performance for different coatings, including polyurethane, polydimethylsiloxane, UV-cured epoxy, poly(acrylic acid) cast from water, and polystyrene cast from toluene. This system provides a simple and practical method for the visual detection of microscopic damage in a wide range of polymer materials under illumination by an appropriate excitation light source.

Subsequently, Lu et al. [13] advanced the work of Robb et al. [20]. They verified the feasibility of a multilayered coating system with three AIEgens (red (2,3-bis[4-(diphenylamino)phenyl]fumaronitrile [BPF]), green (hexaphenylsilole [HPS]), and blue [TPE] emissive colors) loaded with poly(urea–formaldehyde)–polyurethane as shell materials to form double-walled microcapsules. In the event of microcracks, the AIEgen microcapsules in the damaged region ruptured mechanically, followed by the immediate generation of bright fluorescent colors during rapid evaporation of the payload solvent. Therefore, the coating system can distinguish the damage at different depths by adding AIEgens to different coating layers.

CVL was selected as the visual sensing material because of its strong and rapid coloration ability due to its reaction with SiO_2 [21, 22]. The colorless CVL reacts with SiO_2 to cause color development through the reversible opening of the lactone ring [23, 24]. Hu et al. [16] developed a mechanically triggered self-reporting strategy using CVL as an indicator dye because of its excellent coloration ability, which changed from colorless to blue through hydrogen bonding with the hydroxyl groups on the oxide surface. A three-component nanocapsule was prepared in which the CVL was located inside the polymethyl methacrylate shell, and the SiO_2 nanoparticles were located on the capsule surface. Once the coating loaded with this type of nanocapsule was damaged, CVL flowed out and came into contact with SiO_2, producing an intense blue color that indicated mechanical damage.

7.2 Mechanically Responsive Self-reporting Coating Based on DCF

DCF can produce a vibrant color change by reacting with various amino groups as a result of its ring-opening reaction to transform from an acid form to a base form. The DCF was loaded into polyurethane/poly(urea–formaldehyde) (PU/UF) microcapsules, which were then incorporated into shape-memory epoxy polymer (SMEP) coatings. Once the coating was mechanically damaged, DCF was released from the ruptured DCF@PU/UF microcapsules and reacted with the residual amino groups in the polymer matrix to produce a prominent red color. Even with a topcoat of

Table 7.1 The chemicals used in this chapter.

Materials	Specification	Source
Bisphenol A diglycidyl ether	≥85.0% (GC)	Aladdin
Jeffamine D230	—	Aladdin
Resorcinol	AR, 99%	Aladdin
Ethyl phenylacetate	99%	Aladdin
2′,7′-Dichlorofluorescein	—	Sigma-Aldrich
Polyethylene-graft-maleic anhydride	Mw 100 000–500 000	Sigma-Aldrich
Urea	AR, 99%	Macklin
Ammonium chloride	GR, 99.8%	Macklin
Ninhydrin hydrate	98%	Macklin
NaCl	AR, ≥99.5%	Sinopharm
Sodium hydroxide	AR	Sinopharm
Formaldehyde	AR, 37 wt%	Sinopharm
Ethanol	99.5%	Sinopharm
PU prepolymer	Desmodur L75	Shandong
TS01-3 PU varnish	TS01-3	Tianjin

polyurethane varnish or black paint, yellow fluorescence signals of DCF were easily observed upon UV irradiation. For the self-healing performance, the high mobility of the polymer networks endowed the coating with a shape-memory effect to recover the original shape and barrier properties. The content of the amine curing agent was precisely regulated in the SMEP to strengthen both the self-reporting and self-healing performance. This strategy has excellent potential for prolonging the service lifetime of smart coatings.

7.2.1 Materials and Methods

7.2.1.1 Materials

The materials and reagents used in this section are shown in Table 7.1.

7.2.1.2 Instruments

The instruments used in this section are shown in Table 7.2.

7.2.1.3 Preparation of DCF@PU/UF Microcapsules

DCF@PU/UF microcapsules were synthesized using a single batch process, as shown in Figure 7.1. Specifically, the water phase was prepared by adding 1.0 g of urea, 0.1 g of NH_4Cl, and 0.1 g of resorcinol to 60 ml of a 0.5 wt% polyethylene-graft-maleic anhydride (EMA) aqueous solution, and the pH of the solution was adjusted to 3.5 using 0.1 M NaOH. Then, 24 ml of ethyl phenylacetate (EPA) was mixed with 0.8 g of PU and 0.05 g of DCF in an ultrasonic

Table 7.2 The instruments and equipment.

Instrument	Production type	Manufacturer
Drying oven	303-3S	China BaiHui
Magnetic stirrer	XH-C	China labCAN
Vortex oscillator	ZNCL-GS	China Huate
Vacuum freeze dryer	YTLG-10A	China Yetuo
Electrochemical workstation	PARSTAT-2273	America Princeton
SEM	MIRA LMS	Czech Tescan
UV–Vis spectroscopy	U4150	Japan Hitachi
Fourier transform infrared spectroscopy (FTIR)	VERTEX 70r	Germany Bruker
Differential scanning calorimetry (DSC)	DSC3	Switzerland Mettler
Dynamic mechanical analysis (DMA)	Q800	America TA
Optical microscope	DM2700M	Germany Leica
Confocal laser scanning microscope	VK-X200 series	Japan Keyence

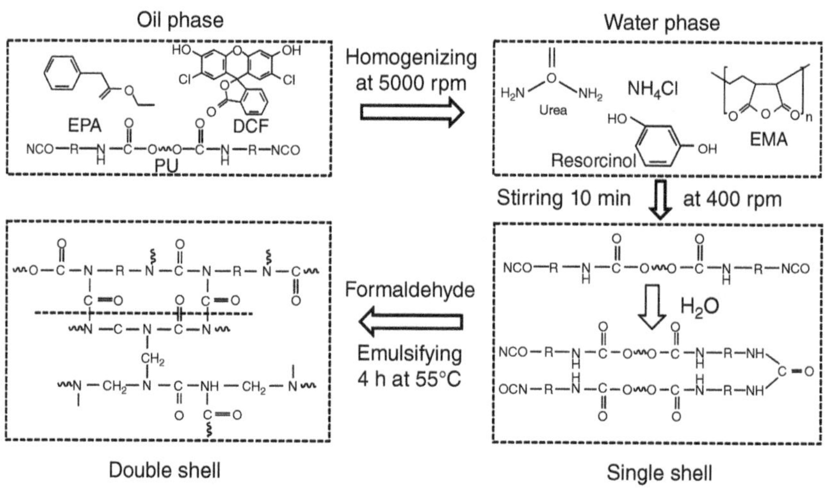

Figure 7.1 The preparation process of DCF@PU/UF microcapsules.

bath for 10 minutes to obtain the oil phase. The oil phase was added to the water phase under mechanical stirring at 5000 rpm for three minutes, and the obtained emulsion was transferred to a flask and stirred at 400 rpm. After 10 minutes, 2.54 ml of a formaldehyde solution was added to the emulsion, and the mixture was heated to 55 °C. The emulsification process lasted four hours. The synthesized emulsions containing the microcapsules were gently rinsed with deionized water to remove excess surfactants. The solution was filtered and dried in a vacuum freeze dryer for 24 hours. Finally, the obtained DCF@PU/UF microcapsules were sieved through a 120-mesh sieve to avoid agglomeration.

Table 7.3 Names of different SMEP samples and the corresponding formulation.

NameFormulation	EP0.5	EP0.6	EP0.7	EP0.8	EP0.9	EP1.0
DGEBA (mol)	1	1	1	1	1	1
D230 (mol)	0.5	0.6	0.7	0.8	0.9	1.0

7.2.1.4 Fabrication of SMEP Coating Containing DCF@PU/UF

Aluminum alloy 2024 (AA2024) plates were sanded with 240 SiC sandpaper and cleaned with acetone. The SMEP coatings consisted of a DGEBA epoxy monomer and D230 amine curing agent mixed in different molar ratios listed in Table 7.3. DGEBA and D230 were mixed using a magnetic stirrer for 20 minutes and vacuum-pumped to remove bubbles. Subsequently, 5 wt% DCF@PU/UF microcapsules were added to EP0.8, EP0.9, and EP1.0 samples, and the obtained mixtures were uniformly mixed by a vortex oscillator and then uniformly coated on the AA2024 substrates. Subsequently, the coatings were cured at 60 °C for 24 hours to obtain the final product. The prepared coatings are DCF@PU/UF-EP0.8, DCF@PU/UF-EP0.9, and DCF@PU/UF-EP1.0. The thicknesses of all the coatings were controlled to be approximately 220 ± 20 μm.

7.2.1.5 Characterization of DCF@PU/UF Microcapsules

The morphologies of the DCF@PU/UF microcapsules were examined using SEM (TESCAN MIRALMS), and their particle size distributions were measured using Nano Measure software. FTIR spectroscopy (VERTEX 70, Bruker) of the DCF, EPA, and DCF@PU/UF microcapsules was performed to check whether DCF was successfully encapsulated.

7.2.1.6 Characterization of Coatings

To obtain the T_g values of epoxy (EP) coatings with different ratios of DGEBA and D230, a DSC experiment was conducted from −10 to 150 °C under nitrogen protection, with a heating rate of 5 °C min^{-1}. A DMA test was conducted using a tensile model (Q800, TA Instruments), which was heated from −10 to 150 °C at a rate of 5 °C min^{-1} in a nitrogen atmosphere. The dimensions of the DSC and DMA samples were $20 \times 5 \times 2$ mm^3. The residual amino group content of the EP samples was determined using the ninhydrin hydrate reaction. Specifically, coating samples were immersed in a 1×10^{-2} M ninhydrin/ethanol solution for 1 hour and then heated at 105 °C for 15 minutes. The color change was recorded using a digital camera, and the transmittance was measured using UV–Vis spectroscopy (U4150, Hitachi). Parallel studies were conducted using a 1×10^{-2} M DCF/EPA solution to determine the reaction of DCF with the residual amino groups in the epoxy resins. The barrier properties of the DCF@PU/UF-EP coatings were assessed via electrochemical impedance spectroscopy (EIS) in the frequency range of 10^5–10^{-2} Hz with a voltage perturbation amplitude of 20 mV. Each sample was tested three times to ensure reproducibility.

7.2.1.7 Characterization of Self-reporting and Self-healing Performance

To demonstrate the self-reporting performance of the DCF@PU/UF-EP coatings, a fresh scratch with a width of ~150 μm was made on the coating surface by a No. 11 scalpel, and then the color change of the scratched coating area was recorded using a digital camera. In addition, microscopic images of the scratched areas were obtained using an optical microscope (OM, Leica DM2700M) in a dark-field model, and the fluorescence signals of the scratched areas were evaluated using a laser confocal microscope (CLSM, Zeiss LSM780). In addition, the effect of PU varnish and PU-based black paint with a thickness of ~100 μm on the self-reporting properties of DCF@PU/UF-EP coatings was investigated.

To evaluate the shape-memory effect of SMEP, the morphologies of scratched coatings were recorded by CLSM (KEYENCE VK-X200 series). Afterward, the coating was placed in an oven at 50 °C for 30 minutes, and the CLSM was again used to record the closure process of the coating scratches. EIS tests were performed on the scratched and healed coatings to confirm the self-healing properties.

7.2.2 Characterization of DCF@PU/UF Microcapsules

Figure 7.2 shows the preparation process for the DCF@PU/UF microcapsules. Briefly, water and oil phases were mixed to obtain single-shell DCF@PU microcapsules, and formaldehyde was added to form the PUF shell. The morphology of the as-prepared DCF@PU/UF microcapsules was examined using SEM, as shown in Figure 7.3. Most of the microcapsules were spherical with a uniform size distribution, and the average particle size was approximately 20 μm.

The microcapsules were well encapsulated, and some small particles grew on the surface of DCF@PU/UF because of the polyurea formaldehyde nanoparticles attached to the PUF shell [25]. As shown in Figure 7.3b, the wall thickness of the DCF@PU/UF microcapsules was approximately 117 nm, ensuring sufficient internal space for DCF encapsulation.

Figure 7.4 presents the FTIR spectra of DCF, EPA, and ruptured DCF@PU/UF microcapsules. For DCF, the absorption peaks at 1735 and 1058 cm^{-1} are related to

Figure 7.2 (a) SEM images and (b) the corresponding size distributions of the DCF@PU/UF microcapsules.

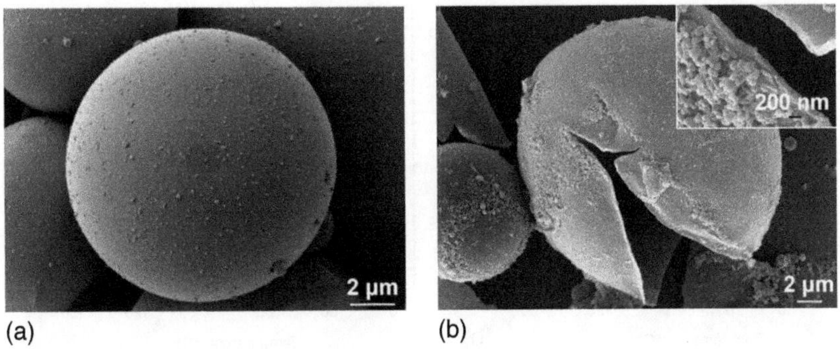

Figure 7.3 (a) The intact microcapsule; (b) the ruptured microcapsule (the inset is the wall thickness).

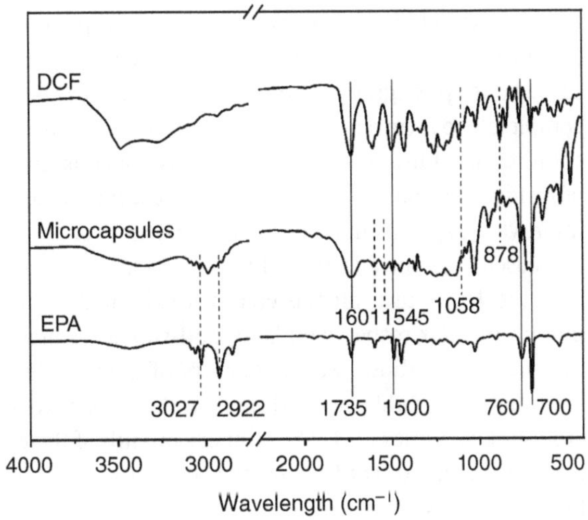

Figure 7.4 FT-IR spectra of DCF, EPA, and ruptured DCF@PU/UF microcapsules.

the stretching vibration of the C=O and the stretching of C—Cl bonds, respectively. Additionally, absorption peaks appear at 1500, 760, and 700 cm^{-1}, corresponding to the benzene ring of DCF. EPA exhibited characteristic FTIR peaks at 3027 and 2922 cm^{-1}, assigned to the aromatic and aliphatic C–H stretching vibrations, respectively [26–30]. These DCF and EPA FTIR peaks were also observed in the spectrum of the ruptured DCF@PU/UF microcapsules, confirming the successful encapsulation of DCF and EPA in the microcapsules. Moreover, the DCF@PU/UF microcapsules exhibited absorption peaks at 1601 and 1545 cm^{-1}, respectively, assigned to the NH–CO and N–H stretching vibrations of PU [31].

7.2.3 Characterization of EP Coatings

As shown in Figure 7.5, the thermodynamic and thermomechanical properties of the different coatings were evaluated via DSC and DMA. From the EP0.5 coating to the EP1.0 coating, the DSC curves indicated a sharp reduction in T_g from 81.2 to

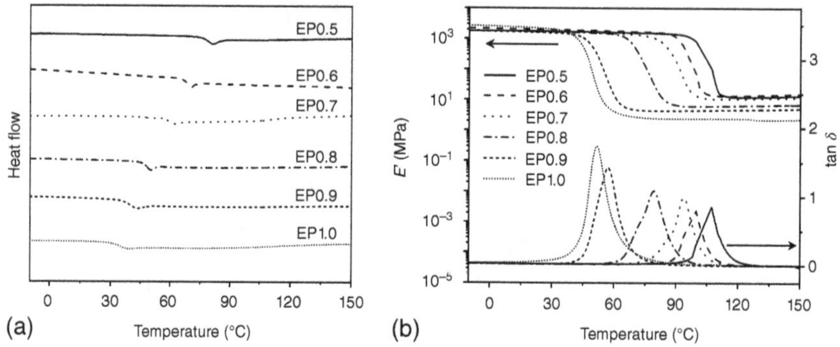

Figure 7.5 (a) DSC and (b) DMA curves of the EP samples.

33.0 °C. All the coatings exhibited a flat rubbery-state plateau, indicating that the samples exhibited elastomeric behavior at temperatures higher than T_g [32]. During the curing of the epoxy coatings, one active hydrogen atom on the amino group consumed one epoxy group. Therefore, for the EP0.5 coating, D230 and DGEBA reacted completely to form a network structure with a high degree of cross-linking. With the continuous addition of D230, D230 was not consumed completely, and the amount of residual amino groups increased; thus, the flexibility of the epoxy network was improved [32]. This indicated that the degree of cross-linking of the coating system decreased with the addition of D230 [33–35].

In the DMA test, all the coatings exhibited relatively constant energy storage moduli in their respective glassy and rubbery states [32, 36–38], and the value of the rubbery modulus was 2–3 orders of magnitude lower than that of the glassy modulus. The rubbery modulus of thermoset systems is proportional to their cross-link densities [35]. The rubbery moduli of the coatings decreased as the D230 content increased, confirming that the cross-linking degree of the epoxy network further decreased.

In this coating system, the amino groups in the epoxy coatings included those that reacted with the epoxy groups and residual unreacted groups. It is known that the reaction between ninhydrin hydrate and amino groups can form diketohydrindylidenediketohydrindamine, exhibiting a purple-colored ninhydrin chromophore (called Ruhemann's purple, λ = 570 nm) [39–42]. Therefore, the presence of residual amino groups in the coatings was confirmed by the ninhydrin hydrate reaction. As shown in Figure 7.6a, all pristine EP samples exhibited a transparent luster. After immersion in a 1×10^{-2} M ninhydrin/ethanol solution, the color of the samples gradually changed. The color of EP0.5 remained almost unchanged, which indicated the complete reaction between D230 and DGEBA and, thus, the absence of residual amino groups.

With an increase in the D230 content of the EP samples, the colors of EP0.6, EP0.7, EP0.8, and EP0.9 darkened, and EP1.0 appeared to be saturated with a distinct Ruhemann's purple color. The color depth can be used as a parameter to determine the residual amino content of the epoxy coatings. In addition, the degree of coloration was quantified using the UV–Vis spectral transmission intensity (Figure 7.6c). The

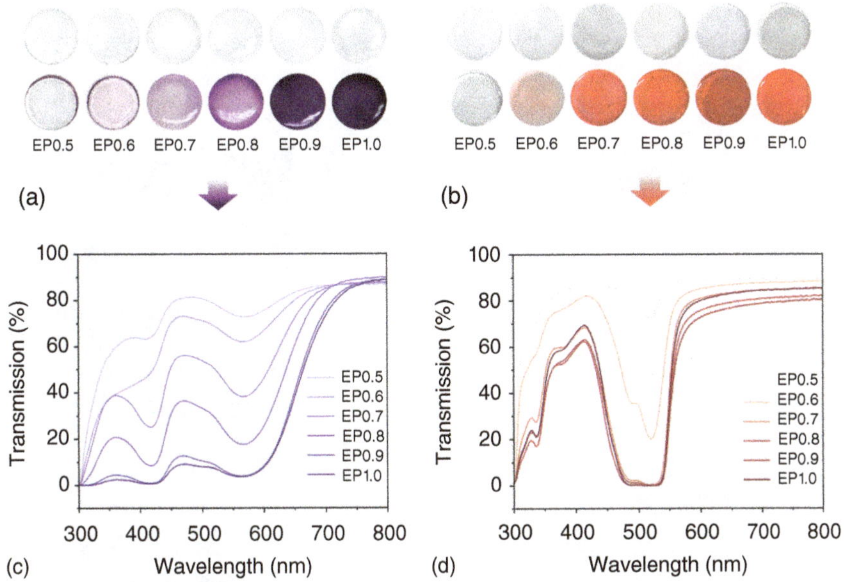

Figure 7.6 The color change and corresponding UV–Vis spectra of EP samples after immersing in 1×10^{-2} M (a,b) ninhydrin/ethanol and (b,d) DCF/EPA solution.

intensity of the transmission peak at 570 nm decreased with an increase in the D230:DGEBA molar ratio, indicating that the content of residual amino groups in the epoxy resin increased with the D230 content in the EP0.5–EP1.0 samples.

Moreover, DCF has high reactivity with various amino groups to produce a vibrant color change, attributed to its ring-opening reaction to transform from an acidic form into a basic form [31, 43]. In other words, DCF can react with the residual amino groups in epoxy coatings, resulting in a color change. Figure 7.6b shows the optical images of the EP samples before and after immersion in a 1×10^{-2} M DCF/EPA solution. After immersion, the samples changed from transparent to light red and then to crimson in the order of EP0.5 to EP1.0. In Figure 7.6d, the reduction in the transmission intensity at 420 nm supports the color-change rule of these samples. It also indicates that more residual amino groups in the coating correspond to better color-rendering performance. According to the visual observations and UV–Vis transmission spectra, EP0.8, EP0.9, and EP1.0 samples have sufficient residual amines to achieve excellent color development efficiency when reacting with DCF.

7.2.4 Self-reporting Performance of DCF@PU/UF-EP Coatings

After confirming the color rendering effect between the residual amino groups and DCF in the EP coatings, the self-reporting performance of the DCF@PU/UF-EP coatings containing 5 wt% microcapsules was evaluated by forming a scratch with a width of ~150 μm on the coating surface, which was recorded by a digital camera and CLSM, using EP0.9 coatings as reference samples. Figure 7.7 shows that the

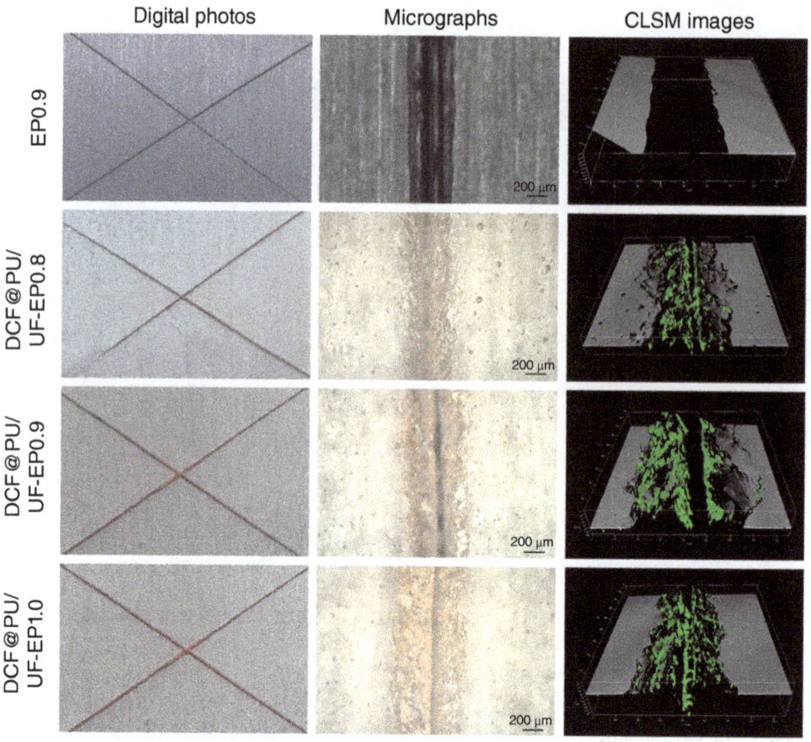

Figure 7.7 Digital photos and corresponding micrographs, fluorescent microscopic CLSM images of EP0.9, DCF@PU/UF-EP0.8, DCF@PU/UF-EP0.9, and DCF@PU/UF-EP1.0 coatings.

EP0.9 coating lacked both color and fluorescence signals to characterize the coating damage. When the DCF@PU/UF-EP coatings were scratched, strong red signals were observed in the scratched areas. This is because the broken DCF@PU/UF microcapsules in the damaged coating area released DCF, which reacted with the residual amino groups in the coating matrix to produce a red color. Thus, mechanically responsive self-reporting was achieved. Notably, the color of the coating scratches deepened as the content of residual amino groups increased in the order of DCF@PU/UF-EP0.8, DCF@PU/UF-EP0.9, and DCF@PU/UF-EP1.0. In particular, the DCF@PU/UF-EP0.9 and DCF@PU/UF-EP1.0 coatings exhibited significant color development, suggesting that there were sufficient amino groups in these two coatings to react with the released DCF and achieve sensitive self-reporting performance. In the microscopic images, the DCF@PU/UF-EP0.8 coating exhibited a light red color in the scratched region, whereas the DCF@PU/UF-EP0.9 and DCF@PU/UF-EP1.0 coatings exhibited distinct red colors around the scratches. In the fluorescence microscopy CLSM images, owing to the fluorescence property of DCF, all three coatings exhibited fluorescence signals in the scratched area, particularly DCF@PU/UF-EP0.9 and DCF@PU/UF-EP1.0, which had strong fluorescence signals.

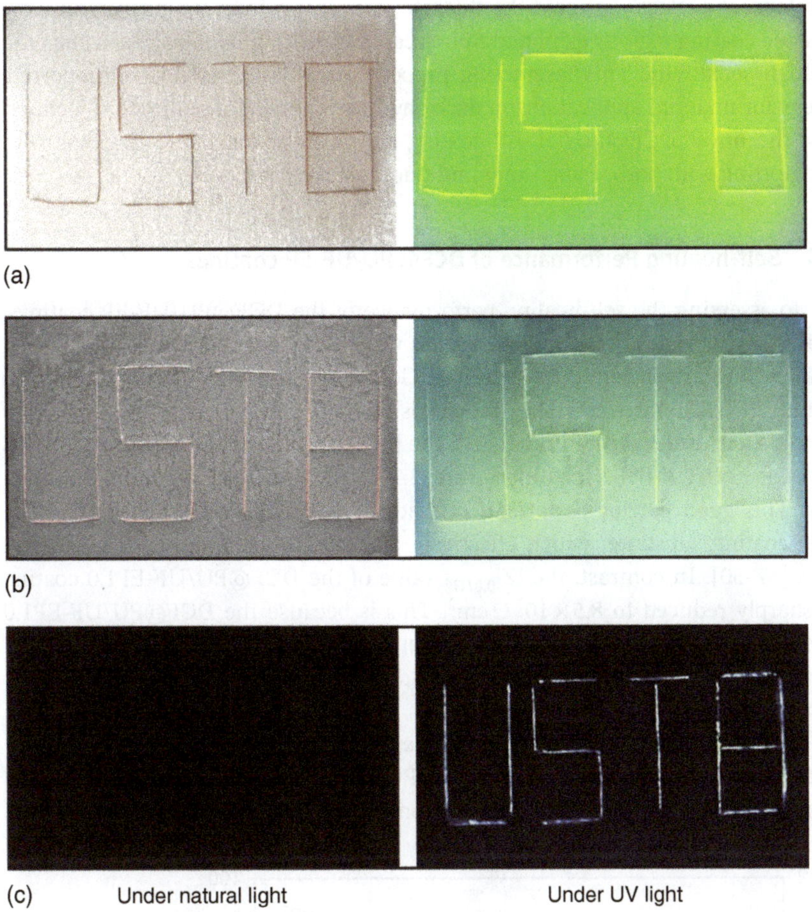

Figure 7.8 The photos of (a) DCF@PU/UF-EP0.9 coating, (b) DCF@PU/UF-EP0.9/PV coating, (c) DCF@PU/UF-EP0.9/PB coating with "USTB" shaped scratches under natural light or UV light irradiation.

To evaluate the practical application ability of DCF@PU/UF-EP coatings, pristine DCF@PU/UF-EP0.9 coatings were separately covered with PU varnish (DCF@PU/UF-EP0.9/PV) and PU-based black paint (DCF@PU/UF-EP0.9/PB). The coating surface was scratched to show the "USTB" logo, and the coating was irradiated with both natural and UV light to evaluate the self-reporting performance. As shown in Figure 7.8, the DCF@PU/UF-EP0.9 coating surface exhibited a clear "USTB" red signal, and even after the surface was covered with ~100 μm thick PU varnish, the red damage signal was clearly observed. Under UV irradiation, the scratched regions of the DCF@PU/UF-EP0.9 and DCF@PU/UF-EP0.9/PV coatings exhibited strong fluorescent signals, which originated from the DCF released from the microcapsules and the aggregation effect.

On the DCF@PU/UF-EP0.9/PB black topcoat coating, it was difficult to observe the coating damage directly with the naked eye. However, under UV irradiation, clear fluorescence was observed in the damaged areas (Figure 7.8c). Because the black topcoat perfectly covered the fluorescence signals from the primer, only the scratched coating regions exhibited fluorescence, resulting in damage sensing with ultrahigh sensitivity. This fascinating property provides a reliable self-reporting strategy for material applications in dark environments and nighttime inspections. Thus, the novel DCF@PU/UF-EP coating can achieve mechanically responsive self-reporting with both visible color and fluorescence responses.

7.2.5 Self-healing Performance of DCF@PU/UF-EP coatings

Prior to assessing the self-healing performance of the DCF@PU/UF-EP coatings, the barrier properties of the intact coatings were tested in a 3.5 wt% NaCl solution and the results are presented in Figure 7.9. Generally, the low-frequency impedance modulus ($|Z|_{0.01Hz}$) reflects the anticorrosion performance of coating systems [44–46]. After three days of immersion, the $|Z|_{0.01Hz}$ values of DCF@PU/UF-EP0.8 and DCF@PU/UF-EP0.9 coatings were 3.7×10^{10} and $5.5 \times 10^{10}\,\Omega\,cm^2$, respectively. This good barrier performance is attributed to the cross-linked structure of the coating substrate, which effectively blocked the penetration of aggressive media [47–50]. In contrast, the $|Z|_{0.01Hz}$ value of the DCF@PU/UF-EP1.0 coating was sharply reduced to $8.5 \times 10^7\,\Omega\,cm^2$. This is because the DCF@PU/UF-EP1.0 coating had excessive residual amino groups, resulting in relatively softness and inferior curing degree of the coating. The Bode phase angle plots also reflected the anticorrosion behavior of the coatings.

As shown in Figure 7.9b, the phase angles of all the coatings started from −90° in the high-frequency region. The DCF@PU/UF-EP0.8 and DCF@PU/UF-EP0.9 coatings had capacitive properties for only one time constant, whereas the

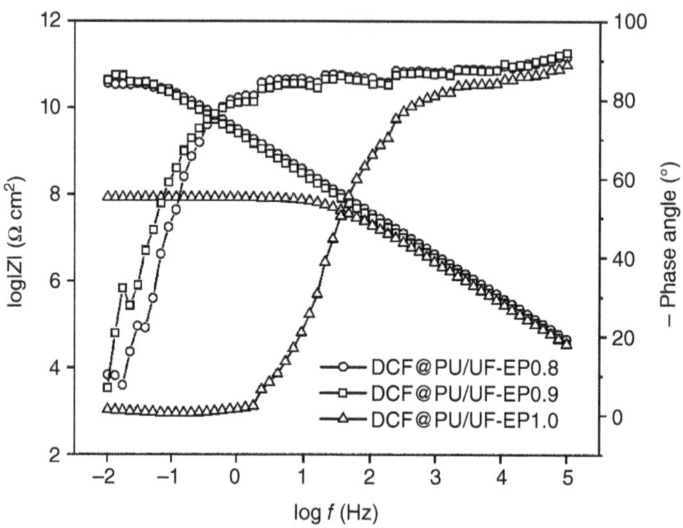

Figure 7.9 Bode magnitude and phase angle plots of DCF@PU/UF-EP0.8, DCF@PU/UF-EP0.9 and DCF@PU/UF-EP1.0 coatings after three days of immersion in 3.5 wt% NaCl solution.

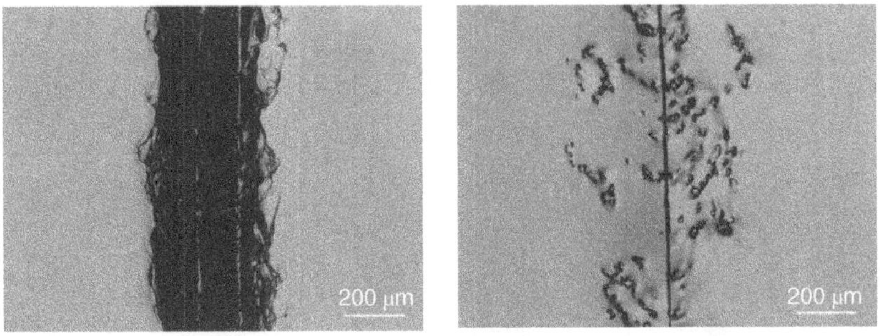

(a) (b)

Figure 7.10 Micrographs and 3D images of DCF@PU/UF-EP0.9 scratch (a) before and (b) after healing.

DCF@PU/UF-EP1.0 coating exhibited resistive behavior for two time constants in the low-frequency region, indicating that the corrosive ions reached the AA2024 substrate surface through microscopic coating defects [51]. Thus, the DCF@PU/UF-EP0.8 and DCF@PU/UF-EP0.9 coatings exhibited better corrosion resistance. Considering the results presented in Figures 7.7–7.9, the DCF@PU/UF-EP0.9 coating with excellent self-reporting performance and good barrier performance was selected for further study of the self-healing performance.

To evaluate the self-healing performance, a ~150 μm wide scratch was made on the DCF@PU/UF-EP0.9 coating surface, and the coating was heated at 50 °C (slightly higher than the T_g of EP0.9 at ~43 °C) for 30 minutes to activate the self-healing effect. The surface morphologies of the coating scratches before and after healing are presented in Figure 7.10. As shown, the coating scratches were significantly narrowed after heating, leaving a small gap with a width of ~15 μm. During the heating process, the shape-memory effect of the epoxy matrix was activated, promoting the movement of the polymer molecular chains and causing significant closure of the coating scratch [52–57]. Some bulges were observed in the healed coating scratch area, which can be explained by the movement of the microcapsules from the scratch area to both sides under stress.

EIS measurements were performed to further examine the self-healing performance of the DCF@PU/UF-EP0.9 coating, using the scratched EP0.9 and unheated DCF@PU/UF-EP0.9 coatings as control samples. Figure 7.11 presents the Nyquist plots of different coatings immersed in a 3.5 wt% NaCl solution over seven days. The EP0.9 coating exhibited the smallest capacitive arc radius, indicating that it had the worst barrier performance during immersion [2, 58–60]. Warburg impedance occurred on the coating, which is attributed to semi-infinite diffusion at the interface between the coating and the metal surface [61–63]. The capacitive arcs were larger for the scratched DCF@PU/UF-EP0.9 coating, which can be explained by the release of substances from the ruptured DCF@PU/UF to fill the coating scratch. The EPA in the microcapsules was insoluble in the aqueous solution, which may have hindered the invasion of corrosive ions and reduced the corrosion activity in the scratched region. The largest capacitive arcs were observed for the healed DCF@PU/UF-EP0.9 coating, indicating its excellent self-healing performance.

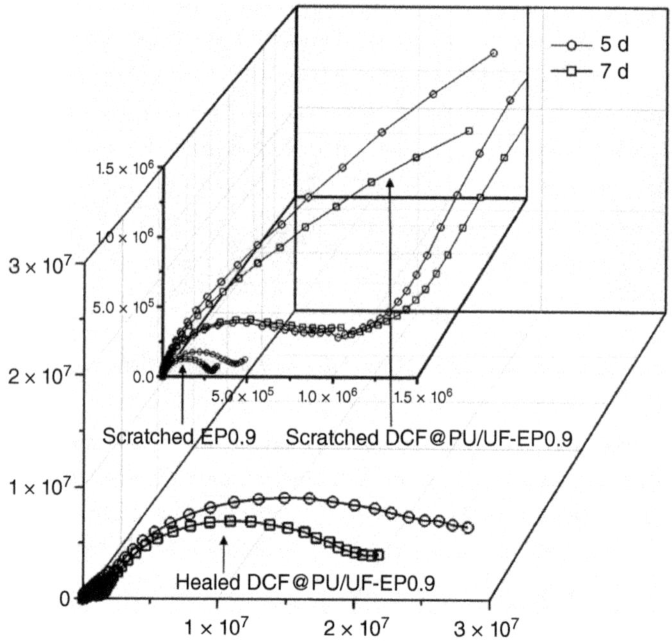

Figure 7.11 Nyquist diagrams of different coatings, during immersion in 3.5 wt% NaCl solution for five and seven days.

Furthermore, the $|Z|_{0.01\text{Hz}}$ values were compared to evaluate the anticorrosion properties of the different coatings, as shown in Figure 7.12. The $|Z|_{0.01\text{Hz}}$ of the EP0.9 coating decreased from 7.2×10^5 to $5.7 \times 10^5\ \Omega\ \text{cm}^2$ during the seven days of immersion, which was attributed to the corrosion activity on the exposed metal surface. The $|Z|_{0.01\text{Hz}}$ of the scratched DCF@PU/UF-EP0.9 coating was maintained at approximately $3.0 \times 10^6\ \Omega\ \text{cm}^2$. Notably, the healed DCF@PU/UF-EP0.9 coating exhibited the highest $|Z|_{0.01\text{Hz}}$ value ($>2.2 \times 10^7\ \Omega\ \text{cm}^2$) throughout the immersion period, indicating that the healed scratch largely prevented the permeation of corrosive media [64–67]. In phase angle plots, the time constant at high frequencies (10^5 Hz) corresponds to the coating response [68–73]. The high-frequency phase angles of the scratched EP0.9 and DCF@PU/UF-EP0.9 coatings were ~10° and ~20°, respectively, whereas this value was almost 90° for the healed DCF@PU/UF-EP0.9 coating. These findings confirmed that the DCF@PU/UF-EP0.9 coating was effectively healed and the corrosion protection capability was significantly recovered.

7.2.6 Dual Self-reporting and Self-healing Mechanisms

Figure 7.13 depicts the dual self-reporting and self-healing mechanisms of the DCF@PU/UF-EP coating. When the coating was damaged, the DCF@PU/UF-EP microcapsules around the damaged area ruptured, and the DCF molecules inside flowed out and contacted the coating base. As DCF reacted with the residual amino groups in the coating, the molecular structure transformed from an acidic form

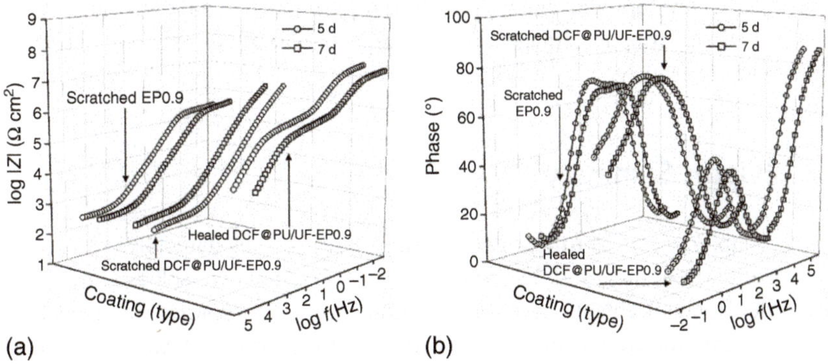

Figure 7.12 (a) Bode magnitude and (b) phase angle plots of scratched EP0.9 coating, scratched DCF@PU/UF-EP0.9 coating, and healed DCF@PU/UF-EP0.9 coating, during immersion in 3.5 wt% NaCl solution for five and seven days.

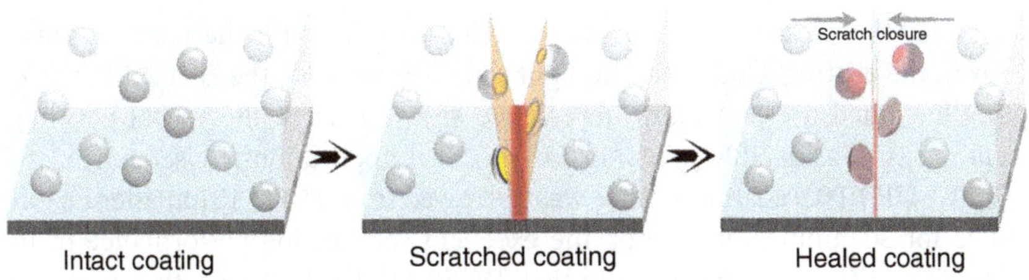

Figure 7.13 The schematic diagram of the self-reporting and self-healing dual functions for the DCF@PU/UF-EP coating.

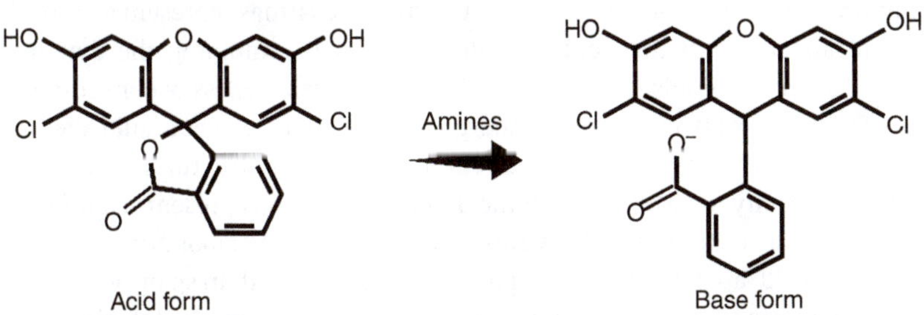

Figure 7.14 The molecular mechanism of the reaction between DCF and amines.

into a basic form, resulting in a color change from light yellow to red (Figure 7.14). A sensitive early warning of coating damage was realized. In addition, because of the fluorescent properties of DCF, a bright yellow fluorescent signal was observed under UV irradiation. In addition, owing to the shape-memory effect of the EP coatings, the motility of the molecular chains of the polymer matrix was enhanced during the heating process, which promoted their movement and rearrangement [74–76]. Consequently, the scratches were significantly narrowed after heating, reducing the exposed metal surface and providing good corrosion protection. By precisely controlling the molecular ratio of the curing agent to the epoxy monomer, coatings with moderate amounts of residual amino groups ensured both the sensitive color-development effect of DCF and an effective shape-memory effect. Thus, when the coating is damaged, the self-reporting and self-repair functions are realized simultaneously.

7.3 Chapter Summary

In this chapter, a dual-functional coating with a mechanically responsive self-reporting capability and shape-memory effect-activated self-healing function was developed. DCF@PU/UF microcapsules were incorporated into the SMEP matrix to endow the composite coating with dual functions. DCF was successfully encapsulated inside double-shell microcapsules (DCF@PU/UF microcapsules), spherical in shape, with an average diameter of 20 μm. With precise adjustment of the molar ratios of the curing agent and epoxy monomer in the epoxy resin, the damaged DCF@PU/UF-EP coatings immediately exhibited a distinct red color owing to the active reaction of the released DCF with the residual amino groups in the coating. Under natural light or UV irradiation, the DCF@PU/UF-EP coating exhibited a clear color/fluorescence signal, even with topcoat coverage, indicating its considerable potential for practical applications. A scratch on the DCF@PU/UF-EP0.9 coating surface was narrowed from 150 to 15 μm after heating at 50 °C for 30 minutes, indicating the excellent self-healing performance of the coating. The $|Z|_{0.01Hz}$ value of a healed DCF@PU/UF-EP0.9 coating was nine times that of an unhealed coating after seven days of immersion in a 3.5 wt% NaCl solution, indicating the recovery of the anticorrosion performance.

Mechanically responsive self-reporting anticorrosive coatings represent a groundbreaking advancement in the field of material science, combining the dynamic properties of smart materials with the durability and protectiveness of conventional coatings. These innovative coatings undergo a color change when subjected to mechanical stress, thereby serving as an early indicator of structural damage or corrosion. This ability to visually signal mechanical alterations presents significant potential for various industries. However, the journey from laboratory development to real-world application is fraught with challenges that span across the material composition, environmental stability, production methodologies, and cost efficiency. Addressing these challenges necessitates a multifaceted approach involving advanced material science, innovative engineering, and environmental considerations.

The following challenges should be considered. (i) Sensitivity and response mechanism. The primary challenge in developing mechanically responsive self-reporting coatings is achieving high degree of sensitivity to mechanical stimuli. This requires a deep understanding of the mechanical response mechanism at the molecular level. The phenomenon, where mechanical stress induces structural or electronic changes leading to a visible color shift, demands materials with precisely engineered properties. (ii) Durability under harsh environmental conditions. Anticorrosive coatings are often exposed to extreme and varying environmental conditions. Ensuring these coatings maintain their mechanical response and protective properties under such conditions is a substantial challenge. External factors such as UV radiation, temperature fluctuations, and exposure to harsh chemicals can degrade the mechanical integrity and color-changing properties of these coatings. (iii) Reversibility and longevity. For practical applications, it is essential for coatings to revert to their original state after the removal of the applied force and to maintain this functionality over multiple cycles. Developing materials capable of withstanding repeated stress without degradation of their mechanical response properties is a critical research area. This aspect of material development not only impacts the functional longevity of the coatings but also affects their sustainability and applicability in various industries. (iv) Cost-effectiveness and scalability of production. Many current mechanically responsive materials are expensive and complex to produce, especially on a large scale. This limitation hinders their practical application. Finding cost-effective and scalable production methods is crucial for the widespread adoption of these smart coatings. Overcoming the barriers to large-scale production involves not only material innovations but also advancements in manufacturing technologies. (v) Environmental impact and sustainability. In an era increasingly focused on sustainability, there is a significant demand for coatings that are both effective and environmentally friendly. This challenge encompasses the need for non-toxic, sustainable materials and processes in the development of mechanically responsive self-reporting coatings. The environmental impact of these materials spans their entire lifecycle, from production to application and eventual disposal or recycling.

For future development aspects, advanced material research and nanotechnology can be applied first. The exploration of new materials, such as stimuli-responsive polymers and nanocomposites, holds the key to enhancing the sensitivity and stability of mechanical-responsive coatings. Nanotechnology, particularly at the forefront of material science research, offers promising avenues for improving the properties of these materials. Nanoscale engineering can lead to coatings with heightened sensitivity to mechanical stress, improved color-changing properties, and enhanced overall performance. Secondly, the mechanical response mechanism should be deeply understood. Gaining a more profound understanding of the molecular mechanisms behind mechanical response is essential for refining the control over the materials' properties. Research should focus on elucidating how molecular and nanoscale structures are altered under mechanical stress and how these changes translate into visible color shifts. This in-depth understanding is crucial for developing materials that can precisely and reliably respond to mechanical alterations. Third, the development of coatings that combine mechanical response with other functionalities, such as UV resistance and enhanced mechanical strength, represents a significant research area. Such multifunctional coatings could offer comprehensive protection for various applications, from industrial machinery to consumer products. The integration of additional functionalities would extend the utility of these coatings, making them suitable for a wider range of applications and environments. In addition, the integration of mechanical responsive self-reporting coatings with advanced sensor technology can lead to the development of intelligent surfaces. These smart surfaces could be capable of real-time monitoring and reporting of structural integrity, thus enabling proactive maintenance strategies. This approach is particularly promising for critical applications in aerospace, automotive, construction, and civil engineering industries. The use of AI technologies and advanced data analytics could further enhance the monitoring capabilities, providing comprehensive insights into the condition and performance of coated structures. Regarding the environmental factors, improving the coatings' resistance to environmental factors such as moisture, salinity, extreme temperatures, and pH levels is essential for ensuring their durability in various applications. This aspect of research is particularly crucial for applications in harsh environmental conditions, such as marine or industrial settings. The development of coatings that can withstand these conditions without compromising their mechanical response sensitivity. Emphasizing green chemistry in the production of mechanochromic materials is crucial. This involves minimizing the use of toxic substances and reducing waste during the manufacturing process. Last but not least, sustainability and environmental protection should be considered. Developing more environmentally friendly manufacturing processes and materials to reduce the environmental impact during production is imperative. Mechanically responsive self-reporting coatings hold great promise for revolutionizing material science and engineering. Their development intersects numerous disciplines, including chemistry, materials science, and engineering. As research progresses, these coatings have the potential to significantly impact various industries by providing enhanced safety, longevity, and performance of materials and structures. The future of this field lies in overcoming current challenges through innovation, interdisciplinary collaboration, and a focus on sustainability.

References

1 Fan, D., Liu, X., Qi, K. et al. (2022). A smart-sensing coating based on dual-emission fluorescent Zr-MOF composite for autonomous warning of coating damage and aluminum corrosion. *Progress in Organic Coatings* 172: 107150. https://doi.org/10.1016/j.porgcoat.2022.107150.

2 Ji, X., Wang, W., Zhao, X. et al. (2022). Poly(dimethyl siloxane) anti-corrosion coating with wide pH-responsive and self-healing performance based on core-shell nanofiber containers. *Journal of Materials Science and Technology* 101: 128–145. https://doi.org/10.1016/j.jmst.2021.06.014.

3 Wang, J., Song, Y., Wang, J. et al. (2017). pH-Responsive polymer coatings for reporting early stages of metal corrosion. *Macromolecular Materials and Engineering* 302: 1700128. https://doi.org/10.1002/mame.201700128.

4 Liu, J., Li, M., Luo, C. et al. (2022). Eco-friendly synthesis of self-reporting robust superhydrophobic coatings with damage sensitive photoluminescence. *Chemical Engineering Journal* 431: 134162. https://doi.org/10.1016/j.cej.2021.134162.

5 Jang, H.G., Jo, J.Y., Park, H. et al. (2023). Mechano-responsive spiropyran microbeads: a facile fabrication strategy for self-reporting materials. *Advanced Materials Technologies* 8: 2200566. https://doi.org/10.1002/admt.202200566.

6 Wang, J., Ma, L., Guo, X. et al. (2022). Two birds with one stone: nanocontainers with synergetic inhibition and corrosion sensing abilities towards intelligent self-healing and self-reporting coating. *Chemical Engineering Journal* 433: 134515. https://doi.org/10.1016/j.cej.2022.134515.

7 Davis, D.A., Hamilton, A., Yang, J. et al. (2009). Force-induced activation of covalent bonds in mechanoresponsive polymeric materials. *Nature* 459: 68–72. https://doi.org/10.1038/nature07970.

8 Lee, T.H., Song, Y.K., Park, S.H. et al. (2018). Dual stimuli responsive self-reporting material for chemical reservoir coating. *Applied Surface Science* 434: 1327–1335. https://doi.org/10.1016/j.apsusc.2017.11.219.

9 Tao, Z., Cui, J., Qiu, H. et al. (2021). Microcapsule/silica dual-fillers for self-healing, self-reporting and corrosion protection properties of waterborne epoxy coatings. *Progress in Organic Coatings* 159: 106394. https://doi.org/10.1016/j.porgcoat.2021.106394.

10 Chen, Z., Scharnagl, N., Zheludkevich, M.L. et al. (2023). Micro/nanocontainer-based intelligent coatings: synthesis, performance and applications – a review. *Chemical Engineering Journal* 451: 138582. https://doi.org/10.1016/j.cej.2022.138582.

11 Guo, Z., Li, T., Chen, T. et al. (2023). Salt-gated releasing effect to open visual self-reporting/self-repairing function of coatings toward corrosion protection. *ACS Sustainable Chemistry & Engineering* 11: 4738–4748. https://doi.org/10.1021/acssuschemeng.2c07185.

12 Chen, S., Han, T., Zhao, Y. et al. (2020). A facile strategy to prepare smart coatings with autonomous self-healing and self-reporting functions. *ACS Applied Materials & Interfaces* 12: 4870–4877. https://doi.org/10.1021/acsami.9b18919.

13 Lu, X., Li, W., Sottos, N.R. et al. (2018). Autonomous damage detection in multilayered coatings via integrated aggregation-induced emission luminogens. *ACS Applied Materials & Interfaces* 10: 40361–40365. https://doi.org/10.1021/acsami.8b16454.

14 Song, Y.K., Lee, T.H., Lee, K.C. et al. (2020). Coating that self-reports cracking and healing using microcapsules loaded with a single AIE fluorophore. *Applied Surface Science* 511: 145556. https://doi.org/10.1016/j.apsusc.2020.145556.

15 Cheng, L., Liu, C., Zhao, H. et al. (2023). Hierarchically self-reporting and self-healing photothermal responsive coatings towards smart corrosion protection. *Chemical Engineering Journal* 467: 143463. https://doi.org/10.1016/j.cej.2023.143463.

16 Hu, M., Peil, S., Xing, Y. et al. (2018). Monitoring crack appearance and healing in coatings with damage self-reporting nanocapsules. *Materials Horizons* 5: 51–58. https://doi.org/10.1039/C7MH00676D.

17 Guo, Y.K., Chen, L., Xu, D.G. et al. (2016). A dual functional epoxy material with autonomous damage indication and self-healing. *RSC Advances* 6: 65067–65071. https://doi.org/10.1039/C6RA13519F.

18 Li, W., Matthews, C.C., Yang, K. et al. (2016). Autonomous indication of mechanical damage in polymeric coatings. *Advanced Materials* 28: 2189–2194. https://doi.org/10.1002/adma.201505214.

19 Liu, C., Cheng, L., Cui, L.Y. et al. (2023). Robust damage warning and healing tracing strategy for anticorrosion coatings on magnesium alloy AZ31enabled by polydopamine fluorophores. *Journal of Materials Science and Technology* 152: 169–180. https://doi.org/10.1016/j.jmst.2022.12.043.

20 Robb, M.J., Li, W., Gergely, R.C.R. et al. (2016). A robust damage-reporting strategy for polymeric materials enabled by aggregation-induced emission. *ACS Central Science* 2: 598–603. https://doi.org/10.1021/acscentsci.6b00198.

21 Vidinejevs, S., Aniskevich, A.N., Gregor, A. et al. (2012). Smart polymeric coatings for damage visualization in substrate materials. *Journal of Intelligent Material Systems and Structures* 23: 1371–1377. https://doi.org/10.1177/1045389X12447289.

22 Zheng, X., Wang, Q., Li, Y. et al. (2020). Fabrication of self-reactive microcapsules as color visual sensing for damage reporting. *Journal of Materials Science* 55: 8861–8867. https://doi.org/10.1007/s10853-020-04668-6.

23 Zhang, H., Wang, Q., Li, Y. et al. (2022). Preparation and characterization of damage indication and self-healing microcapsules for surface micro-cracks in mortar coating. *Journal of Applied Polymer Science* 139: 52198. https://doi.org/10.1002/app.52198.

24 Li, Y., Wang, Q., Zheng, X. et al. (2020). Microcapsule encapsulated with leuco dye as a visual sensor for concrete damage indication via color variation. *RSC Advances* 10: 1226–1231. https://doi.org/10.1039/C9RA09492J.

25 Wu, G., An, J., Tang, X.Z. et al. (2014). A versatile approach towards multifunctional robust microcapsules with tunable, restorable, and solvent-proof superhydrophobicity for self-healing and self-cleaning coatings. *Advanced Functional Materials* 24: 6751–6761. https://doi.org/10.1002/adfm.201401473.

26 Han, T., Chen, S., Wang, X. et al. (2022). Autonomous visualization of damage in polymers by metal-free polymerizations of microencapsulated activated alkynes. *Advanced Science* 9: 2105395. https://doi.org/10.1002/advs.202105395.

27 Geng, W., Nakajima, T., Takanashi, H. et al. (2009). Analysis of carboxyl group in coal and coal aromaticity by Fourier transform infrared (FT-IR) spectrometry. *Fuel* 88: 139–144. https://doi.org/10.1016/j.fuel.2008.07.027.

28 Dubis, E.N., Dubis, A.T., and Popławski, J. (2001). Determination of the aromatic compounds in plant cuticular waxes using FT-IR spectroscopy. *Journal of Molecular Structure* 596: 83–88. https://doi.org/10.1016/S0022-2860(01)00691-3.

29 Sutcu, H., Toroglu, I., and Piskin, S. (2005). Structural characterization of oil component of high temperature pyrolysis tars. *Energy Sources* 27: 521–534. https://doi.org/10.1080/00908310490449234.

30 Russo, C., Stanzione, F., Tregrossi, A. et al. (2014). Infrared spectroscopy of some carbon-based materials relevant in combustion: qualitative and quantitative analysis of hydrogen. *Carbon* 74: 127–138. https://doi.org/10.1016/j.carbon.2014.03.014.

31 Chen, Y., Li, W., Luo, J. et al. (2021). Robust damage-reporting strategy enabled by dual-compartment microcapsules. *ACS Applied Materials & Interfaces* 13: 14518–14529. https://doi.org/10.1021/acsami.0c20276.

32 Liu, T., Zhao, H., Zhang, D. et al. (2021). Ultrafast and high-efficient self-healing epoxy coatings with active multiple hydrogen bonds for corrosion protection. *Corrosion Science* 187: 109485. https://doi.org/10.1016/j.corsci.2021.109485.

33 Chen, J., Fang, L., Xu, Z. et al. (2016). Self-healing epoxy coatings curing with varied ratios of diamine and monoamine triggered via near-infrared light. *Progress in Organic Coatings* 101: 543–552. https://doi.org/10.1016/j.porgcoat.2016.09.020.

34 Fang, L., Chen, J., Zou, Y. et al. (2017). Self-healing epoxy coatings via focused sunlight based on photothermal effect. *Macromolecular Materials and Engineering* 302: 1700059. https://doi.org/10.1002/mame.201700059.

35 Xie, T. and Rousseau, I.A. (2009). Facile tailoring of thermal transition temperatures of epoxy shape memory polymers. *Polymer* 50: 1852–1856. https://doi.org/10.1016/j.polymer.2009.02.035.

36 Forrest, J.A., Dalnoki-Veress, K., Stevens, J.R. et al. (1996). Effect of free surfaces on the glass transition temperature of thin polymer films. *Physical Review Letters* 77: https://doi.org/10.1103/PhysRevLett.77.2002.

37 Chen, J., Li, J., Xu, L. et al. (2019). The glass-transition temperature of supported PMMA thin films with hydrogen bond/plasmonic interface. *Polymers* 11: 601. https://doi.org/10.3390/polym11040601.

38 White, R.P. and Lipson, J.E.G. (2011). Thermodynamic treatment of polymer thin-film glasses. *Physical Review E* 84: 041801. https://doi.org/10.1103/PhysRevE.84.041801.

39 Lamothe, P.J. and McCormick, P.G. (1972). Influence of acidity on the reaction of ninhydrin with amino acids. *Analytical Chemistry* 1972 (44): 821–825. https://doi.org/10.1021/ac60312a003.

40 Lamothe, P.J. and McCormick, P.G. (1973). Role of hydrindantin in the determination of amino acids using ninhydrin. *Analytical Chemistry* 45: 1906–1911. https://doi.org/10.1021/ac60333a021.

41 McCaldin, D.J. (1960). The chemistry of ninhydrin. *Chemical Reviews* 60: 39–51. https://doi.org/10.1021/cr60203a004.

42 Schertz, T.D., Reiter, R.C., and Stevenson, C.D. (2001). Zwitterion radicals and anion radicals from electron transfer and solvent condensation with the fingerprint developing agent ninhydrin. *Journal of Organic Chemistry* 66: 7596–7603. https://doi.org/10.1021/jo0102720.

43 Ma, L., Ren, C., Wang, J. et al. (2021). Self-reporting coatings for autonomous detection of coating damage and metal corrosion: a review. *Chemical Engineering Journal* 421: 127854. https://doi.org/10.1016/j.cej.2020.127854.

44 Ding, J., Liu, P., Zhou, M. et al. (2020). Nafion-endowed graphene super-anticorrosion performance. *ACS Sustainable Chemistry & Engineering* 8: 15344–15353. https://doi.org/10.1021/acssuschemeng.0c05679.

45 Ding, J., Zhao, H., Ji, D. et al. (2019). Achieving long-term anticorrosion via the inhibition of graphene's electrical activity. *Journal of Materials Chemistry A* 7: 2864–2874. https://doi.org/10.1039/C8TA10337B.

46 Zhao, H., Ding, J., and Yu, H. (2018). The efficient exfoliation and dispersion of hBN nanoplatelets: advanced application to waterborne anticorrosion coatings. *New Journal of Chemistry* 42: 14433–14443. https://doi.org/10.1039/C8NJ03113D.

47 Ding, J., Zhao, H., and Yu, H. (2022). Bio-inspired multifunctional graphene–epoxy anticorrosion coatings by low-defect engineered graphene. *ACS Nano* 16: 710–720. https://doi.org/10.1021/acsnano.1c08228.

48 Kaur, H., Sharma, J., Jindal, D. et al. (2018). Crosslinked polymer doped binary coatings for corrosion protection. *Progress in Organic Coatings* 125: 32–39. https://doi.org/10.1016/j.porgcoat.2018.08.026.

49 Pathania, A., Arya, R.K., and Ahuja, S. (2017). Crosslinked polymeric coatings: preparation, characterization, and diffusion studies. *Progress in Organic Coatings* 105: 149–162. https://doi.org/10.1016/j.porgcoat.2016.12.023.

50 Zhang, J., Zhao, X., Wei, J. et al. (2021). Superhydrophobic coatings with photothermal self-healing chemical composition and microstructure for efficient corrosion protection of magnesium alloy. *Langmuir* 37: 13527–13536. https://doi.org/10.1021/acs.langmuir.1c02355.

51 Ding, J., Zhao, H., Zhao, X. et al. (2019). How semiconductor transition metal dichalcogenides replaced graphene for enhancing anticorrosion. *Journal of Materials Chemistry A* 7: 13511–13521. https://doi.org/10.1039/C9TA04033A.

52 Huang, Y., Deng, L., Ju, P. et al. (2018). Triple-action self-healing protective coatings based on shape memory polymers containing dual-function microspheres. *ACS Applied Materials & Interfaces* 10: 23369–23379. https://doi.org/10.1021/acsami.8b06985.

53 Wang, B., Liu, J., Liu, M. et al. (2023). Preparation and corrosion resistance of shape memory self-healing coatings responsive to near-infrared light. *Polymer Testing* 126: 108146. https://doi.org/10.1016/j.polymertesting.2023.108146.

54 Jacobson, N.D. and Iroh, J. (2021). Shape memory corrosion-resistant polymeric materials. *International Journal of Polymeric Science* 2021: 1–18. https://doi.org/10.1155/2021/5558457.

55 Qian, H., Xu, D., Du, C. et al. (2017). Dual-action smart coatings with a self-healing superhydrophobic surface and anti-corrosion properties. *Journal of Materials Chemistry A* 5: 2355–2364. https://doi.org/10.1039/C6TA10903A.

56 Wang, L., Deng, L., Zhang, D. et al. (2016). Shape memory composite (SMC) self-healing coatings for corrosion protection. *Progress in Organic Coatings* 97: 261–268. https://doi.org/10.1016/j.porgcoat.2016.04.041.

57 Dong, Y., Gong, M., Huang, D. et al. (2019). Shape memory, self-healing property, and NIR photothermal effect of epoxy resin coating with polydopamine@polypyrrole nanoparticles. *Progress in Organic Coatings* 136: 105232. https://doi.org/10.1016/j.porgcoat.2019.105232.

58 Liu, T., Wei, J., Ma, L. et al. (2021). Effect of polyaniline-based plate on the anticorrosion performance of epoxy coating. *Progress in Organic Coatings* 151: 106109. https://doi.org/10.1016/j.porgcoat.2020.106109.

59 Dong, Y., Wu, K., Yin, Y. et al. (2021). Shape memory self-healing coating based on photothermal effect of PPy@PDA nanoparticles. *Synthetic Metals* 280: 116869. https://doi.org/10.1016/j.synthmet.2021.116869.

60 Wang, H., Xu, J., Du, X. et al. (2022). Stretchable and self-healing polyurethane coating with synergistic anticorrosion effect for the corrosion protection of stainless steels. *Progress in Organic Coatings* 164: 106672. https://doi.org/10.1016/j.porgcoat.2021.106672.

61 Wang, T., Wang, W., Feng, H. et al. (2022). Photothermal nanofiller-based polydimethylsiloxane anticorrosion coating with multiple cyclic self-healing and long-term self-healing performance. *Chemical Engineering Journal* 446: 137077. https://doi.org/10.1016/j.cej.2022.137077.

62 Yang, J., Wang, D., Wang, J. et al. (2022). Corrosion resistance and near-infrared light induced self-healing behavior of polycaprolactone coating with MIL-53@TA on magnesium alloy. *Applied Surface Science* 585: 152729. https://doi.org/10.1016/j.apsusc.2022.152729.

63 Ma, Y., Yang, Y., Cao, D. et al. (2021). Facile repair of anti-corrosion polymeric composite coatings based on light triggered self-healing. *Macromolecular Materials and Engineering* 306: 2100106. https://doi.org/10.1002/mame.202100106.

64 Huangfu, H., Guo, X., Li, N. et al. (2023). A smart composite coating with self-reporting and self-healing functions to enhance corrosion protection for magnesium alloys. *Progress in Organic Coatings* 181: 107598. https://doi.org/10.1016/j.porgcoat.2023.107598.

65 Li, W., Tao, J., Chen, Y. et al. (2023). Porous microspheres with corrosion sensing and active protecting abilities towards intelligent self-reporting and anti-corrosion coating. *Progress in Organic Coatings* 178: 107468. https://doi.org/10.1016/j.porgcoat.2023.107468.

66 Cai, Z., Li, C., Zhang, D. et al. (2022). An anti-corrosion coating with self-healing function of polyurethane modified by lipoic acid. *Journal of Coatings Technology and Research* https://doi.org/10.1007/s11998-022-00703-0.

67 Li, B., Xue, S., Mu, P. et al. (2022). Robust self-healing graphene oxide-based superhydrophobic coatings for efficient corrosion protection of magnesium alloys. *ACS Applied Materials & Interfaces* 14: 30192–30204. https://doi.org/10.1021/acsami.2c06447.

68 Wang, J., Ma, L., Huang, Y. et al. (2022). Photothermally activated self-healing protective coating based on the "close and seal" dual-action mechanisms. *Composites Part B: Engineering* 231: 109574. https://doi.org/10.1016/j.compositesb.2021.109574.

69 Meng, L., Xu, F., Zhang, M. et al. (2023). A novel anticorrosive performance coating through introducing epoxidized hydroxyl terminated polybutadiene. *Progress in Organic Coatings* 175: 107334. https://doi.org/10.1016/j.porgcoat.2022.107334.

70 Huang, Y., Wang, P., Tan, W. et al. (2022). Photothermal and pH dual-responsive self-healing coating for smart corrosion protection. *Journal of Materials Science and Technology* 107: 34–42. https://doi.org/10.1016/j.jmst.2021.08.044.

71 Ma, L., Wang, J., Zhang, D. et al. (2021). Dual-action self-healing protective coatings with photothermal responsive corrosion inhibitor nanocontainers. *Chemical Engineering Journal* 404: 127118. https://doi.org/10.1016/j.cej.2020.127118.

72 Cao, L., Wang, W., Li, Q. et al. (2022). Three-dimensional nanofibers network multifunctional material for photothermal self-healing protective coating. *Chemical Engineering Journal* 440: 134943. https://doi.org/10.1016/j.cej.2022.134943.

73 Liu, C., Li, J., Jin, Z. et al. (2019). Synthesis of graphene-epoxy nanocomposites with the capability to self-heal underwater for materials protection. *Composites Communications* 15: 155–161. https://doi.org/10.1016/j.coco.2019.07.011.

74 Bai, Y., Zhang, J., Wen, D. et al. (2020). A reconfigurable, self-healing and near infrared light responsive thermoset shape memory polymer. *Composites Science and Technology* 187: 107940. https://doi.org/10.1016/j.compscitech.2019.107940.

75 Cheng, L., Liu, C., Zhao, H. et al. (2021). Photothermal-triggered shape memory coatings with active repairing and corrosion sensing properties. *Journal of Materials Chemistry A* 9: 22509–22521. https://doi.org/10.1039/D1TA05761H.

76 Dong, Y., Geng, C., Liu, C. et al. (2020). Near-infrared light photothermally induced shape memory and self-healing effects of epoxy resin coating with polyaniline nanofibers. *Synthetic Metals* 266: 116417. https://doi.org/10.1016/j.synthmet.2020.116417.

8

Development of Aging-Resistant Coating with Self-healing and Self-reporting Properties

8.1 Introduction

Aging behavior of coatings refers to the degradation of a coating's polymer chain segments under the influence of both internal and external factors, leading to a decline in its overall performance. When aging behavior advances to a certain extent, the coating becomes functionally useless. There are numerous external elements, which contribute to the coating aging, including temperature, exposure to environmental light, oxidation, chemical media, biological effects (such as those from microorganisms), and changes in atmospheric climate. Aging is typically a prolonged process, often seen as a series of reactions involving free groups. As coatings are directly exposed to the external environment, they are subject to the combined effects of oxygen, light, and heat. These elements can react with the molecular chains of the coating's main components, leading to their decomposition and the eventual aging of the coating. In atmospheric conditions, the presence of oxygen, light, and heat can damage the primary film-forming substances of the coating. This damage results in the destruction and decomposition of molecular chains, thereby causing the coating to age and deteriorate. This process highlights the importance of considering environmental factors in the design and formulation of coatings, especially for applications where long-term durability is essential.

The aging of organic coatings is significantly influenced by environmental factors such as light, temperature, water, and chemical media, which are explained as follows: (i) *Light*—sunlight, particularly its UV component, plays a critical role in the aging process of coatings in the atmospheric environment. Despite constituting only about 5% of the solar spectrum, UV light is more energetic than other spectra and can easily break the chemical bonds in polymers. When coatings are exposed to the atmosphere, oxygen acts as a natural catalyst. In this context, UV light initiates not only photodecomposition of the coating but also photooxidation. This results in the main chain breakage in the coating. During this reaction process, intermediate products such as peroxides and hydrophilic small molecules are formed, leading to the aging and eventual degradation of the coating. The underlying principle of UV-induced aging and degradation is photochemical reactions occurring during radiation exposure. The key mechanism involves high-energy photons in UV radiation that break certain chemical bonds, particularly those with relatively low

Smart Protective Coatings for Corrosion Control, First Edition. Lingwei Ma and Dawei Zhang.

bond energy, resulting in the formation of free radicals. Owing to the instability of free radicals, they often further react with other molecules, producing oxygen or peroxyl free radicals. These radicals instigate subsequent photochemical reactions until the unstable radicals combine to form stable compounds. The main consequence of these photochemical reactions is breaking of molecular chemical bonds, followed by the generation of free radicals that react with other molecules, eventually leading to new compounds. This process culminates in the degradation, failure, and deterioration of organic coatings. (ii) *Temperature*—Temperature plays a significant role in the aging of organic anticorrosion coatings, particularly in high temperatures. The primary mechanism behind this is that as the temperature increases, molecular thermal motion accelerates, leading to the degradation and cross-linking of some polymers within the coating. In addition, thermal expansion and contraction effects can cause the coating to expand or contract. This physical change can lead to damage and further accelerate the aging process of the coating. (iii) *Water*—The impact of water on the surface of organic anticorrosive coatings is gradual but significant over time. With prolonged exposure, water molecules can penetrate the coating through surface defects, such as pores and cracks. This infiltration leads to the expansion or contraction of the organic coating, altering its stress and eventually causing stress cracks and detachment. Moreover, water infiltration can cause chemical reactions within the coating. It may interact with soluble additives, leading to the extraction, migration, and dissolution, thereby adversely affecting the coating's performance. In addition, water on the coating surface can remove pollutants, exposing the coating to UV radiation in natural environment and exacerbating the aging process. Moisture also affects the coating's aging through volumetric changes. Natural hydration or dehydration processes can cause the coating to expand or contract, leading to cracking and detachment. During alternating dry and wet periods, significant stress is generated on the coating. When this stress exceeds a certain threshold, it can cause the coating to detach and lose its protective function. This expansion cycle during wet periods and contraction during dry periods underlines the importance of considering moisture resistance in the design and application of organic anticorrosive coatings. (iv) *Chemical media*—Various chemical media significantly influence the aging of organic anticorrosion coatings. This category encompasses not only basic chemicals such as acids, bases, and salts but also more complex compounds such as organic acids, different organic solvents, and chlorine gas. Over time, these substances can penetrate into the coating's interior and react with its constituents. When dissolved in aqueous solutions, acids, bases, salts, and certain gases ionize into small radius ions. These ions can easily navigate through microscopic defects in the anticorrosion coating and diffuse toward the interface between the coating and metal substrate. They may form microscopic corrosion galvanic cells and initiate reactions with the metal. This interaction can lead to the corrosion of the metal substrate beneath the coating, eventually causing physical damage to the coating, manifested as bubbling and detachment. In contrast, the destructive impact of organic solvents and organic acids on coatings is not due to the penetration. Instead, it occurs through the migration of their macromolecules within the coating system, typically around the coating's pores. When these corrosive chemical media reach the interface between the coating and the metal substrate through the microscopic defects of coatings, they form microscopic corrosion galvanic cells with the metal. This interaction results in the corrosion of the metal substrate and subsequent damage to the coating, such as bubbling, cracking, and ultimately detachment.

Incorporating antioxidants into organic coatings is a key strategy to delay oxidation and minimize the risk of aging failure. Traditionally, many antioxidants used in coatings are derived from petrochemical sources. However, these substances often raise environmental concerns and pose toxic side effects, which significantly limit their future applicability. In light of growing environmental awareness and the need to conserve fossil energy, there is a pressing demand for nontoxic and eco-friendly antioxidants. This shift toward sustainable practices has catalyzed recent research into nontoxic and eco-friendly antioxidants, along with nanofillers. These new developments aim to provide the same or higher levels of protection against oxidation and aging as their petrochemical counterparts, but without the detrimental environmental and health impacts.

Natural polyphenols, predominantly derived from plants, are abundant in nature, ranking second only to sugars in terms of their prevalence. Similar to proteins, nucleic acids, fats, and sugars, polyphenols are vital natural compounds. Polyphenols are well-known for their antioxidant properties. Almost all plants produce these substances as secondary metabolites, offering protection against UV radiation and serving as a defense mechanism against consumption by other plants and animals. The effectiveness of polyphenols as antioxidants is deeply linked to their chemical properties. One key reason for their efficacy as nucleophiles lies in the hydroxyl groups they contain, which exhibit strong electron donation capabilities. However, these groups are also prone to losing electrons and undergoing oxidation. In addition, the hydrogen atoms in the phenolic hydroxyl groups play a critical role in biochemical reactions. In the presence of a base, these hydrogen atoms can be released as protons or hydrogen free radicals. Subsequently, the resulting phenol anions and free radicals can transform into each other through electron transfer. Owing to the resonance effect, phenol-free radicals are relatively stable and can combine with other free radicals, effectively scavenging them. Further oxidation of phenol radicals and anions can lead to the formation of *o*-quinone structures, known to be highly active electrophilic agents. These agents are readily attacked by external nucleophilic reactants, effectively reversing the reactivity of phenol. This intricate chemical behavior underscores the potential of natural polyphenols as powerful antioxidants in various applications. When integrated into coatings, they hold the promise to significantly enhance the anti-aging performance by neutralizing free radicals that cause degradation of the coating materials, thereby prolonging the life span of the coating.

This chapter describes two types of corrosion protection coating with aging-resistant, self-reporting, and self-healing properties. Tannic acid (TA) was incorporated into mesoporous silica nanocontainers (TA-MSNs), and then the TA-MSNs nanofillers were loaded in an epoxy matrix. TA performs three important roles, that is, radical scavenging, color development, and corrosion inhibition [1]. The TA-MSNs nanomaterial was prepared to alleviate coating degradation by scavenging the radicals generated during UV irradiation. Compared with the blank coating, those containing TA-MSNs exhibited much less degradation in terms of surface morphology, wettability, and glossiness and maintained good barrier properties. Moreover, once the coating was damaged, TA was released and reacted with Fe^{3+} ions to form a chelate that has two functions: its visible black color indicates the onset of corrosion, and it also inhibits corrosion propagation by causing significant self-healing in the coating. Moreover, tea polyphenol (TP), with the similar characteristics compared with TA, was selected as an antioxidant substance and incorporated into MSNs to design another aging-resistant coating. The effect of TP-MSNs filler amount on the anticorrosion and aging-resistant properties of the coating and the positive effect of antioxidant filler on the aging resistance of the coating were studied.

8.2 Aging-Resistant, Self-healing, and Self-reporting Smart Coatings Based on TA Additives

In recent years, TA has received extensive attention as an eco-friendly corrosion inhibitor owing to its nontoxic and eco-friendly nature, good water solubility, and wide range of sources. It has been applied as a corrosion inhibitor for iron and steel materials by interacting with the metal surface to form a complex film layer. TA also reacts with Fe^{3+} to form a blue-black precipitate, which can be used as a color indicator for corrosion-induced self-reporting property [2]. In addition, TA is considered a promising green radical scavenger owing to its abundant *o*-phenylphenol groups, which can donate hydrogen atoms to the free radicals generated during aging and thus stop free radical chain reactions [3].

Zang and Ma [4] compounded TA with sodium silicate, zinc dihydrogen phosphate, and ethanolamine and applied the mixture to protect iron cultural relics to avoid the adverse effects of traditional acid corrosion inhibitors. Xu et al. [5] studied the influence of TA concentration on the corrosion resistance of carbon steel. They demonstrated that increasing the amount of TA first increased and then somewhat decreased the corrosion inhibition. There are currently two explanations for the corrosion inhibition mechanism of TA. One is that TA can interact with metal ions to form a complex film layer, hindering the direct contact between the metal substrate and the corrosive medium [6]. The other is that TA itself forms a dense and stable adsorption film on the metal surface through physicochemical adsorption to inhibit the aggravation of corrosion reactions [7].

As shown in Figure 8.1, TA contains numerous aromatic rings and phenolic hydroxyl groups. In addition to inhibiting metal corrosion, the phenolic hydroxyl group of TA also reacts with Fe^{3+} produced during corrosion to form a blue-black chelate, which can function as a color indicator of corrosion initiation. This section describes the color development, corrosion inhibition, and radical scavenging properties of TA. The mechanism of color development was investigated by mixing aqueous solutions of Fe^{3+} and TA at different concentrations, and the sensitivity of color development was also studied. The protective effects of TA on carbon steel

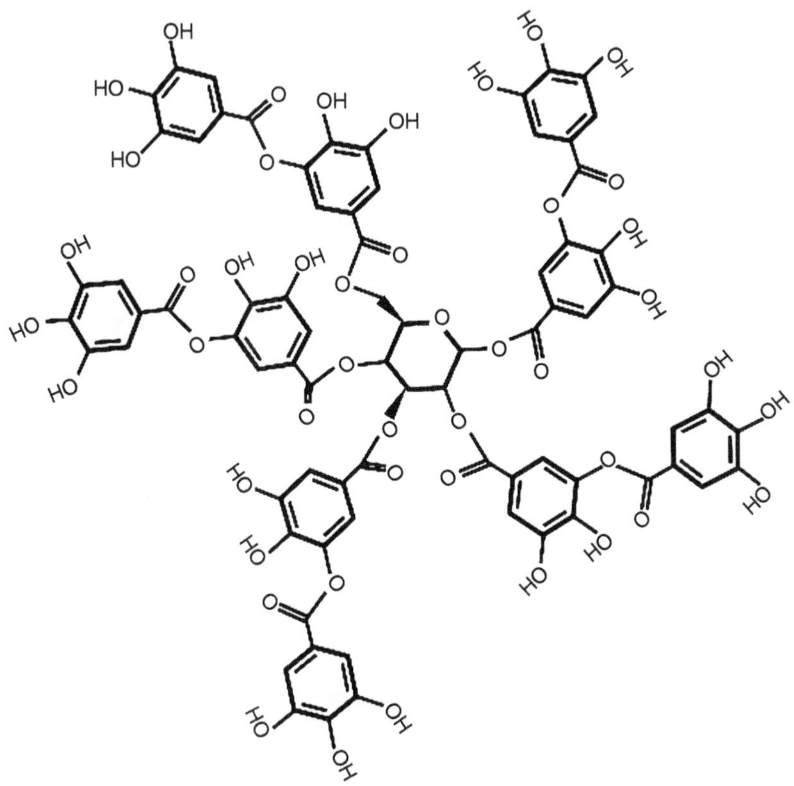

Figure 8.1 Schematic diagram of the chemical structure of TA.

substrate were characterized by electrochemical tests and surface analyses. The corrosion inhibition mechanism was also investigated [8, 9].

8.2.1 Materials and Methods

8.2.1.1 Materials

The materials and reagents used in this section are shown in Table 8.1.

8.2.1.2 Instruments

The instruments used in this section are shown in Table 8.2.

8.2.1.3 Coloration Test of TA

First, we mixed Fe^{3+} aqueous solutions at different concentrations with TA aqueous solution to evaluate the chromogenic properties of TA. Specifically, TA (0.1 g) was weighed and dissolved in 100 ml of deionized water. After complete dispersion, a pipette was used to transfer multiple sets of 2 ml TA solution into sample bottles. Equal volumes of Fe^{3+} solutions of different concentrations ($0, 1 \times 10^{-4}, 2 \times 10^{-4}, 4 \times 10^{-4}, 6 \times 10^{-4}$, and 8×10^{-4} M) were added to observe the color change of the

Table 8.1 Experimental materials and reagents.

Materials	Purity	Company
Tannic acid (TA)	Analytical purity	Alfa Aesar
Sodium chloride (NaCl)	Analytical purity	Sinopharm Group
Hydrochloric acid (HCl)	Analytical purity	Sinopharm Group
Hexamethylenetetramine	Analytical purity	Sinopharm Group
Ferric chloride ($FeCl_3$)	Analytical purity	Aladdin Biochemical Technology Co., LTD
Tetraethyl orthosilicate (TEOS)	Analytical purity	Sinopharm Group
Ammonium hydroxide	Analytical purity	Sinopharm Group
Anhydrous ethanol	Analytical purity	Sinopharm Group
Sodium hydroxide (NaOH)	Analytical purity	Aladdin Biochemical Technology Co., LTD
Ferric chloride ($FeCl_3$)	Analytical purity	Aladdin Biochemical Technology Co., LTD
Acetone	Analytical purity	Sinopharm Group
Bisphenol A diglycidyl ether (DGEBA)	Analytical purity	Sigma Corporation
Neopentyl glycol diglycidyl ether (NGDE)	Analytical purity	Aladdin Biochemical Technology Co., LTD
Polyetheramine (D230)	Analytical purity	Aladdin Biochemical Technology Co., LTD
Q235 mild steel	$40 \times 30 \times 1\,mm^3$	Henderson Steel

solutions, and photos were taken at the same time to record the visual color difference. UV–Vis absorption spectra of the mixed solutions were also collected to reveal the mechanism of color development and color-response sensitivity of TA.

8.2.1.4 Corrosion Inhibition Performance of TA

The corrosion inhibition performance of TA on Q235 carbon steel substrate was investigated using EIS. The electrolyte was 3.5 wt% NaCl aqueous solution with or without 1 wt% TA. After immersing carbon steel in the electrolyte for 72 hours, the formed corrosion product was removed by immersing for 10 minutes at room temperature in a 100-ml solution containing concentrated hydrochloric acid (50 ml), hexamethylenetetramine (0.35 g), and deionized water. Then, excess solution on the surface was wiped off, and the corrosion morphology and surface roughness were observed using CLSM.

8.2.1.5 Preparation of TA-MSNs

TA-MSNs with dual functions of color development and corrosion inhibition were prepared by a one-step method. Specifically, TA (544 mg) was weighed and added to a beaker containing 50 ml of anhydrous ethanol and fully dissolved. Concentrated

Table 8.2 The instruments and equipment.

Instrument	Production type	Manufacturer
UV/Vis/NIR Spectrometer	Lambda 750	PerkinElmer, USA
Electrochemical workstation	PARSTAT-2273	Princeton, USA
Laser confocal microscope	VK-X2000 series	Japan KEYENCE
Electronic balance	ME204	Switzerland Mettler Toledo
Magnetic stirrer	JEM-2100F	Japan JEOL
Centrifuges	Allegra X-30	Beckman, USA
pH meter	PHS-3C	China Thunderbolt
Vacuum freeze dryer	Lab-1D-50	China Bo Medical Kang
Box-type heating furnace	ksl-1200x-5l-ul	China Cogent
Magnetic stirrer	JEM-2100F	JEOL, Japan
Constant temperature blast drying oven	303-3S	China Baihui
Salt spray corrosion tester	LRHS-108-RY	China Forestry Frequency (CFF)
Electrochemical workstation	PARSTAT-2273	Princeton, USA
Laser confocal microscope	VK-X2000 series	Japan KEYENCE

ammonium hydroxide (25 ml) was added into the ethanol solution and vigorously stirred for one minute to ensure uniform dispersion. 0.3 ml of TEOS was then added dropwise as the silicon source, and the mouth of the beaker was covered with cling film. The solution became turbid, indicating the formation of MSNs. After vigorously stirring the mixture for two hours, the reaction was complete. The suspension was centrifuged (11 000 rpm, 10 minutes) to separate the nanocontainers, which were washed three times with deionized water. Finally, the product was lyophilized for 24 hours and ground to obtain the TA-MSNs. As a control, empty nanocontainers (denoted as MSNs) were prepared by calcinating TA-MSNs in a box-type furnace (heating to 500 °C at 2 °C $minute^{-1}$ and held for four hours) to remove the encapsulated TA. Figure 8.2 is a flowchart for the preparation of TA-MSNs.

8.2.1.6 Coloration Test of TA-MSNs

The color performance of TA-MSNs was evaluated using the complexation reaction between TA molecules and Fe^{3+}. TA-MSNs (0.2 g) were weighed and completely dispersed in 1 l of deionized water. After complete dispersion, a pipette was immediately used to transfer multiple sets of 2 ml dispersion. The solution (2 ml) was placed in the sample bottle and added with 2 ml of Fe^{3+} solution ($0, 1\times10^{-4}, 2\times10^{-4}, 4\times10^{-4}, 6\times10^{-4}, 8\times10^{-4}$ M). The color development was then observed using optical images and UV–Vis spectra.

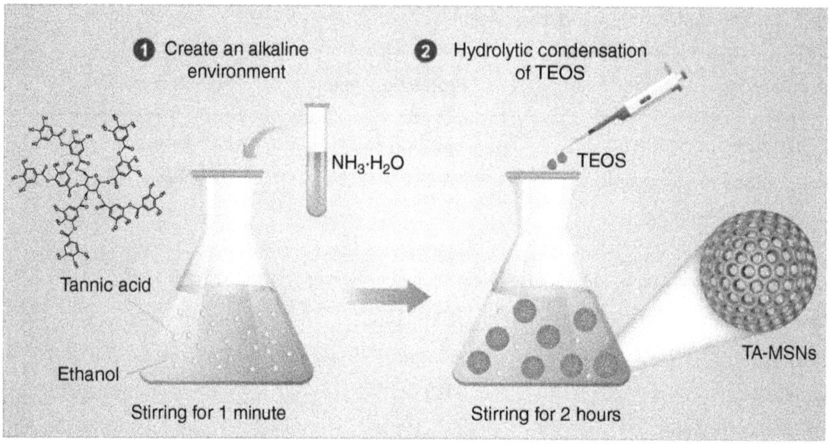

Figure 8.2 Schematic diagram of TA-MSNs prepared by one-step synthesis.

8.2.1.7 Corrosion Inhibition Test of TA-MSNs

The corrosion inhibition performance of TA-MSNs on carbon steel substrate was investigated by EIS and polarization curves. The electrolyte solutions contained 3.5 wt% NaCl aqueous solution and 3.5 wt% NaCl aqueous solution spiked with $2\,g\,l^{-1}$ of TA-MSNs, respectively. The surface morphology of carbon steel was observed by laser confocal microscope to investigate the corrosion inhibition performance of TA-MSNs. The carbon steel was immersed in 3.5 wt% NaCl and 3.5 wt% NaCl with $2\,g\,l^{-1}$ TA-MSNs, and the corrosion products were removed after 24 hours. The specimens were immersed in the cleaning solution for 10 minutes at room temperature, and then the CLSM was used to observe the surface morphology and roughness of the carbon steel after removing the surface corrosion products. 50 ml of concentrated hydrochloric acid and 0.35 g of hexamethylenetetramine were added into deionized water to form a 100-ml solution of cleaning formula.

8.2.1.8 Preparation of TA-MSNs-Based Coatings

Q235 carbon steel was selected as the metal substrate. This mild steel was sanded with 150- and 240-grit sandpapers until a metallic sheen appeared, followed by ultrasonic cleaning in anhydrous ethanol for 10 minutes. The epoxy matrix consisted of bisphenol A diglycidyl ether and neopentyl glycol diglycidyl ether, and D230 as curing agent in a molar ratio of 1 : 1 : 1. First, the two resin monomers were mixed, and then TA-MSNs (2.5, 5, 7.5, and 10 wt%) were added into the mixed solution to be fully dispersed. Next, D230 was added, the mixture was fully stirred, and then a small amount of the mixture was pipetted onto the cleaned carbon steel and uniformly spread using a scraping rod. The coated samples were stored at room temperature for 12 hours and then heated in an oven at 55 °C for 24 hours. According to the TA-MSNs ratio, the coatings were designated as 2.5, 5, 7.5, and 10 wt% TA-MSNs.

Pure epoxy coating without TA-MSNs was also prepared as control using the same procedure.

8.2.1.9 Coloration Properties of Coatings

To test the corrosion-sensing properties of coatings with TA-MSNs, salt spray tests were used to investigate the color development properties of the damaged coatings with different levels of TA-MSNs content. Prior to the test, a scratch was created on the coating surface using a scalpel to expose the metal substrate. The scratched coating samples were put into a salt spray chamber for the salt spray test with 5.0 wt% NaCl solution, with an interval of 15 seconds for every one minute of spraying. The ambient temperature was controlled at 25 °C, and the samples were taken out and photographed every once in a while. The color change of the scratched area of broken coatings added with different contents of fillers was compared with the color change of the scratched area over time. The coloration performance of the coating was evaluated by measuring the color change at the scratch of the coating [10, 11].

8.2.1.10 Self-healing Property of Coatings

The corrosion-resistant performance is an important indicator of the overall performance of coatings and closely related to their service life. A scratch with a width of about 30 μm and a length of 5 mm was made on the coating surface to create a damaged coating. EIS was conducted to investigate the corrosion-resistant performance of the intact coatings with different loading amounts of TA-MSNs and the self-healing performance of the damaged coatings. After the EIS measurement, a scalpel was used to peel off the coating and expose the metal substrate. Then, CLSM was used to observe the surface morphology and roughness of the corroded metal substrate to assess the barrier performance.

8.2.1.11 Aging-Resistant Property of Coatings

An accelerated-aging testing machine (SUN-II, Shanghai Yiheng Instrument Co., Ltd.) was used to assess the aging resistance. Coated samples were exposed to cyclic UV radiation (λ = 300–400 nm, power density = 65 W m^{-2}, two hours of irradiation followed by two hours of dark storage). After 192 and 384 hours of accelerated aging, the coating samples were removed from the test chamber for surface characterization.

Morphological evolution of the coatings was recorded using a digital camera. The surface morphology and roughness of the coatings were characterized using CLSM (VK-X, Keyence). FTIR spectra of the coatings were collected before the aging test and after 384 hours of aging test to evaluate molecular conformation in the coatings. The water contact angle (WCA) was measured using a Dataphysics OCA20 instrument by depositing 5 μl water droplets on the sample surface. According to the Chinese national standards (GB/T 9574-2007), the 60° gloss of coatings was measured using a WGG60-Y4 gloss meter (KSJ Photoelectric Instruments Co., Ltd.). After 384 hours of aging, EIS tests were carried out on the coatings to evaluate their aging resistance.

8.2.2 Color Development and Corrosion Inhibition Properties of TA

8.2.2.1 Chromogenic Properties of TA

The complexation behavior and color development properties of TA molecules with Fe^{3+} were investigated. An aqueous solution of 5×10^{-4} M TA was mixed with equal volumes of different concentrations of Fe^{3+} aqueous solutions (from left to right, 1×10^{-4}, 5×10^{-4}, 1×10^{-3}, 5×10^{-3}, and 1×10^{-2} M, respectively), and the colors of solutions after mixing are shown in Figure 8.3. Owing to the low concentration of TA aqueous solution, 5×10^{-4} M TA solution was nearly colorless. After adding different concentrations of Fe^{3+} solution, the color of the mixed solution gradually changed from colorless to blue-black, due to the complexation of TA molecules with Fe^{3+}. When the added Fe^{3+} solution reached 5×10^{-3} M, the deposited TA–Fe^{3+} complex appeared on the bottom layer of the mixed solution, and the amount of deposits gradually increased with the concentration of Fe^{3+} ions. The blue-black color of the TA–Fe^{3+} complex made TA an ideal indicator to detect the formation of Fe^{3+} ions during the corrosion reaction of the iron and steel substrate.

The complexation reaction between TA and Fe^{3+} was further investigated using UV–Vis spectroscopy. The UV–Vis spectra of TA solution (1×10^{-5} M) mixed with Fe^{3+} solutions at different concentrations (0, 1×10^{-4}, 2×10^{-4}, and 3×10^{-4} M) are shown in Figure 8.4. On adding Fe^{3+}, a new absorption peak appeared at 570 nm due to the formation of TA–Fe^{3+} complex, and the corresponding absorbance increased with the Fe^{3+} concentration. Because this characteristic peak was observed at the lowest Fe^{3+} concentration (1×10^{-4} M), TA has a high sensitivity to Fe^{3+} generated during local corrosion.

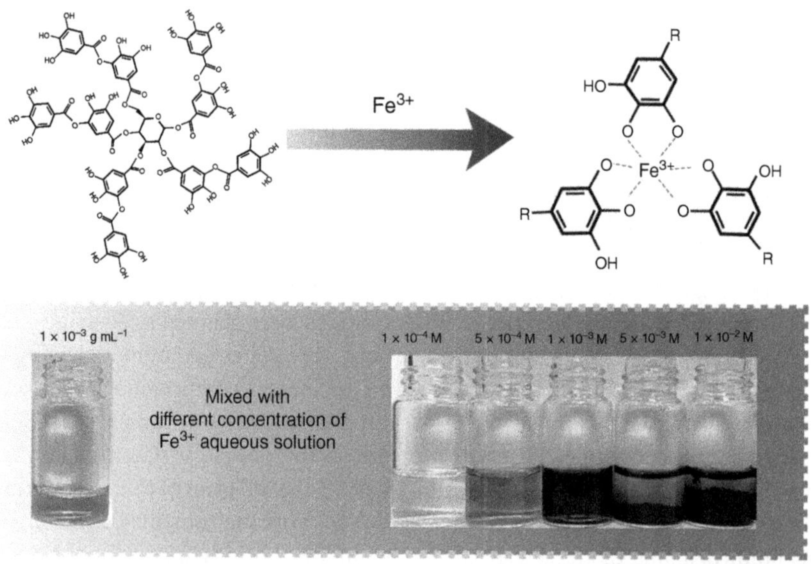

Figure 8.3 Comparison of the color of 5×10^{-4} M TA with equal volumes of Fe^{3+} solutions of different concentrations.

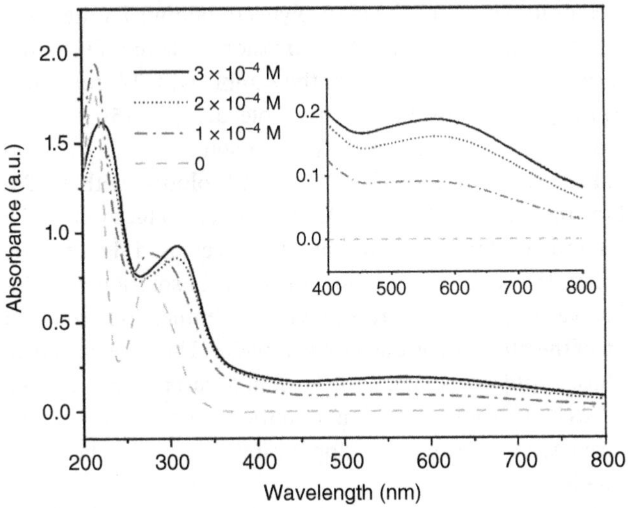

Figure 8.4 UV–Vis spectra of TA complexed with different concentrations of Fe^{3+} aqueous solution.

8.2.2.2 Corrosion Inhibition Performance of TA

The effect of TA molecules on the corrosion inhibition properties of Q235 metal substrates was investigated. Q235 steels were immersed in 3.5 wt% NaCl solution without and with 1.0 wt% TA and the corrosion inhibition effect of TA on the carbon steel substrate was tested by EIS. Figure 8.5 shows the Nyquist plots of Q235 carbon steel after immersion for different times in 3.5 wt% NaCl solution with and without TA. When the solution contained no TA, the arc resistance had a small radius that decreased with the immersion time, proving that the corrosion reaction became more serious with time. In contrast, the radius of arc resistance for the samples immersed in TA-containing electrolyte was significantly larger than those immersed in pure saline; and this radius increased with immersion time, indicating that TA had a certain inhibition effect on corrosion. After immersion for 48 hours, the radius

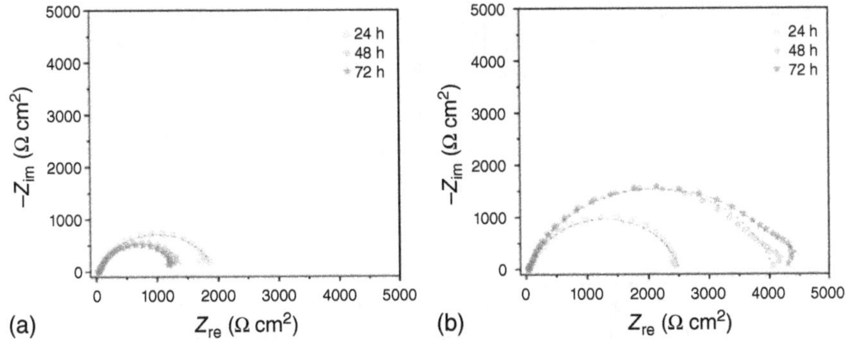

Figure 8.5 Nyquist plots of Q235 mild steel immersed in different solutions: (a) 3.5 wt% NaCl, (b) 3.5 wt% NaCl + 1.0 wt% TA.

of arc resistance was three times larger for the electrolyte containing TA than that without it, meaning that TA significantly inhibited corrosion in saline. Even after immersing for 72 hours, the arc resistance radius of the sample with TA continued to grow slowly, reflecting the gradual adsorption of TA molecules [12–15].

The Bode plots of Q235 mild steel immersed in different solutions are shown in Figure 8.6. Panels (a) and (c) show the case of 3.5 wt% NaCl solution without TA, and panels (b) and (d) show the case with 1.0 wt% TA. The $|Z|_{0.01\,\text{Hz}}$ values were quite low when the sample was immersed in pure saline, and they decreased significantly with immersion time. When the sample was immersed in NaCl solution containing TA, the $|Z|_{0.01\,\text{Hz}}$ value was higher and increased with the immersion time, and the mid-frequency region of the phase angle diagram was wider. These results prove that TA in the solution phase could slow down the corrosion of carbon steel [16–19]. In conclusion, TA molecules had obvious corrosion inhibition effect on carbon steel substrate, and it can be used as a promising corrosion inhibitor to inhibit the corrosion reaction of carbon steels.

The equivalent circuit shown in Figure 8.7 was used to fit the EIS data. R_s is the solution resistance, CPE_f is the film capacitance, R_f is the film resistance, CPE_{dl} is the double-layer capacitance, and R_{ct} is the charge transfer resistance. Considering the nonuniformity of the surface and the nonideal characteristics of the capacitance,

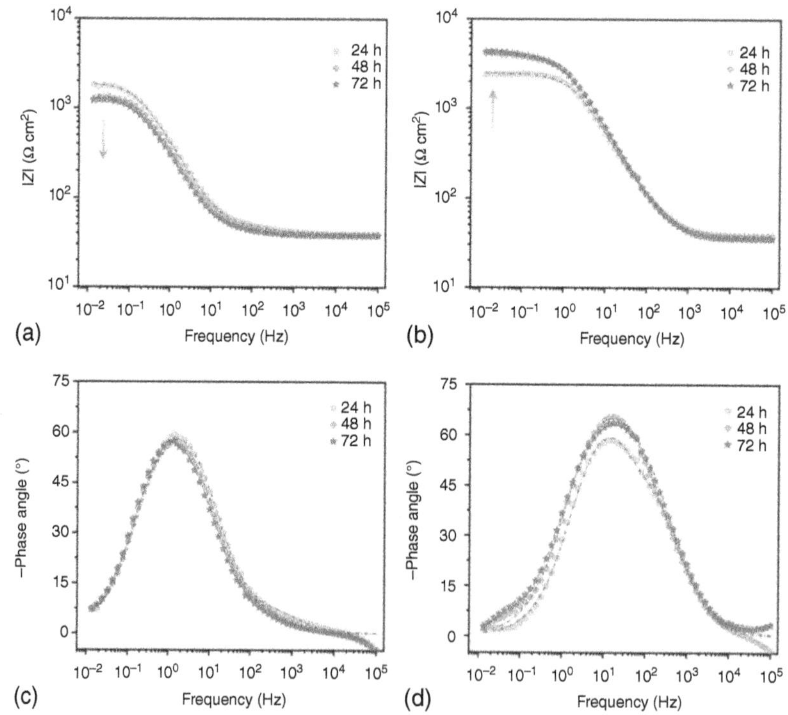

Figure 8.6 Bode diagram of Q235 mild steel immersed in different solutions: (a,c) 3.5 wt% NaCl; (b,d) 3.5 wt% NaCl + 1.0 wt% TA.

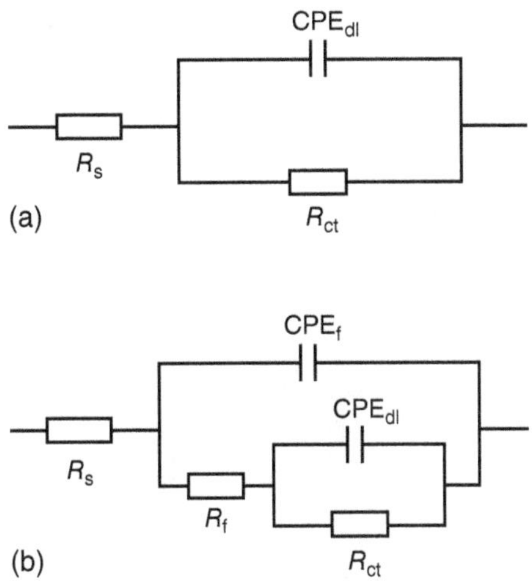

Figure 8.7 EIS equivalent fitting circuit for different samples: (a) Q235 mild steel immersed in 3.5 wt% NaCl solution, and (b) Q235 mild steel immersed in 3.5 wt% NaCl + 1.0 wt% TA solution.

Table 8.3 EIS electrochemical fitting parameters for different samples.

Sample	Time (h)	R_s (Ω cm^2)	CPE$_f$ (Ω^{-1} cm^{-2} S^n)	n_f	R_f (Ω cm^2)	CPE$_{dl}$ (Ω^{-1} cm^{-2} S^n)	n_{dl}	R_{ct} (Ω cm^2)
3.5 wt% NaCl	24	39.84	—	—	—	5.39×10^{-5}	0.80	1963
	48	38.04	—	—	—	6.51×10^{-5}	0.82	1445
	72	37.43	—	—	—	7.60×10^{-5}	0.81	1342
3.5 wt% NaCl + 1.0 wt% TA	24	37.44	7.12×10^{-6}	1.00	40.71	4.16×10^{-5}	0.80	2474
	48	38.25	9.06×10^{-6}	1.00	45.26	4.22×10^{-5}	0.79	3980
	72	38.08	1.28×10^{-5}	0.96	45.95	4.69×10^{-5}	0.78	4310

a phase angle element (CPE) was used instead of pure capacitance [20, 21]. Table 8.3 shows the parameters obtained by fitting the EIS data.

R_{ct} reflects the difficulty of charge transfer, and it can be used to measure the corrosion resistance of a material. In general, the larger the value of R_{ct}, the more difficult the transfer of electrons to induce corrosion [22–24]. The variation of R_{ct} value is shown in Figure 8.8. During the immersion process, R_{ct} of the carbon steel samples gradually decreased with the immersion time in pure saline solution, whereas it increased continuously during immersion in saline solution with added TA. After 72 hours, the latter R_{ct} value was approximately 3.2 times higher than the former one because TA and Fe^{3+} formed a stable complex that served as a protective layer against corrosion of the carbon steel substrate.

To more intuitively demonstrate the corrosion inhibition performance of TA, the corrosion product formed after 72 hours of immersion was removed from the Q235 steel substrate, and the micromorphology of the corroded sample surface

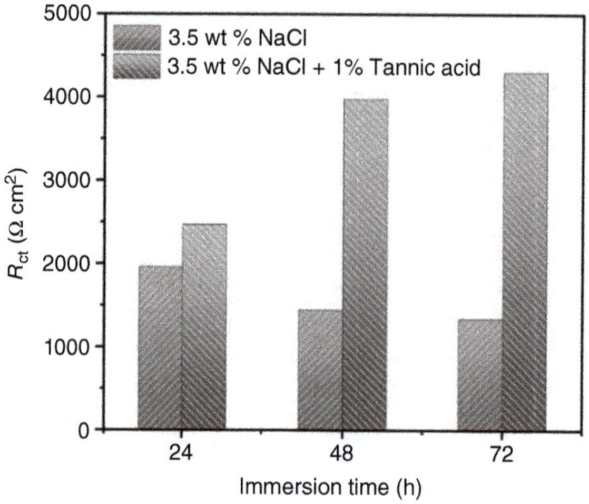

Figure 8.8 Variation of R_{ct} values of samples immersed in different solutions.

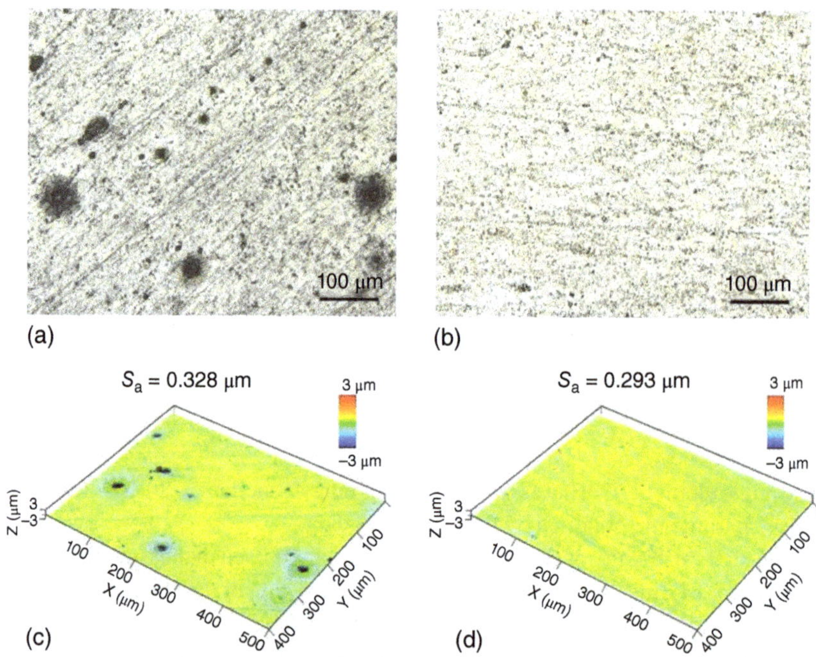

Figure 8.9 CLSM images of Q235 carbon steel after immersion in different solutions: (a,c) 3.5 wt% NaCl; (b,d) 3.5 wt% NaCl + 1.0 wt% TA.

was observed using CLSM. In Figure 8.9, Panels (a) and (c) show the surface morphology and 3D image of the sample immersed in 3.5 wt% NaCl solution, and Panels (b) and (d) show those of the sample immersed in 3.5 wt% NaCl + 1.0 wt% TA solution, respectively. The surface after immersion in pure saline solution showed obvious corrosion with multiple pits of various sizes, and surface roughness was measured as 0.328 μm. In contrast, there were almost no obvious surface corrosion defects on the sample immersed in NaCl + TA solution, and the surface roughness was relatively lower (0.293 μm). Therefore, TA had a certain inhibition effect on the corrosion of carbon steel in saline solution.

The corrosion inhibition mechanism of TA on carbon steel can be explained as follows. Fe^{3+} produced by corrosion formed stable, insoluble complexes with TA in the solution. These complexes were deposited on the surface of the substrate, producing a protective film that hindered the intrusion of corrosive media, reducing the corrosion rate and improving the corrosion resistance [25, 26].

The TA–Fe^{3+} complexes (denoted as $(Fe^{3+})_n$tannate) could be formed in three pathways [27–29]. In the first pathway, TA combines with Fe^{2+} to form $(Fe^{2+})_n$tannate, which is easily oxidized to $(Fe^{3+})_n$tannate. In the second pathway, TA directly interacts with Fe^{3+} to form $(Fe^{3+})_n$tannate. In the third pathway, Fe^{3+} is reduced by TA to Fe^{2+}, forms $(Fe^{2+})_n$tannate with TA, and then $(Fe^{2+})_n$tannate reacts with oxygen to produce $(Fe^{3+})_n$tannate. The reaction mechanism is shown in Eqs. (8.1)–(8.8).

$$\mathrm{Fe} \rightarrow \mathrm{Fe}^{2+} + 2\mathrm{e}^- \tag{8.1}$$

$$\mathrm{Fe}^{2+} + \frac{1}{4}\mathrm{O}_2 + \mathrm{H}^+ \rightarrow \mathrm{Fe}^{3+} + \frac{1}{2}\mathrm{H}_2\mathrm{O} \tag{8.2}$$

$$2\mathrm{Fe}^{3+} + \mathrm{Fe} \rightarrow 3\mathrm{Fe}^{2+} \tag{8.3}$$

$$\text{Tannins} \rightarrow \text{tannate} + n\mathrm{e}^- + n\mathrm{H}^+ \tag{8.4}$$

$$\text{Tannate} + n\,\mathrm{Fe}^{3+} \rightarrow (\mathrm{Fe}^{3+})_n\text{tannate} \tag{8.5}$$

$$\text{Tannate} + n\,\mathrm{Fe}^{2+} \rightarrow (\mathrm{Fe}^{2+})_n\text{tannate} \tag{8.6}$$

$$\text{Tannins} + n\,\mathrm{Fe}^{3+} \rightarrow (\mathrm{Fe}^{2+})_n\text{tannate} + n\mathrm{H}^+ \tag{8.7}$$

$$(\mathrm{Fe}^{2+})_n\text{tannate} + \frac{1}{4}\mathrm{O}_2 + n\mathrm{H}^+ \rightarrow (\mathrm{Fe}^{3+})_n\text{tannate} + \frac{1}{2n}\mathrm{H}_2\mathrm{O} \tag{8.8}$$

According to the results described earlier, TA shows a highly sensitive color change in response to Fe^{3+} and performs well in inhibiting the corrosion of carbon steel. Therefore, it can function as both an early corrosion indicator and a corrosion inhibitor for iron and steel materials. The color change due to the complexation reaction between TA and Fe^{3+} was investigated by visual observation and UV–Vis spectroscopy. The formed TA–Fe^{3+} complex was blue-black in color and insoluble, making it obvious to the naked eye that the corrosion had occurred. In the UV–Vis spectra, characteristic peaks of the TA–Fe^{3+} complex were observed even when the Fe^{3+} concentration was as low as 1×10^{-4} M, indicating that TA is sensitive for detecting Fe^{3+} generated during the corrosion of carbon steel. The corrosion inhibition performance of TA was characterized using EIS and morphological observations. When the formed TA–Fe^{3+} complex was deposited on the carbon steel surface, a protective film was formed to hinder the intrusion of corrosive media. As a result, the corrosion rate was reduced, and the substrate displayed a higher corrosion resistance. Results showed that 1.0 wt% free TA has good performance in corrosion inhibition.

8.2.3 Characterization and Property of TA-MSNs

As TA has dual properties of color development and corrosion inhibition by complexing with Fe^{3+}, we loaded TA as an indicator and a corrosion inhibitor into silicate nanocontainers for use in coatings. To avoid the reaction between TA and the coating matrix and to ensure the effective storage and controlled release of TA, this section used TA as the encapsulating substance and used the hydrolysis and condensation of ethyl orthosilicate in alkaline environment to produce TA-loaded mesoporous silica (TA-MSNs) in one step.

8.2.3.1 Morphology and Structure of TA-MSNs

TA-MSNs were prepared using a one-step synthesis method, and TA was used as the pore-forming agent. Under alkaline conditions, TEOS hydrolyzed and condensed on the surface of TA, forming a mesoporous silica framework. This allows TA to be directly loaded into the silica framework and become effectively encapsulated. Figure 8.10 shows the SEM and TEM morphologies of TA-MSNs. In the SEM images, TA-MSNs prepared by one-step synthesis method were spherical particles with a diameter of approximately 200 nm, with uniform size. In the TEM images, these particles clearly exhibit a mesoporous structure that facilitates TA loading.

Figure 8.11 shows the TEM images and EDS data of TA-MSNs. In the EDS elemental distribution, the distribution of Si corresponds to the mesoporous SiO_2, the distribution of C corresponds to the loaded TA molecule, and the O contributes from both the mesoporous SiO_2 and the TA molecules. The distributions of C, O, and Si elements overlap strongly, confirming that TA was successfully loaded into the nanocontainers and uniformly distributed inside.

The loading behavior of TA in the nanocontainers was further analyzed using FT-IR spectroscopy. The FT-IR spectra of TA-MSNs, MSNs (after removing TA by calcination), and pure TA are shown in Figure 8.12 to qualitatively compare the chemical bonds and functional groups in these samples. The TA-MSNs and MSNs samples had obvious absorption peaks at 802 and 1080 cm^{-1}, where 802 cm^{-1} was the symmetric stretching vibration absorption peak of the Si—O—Si bond of SiO_2, and 1080 cm^{-1} was the asymmetric vibration absorption peak of the Si—O—Si bond

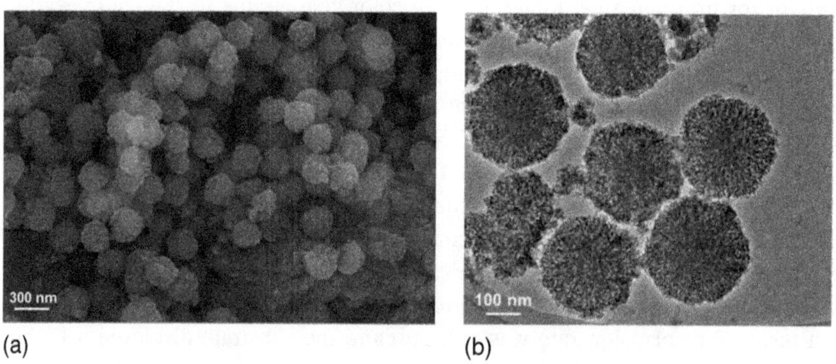

Figure 8.10 SEM image and TEM image of TA-MSNs.

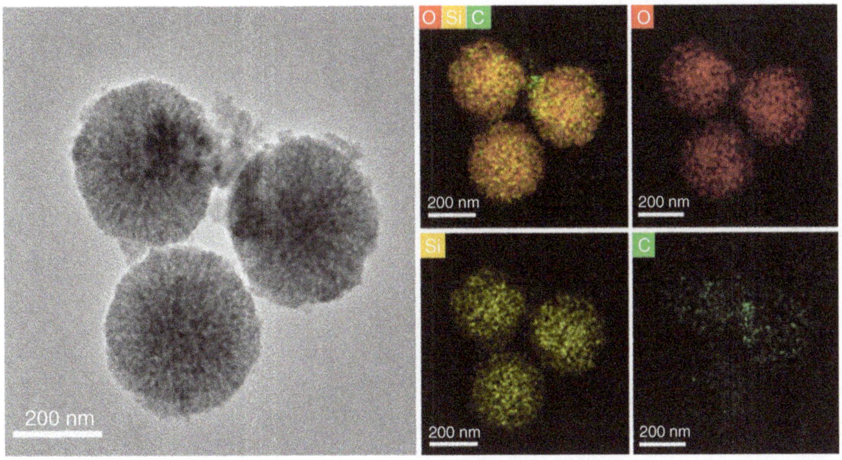

Figure 8.11 TEM image and EDS results of TA-MSNs.

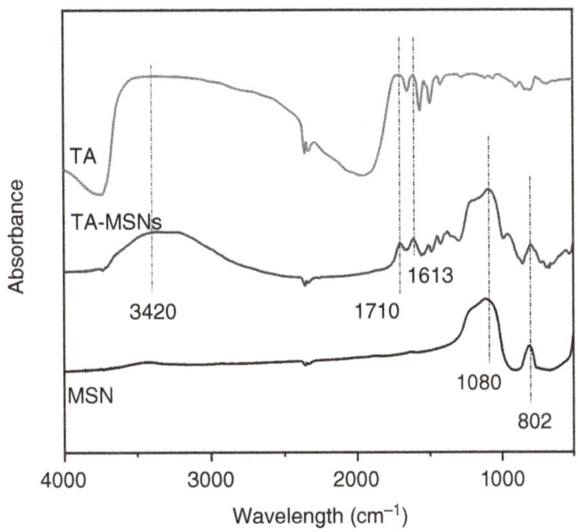

Figure 8.12 FT-IR of TA-MSNs, MSNs, and TA.

of SiO_2 proving the formation of mesoporous SiO_2. TA-MSNs and TA samples had obvious absorption peaks at 1613 and 1710 cm^{-1}, with 1613 cm^{-1} being the absorption peak of the benzene ring from TA and 1710 cm^{-1} being the stretching vibration absorption peak of C=O bonds, indicating that TA molecules were successfully loaded into mesoporous SiO_2 [30].

To further determine the loading of TA in the MSNs, thermogravimetric experiments were conducted on TA-MSNs, MSNs, and TA using a synchronous thermal analyzer [31–33], and their curves are shown in Figure 8.13. The main decomposition temperature of TA was between 200 and 550 °C. The loading rate of TA in TA-MSNs can be calculated using the following formula [34].

$$Wt_{(TA)} = \frac{Wt_{(TA-MSNs,550°C)} - Wt_{(TA-MSNs,200°C)}}{Wt_{(TA,550°C)} - Wt_{(TA,200°C)}} \times 100\% \tag{8.9}$$

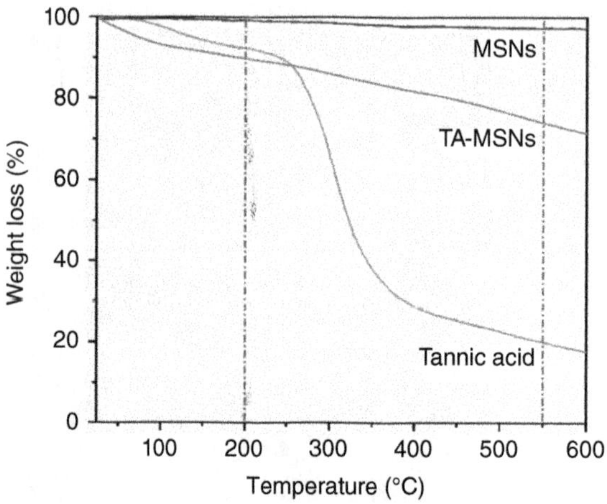

Figure 8.13 Thermogravimetric curves of TA-MSNs, MSNs, and TA.

From the thermogravimetric curves, it can be seen that the mass loss in the range of 200–550 °C was about 72% for TA, about 15% for TA-MSNs, and only about 1% for MSNs; and the difference in mass loss between MSNs and TA-MSNs was mainly due to the successful loading of TA molecules into TA-MSNs. The TA loading in TA-MSNs was calculated to be approximately 21%.

The pore structure in TA-MSNs was investigated by nitrogen isothermal adsorption–desorption experiments using a physical adsorbent apparatus. The obtained adsorption–desorption curves and pore size distributions of TA-MSNs and MSNs are shown in Figure 8.14. The nitrogen adsorption capacity of TA-MSNs was very low, and the BET-specific surface area was 21.3 $m^2\ g^{-1}$, because the mesoporous pore channels were occupied by the loaded TA molecules. However, after high-temperature calcination, the BET-specific surface area of MSNs increased greatly to 375.5 $m^2\ g^{-1}$, the pore volume was 0.46 $cm^3\ g^{-1}$, and the average pore diameter was 5.32 nm. These results proved that TA loaded in TA-MSNs was completely decomposed after calcination leaving a structure with mesopores. From

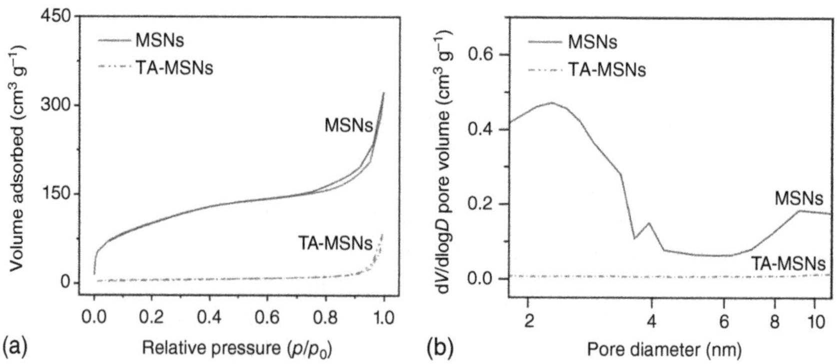

Figure 8.14 (a) Nitrogen isothermal adsorption–desorption curves and (b) BJH pore size distribution of TA-MSNs and MSNs.

Figure 8.14, TA-MSNs had rich mesoporous structure, which ensured the effective loading of TA molecules in the mesoporous silica pores.

8.2.3.2 Release Behavior of TA-MSNs

The release rate of TA from TA-MSNs at different pH values was investigated using UV–Vis absorption spectroscopy [35–38]. First, we need standard curves that relate the optical absorption to the TA concentration. Figure 8.15 shows the UV–Vis absorption spectra of 3.5 wt% NaCl solutions containing different concentrations of TA at pH 3, 7, and 11. The spectra in acidic and neutral solutions are similar, with the characteristic absorption peak of TA located at 274 nm. In the alkaline solution, this peak was shifted to 284 nm. Moreover, the relationship between the concentration of TA and the UV–Vis absorption intensities is plotted, and a correlation scatter plot is drawn in Figure 8.16. Regression analysis was performed to determine the correlation between these two variables, which could be used to determine the TA concentration in unknown solutions. The fitting lines were added to the plot for 3.5 wt% NaCl solutions at different pH values.

To study the TA release performance of TA-MSNs under different pH conditions, the effect of pH on the TA release behavior was investigated using solution's pH value as a variable condition. 0.01 g of TA-MSNs was dispersed in 2 ml of 3.5 wt% NaCl

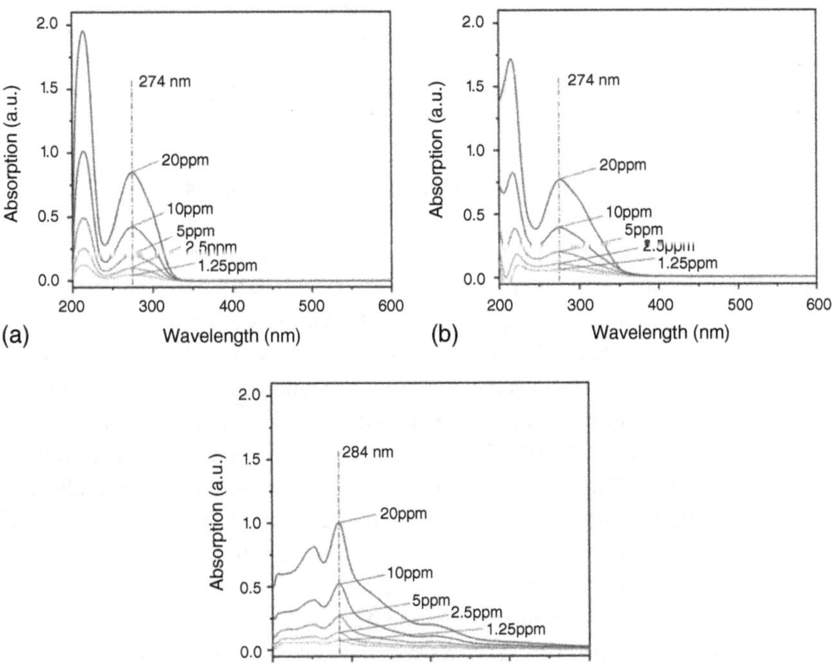

Figure 8.15 UV–Vis absorption spectra of TA in 3.5 wt% NaCl solution at different pH: (a) pH = 3, (b) pH = 7, (c) pH = 11.

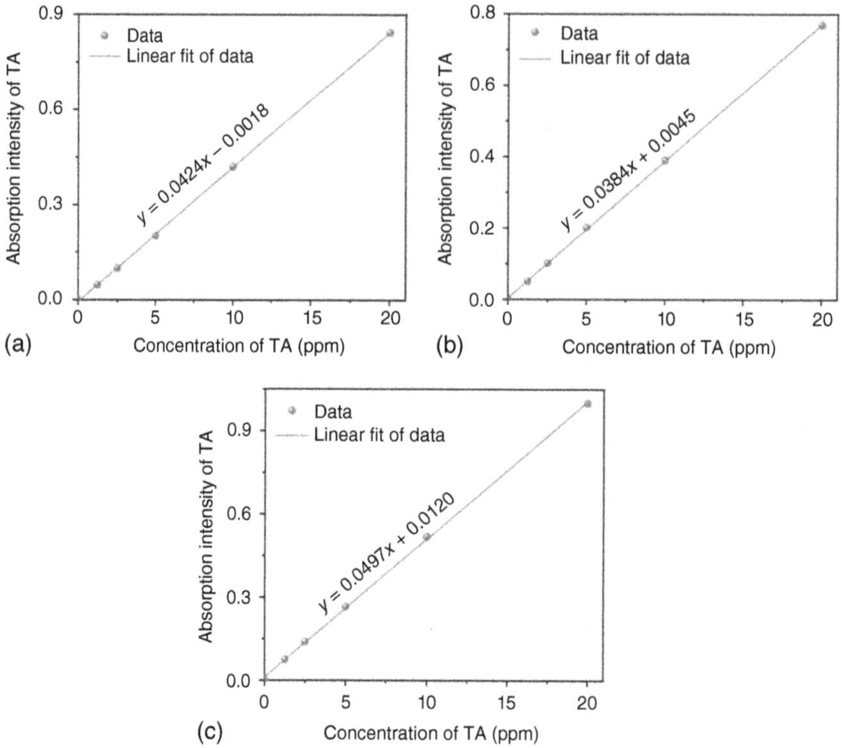

Figure 8.16 UV–Vis absorption standard curve of TA in 3.5 wt% NaCl solution: (a) pH = 3, (b) pH = 7, (c) pH = 11.

solutions at pH 3, 7, and 11 and then added into the dialysis bag. 48 ml of NaCl solution was weighed, the dialysis bag was immersed in the NaCl solution, and the pH inside and outside the bag was kept constant. The solution was stirred at 400 rpm on a magnetic stirrer. At regular intervals, a suitable amount of the solution was sucked up from the beaker with a pipette gun and transferred to a cuvette for UV–Vis spectroscopy measurement. According to the characteristic UV–Vis absorption peaks of TA at different pH values, the corresponding peaks were recorded, and the solution was poured back into the centrifuge tube immediately after the test [39, 40].

Three conditions, pH = 3, pH = 7, and pH = 11, were selected respectively as shown in Figure 8.17. The release rate of TA molecules varied greatly under different pH conditions, with the fastest release rate at pH = 3, followed by pH = 7, and the slowest release rate at pH = 11. With the prolongation of the release time, the TA molecules gradually reached a plateau. At 600 minutes, the TA release rate was about 97% at pH = 3, 60% at pH = 7, and 49% at pH = 11. The results showed that the release of TA from the TA-MSNs was the fastest at pH = 3, the slowest at pH = 11, and reached a plateau with the increase of release time. From the aforementioned analysis, it can be concluded that the release of TA from TA-MSNs was affected by the pH value in the solution, and the fastest release was under acidic conditions.

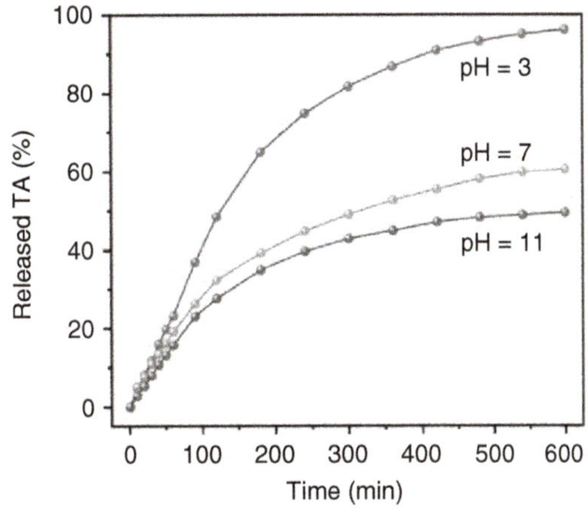

Figure 8.17 Release of TA-MSNs under different pH conditions.

8.2.3.3 Coloration Properties of TA-MSNs

This section explores the coloration performance of TA-MSNs. TA MSNs dispersion was prepared with a concentration of $0.2\,g\,l^{-1}$, and Fe^{3+}solutions of equal volume and different concentrations were immediately added. The color development degree of the solution was observed to characterize the color development performance of TA-MSNs [41, 42]. Figure 8.18 shows the color comparison between TA-MSNs dispersion and Fe^{3+} ion solutions at different concentrations. The $0.2\,g\,l^{-1}$ TA-MSNs dispersion exhibited a very light yellow color. After adding different concentrations of Fe^{3+} solutions ($0, 1 \times 10^{-4}, 2 \times 10^{-4}, 4 \times 10^{-4}, 6 \times 10^{-4}, 8 \times 10^{-4}$ M from left to right), the color of mixed solution quickly turned black, indicating that TA molecules in TA-MSNs were rapidly released and complexed with Fe^{3+}. The excellent color response speed of TA-MSNs was varied. After adding Fe^{3+} solution with 1×10^{-4} M, a visible black color can be observed in the mixed solution, indicating that TA-MSNs had excellent color sensitivity. At higher Fe^{3+}concentrations, the black color became gradually deeper.

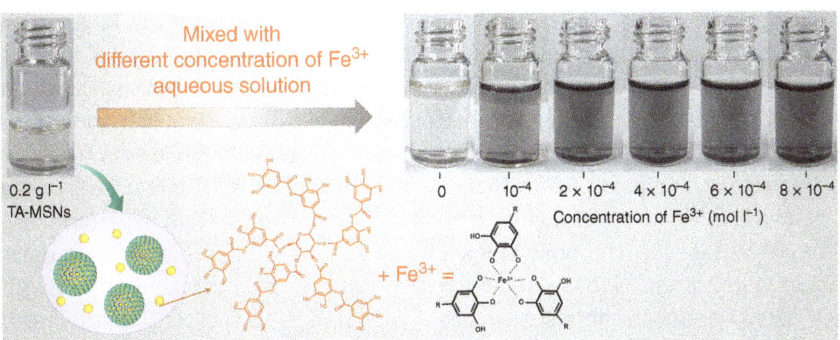

Figure 8.18 Color comparison of TA-MSNs with different concentrations of Fe^{3+} in aqueous solution.

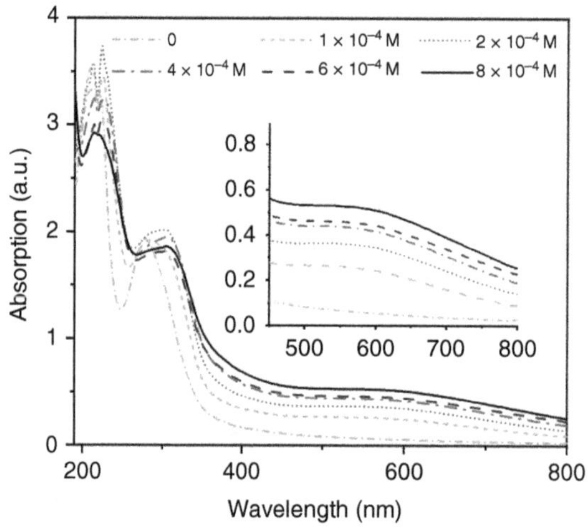

Figure 8.19 UV–Vis absorption spectra of TA-MSNs in mixed solutions with different concentrations of Fe^{3+}.

To illustrate the high sensitivity of TA-MSNs to localized corrosion, UV–Vis spectroscopy was used to investigate the UV–Vis absorption intensity of the aforementioned mixtures at different concentrations. Figure 8.19 shows the UV–Vis absorption spectra of the mixed solutions of TA-MSNs and different concentrations of Fe^{3+}. It can be seen that with the addition of Fe^{3+}, the spectra showed a distinct peak at 570 nm, which represented the formation of TA–Fe^{3+} complex. When the concentration of Fe^{3+} reached 1×10^{-4} M, a distinct characteristic peak of TA–Fe^{3+} complex can be observed, indicating that the TA-MSNs had a high sensitivity [43–45]. Therefore, TA-MSNs had excellent color response speed and color development sensitivity, which can be used as an indicator for the occurrence of carbon steel corrosion.

8.2.3.4 Corrosion Inhibition Performance of TA-MSNs

In this section, the corrosion inhibition properties of TA-MSNs on carbon steel substrates are further investigated. Q235 carbon steel was immersed in 3.5 wt% NaCl solution and 3.5 wt% NaCl solution together with $2.0\,g\,l^{-1}$ TA-MSNs. Figure 8.20 shows the Nyquist and Bode plots of Q235 carbon steel immersed in different solutions. The Nyquist plots show that the arc radius of capacitive reactance of the samples immersed in the TA-MSNs solution was significantly larger than that of the samples immersed in the pure NaCl solution, which indicates that the filler can inhibit the corrosion reaction of the substrate. With the extension of immersion time, the arc radius of capacitive reactance of the sample in pure NaCl solution decreased. It indicated that the corrosion reaction of carbon steel substrate was increasing, compared with which the arc radius of capacitive reactance of the sample with filler increased slightly. It verified the corrosion inhibition performance of TA-MSNs [46–48]. The Bode plot shows that the $|Z|_{0.01\,Hz}$ value of the samples immersed in 3.5 wt% NaCl solution was relatively low and decreased further with time up to 72 hours. In contrast, the sample immersed in 3.5 wt% NaCl solution with $2\,g\,l^{-1}$ TA-MSNs had a higher $|Z|_{0.01\,Hz}$ value and rose slightly with the extension of time up to 72 hours. This is attributed to the release of TA molecules from TA-MSNs, which reacted with the Fe^{3+} ions formed on the steel surface to form a protective

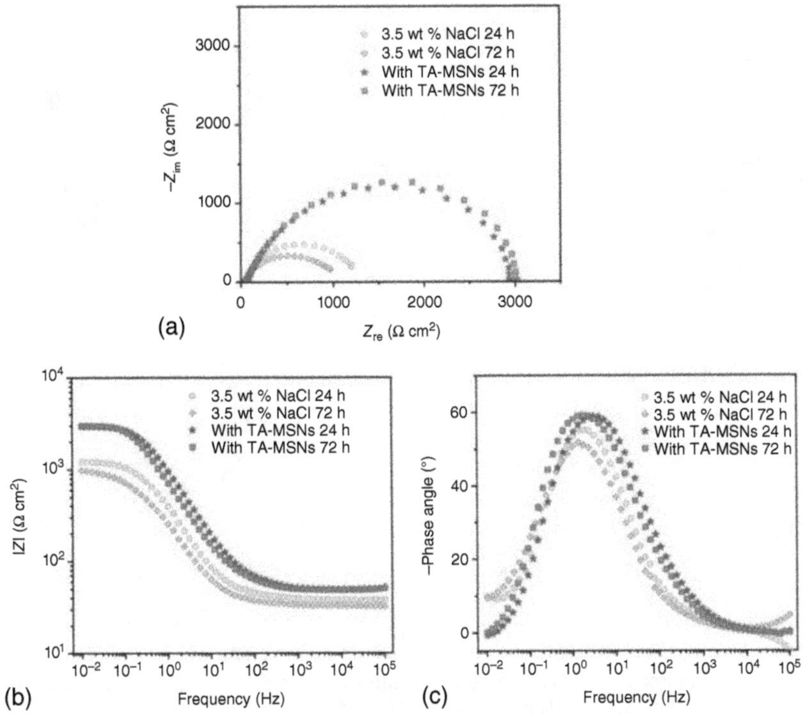

Figure 8.20 (a) Nyquist and (b, c)Bode plots of Q235 carbon steel immersed in 3.5 wt% NaCl solution and 3.5 wt% NaCl solution together with 2.0 g l^{-1} TA-MSNs.

layer [49, 50]. It could inhibit the corrosion of the metal against corrosion. The aforementioned analysis shows that TA-MSNs in solution can inhibit corrosion reaction and improve corrosion resistance of steel surfaces.

Then, the surface morphology and roughness of the samples after 72 hours of immersion were observed by CLSM. Figure 8.21 shows the CLSM images of Q235 carbon steel after immersion. In Figure 8.21, panels a and c are the samples immersed in 3.5 wt% NaCl solution, and Figure 8.21 panels b and d are the samples immersed in 3.5 wt% NaCl solution with 2.0 g l^{-1} TA-MSNs. It is clear that no obvious corrosion pits were found in the samples immersed in the solution with TA-MSNs, and the surface roughness (S_a) was as low as 0.285 μm, which indicates that the TA-MSNs had the ability to protect the metal substrate from corrosion. On the contrary, carbon steel without TA-MSNs showed obvious corrosion pits on the surface, and the surface roughness increased to 0.312 μm, indicating that more severe corrosion occurred on the metal surface [51, 52]. This can be explained by the fact that the TA molecules in TA-MSNs, after being released into the saline solution, complexed with Fe^{3+} to form a protective film layer deposited on the metal surface to achieve effective corrosion protection [53–55].

Figure 8.22 shows the results of polarization curves for samples immersed in 3.5 wt% NaCl solution with and without TA-MSNs. The corrosion current density in pure NaCl solution was about 8×10^{-6} A cm^{-2}, and adding TA-MSNs to saline

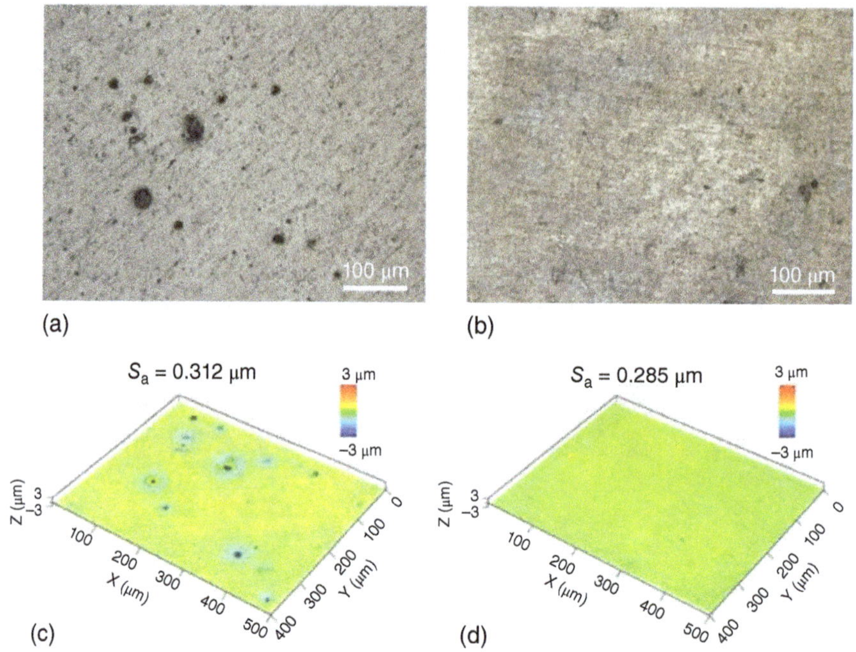

Figure 8.21 CLSM images of Q235 carbon steel after immersion in different solutions: (a,c) 3.5 wt% NaCl; (b,d) 3.5 wt% NaCl + 2.0 g l^{-1} TA-MSNs.

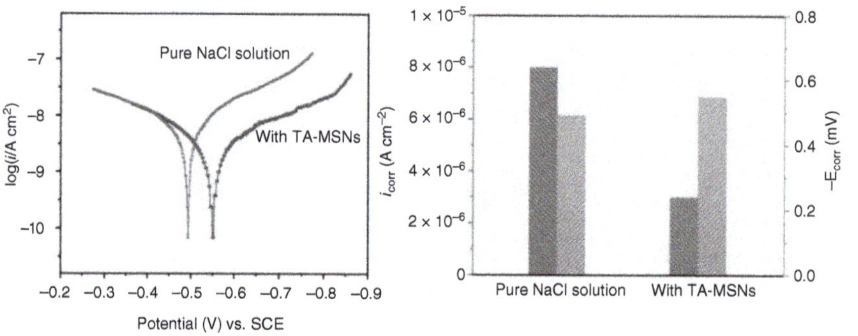

Figure 8.22 (a) Polarization curves and (b) corresponding electrochemical parameters of carbon steels in different solutions.

solution successfully reduced the corrosion current density to $3 \times 10^{-6}\,A\,cm^{-2}$. The corresponding corrosion inhibition efficiency (62.5%) indicates substantial corrosion inhibition effect from TA-MSNs in solution on carbon steel in neutral environments [45, 56, 57].

8.2.3.5 Radical Scavenging Activity of TA-MSNs

Figure 8.23 shows the scavenging activity of different concentrations of TA-MSNs when interacting with 2,2-diphenyl-1-picrylhydrazyl (DPPH) free radicals, with TA and MSNs as control samples. When the concentration of TA was only 4 ppm,

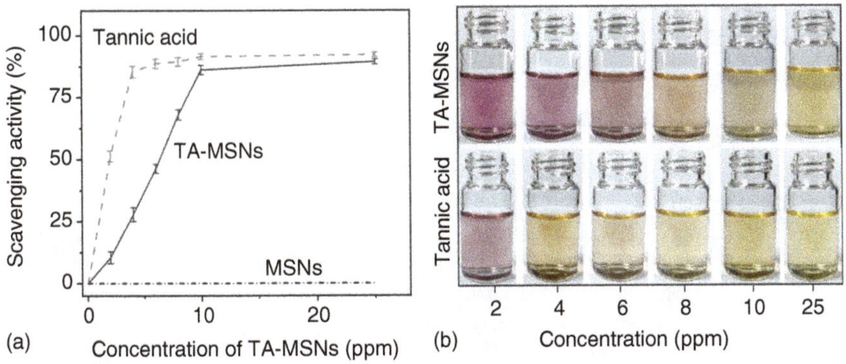

Figure 8.23 (a) DPPH radical scavenging activity of MSN, TA-MSNs, and TA at different concentrations; (b) digital images of DPPH reacted with different concentrations of TA-MSNs and TA.

TA molecules already showed obvious free radical scavenging activity within 30 minutes of reaction. Benefiting from the efficient loading of TA in the TA-MSNs, TA-MSNs also showed significant DPPH radical scavenging activity at a concentration of 10 ppm. In contrast, for MSNs nanostructures, no radical scavenging activity was shown due to the complete removal of the internal TA molecules. Figure 8.23b shows a digital photograph of reaction mixtures containing DPPH and different concentrations of TA and TA-MSNs. With the increase in the concentration of additives, the color of the mixture containing TA-MSNs and TA molecules changed from dark purple to light purple and finally to yellow, indicating that these substances have good radical scavenging ability.

8.2.4 Characterization and Property of TA-MSNs/Epoxy Coating

According to previous results, it is demonstrated that TA possessed both color development and corrosion inhibition properties, and the TA-MSNs were successfully prepared in a one-step method with excellent color development and corrosion inhibition properties. These properties were prepared using a facile one-step method. In this section, TA-MSNs are added to the epoxy matrix to investigate the effects of filler addition on the anticorrosion effect and aging-resistant properties of intact coatings, as well as the effect of filler addition on the self-reporting and self-healing properties on composite coatings.

8.2.4.1 Self-reporting Property of Coatings

In this section, we investigate the coloration performance of TA-MSNs coating and the influence of the amount of nanofiller on the self-reporting performance of coatings. The color performances of TA molecules and TA-MSNs have been demonstrated in previous sections. Therefore, 2.5, 5, 7.5, and 10 wt% of TA-MSNs nanofillers were added into the coating, and the color change at the coating scratch was observed. Figure 8.24 presents the color change of pure epoxy coating and 2.5, 5, 7.5, and 10 wt% TA-MSNs containing coatings during salt spray. As shown in the

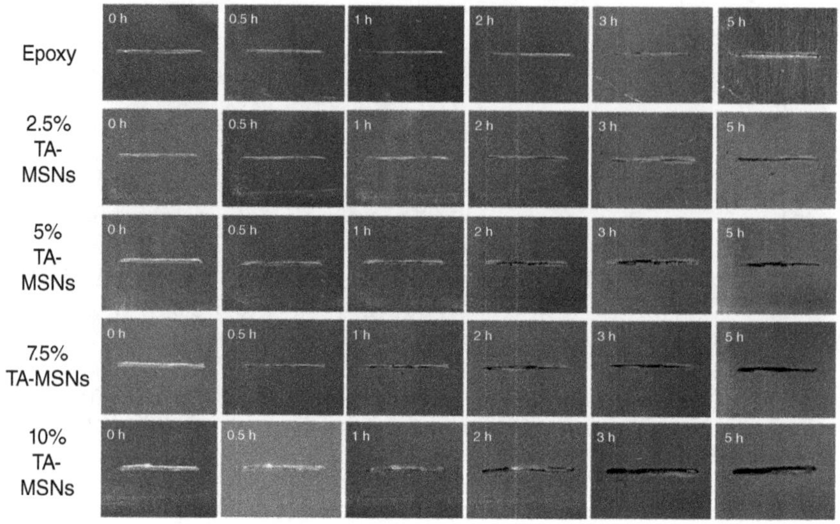

Figure 8.24 Corrosion self-reporting pictures of coatings with different contents of TS-MSNs during salt spray process.

figure, no color change was observed in the scratched area of the pure epoxy coating after five hours of salt spraying. In contrast, TA-MSNs coatings exhibited intuitive color changes after a period of salt spraying; an obvious black color was observed at the scratch, and the color intensity increased with the extension of the salt spray time. The color changes confirmed the formation of Fe^{3+} during the corrosion process. Fe^{3+} ions can effectively combine with the TA molecules released by the TA-MSNs to induce a coloration effect during the corrosion initiation process. A black color was observed in the scratched area of the coating containing 2.5 wt% TA-MSNs after three hours of salt spraying but the coloration effect was not obvious [44, 58–61]. This can be explained by the fact that the TA-MSNs nanofillers did not provide sufficient TA molecules to react with Fe^{3+} per unit area. With an increase in the filler content, the density of TA-MSNs increased, resulting in higher color intensity. When the filler content of the coating was increased to 5 and 10 wt%, the visible black color was observed after one hour of salt spraying, and the color intensity increased during the salt spraying. This indicated that the content of TA-MSNs played a significant role in early corrosion warning. Compared with the pure epoxy coating, the degree of peeling of the TA-MSNs coating was significantly reduced after five hours of salt spraying. Thus, the TA-MSNs coating can detect Fe^{3+} ions associated with anodic dissolution and achieve self-reporting during the corrosion initiation stage. The obvious color response is helpful for warning about coating damage and facilitating timely coating maintenance and repair [62, 63].

8.2.4.2 Corrosion Resistance of Coatings

Thereafter, the influence of amount of TA-MSNs nanofiller added on the corrosion protection ability of the coatings was investigated via electrochemical tests and CLSM observation. Figure 8.25 shows the EIS results of the epoxy coatings with

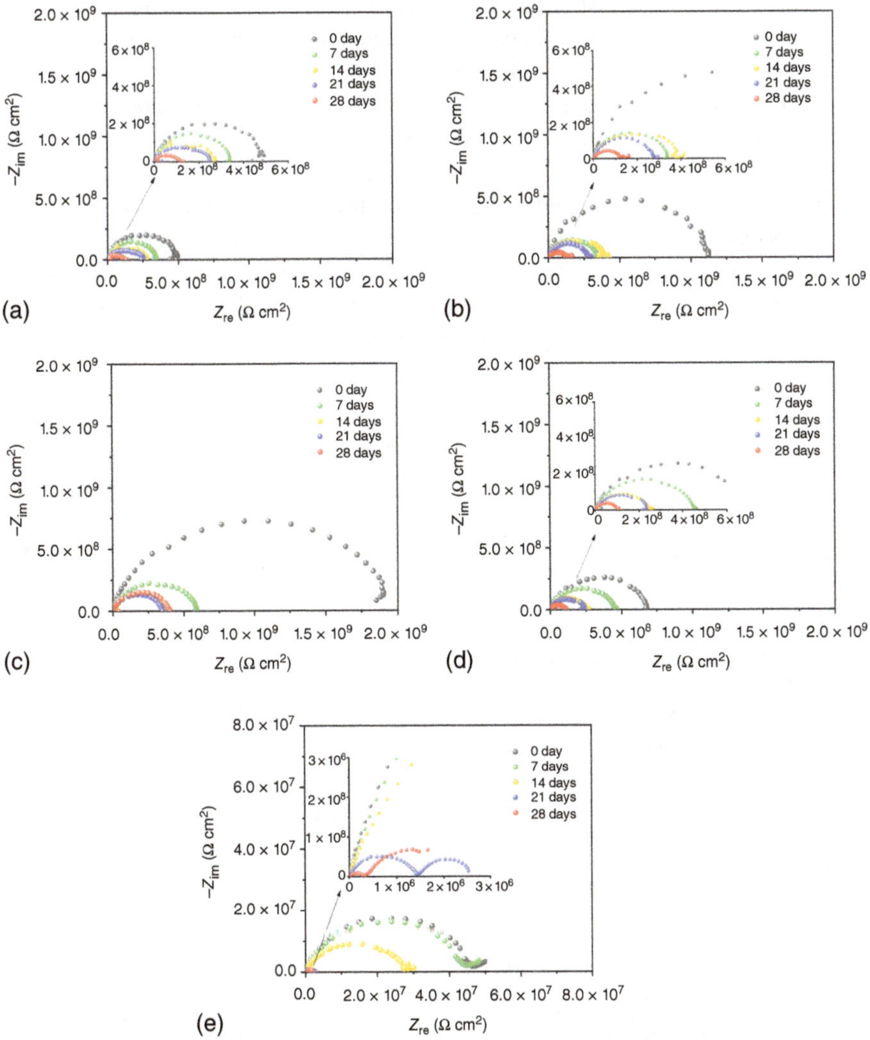

Figure 8.25 Nyquist diagram of intact coatings with different contents of TA-MSNs during the immersion process: (a) epoxy coating, (b) 2.5 wt% TA-MSNs coating, (c) 5 wt% TA-MSNs coating, (d) 7.5 wt% TA-MSNs coating, (e) 10 wt% TA-MSNs coating.

different nanofiller contents during immersion in 3.5 wt% NaCl solution for 1, 7, 14, 21, and 28 days. After immersion for 28 days, the pure epoxy coating and the epoxy coating with 2.5, 5, and 7.5 wt% TA-MSNs exhibited only one capacitive reactance arc, indicating that they had good barrier performance. However, for the epoxy coating with 10 wt% TA-MSNs, a second capacity-reactance arc appeared on the 1st day of immersion, indicating that corrosion of the metal substrate had begun. Compared with the epoxy coatings with 2.5, 5, and 7.5 wt% TA-MSNs, the diameter of the capacitive reactance arc for the pure epoxy coating was smaller and decreased rapidly during the test period. In contrast, the addition of 2.5, 5, and

7.5 wt% TA-MSNs yielded larger arc diameters on the first immersion day, with 5 wt% TA-MSNs having the largest arc diameter. The diameters of the capacitive reactance of the three coatings decreased with an increase in the immersion time [64, 65]. Notably, the coating with 5 wt% TA-MSNs nanofillers exhibited a relatively stable state and had the best corrosion resistance after being immersed in 3.5 wt% NaCl solution for 7–28 days.

Figure 8.26 shows the Bode diagrams of the intact coatings with different TA-MSNs contents during the immersion process. The phase angle diagram can reflect the corrosion degree of coatings, from which the breakpoint frequency at a phase angle of −45° (f_b) is an effective parameter to verify the protective properties of coatings, which are closely related to their microscopic hierarchical regions.

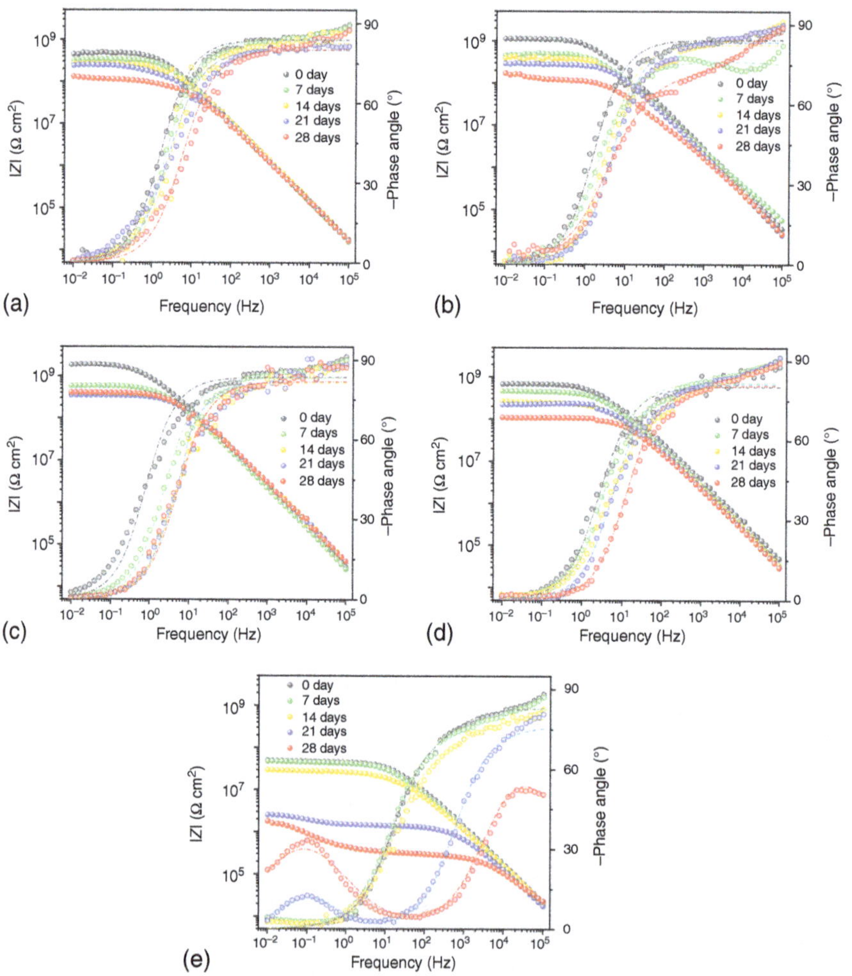

Figure 8.26 Bode diagram of intact epoxy coating with different contents of TA-MSNs in the immersion process: (a) epoxy coating, (b) 2.5 wt% TA-MSNs coating, (c) 5 wt% TA-MSNs coating, (d) 7.5 wt% TA-MSNs coating, (e) 10 wt% TA-MSNs coating.

As shown in Figure 8.26, with the extension of the immersion time, the f_b values of all the coatings moved toward higher frequencies. Compared with the pure epoxy coating, the f_b value of the epoxy coating with 5 wt% TA-MSNs was higher. The most stable f_b value exhibited only a slight increase in the TA-MSNs, indicating that the addition of TA-MSNs inhibited the penetration of the corrosive medium and stripping of the coating. However, the f_b value of 10 wt% TA-MSNs epoxy coating moved to a higher frequency with two time constants, indicating that the coating almost lost its barrier effect, and the corrosion reaction between the coating and metal substrate was strongly intensified.

The $|Z|_{0.01\,Hz}$ value changes of the coatings with different filler contents were analyzed as shown in Figure 8.27. The initial impedance on the first immersion day was $4.5 \times 10^8\ \Omega\,cm^2$. After 28 days, the $|Z|_{0.01\,Hz}$ value was reduced to $1.3 \times 10^8\ \Omega\,cm^2$, which indicated that the coating could not completely hinder the electrolyte penetration and that the barrier performance gradually deteriorated. With the addition of TA-MSNs nanofillers, the initial $|Z|_{0.01\,Hz}$ values of 2.5, 5, and 7.5 wt% TA-MSNs epoxy coatings were higher than that of the pure epoxy coating, whereas the initial $|Z|_{0.01\,Hz}$ value of the 10 wt% TA-MSNs epoxy coating was lower than that of the pure epoxy coating. It was proven that adding appropriate nanofillers can compensate for the defects of the epoxy matrix to a certain extent, thereby improving the compactness of the coating, and the slow release of TA molecules in the nanofiller inhibits the development of the corrosion reaction [66–68]. However, an excessive amount of filler produces more corrosion channels at the interface between the filler and coating, which degrades the barrier performance of the coating. After 28 days, the impedance values of the 2.5 and 7.5 wt% TA-MSNs epoxy coatings were close to those of pure epoxy coatings, whereas the $|Z|_{0.01\,Hz}$ values of the 5 wt% TA-MSNs epoxy coatings remained at higher levels, and the impedance

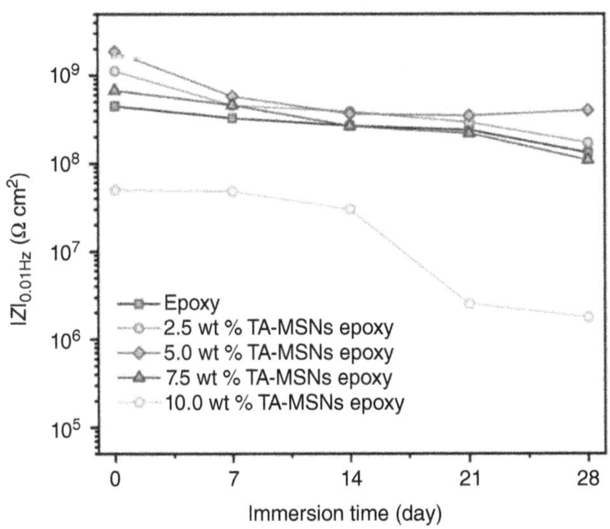

Figure 8.27 The $|Z|_{0.01\,Hz}$ changes of intact coatings with different filler content during immersion time.

values of the 10 wt% TA-MSNs epoxy coatings were far lower than those of the other coatings [69]. It was proven that an excessive amount of additives provided more channels for the invasion of corrosive media. An epoxy coating with 5 wt% TA-MSNs maintained the highest $|Z|_{0.01\ \mathrm{Hz}}$ value throughout the immersion process. On the 1st day of immersion, the $|Z|_{0.01\ \mathrm{Hz}}$ value was as high as $1.9 \times 10^9\ \Omega\,\mathrm{cm}^2$. It was reduced to $5.8 \times 10^8\ \Omega\,\mathrm{cm}^2$ on the 7th day, then remained relatively stable and was $4.0 \times 10^8\ \Omega\,\mathrm{cm}^2$ on the 28th day. The Nyquist and Bode diagrams indicated that up to 7.5 wt% TA-MSNs can be added to the epoxy coating without affecting its protective performance, and TA-MSNs had the best barrier performance when the addition amount was 5 wt%.

The EIS results for the coatings with different filler contents were fitted to the Nyquist and Bode curves. The equivalent circuit used to fit the data is shown in Figure 8.28. Here, R_s represents the solution resistance, CPE_c represents the coating capacitance, R_{pore} represents the pore resistance, CPE_{dl} represents the double-layer capacitance, and R_{ct} represents the charge-transfer resistance [70, 71].

Tables 8.4–8.8 explain the EIS fitting parameters of pure epoxy coating, 2.5 wt% TA-MSNs epoxy coating, 5 wt% TA-MSNs epoxy coating, 7.5 wt% TA-MSNs epoxy coating, and 10 wt% TA-MSNs epoxy coating, respectively. In general, R_{pore} characterizes the barrier performance of a coating. Figure 8.29 presents the R_{pore} values of coatings with different filling amounts. As shown, the R_{pore} values of all the coatings decreased with the extension of the immersion time. The R_{pore} values increased in the following order: 10 wt% TA-MSNs epoxy coating < pure epoxy coating < 7.5 wt% TA-MSNs epoxy coating < 2.5 wt% TA-MSNs epoxy coating < 5 wt% TA-MSNs epoxy coating. This indicated that the TA-MSNs nanofiller had a positive effect on the protective performance of the coating when the filler content was <7.5 wt%. When the filler content was 10 wt%, the excessive filler content caused the agglomeration problem ascribed to the poor dispersion of fillers in the coating. This results in more pores and microcracks and accelerates the corrosion process. The R_{pore} values of the 5 wt% TA-MSNs epoxy coating remained the highest and most stable, indicating excellent corrosion protection performance.

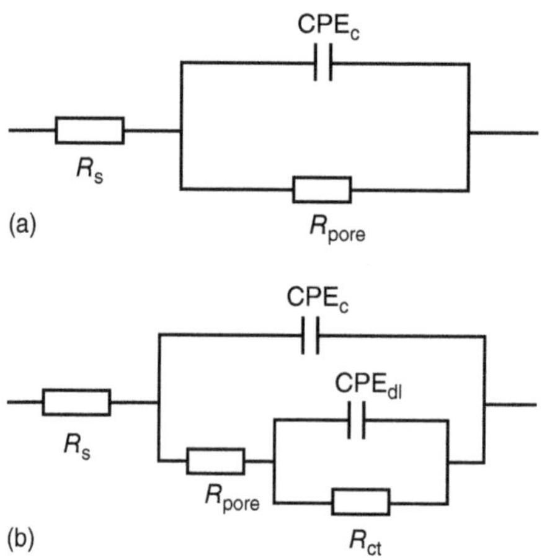

Figure 8.28 Equivalent circuit diagram of EIS fitting: (a) pure epoxy coating, 2.5 wt% TA-MSNs coating, 5 wt% TA-MSNs coating, and 7.5 wt% TA-MSNs coating; (b) 10 wt% TA-MSNs coating.

Table 8.4 EIS electrochemical fitting parameters of pure epoxy coating during immersion in 3.5 wt% NaCl solution for different time.

Immersion time (day)	CPEc (Ω^{-1} cm^{-2} S^n)	n_c	R_{pore} (Ω cm^2)
0	2.27×10^{-10}	0.93	4.73×10^8
7	2.40×10^{-10}	0.93	3.35×10^8
14	2.47×10^{-10}	0.92	2.58×10^8
21	3.58×10^{-10}	0.89	2.28×10^8
28	3.60×10^{-10}	0.90	1.10×10^8

Table 8.5 EIS electrochemical fitting parameters of 2.5 wt% TA-MSNs coating during immersion in 3.5 wt% NaCl solution for different time.

Immersion time (day)	CPEc (Ω^{-1} cm^{-2} S^n)	n_c	R_{pore} (Ω cm^2)
0	9.87×10^{-11}	0.94	1.10×10^9
7	2.26×10^{-10}	0.85	5.02×10^8
14	2.47×10^{-10}	0.92	2.58×10^8
21	3.58×10^{-10}	0.89	2.28×10^8
28	3.60×10^{-10}	0.90	1.10×10^8

Table 8.6 EIS electrochemical fitting parameters of 5 wt% TA-MSNs coating during immersion in 3.5 wt% NaCl solution for different time.

Immersion time (day)	CPEc (Ω^{-1} cm^{-2} S^n)	n_c	R_{pore} (Ω cm^2)
0	1.23×10^{-10}	0.93	1.81×10^9
7	1.43×10^{-10}	0.92	5.67×10^8
14	1.26×10^{-10}	0.90	3.56×10^8
21	1.21×10^{-10}	0.91	3.41×10^8
28	1.20×10^{-10}	0.91	3.86×10^8

The degree of corrosion of the carbon steel surfaces beneath the coatings was examined using CLSM after 28 days of immersion. Figure 8.30 shows the surface morphology of carbon steel covered by coatings with different filler contents. Scratches treated with sandpaper could be clearly observed on the metal surface below the pure epoxy coating and the epoxy coating with 2.5, 5, and 7.5 wt% TA-MSNs, and no corrosion product was observed; thus confirming that the surface of the carbon steel was essentially not corroded, and the coating effectively prevented the invasion of corrosive media. However, scratches treated with sandpaper on the metal surface for the 10 wt% TA-MSNs epoxy coating were corroded, and

Table 8.7 EIS electrochemical fitting parameters of 7.5 wt% TA-MSNs coating during immersion in 3.5 wt% NaCl solution for different time.

Immersion time (day)	CPEc (Ω^{-1} cm^{-2} S^n)	n_c	R_{pore} (Ω cm^2)
0	1.20×10^{-10}	0.89	6.58×10^8
7	1.37×10^{-10}	0.92	4.38×10^8
14	1.97×10^{-10}	0.88	2.43×10^8
21	1.67×10^{-10}	0.90	2.24×10^8
28	1.86×10^{-10}	0.90	1.05×10^8

Table 8.8 EIS electrochemical fitting parameters of 10 wt% TA-MSNs during immersion in 3.5 wt% NaCl solution for different time.

Immersion time (day)	CPEc (Ω^{-1} cm^{-2} S^n)	n_c	R_{pore} (Ω cm^2)	CPEct (Ω^{-1} cm^{-2} S^n)	n_{ct}	R_{ct}(Ω cm^2)
0	2.13×10^{-10}	0.92	1.08×10^7	5.42×10^{-10}	0.70	3.67×10^7
7	2.66×10^{-10}	0.91	8.19×10^6	7.06×10^{-10}	0.67	3.77×10^7
14	3.65×10^{-10}	0.88	5.71×10^6	6.85×10^{-10}	0.74	2.20×10^7
21	6.81×10^{-10}	0.85	1.40×10^6	1.42×10^{-6}	0.74	1.26×10^6
28	1.71×10^{-9}	0.77	3.00×10^5	1.68×10^{-6}	0.70	2.23×10^6

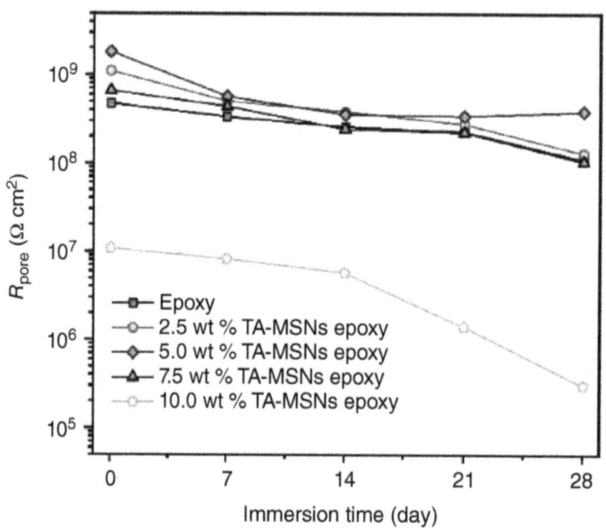

Figure 8.29 R_{pore} values of intact coatings with different filler contents during immersion time.

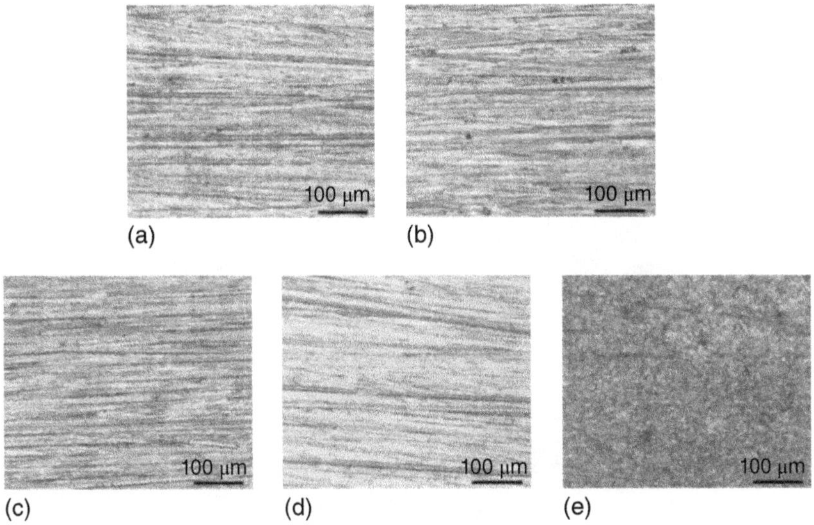

Figure 8.30 Surface morphology of carbon steel after immersion of intact coatings with different filler contents: (a) epoxy coating, (b) 2.5 wt% TA-MSNs coating, (c) 5 wt% TA-MSNs coating, (d) 7.5 wt% TA-MSNs coating, (e) 10 wt% TA-MSNs coating.

a large amount of corrosion products accumulated, indicating that the barrier performance of the coating failed and that the substrate was corroded by the corrosive medium. This is attributed to the addition of excess filler, which resulted in more defects owing to poor dispersion.

Figure 8.31 presents the surface roughness of carbon steel covered by coatings with different filler contents after immersion for 28 days. As shown in Figure 8.31a–d, the surface roughness values of the pure epoxy coating and the epoxy coatings with 2.5, 5, and 7.5 wt% TA-MSNs samples were similar, scratches were observed on the surface, and no corrosion products were generated, indicating that the surface of the carbon steel was not corroded. The scratch marks on the surface of the 10 wt% TA-MSNs epoxy coating (Figure 8.31e) were no longer obvious, and the surface roughness was reduced, possibly owing to the corrosion of the carbon steel surface by the corrosive medium and the formation of corrosion products.

8.2.4.3 Self-healing Property of Damaged Coatings

According to the analysis in the previous section, adding appropriate TA-MSNs nanofillers can improve the barrier properties of the complete coating, and the coating with 5 wt% filler had the optimal protective performance. In this section, the influence of the amount of added TA-MSNs nanofillers on the self-healing ability of the damaged coating is analyzed. Figure 8.32 shows the Nyquist plots of the scratched pure epoxy coating, 2.5 wt% TA-MSNs epoxy coating, 5 wt% TA-MSNs epoxy coating, 7.5 wt% TA-MSNs epoxy coating, and 10 wt% TA-MSNs epoxy coating during immersion in 3.5 wt% NaCl solutions after 1, 3, 5, 7, 14, 21, and 28 days. The diameter of the capacitive reactance arc of the pure epoxy coating and the 10 wt% TA-MSNs epoxy coating decreased with the extension of the immersion

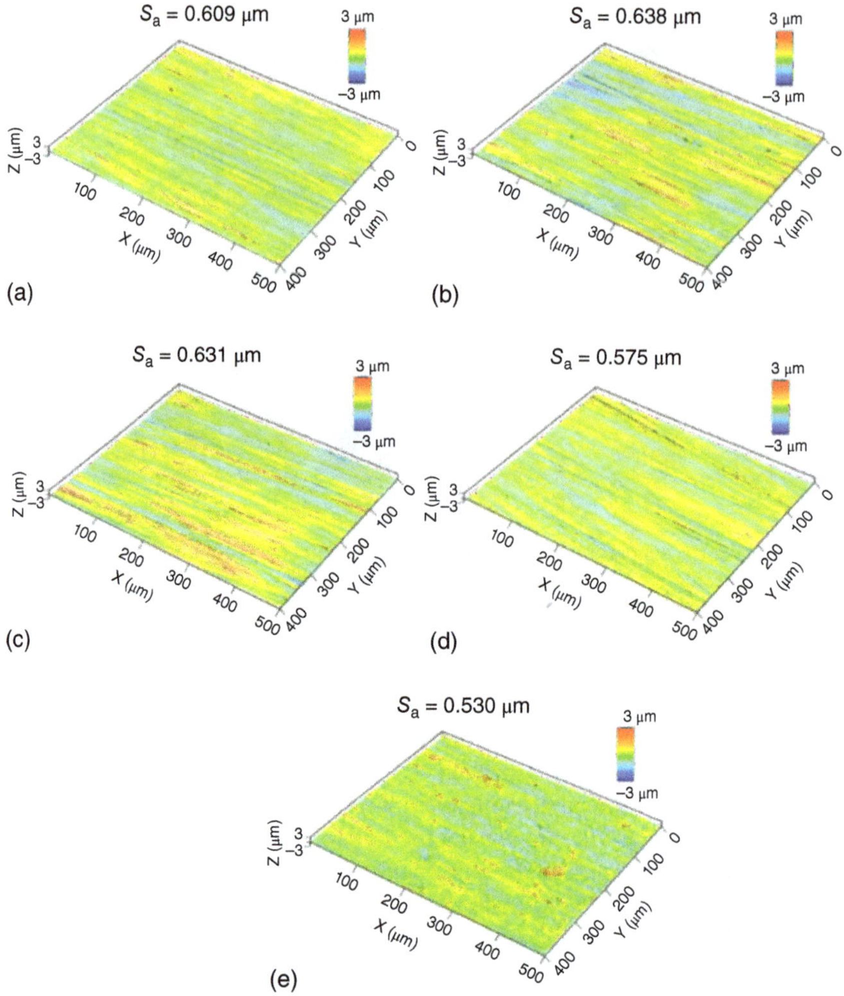

Figure 8.31 Surface roughness of carbon steel under intact coatings with different filler contents: (a) epoxy coating, (b) 2.5 wt% TA-MSNs coating, (c) 5 wt% TA-MSNs coating, (d) 7.5 wt% TA-MSNs coating, (e) 10 wt% TA-MSNs coating.

time, indicating that the barrier property of the coating decreased and that the corrosion of the metal substrate was intensified. The diameter of the capacitive reactance arc for the 2.5, 5, and 7.5 wt% TA-MSNs epoxy coatings first decreased and then increased, owing to the gradual release of TA molecules from TA-MSNs in the coating, which reacted with the Fe^{3+} ions in the carbon steel matrix generated during corrosion. The TA molecules and Fe^{3+} ions formed a complex layer on the metal surface, effectively inhibiting the aggravation of corrosion in the damaged area of the coating [72–74]. As indicated by the Nyquist diagram of the 5 wt% TA-MSNs epoxy coating, with the extension of the immersion time, two capacitive reactance arcs appeared at the scratch, indicating that TA molecules were released

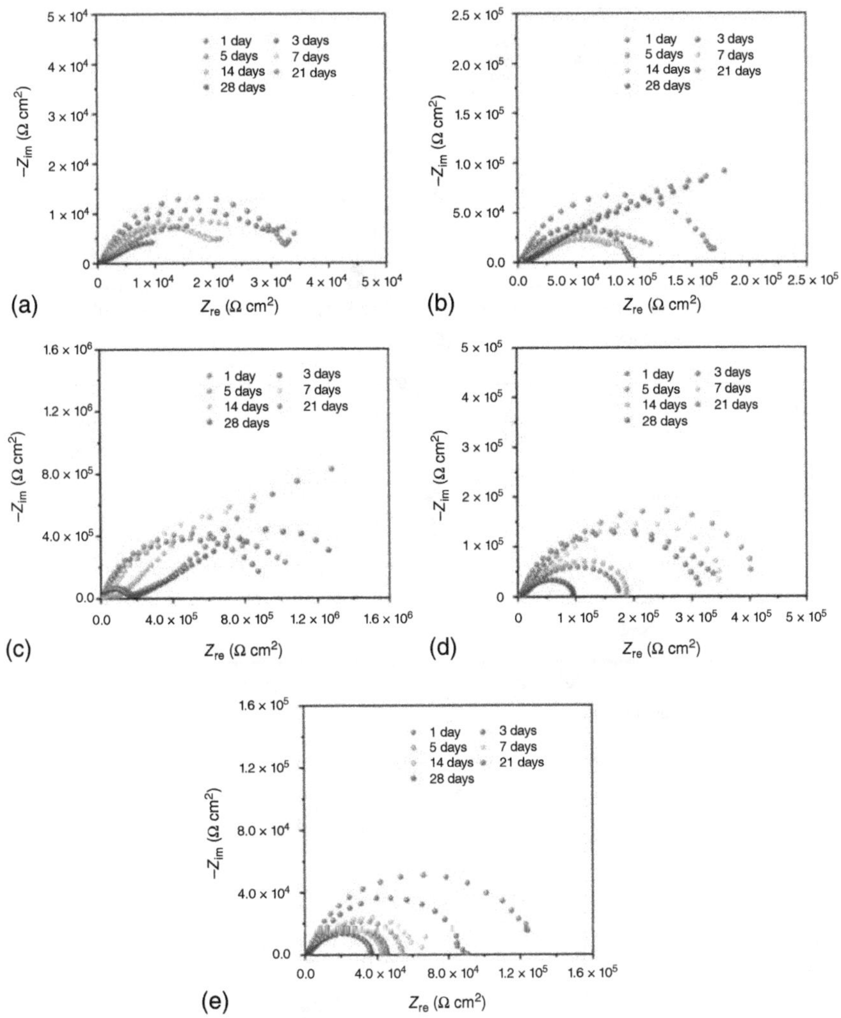

Figure 8.32 Nyquist diagram of scratched coatings with different contents of TA-MSNs: (a) Epoxy coating, (b) 2.5 wt% TA-MSNs coating, (c) 5 wt% TA-MSNs coating, (d) 7.5 wt% TA-MSNs coating, (e) 10 wt% TA-MSNs coating.

from TA-MSNs nanofillers and formed a protective film of a TA–Fe^{3+} complex on the surface of the carbon steel matrix [75, 76].

The self-healing performance of the TA-MSNs coatings was verified by the phase angle diagram shown in Figure 8.33. For the coating systems, the high frequency (10^4–10^5 Hz) reflects the coating properties, the middle frequency is related to the corrosion products, and the low frequency reflects the corrosion activity at the interface between the carbon steel and the coating. A new time constant for the scratched pure epoxy coating was gradually generated during immersion, which was primarily caused by the accumulation of corrosion products. In contrast, the high-frequency phase angle of the scratch coating with the addition of 2.5 and 5 wt%

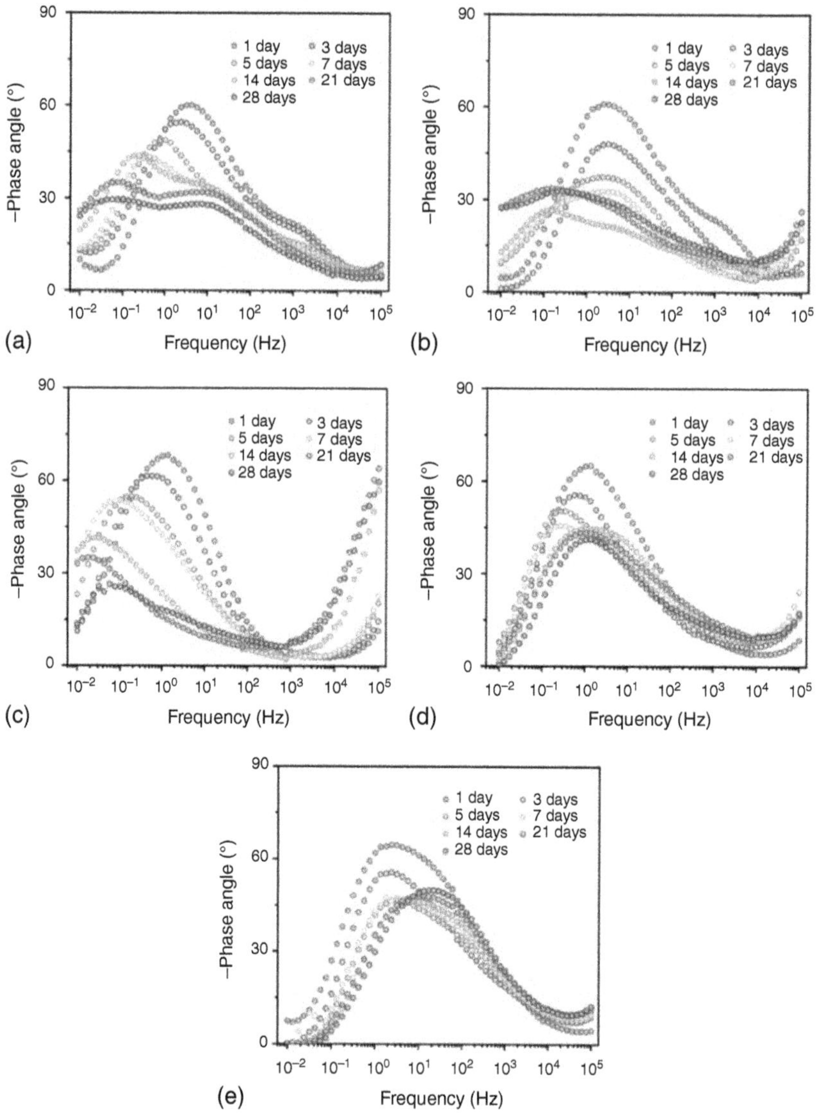

Figure 8.33 Bode phase angle of damaged coatings with different contents of TA-MSNs during immersion. (a) Pure epoxy coating, (b) 2.5 wt% TA-MSNS epoxy coating, (c) 5 wt% TA-MSNs coating, (d) 7.5 wt% TA-MSNs coating, (e) 10 wt% TA-MSNs coating.

TA-MSNs nanofillers gradually increased and with 5 wt% TA-MSNs nanofiller significantly increased, indicating that a relatively dense film layer was formed in the exposed scratch area on the coating. Thus, the corrosion reaction was delayed, and the self-healing effect was evident [77–79].

Figure 8.34 shows the Bode impedance moduli of the damaged coatings with different TA-MSN contents in the immersion process. All the coatings with the addition of the TA-MSNs nanofillers exhibited higher $|Z|_{0.01\,\mathrm{Hz}}$ values than the pure

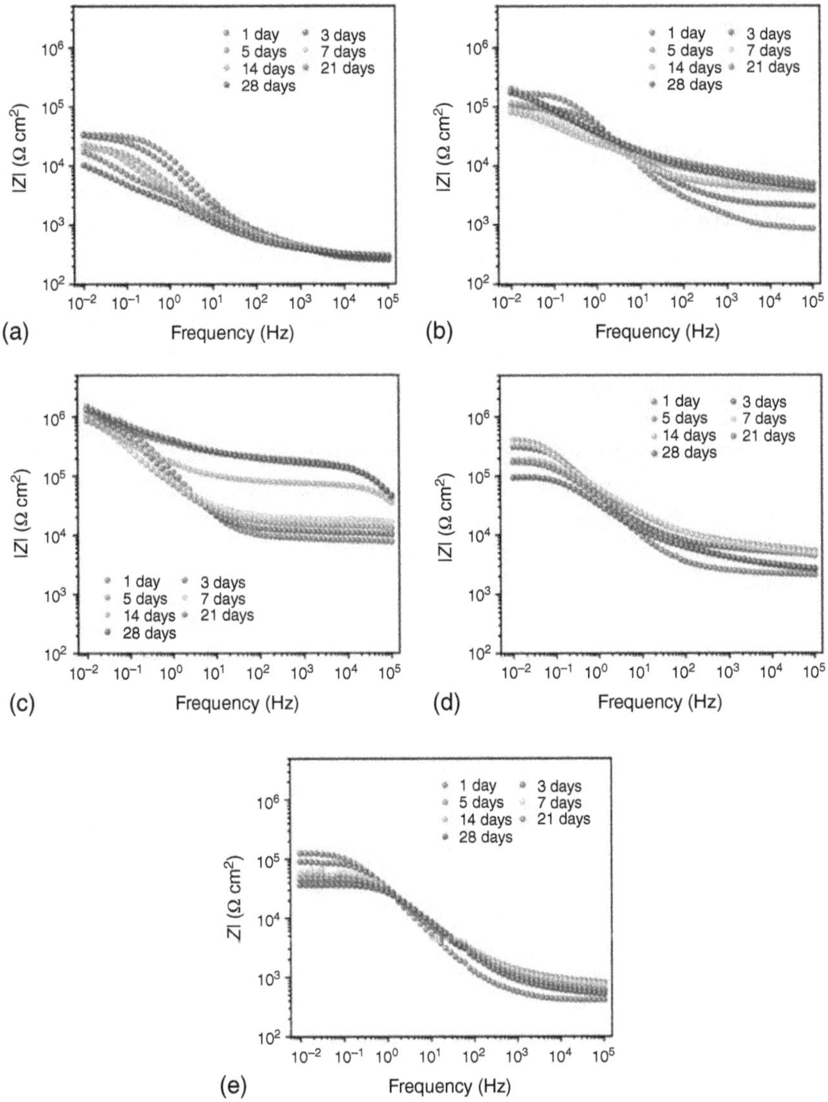

Figure 8.34 Bode impedance diagram of damaged coatings with different contents of TA-MSNs. (a) Epoxy coating, (b) 2.5 wt% TA-MSNs coating, (c) 5 wt% TA-MSNs coating, (d) 7.5 wt% TA-MSNs coating, (e) 10 wt% TA-MSNs coating.

epoxy coating. After 1 day of immersion, the $|Z|_{0.01\,\text{Hz}}$ value of the pure epoxy coating was $3.4 \times 10^4\,\Omega\,\text{cm}^2$, that of the 2.5 wt% TA-MSNs coating was $1.7{\times}10^5\,\Omega\,\text{cm}^2$, 5.0 wt% TA-MSNs coating was $1.1 \times 10^6\,\Omega\,\text{cm}^2$, 7.5 wt% TA-MSNs coating was $3.4 \times 10^5\,\Omega\,\text{cm}^2$, and 10 wt% TA-MSNs coating was $1.3 \times 10^5\,\Omega\,\text{cm}^2$. The 5.0 wt% TA-MSNs coating exhibited the highest $|Z|_{0.01\,\text{Hz}}$ value, which was two orders of magnitude higher than that of the pure epoxy coating. This can be explained by the fact that when the coating is damaged, TA molecules are released from the

TA-MSNs nanofillers, complexed with the Fe^{3+} ions of the carbon steel matrix, and deposited on the steel surface to form a protective film, effectively inhibiting the aggravation of corrosion in the damaged area of the coating. With an increase in the immersion time, the impedance value of the pure epoxy coating decreased continuously because of the continuous invasion of Cl^-, H_2O, O_2, and other corrosive ions, which aggravated the corrosion of the damaged area. However, the low-frequency impedance value of the coating with nanofillers did not increase or decrease significantly. As the TA-MSNs kept releasing TA molecules in the damaged area, and these molecules continuously deposited on the surface and hindered the aggravation of corrosion activities. After 28 days of immersion, the impedance value of the pure epoxy coating was reduced to $1.0 \times 10^4\,\Omega\,cm^2$, whereas the impedance value of the coating with nanofillers exhibited a smaller reduction. The 5 wt% TA-MSNs coating still maintained a high-impedance value of $1.3 \times 10^6\,\Omega\,cm^2$, exhibiting excellent self-healing performance. In conclusion, TA-MSNs nanofillers can reduce the corrosion activity of the metal substrate in the damaged area to a certain extent and improve the protective performance of the coating and the damaged coating with 5 wt% TA-MSNs nanofillers had the best self-healing effect [80–82].

We further explored the $|Z|_{0.01\,Hz}$ changes of different coatings during immersion, as shown in Figure 8.35. The impedance values of the TA-MSNs nanofillers were higher than those of the pure epoxy coatings. The $|Z|_{0.01\,Hz}$ value of the pure epoxy coatings was $10^4\,\Omega\,cm^2$. However, the $|Z|_{0.01\,Hz}$ values of the 2.5, 7.5, and 10 wt% TA-MSNs coatings were $>10^5\,\Omega\,cm^2$, and the impedance value of the 5 wt% TA-MSNs coating was $10^6\,\Omega\,cm^2$. This indicates that the TA-MSNs nanofillers can slow the corrosion of the metal substrate and improve the self-healing effect of the coating. The amount of nanofiller significantly influenced the self-healing effect of the coating. The impedance modulus of the 5 wt% TA-MSNs coating was the highest, the impedance of 7.5 wt% TA-MSNs coating was higher than that of the 2.5 wt% TA-MSNs coating, and the impedance of the 10 wt% TA-MSNs

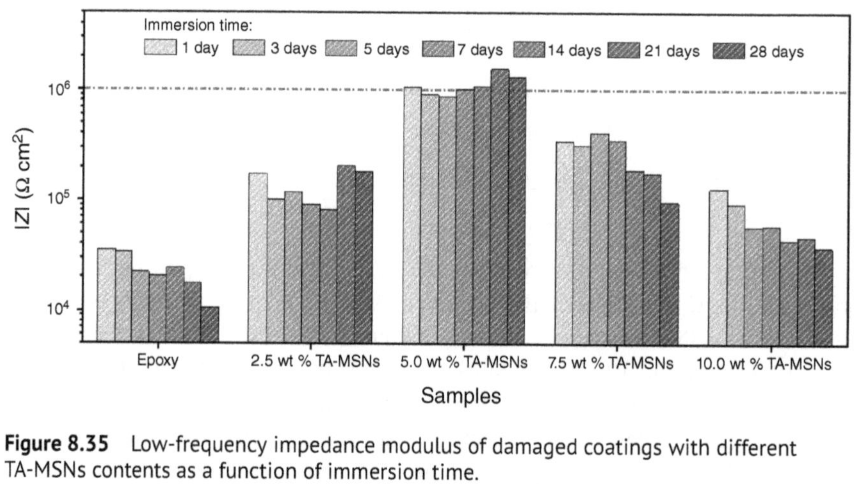

Figure 8.35 Low-frequency impedance modulus of damaged coatings with different TA-MSNs contents as a function of immersion time.

coating was the lowest [83–86]. This is because within a certain range of nanofiller amounts, with a larger nanofiller amount, more TA molecules were released, and the corrosion inhibition effect was stronger. The self-healing effect was optimal when the amount of TA-MSNs was 5 wt%. However, too much nanofiller will cause agglomeration owing to the high-nanofiller content, and more corrosion channels will be generated at the interface between the filler and the coating, degrading the self-healing performance of the coating.

The immersion time of a single coating sample was used as a variable in the analysis. As the immersion time increased, the $|Z|_{0.01\,Hz}$ values of the pure epoxy coating and 10 wt% TA-MSNs coating decreased. This is because the corrosive medium in the immersion process invaded ceaselessly, causing damage to the coating regions with intensified corrosion. The $|Z|_{0.01\,Hz}$ values of the 2.5, 5, 7.5 wt% TA-MSNs coatings appeared to decline at the beginning of immersion. The $|Z|_{0.01\,Hz}$ values for the 5 wt% TA-MSNs coating then increased, indicating that the coating with the proper nanofiller content had both the barrier and the corrosion inhibition properties. This explains why the impedance modulus of the 10 wt% TA-MSNs coating exhibited a decreasing trend. Owing to excessive filler agglomeration in the coating, the protective properties of the coating deteriorated. Even with more TA corrosion inhibitors in the coating, the self-healing ability cannot be guaranteed [87–89].

After immersing in 3.5 wt% NaCl solution for 28 days, the surface morphology at the scratched coating areas was examined via CLSM. As shown in Figure 8.36, the peeling of the pure epoxy coating and 10 wt% TA-MSNs coating near the scratch was severe, and the width of the corrosion area reached to 400 and 500 μm, respectively. The invasion of the corrosion medium was evident, and a large amount of corrosion products was observed at the scratch. In contrast, the peeling of the 2.5 and 7.5 wt% TA-MSNs coatings near the scratch was effectively inhibited, and the

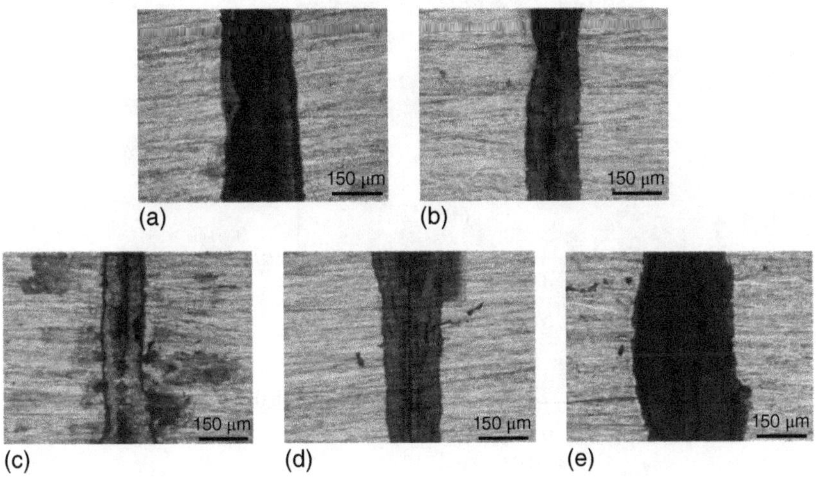

Figure 8.36 Surface morphology of damaged coating with different TA-MSNs filler contents after immersion for 28 days. (a) Epoxy coating, (b) 2.5 wt% TA-MSNs coating, (c) 5 wt% TA-MSNs coating, (d) 7.5 wt% TA-MSNs coating, (e) 10 wt% TA-MSNs coating.

amount of corrosion products in the scratch area was far smaller than those for the pure epoxy coatings. The 5 wt% TA-MSNs coating had a lower degree of peeling in the scratch area (<200 μm), and almost no corrosion products were formed inside the scratch and in the surrounding area.

8.2.4.4 Aging Resistance of Coatings

The aging resistance of the 5 wt% TA-MSNs coating was evaluated via an accelerated aging test in comparison with the blank epoxy resin and 5 wt% MSNs coating. Figure 8.37a shows macroscopic images of the different coatings before and after aging. All the coatings exhibited a smooth and clean surface morphology before the aging test. Compared with the blank epoxy and 5 wt% MSNs coatings, the original 5 wt% TA-MSNs coating appeared slightly yellow owing to the natural color of TA. With the extension of the aging time, obvious white hydrophilic and low molecular weight aging products appeared on the blank epoxy resin and the 5 wt% MSNs coating. In addition, the blank epoxy coating turned yellow, which is attributed to the formation of a chromogenic group due to the degradation of the epoxy resin. In contrast, the 5 wt% TA-MSNs coating not only exhibited little change in surface morphology but also had no change in the coating color appearance. The microscale morphologies and surface roughness values (S_a) of the different weathered coatings were analyzed using CLSM. As shown in Figure 8.37b$_1$, a rough and uniformly distributed bulge was observed on the blank epoxy coating with an S_a value of 0.449 μm, which may have been due to aging products and microscale defects generated during the degradation of the epoxy matrix. The aging products of the 5 wt% MSNs coating in Figure 8.37b$_2$ were only locally produced, and the S_a value (0.329 μm) was slightly lower than that of the blank epoxy resin. In contrast, the 5 wt% TA-MSNs coating had

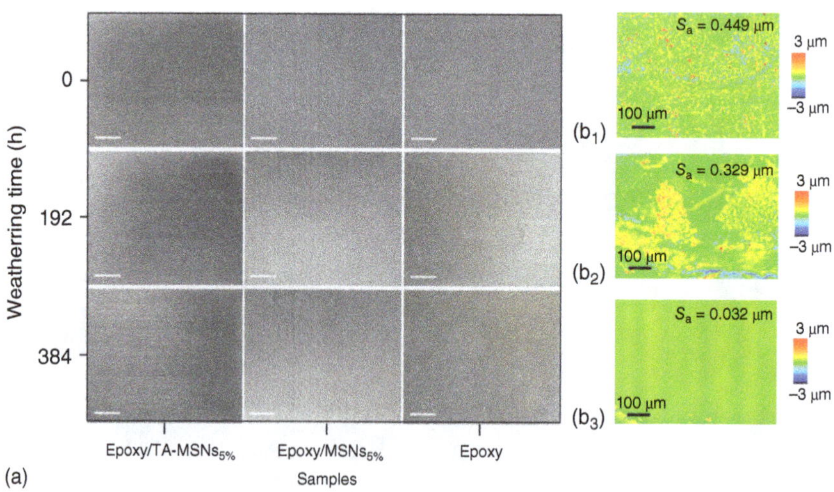

Figure 8.37 Digital pictures of different coatings during aging test (scale bar: 5 mm); the surface morphology and S_a of (b$_1$) epoxy coating, (b$_2$) 5 wt% MSNs coating, and (b$_3$) 5 wt% TA-MSNs coating after 384 hours of aging test.

a relatively smooth surface with no aging products (Figure 8.37b$_3$), and the S_a value was as low as 0.032 µm [90–93].

8.3 Aging-Resistant Anticorrosion Coatings Based on TP Additives

In this section, a kind of aging-resistant anticorrosion based on antioxidation nanofillers is prepared to cope with the aging of coating in complex service environments. TP with free radical scavenging ability was encapsulated in MSNs, and the TP-MSNs nanoparticles were dispersed in epoxy coating [94, 95]. First, properties of different antioxidants were investigated, and TP molecules with the best free radical scavenging ability were selected to prepare nanofillers. The morphology and structure of the TP-MSNs nanofillers were observed by SEM, TEM, FTIR. Thermogravimetric analysis was used to determine the drug-loading rate of TP in TP-MSNs nanofillers. The prepared TP-MSNs were dispersed into the epoxy coating to prepare the composite coating with aging resistance. The effect of the amount of nanofillers on the physical barrier properties of the coating was studied by EIS measurement. WCA and EIS tests were used to study the effect of nanofiller content on the aging-resistant properties of the coatings, and the positive role of TP antioxidants in improving the aging resistance of coatings was illustrated.

8.3.1 Materials and Methods

8.3.1.1 Materials

The materials and reagents used in this section are shown in Table 8.9.

8.3.1.2 Instruments

The instruments used in this section are shown in Table 8.10.

8.3.1.3 Selection of Antioxidant

Plant polyphenols are excellent antioxidants, which can effectively trap and remove free radicals. In this study, plant polyphenols were selected as antioxidant substances, which were loaded with nanocapsules and placed in epoxy coatings, to improve the aging resistance of organic coatings and extend their service life.

Table 8.9 Experimental materials and reagents.

Materials	Purity	Company
Tea polyphenols ($C_{17}H_{19}N_3O$)	Analytically pure	Source leaf organisms
Anhydrous ethanol (C_2H_5OH)	Analytically pure	Sinopharm Group
Tetraethyl silicate (TEOS)	Analytically pure	Aladdin Inc.
Hydrochloric acid (HCl)	Analytically pure	Sinopharm Group

Table 8.10 The instruments and equipment.

Instrument	Production type	Manufacturer
SEM	JEM-2100F	Japan JEOL
FTIR	Vector 22	Bruker, Germany
UV–Vis spectrometer	U-3900/3900H	Hitachi, Japan
Electronic balance	ME204	Switzerland Mettler Toledo
Electrochemical workstation	PARSTAT-2273	Princeton, USA
Constant temperature blast drying oven	303-3S	China Baihui
Xenon lamp aging and aging test chamber	SUN-II	Shanghai Yihang, China

First, the antioxidant properties of four polyphenols, TA, TP, chlorogenic acid, and caffeic acid, were compared. Different concentrations of these phenols, 10, 5, 2.5, and 1.25 ppm, were prepared; and 450 μl of the above solutions were extracted with a pipette gun and placed on the 48-well plate. Each solution was provided with three parallel samples, and then 450 μl of DPPH reagent (including 0.04 mg ml^{-1} of DPPH ethanol solution) was added and covered with tin foil to observe the color change after 30 minutes.

8.3.1.4 Preparation of TP-MSNs

The TP-MSNs nanostructures were prepared by hydrolysis and condensation of TEOS under alkaline conditions. TEOS was selected as the silicon source material of mesoporous microcapsules, and TP was used as both the loading material and the pore-causing agent of mesoporous microcapsules. First, 1 g of TP was dispersed into 100 ml of anhydrous ethanol, and then 50 ml of concentrated ammonia was injected and poured into a flask under magnetic stirring at 550 rpm for one minute. Next, 0.6 ml TEOS was added to the flask and stirred for another two hours (covered with plastic wrap). Finally, the reaction suspension was centrifuged at 6000 rpm for two minutes, washed twice with DI water, and dried in a freeze-dryer for 24 hours. Finally, the TP-MSNs antioxidant nanofiller was obtained. As a comparison, the control group was prepared by adjusting the additional amount of TP to 0.5 and 1.5 g, and the preparation of TP-MSNs was the same. These three samples are denoted as TP-MSN0.5, TP-MSN1.0, and TP-MSN1.5, respectively.

8.3.1.5 Characterization of TP-MSNs

Using FTIR, the infrared spectrum of TP-MSNs was measured, and the chemical composition of the sample was characterized. UV–Vis spectrometer was used to test the absorption spectra of different samples and analyze the scavenging efficiency of DPPH free radicals in samples with different antioxidant dosages. The antioxidant properties of TP-MSN0.5, TP-MSN1.0, and TP-MSN1.5 samples were characterized by DPPH free radical scavenging experiment [96, 97]. A certain amount of DPPH was weighed and solution of 0.04 g l^{-1} was prepared with anhydrous ethanol. 2 ml of TP-MSN0.5, TP-MSN1.0, and TP-MSN1.5 solutions with different concentrations (12.5, 25, 50, 100, 150, 200 ppm) were mixed with 2 ml of DPPH solution. The UV–Vis absorption intensity of the mixture was measured at 517 nm. SEM, TEM, and EDS were used to observe the microstructure of TP-MSN.

8.3.1.6 Preparation of TP-MSNs-Based Anti-aging Coating

Q235 steel substrates were polished with No. 150 and No. 240 sandpapers until they showed metallic luster. 1, 2, and 3 wt% TP-MSNs were loaded into the epoxy coating. After curing at room temperature for 12 hours, and then curing in the oven at 55 °C for 24 hours, the composite coatings were prepared. Epoxy coatings without nanofillers were prepared using the same method. The thicknesses of all coatings were controlled at ~80 μm.

8.3.1.7 Characterization of TP-MSNs-Based Anti-aging Coating

Immersion experiments were conducted on intact coatings with different contents of TP-MSNs, and EIS was conducted to compare the protective effect and failure behavior of different coatings so as to analyze the key role and synergistic effect of TP and MSNs on the corrosion and aging resistances of coatings. For the UV aging test, xenon lamp aging-resistance test chamber was used to accelerate UV aging of coatings with different nanofiller additions. The total accelerated aging time was 384 hours. The irradiation wavelength was 300–400 nm, the average irradiation intensity of the lamp was 65 W cm^{-2}, and the irradiation time was two hours (102 minutes of UV illumination and 18 minutes of UV illumination + salt spraying), followed by darkness for 2 hours in each cycle. The cyclic irradiation lasted 96 times. At different predetermined times, the sample was taken out, and the variation of coating properties was recorded. The UV-aging resistance of coatings was evaluated by measuring the WCAs, surface morphology, and electrochemical data of coatings [98, 99].

8.3.2 Preparation and Oxidation Resistance of TP-MSNs

8.3.2.1 Oxidation Resistance of Different Polyphenols

DPPH free radical scavenging experiment and colorimetric analysis were carried out for different concentrations of TA, chlorogenic acid, caffeic acid, and TPs. Different substances were prepared in same concentrations. Figure 8.38 illustrates four kinds of polyphenols with a certain free radical scavenging ability, and this effect gradually improved with the increase in polyphenol concentrations. When the free radicals were scavenged, the color of DPPH changed from purple to yellow. Among these substances, at the lowest concentration of 1.25 ppm, the purple color of DPPH was the lightest when mixed with TP, and at the concentration of 2.5 ppm, the solution of DPPH-TP had a trend of turning yellow. These results showed that the TP possessed the best performance for free radical scavenging.

To further verify the best free radical scavenging ability of TP, UV–Vis spectra were applied to monitor the absorbance intensity changes at 400–800 nm, as shown in Figure 8.39. According to the UV–Vis peak of DPPH at 521 nm, the absorption intensity decreased most significantly when interacting with TP (see Figure 8.39b). It

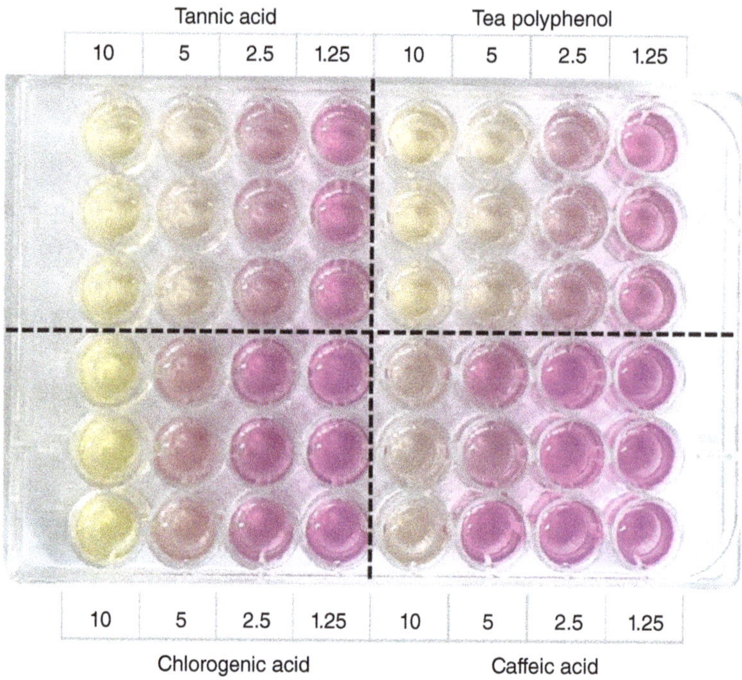

Figure 8.38 DPPH free radical scavenging ability of polyphenol at different concentrations.

demonstrates that TP molecules had the strongest free radical scavenging ability and best antioxidant ability and were selected as the antioxidants to prepare nanofillers in the following sections.

8.3.2.2 Morphology and Oxidation Resistance of TP-MSNs

The morphologies of TP-MSNs were observed by SEM and TEM. As shown in Figure 8.40, TP-MSN0.5 and TP-MSN1.0 had uniform monodisperse morphology with diameters ranging from 300 to 400 nm, whereas TP-MSN1.5 had uneven surface roughness with different sizes. According to the TEM images, the pores on TP-MSN0.5 were almost invisible, the pores on TP-MSN1.0 were large and clearly visible, and the pores on TP-MSN1.5 were too large that the structure was almost destroyed, leading to premature leakage of TP during the preparation process.

To further determine the loading property of TP in TP-MSNs, TP-MSN0.5, TP-MSN1.0, TP-MSN1.5, and TP molecules are qualitatively characterized by FTIR. Figure 8.41 shows that the characteristic FTIR peaks of TP-MSN0.5, TP-MSN1.0, and TP-MSN1.5 were located at 790 and 1080 cm^{-1}, corresponding to the Si—O—Si symmetric stretching and Si—O—Si asymmetric vibration, respectively. It proved the formation of mesoporous silica structure. The FTIR peaks at 1613, 1530, and 1453 cm^{-1} were caused by the C=C skeleton vibration in the benzene torus of TP, which proved that TP molecules were successfully loaded into mesoporous silica.

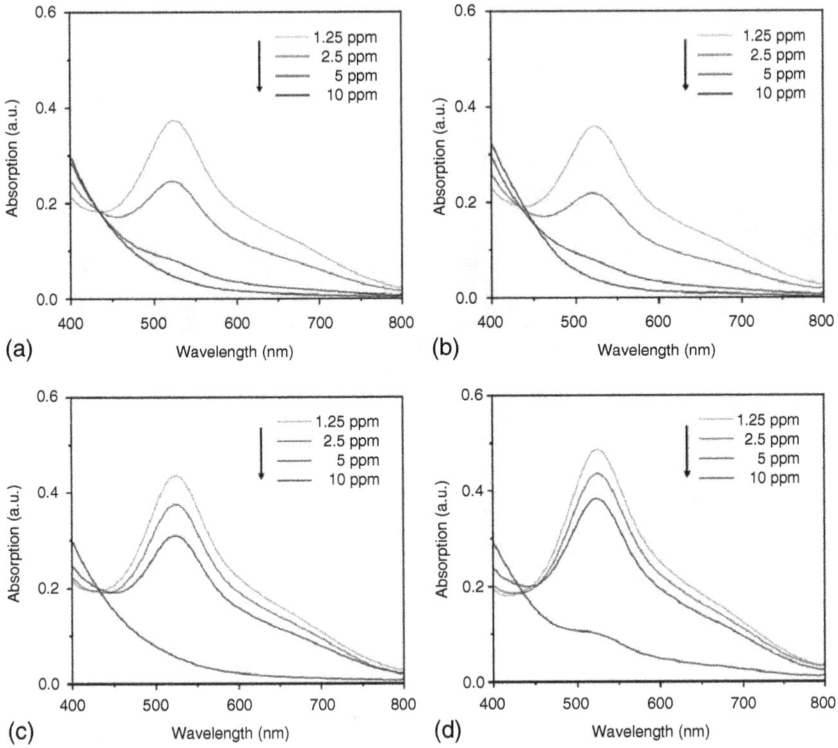

Figure 8.39 UV–Vis spectra of DPPH when mixing with different concentration of (a) TA, (b) TP, (c) chlorogenic acid, and (d) caffeic acid.

In addition, the characteristic absorption peak of TP in TP-MSN0.5 sample was weak, indicating that the loading rate of TP in the TP-MSN0.5 sample was low.

The DPPH free radical scavenging ability of different concentrations of TP-MSN0.5, TP-MSN1.0, and TP-MSN1.5 was analyzed by colorimetric analysis. As shown in Figure 8.42, under the same concentration, TP-MSN1.0 had the strongest antioxidant capacity, whereas TP-MSN0.5 had the weakest antioxidant capacity. For the same sample under different concentrations, with the increase of concentrations of TP-MSN0.5, the color of DPPH changed from dark purple to light purple, indicating a relative weak free radical scavenging ability. As for TP-MSN1.0 and TP-MSN1.5, the color of DPPH changed from dark purple to yellow with the increase of TP-MSNs concentration.

To further compare the antioxidant capacity of the three TP-MSNs samples, the scavenging efficiencies of free radicals are plotted in Figure 8.43 [100, 101]. It shows that the free radical scavenging rate of the three samples increased with the increase of TP-MSNs concentration, and the free radical scavenging ability was in the following order: TP-MSN1.0 > TP-MSN1.5 > TP-MSN0.5. When the concentration of TP-MSNs was between 0 and 25 ppm, the free radical scavenging rate increased rapidly, and then the scavenging activity became weaker with the further increase

Figure 8.40 SEM images of (a_1) TP-MSN0.5, (b_1) TP-MSN1.0, (c_1) TP-MSN1.5; TEM images of (a_2) TP-MSN0.5, (b_2) TP-MSN1.0, and (c_2) TP-MSN1.5.

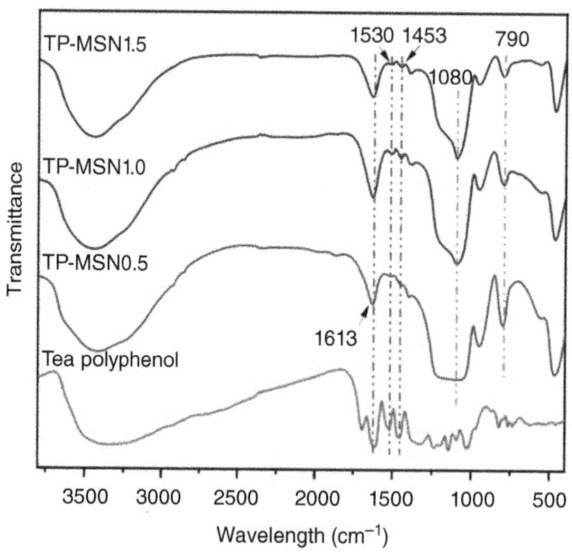

Figure 8.41 FTIR spectra of TP-MSNs, TP-MSN0.5, TP-MSN1.0, TP-MSN1.5, and TP [100].

Figure 8.42 Colorimetric photos of DPPH free radical scavenging mixed with TP-MSN0.5, TP-MSN1.0, and TP-MSN1.5 at different concentrations [100].

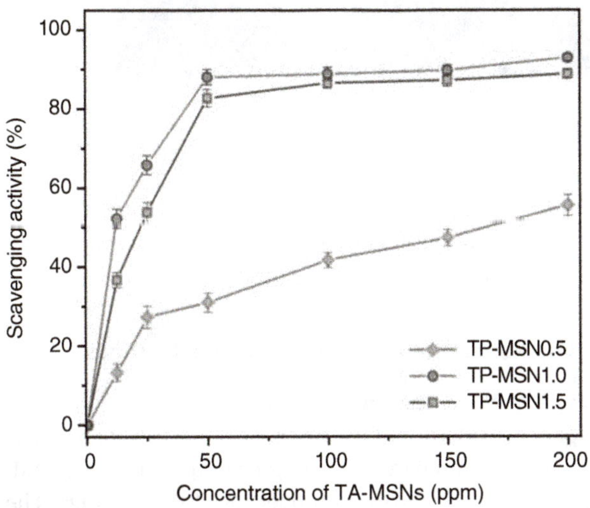

Figure 8.43 Scavenging activity of TP-MSN0.5, TP-MSN1.0, and TP-MSN1.5 at different concentrations [100].

of TP-MSNs concentration. The maximum scavenging activity reached more than 90% when the TP-MSN1.0 concentration was 100 ppm or higher.

Through the above tests, it is concluded that TP-MSN1.0 has the best antioxidant capacity. EDS spectrum analysis was performed on the TP-MSN1.0 sample, as shown in Figure 8.44. The distribution of Si and O corresponded to the mesoporous

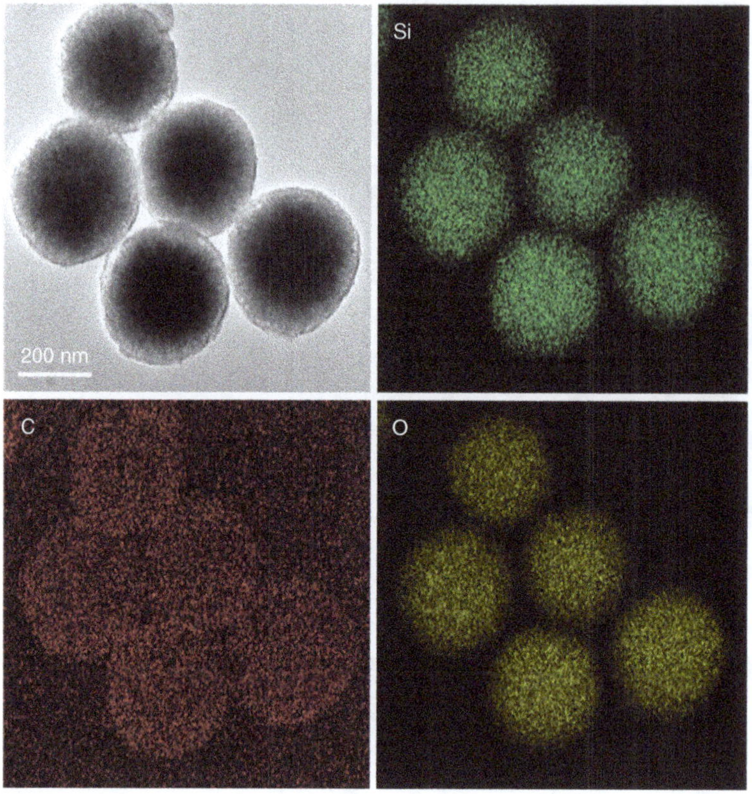

Figure 8.44 TEM image and EDS spectra of TP-MSN1.0.

SiO_2 structure, and the distribution of C indicated that TP molecules were successfully and uniformly coated into TP-MSN s. In summary, it can be concluded that TP-MSN1.0 has the best antioxidant performance, and it is selected for the subsequent preparation and characterization of anti-aging coatings.

8.3.3 Preparation and Aging Resistance of TP-MSNs-Loaded Coatings

8.3.3.1 Corrosion Resistance of TP-MSNs-Loaded Coatings

After characterizing the oxidation resistance of the prepared TP-MSNs nanofillers, they were mixed with epoxy resin to prepare anti-aging coatings. First, we studied the effect of filler amount on the anticorrosion properties of the coatings. The coatings with different filler contents were immersed in 3.5 wt% NaCl solution for different durations. Figure 8.45 shows the Nyquist diagrams for different coatings after immersion. At the initial stage, the arc radius of capacitive reactance of coatings with 2 wt% filler content was the largest, followed by coatings with 1 wt% filler content and 3 wt% filler content, and the arc radius of capacitive reactance of the blank epoxy coating was the smallest. The best corrosion resistance property of the coating containing 2 wt% TP-MSNs nanofillers can be explained by the fact that addition of the fillers in proper amounts can fill the micropores in the epoxy resin, reduce the

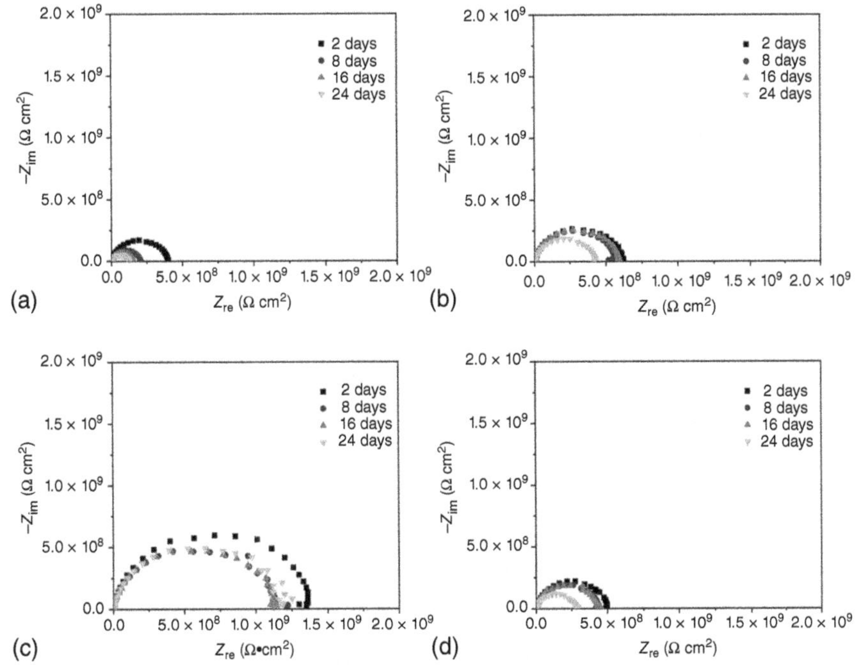

Figure 8.45 Nyquist diagrams for different coatings after immersion in 3.5 wt% NaCl solution for different period: (a) blank epoxy coating, (b) epoxy coating containing 1 wt% TP-MSNs, (c) epoxy coating containing 2 wt% TP-MSNs, (d) epoxy coating containing 3 wt% TP-MSNs.

intrusion path of the corrosive media, and thus slowing the intrusion of corrosive substances. The arc radii of capacitive reactance of the coatings with 1 and 3 wt% TP-MSNs nanofillers were smaller compared with the coating containing 2 wt% TP-MSNs. The addition of insufficient or excessive nanofillers may affect the barrier properties of the coating. When the filler content is insufficient, it cannot fully compensate for the gaps and defects in the coating, whereas when the nanofiller is excessive, agglomeration phenomenon happens in the coating, making the polymer matrix with more defects that provide more channels for the diffusion of corrosive media [102, 103].

Figure 8.46 shows the Bode diagram of the above coatings. The impedance diagrams of all coatings show capacitance characteristics in the high-frequency region and reach a plateau in the low-frequency region. With the extension of immersion time in 3.5 wt% NaCl solution, the frequency range occupied by the impedance behavior gradually became wider for the blank epoxy coating sample, and the impedance modulus decreased gradually, which proves that the protective ability of coatings was gradually decreased. Within 24 days of immersion, the impedance diagram of the coating with 2 wt% TP-MSNs nanofillers remained stable, showing excellent protection performance. The protective properties of coatings with 1 and 3 wt% nanofillers decreased to a certain degree, but their barrier properties were still better than the blank epoxy coating. The variation of the $|Z|_{0.01\,\text{Hz}}$ values of different

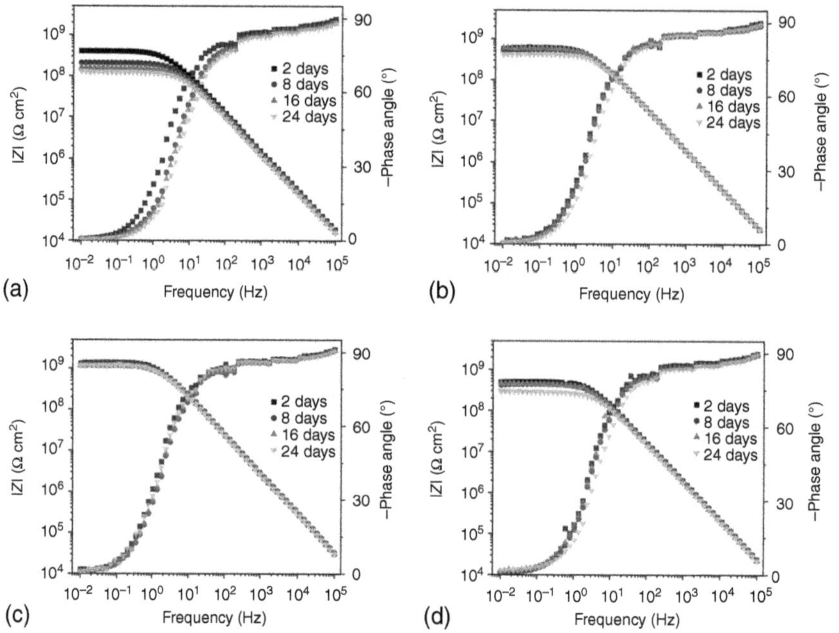

Figure 8.46 Bode plots for different coatings after immersion in 3.5 wt% NaCl solution for different periods: (a) blank epoxy coating, (b) epoxy coating containing 1 wt% TP-MSNs, (c) epoxy coating containing 2 wt% TP-MSNs, (d) epoxy coating containing 3 wt% TP-MSNs.

coatings is analyzed. After immersion for 2 days, the $|Z|_{0.01\,Hz}$ value of blank epoxy coating was $4.0 \times 108\,\Omega\,cm^2$ and declined further with the increase of immersion time. After 24 days, the $|Z|_{0.01\,Hz}$ value was as low as $1.3 \times 10^8\,\Omega\,cm^2$. The $|Z|_{0.01\,Hz}$ values of coatings with 1 and 3 wt% of TP-MSNs were slightly higher than that of blank epoxy coating after 2 days, which were $5.7 \times 10^8\,\Omega\,cm^2$ and $4.9 \times 10^8\,\Omega\,cm^2$, respectively. The $|Z|_{0.01\,Hz}$ values decreased with the prolongation of corrosion time, which were 4.2×10^8 and $2.9 \times 10^8\,\Omega\,cm^2$ after 24 days of immersion. In contrast, the $|Z|_{0.01\,Hz}$ values of the coating with 2 wt% TP-MSNs nanofillers were higher than $10^9\,\Omega\,cm^2$, and changed little during the whole immersion process, which again indicates that the coating with 2 wt% nanofillers exhibited the best corrosion-resistant property and maintain a stable anticorrosion effect in a certain period.

8.3.3.2 Aging Resistance of TP-MSNs-Loaded Coatings

Based on the anticorrosion properties of the coatings with different contents of TP-MSNs, the UV-aging-resistant property was studied on the coating with 2 wt% of nanofillers. The blank epoxy coating and the coating with 2% MSNs (without TP) were used as the control samples, and the UV-aging accelerated test was carried out by measuring the WCAs and characterizing the surface morphology on the coating surfaces. The changes of WCAs on coatings after 192 and 384 hours of UV irradiation are shown in Figure 8.47. Before UV irradiation, the WCAs of the three coatings were similar at approximately 67°. After 384 hours of UV

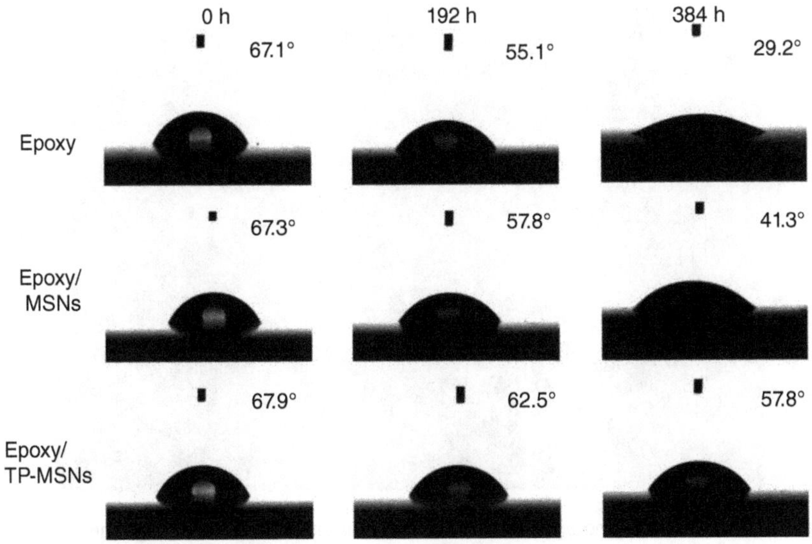

Figure 8.47 WCAs of three coatings after different UV aging time.

aging, the WCAs decreased to 29.2°, 42.5°, and 56.8° for blank epoxy coating, 2% MSNs containing coating, and 2% TP-MSNs containing coating, respectively. The decline trend was the largest for the blank epoxy coating and the slightest for the 2 wt% TP-MSNs-based coating. The results show that the coating containing 2 wt% TP-MSNs had the most stable aging resistance and effectively inhibit the harmful effects of UV radiation on the coating aging performance. In addition, by comparing with the result of epoxy/MSNs sample, the effect of MSN on the aging resistance was excluded, indicating that TP molecules played a major role in the aging-resistant property [104, 105].

The surface morphologies of the three coatings (epoxy, epoxy/MSNs, epoxy/TP-MSNs) were observed to reflect the aging degree of different coatings. From Figure 8.48, it shows that when the UV aging time reached 192 hours, the surface of the blank epoxy coating illustrated obvious yellow degeneration, which was due to the destruction of the cross-linked network of the coating by UV irradiation and the promotion of the generation of carbonyl chromophore. After 384 hours, the aging behavior became more serious, and the surface of the coating was silted and denuded. The exopy/MSNs coating had no yellowing phenomenon after 192 hours, and the aging phenomenon was more serious with a powdery denudation phenomenon after 384 hours. For the epoxy/TP-MSNs, neither yellowing nor powder denudation was observed on the coating after 384 hours of UV irradiation. This shows that the addition of antioxidation nanofillers can significantly improve the aging-resistant property of the coating.

EIS measurement was conducted on the three coatings after 384 hours of UV irradiation. As shown in Figure 8.49, the epoxy/TP-MSNs coating had the largest capacitive reactance arc, followed by epoxy/MSNs and epoxy coating. The $|Z|_{0.01\,\text{Hz}}$ values were 8.9×10^7, 3.5×10^7, and $6.4\times10^6\ \Omega\,\text{cm}^2$ for epoxy/TP-MSNs,

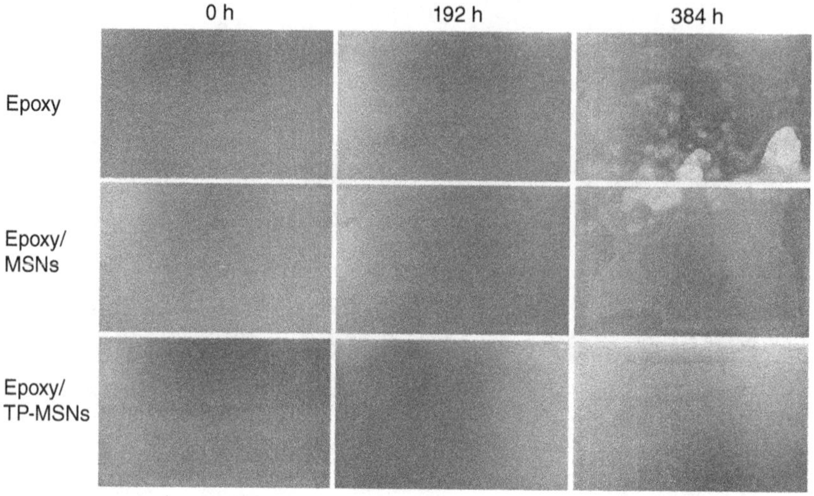

Figure 8.48 Surface morphologies of three coatings after different UV aging time.

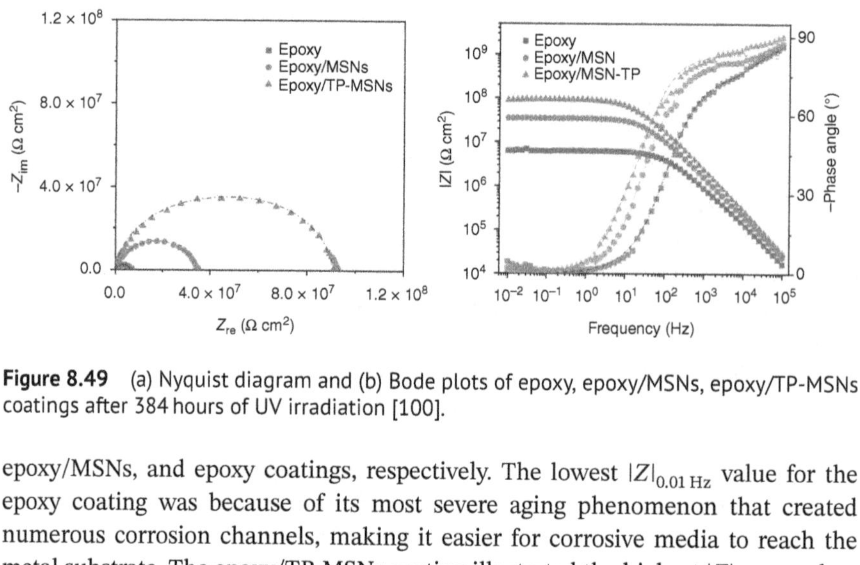

Figure 8.49 (a) Nyquist diagram and (b) Bode plots of epoxy, epoxy/MSNs, epoxy/TP-MSNs coatings after 384 hours of UV irradiation [100].

epoxy/MSNs, and epoxy coatings, respectively. The lowest $|Z|_{0.01\ \mathrm{Hz}}$ value for the epoxy coating was because of its most severe aging phenomenon that created numerous corrosion channels, making it easier for corrosive media to reach the metal substrate. The epoxy/TP-MSNs coating illustrated the highest $|Z|_{0.01\ \mathrm{Hz}}$ value, which was an order of magnitude higher than the blank epoxy coating and twice higher than the epoxy/MSN coating. The excellent antioxidant effect of TP plays a significant role in the aging resistance of the composite coating.

Under the synergistic effect of environmental factors such as UV irradiation, moisture, and oxygen, the surface of organic protective coatings undergoes significant degradation. This process not only leads to the generation of small soluble products but also results in the formation of hydrophilic groups. The increased hydrophilicity of the coating surface is indicative of poor UV aging resistance. The presence of these hydrophilic groups in the coating enhances its water absorption capabilities. Consequently, this can lead to the formation of permeable microvoids, contributing to defects such as pores and cracks in the coating structure. Moreover, in environments that alternate between dry and wet conditions, these defects can manifest as surface foaming on the coatings. Similarly, when exposed to cycles of light and darkness, these environmental stresses can cause fluctuations in the surface tension of the coating. Such variations in tension can result in the coating cracks, leading to a significant decrease in its aging resistance. Understanding and addressing these vulnerabilities are critical for extending the life span and maintaining the integrity of protective coatings in various applications.

8.4 Chapter Summary

TA-MSNs were prepared via a one-step synthesis method using TA molecules as the pore-generation material. The TA-MSNs had a near-spherical morphology, with a diameter of approximately 200 nm and uniform size, and the rich mesoporous structure of the TA-MSNs provided nanofillers with a high TA loading rate of approximately 21%. The release of TA from TA-MSNs was investigated via UV–Vis spectroscopy under different pH environments. The release of TA from the TA-MSNs nanofillers was affected by the pH of the solutions, and the fastest release was observed in acidic environments. In addition, the color-development performance of TA-MSNs was investigated. They exhibited excellent color response speed and color development sensitivity, which is attributed to the high TA loading and release ability of the nanofillers, proving that they can be used as an indicator of corrosion on carbon steel.

The corrosion inhibition performance of the TA-MSNs on the steel substrates was tested using EIS. TA molecules loaded with nanofillers can complex with Fe^{3+} ions on the steel surface and be adsorbed onto the sample surface to form a protective film, thereby inhibiting the corrosion reaction on the steel surface and improving the corrosion resistance of the steel. The prepared TA-MSNs nanofillers exhibited excellent coloration and corrosion inhibition properties, laying the foundation for the next step in preparing TA-MSNs-based self-reporting and self-healing functional coatings. Moreover, the TA-MSNs nanofillers exhibited excellent DPPH free radical scavenging activity, which is ascribed to the abundant o-phenylphenol groups of TA. The abundant o-phenylphenol groups of the TA molecules can donate hydrogen atoms to the free radicals of the coatings generated under aging conditions, stopping the free radical chain reactions in the polymer resin [106–109].

After the TA-MSNs nanofillers were incorporated into the epoxy coatings, the self-reporting, barrier, self-healing, and aging-resistant properties of the coatings with different contents of TA-MSNs were investigated. According to the salt spray experiment, 2.5 wt% TA-MSNs fillers endowed the coating with obvious coloration, and the self-reporting effect of the coating was positively correlated with the filler content of the coating. This can be explained by the fact that with an increase in the amount of TA-MSNs nanofiller, the density of TA-MSNs per unit area increased, providing more TA molecules to react with Fe^{3+} ions, and the early self-reporting effect of the coating was improved. The barrier properties of the intact coatings were evaluated. When the amount of TA-MSNs nanofillers was 2.5, 5, and 7.5 wt%, the barrier performance of the coating improved with the increasing nanofiller content. However, when the nanofiller content was 10 wt%, the barrier performance of the coating was degraded because the agglomeration caused by excessive filler addition caused more defects and pores in the coating. According to the results of the electrochemical tests and surface analyses, it was concluded that the coating exhibited the best barrier properties and self-healing performance when the filler content was 5 wt%. After immersion in 3.5 wt% NaCl solution for 28 days, the low-frequency impedance value of the intact 5 wt% TA-MSNs coating was $4.0\times10^8\ \Omega\,cm^2$, which was approximately three times that of the pure epoxy coating without nanofillers. Regarding the self-healing performance, after 28 days of immersion, owing to the gradual release of TA corrosion inhibitors from the coating, the $|Z|_{0.01\ Hz}$ value of the damaged-then-healed 5 wt% TA-MSNs coating was $1.3 \times 10^6\ \Omega\,cm^2$, which was 2 orders of magnitude higher than that of the damaged pure epoxy coating. In summary, the 5 wt% TA-MSNs coating had both an apparent self-reporting and good self-healing effects. The TA molecules in the TA-MSNs complexed with Fe^{3+} ions generated by the steel substrate to form a protective film on the metal surface of the defect, which inhibited the intensification of the corrosion reaction. Finally, regarding aging-resistant behavior, the introduction of TA-MSNs can mitigate the degradation of coatings by removing free radicals generated by UV irradiation. Compared with the blank coating, the coating containing 5 wt% TA-MSNs nanofillers exhibited far less degradation in the surface morphology, wettability, and gloss and maintained a better barrier property after 384 hours of the accelerated aging test [110, 111].

Natural polyphenols, derived from plants, offer a sustainable and eco-friendly alternative to synthetic chemicals in coatings. Their use aligns with the global trend toward green chemistry and the reduction of environmental footprints. The incorporation of natural polyphenols into anti-aging coatings represents a fascinating and growing field of research, blending the realms of natural product chemistry and material science. In addition to TA and TP, other types of polyphenol substances, such as theaflavin, caffeic acid, gallic acid, and ellagic acid, have considerable promise as nanofillers in smart coatings for achieving self-reporting, self-healing, and aging resistance. The corrosion inhibition properties of different types of polyphenol substances on different types of metal substrates, such as carbon steel, copper, aluminum alloys, and magnesium alloys, should be investigated, and the corrosion inhibition mechanism should be clarified. In addition, the complex behavior of polyphenol substances with different types of metal ions must be characterized so that self-reporting can be achieved on specific metal surfaces. Aging-resistant properties are important in practical outdoor applications. An efficient strategy to improve the aging resistance of organic coatings is the introduction of radical scavengers, such as hindered amines and hindered phenols. However, these synthetic radical scavengers are potentially harmful to human health and the environment; therefore, the development of eco-friendly radical scavengers is of considerable research significance. Recently, various eco-friendly radical scavengers have been introduced, including graphene, carbon dots, $Ti_3C_2T_x$, MXenes, and a series of bio-based additives such as chitosan, phytic acid, and plant polyphenols.

The free radical removal abilities of different polyphenol substances should be compared, and the mechanism of aging resistance can provide more options for optimization. In conclusion, eco-friendly materials have tremendous potential for application as anticorrosive coatings. These substances should be further developed, and their unique properties should be exploited to achieve an intelligent response to the external environment and long-term corrosion protection effects. Furthermore, improving the methods for testing and predicting the performance of coatings is crucial. The use of advanced simulation tools and accelerated aging tests helps in the faster development of more effective coatings. To make the use of polyphenols in coatings more feasible, advances in the extraction of these compounds from natural sources, as well as their synthesis, are necessary. Efficient and cost-effective methods for obtaining high-purity polyphenols will be crucial. In summary, the use of natural polyphenols in anti-aging coatings is a promising area of development, with a focus on sustainability, enhanced protective properties, and the potential for innovative functionalities such as self-healing. This field is likely to see significant growth as research continues to unlock the potential of these natural compounds in advanced coating applications.

References

1 Hou, C., Wang, Q., Song, W. et al. (2023). Chlorine resistance of tannic acid anticorrosive coating on galvanized steel in simulated concrete pore solution. *Construction and Building Materials* 401: 132657. https://doi.org/10.1016/j.conbuildmat.2023.132657.

2 Zhang, J., Zu, F., Song, W. et al. (2013). Corrosion inhibition mechanism of imidazoline-based dissymmetric bis-quaternary ammonium salts with different hydrophobic chain length on Q235 steel in 1 m HCl solution. *Journal of Surfactants and Detergents* 16 (4): 559–569. https://doi.org/10.1007/s11743-012-1390-8.

3 Qian, B., Hou, B., and Zheng, M. (2013). The inhibition effect of tannic acid on mild steel corrosion in seawater wet/dry cyclic conditions. *Corrosion Science* 72: 1–9. https://doi.org/10.1016/j.corsci.2013.01.040.

4 Zhang, Z.G. and Ma, Q.L. (2017). Application of tannic acid compound corrosion inhibitor in iron cultural relics. *Surface Technology* 46 (2): 27–32. https://doi.org/10.16490/j.cnki.issn.1001-3660.2017.02.004.

5 Xu, W., Han, E., and Wang, Z. (2019). Effect of tannic acid on corrosion behavior of carbon steel in nacl solution. *Journal of Materials Science and Technology* 35 (1): 64–75. https://doi.org/10.1016/j.jmst.2018.09.001.

6 Dong, Z.H., Shi, W., Zhang, G.A. et al. (2011). The role of inhibitors on the repassivation of pitting corrosion of carbon steel in synthetic carbonated concrete pore solution. *Electrochimica Acta* 56 (17): 5890–5897. https://doi.org/10.1016/j.electacta.2011.04.120.

7 Kusmierek, E. and Chrzescijanska, E. (2015). Tannic acid as corrosion inhibitor for metals and alloys. *Materials and Corrosion* 66 (2): 169–174. https://doi.org/10.1002/maco.201307277.

8 Chang, J., Wang, Z., Han, E. et al. (2021). Corrosion resistance of tannic acid, D-limonene and nano-ZrO_2 modified epoxy coatings in acid corrosion environments. *Journal of Materials Science and Technology* 65: 137–150. https://doi.org/10.1016/j.jmst.2020.03.081.

9 Kaghazchi, L., Naderi, R., Ramezanzadeh, B., and B. (2020). Construction of a high-performance anti-corrosion film based on the green tannic acid molecules and zinc cations on steel: electrochemical/surface investigations. *Construction and Building Materials* 262: 120861. https://doi.org/10.1016/j.conbuildmat.2020.120861.

10 Matsumoto, A., Takashima, T., and Murakami, M. (2023). Nanoparticle-releasing xerogelfilms prepared by a one-step method. *Results in Materials* 20: 100467. https://doi.org/10.1016/j.rinma.2023.100467.

11 Verma, G. (2019). Aging, salt spray corrosion and mar resistance mechanism of clay (nano-platelet) reinforced polyurethane nanocomposite coatings. *Progress in Organic Coatings* 129: 260–270. https://doi.org/10.1016/j.porgcoat.2019.01.028.

12 Liu, H., Zhu, Z.J., Hu, J. et al. (2022). Inhibition of Q235 corrosion in sodium chloride solution by chitosan derivative and its synergistic effect with zno. *Carbohydrate Polymers* 296: 119936. https://doi.org/10.1016/j.carbpol.2022.119936.

13 Zhang, Q.H., Li, Y.Y., Zhu, G.Y. et al. (2021). In-depth insight into the synergistic inhibition mechanism of s-benzyl-l-cysteine and thiourea on the corrosion of carbon steel in the CO_2-saturated oilfield produced water. *Corrosion Science* 192: 109807. https://doi.org/10.1016/j.corsci.2021.109807.

14 Anadebe, V.C., Chukwuike, V.I., Porsia, A.J. et al. (2023). Insight into the corrosion inhibition of Ce-MOF on copper in aqueous chloride environment: from experimental validation to molecular-level prediction. *Applied Surface Science Advances* 18: 100526. https://doi.org/10.1016/j.apsadv.2023.100526.

15 Vaghefinazari, B., Lamaka, S.V., Gazenbiller, E. et al. (2023). Corrosion inhibition of decylphosphonate on bare and PEO-coated Mg alloy. *Corrosion Science* 226: 111651. https://doi.org/10.1016/j.corsci.2023.111651.

16 Sriraman, K.R., Brahimi, S., Szpunar, J.A. et al. (2013). Characterization of corrosion resistance of electrodeposited Zn–Ni Zn and Cd coatings. *Electrochimica Acta* 105: 314–323. https://doi.org/10.1016/j.electacta.2013.05.010.

17 Zhao, S., You, Z., Zhang, X. et al. (2021). Magnetic field effects on the corrosion and electrochemical corrosion of $Fe_{83}Ga_{17}$ alloy. *Materials Characterization* 174: 110994. https://doi.org/10.1016/j.matchar.2021.110994.

18 Zhao, Y., Zhang, T., Xiong, H. et al. (2021). Bridge for the thermodynamics and kinetics of electrochemical corrosion: modeling on dissolution, ionization, diffusion and deposition in metal/solution interface. *Corrosion Science* 191: 109763. https://doi.org/10.1016/j.corsci.2021.109763.

19 Ibrahim, T.H., Chehade, Y., and Zour, M.A. (2011). Corrosion inhibition of mild steel using potato peel extract in 2M HCl solution. *International Journal of*

Electrochemical Science 6: 6542–6556. https://doi.org/10.1016/S1452-3981(23)19699-4.

20 Zhang, Z., Lu, S., Lv, W. et al. (2023). Enhanced corrosion resistance and bio-functionality of Zn-Al layered double hydroxide coating on micro-arc oxidized zk60 mg alloy viaion exchange. *Materials Chemistry and Physics* 299: 127482. https://doi.org/10.1016/j.matc-hemphys.2023.127482.

21 Zhang, Q., Shen, M., Liu, X. et al. (2023). Construction and utilization of rare earth complexes as efficient corrosion inhibitor in epoxy coating. *Progress in Organic Coatings* 186: 108024. https://doi.org/10.1016/j.porgcoat.2023.108024.

22 Zhang, G., Khanlari, K., Huang, S. et al. (2023). Dual-structured oxide coatings with enhanced wear and corrosion resistance prepared by plasma electrolytic oxidation on Ti-Nb-Ta-Zr-Hf high-entropy alloy. *Surface and Coatings Technology* 456: 129254. https://doi.org/10.1016/j.surfcoat.2023.129254.

23 Stratmann, M. (1990). The atmospheric corrosion of iron-a discussion of the physico-chemical fundamentals of this omnipresent corrosion process invited review. *Berichte der Bunsengesellschaft für Physikalische Chemie* 94 (6): 626–639. https://doi.org/10.1002/bbpc.19900940603.

24 Rahim, A.A., Rocca, E., Steinmetz, J. et al. (2008). Antioxidant activities of mangrove rhizophora apiculata bark extracts. *Food Chemistry* 107 (1): 200–207. https://doi.org/10.1016/j.foodchem.2007.08.005.

25 Obot, I.B. and Madhankumar, A. (2015). Enhanced corrosion inhibition effect of tannic acid in the presence of gallic acid at mild steel/HCl acid solution interface. *Journal of Industrial and Engineering Chemistry* 25: 105–111. https://doi.org/10.1016/j.jiec.2014.10.019.

26 Liu, Y., Mo, Y., Zhong, H. et al. (2023). Construction of Al-BTA passivation film on the surface of electrolytic copper foil and study of corrosion resistance mechanism. *Colloids and Surfaces A: Physicochemical and Engineering Aspects* 676: 132179. https://doi.org/10.1016/j.colsurfa.2023.132179.

27 Santos, L.R.L., Leal, D.A., Marino, C.E.B. et al. (2022). pH-triggered release of tannic acid as green corrosion inhibitor for carbon steel. *Materials Today Communications* 31: 103730. https://doi.org/10.1016/j.mtcomm.2022.103730.

28 Chen, S., Wang, S.C., Suo, Y. et al. (2022). Inhibition effect of tannic acid and sodium molybdate for the flow corrosion of 304 stainless steel on 90°elbow. *Journal of Materials Research and Technology* 20: 2408–2420. https://doi.org/10.1016/j.jmrt.2022.07.191.

29 Hao, Y., Sani, L.A., Ge, T. et al. (2017). The synergistic inhibition behaviour of tannic acid and iodide ions on mild steel in H_2SO_4 solutions. *Corrosion Science* 123: 158–169. https://doi.org/10.1016/j.corsci.2017.05.001.

30 Usman, B.J., Umoren, S.A., and Gasem, Z.M. (2017). Inhibition of API 5L X60 steel corrosion in CO_2-saturated 3.5% NaCl solution by tannic acid and synergistic effect of KI additive. *Journal of Molecular Liquids* 237: 146–156. https://doi.org/10.1016/j.molliq.2017.04.064.

31 Adamczyk, A. and Rokita, M. (2016). The structural studies of ag containing TiO_2–SiO_2 gels and thin films deposited on steel. *Journal of Molecular Structure* 1114: 171–180. https://doi.org/10.1016/j.molstruc.2016.02.054.

32 Rao, C.J., Ningshen, S., Mallika, C. et al. (2018). Molten salt corrosion behavior of structural materials in LiCl-KCl-UCl_3 by thermogravimetric study. *Journal of Nuclear Materials* 501: 189–199. https://doi.org/10.1016/j.jnucmat.2018.01.012.

33 Kourtidou, D., Chaliampalias, D., Vogiatzis, C. et al. (2019). Deposition of Ni-Al coatings by pack cementation and corrosion resistance in high temperature and marine environments. *Corrosion Science* 148: 12–23. https://doi.org/10.1016/j.corsci.2018.11.003.

34 Koszelow, D., Makowska, M., Marone, F. et al. (2021). High temperature corrosion evaluation and lifetime prediction of porous $Fe_{22}Cr$ stainless steel in air in temperature range 700–900 °C. *Corrosion Science* 189: 109589. https://doi.org/10.1016/j.corsci.2021.109589.

35 Galvão, T.L.P., Sousa, I., Wilhelm, M. et al. (2018). Improving the functionality and performance of AA2024 corrosion sensing coatings with nanocontainers. *Chemical Engineering Journal* 341: 526–538. https://doi.org/10.1016/j.cej.2018.02.061.

36 Li, J., Jiang, Z., Gan, L. et al. (2018). Functionalized graphene/polymer composite coatings for autonomous early-warning of steel corrosion. *Composites Communications* 9: 6–10. https://doi.org/10.1016/j.coco.2018.04.002.

37 Lee, T.H., Song, Y.K., Park, S.H. et al. (2018). Dual stimuli responsive self-reporting materi-al for chemical reservoir coating. *Applied Surface Science* 434: 1327–1335. https://doi.org/10.1016/j.apsusc.2017.11.219.

38 Tian, J., Yu, J., and Sun, X. (2008). Chitosan microspheres as candidate plasmid vaccine carrier for oral immunisation of Japanese flounder (Paralichthys olivaceus). *Veterinary Immunology and Immunopathology* 126 (3): 220–229. https://doi.org/10.1016/j.vetimm.2008.07.002.

39 Maia, F., Tedim, J., Bastos, A.C. et al. (2013). Nanocontainer-based corrosion sensing coating. *Nanotechnology* 24 (41): 415502. https://doi.org/10.1088/0957-4484/24/41/415502.

40 Mata, D., Scharnagl, N., Lamaka, S.V. et al. (2018). Validating the early corrosion sensing functionality in poly (ether imide) coatings for enhanced protection of magnesium alloy AZ31. *Corrosion Science* 140: 307–320. https://doi.org/10.1016/j.corsci.2018.05.034.

41 Li, W., Matthews, C.C., Yang, K. et al. (2016). Autonomous indication of mechanical damage in polymeric coatings. *Advanced Materials* 28 (11): 2189–2194. https://doi.org/10.1002/adma.201505214.

42 Zheng, X., Wang, Q., Li, Y. et al. (2020). Fabrication of self-reactive microcapsules as color visual sensing for damage reporting. *Journal of Materials Science* 55 (21): 8861–8867. https://doi.org/10.1007/s10853-020-04668-6.

43 Davis, D.A., Hamilton, A., Yang, J. et al. (2009). Force-induced activation of covalent bonds in mechanoresponsive polymeric materials. *Nature* 459 (7243): 68–72. https://doi.org/10.1038/nature07970.

44 Lu, X., Li, W., Sottos, N.R. et al. (2018). Autonomous damage detection in multilayered coatings via integrated aggregation-induced emission luminogens. *ACS Applied Materials & Interfaces* 10 (47): 40361–40365. https://doi.org/10.1021/acsami.8b16454.

45 Laleh, M., Pathiranam, M., and Tan, M.Y. (2023). Site-specific local polarisation curve measurements for probing localised corrosion and inhibition. *Corrosion Science* 214: 111019. https://doi.org/10.1016/j.corsci.2023.111019.

46 Lu, J.C., Wang, Z.B., Hu, H.X. et al. (2023). Understanding localized corrosion mechanism of 90/10 copper-nickel alloy in flowing NaCl solution induced by partial coverage of corrosion products films. *Corrosion Science* 227: 111716. https://doi.org/10.1016/j.corsci.2023.111716.

47 Mei, D., Li, Y., Tian, Y. et al. (2023). The effect of selected corrosion inhibitors on localized corrosion of magnesium alloy: the expanded understanding of "inhibition efficiency". *Corrosion Science* 226: 111650. https://doi.org/10.1016/j.corsci.2023.111650.

48 Zou, J., Guan, J., Wang, X. et al. (2023). Corrosion and wear resistance improvements in NiCu alloys through flame-grown honeycomb carbon and CVD of graphene coatings. *Surface and Coatings Technology* 473: 130040. https://doi.org/10.1016/j.surfcoat.2023.130040.

49 Zhang, J., Zhan, F., and Ai-Zaqri, N. (2023). Corrosion inhibition performance of *Punica granatum* L extract on Q235 steel in H_2SO_4 environment: electrochemical, morphological, and quantum chemical studies. *International Journal of Electrochemical Science* 18 (10): 100267. https://doi.org/10.1016/j.ijoes.2023.100267.

50 Liao, W., Liu, H., Jin, Z. et al. (2023). Synergistic inhibition effect of ultraviolet irra-diation and benzalkonium chloride on the corrosion of 316L stainless steel caused by *Aspergillus terreus*. *Bioelectrochemistry* 153: 108485. https://doi.org/10.1016/j.bioelechem.2023.108485.

51 Leiva-García, R., García-Antón, J., and Muñoz-Portero, M.J. (2010). Contribution to the elucidation of corrosion initiation through confocal laser scanning microscopy (CLSM). *Corrosion Science* 52 (6): 2133–2142. https://doi.org/10.1016/j.corsci.2010.02.034.

52 Du, P., Deng, S., Du, G. et al. (2023). Synergistic inhibition effect of Mikania micrantha extract with potassium iodide on the corrosion of cold rolled steel in methanesulfonic acid solution. *Corrosion Science* 220: 111296. https://doi.org/10.1016/j.corsci.2023.111296.

53 Baer, E., Koutsky, J.A., and Walton, A.G. (1965). The nucleation of water vapor on surface films formed by adsorption on metal surfaces. *Surface Science* 3 (3): 280–286. https://doi.org/10.1016/0039-6028(65)90051-8.

54 Yin, Y., Li, H., Pan, S. et al. (2022). Electrochemical behaviour of passivation film formed on SLM-fabricated Hastelloy X superalloy surface in 10 wt% $NaNO_3$ solution. *Corrosion Science* 206: 110494. https://doi.org/10.1016/j.corsci.2022.110494.

55 Wu, J., Wu, L., Yao, W. et al. (2023). Preparation and corrosion behavior of Mg(Li)-M layered double hydroxides films on the surface of magnesium-lithium alloy. *Corrosion Science* 224: 111511. https://doi.org/10.1016/j.corsci.2023.111511.

56 Fayard, L., Bizouard, V., Feynerol, V. et al. (2023). Equivalent electrical model for high current density zinc electrodeposition: investigation through polarization curves and electrochemical impedance spectroscopy. *International Journal of Electrochemical Science* 18 (7): 100185. https://doi.org/10.1016/j.ijoes.2023.100185.

57 Morales, D.M. and Arteaga, C.C. (2016). Determination of the corrosion resistance of SS-304 in synthetic seawater at two temperatures using electrochemical noise and polarization curves. *International Journal of Electrochemical Science* 11 (10): 8683–8696. https://doi.org/10.20964/2016.10.28.

58 Nevejans, S., Ballard, N., Fernández, M. et al. (2019). The challenges of obtaining mechanical strength in self-healing polymers containing dynamic covalent bonds. *Polymer* 179: 121670. https://doi.org/10.1016/j.polymer.2019.121670.

59 Augustyniak, A., Tsavalas, J., and Ming, W. (2009). Early detection of steel corrosion via "turn-on" fluorescence in smart epoxy coatings. *ACS Applied Materials and Interfaces* 1 (11): 2618–2623. https://doi.org/10.1021/am900527s.

60 Hu, M., Peil, S., Xing, Y. et al. (2018). Monitoring crack appearance and healing incoatings with damage self-reporting nanocapsules. *Materials Horizons* 5 (1): 51–58. https://doi.org/10.1039/C7MH00676D.

61 Zhang, H., Zhang, X., Bao, C. et al. (2019). A skin-inspired, fully autonomous self-warning and self-repairing polymeric material under damaging events. *Chemistry of Materials* 31 (7): 2611–2618. https://doi.org/10.1021/acs.chemmater.9b00398.

62 Chen, S., Han, T., Zhao, Y. et al. (2020). A facile strategy to prepare smart coatings with autonomous self-healing and self-reporting functions. *ACS Applied Materials and Interfaces* 12 (4): 4870–4877. https://doi.org/10.1021/acsami.9b18919.

63 Feng, C.G., Sun, Y.R., Li, M.Y. et al. (2012). Infrared spectroscopy of synthesis and adsorption properties of aminophosphonic acid chelating fibers. *Spectroscopy and Spectral Analysis* 32 (12): 5. https://doi.org/10.3964/j.issn.1000-0593(2012)12-3188-05.

64 Wang, J., Song, X., Wang, J. et al. (2019). Smart-sensing polymer coatings with autonomously reporting corrosion dynamics of self-healing systems. *Advanced Materials Interfaces* 6 (10): 1900055. https://doi.org/10.1002/admi.201900055.

65 Liu, T., Zhao, H., Zhang, D. et al. (2021). Ultrafast and high-efficient self-healing epoxy coatings with active multiple hydrogen bonds for corrosion protection. *Corrosion Science* 187: 109485. https://doi.org/10.1016/j.corsci.2021.109485.

66 Plawecka, M., Snihirova, D., Martins, B. et al. (2014). Self healing ability of inhibitor-containing nanocapsules loaded in epoxy coatings applied on aluminium 5083 and galvanneal substrates. *Electrochimica Acta* 140: 282–293. https://doi.org/10.1016/j.electacta.2014.04.035.

67 Cano, E., Lafuente, D., and Bastidas, D.M. (2010). Use of EIS for the evaluation of the protective properties of coatings for metallic cultural heritage: a review. *Journal of Solid State Electrochemistry* 14 (3): 381–391. https://doi.org/10.1007/s10008-009-0902-6.

68 Ramezanzadeh, B., Niroumandrad, S., Ahmadi, A. et al. (2016). Enhancement of barrier and corrosion protection performance of an epoxy coating through wet transfer of amino functionalized graphene oxide. *Corrosion Science* 103: 283–304. https://doi.org/10.1016/j.corsci.2015.11.033.

69 Cheng, L., Liu, C., Wu, H. et al. (2021). A mussel-inspired delivery system for enhancing self-healing property of epoxy coatings. *Journal of Materials Science and Technology* 80: 36–49. https://doi.org/10.1016/j.jmst.2020.10.075.

70 Zeng, Y., Kang, L., Wu, Y. et al. (2022). Melamine modified carbon dots as high effective corrosion inhibitor for Q235 carbon steel in neutral 3.5 wt% NaCl solution. *Journal of Molecular Liquids* 349: 118108. https://doi.org/10.1016/j.molliq.2021.118108.

71 Wang, Y., Wang, J., Ma, L. et al. (2021). Qualitative and quantitative detection of corrosion inhibitors using surface-enhanced Raman scattering coupled with multivariate analysis. *Applied Surface Science* 568: 150967. https://doi.org/10.1016/j.apsusc.2021.150967.

72 Saraswathy, V., Muralidharan, S., Kalyanasundaram, R.M. et al. (2001). Evaluation of a composite corrosion-inhibiting admixture and its performance in concrete under macrocell corrosion conditions. *Cement and Concrete Research* 31 (5): 789–794. https://doi.org/10.1016/S0008-8846(01)00468-9.

73 Tan, Y.J., Bailey, S., and Kinsella, B. (2001). Mapping non-uniform corrosion using the wire beam electrode method. II. Crevice corrosion and crevice corrosion exemption. *Corrosion Science* 43 (10): 1919–1929. https://doi.org/10.1016/S0010-938X(00)00191-8.

74 Umoren, S.A., Obot, I.B., Ebenso, E.E. et al. (2006). Gum arabic as a potential corrosion inhibitor for aluminium in alkaline medium and its adsorption characteristics. *Anti-Corrosion Methods and Materials* 53 (5): 277–282. https://doi.org/10.1108/00035590610692554.

75 Krishnegowda, P.M., Venkatesha, V.T., Krishnegowda, P.K.M. et al. (2013). Acalypha torta leaf extract as green corrosion inhibitor for mild steel in hydrochloric acid solution. *Industrial and Engineering Chemistry Research* 52 (2): 722–728. https://doi.org/10.1021/ie3018862.

76 Zhong, F., He, Y., Wang, P. et al. (2022). Novel pH-responsive self-healing anti-corrosion coating with high barrier and corrosion inhibitor loading based on reduced graphene oxide loaded zeolite imidazole framework. *Colloids and Surfaces A: Physicochemical and Engineering Aspects* 642: 128641. https://doi.org/10.1016/j.colsurfa.2022.128641.

77 Hamdy, A.S., Doench, I.M., and Hwald, H. (2011). Smart self-healing anti-corrosion vana-dia coating for magnesium alloys. *Progress in Organic Coatings* 72 (3): 387–393. https://doi.org/10.1016/j.porgcoat.2011.05.011.

78 Yang, C., Xu, W., Meng, X. et al. (2021). A pH-responsive hydrophilic controlled release system based on ZIF-8 for self-healing anticorrosion application. *Chemical Engineering Journal* 415 (32): 128985. https://doi.org/10.1016/j.cej.2021.128985.

79 Cui, X., Yan, Y., Huang, J. et al. (2022). A substrate-independent isocyanate-modified polydimethylsiloxane coating harvesting mechanical durability, self-healing ability and low surface energy with anti-corrosion/biofouling potential. *Applied Surface Science* 3 (30): 579. https://doi.org/10.1016/j.apsusc.2021.152186.

80 Dhouibi, L., Triki, E., and Raharinaivo, A. (2002). The application of electrochemical impedance spectroscopy to determine the long-term effectiveness of corrosion inhibitors for steel in concrete. *Cement and Concrete Composites* 24 (1): 35–43. https://doi.org/10.1016/S0958-9465(01)00062-2.

81 Darowicki, K., Krakowiak, S., and Lepski, P. (2004). Evaluation of pitting corrosion by means of dynamic electrochemical impedance spectroscopy. *Electrochimica Acta* 49 (17–18): 2909–2918. https://doi.org/10.1016/j.electacta.2004.01.070.

82 Angelini, E., Grassini, S., Rosalbino, F. et al. (2003). Electrochemical impedance spectroscopy evaluation of the corrosion behaviour of Mg alloy coated with PECVD organosilicon thin film. *Progress in Organic Coatings* 46 (2): 107–111. https://doi.org/10.1016/S0300-9440(02)00217-5.

83 Zyilmaz, A.T., Erbil, M., and Yazici, B. (2004). Investigation of corrosion behaviour of stainless steel coated with polyaniline via electrochemical impedance spectroscopy. *Progress in Organic Coatings* 51 (1): 47–54. https://doi.org/10.1016/j.porgcoat.2004.06.001.

84 üttner, K., Lorenz, W.J., and Paatsch, W. (1989). The role of surface inhomogeneities in corrosion processes-electrochemical impedance spectroscopy (EIS) on different aluminium oxide films. *Corrosion Science* 29 (2–3): 279–288. https://doi.org/10.1016/0010-938X(89)90036-X.

85 Rout, T.K. (2007). Electrochemical impedance spectroscopy study on multi-layered coated steel sheets. *Corrosion Science* 49 (2): 794–817. https://doi.org/10.1016/j.corsci.2006.06.008.

86 Gu, X.N., Li, N., Zhou, W.R. et al. (2011). Corrosion resistance and surface biocompatibility of a microarc oxidation coating on a Mg–Ca alloy. *Acta Biomaterialia* 7 (4): 1880–1889. https://doi.org/10.1016/j.actbio.2010.11.034.

87 Meng, F., Liu, L., Tian, W. et al. (2015). The influence of the chemically bonded interface between fillers and binder on the failure behaviour of an epoxy coating under marine alternating hydrostatic pressure. *Corrosion Science* 2015 (101): 139–154. https://doi.org/10.1016/j.corsci.2015.09.011.

88 Qiao, Y.X., Zheng, Y.G., Okafor, P.C. et al. (2009). Electrochemical behaviour of high nitrogen bearing stainless steel in acidic chloride solution: effects of oxygen, acid concentration and surface roughness. *Electrochimica Acta* 54 (8): 2298–2304. https://doi.org/10.1016/j.electacta.2008.10.038.

89 Di, Q.F., Piazza, S., and Sunseri, C. (1995). Photoelectrochemistry in corrosion studies: achievements and perspectives. *Materials Science Forum* 192–194: 633–648. https://doi.org/10.4028/www.scientific.net/MSF.192-194.633.

90 Crolet, J.L. (1993). The electrochemistry of corrosion beneath corrosion deposits. *Journal of Materials Science* 28 (10): 2577–2588. https://doi.org/10.1007/BF00356193.

91 Pang, X., Zhou, F., Li, B. et al. (2021). Optical thermostability and weatherability of TiN/TiC-Ni/Mo cermet-based spectral selective absorbing coating by laser cladding. *Optical Materials* 117: 111195. https://doi.org/10.1016/j.optmat.2021.111195.

92 Park, H.S., Yang, I.M., Wu, J.P. et al. (2001). Synthesis of silicone–acrylic resins and their applications to superweatherable coatings. *Journal of Applied Polymer Science* 81 (7): https://doi.org/10.1002/app.1592.

93 Perrin, F.X., Merlatti, C., Aragon, E. et al. (2009). Degradation study of polymer coating: improvement in coating weatherability testing and coating failure prediction. *Progress in Organic Coatings* 64 (4): 466–473. https://doi.org/10.1016/j.porgcoat.2008.08.015.

94 Ma, L., Wang, X., Liu, X.X. et al. (2023). Study on the inhibition of coal spontaneous combustion by superabsorbent antioxidant inhibitors. *Combustion Science and Technology* https://doi.org/10.1080/00102202.2023.2270150.

95 Xu, C., Zhou, S., Song, H. et al. (2023). Green tea polyphenols-derived hybrid materials in manufacturing, environment, food and healthcare. *Nano Today* 52: 101990. https://doi.org/10.1016/j.nantod.2023.101990.

96 Yan, M., Wang, Y., Wang, C. et al. (2024). Whey protein isolate-resveratrol complex as a radical scavenging foaming ingredient with increased ultraviolet stability. *Food Chemistry* 434: 137519. https://doi.org/10.1016/j.foodchem.2023.137519.

97 Jeleń, M., Ying, P.T.C., Hao, Y.J. et al. (2024). In vitro study of antioxidant, antigylycation, sugar hydrolysis enzyme inhibitory effect and molecular in silico docking study of angularly condensed diquinothiazines. *Journal of Molecular Structure* 1296 (2): 136856. https://doi.org/10.1016/j.molstruc.2023.136856.

98 Wang, J., Ma, L., Ding, X. et al. (2023). Tea polyphenol radical scavenger loaded UV absorber for corrosion resistant and weathering resistant epoxy coating fabrication. *Progress in Organic Coatings* 180: 107553. https://doi.org/10.1016/j.porgcoat.2023.107553.

99 Mahltig, B., Böttcher, H., Rauch, K. et al. (2005). Optimized UV protecting coatings by combination of organic and inorganic UV absorbers. *Thin Solid Films* 485 (1–2): 108–114. https://doi.org/10.1016/j.tsf.2005.03.056.

100 Liu, C., Wang, W.Z., Song, M.T. et al. (2021). Radical scavenging efficiency of flavonoids increased by calcium(II) binding: structure-activity relationship. *ChemistrySelect* 6 (38): 10323–10323. https://doi.org/10.1002/slct.202101560.

101 Gaudel, N., Bémer, D., and Gérardin, F. (2021). A parametric study of aerosol particle scavenging efficiency by spray droplets. *Journal of Aerosol Science* 159: 105843. https://doi.org/10.1016/j.jaerosci.2021.105843.

102 Sun, X., Gu, S., Wang, L. et al. (2023). Multifunctional liquid-like magnetic nanofluids mediated coating with anticorrosion and self-healing performance. *Journal of Colloid and Interface Science* 654: 25–35. https://doi.org/10.1016/j.jcis.2023.09.182.

103 Sun, C., Liu, C., Wang, Y. et al. (2023). Corrosion sensing and self-healing composite coatings on magnesium alloy AZ31 enabled by stimuli-responsive microcapsules loaded with phenolphthalein and epoxy resin. *Progress in Organic Coatings* 186: 108043. https://doi.org/10.1016/j.porgcoat.2023.108043.

104 Chen, Y., Liu, R., and Luo, J. (2023). Improvement of anti-aging property of UV-curable coatings with silica-coated TiO_2. *Progress in Organic Coatings* 179: 107479. https://doi.org/10.1016/j.porgcoat.2023.107479.

105 Ma, Y., Chen, R., Fei, G. et al. (2022). Enhanced anti-aging and anti-corrosion performance of waterborne epoxy coating layers over the dual effects of g-C_3N_4 photocatalysis. *Journal of Applied Polymer Science* 139 (24): 52356. https://doi.org/10.1002/app.52356.

106 Rahim, A.A., Rocca, E., Steinmetz, J. et al. (2008). Inhibitive action of mangrove tannins and phosphoric acid on pre-rusted steel via electrochemical methods. *Corrosion Science* 50 (6): 1546–1550. https://doi.org/10.1016/j.corsci.2008.02.013.

107 Oki, M., Charles, E., Alaka, C. et al. (2011). Corrosion inhibition of mild steel in hydrochloric acid by Tannins Rom Rhizophora Racemosa. *Materials Sciences and Applications* 2 (6): https://doi.org/10.4236/msa.2011.26079.

108 Ashassi-Sorkhabi, H. and Es'Haghi, M. (2009). Corrosion inhibition of mild steel in hydrochloric acid by betanin as a green inhibitor. *Journal of Solid State Electrochemistry* 13 (8): 1297–1301. https://doi.org/10.1007/s10008-008-0673-5.

109 Kaddouri, M., Rekkab, S., Bouklah, M. et al. (2013). Experimental study of inhibition of corrosion of mild steel in 1 M HCl solution by two newly synthesized calixarene derivatives. *Research on Chemical Intermediates* 39 (8): 3649–3667. https://doi.org/10.1007/s11164-012-0869-2.

110 Yu, D., Wen, S., Yang, J. et al. (2017). RGO modified ZnAl-LDH as epoxy nanostructure filler: a novel synthetic approach to anticorrosive waterborne coating. *Surface and Coatings Technology* 326: https://doi.org/10.1016/j.surfcoat.2017.07.053.

111 Jeyasubramanian, K., Benitha, V.S., and Parkavi, V. (2019). Nano iron oxide dispersed alkyd coating as an efficient anticorrosive coating for industrial structures. *Progress in Organic Coatings* 132: 76–85. https://doi.org/10.1016/j.porgcoat.2019.03.023.

9

Development of Superhydrophobic Coating with Good Mechanical Durability and Self-healing Effect

9.1 Introduction

In recent years, inspired by "biological bionics," more and more researchers have begun to pay attention to superhydrophobic surfaces. The superhydrophobic surface has been widely used in anti-icing, anti-fouling, drag reduction, and other fields, and its unique superhydrophobic performance also plays an important role in reducing the corrosion of metal materials in the atmosphere, ocean, soil, and even industrial environment. Therefore, the development of a series of superhydrophobic surface coatings with comprehensive properties such as self-cleaning [1], anticorrosion, and wear resistance, and its construction technology is an important research direction and has broad development prospects.

At present, more and more superhydrophobic surfaces with micro-roughness have been applied in the field of metal corrosion protection, but the key mechanism of these microstructures in improving the physical barrier effect and inhibiting the development process of corrosion initiation has not been deeply discussed [2]. It is generally believed that for the superhydrophobic surface exposed to the atmosphere, its micron and nanostructure can capture a large amount of air, greatly reducing the contact area between water droplets and the surface, making it difficult to stay on the coating surface, and shortening the contact time between water and the coating surface. For the superhydrophobic surface immersed in a water environment, the gas trapped by the microstructure forms a gas film, which increases the physical barrier effect of the superhydrophobic surface structure on water and other corrosive media to a certain extent [3]. The biomimetic superhydrophobic surface is also applied to ice protection. The ice-coating process of the material surface is very complicated and is not only related to the environment but also involves the phase transition process of water [4]. By reducing the contact area between the water droplets and the material, the superhydrophobic surface reduces the heat transfer rate and delays the freezing time of the water droplets. In the process of icing, it delays the formation of heterogeneous ice nuclei, reduces the generation of surface fog/frost, prevents the transformation of water droplets from Cassie state to Wenzel state on the superhydrophobic surface, increases the probability of water droplets sliding on the surface, reduces the adhesion strength between the ice-solid interface, and has an anti-icing effect [5].

Smart Protective Coatings for Corrosion Control, First Edition. Lingwei Ma and Dawei Zhang.

Due to the lack of basic research on the mechanism of superhydrophobic coating and its main influencing factors [6], there are many problems in the practical application of superhydrophobic coating. For example, most of the methods for preparing superhydrophobic surfaces can only be performed in the laboratory, the process is complicated and cumbersome, the acquisition cost of raw materials is high, and the mechanical strength of the prepared superhydrophobic surfaces is low. The durability of superhydrophobic surface coating is very poor, and the structural stability of superhydrophobic surface is very low. Second, the superhydrophobic coating cannot achieve large-area spraying construction, resulting in the superhydrophobic coating being difficult to achieve in large-scale engineering applications [7]. Therefore, under the premise of reducing the cost and simplifying the construction process, the development of superhydrophobic coatings with good stability and high surface mechanical strength is also an extremely important research direction. Therefore, it is urgent to develop new coating systems and exploit new preparation methods to achieve simple, large-scale preparation and construction of superhydrophobic coatings [8].

In this chapter, a series of new coatings were prepared by a facile spraying process using epoxy resin as matrix, nano-silica ($nSiO_2$) particles, carbon nanotubes (CNTs), and corrosion inhibitor containing CNTs/layered bimetallic oxide (CNTs/LDH/VO_3^-) nanoparticles as additives. Based on the analysis and study of the superhydrophobic mechanism of the surface coating, the influence of the hybrid particles on the performance of the superhydrophobic coating was studied [9], and the action mechanism of $nSiO_2$ and CNTs on superhydrophobic coating was clarified. On the basis of the above research, the proportion of resin and curing agent is adjusted to make the coating have a shape memory effect, and the superhydrophobic coating with self-healing property is prepared. Further, corrosion inhibitors are added to the coating to enhance the corrosion resistance of the coating [10]. Therefore, a series of new superhydrophobic coatings with excellent properties that can be sprayed in large areas are developed, which provides a basis for the engineering practice of preparing large-scale superhydrophobic coatings by spraying technology [11].

9.2 Preparation and Evaluation of Modified Nano SiO_2/Epoxy Resin Superhydrophobic Coating

$nSiO_2$ particles have the advantages of high hardness, stable chemical properties, high-temperature resistance, good insulation, and good wear resistance, which help improve the mechanical properties and thermal stability of resin materials [12]. However, as a kind of inorganic filler, the compatibility of $nSiO_2$ particles with polymer is poor, resulting in weak interface bonding ability. Due to the rich silica hydroxyl group on the surface of $nSiO_2$, the interface bonding force between $nSiO_2$ and polymer can be improved by chemical modification to meet the different needs of practical applications [13].

The commonly used modifiers for $nSiO_2$ are silane coupling agent, titanate coupling agent, hyperdispersant, and so on. Research shows that after modification by silane coupling agents, organic connections can be formed between SiO_2 nanoparticles and polymer interface, and the interface binding force is significantly enhanced. Yan et al. used the Stober method to modify $nSiO_2$ with dichlorodimethylsilane, and the dispersion of modified $nSiO_2$ particles in epoxy resin was more uniform [14]. By changing the content of the modifier, the surface wettability of $nSiO_2$ particles can be regulated, so that it has hydrophobic properties, and the water contact angle (WCA) can reach a maximum of 146.1°. Anjum et al. [15] used the double continuous phase emulsion method/double phase stratification method to synthesize $nSiO_2$ particles with wrinkled surface. The coating with superhydrophobic properties was successfully prepared by adding the synthesized $nSiO_2$ particles into different resins [16]. However, the preparation process of the superhydrophobic coating is more complicated, the production cost is high, and a large number of solvents are required, which is easy to cause serious environmental pollution.

In order to optimize the modification efficiency of $nSiO_2$ particles, silane coupling agent KH560 was used to modify the surface of $nSiO_2$. Compared with KH550 and KH570, KH560 has a higher grafting rate, the least amount required, and the modification process is simple. After that, the coating with superhydrophobic properties is prepared by a convenient one-step spraying method [17].

9.2.1 Materials and Methods

9.2.1.1 Materials

The materials and reagents used in this section are shown in Table 9.1.

9.2.1.2 Modification of $nSiO_2$

3.0 g $nSiO_2$ powder was added into 100 ml anhydrous ethanol and was under ultrasonic dispersion for 30 minutes, to obtain uniformly dispersed $nSiO_2$/ethanol

Table 9.1 Experimental materials and reagents.

Materials	Purity	Company
SiO_2 particles	Nanosize	Guihuang Metal Materials Co., LTD
KH560	Analytical purity	Sinopharm Group
Ethyl alcohol	Analytical purity	Sinopharm Group
Methylbenzene	Analytical purity	Sinopharm Group
Acetone	Analytical purity	Sinopharm Group
Bisphenol A diglycidyl ether (DGEBA)	Analytical purity	Sigma Corporation
Polyetheramine (D230)	Analytical purity	Aladdin, Inc.

suspension. Then, 0.015 g of KH560 silane coupling agent and 10 ml of toluene were added to the $nSiO_2$/ethanol suspension, which was first treated with ultrasound for five minutes, and then stirred at 80 °C for two hours with a magnetic agitator. Finally, the mixture was poured into the sand core funnel, filtered by vacuum negative pressure with deionized water 3 times, and washed with acetone several times. The obtained modified $nSiO_2$ filter was dried at 80 °C and placed in a drying oven for use.

9.2.1.3 Preparation of Modified $nSiO_2$/Epoxy Resin Composite Coating

240 grits of sandpaper was used to sand the aluminum sheet, rinsed with alcohol, and deionized water, and dried with hot air. 1.875 g of bisphenol A diglycidyl ether was added into 50 ml of acetone to each beaker to reduce the viscosity of the epoxy resin. Then, 0.25, 0.50, and 0.75 g modified $nSiO_2$ were added to the beaker respectively, and were ultrasonically dispersed at room temperature for one hour to make them uniformly dispersed in the epoxy resin. Then, 0.625 g D230 was weighed and added into the above uniformly dispersed $nSiO_2$-epoxy dispersion solution. After ultrasonic treatment for 30 minutes at room temperature, the mixed solution was sprayed on the prepared aluminum sheet with a spray gun three times. After curing in a 60 °C incubator for 48 hours, the modified $nSiO_2$/epoxy composite coating with mass fractions of 10, 20, and 30 wt% was obtained.

9.2.1.4 Structure Analysis of Modified $nSiO_2$

The unmodified and modified $nSiO_2$ were characterized and analyzed by Fourier infrared spectroscopy (FTIR, VERTEX 70) with a measurement range of 4000–400 cm^{-1}, and a resolution of 4 cm^{-1}. The morphology of unmodified and modified $nSiO_2$ was observed by transmission electron microscopy (TEM, JOEL2100).

9.2.1.5 Hydrophobic Property of Modified $nSiO_2$/Epoxy Coating

To verify the superhydrophobic properties of the coating, the WCA and rolling angle of the coating surface were measured at room temperature using a dataphysics OCA20 contact angle measurement instrument with a drop volume of 6 μl. By tilting the sample table, the smallest angle at which water droplets roll off the sample surface is defined as the rolling angle. In order to reduce the experimental error, five locations on each sample surface were selected for testing, and the average of five measured values was taken as the final data.

9.2.1.6 Mechanical Durability and Stability Analysis of Coating

According to the international standard ASTM D3359-09, the mechanical durability of $nSiO_2$/epoxy coating was evaluated by tape stripping test. During the test, the 3M tape was pressed against the surface of the sample, and a pressure of 30 KPa was used to ensure good contact between the tape and the surface. The tape was removed from the coating surface, the WCAs were measured, and the tape was pressed back onto the sample surface. The process was repeated until the coating surface lost its superhydrophobicity. In order to characterize the chemical stability of the coating,

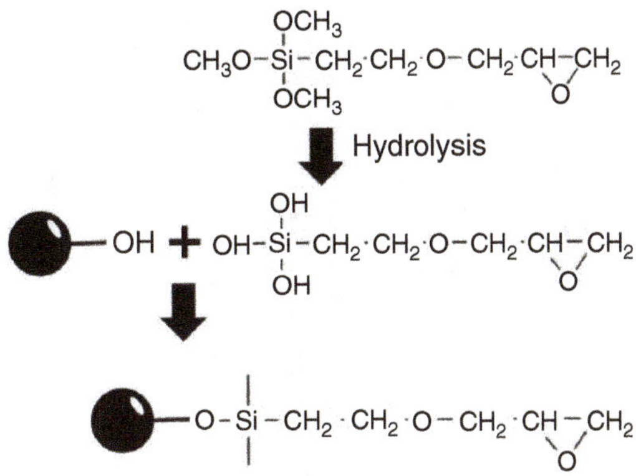

Figure 9.1 Chemical modification of $nSiO_2$ using KH560.

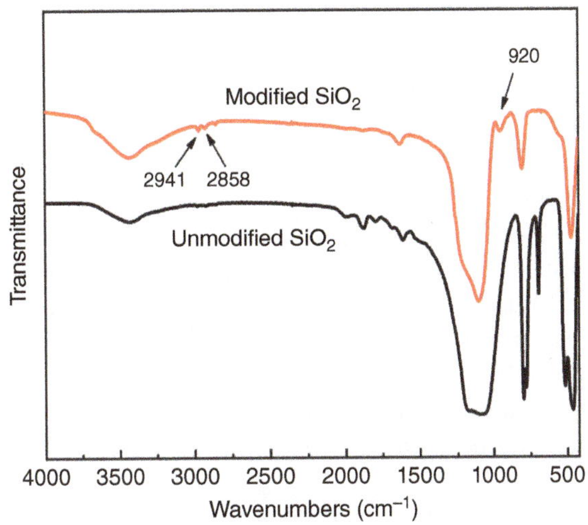

Figure 9.2 FTIR spectra of $nSiO_2$ before and after KH560 modification.

acid–base solutions with pH values of 3, 5, 7, 9, 11, and 13 were prepared, and these acid–base solutions were dropped on the surface of the material to determine the liquid contact angle and rolling angle of the samples.

9.2.2 Characterization of $nSiO_2$ and $nSiO_2$/Epoxy Coating

9.2.2.1 Characterization of $nSiO_2$

The surface of $nSiO_2$ particles was modified by the KH560 silane coupling agent with the epoxy group. The epoxy group was grafted to the surface of $nSiO_2$ by the condensation of the silicone hydroxyl group on $nSiO_2$ and the silicone oxygen bond generated by the hydrolysis of the silane coupling agent to achieve the effect of the modification. The reaction mechanism is shown in Figure 9.1.

Figure 9.2 shows the FTIR spectra of $nSiO_2$ before and after KH560 modification. 3550 cm^{-1} is the Si–OH stretching vibration absorption peak of the $nSiO_2$ surface, 1082 cm^{-1} is the antisymmetric absorption peak of Si–O–Si, and 468 cm^{-1} is the bending vibration absorption peak of Si–O–Si. After KH560 surface modification, the absorption peak of $nSiO_2$ showed obvious changes, the absorption peak

Figure 9.3 TEM images of (a) unmodified $nSiO_2$ and (b) modified $nSiO_2$ and the dispersion photos in water.

of the epoxy group appeared at $920\,cm^{-1}$, and the absorption peak of the methyl group appeared at $2941\,cm^{-1}$. Methylene absorption peak appeared at $2858\,cm^{-1}$ [18]. From the above analysis, it can be seen that the epoxy group is successfully modified on the surface of $nSiO_2$ after modification by the KH560 silane coupling agent.

Figure 9.3 shows the TEM images of unmodified and modified $nSiO_2$ and the dispersion photos in water respectively. It can be seen that the diameter of $nSiO_2$ particles before and after modification was about 25 nm, and the diameter did not show much change. However, it can be seen from the illustration that the unmodified $nSiO_2$ particles sunk to the bottom of the bottle after standing for half an hour, while the modified $nSiO_2$ particles floated on the aqueous solution [19], indicating that the modified $nSiO_2$ particles had hydrophobic properties.

9.2.2.2 Structure and Hydrophobicity of $nSiO_2$/Epoxy Coating

A scanning electron microscope (SEM) was used to observe the surface morphology of the $nSiO_2$/epoxy resin composite coating. In Figure 9.4, the content of modified $nSiO_2$ was 10, 20, and 30 wt%, respectively. As can be seen from the figure, when the $nSiO_2$ content was 10 wt%, the coating surface was relatively smooth. When the content of $nSiO_2$ increased to 20 wt%, the surface of the coating showed obvious fluctuation and rough structure [20]. When the content of $nSiO_2$ further increased to 30 wt%, the coating surface appeared as a small spherical structure, and the sample

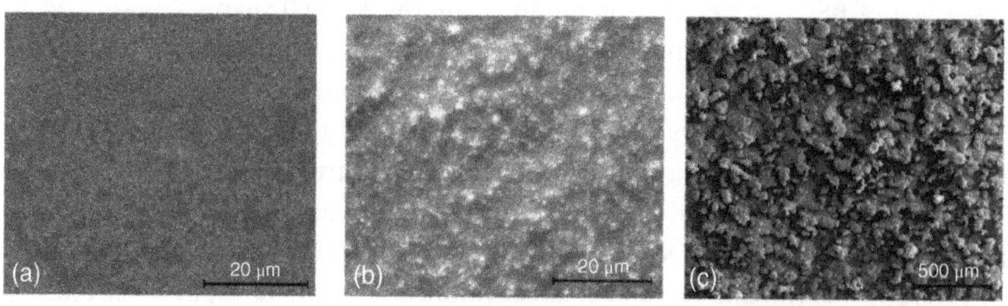

Figure 9.4 Surface morphology of $nSiO_2$/epoxy resin composite coating containing (a) 10, (b) 20, and (c) 30 wt% $nSiO_2$.

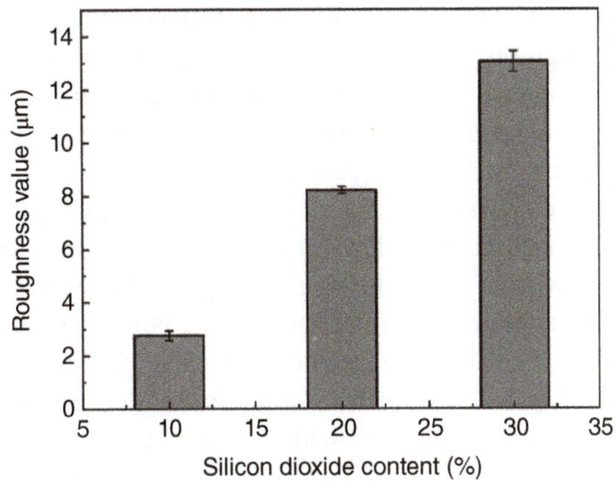

Figure 9.5 Surface rough of $nSiO_2$/epoxy resin composite coating containing 10, 20, and 30 wt% $nSiO_2$.

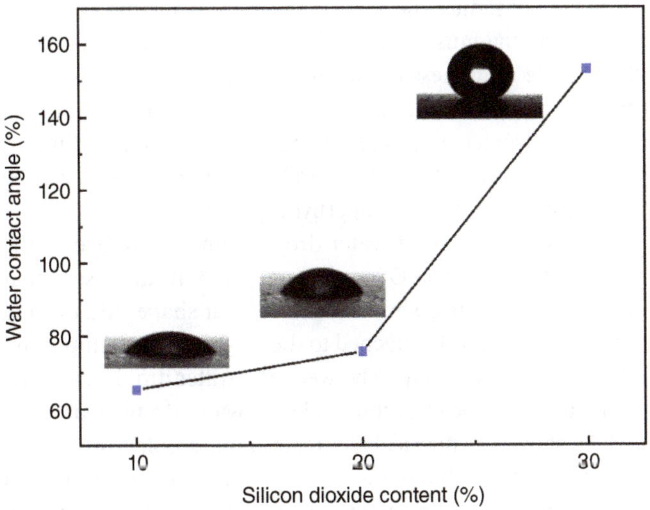

Figure 9.6 WCAs of $nSiO_2$/epoxy resin composite coating containing 10, 20, and 30 wt% $nSiO_2$.

surface became rougher. Figure 9.5 shows the surface roughness histogram of the three coatings. As can be seen from the figure, with the increase of $nSiO_2$ content, the surface granular structure increased, making the surface roughness of the coating increase from 2.75 to 8.20 μm and 13.04 μm. The above results were in agreement with those obtained by SEM observation

Figure 9.6 shows the WCA results of different $nSiO_2$/epoxy composite coatings. As the modified $nSiO_2$ content increased from 10 to 20 wt% and 30 wt%, the water droplets showed different morphologies on the coating surface. When the $nSiO_2$ content was 10 wt%, the water droplets were spread on the surface of the coating, and the contact area between the water droplets and the coating was large, and the WCA was 65.2°. When the $nSiO_2$ content was 20 wt%, the water droplets gradually arched, and the contact area between the water droplets and the coating decreased with the WCA of 72.3°. When the SiO_2 content was 30 wt%, the water droplets existed as balls on the surface of the coating, and the WCA was 152.4°, at which time the coating had reached the requirement of superhydrophobic. The above results show that suitable surface roughness is one of the necessary conditions for constructing superhydrophobic structures.

The wettability of a solid surface is closely related to its microstructure and chemical composition [21]. When the $nSiO_2$ content was 10 wt%, the wetting behavior of the composite coating was in the Wenzel state, and the contact area of the liquid/solid interface was the largest. As the content of $nSiO_2$ increased to 20 wt% and the surface energy of $nSiO_2$ grafted by the epoxy group was lower, the wetting behavior of the composite coating gradually changed from Wenzel state to Cassie–Baxter state, which reduced the contact area between the liquid and the coating. According to the Cassie–Baxter model, superhydrophobicity is based on the simultaneous action of two key factors – surface roughness and low surface energy [22]. Therefore, in order to further reduce the contact area between liquid and solid, the surface roughness of the coating must be improved. When the content of $nSiO_2$ increased to 30 wt%, the surface roughness of the coating was relatively high, and the water droplets formed an air barrier with the coating surface. A large amount of air retention caused the liquid droplets to produce a multistage and discontinuous three-phase boundary at the solid/liquid interface, and the liquid droplets can be suspended on the coating surface to achieve a superhydrophobic effect.

Figure 9.7a shows the physical picture of water droplets on the surface of the composite epoxy coating with a modified $nSiO_2$ content of 10 wt%. It can be seen that the water drops spread out on the coating, showing an irregular shape. After tilting the sample, the water droplets were firmly attached to the surface of the sample and did not slip off. This is because the contact area between the water droplets and the coating was large, and the larger surface tension made the water droplets produce strong adhesion. Figure 9.7b shows the physical picture of water droplets on the surface of the composite epoxy coating with modified $nSiO_2$ content of 30 wt%. Due to the better hydrophobic performance of the coating, the water droplets appeared as spheres on the coating. Due to the small contact area between the water droplets

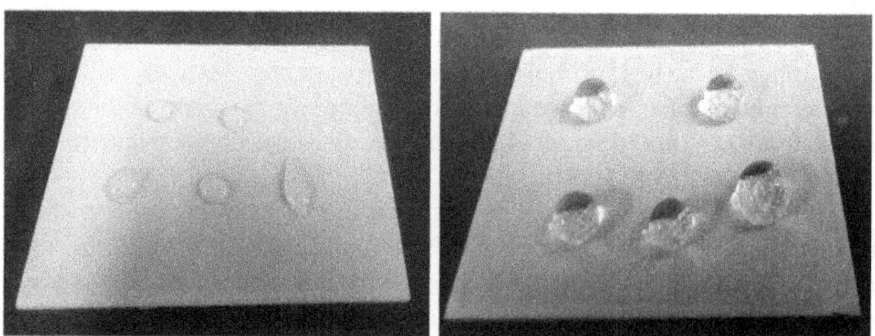

(a) (b)

Figure 9.7 The hydrophobic effect of the coating: (a) modified composite epoxy coating with $nSiO_2$ content of 10 wt%; (b) modified composite epoxy coating with $nSiO_2$ content of 30 wt%.

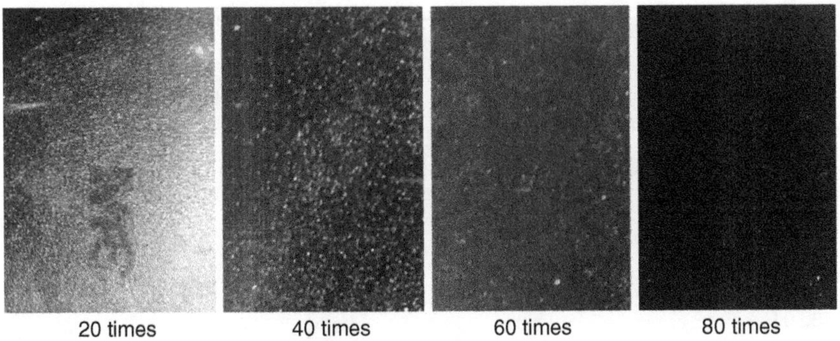

Figure 9.8 The macroscale morphology of $nSiO_2$/epoxy coating with $nSiO_2$ content of 30 wt% after different tape stripping times.

and the coating on the superhydrophobic surface and the small surface tension, the water droplets were easy to slip off the coating surface when the sample surface was tilted.

9.2.2.3 Stability and Anti-icing Ability of $nSiO_2$/Epoxy Coating

At present, one of the main problems facing superhydrophobic coatings is poor mechanical durability, which limits the practical application of superhydrophobic coatings. In this experiment, the mechanical durability of superhydrophobic coated surfaces was evaluated using tape stripping tests. Figure 9.8 shows the macro morphology of the modified superhydrophobic coating with $nSiO_2$ content of 30 wt% after different tape stripping times. As can be seen from the figure, the white particles on the surface of the sample decreased and the coating color gradually deepened with an increase in the tape stripping times, indicating that the $nSiO_2$ particles on the surface layer were gradually consumed. Figure 9.9 shows the variation of WCAs of the coating after different tape stripping times. The WCAs of the composite coating were reduced to 149.5° after tape stripping 80 times, but it still had a good hydrophobic performance. This is because the epoxy resin reacts with the curing agent to form a highly cross-linked three-dimensional network structure [23], and the modified $nSiO_2$ is combined with the epoxy matrix as a filler, resulting in a strong binding force. Although the surface $nSiO_2$ particles were consumed, the resin matrix can fix the $nSiO_2$ particles embedded in it, maintaining the microscopic rough structure of the coating surface and having good mechanical durability.

In addition to mechanical durability, it is also necessary to investigate whether the wettability of the coating is affected by the acidity and alkalinity of the solution. The degree of acid and alkali resistance [24] of the coating was mainly assessed by exposure to solutions of different pH values. During the test, droplets with pH values of 3, 5, 7, 9, 11, and 13 were added to the coating for 10 seconds and the WCAs were measured. Figure 9.10 shows the changing trend of static WCAs and the rolling angle of droplets at different pH values on the surface of the epoxy coating with modified $nSiO_2$ content of 30 wt%. With the increase in pH value, the static WCAs gradually increased, and the rolling angle gradually decreased. The WCAs were greater than 150°, and the rolling angle was less than 10°, indicating that the coating maintained superhydrophobic properties in contact with solutions of different pH values. The

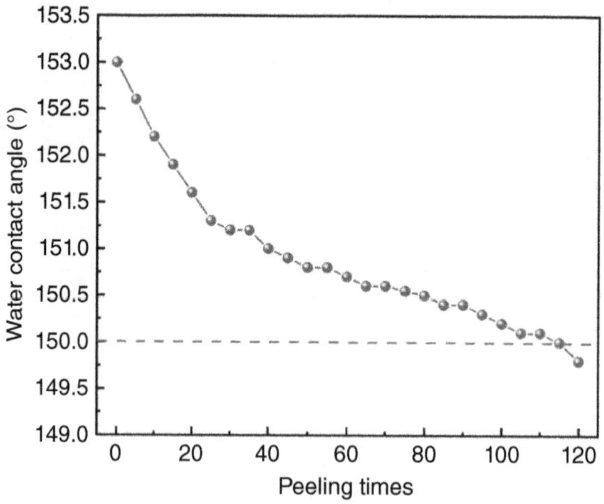

Figure 9.9 WCAs of $nSiO_2$/epoxy coating with $nSiO_2$ content of 30 wt% after different tape stripping times.

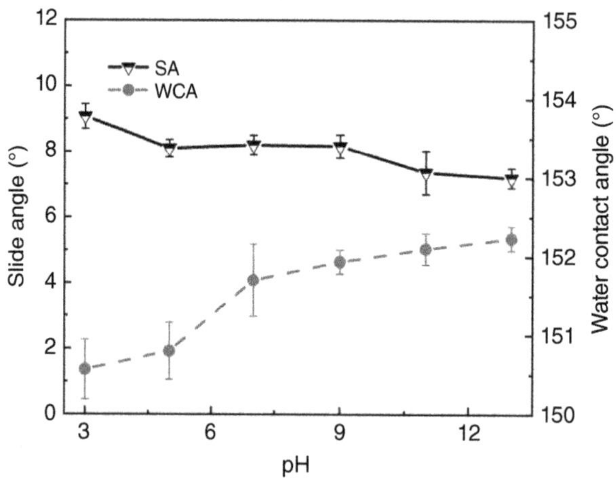

Figure 9.10 WCAs and rolling angles of $nSiO_2$/epoxy coating with $nSiO_2$ content of 30 wt% in solutions at different pH values.

superhydrophobic coating has excellent acid and alkali resistance mainly due to the formation of three-dimensional stable molecular chains after epoxy resin curing, which has strong corrosion resistance to acid and alkali solution. At the same time, the superhydrophobic coating reduces the contact area between the coating surface and the acid–base solution, which further reduces the influence of pH change in solutions on the structure and properties of the coating [25].

9.3 Preparation and Evaluation of CNTs/Epoxy Resin Superhydrophobic Coating

Through the research in the previous section, it was found that the superhydrophobic surface of the $nSiO_2$ particles accumulated has a certain wear resistance, but the nanoparticles would still fall off obviously with the increase of stripping times, and the mechanical friction property of the coating was unstable, which restricted the practicability of the superhydrophobic coating. Therefore, it is necessary to further improve the mechanical durability [26] of the coating while ensuring its hydrophobic performance.

CNTs have the characteristics of flexibility and lubricity, which can effectively reduce the friction coefficient. It has been experimentally proven that the composite coating prepared by CNTs as raw material has excellent wear resistance. The material prepared by the combination of CNTs, and polymer resin matrix has superhydrophobicity, and the micro-nanostructure on the surface also has high mechanical durability [27]. However, many organic chemical reagents such as sodium stearate, hexafluorobisphenol A, and 4,4-difluorobenzophenone [28] need to be added during the preparation process, resulting in more harmful gases, serious environmental pollution, and complex preparation process.

In this section, CNTs were used as raw materials, acetone was added to promote dispersion, and CNTs/epoxy resin composite coatings were prepared by a simple and effective one-step spraying method. The superhydrophobic surface exhibited layered micro-nanostructure on the coating surface. By changing the content of CNTs, the microstructure of the composite coating was regulated, and the hydrophobicity and wear resistance of the coating surface were improved.

9.3.1 Materials and Methods

9.3.1.1 Materials

The materials and reagents used in this section are shown in Table 9.2.

9.3.1.2 Preparation of CNTs/Epoxy Resin Composite Coating

The Q235 steel plate ($40 \times 60\,mm^2$) was used as the metal substrate, which was polished with 240 mesh sandpaper, and then cleaned with acetone in the ultrasonic cleaning machine. CNTs of different masses (0.125, 0.25, 0.5, and 0.75 g) were dispersed in 1.875 g bisphenol A diglycidyl ether, and 40 ml acetone solution was added for ultrasonic dispersion. After ultrasound for 45 minutes, 0.625 g D230 curing agent was added for mechanical stirring. The mass fractions of CNTs relative

Table 9.2 Experimental materials and reagents.

Materials	Purity	Company
Multiwalled carbon nanotube	—	Suzhou Carbon Feng Technology Co., LTD
Acetone	Analytical purity	Sinopharm Group
Bisphenol A diglycidyl ether (DGEBA)	Analytical purity	Sigma Corporation
Polyetheramine (D230)	Analytical purity	Aladdin, Inc.

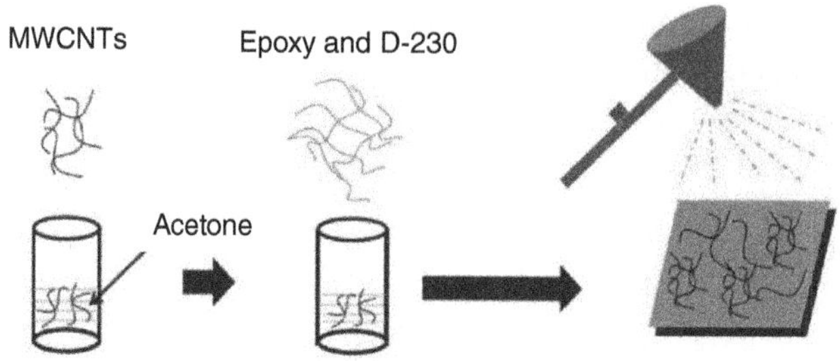

Figure 9.11 Preparation diagram of CNTs/epoxy coating.

to epoxy resin were 5, 10, 20, and 30 wt%, respectively. The mixture was sprayed three times on the steel surface with a 0.5 MPa air pressure gun. Finally, after the samples were cured in a 60 °C incubator for 48 hours, the CNTs/epoxy composite coating was obtained, and the coating thickness was 100 ± 10 μm. The entire preparation process is shown in Figure 9.11.

9.3.1.3 Surface Characterization of CNTs/Epoxy Composite Coatings

The morphology of CNTs was observed by TEM, JOEL2100. SEM (SU1510) was used to observe the morphology and structure of the coating surface. The surface roughness of the coating was measured and analyzed by laser confocal scanning microscope (CLSM, KEYENCE VK-X). The hydrophobicity of the coating surface was measured with a dataphysics OCA20 contact angle measurement instrument with a drop volume of 6 μl. By tilting the sample table, the smallest angle at which water droplets roll off the sample surface is the rolling angle. In order to reduce the experimental error, five locations on each sample surface were selected for testing, and the average of five measured values was taken as the final data.

9.3.1.4 Anti-icing Property of CNTs/Epoxy Composite Coatings

The anti-icing properties of the coating surface were studied by dynamic water drop test and static icing test [29]. The dynamic water drop icing test was carried out in a self-made ice chamber. The schematic diagram of the experimental device is shown in Figure 9.12. The working temperature of the ice chamber was set at −10 °C. The water temperature in the funnel above the device was stabilized at 0 °C to form a mixture of ice water. The prepared coating sample was placed at the bottom of the ice chamber at a tilt of 30°, and the distance from the funnel to the sample was 15 cm. The sample was taken out after being placed in the ice chamber for 15 minutes, and the image of the coating surface was taken with a digital camera. In the static icing test [30], the sample was placed on the cold table, the temperature was set to −10 °C, and the water droplets were placed on different sample surfaces to show the icing behavior of water droplets on the coating surface.

9.3.1.5 Mechanical Durability Test

According to the international standard ASTM D3359-09, the mechanical durability of the CNTs/epoxy composite coating was evaluated by tape stripping test [31].

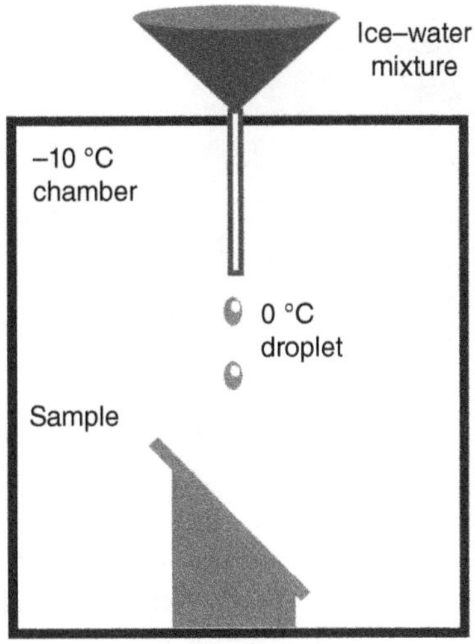

Figure 9.12 Self-made dripping test diagram of ice chamber.

During the test, the surface of the sample was pressed with 3M tape, and a pressure of 30 KPa was used to ensure good contact between the tape and the surface. The tape was removed from the coating surface, the WCAs were measured, and the tape was pressed back onto the sample surface. The process was repeated until the coating surface lost its superhydrophobicity. The WCA of the coating surface was recorded for different times of peeling.

9.3.1.6 Electrochemical Analysis

The coatings were tested by electrochemical impedance spectroscopy (EIS) using the PARSTAT 2273 electrochemical workstation in 3.5 wt% NaCl solution at room temperature. The open circuit potential was measured before the impedance spectrum test, the test time was 30 minutes, and the electrochemical impedance spectrum test was carried out when the whole system reached a steady state. The test frequency range of the impedance spectrum was 10^5–10^{-2} Hz, and the disturbance potential was 20 mV.

9.3.2 Characterization of CNTs/Epoxy Coating

9.3.2.1 Surface Characterization of CNTs/Epoxy Coating

Figure 9.13 shows the TEM image of CNTs. The CNTs were disordered and had a diameter of about 15 nm. Different qualities of CNTs were added to the epoxy coating, Figure 9.14 shows the surface morphology of the CNTs/epoxy composite coating with CNTs content of 5, 10, 20, and 30 wt%. It can be seen that the surface morphology of the coating was closely related to the content of CNTs. When the CNTs content was 5 wt%, the epoxy resin and CNTs aggregated, forming tiny bumps on the surface of the composite coating, showing a rough microstructure. Most of the aggregates were sparsely dispersed, with an average diameter of about 10 μm, and only a few protrusions with a diameter of 30 μm. When the content of

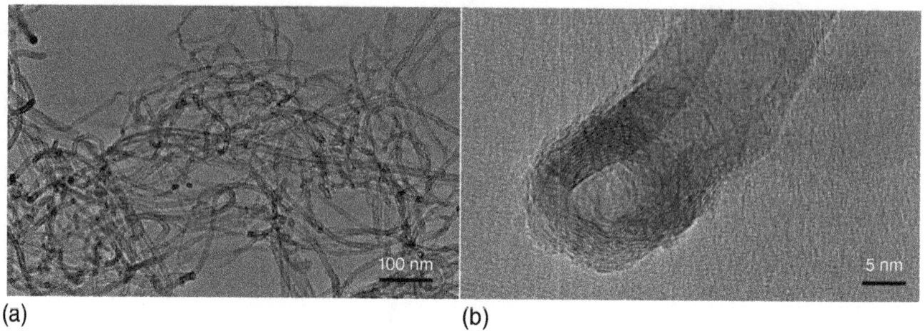

Figure 9.13 TEM images of CNTs.

CNTs increased to 10 wt%, the density of aggregate particles on the coating surface increased, and the average size reached 25 μm with uniform size. When the CNTs content was 20 wt%, some large aggregates began to appear on the surface, and the average size of the aggregates increased to ~75 μm. When the CNTs content was further increased to 30 wt%, the aggregate size of the coating surface was about 100 μm, and the coating had a denser micro/nano microstructure, similar to that of the lotus leaf surface.

In order to further observe the changes in the surface topography of coatings, CLSM was used and the results are shown in Figure 9.15. When the content of CNTs increased from 5 to 30 wt%, the protrusions on the surface of the sample gradually became larger, and the depth between the protrusions at adjacent sites increased from 20 to 100 μm. As shown in Figure 9.16, with the increase of CNT content, the arithmetical mean of coating surface roughness (R_a) increased from 2.98 ± 0.51 to 25.45 ± 0.44 μm. It can be seen from the above results that a higher CNT content would form a larger aggregate, and further roughen the coating surface to form a micro/nanocomposite microstructure.

9.3.2.2 Surface Hydrophobicity of CNTs/Epoxy Coatings with Different CNT Content

As the surface roughness increased, more pores were created between the coated surface and the water droplets to trap air, forming a physical barrier to reduce the contact between the water droplets and the surface. Figure 9.17 shows the change of WCAs on the coating surface with the increase of CNTs content. The surface of the composite coating with 5 wt% CNTs content had a certain hydrophobicity, and the WCAs were about 89.3°. With the increase of CNTs content to 10 and 20 wt%, the hydrophobic angle of the coating gradually increased to 118.2° and 144.9°, respectively. When the CNTs content increased to 30 wt%, the WCA of the coating surface reached 154.1°. Figure 9.18 shows the surface wettability test for an epoxy coating containing 30 wt% CNTs. When the sample inclination angle was 7°, 10 μl water droplets easily rolled off the coating surface, indicating that the CNTs/epoxy composite coating containing 30 wt% CNTs satisfied the requirements of the superhydrophobic surface.

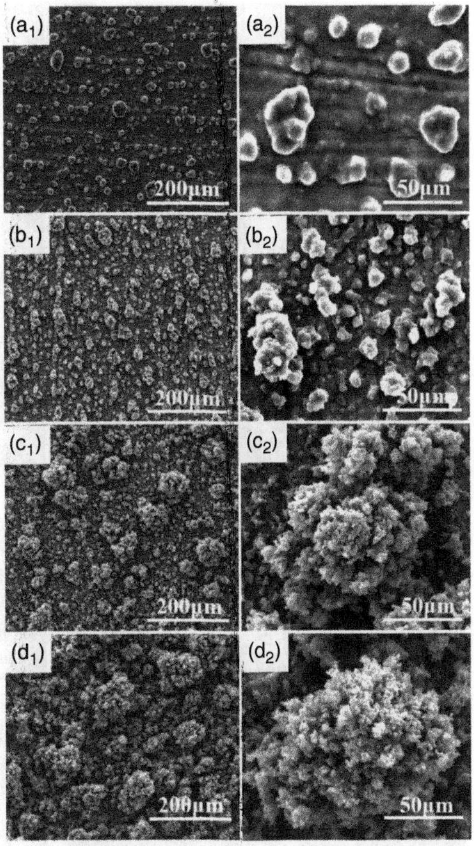

Figure 9.14 The surface morphology of CNTs/epoxy composite coatings with different CNTs content: (a) 5; (b) 10; (c) 20; (d) 30 wt%.

9.3.2.3 Anti-Icing Properties of CNTs/Epoxy Composite Coatings

The anti-icing properties of material surfaces are of great significance to the safe and economical operation of power generation, machinery, and transportation systems. The superhydrophobic surface has a certain anti-icing effect because water droplets are not easy to stay on the surface of materials. In this experiment, a self-made anti-icing device was used to characterize the anti-icing performance of CNTs/epoxy composite coating. First, the coating sample was placed at the bottom of the ice chamber at a tilt angle of 30°, and 0 °C water droplets were added from the top of the ice chamber. Figure 9.19 shows the surface morphology of the composite coating with different CNTs content after 30 minutes of dynamic icing. It can be seen that the coating surface with CNTs content of 5, 10, and 20 wt% had white solid ice adhesion, while the coating surface with 30 wt% CNTs had no solid ice. This is mainly due to the different hydrophobicity of the coating. When the content of CNTs was small, the hydrophobicity of the coating surface was poor, so the rolling of water droplets on the coating surface was difficult, leading to some water droplets staying on the surface. In contrast, the surface of the composite coating with 30 wt% CNTs content exhibited the superhydrophobic property, and the water droplets easily rolled off the coating surface [32].

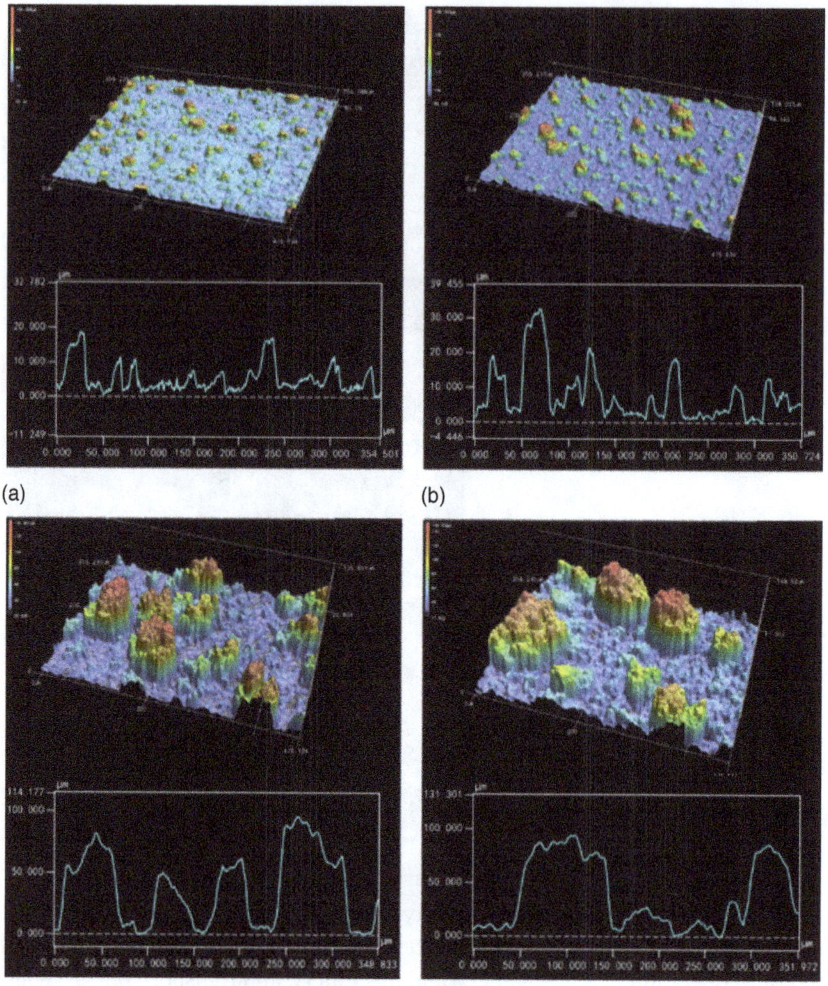

(a) (b) (c) (d)

Figure 9.15 The surface topography of CNTs/epoxy composite coatings with different CNTs content: (a) 5; (b) 10; (c) 20; (d) 30 wt%.

To further illustrate the anti-icing performance of different coatings, Figure 9.20 shows the results of static icing tests [33] to simulate the icing morphology of water droplets on the surface. When the CNTs contents were 5 and 10 wt%, the water droplets showed ellipsoid on the coating surface, and there was a large solid–liquid contact area between the surface and the water droplets. As the CNTs content increased to 20 and 30 wt%, the water droplets were nearly spherical on the coating surface [34]. For the coating with 30 wt% CNT, a tilting angle of 45° led to the rapid falling off of ice particles from the coating surface. The results show that the superhydrophobicity reduced the adhesion of ice particles on the coating surface [35], and effectively improved the anti-ice performance of the coating.

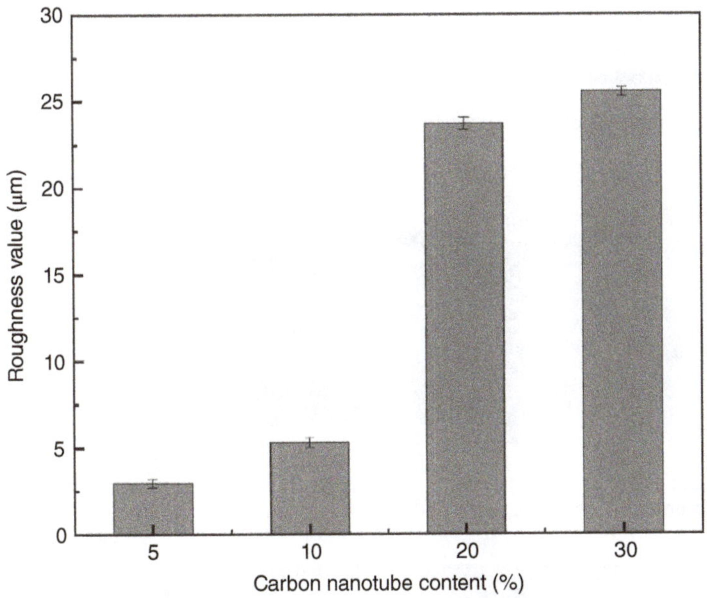

Figure 9.16 Surface roughness values of CNTs/epoxy composite coatings with different CNTs content.

Figure 9.17 Water contact angles of different CNTs/epoxy composite coatings.

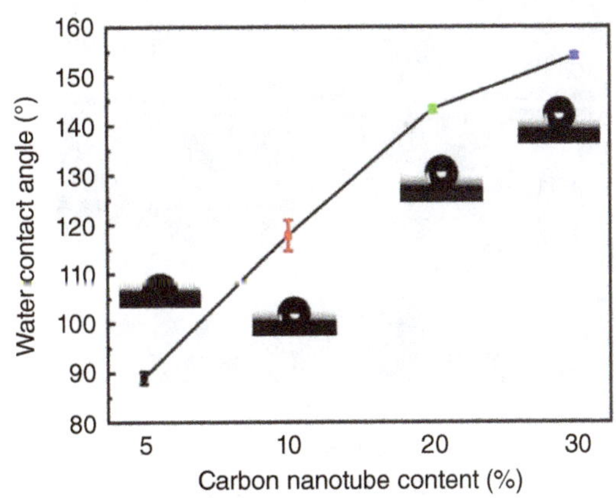

9.3.2.4 Mechanical Durability of CNTs/Epoxy Composite Coatings

Superhydrophobic surfaces often face the problem of poor mechanical durability during use, which greatly limits the wide application of superhydrophobic materials. At present, most of the superhydrophobic coatings may damage their microstructure during use even if they are slightly pressed by fingers or slightly rubbed [36, 37]. In this experiment, the mechanical durability of the superhydrophobic coating was evaluated by the tape stripping method. Figure 9.21a shows the WCAs of the superhydrophobic coating surface with CNTs content of 30 wt% after the tape stripping test. The WCAs began to decrease to below 150° after 100 times of tape stripping [38]. As can be seen from Figure 9.21b,c, after the tape

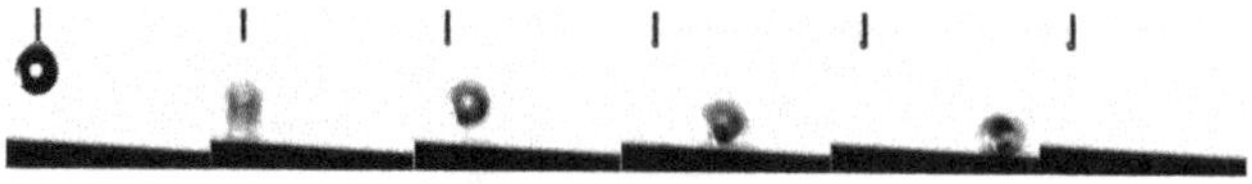

Figure 9.18 Video snapshot of 10 μl water droplets rolling on a 30 wt% CNTs/epoxy composite coating.

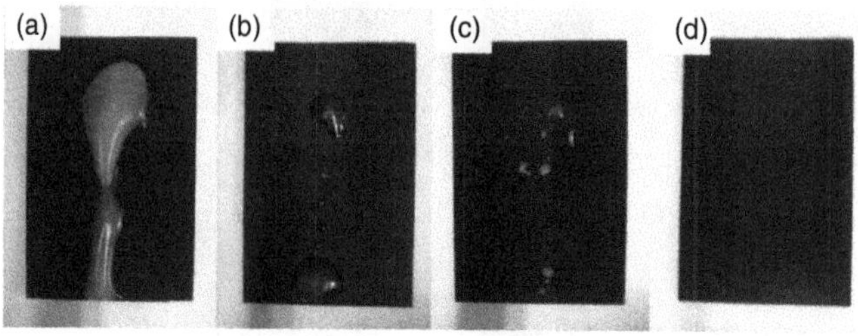

Figure 9.19 Photos of dynamic icing of water droplets on CNTs/epoxy coatings with different CNTs content: (a) 5, (b) 10, (c) 20, (d) 30 wt%.

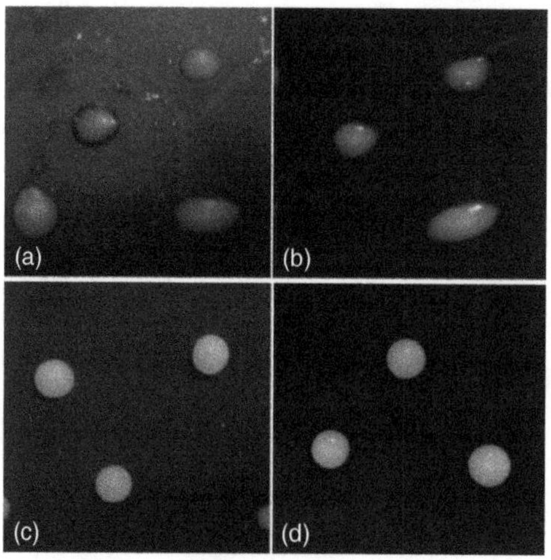

Figure 9.20 Static icing images of water droplets on CNTs/epoxy coatings with different CNTs content: (a) 5; (b) 10; (c) 20; (d) 30 wt%.

stripping test [39], the surface of the superhydrophobic coating still retained a good micro-nanostructure. This coating exhibited better mechanical durability compared with the $nSiO_2$/epoxy resin coatings in Section 9.2, which indicates that the superhydrophobic coating containing CNTs could offer better mechanical durability.

CNTs/epoxy composite coating has good mechanical durability for the following reasons – First of all, compared with most micro/nano superhydrophobic structures, the epoxy resin content of the prepared coating reached 70 wt%, and a stable three-dimensional cross-linked network [40] structure was formed after curing, which can firmly wrap CNTs in the epoxy resin and improve the mechanical durability of the coating. Second, CNTs had a winding agglomeration in the epoxy resin, thus forming a micro/nano coexistence structure. When the surface was damaged by the adhesive stripping test, part of the nanostructure was damaged,

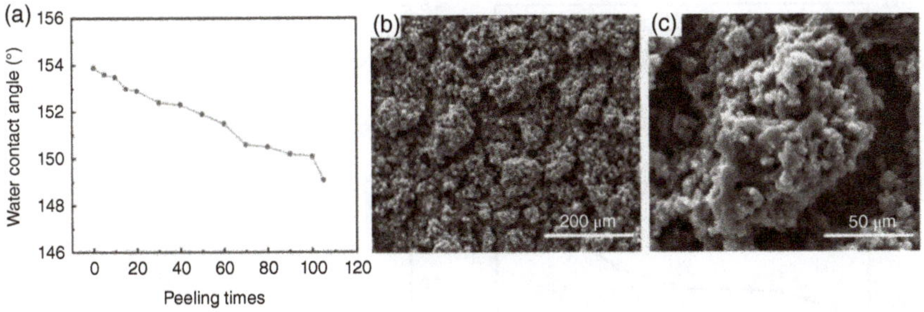

Figure 9.21 (a) The change of WCA of 30 wt% CNTs/epoxy composite coating during tape stripping test; (b,c) SEM image of the surface microstructure of the coating after 100 times of tape stripping.

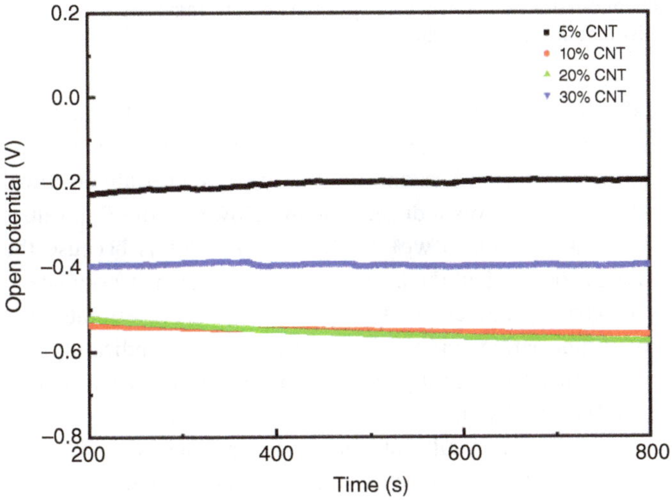

Figure 9.22 E_{ocp} values of epoxy composite coatings with different CNTs contents immersed in 3.5 wt% NaCl solutions.

but the micron structure of the surface still existed, so that the superhydrophobic function was retained. Finally, CNTs had a high Young's modulus and tensile strength [41], and the high entanglement of aggregates also enhanced the stripping resistance of the surface microstructure [42].

9.3.2.5 Electrochemical Test of CNTs/Epoxy Composite Coatings

The effect of CNTs content in 3.5 wt% NaCl solution on the corrosion resistance of CNTs/epoxy composite coating was studied by electrochemical experiments. The open circuit potential (OCP) value can reflect the sensitivity and corrosion tendency of the corrosion process from the thermodynamic point of view. The more positive the open circuit potential, the lower the corrosion tendency and sensitivity. Figure 9.22 shows the open circuit potential (E_{ocp}) values of different coating surfaces [43]. The E_{ocp} value of the composite coating with 5 wt%CNT content was −0.18 V VS_{SCE}, and when the CNT content increased from 5 to 20 wt%,

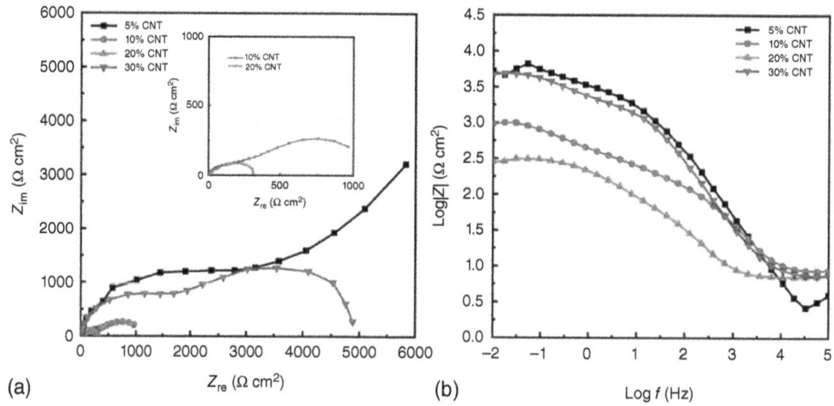

Figure 9.23 EIS results of different CNTs/epoxy composite coatings immersed in 3.5 wt% NaCl solution: (a) Nyquist diagram, (b) Bode plots.

the E_{ocp} value gradually decreased to −0.58 V VS_{SCE}. This is due to the moderate hydrophobic properties of the coating surface; the solution would dip into the depression position of the coating surface, and the coating had a certain electrical conductivity so that the E_{ocp} value was reduced [44–46]. However, the E_{ocp} values were increased for the coating with 30 wt% CNTs content. This is because the coating with 30 wt% CNTs content had the best superhydrophobic performance to form an air film on the coating surface. As a good insulator, air deteriorated the electrical conductivity of the coating, and the E_{ocp} value increased, indicating that the superhydrophobic coating significantly reduced the corrosion sensitivity of carbon steel substrate in NaCl solution.

Figure 9.23 shows the EIS test results of composite coatings with different contents of CNTs. Figure 9.23a shows the Nyquist diagram of the coatings, which shows two arcs of capacitive resistance. The arc resistance of the high-frequency region reflects the barrier effect of the coating. The corrosion-resistant effect of the coating came from the barrier effect of the coating itself and the air film in the rough microstructure of the coating surface. The second arc of capacitive reactance observed in the low-frequency region was usually related to the electrochemical reaction occurring at the coating/interface. With the increase of CNTs content, the arc diameter in the Nyquist plot decreased first and then increased, and reached the minimum in coatings with CNTs content of 20 wt%. As can be seen from Figure 9.23b, the low-frequency impedance modulus of the coating ($|Z|_{0.01\ Hz}$) showed the same changing trend with the increase of CNTs content. $|Z|_{0.01\ Hz}$ value is often used for semiquantitative evaluation of barrier properties of coatings – the higher the $|Z|_{0.01\ Hz}$ value, the better the corrosion resistance of the material. When the CNTs content was 5 wt%, CNTs were uniformly dispersed in the coating, and no effective conductive connection had been formed, and it was only used as a barrier material to prevent contact between the corrosive medium and the metal material. With the increase of CNTs content to 10 and 20 wt%, CNTs were overlapping with each other, which makes it easy to form a highly conductive network, resulting in

Table 9.3 Electrical resistance of different CNTs/epoxy composite coatings.

CNTs content (wt%)	5%	10%	20%	30%
Electrical resistance (KΩ)	4.48 ± 0.76	0.74 ± 0.08	0.16 ± 0.03	0.06 ± 0.01

a rapid decline of $|Z|_{0.01\,Hz}$ value. When the CNTs content was further increased to 30 wt%, the coating surface formed a superhydrophobic structure, and the large surface roughness made the coating surface with a large amount of nonconductive air, which inhibited the corrosive medium from reaching the metal substrate surface.

Table 9.3 shows the electrical resistance of the coatings. During measurement, the distance between the coating surface and the probe of the multimeter was kept at 1 cm. When the CNTs content increased from 5 to 30 wt%, the coating resistance decreased sharply from 4.48 ± 0.76 to 0.06 ± 0.01 KΩ. This shows that the composite coating has essentially become an excellent conductor due to its high CNTs content, rather than the traditional protective coating.

9.4 Shape Memory Self-healing Superhydrophobic Coating Based on SiO_2–CNT Hybrids

Most of the superhydrophobic coatings have excellent waterproof [47], self-cleaning, and anti-fouling [48] properties, so they have important applications in the fields of solar cells, display screens, lenses, goggles, windshields, and so on. However, the outdoor use of superhydrophobic coatings is still greatly limited, and it is highly susceptible to human scratches or material friction, resulting in a significant reduction in surface hydrophobicity and protective ability.

In order to improve the service life of superhydrophobic coatings, four ways to achieve self-healing superhydrophobic coatings are proposed in the literature review. For example, the superhydrophobic coating can store low surface energy substances [49] in the coating, and use external humidity, temperature, organic solvents, light, and other stimuli to migrate low surface energy substances to the coating surface, which could restore the hydrophobic property of the coating surface. Methods for storing low surface energy substances in coatings include layer self-assembly, ultraviolet curing [50], and Pickering emulsion polymerization. The preparation process is complex and time-consuming, and most of them are not suitable for industrial production. In addition, the superhydrophobic coating can also realize the self-healing effect of the hydrophobic ability of the coating surface by restoring the microstructure and morphology of the coating surface [51]. For example, Qian et al. [52] prepared a self-healing superhydrophobic coating with temperature-sensitive shape memory effect by changing the flexibility of the molecular chain by adjusting the amount of the toughening agent in the epoxy resin. When heated at a temperature higher than 60 °C, the shape memory effect of polymer resin is triggered, causing the surface topography of the damaged coating to recover and the scratch defect in the coating to close.

According to previous studies, when the additive content was 30 wt% relative to the resin system, the superhydrophobic properties, mechanical properties, and anti-icing properties of the composite coatings achieved the best performance. In order to achieve the self-healing effect of the superhydrophobic coating, this section incorporated SiO_2–CNT hybrid in the shape memory epoxy coating, and the ratio of flexible curing agent and epoxy resin was adjusted to make the resin system with better flexibility, so as to achieve the self-healing ability based on the shape memory effect.

9.4.1 Materials and Methods

9.4.1.1 Materials

The materials and reagents used in this section are shown in Table 9.4.

9.4.1.2 Preparation of SiO_2–CNT Hybrid/Epoxy Shape Memory Coating

Two parts of E-51 epoxy resin with a mass of 1.875 g were placed into a beaker and labeled as E-51A and E-51B, respectively. To reduce the viscosity of the epoxy resin, 30 ml of acetone solution was added to each beaker. Then, 0.9 g of SiO_2–CNT hybrid was added to beaker A, and 1.125 g of SiO_2–CNT hybrid was added to beaker B for ultrasonic dispersion for one hour, so that the hybrid was evenly dispersed in the epoxy resin. According to the mass ratio of epoxy resin to curing agent of 1 : 0.6 and 1 : 1.0, 1.125 g of curing agent polyether amine D-400 was added to beaker A, and 1.875 g of curing agent polyether amine D-400 was added to beaker B, and the beaker was placed in the ultrasonic dispersion instrument for 15 minutes to obtain a uniformly dispersed hybrid suspension. Using a spray gun, the suspension-containing hybrid was sprayed three times on the pre-polished aluminum sheet. After curing in a 60 °C incubator for 48 hours, the hybrid/epoxy resin coatings with different epoxy resin ratios were obtained. In this work, E106 coating and E110 coating were used to represent the coatings prepared by the ratio of epoxy resin and curing agent 1 : 0.6 and 1 : 1.0, respectively.

Table 9.4 Experimental materials and reagents.

Materials	Purity	Company
SiO_2 particles	Nanosize	Guihuang Metal Materials Co., LTD
Multiwalled carbon nanotube	—	Suzhou Carbon Feng Technology Co., LTD
KH560	Analytical purity	Sinopharm Group
Acetone	Analytical purity	Sinopharm Group
E-51 epoxy resin	Analytical purity	Aladdin, Inc.
Polyetheramine (D400)	Analytical purity	Aladdin, Inc.

9.4.1.3 Glass Transition Temperature (T_g) Test of Coatings

The heat release rate of coatings was measured by Q2000 V24.11 build differential scanning calorimeter (DSC). The nitrogen flow rate was 50 ml min^{-1}, the temperature range was −20–120 °C, and the heating rate was 10 °C min^{-1}.

9.4.1.4 Surface Characterization of Coatings

SEM (SU1510 system) was used to observe the microscopic morphology of coatings. The acceleration voltage selected for the test was 10 kV. Before the test, gold was sprayed on the sample surface to improve the electrical conductivity. The surface roughness of the coating was analyzed and measured by CLSM (KEYENCE VK-X).

To verify the superhydrophobic property of coatings, the WCA and rolling angle of the coating surface were measured at room temperature using a dataphysics OCA20 contact angle measuring instrument with a drop volume of 6 μl. By tilting the sample, the smallest angle at which water droplets rolled off the sample surface was defined as the rolling angle. In order to reduce the experimental error, five locations on each sample surface were selected for testing, and the average of five measured values was taken as the final data.

9.4.1.5 Electrothermal Performance Test

A voltage of 6 V was applied on both sides of the coating, the temperature of the coating surface was measured using a point thermometer, and the temperature change was recorded. After the coating surface temperature stabilized, the WCA and surface temperature of the coatings were measured by dataphysics OCA20 measuring instrument every one hour.

9.4.1.6 Surface Mechanical Durability

The coating surface was rubbed with a tribometer (HD8837 model), with a weight load of 200 g to ensure sufficient friction between the rubber and the coating surface until the superhydrophobic properties of the coating failed.

9.4.1.7 Electrochemical Analysis

The corrosion-resistant properties of coatings were tested by EIS in 3.5 wt% NaCl solution at room temperature using the PARSTAT 2273 electrochemical workstation. The test frequency ranges from 10^5 to 10^{-2} Hz, and the disturbance potential was 20 mV.

9.4.2 Characterization of SiO_2–CNTs/Epoxy Composite Coating

9.4.2.1 T_g of SiO_2–CNTs/Epoxy Composite Coating

The epoxy resin system in this study was heated above the T_g to make it with a highly elastic state and trigger the shape memory effect to achieve self-healing. Figure 9.24 shows the DSC curves of pure epoxy resin with different proportions and resin systems with SiO_2–CNTs hybrid in the temperature range of −20 to 120 °C obtained by differential scanning calorimetry. As can be seen from the figure, the T_g of the coating with a ratio of 1 : 0.6 was 28 °C, and the T_g of the coating with

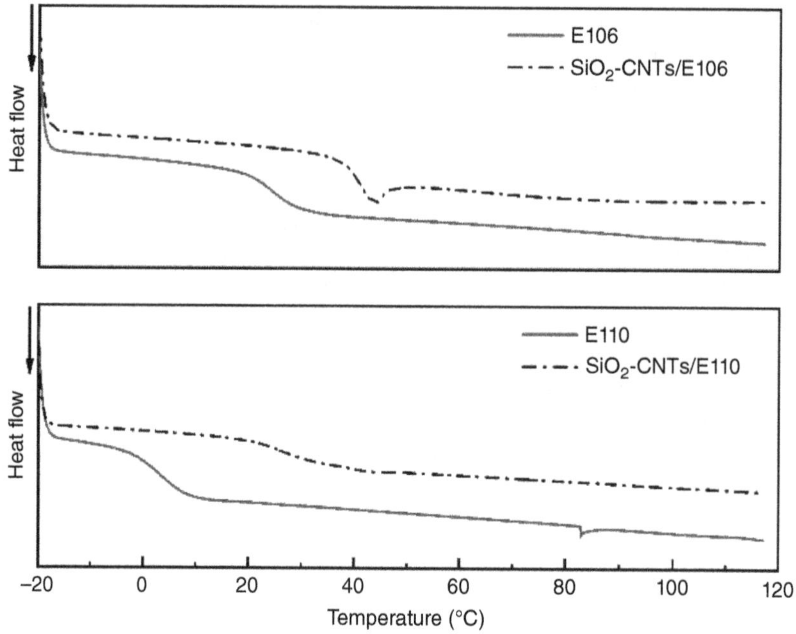

Figure 9.24 DSC images of E106 and E110 based coating systems.

the addition of SiO_2–CNTs hybrids was 40 °C. The T_g of the coating with the ratio of 1 : 1.0 was 10 °C, and the T_g of the coating with the addition of SiO_2–CNTs hybrid was 25 °C. It is concluded that the T_g of the coating decreased with the increase of curing agent content. This is because the curing agent polyether amine contained flexible chain segments. As the temperature increased or decreased, the long chain structure of the flexible chain segment could shrink, stretch, or rotate freely, and the flexible chain segment was combined with the epoxy resin through the bonding of chemical bonds to form a loose interphase network structure, and the chain segment was more likely to migrate when the temperature exceeded the T_g value. With the increase of the content of flexible curing agent, the toughness of the resin system increases, and the T_g decreases. The T_g of resin systems with different proportions showed an increasing trend after the addition of SiO_2–CNTs hybrid. The SiO_2–CNTs hybrid contained CNTs with high rigidity. When CNTs were added to the epoxy resin, it became an obstacle to the movement of the molecular chain of the epoxy resin to a certain extent and reduced the migration ability of the molecular chain segment of the resin, in which way the T_g value was increased [53].

9.4.2.2 Surface Hydrophobicity of SiO_2–CNTs/Epoxy Composite Coating

Figure 9.25 shows the WCA results when the coatings were in an intact, scratched, and healed state. The coatings were scratched using a scalpel with a scratch width of approximately 50 μm and were healed by heating in an oven above the T_g temperatures. Figure 9.25a–c shows the WCAs of the E106 coating. The WCA of the intact coating was 159.4°, the WCA of the scratched coating was 131.9°, and the WCA of

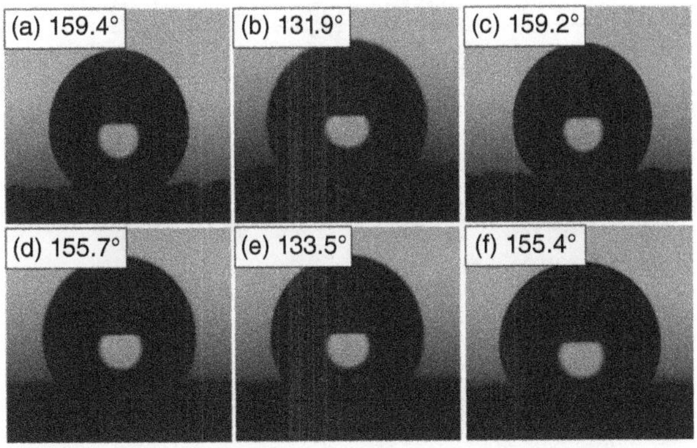

Figure 9.25 The WCAs of (a) the unscratched E106 coating, (b) the scratched E106 coating, and (c) the healed E106 coating; the WCAs of (d) the intact E110 coating, (e) the scratched E110 coating, and (f) the healed E110 coating.

the healed coating was 159.2°. Figure 9.25d–f shows the WCAs of the E110 coating, which had the same variation trend as the WCAs of the E106 coating. The WCA of the intact coating was 155.7°, the WCA of the scratched coating was 133.5°, and the WCA of the healed coating was 155.4°. It shows that the superhydrophobic properties of the healed coatings can be completely restored. Second, the WCAs of E110 coating in both intact and scratched conditions were lower than those of the E106 coating, which was due to the increase in the content of curing agent, the increase in the dispersion degree of SiO_2–CNTs hybrid in the resin system, and the decrease in surface roughness.

Figure 9.26 shows WCAs variation of the two coatings during 10 cycles of scratch damage and healing. When the coating surface was scratched, the hydrophobicity decreased, and it returned to the superhydrophobic state after thermal healing. After multiple cycles of scratch damage and healing, the coating still maintained the superhydrophobic property. It indicates that the toughening of epoxy resin could not only improve the toughness of the material but could also reduce the damage of

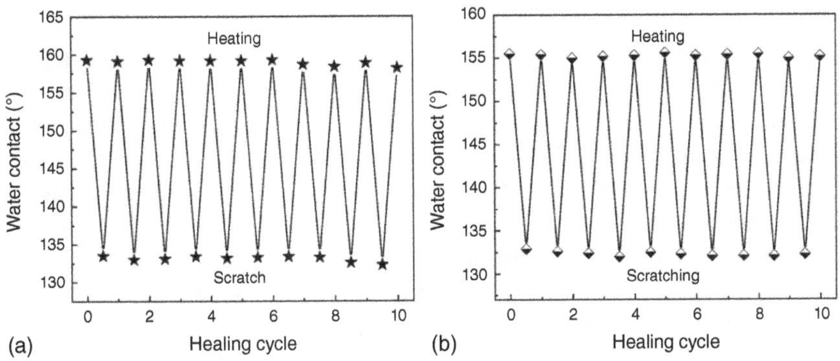

Figure 9.26 WCAs of (a) E106 coating, and (b) E110 coating during multiple scratch damage and healing cycles.

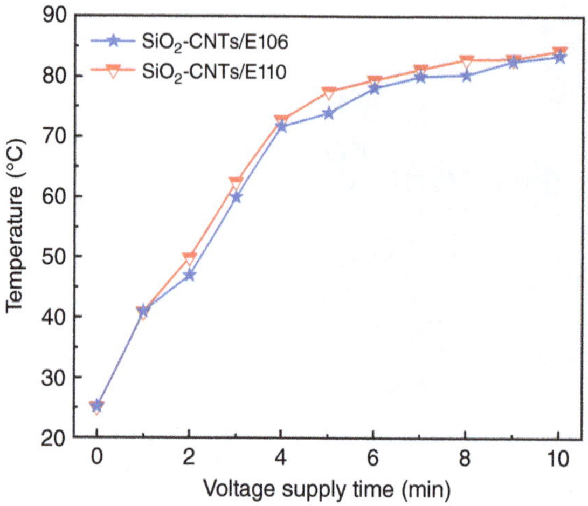

Figure 9.27 Surface temperature curves of the two coatings with SiO_2-CNTs hybrid during voltage application.

the material through the introduction of flexible chain segments, in which way the service life of the superhydrophobic coating was extended.

The healing mechanism of scratched coatings can be explained as follows – when the coating is destroyed, the surface temperature is heated to be higher than the T_g of the resin matrix. With the increase in temperature, the polymer resin becomes soft and returns to a highly elastic state. The rebound of the resin near the damaged area makes the scratch significantly smaller and covers the underneath metal substrate. The SiO_2–CNTs hybrid in the polymer resin moves together with the resin, so that the micro-morphology and roughness of the defect are restored, and the superhydrophobic property of the coating surface is recovered.

9.4.2.3 Electrothermal Property of SiO_2–CNTs Hybrid/Epoxy Coatings

In the previous section, we proved that the CNTs can convert electrical energy into heat energy to achieve the effect of heating coatings, which can activate the shape memory effect of damaged epoxy coating. A voltage of 6 V was applied to the E106 and E110 coatings. Figure 9.27 shows the temperature change of the coating surfaces with the addition of SiO_2–CNTs hybrid. After five min of voltage supply, the temperature of the coating surfaces rapidly increased to 70 °C, and then slowly reached 82 °C and tended to equilibrium with further application of voltage [54]. The temperature of the SiO_2–CNTs/E106 coating was slightly higher than that of the SiO_2–CNTs/E110 coating. This is due to the fact that although the mass of SiO_2–CNTs hybrid added to the epoxy resin was the same, the proportion of resin and curing agent was different so that the percentage of SiO_2–CNTs hybrid in the E110 coating was relatively low, resulting in a decrease in the temperature of the surface of the SiO_2–CNTs/E110 coating.

A voltage of 6 V was applied to the coatings, and the WCAs of the coatings were measured to test the stability of the superhydrophobic ability. The test results are shown in Figure 9.28. When the constant voltage current was applied to the coating surface for 10 hours, the WCAs of the coating surface were measured every one hour. The superhydrophobicity and maximum temperature of the two coatings

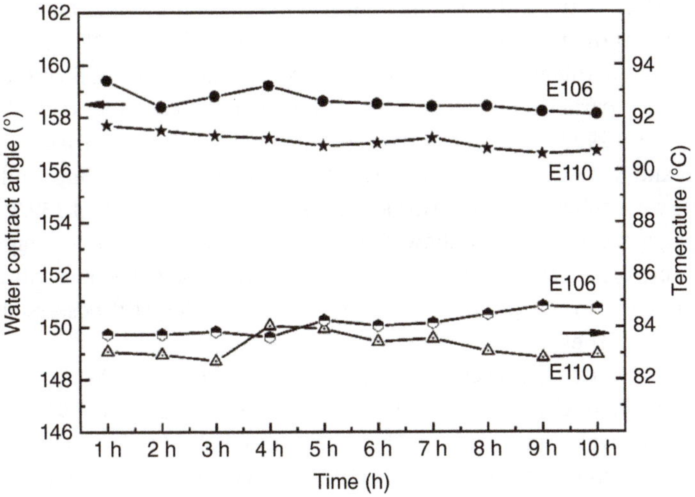

Figure 9.28 Variation of the WCAs and coating surface temperatures during long-term voltage application.

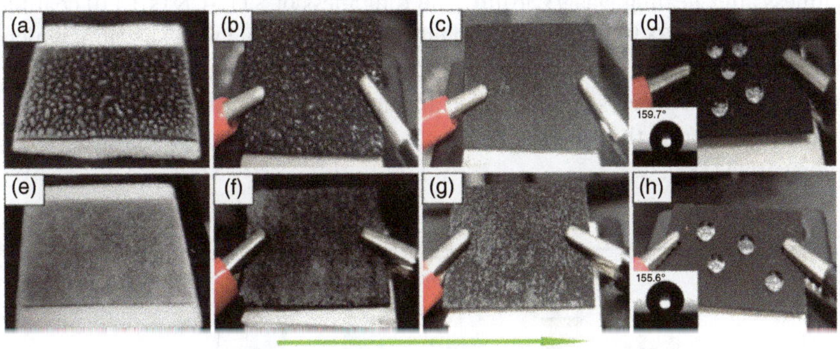

Figure 9.29 Melting process of coating surfaces under voltage application: (a) initial icing state of SiO_2-CNTs/E106 coating; (b) thawing state of SiO_2-CNTs/E106 coating after 80 s of voltage apply; (c) thawing state of SiO_2-CNTs/E106 coating after 100 s of voltage application; (d) the state of water droplets on the SiO_2-CNTs/E106 coating after 200 s of voltage application; (e) initial icing state of SiO_2-CNTs/E110 coating; (f) thawing state of SiO_2-CNTs/E110 coating after 80 s of voltage application; (g) thawing state of SiO_2-CNTs/E110 coating after 100 s of voltage application; (h) the state of water droplets on the SiO_2-CNTs/E110 coating after 200 s of voltage application.

remained unchanged, indicating that the electrothermal conversion capacity and superhydrophobic capacity of the coating were stable and could be used for deicing applications.

Figure 9.29 shows the melting of ice attached to the coating surfaces. Figure 9.29a shows the icing of the SiO_2–CNTs/E106 coating surface after five min on the cold table at −5 °C. Figure 9.29b,c show the melting of surface ice particles when a voltage of 6 V was applied for 80 and 100 seconds. It can be seen that after 100 s, the ice particles on the surface of the coating completely melted. Figure 9.29e shows the icing on the surface of the E110 coating after five minutes on the cold table at −5 °C. Figure 9.29f,g show the melting of ice on the surface of SiO_2–CNTs/E110 coating after applying a voltage of 6 V for 80 and 100 seconds. After applying a voltage of 100, the ice particles on the surface of the coating completely melted. As shown in Figure 9.29d,h, 6 μl water droplets were added to the surface of SiO_2–CNTs/E106 and SiO_2–CNTs/E110 coatings after ice melting and drying, and the water droplets were spherical on the surface of the coating, with WCA of 159.7° and 155.6°, respectively. The results show that the surface of the two coatings still maintained micro/nanostructure after ice melting, the applied voltage did not affect the surface morphology of the coatings, so the coating still possessed superhydrophobic properties.

When the coatings are damaged and electrically healed, the WCAs of SiO_2–CNTs/E106 and SiO_2–CNTs/E110 coatings changed from 133.2° to 158.9° and 133.4° to 155.6°, respectively, indicating that the superhydrophobic properties of the two coatings were recovered. Since the change of WCAs of coatings is related to the roughness of the coating, the surface roughness of the two coatings before and after electrical healing was measured by CLSM, and the results are shown in Figure 9.30. The roughness of SiO_2–CNTs/E106 coating before healing was 75.24 μm, and the roughness after healing was 93.56 μm. The roughness of SiO_2–CNTs/E110 coating before healing was 78.91 μm, and the roughness after healing was 94.82 μm, which showed consistent values compared with the coatings without scratches [54]. It shows that the SiO_2–CNTs hybrid, which affected the roughness of the coating, flew with the resin during the rebounding process, so as to restore the roughness of the coating surface [55].

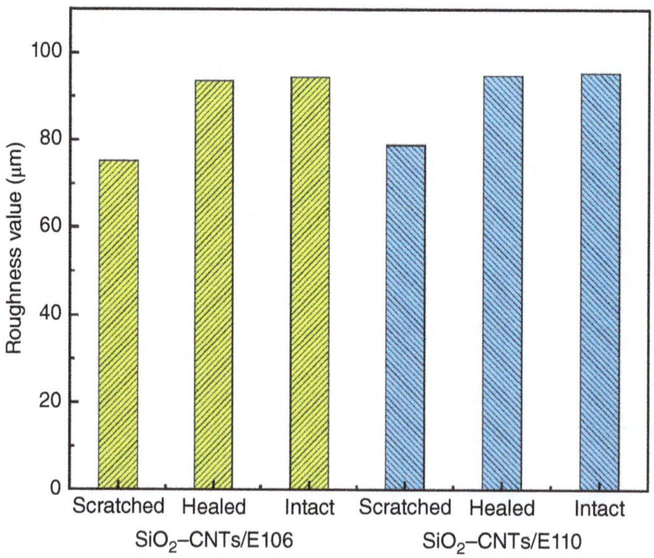

Figure 9.30 Surface roughness and WCAs of the two coatings before and after electrical healing.

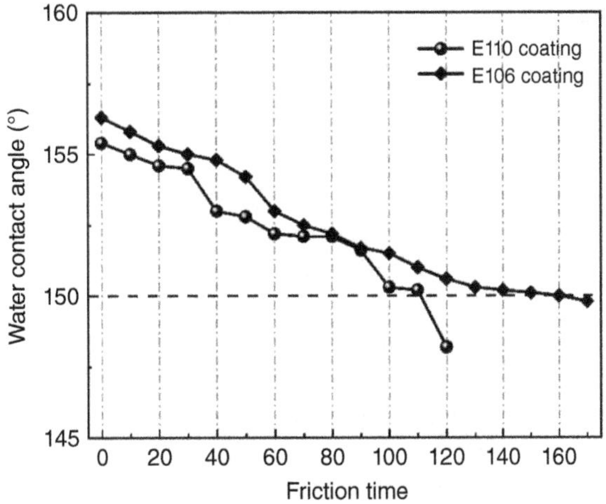

Figure 9.31 WCAs variation of SiO_2–CNTs/E106 and SiO_2–CNTs/E110 coatings after different friction times.

9.4.2.4 Mechanical Durability of SiO_2–CNTs Hybrid/Epoxy Coatings

Figure 9.31 shows the WCA change of SiO_2–CNTs/E106 and SiO_2–CNTs/E110 coatings after different friction times. The WCAs of coatings changed a little in the early friction stage and decreased in the later friction stage. This is because at the early stage of the friction process, the SiO_2–CNTs hybrid on the coating surface was firmly kept, and the hybrid that affected the coating roughness moved with the resin during the resin rebound process [56]. However, with the increase of friction times, the SiO_2–CNTs hybrid on the coating surface began to get loose, so the WCAs reduced. After 120 times of surface friction on the SiO_2–CNTs/E110 coating, the WCA decreased to 148.2°, and the coating surface no longer retained the superhydrophobic state. After 170 times of friction on the surface of SiO_2–CNTs/E106 coating, the WCA of the coating was 149.8°. Therefore, the friction resistance of SiO_2–CNTs/E106 coating was better than that of the SiO_2–CNTs/E110 coating, which was due to the higher content of curing agent polyether amine D400 flexible chain segment in the SiO_2–CNTs/E110 coating [57].

First of all, the additional amount of curing agent in the E110 coating was relatively large, and after reacting with the epoxy resin, excess curing agent did not participate in the reaction [58]. When the external temperature reached the T_g of the resin system, it was conducive to the movement of chain segments, and the resin with shape memory effect began to deform, thus achieving the self-healing effect of the coating [59]. The SiO_2–CNTs/hybrid in the resin also moved with the resin, generally recovering both the surface roughness and WCA property of the composite coating. On the other hand, the additional amount of curing agent in the E106 coating was relatively less, therefore there were fewer flexible chain segments after coating curing and fewer chain segments that could move freely in the resin system [60]. After curing, the epoxy resin formed a three-dimensional network structure, which was not conducive to the rebound of the resin system. When the temperature reached the T_g of the coating, the chain segments moved and the coating began to rebound. Second, the curing agent affected the T_g of the resin system, which led to the T_g of the E110 coating being higher than that of the E106 coating. In addition, due to the high curing agent amount in the E110-based coating system, there were more flexible chain segments in the coating and less stable three-dimensional network structure, which reduced the wear resistance of the coating [61]. The stable three-dimensional network structure of the E106-based composite coating after curing occupied a higher proportion. The rigid benzene ring structure of the epoxy resin structure [62] was also conducive to the wear resistance of the coating.

9.4.2.5 Electrochemical Test of SiO_2–CNTs Hybrid/Epoxy Coatings

SiO_2–CNTs/E106 coating and SiO_2–CNTs/E110 coating were scratched and thermally healed at a temperature above the T_g value. The scratched SiO_2–CNTs/E106 coating was used as the reference sample. Figure 9.32 shows the Nyquist and Bode plots of the EIS test results of three coatings after different immersion times in 3.5 wt% NaCl solution. For the healed SiO_2–CNTs/E106 coating, due to the restoration of superhydrophobic property, the impedance in the low-frequency region was relatively high at the early stage of immersion, and the value of $|Z|_{0.01\,\text{Hz}}$ was about $8.7 \times 10^5\,\Omega\,\text{cm}^2$. With the extension of immersion time, the air layer stored on the superhydrophobic surface gradually escaped and the impedance decreased accordingly [63] and gently from the third day to the seventh day of immersion time. This indicates that after the coating lost the air layer on the superhydrophobic surface, the healed coating still had the barrier performance between corrosive medium and metal substrate. Figure 9.32c,d shows the EIS measurement results after the healed SiO_2–CNTs/E110 coating was immersed in 3.5 wt% NaCl solution for different times. After healing, the superhydrophobic property of the coating recovered, and the initial $|Z|_{0.01\,\text{Hz}}$ value was as high as $5.7 \times 10^5\,\Omega\,\text{cm}^2$. With the extension of immersion time, the superhydrophobic property was lost, and the impedance decreased accordingly. After the third day of immersion, the declining trend of the $|Z|_{0.01\,\text{Hz}}$ became stabilized, indicating that the healed coating could effectively delay the infiltration of corrosive media [64].

Figure 9.32e,f show the EIS measurement results after the unhealed SiO_2–CNTs/E106 coating was immersed in 3.5 wt% NaCl solution for different times. With the extension of immersion time, the radius of the impedance arc showed a decreasing trend. The $|Z|_{0.01\,\text{Hz}}$ value was about $2 \times 10^5\,\Omega\,\text{cm}^2$ after the first day of immersion, lower than the healed SiO_2–CNTs/E106 coating and SiO_2–CNTs/E110 coating. After immersion for seven days, the $|Z|_{0.01\,\text{Hz}}$ value was only $1.0^3 \times 10^5\,\Omega\,\text{cm}^2$, which was lower than the two healed coatings. It can be seen from the above results that it is difficult to form a continuous protective gas layer at the scratch of the coating, and the solution penetrates through the scratch, resulting in corrosion of the metal matrix [65].

Figure 9.32g shows the variation trend of the $|Z|_{0.01\,\text{Hz}}$ values of the three coatings with the extension of immersion time. As can be seen from the figure, the impedance values of both healed and scratched coatings showed a decreasing trend with the extension of immersion time, but the impedance values of healed coatings were higher than that of damaged coating [66], and the corrosion resistance of healed

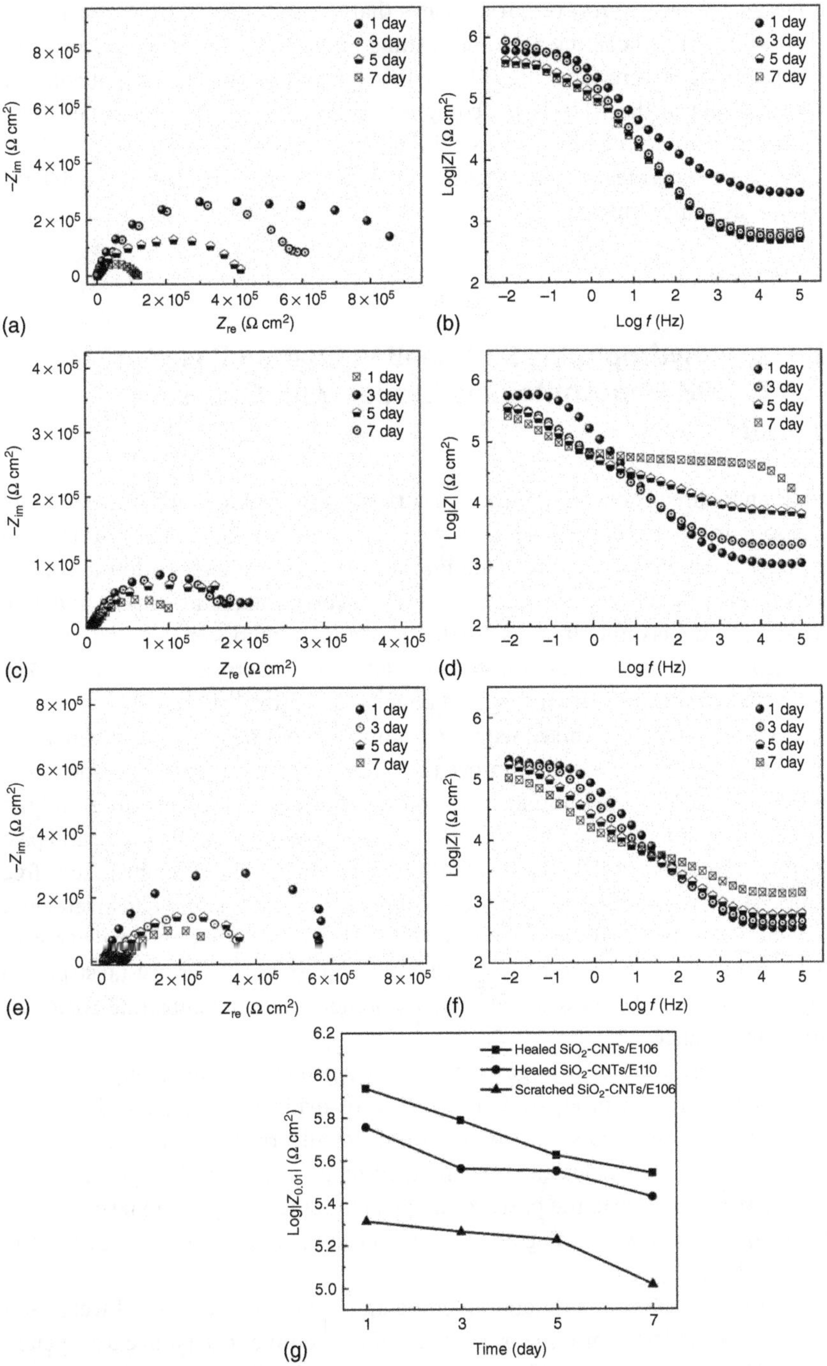

Figure 9.32 Nyquist and Bode plots of different coatings after immersion in 3.5 wt% NaCl solution for different time: (a,b) healed SiO_2–CNTs/E106 coating; (c,d) healed SiO_2–CNTs/E110 coating; (e,f) scratched SiO_2–CNTs/E106 coating; (g) Comparison of $|Z|_{0.01\,Hz}$ values of three coatings.

coatings significantly improved. It indicates that the SiO_2–CNTs/E106 coating and the SiO_2–CNTs/E110 coating with self-healing properties have good anticorrosion effects [67]. There were fewer flexible chain segments in the E106-based composite coating than in the E110-based composite coating, and the three-dimensional network structure formed by the cross-linking was denser in the SiO_2–CNTs/E106 coating, and the permeability resistance to NaCl solution was better to provide the corrosion resistant property [68].

9.5 Superhydrophobic Self-healing Coating Based on CNTs/LDH Nanocontainers Loaded with Corrosion Inhibitors

In addition to directly covering the metal surface with organic coating, the addition of corrosion inhibitors can further improve the corrosion protection performance of coatings [69]. However, the direct addition of corrosion inhibitors into the coating may cause coating defects, forming a channel for corrosion media to penetrate into the coating, and it is difficult to release the corrosion inhibitors on demand. In addition to loading corrosion inhibitors into capsules [70], fibers, and nanotubes [71], they can also be inserted into the layered bimetallic hydroxide [72] (LDH), and then the LDH structure can be mixed with the coating. Benefiting from the lamellar texture of LDH structure, it could improve the barrier property of coatings due to the labyrinth effect [73]. In addition, the capturing ability of corrosive anions and the release of corrosion inhibitors from LDH can be used to improve the active and passive protection effect of the coatings [74]. Buchheit found that the chromium-free organic coating with corrosion inhibitors loaded into LDH structure had an efficient active protection effect on the aluminum alloy matrix [75]. When the loaded corrosion inhibitor $V_{10}O^{6-}{}_{28}$ was exchanged with chloride ions [76], the crystal structure changed, which can be used to determine whether the metal substrate corrosion occurred in advance.

Due to the presence of a large number of oxygen-containing groups, such as hydroxyl and carboxyl groups on the surface of oxidized CNTs, the dispersion of oxidized CNTs in resins can be improved. Small organic molecules, organic polymers, and nanoparticles can be combined with oxidized CNTs. For example, LDH nanoparticles, due to the positive charge of the surface layer, can be combined with oxidized CNTs with a negative charge on the surface through electrostatic interaction [77].

In this section, the coprecipitation method and ion exchange method were used to prepare CNTs/LDH composite by changing the amount of LDH and carboxylated CNTs. FTIR and XPS were used to analyze the CNTs loaded with LDH nanoparticles, and the surface was observed by SEM and TEM. The release of VO^{3-} corrosion inhibitor was characterized by UV–Vis spectroscopy. The corrosion resistance of the coating in 3.5 wt% NaCl solution was tested by the EIS [78].

9.5.1 Materials and Methods

9.5.1.1 Materials

The materials and reagents used in this section are shown in Table 9.5.

9.5.1.2 Preparation of CNTs/LDH/VO_3^-

The carboxylation process of CNTs. First, 3 g of CNTs was put in a vacuum drying oven at 100 °C for 12 hours. 10 ml of concentrated sulfuric acid and 10 ml of concentrated nitric acid were added into the three-necked flask, which was kept in the water bath condition at 80 °C. The dried CNTs were added into the three-necked flask with a condensing return pipe at the mouth of the flask, and the mixture was magnetically stirred for 24 hours. After the reaction, the flask was cooled to room temperature, and the mixture was transferred back to a beaker filled with deionized water. When the black CNTs settled to the bottom of the beaker, the filter was pumped with a microporous filter membrane, cleaned with ethanol five times, and the obtained filter was placed in a vacuum drying oven and dried at 60 °C for 24 hours to finally obtain carboxylated CNTs (CNTs–COOH).

The preparation process of LDH is as follows – 0.1 M $Mg(NO_3)_2 \cdot 6H_2O$ and 0.05 M $Al(NO_3)_3 \cdot 9H_2O$ were added into the beaker. In addition, 0.4 M $NaNO_3$ and 2 M NaOH were added slowly to maintain the pH of the solution at 10 ± 0.3. After magnetically stirring for 30 minutes, the suspension was made to react in a condensation reflux unit at 80 °C for 24 hours, and the reaction process was carried out in an argon atmosphere. Then, the obtained substance was dried in a vacuum freeze-drying system for two hours to obtain LDH products.

The preparation process of CNTs/LDH is as follows – three beakers containing 100 ml deionized water were added with 100 mg CNTs–COOH powder, and ultrasonic dispersion at 100 W power was performed to obtain a uniform CNTs–COOH suspension. Then 50, 100, and 200 mg LDH were added according to the proportions of CNTs and LDH 1 : 0.5, 1 : 1, and 1 : 2, respectively. After magnetic stirring

Table 9.5 Experimental materials and reagents.

Materials	Purity	Company
Multiwalled carbon nanotube	—	Suzhou Carbon Feng Technology Co., LTD
Magnesium nitrate	Analytical purity	Sinopharm Group
Aluminium nitrate	Analytical purity	Sinopharm Group
Sodium nitrate	Analytical purity	Sinopharm Group
Sodium metavanadate	Analytical purity	Sinopharm Group
NaOH	Analytical purity	Sinopharm Group
Ethyl alcohol	Analytical purity	Sinopharm Group
E-51 epoxy resin	Analytical purity	Aladdin, Inc.
Polyetheramine (D400)	Analytical purity	Aladdin, Inc.

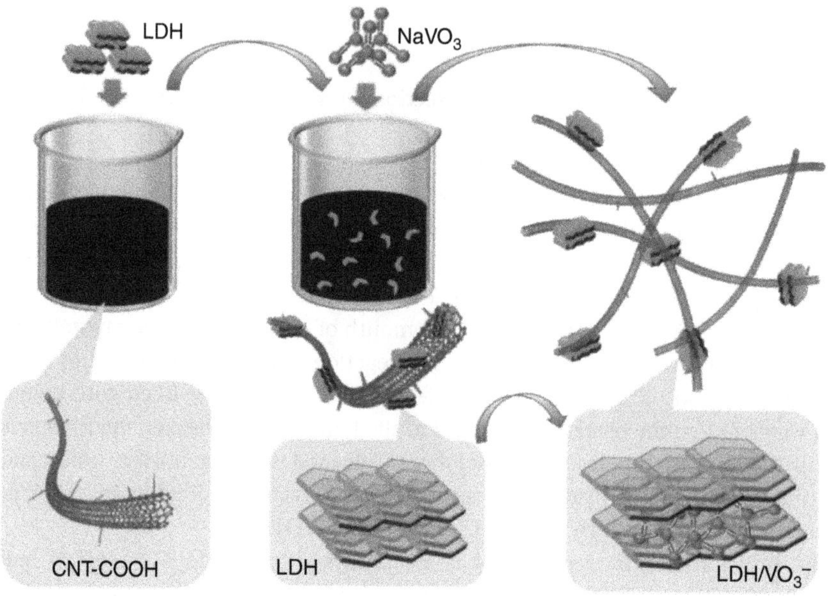

Figure 9.33 Preparation process of CNTs/LDH/VO_3^-.

for 30 minutes, the suspension was made to react in the condensing reflux unit at 80 °C for 24 hours, and then washed with deionized water 4 times. After drying in a vacuum freeze-drying system for two hours, three kinds of CNTs/LDH products were obtained, which were named CNTs/0.5LDH, CNTs/1LDH, and CNTs/2LDH, respectively.

The preparation process of CNTs/LDH/VO_3^- is as follows – CNTs/LDH intercalated with VO_3^- corrosion inhibitor was prepared by cation exchange reaction. The three proportions of CNTs/LDH were weighed by 3 g respectively and added into 0.1 M $NaVO_3$ solution. The cation exchange reaction was completely carried out by magnetic stirring at 40 °C for 24 hours. To preserve the pH of the solution at 8.4 ± 0.3, a solution of 0.2 M NaOH was added, followed by a centrifuge to remove the unreacted $NaVO_3$ and repeated cleaning with deionized water to achieve a pH of 7.0. Finally, CNTs/LDH containing $NaVO_3$ corrosion inhibitors was obtained by vacuum drying. The synthesis process of CNTs/LDH/VO_3^- is shown in Figure 9.33. The prepared samples with different proportions of CNTs and LDH were named CNT/0.5LDH/VO_3^-, CNT/1LDH/VO_3^-, and CNT/2LDH/VO_3^-, respectively.

9.5.1.3 Preparation of CNTs/LDH/VO_3^- Epoxy Coating

Four parts of bisphenol A diglycidyl ether with a mass of 2 g were weighed, and 0.6 g of three different types of CNTs/LDH/VO_3^- were prepared and added to the epoxy resin (the ratio of bisphenol A diglycidyl ether to curing agent used here was 1 : 0.6). Then, 50 ml anhydrous ethanol solution was added, and the two were mixed uniformly by ultrasonic shock for 30 minutes. The anhydrous ethanol was steamed out of the mixture using a rotary evaporator to obtain a uniformly dispersed suspension. Next, 1.2 g polyether amine D-400 in acetone solution was added to reduce the viscosity of the resin, and the resin was ultrasonically dispersed for 30 minutes. The coating was sprayed onto an aluminum sheet using a spray gun three times and cured in a 60 °C oven for 48 hours to obtain three different types of CNTs/LDH/VO_3^- coating and CNTs/LDH coating. One coating was added with 0.6 g CNT/0.5LDH without corrosion inhibitor as the reference sample. These coatings are named CNTs/0.5LDH/VO_3^- coating, CNTs/1LDH/VO_3^- coating, CNT/2LDH/VO_3^- coating, and CNT/0.5LDH coating.

9.5.1.4 Characterization of CNTs/LDH/VO_3^- Epoxy Coating

Fourier infrared spectroscopy (FTIR, VERTEX 70) was used to measure the chemical structure of CNTs–COOH, LDH, and CNTs/LDH/VO_3^- in the range of 4000–400 cm^{-1}. The loading of LDH nanoparticles on CNTs was investigated by X-ray photoelectron spectroscopy (XPS, Thermo SCIENTIFIC ESCALAB 250Xi). SEM (SU1510 system) was used to observe the microstructure of CNTs, LDH, and CNTs/LDH composites. An energy dispersive X-ray spectrometer (EDS, ZEISS, SUPRA 55) was used for elemental analysis of a region with an area of $2 \times 1.6\,\mu m^2$. The morphologies of CNTs, LDH, and CNTs/LDH composites were analyzed by TEM (JEOL 2100). Before observation, the sample was ultrasonically dispersed in anhydrous ethanol and dropped on a clean silicon wafer for air drying.

9.5.1.5 Water Contact Angle Measurement

To verify the superhydrophobic property of coatings, the WCAs and rolling angle of the coating surface were measured at room temperature using a dataphysics OCA20 contact angle measuring instrument with a drop volume of 6 μl. By tilting the sample surface, the smallest angle at which the water droplets rolled off the sample surface is defined as the rolling angle. In order to reduce the experimental error, five locations on each sample surface were selected for testing, and the average of five measured values was taken as the final data.

9.5.1.6 Release Behavior of Corrosion Inhibitors From CNTs/LDH/VO_3^-

An ultraviolet-visible spectrophotometer (UV–Vis, Agilent 7700) was used to characterize the release of corrosion inhibitors in CNTs/LDH/VO_3^- complexes. The Zeta potential values of CNTs–COOH, LDH, and different proportions of CNTs/LDH in water were measured by Zeta potential analyzer (Zatasizer Nano ZS90).

9.5.1.7 Electrochemical Analysis

The corrosion-resistant properties of coatings were tested by EIS in 3.5 wt% NaCl solution at room temperature using the PARSTAT 2273 electrochemical workstation. The test frequency ranges from 10^5 to 10^{-2} Hz, and the disturbance potential is 20 mV.

9.5.2 Characterization of CNTs/LDH/VO_3^- Epoxy Composite Coating

9.5.2.1 Characterization of CNTs/LDH/VO_3^-

Figure 9.34 shows the Zeta potentials of CNTs–COOH, LDH, and different proportions of CNTs/LDH in water. The Zeta potential of CNTs–COOH was negative

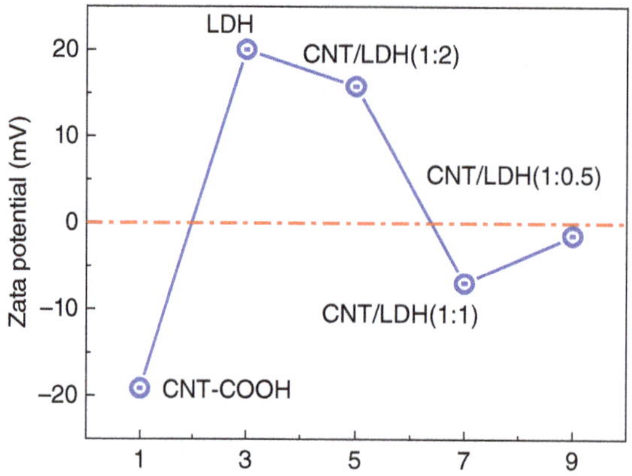

Figure 9.34 Zeta potential of CNTs–COOH, LDH, and CNTs/LDH with different proportions of CNTs and LDH.

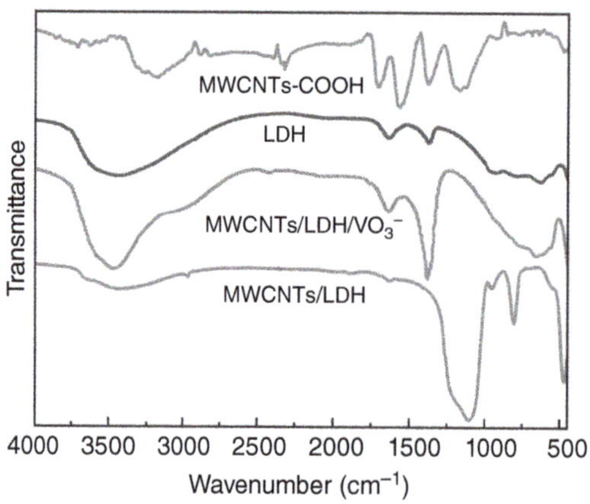

Figure 9.35 FTIR spectra of CNTs–COOH, LDH, CNTs/LDH, and CNTs/LDH/VO_3^-.

(−21.02 ± 1.15 mV) due to the presence of a large number of oxygen-containing functional groups on the surface of carboxylated CNTs. The Zeta potential of LDHs was positive (20.05 ± 1.05 mV). After mixing CNTs–COOH and LDH at the ratio of 1 : 2, the potential value became 15.82 ± 1.71 mV, indicating that the CNTs–COOH with negative potential was adsorbed with the LDH surface with the positive potential. When CNTs–COOH and LDH were mixed at a ratio of 1 : 1, the potential value moved to −6.91 ± 1.20 mV. When the ratio of CNT to LDH was 1 : 0.5, the potential value was −1.4 ± 1.54 mV, indicating that the negative charge on the surface of CNTs–COOH was almost neutralized by the positively charged LDH. The electrostatic interaction between LDH and CNTs–COOH was demonstrated by the Zeta potential measurement.

Figure 9.35 shows the FTIR test results of CNTs–COOH, LDH, CNTs/LDH, and CNTs/LDH/VO_3^-. For the CNTs–COOH sample, 1714 cm^{-1} is the stretching vibration peak of the carbon-oxygen double bond (C=O) on the carboxyl group, 3178 cm^{-1} is the absorption peak caused by the hydroxyl stretching vibration, 1565 cm^{-1} is the characteristic absorption peak of C—H bond, and 1184 cm^{-1} is the stretching vibration absorption peak of C—O [79]. The absorption peaks at 1631 and 1370 cm^{-1} are the stretching vibration peaks of LDH interlayer water molecules and hydroxyl groups on the lamellar [80]. The absorption peaks in the FTIR spectral range from 400 to 900 cm^{-1} were caused by the lattice vibrations of metal–oxygen atoms (M—O) and oxygen-metal-oxygen atoms (O—M—O) [81] on the LDHs lamination. In the FTIR spectrum of CNTs/LDH, the absorption peak at 3473 cm^{-1} was caused by the contraction vibration of hydroxyl group between LDH laminates and interlayer water molecules [82], and the absorption peak at 1380 cm^{-1} was caused by the stretching vibration of NO_3^- [83]. In addition, the carbon–oxygen double bond (C=O) peak at 1714 cm^{-1} of CNTs-COOH disappeared, indicating that the active group of CNTs-COOH successfully combined with LDH. For the FTIR spectrum of CNTs/LDH/VO_3^-, new characteristic absorption peaks appeared at 1105, 951, and 802 cm^{-1}, which was due to the loading of $NaVO_3$ corrosion inhibitors. VO_3^- could exchange with NO_3^- and intercalate in the LDH structure. It indicates that the CNTs/LDH/VO_3^- structure as coating filler has been successfully prepared.

The chemical bond between CNTs-COOH and CNTs/LDH/VO_3^- was analyzed by XPS, and the results are shown in Figure 9.36. It can be seen from the figure that compared with the elements from the XPS spectrum of CNTs-COOH, the elements of C, O, Mg, Al, and V are shown on the CNTs/LDH/VO_3^- structure. The three peaks of the O 1s spectrum in Figure 9.36b were observed at 534, 533.1, and 532.6 eV, which are attributed to the O=C—OH, C=O, and C—OH bonds [84, 85], respectively. In Figure 9.36c, the peaks of high-resolution C 1s spectra at 285.6, 289.5, 285, and 284.8 were C—OH, O=C—OH, C=C, and C—C bonds [86], respectively. In Figure 9.36d–f, the Al 2s, Mg 1s, and V 2s spectra were centered at 1304.4, 75.8, and 517, which corresponded to Al—O, Mg—O, and V—O bonds [87], respectively. The XPS results indicated that the corrosion inhibitors have been loaded into the CNTs/LDH structure.

Figure 9.37 shows TEM images of CNTs, LDH, and CNTs/LDH/VO_3^-. Figure 9.37a shows the TEM image of CNTs-COOH, and Figure 9.37b shows the TEM image of LDH. It can be seen from the figure that LDH was an irregular hexagon [88]. Due to the strong force between LDH, agglomeration will occur when there is excessive LDH. Figure 9.37c shows that the CNTs were entwined, and the LDH was inserted between the CNTs.

In order to obtain the standard release curve of $NaVO_3$ corrosion inhibitors, 2, 5, 10, 20, and 50 ppm of $NaVO_3$ were dissolved in 0.05 M NaCl solution respectively. The UV–Vis spectra of $NaVO_3$ was characterized between 190 and 500 nm, and shown in Figure 9.38a. The characteristic absorption peak of $NaVO_3$ was at 265 nm. According to Figure 9.38b, the UV–Vis absorbance intensities at different $NaVO_3$ concentrations were obtained, and its standard release curve was obtained. The linear fitting formula of the standard release curve was obtained as $y = 0.01182x - 0.00313$.

50 mg CNTs/LDH/VO_3^- was placed in a dialysis bag and 50 ml 0.05 M NaCl solution was added. After 5, 10, 15, 25, 30, and 35 hours of time intervals, the solution samples were taken out and placed in a colorimetric dish, and the

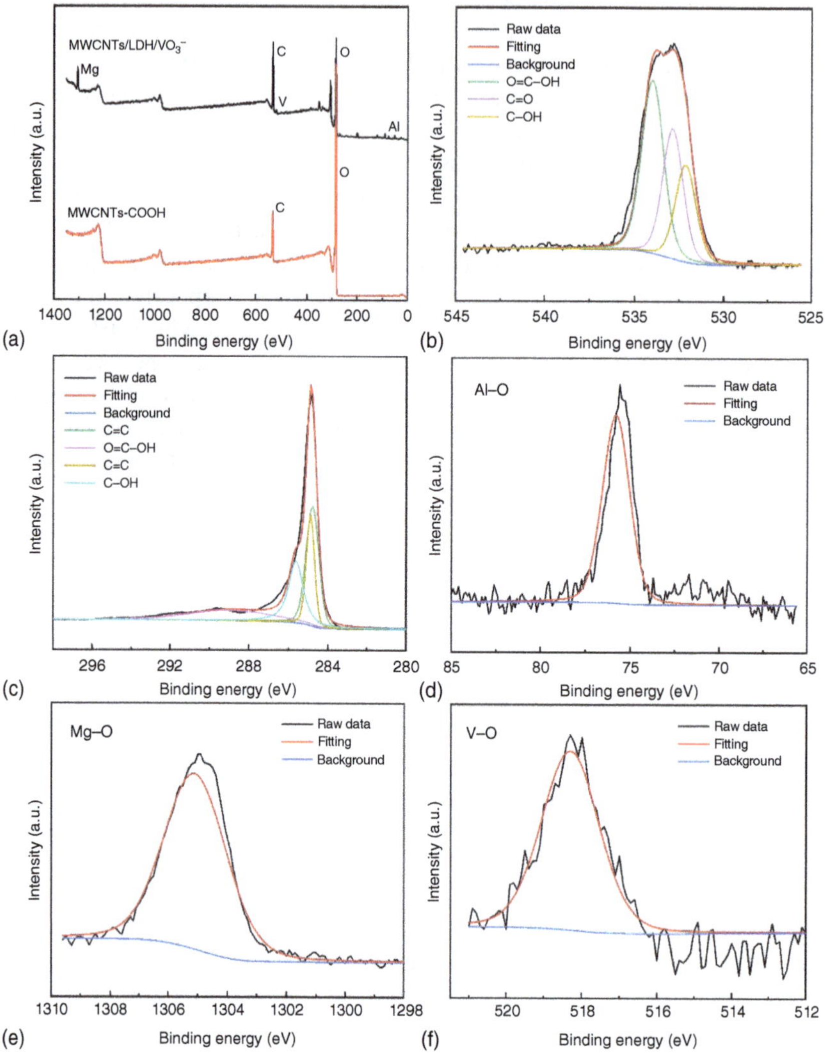

Figure 9.36 FTIR spectra of CNTs–COOH and CNTs/LDH/VO_3^-.

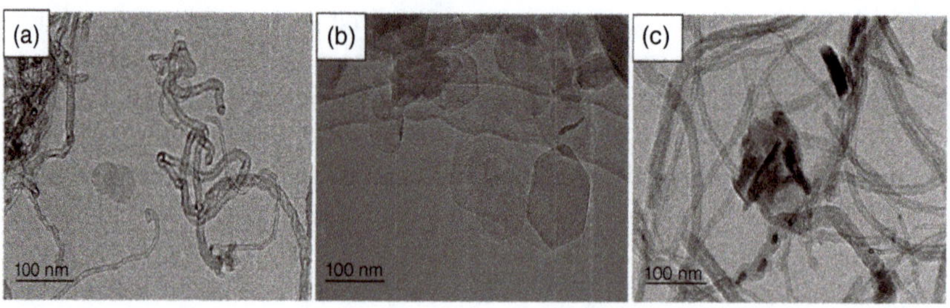

Figure 9.37 TEM images of (a) CNTs–COOH; (b) LDH; (c) CNTs/LDH/VO_3^-.

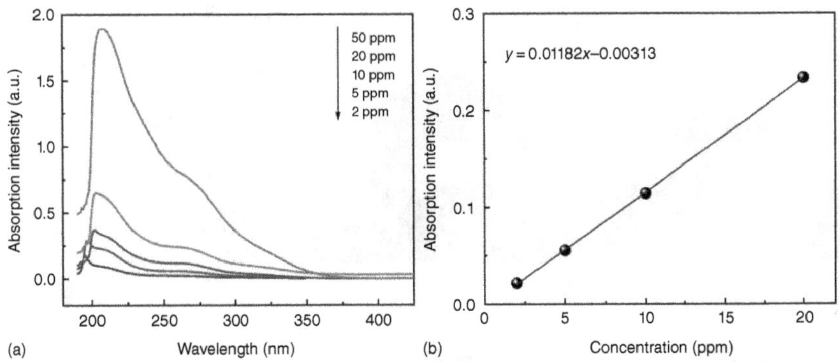

Figure 9.38 (a) UV–Vis spectra of different concentrations of $NaVO_3$ dissolved in 0.05 M NaCl solution; (b) the relationship between UV–Vis intensities and $NaVO_3$ concentrations.

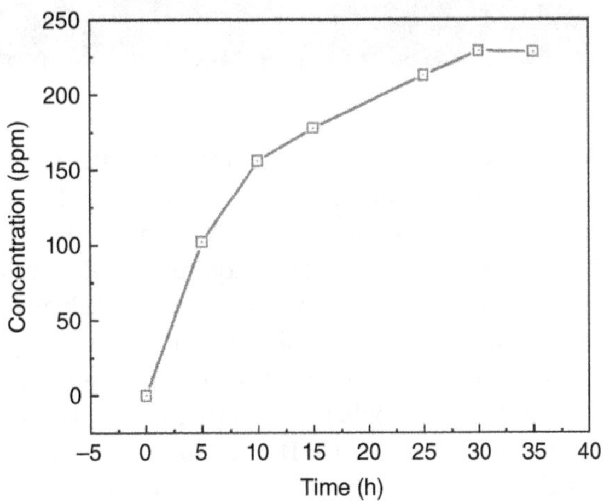

Figure 9.39 Release behavior of $NaVO_3$ from CNTs/LDH/VO_3^- containers during immersion in 0.05 M NaCl solution.

concentrations of VO_3^- in the solution were measured by the UV–Vis spectrophotometer [89]. The UV–Vis intensities at 265 nm after different immersion times were substituted into the standard curve to obtain the release of VO_3^-. Figure 9.39 shows the LDH release behavior of $NaVO_3$ within 50 hours. After immersion for five hours, the ion exchange between VO_3^- and Cl^- caused the rapid release of $NaVO_3$ to 101.97 ppm. When the immersion time was 15 hours, the concentration of VO_3^- ions in the solution increased rapidly to reach 119 ppm, and then the release rate slowed down, reaching the maximum concentration of VO_3^- ions at 30 hours. The above experimental results show that the corrosion inhibitor will be released from LDH in the presence of corrosive medium [90], and the loading rate of the corrosion inhibitors in LDH was 22.9%.

9.5.2.2 Superhydrophobic Property of CNTs/LDH/VO_3^- Based Coatings

The above results show that the CNTs/LDH/VO_3^- structure has been successfully prepared. Then 0.6 g of CNTs/LDH/VO_3^- was mixed into the epoxy resin to verify its hydrophobic property on the coating surface. Figure 9.40 shows the WCAs of

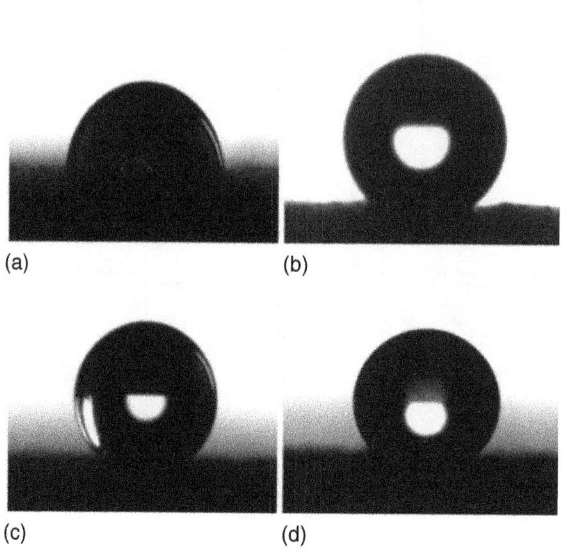

Figure 9.40 WCAs of different coatings: (a) CNT/2LDH/VO_3^- loaded epoxy coating; (b) CNT/1LDH/VO_3^- loaded epoxy coating; (c) CNT/0.5LDH/VO_3^- loaded epoxy coating; and (d) CNT/0.5LDH loaded epoxy coating.

coatings with different ratios of CNT to LDH. The WCA of the CNT/2LDH/VO_3^- loaded epoxy coating was 90.4°, the WCA of the CNT/1LDH/VO_3^- loaded epoxy coating was 142.6°, and the WCA of the CNT/0.5LDH/VO_3^- loaded epoxy coating was 154.2°, which met the requirement of superhydrophobic performance. The superhydrophobic property of the coating gradually decreased with the increase of the ratio of CNT to LDH, which was due to the existence of some water molecules in the form of crystallization between LDH layers. Therefore, excessive LDH content will affect the hydrophobic properties of the material [91]. Figure 9.40d shows that the WCA of the CNT/0.5LDH coating without corrosion inhibitors was 152.7°, in which case the coating also met the superhydrophobic performance.

9.5.2.3 Electrochemical Performance of CNTs/LDH/VO_3^- Based Coatings

As demonstrated in Section 9.3, the epoxy resin (E106) with E-51 : D-400 = 1 : 0.6 exhibited a good shape memory effect. The CNT/0.5LDH/VO_3^- structures were loaded into the E106 resin to prepare the superhydrophobic coating, using bare E106 coating and CNT/0.5LDH epoxy coating as control samples. The coatings were scratched using a scalpel with a scratch width of approximately 50 μm, and were healed by heating in an oven at 40 °C. Figure 9.41 shows the Nyquist and Bode plots of the three coatings at different immersion times in 3.5 wt% NaCl solutions.

Figure 9.41a,b show the variation of the barrier properties of E106 coatings after immersion in 3.5 wt% NaCl solution for different times. With the extension of time, the impedance arc radius of E106 coating decreased gradually, and the $|Z|_{0.01\ Hz}$ value decreased from $8.7 \times 10^5\ \Omega\ cm^2$ on the first day to $1.4 \times 10^5\ \Omega\ cm^2$ on the sixth day, and the protective performance of the coating declined. Figure 9.41c,d show the corrosion protection performance of CNT/0.5LDH coating after immersing in 3.5 wt% NaCl solution for different times. The $|Z|_{0.01\ Hz}$ value decreased from $8.04 \times 10^6\ \Omega\ cm^2$ on the first day to $2.98 \times 10^6\ \Omega\ cm^2$ on the sixth day. Compared with

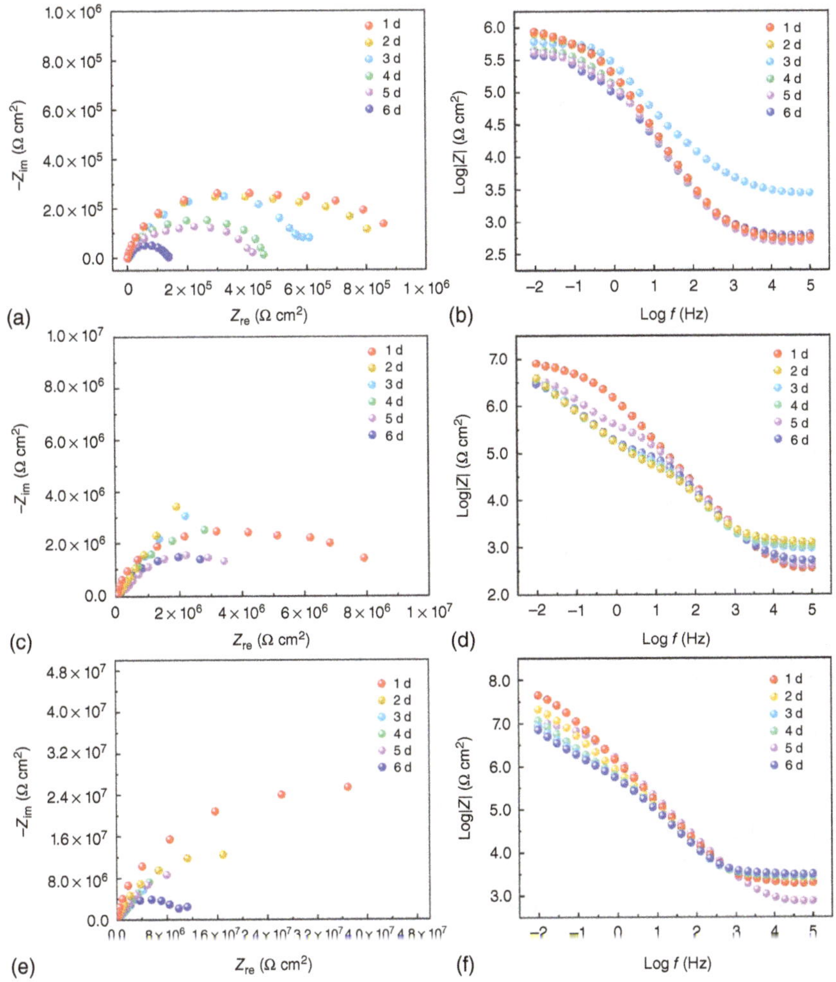

Figure 9.41 Nyquist and Bode diagrams of (a,b) E106 coating, (c,d) CNT/0.5LDH coating, and (e,f) CNT/0.5LDH/VO_3^- coating after different immersion times in 3.5 wt% NaCl solutions.

the E106 coating, the $|Z|_{0.01\,Hz}$ value was one order of magnitude higher, indicating that the CNT/0.5LDH coating can more effectively avoid the substrate from the influence of corrosive media. Figure 9.41e,f shows the change of barrier property of CNT/0.5LDH/VO_3^- coating after immersion in 3.5 wt% NaCl solution for different times. With the extension of immersion time, the radius of the impedance arc also decreased gradually. As can be seen from the Bode plots, the initial $|Z|_{0.01\,Hz}$ value of the coating was $4.48\times10^7\,\Omega\,cm^2$, and when the immersion time prolonged to the third day, the $|Z|_{0.01\,Hz}$ value of the coating quickly dropped to $8.9\times10^6\,\Omega\,cm^2$. Surprisingly, with a further increase in the immersion time, the $|Z|_{0.01\,Hz}$ value of the coating increased slightly. After six days, the $|Z|_{0.01\,Hz}$ value decreased again to $7.08\times10^6\,\Omega\,cm^2$. The variation trend of $|Z|_{0.01\,Hz}$ values of the three coatings after

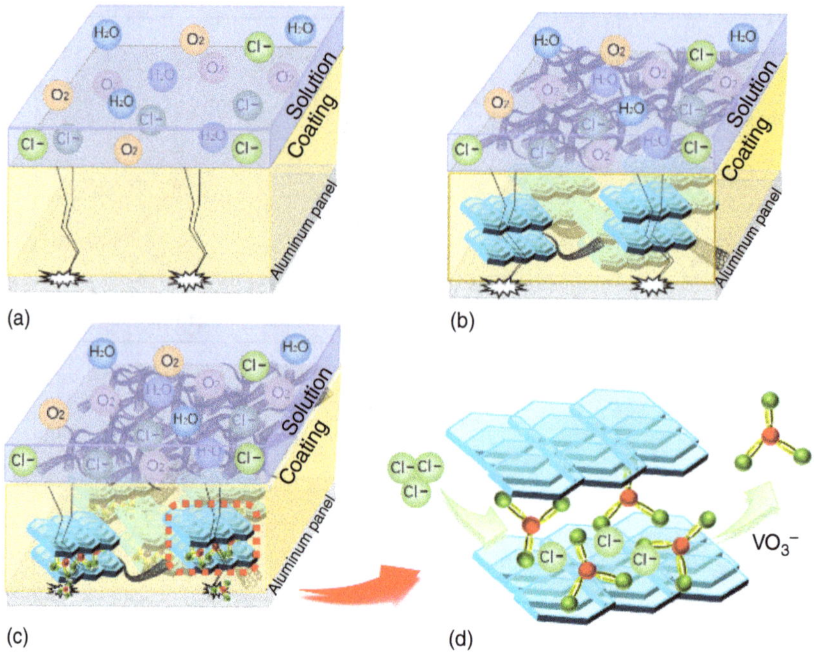

Figure 9.42 Corrosion protection mechanism of (a) epoxy coating, (b) CNTs/LDH epoxy coating, and (c) CNTs/LDH/VO_3^- loaded epoxy coating; (d) anion exchange mechanism of corrosion inhibitors within LDH interlayer.

immersion in 3.5 wt% NaCl solution for different times was summarized. It is noted that the $|Z|_{0.01\,Hz}$ value of the CNT/0.5LDH/VO_3^- coating was always higher than that of the CNT/0.5LDH coating and E106 coating, demonstrating better corrosion resistance property thanks to the shape memory effect of E106 resin and the release of corrosion inhibitors.

9.5.3 Corrosion Protection Mechanism of CNTs/LDH/VO_3^- Based Coating

Figure 9.42 shows the corrosion protection mechanism of pure epoxy resin and epoxy resin with corrosion inhibitors. As shown in Figure 9.42a, for pure epoxy resin, there are pores, voids, and other defects in the coating after long-term immersion in corrosive media. Chloride ions, oxygen, and water in the solution will easily reach the metal substrate through the coating, and a corrosion reaction will occur on the exposed metal surface. Figure 9.42b shows the degradation process of the CNTs/LDH-loaded epoxy coating. Because CNTs possess a hydrophobic property and can intertwine with each other on the coating surface to form nanostructures [92], the roughness of the coating surface increases to strengthen the superhydrophobic property of the coating. The superhydrophobic effect of CNTs can form an air layer on the coating surface, which can block the penetration of O_2 and Cl^- from the corrosive solution to the coating, thus delaying metal corrosion.

However, with the extension of immersion time, microscale damages are generated on the coating. At this time, LDH's close packing on the metal surface can act as a barrier to seal the defects on the metal surface, thereby improving the corrosion resistance of metals [93].

Figure 9.42c,d show the corrosion protection mechanism of CNTs/LDH/VO_3^- loaded coating. In addition to the role of CNTs and LDH in protecting the metal substrate in the coating, the incorporated corrosion inhibitors also perform an active corrosion protection role. Since the anion corrosion inhibitors cannot directly act on the metal surface, LDH is used to load the corrosion inhibitor $NaVO_3$. Corrosive anions in the solution and the variation of pH value can promote the release of corrosion inhibitors and capture corrosive anions in the coating [94]. In addition, in the process of capturing Cl^- in the solution, LDH will consume H^+ in the solution and increase the localized pH value, thereby neutralizing the acidic electrolyte generated in the corrosion zone to promote the re-passivation of the metal and improve the corrosion resistance of the substrate.

9.6 Chapter Summary

In this chapter, different kinds of superhydrophobic coatings were prepared to strengthen the anti-icing, anticorrosion, mechanical stability, and self-healing properties. First, the silane coupling agent KH560 was used to modify SiO_2 nanoparticles, and a one-step spraying method was used to prepare composite coatings with different SiO_2 particle contents. When the content of modified SiO_2 was 30 wt%, the coating achieved superhydrophobicity, and the WCA reached 152.4°. The coating maintained its good superhydrophobic performance within 80 times of tape-peeling tests, which was because the resin matrix can fix the $nSiO_2$ particles to maintain the microscopic rough structure on the coating surface, exhibiting good mechanical properties. The SiO_2/epoxy composite coating showed excellent chemical stability in solutions of different pH values. The superhydrophobic structure made the coating with good anti-icing properties.

To improve the wear resistance of the coating, a CNT/epoxy composite coating was prepared. When the CNTs content was 30 wt%, the coating exhibited superhydrophobic performance. Because CNTs were entangled and agglomerated in the resin to form micro/nanostructures, the coating could maintain the superhydrophobic property after 100 times of tape-peeling tests, showing excellent mechanical durability. The EIS results demonstrated that the superhydrophobic composite coating with 30 wt% CNTs content possessed excellent corrosion resistance due to the air trapped in the micro-nanostructure on the surface, which could effectively isolate the corrosive medium from contacting the coating surface.

In order to realize the self-healing performance of the superhydrophobic coating, the SiO_2–CNT hybrid was added to the shape memory epoxy coating. By adjusting the ratio of the soft segment and the hard segment in the resin system, the shape memory effect of the coating was realized. When the ratio of epoxy resin to curing agent was 1 : 0.6, the content of flexible segments in the coating was lower, and the T_g reached 40 °C; when the ratio of epoxy resin to curing agent was 1 : 1.0, the coating contained more flexible segments, and the T_g was 25 °C. By heating the two coatings to above T_g or using the electrothermal effect of CNTs in the SiO_2–CNT hybrid, the shape memory effect of the coating was triggered, thereby restoring the microstructure, superhydrophobicity, and corrosion resistance of the coating.

Finally, CNTs/LDH nanoparticles were prepared using the electrostatic adsorption effect of carboxylated CNTs and LDH, which were then loaded with $NaVO_3$ corrosion inhibitor. CNTs/LDH/VO_3^- particles were loaded into the coating to prepare a superhydrophobic self-healing coating containing corrosion inhibitors. The loading ratio of corrosion inhibitors in the nanoparticles reached 22.9%. The EIS results showed that the coating containing corrosion inhibitors had the best anti-corrosion performance. This is because when the corrosive media penetrated into the coating, Cl^- ions could exchange with the corrosion inhibitors in LDH to accelerate the release of VO_3^- and promote the formation of a passivation film on the metal surface, thereby improving the protective function of the coating on the metal substrate.

Superhydrophobic coatings have garnered significant attention in recent years due to their potential applications in various fields such as self-cleaning surfaces, anti-icing, anticorrosion, and water harvesting. However, despite the progress, there are still several challenges and exciting future directions in this area of research. (i) Durability and mechanical stability. One of the primary challenges in the development of superhydrophobic coatings is their durability under various conditions. Many existing superhydrophobic surfaces can easily lose their water-repellent properties when subjected to mechanical abrasion, chemical attack, or UV radiation. Enhancing the mechanical strength and long-term durability of these coatings without compromising their superhydrophobicity remains a significant challenge. (ii) Scalability and cost-effectiveness. Another challenge is the scalability of the manufacturing process for superhydrophobic coatings. Many methods currently used are not feasible for large-scale production or are too costly. There is a need for cost-effective, scalable manufacturing processes that can produce superhydrophobic coatings on a commercial scale. (iii) Environmental impact. The environmental impact of the materials and processes used to create superhydrophobic coatings is also a concern. Some materials may be toxic or nonenvironment friendly, posing challenges to sustainable development. Research is needed to develop eco-friendly materials and green synthesis methods. (iv) Performance under extreme conditions. Superhydrophobic coatings often fail under extreme conditions like high temperature, pressure, or corrosive environments. Developing coatings that can maintain their properties under such conditions is a key research area. (v) Multifunctionality. There is a growing interest in developing multifunctional superhydrophobic coatings that not only repel water but also have other properties such as antimicrobial activity, self-healing ability, or conductivity. Integrating these functionalities without compromising superhydrophobicity is complex.

Therefore, the future directions of superhydrophobic coatings are as follows: (i) Advanced materials. Research into new materials like graphene, CNTs, and advanced polymers may offer solutions for enhancing durability and multifunctionality. Nanomaterials, in particular, hold promise for creating more robust and versatile coatings. (ii) Smart and responsive coatings. The development of smart superhydrophobic coatings that can change their properties in response to external stimuli (like temperature, pH, or light) is an exciting future direction. Such materials can find applications in sensors, controlled release systems, and smart windows. (iii) Bioinspired designs. Mimicking natural superhydrophobic surfaces, such as lotus leaves or butterfly wings, can inspire the development of innovative coating designs. These bioinspired approaches can lead to more efficient and environmentally friendly solutions. (iv) Multifunctional coatings. Combining superhydrophobic features with other surface functionalities (like oleophobicity, photothermal properties, or electrical conductivity) to create hybrid coatings can expand the application scope of these materials. (v) Application development. There is tremendous potential in expanding the applications of superhydrophobic coatings to new fields such as renewable energy (e.g. enhancing the efficiency of solar panels), biomedical devices, and the automotive and aerospace industries. In summary, while superhydrophobic coatings have shown great promise, overcoming the current challenges is critical. Future research focusing on different aspects can pave the way for innovative applications and sustainable solutions in various industries.

References

1 Hou, X., Cheng, Q., and Wang, H. (2024). Self-cleaning paper-based microfluidic biosensor employing DNAzyme and semiconducting single-walled carbon nanotube for copper ion detection. *Bioelectrochemistry* 155: 108602. https://doi.org/10.1016/j.bioelechem.2023.108602.

2 Li, J., Yang, Y., Gao, Q. et al. (2023). Preparation and characterisation of CeSt3-modified Zn–Ni–B4C composite superhydrophobic coatings. *Applied Surface Science* 641: 158489. https://doi.org/10.1016/j.apsusc.2023.158489.

3 Liu, H., Liu, D., Li, P. et al. (2023). Effect of superhydrophobic surface on the surface trap distribution of silicone rubber composites. *Materials Letters* 347: 134588. https://doi.org/10.1016/j.matlet.2023.134588.

4 Han, S., Choi, M.Y., Kumar, P. et al. (2010). Phase transitions in confined water nanofilms. *Nature Physics* 6: 685–689. https://doi.org/10.1038/nphys1708.

5 Wang, X. and Lin, Z. (2023). A novel high-performance coating with hybrid nanofiller reinforcement for superior self-cleaning, anti-icing, and corrosion resistance properties. *Journal of Building Engineering* 80: 107993. https://doi.org/10.1016/j.jobe.2023.107993.

6 Yang, Q., Cao, J., Ding, R. et al. (2023). The synthesis and mechanism of superhydrophobic coatings with multifunctional properties on aluminum alloys surface: a review. *Progress in Organic Coatings* 183: 107786. https://doi.org/10.1016/j.porgcoat.2023.107786.

7 Zhao, Q., Sun, Q., Xin, S. et al. (2022). High-strength titanium alloys for aerospace engineering applications: a review on melting-forging process. *Materials Science and Engineering A* 845: 143260. https://doi.org/10.1016/j.msea.2022.143260.

8 Li, X., Wang, J., Yi, G. et al. (2024). From waste plastic to artificial lotus leaf: Upcycling waste polypropylene to superhydrophobic spheres with hierarchical micro/nanostructure. *Applied Catalysis B: Environmental* 342: 123378. https://doi.org/10.1016/j.apcatb.2023.123378.

9 Lou, D., Li, T., Liang, E. et al. (2021). Superhydrophobic/superhydrophilic hybrid copper surface enhanced micro heat pipe by using laser selective texturing. *ECS Journal of Solid State Science and Technology* 10: https://doi.org/10.1149/2162-8777/ac3772.

10 Yıldız, R., Arslanhan, S., Döner, A. et al. (2024). Corrosion behavior of mild steel in 1 M HCl with *Cyclotrichium niveum* as a green inhibitor. *Materials Chemistry and Physics* 312: 128654. https://doi.org/10.1016/j.matchemphys.2023.128654.

11 Zhang, L., Zhou, A.G., Sun, B.R. et al. (2021). Functional and versatile superhydrophobic coatings via stoichiometric silanization. *Nature Communications* 12: 982. https://doi.org/10.1038/s41467-021-21219-y.

12 Lin, T.W., Salzmann, C.G., Shao, L.D. et al. (2009). Polyethylene glycol grafting and attachment of encapsulated magnetic iron oxide silica nanoparticles onto chlorosilanized single-wall carbon nanotubes. *Carbon* 47: 1415–1420. https://doi.org/10.1016/j.carbon.2008.11.034.

13 Karimian, A., Norouzi, M., Ebrahimnia, A. et al. (2024). Fe_3O_4@SiO_2@APTES@MPIB-Mn(II) as an eco-friendly and magnetically recyclable nano catalyst for the green synthesis of various xanthene derivatives. *Journal of Molecular Structure* 1297: 19. https://doi.org/10.1016/j.molstruc.2023.137014.

14 Yan, Y., Cai, Y., Liu, X. et al. (2020). Hydrophobic modification on the surface of SiO_2 nanoparticle: wettability control. *Langmuir* 36: 14924–14932. https://doi.org/10.1016/j.polymer.2018.04.011.

15 Anjum, A.S., Sun, K.C., Ali, M. et al. (2020). Fabrication of coral-reef structured nano silica for self-cleaning and super-hydrophobic textile applications. *Chemical Engineering Journal* 401: https://doi.org/10.1016/j.cej.2020.125859.

16 Swisher, J.H., Jibril, L., Petrosko, S.H. et al. (2022). Nanoreactors for particle synthesis. *Nature Reviews Materials* 7: 428–448. https://doi.org/10.1038/s41578-021-00402-z.

17 Chen, Q., Fang, M., Guo, R. et al. (2023). Multi-functional and durable anti-corrosion coatings with hydrophobic, freeze time retardation and photothermal properties by means of a simple spraying method. *Colloids and Surfaces A: Physicochemical and Engineering Aspects* 679: 132549. https://doi.org/10.1016/j.colsurfa.2023.132549.

18 Sun, S., Li, C., Zhang, L. et al. (2006). Effects of surface modification of fumed silica on interfacial structures and mechanical properties of poly(vinyl chloride) composites. *European Polymer Journal* 42: 1643–1652. https://doi.org/10.1016/j.eurpolymj.2006.01.012.

19 Kumara, G.R.A., Pitawala, H.M.G.T.A., Karunarathne, B. et al. (2021). Development of a chemical-free floatation technology for the purification of vein graphite and characterization of the products. *Scientific Reports* 11: 22713. https://doi.org/10.1038/s41598-021-02101-9.

20 Liu, Y., Zou, S., He, Y. et al. (2021). Influence of fractal surface roughness on multiphase flow behavior: lattice Boltzmann simulation. *International Journal of Multiphase Flow* 134: 103497. https://doi.org/10.1016/j.ijmultiphaseflow.2020.103497.

21 Song, J., Li, Y., Xu, W. et al. (2019). Inexpensive and non-fluorinated superhydrophobic concrete coating for anti-icing and anti-corrosion. *Journal of Colloid and Interface Science* 541: 86–92. https://doi.org/10.1016/j.jcis.2019.01.014.

22 Montemor, M.F. (2014). Functional and smart coatings for corrosion protection: a review of recent advances. *Surface and Coatings Technology* 258: 17–37. https://doi.org/10.1016/j.surfcoat.2014.06.031.

23 Jin, K., Luo, H., Wang, Z. et al. (2020). Composition optimization of a high-performance epoxy resin based on molecular dynamics and machine learning. *Materials & Design* 194: https://doi.org/10.1016/j.matdes.2020.108932.

24 Krasnyi, B.L., Tarasovskii, V.P., Rakhmanova, E.V. et al. (2004). Chemical resistance of ceramic materials in acids and alkalis. *Glass and Ceramics* 61: 337–339. https://doi.org/10.1023/B:GLAC.0000048706.05368.71.

25 Chen, C., Yu, M., Zhan, Z. et al. (2023). Effect of pH on the structure and corrosion protection properties of sol-gel coatings. *Corrosion Science* 212: 110955. https://doi.org/10.1016/j.corsci.2022.110955.

26 Wang, L., Wei, R., Luo, Y. et al. (2024). Construction of alternating multilayer films with stable absorption-dominated electromagnetic shielding performance and reinforced mechanical properties via interface engineering. *Composites Part A: Applied Science and Manufacturing* 176: 107862. https://doi.org/10.1016/j.compositesa.2023.107862.

27 Mittal, G., Dhand, V., Rhee, K.Y. et al. (2015). A review on carbon nanotubes and graphene as fillers in reinforced polymer nanocomposites. *Journal of Industrial and Engineering Chemistry* 21: 11–25. https://doi.org/10.1016/j.jiec.2014.03.022.

28 Adams, D.J., Clark, J.H., and McFarland, H. (1998). The formation of 4,4′-difluorobenzophenone from 4,4′-dinitrodiphenylmethane. *Journal of Fluorine Chemistry* 92: 127–129. https://doi.org/10.1016/S0022-1139(98)00274-7.

29 Xuan, S., Yin, H., Li, G. et al. (2023). *Trifolium repens* L.-like periodic micronano structured superhydrophobic surface with ultralow ice adhesion for efficient anti-icing/deicing. *ACS Nano* 17: 21749–21760. https://doi.org/10.1021/acsnano.3c07385.

30 Liu, X., Chen, H., Zhao, Z. et al. (2019). Slippery liquid-infused porous electric heating coating for anti-icing and de-icing applications. *Surface and Coatings Technology* 374: 889–896. https://doi.org/10.1016/j.surfcoat.2019.06.077.

31 Liu, M., Wu, H., Wu, Y. et al. (2022). The weakly negative permittivity with low-frequency-dispersion behavior in percolative carbon nanotubes/epoxy nanocomposites at radio-frequency range. *Advanced Composites and Hybrid Materials* 5: 2021–2030. https://doi.org/10.1007/s42114-022-00541-z.

32 Valsesia, A., Desmet, C., Ojea-Jiménez, I. et al. (2018). Direct quantification of nanoparticle surface hydrophobicity. *Communications Chemistry* 1: 53. https://doi.org/10.1038/s42004-018-0054-7.

33 Tian, Z., Wang, L., Zhu, D. et al. (2023). Passive anti-icing performances of the same superhydrophobic surfaces under static freezing, dynamic supercooled-droplet impinging, and icing wind tunnel tests. *ACS Applied Materials & Interfaces* 15: 6013–6024. https://doi.org/10.1021/acsami.2c15253.

34 Li, F., Liu, G., Liu, F. et al. (2023). A review of self-cleaning photocatalytic surface: effect of surface characteristics on photocatalytic activity for NO. *Environmental Pollution* 327: 121580. https://doi.org/10.1016/j.envpol.2023.121580.

35 Huang, W., Huang, J., Guo, Z. et al. (2022). Icephobic/anti-icing properties of superhydrophobic surfaces. *Advances in Colloid and Interface Science* 304: 102658. https://doi.org/10.1016/j.cis.2022.102658.

36 Li, W. and Kang, Z. (2014). Fabrication of corrosion resistant superhydrophobic surface with self-cleaning property on magnesium alloy and its mechanical stability. *Surface and Coatings Technology* 253: 205–213. https://doi.org/10.1016/j.surfcoat.2014.05.038.

37 Mohamed, A.M.A., Abdullah, A.M., and Younan, N.A. (2015). Corrosion behavior of superhydrophobic surfaces: a review. *Arabian Journal of Chemistry* 8: 749–765. https://doi.org/10.1016/j.arabjc.2014.03.006.

38 Wang, N., Xiong, D., Deng, Y. et al. (2015). Mechanically robust superhydrophobic steel surface with anti-icing, UV-durability, and corrosion resistance properties. *ACS Applied Materials & Interfaces* 7: 6260–6272. https://doi.org/10.1021/acsami.5b00558.

39 van Voorst Vader, P.C., Lier, J.G., Woest, T.E. et al. (1991). Patch tests with house dust mite antigens in atopic dermatitis patients: methodological problems. *Acta Dermato-Venereologica* 71: 301–305.

40 Park, S.-R., Jung, Y.-C., Shin, W.-K. et al. (2017). Cross-linked fibrous composite separator for high performance lithium-ion batteries with enhanced safety. *Journal of Membrane Science* 527: 129–136. https://doi.org/10.1016/j.memsci.2017.01.015.

41 Ndumia, J.N., Kang, M., Gbenontin, B.V. et al. (2023). Investigation of tensile fracture mechanism of Fe-based coatings deposited on different substrates by arc spraying. *Materials Today Communications* 37: 107339. https://doi.org/10.1016/j.mtcomm.2023.107339.

42 Tyson Bryan, M., Abu Al-Rub Rashid, K., Yazdanbakhsh, A. et al. (2011). Carbon nanotubes and carbon nanofibers for enhancing the mechanical properties of nanocomposite cementitious materials. *Journal of Materials in Civil Engineering* 23: 1028–1035. https://doi.org/10.1061/(ASCE)MT.1943-5533.0000266.

43 Suzana, A.F., Ferreira, E.A., Benedetti, A.V. et al. (2016). Corrosion protection of chromium-coated steel by hybrid sol-gel coatings. *Surface and Coatings Technology* 299: 71–80. https://doi.org/10.1016/j.surfcoat.2016.04.075.

44 Khun, N.W., Troconis, B.C.R., and Frankel, G.S. (2014). Effects of carbon nanotube content on adhesion strength and wear and corrosion resistance of epoxy composite coatings on AA2024-T3. *Progress in Organic Coatings* 77: 72–80. https://doi.org/10.1016/j.porgcoat.2013.08.003.

45 Khun, N.W. and Frankel, G.S. (2016). Cathodic delamination of polyurethane/multiwalled carbon nanotube composite coatings from steel substrates. *Progress in Organic Coatings* 99: 55–60. https://doi.org/10.1016/j.porgcoat.2016.05.002.

46 Song, D., Yin, Z., Liu, F. et al. (2017). Effect of carbon nanotubes on the corrosion resistance of water-borne acrylic coatings. *Progress in Organic Coatings* 110: 182–186. https://doi.org/10.1016/j.porgcoat.2017.04.043.

47 Zhang, M., Zhang, B., Li, X. et al. (2022). Progress in research of self-healing superhydrophobic coatings. *Polymer Materials Science & Engineering* 38: 168–175.

48 Dong, W., Li, B., Wei, J. et al. (2021). Environmentally friendly, durable and transparent anti-fouling coatings applicable onto various substrates. *Journal of Colloid and Interface Science* 591: 429–439. https://doi.org/10.1016/j.jcis.2021.02.014.

49 Hui, W., Chao, G., Jiangang, F. et al. (2011). Adsorption behavior of weak hydrophilic substances on low-energy surface in aqueous medium. *Applied Surface Science* 257: 7959–7967. https://doi.org/10.1016/j.apsusc.2011.03.158.

50 Qiang, S., Chen, K., Yin, Y. et al. (2017). Robust UV-cured superhydrophobic cotton fabric surfaces with self-healing ability. *Materials & Design* 116: 395–402. https://doi.org/10.1016/j.matdes.2016.11.099.

51 Davydovich, D. and Urban, M.W. (2020). Water accelerated self-healing of hydrophobic copolymers. *Nature Communications* 11: 5743. https://doi.org/10.1038/s41467-020-19405-5.

52 Qian, H., Xu, D., Du, C. et al. (2017). Dual-action smart coatings with a self-healing superhydrophobic surface and anti-corrosion properties. *Journal of Materials Chemistry A* 5: 2355–2364. https://doi.org/10.1039/c6ta10903a.

53 Grady, B.P., Paul, A., Peters, J.E. et al. (2009). Glass transition behavior of single-walled carbon nanotube–polystyrene composites. *Macromolecules* 42: 6152–6158. https://doi.org/10.1021/ma900375g.

54 Zare, Y. and Rhee, K.Y. (2019). Modeling of viscosity and complex modulus for poly (lactic acid)/poly (ethylene oxide)/carbon nanotubes nanocomposites assuming yield stress and network breaking time. *Composites Part B: Engineering* 156: 100–107. https://doi.org/10.1016/j.compositesb.2018.08.058.

55 Wan, T. and Stylios, G.K. (2017). Effects of coating process on the surface roughness of coated fabrics. *The Journal of The Textile Institute* 108: 712–719. https://doi.org/10.1080/00405000.2016.1180951.

56 Shen, X., Mao, T., Li, C. et al. (2023). Durable superhydrophobic coatings based on CNTs-SiO_2 gel hybrids for anti-corrosion and thermal insulation. *Progress in Organic Coatings* 181: 107602. https://doi.org/10.1016/j.porgcoat.2023.107602.

57 Zhang, F., Xu, D., Zhang, D. et al. (2021). A durable and photothermal superhydrophobic coating with entwinned CNTs-SiO2 hybrids for anti-icing applications. *Chemical Engineering Journal* 423: 130238. https://doi.org/10.1016/j.cej.2021.130238.

58 Liu, Y., Cao, X., Shi, J. et al. (2022). A superhydrophobic TPU/CNTs@SiO_2 coating with excellent mechanical durability and chemical stability for sustainable anti-fouling and anti-corrosion. *Chemical Engineering Journal* 434: 134605. https://doi.org/10.1016/j.cej.2022.134605.

59 Zhang, F., Ju, P., Pan, M. et al. (2018). Self-healing mechanisms in smart protective coatings: a review. *Corrosion Science* 144: 74–88. https://doi.org/10.1016/j.corsci.2018.08.005.

60 Bailey, E.J. and Winey, K.I. (2020). Dynamics of polymer segments, polymer chains, and nanoparticles in polymer nanocomposite melts: a review. *Progress in Polymer Science* 105: 101242. https://doi.org/10.1016/j.progpolymsci.2020.101242.

61 Beake, B.D. (2022). The influence of the H/E ratio on wear resistance of coating systems – insights from small-scale testing. *Surface and Coatings Technology* 442: 128272. https://doi.org/10.1016/j.surfcoat.2022.128272.

62 Hu, Y., Tong, S., Sha, Y. et al. (2023). Cardanol-based epoxy vitrimer/carbon fiber composites with integrated mechanical, self-healing, reprocessable, and welding properties and degradability. *Chemical Engineering Journal* 471: 144633. https://doi.org/10.1016/j.cej.2023.144633.

63 Zhong, Y., Hu, J., Zhang, Y. et al. (2018). The one-step electroposition of superhydrophobic surface on AZ31 magnesium alloy and its time-dependence corrosion resistance in NaCl solution. *Applied Surface Science* 427: 1193–1201. https://doi.org/10.1016/j.apsusc.2017.08.103.

64 Vakarelski, I.U., Chan, D.Y.C., Marston, J.O. et al. (2013). Dynamic air layer on textured superhydrophobic surfaces. *Langmuir* 29: 11074–11081. https://doi.org/10.1021/la402306c.

65 Ma, L., Wang, J., Zhang, D. et al. (2021). Dual-action self-healing protective coatings with photothermal responsive corrosion inhibitor nanocontainers. *Chemical Engineering Journal* 404: 127118. https://doi.org/10.1016/j.cej.2020.127118.

66 Huang, Y., Wang, P., Tan, W. et al. (2022). Photothermal and pH dual-responsive self-healing coating for smart corrosion protection. *Journal of Materials Science and Technology* 107: 34–42. https://doi.org/10.1016/j.jmst.2021.08.044.

67 He, S., Gao, Y., Wu, C. et al. (2023). A self-healing and anticorrosion epoxy coating based on the novel polymer filler containing a side-linked grafting 2-mercaptobenzothiazole. *Journal of Materials Research and Technology* 26: 1107–1121. https://doi.org/10.1016/j.jmrt.2023.07.264.

68 Chen, Z., Yu, J., Nong, Y. et al. (2023). Beyond time: enhancing corrosion resistance of geopolymer concrete and BFRP bars in seawater. *Composite Structures* 322: 117439. https://doi.org/10.1016/j.compstruct.2023.117439.

69 Honarvar Nazari, M., Zhang, Y., Mahmoodi, A. et al. (2022). Nanocomposite organic coatings for corrosion protection of metals: a review of recent advances. *Progress in Organic Coatings* 162: 106573. https://doi.org/10.1016/j.porgcoat.2021.106573.

70 Zhou, H., Zhang, C.L., Zhang, X. et al. (2023). Image classification based on quaternion-valued capsule network. *Applied Intelligence* 53: 5587–5606. https://doi.org/10.1007/s10489-022-03849-x.

71 Divon, Y., Levi, R., Garel, J. et al. (2017). Torsional resonators based on inorganic nanotubes. *Nano Letters* 17: 28–35. https://doi.org/10.1021/acs.nanolett.6b03012.

72 Wu, K., Shi, L., Wang, Z. et al. (2023). A general strategy to generate oxygen vacancies in bimetallic layered double hydroxides for water oxidation. *Chemical Communications* 59: 3138–3141. https://doi.org/10.1039/d3cc00096f.

73 Chen, C., Xiao, G., He, Y. et al. (2020). Bio-inspired superior barrier self-healing coating: self-assemble of graphene oxide and polydopamine-coated halloysite nanotubes for enhancing corrosion resistance of waterborne epoxy coating. *Progress in Organic Coatings* 139: 105402. https://doi.org/10.1016/j.porgcoat.2019.105402.

74 Cao, Y., Zheng, D., Zhang, F. et al. (2022). Layered double hydroxide (LDH) for multi-functionalized corrosion protection of metals: a review. *Journal of Materials Science and Technology* 102: 232–263. https://doi.org/10.1016/j.jmst.2021.05.078.

75 Akbari, T., Ansari, A., and Rahimi Pishbijari, M. (2023). Influence of aluminum alloys on protection performance of metal matrix composite armor reinforced with ceramic particles under ballistic impact. *Ceramics International* 49: 30937–30950. https://doi.org/10.1016/j.ceramint.2023.07.046.

76 Zhou, L., Cai, Y., and Ma, C. (2023). Experimental study and numerical analysis of chloride ion diffusion in hydrotalcite concrete in chloride salt environment. *Materials*.

77 Koromyslov, V.A. and Grigor'ev, A.I. (2002). On the polarization interaction between two closely spaced conducting spheres in a uniform electrostatic field. *Technical Physics* 47: 1214–1218. https://doi.org/10.1134/1.1514798.

78 Jaegle, M., Pernau, H.F., Pfützner, M. et al. (2016). Thermal-electrical impedance spectroscopy for fluid characterisation. *Procedia Engineering* 168: 770–773. https://doi.org/10.1016/j.proeng.2016.11.276.

79 Vüllers, F., Gomard, G., Preinfalk, J.B. et al. (2016). Bioinspired superhydrophobic highly transmissive films for optical applications. *Small* 12: 6144–6152. https://doi.org/10.1002/smll.201601443.

80 Ahmed, A.A.A., Talib, Z.A., Hussein, M.Z.B. et al. (2012). Improvement of the crystallinity and photocatalytic property of zinc oxide as calcination product of Zn–Al layered double hydroxide. *Journal of Alloys and Compounds* 539: 154–160. https://doi.org/10.1016/j.jallcom.2012.05.093.

81 Zaghouane-Boudiaf, H., Boutahala, M., and Arab, L. (2012). Removal of methyl orange from aqueous solution by uncalcined and calcined MgNiAl layered double hydroxides (LDHs). *Chemical Engineering Journal* 187: 142–149. https://doi.org/10.1016/j.cej.2012.01.112.

82 Xue, B., Yu, M., Liu, J. et al. (2017). Synthesis of inhibitor nanocontainers with two-dimensional structure and their anticorrosion action in sol-gel coating on AA2024-T3 aluminum alloy. *Journal of the Electrochemical Society* 164: C641–C652. https://doi.org/10.1149/2.0851712jes.

83 Poznyak, S.K., Tedim, J., Rodrigues, L.M. et al. (2009). Novel inorganic host layered double hydroxides intercalated with guest organic inhibitors for anticorrosion applications. *ACS Applied Materials & Interfaces* 1: 2353–2362. https://doi.org/10.1021/am900495r.

84 Cordier, P., Tournilhac, F., Soulié-Ziakovic, C. et al. (2008). Self-healing and thermoreversible rubber from supramolecular assembly. *Nature* 451: 977–980. https://doi.org/10.1038/nature06669.

85 Al-Gaashani, R., Najjar, A., Zakaria, Y. et al. (2019). XPS and structural studies of high quality graphene oxide and reduced graphene oxide prepared by different chemical oxidation methods. *Ceramics International* 45: 14439–14448. https://doi.org/10.1016/j.ceramint.2019.04.165.

86 Wang, Z., Dong, Y., Li, H. et al. (2014). Enhancing lithium–sulphur battery performance by strongly binding the discharge products on amino-functionalized reduced graphene oxide. *Nature Communications* 5: 5002. https://doi.org/10.1038/ncomms6002.

87 Iannuzzi, M. and Frankel, G.S. (2007). Mechanisms of corrosion inhibition of AA2024-T3 by vanadates. *Corrosion Science* 49: 2371–2391. https://doi.org/10.1016/j.corsci.2006.10.027.

88 Shulha, T., Serdechnova, M., Lamaka, S.V. et al. (2023). Corrosion inhibitors intercalated into layered double hydroxides prepared in situ on AZ91 magnesium alloys: structure and protection ability. *ACS Applied Materials & Interfaces* 15: 6098–6112. https://doi.org/10.1021/acsami.2c18675.

89 Zhao, X., Zhao, K., and Sun, P. (2019). A method to describe the shapes of UV–vis absorbance spectra during the aggregation process of conjugated polymer solutions quantitatively. *Chemical Physics Letters* 730: 582–586. https://doi.org/10.1016/j.cplett.2019.06.030.

90 Yang, H., Xiong, C., Liu, X. et al. (2021). Application of layered double hydroxides (LDHs) in corrosion resistance of reinforced concrete-state of the art. *Construction and Building Materials* 307: 124991. https://doi.org/10.1016/j.conbuildmat.2021.124991.

91 Ge, L., Shao, B., Liang, Q. et al. (2022). Layered double hydroxide based materials applied in persulfate based advanced oxidation processes: property, mechanism, application and perspectives. *Journal of Hazardous Materials* 424: 127612. https://doi.org/10.1016/j.jhazmat.2021.127612.

92 Esmaeilzadeh, H., Su, J., Charmchi, M. et al. (2018). Effect of hydrophobicity on the water flow in carbon nanotube—a molecular dynamics study. *Theoretical and Applied Mechanics Letters* 8: 284–290. https://doi.org/10.1016/j.taml.2018.04.007.

93 Jegdić, B., Bobić, B., Radojković, B. et al. (2019). Corrosion resistance of welded joints of X5CrNi18-10 stainless steel. *Journal of Materials Processing Technology* 266: 579–587. https://doi.org/10.1016/j.jmatprotec.2018.11.029.

94 Tedim, J., Bastos, A.C., Kallip, S. et al. (2016). Corrosion protection of AA2024-T3 by LDH conversion films. Analysis of SVET results. *Electrochimica Acta* 210: 215–224. https://doi.org/10.1016/j.electacta.2016.05.134.

10

Investigation of the Self-healing Performance of Coatings Using Atmospheric Corrosion Monitor Technique

10.1 Introduction

Over the years, the application of organic coatings has been the most mature method used globally for anticorrosion applications. However, due to the porosity and permeability of organic coatings, water, dissolved oxygen, and electrolyte ions can pass through to the metal surface, causing corrosion at the metal–coating interface and leading to the failure of the coating [1–3]. Adding corrosion inhibitors to the coating can effectively enhance the protective performance and extend its service life [4–7].

However, due to the lack of a deep understanding of the coating failure mechanism, it is challenging to accurately assess the protective performance and service life of coatings on the metal substrate. Currently, the electrochemical impedance spectroscopy (EIS) method is the most commonly used method to study the changes in protective performance during the coating failure process [8–10]. By measuring electrochemical parameters such as coating capacitance and resistance, the failure process of the coating system can be studied. However, this method has disadvantages such as a long data accumulation period due to prolonged simulation of the failure experiment, instability during the accumulation process, and certain latencies in data as well as the hypothetical nature of fitting circuit analysis that cannot timely and intuitively reflect changes in the coating system. In addition, due to the fragmented nature of EIS data over time, EIS cannot reflect the process of the entire service life and the failure behavior of coating systems. Moreover, due to the limitations of outdoor atmospheric environments, indoor accelerated corrosion environmental studies on the failure process of anticorrosion coatings are more common, while studies on the failure process of anticorrosion coatings in outdoor atmospheric environments are less frequent and subject to various limitations [11–13]. Therefore, these studies may not accurately reflect the actual application conditions of coatings.

Therefore, monitoring the effectiveness and aging failure process of anticorrosion coatings in atmospheric environments is crucial. This can provide theoretical support for the application of coatings in engineering and experimental evidence for the service life of anticorrosion coatings. This also helps accumulate environmental corrosion data and establish a life assessment system for anticorrosion coatings. Currently, the required data to assess the corrosion degree and environmental parameters mainly come from traditional electrochemical tests, which are fragmented in

Smart Protective Coatings for Corrosion Control, First Edition. Lingwei Ma and Dawei Zhang.

time and space and cannot timely and effectively reflect changes in the environment and the effectiveness and failure process of anticorrosion coatings. Using the atmospheric corrosion monitor (ACM) technique, the corrosion current or impedance information under the thin liquid film of the atmospheric sensor probe surface can be monitored, so that the real-time protection behavior of coatings in environments can be monitored and recorded [14–17].

For self-healing coating systems, the release of corrosion inhibitors and the recovery of the barrier property of the coating are a continuous process and are dramatically influenced by environmental factors [18–20]. Therefore, it is possible to apply a self-healing coating on the surface of the ACM probe and equip it with basic environmental parameters sensors such as temperature and humidity, to monitor the protection, failure, and self-healing process of coatings in real-time. This helps further understand and analyze the failure process and self-healing mechanism of the anticorrosion coating. Currently, there are limited studies about the application of ACM technology in the research of self-healing coatings, and there is a lack of systematic research work in the field of coating research using the ACM technique.

This chapter selects a self-healing anticorrosion coating doped with corrosion inhibitors and uses ACM technology to collect the dynamic corrosion current data of the self-healing coating in different accelerated corrosion environments. Combined with macroscopic electrochemical tests, morphological analysis, and chemical composition analysis, this study deeply analyzes the self-healing mechanism and failure mechanism of the self-healing coating. It provides theoretical support and experimental evidence for the application of self-healing coatings in engineering. It also verifies the reliability, practicality, and sensitivity of the ACM technology in the study of coating systems, offering a new research idea and method for the study of corrosion protection of coatings.

10.2 Characterization Methods for Corrosion Protection Coatings

10.2.1 Experimental Characterization Methods for Corrosion Protection Coatings

10.2.1.1 Salt Spray Test

The salt spray test can be used to assess the corrosion resistance and anti-permeability of coatings by simulating atmospheric corrosion, thereby allowing modification of the coating formula based on the experimental results [21–23]. Depending on the experimental conditions, it can be divided into neutral salt spray, acid salt spray, copper-accelerated acetic acid salt spray, and alternating salt spray tests. The harsher the conditions, the faster the accelerated corrosion rate. Chloride ions in the salt fog have a high penetration ability to metal oxides and rust products, which will continuously corrode the metal substrate. The spray system ensures a continuously renewed thin liquid film on the surface of coupons, containing a large amount of oxygen, so the cathodic depolarization will continuously proceed.

Therefore, only several hours are needed to simulate the corrosion reactions in atmospheric conditions for one year. This benefits rapid research and development of anticorrosion coating systems or surface treatment processes.

10.2.1.2 Immersion Test

The corrosive medium is prepared by adding refined sea salt into distilled water to obtain a saltwater solution at certain concentrations. Coating coupons are added to the saltwater solution for a certain period to observe the failure process of coatings. The temperature is usually maintained at 25 °C. The solution contains a large number of chloride ions, which have strong permeability to the coating systems. The immersion test can quickly cause defects or blistering, allowing rapid evaluation of coating performance [24–26].

10.2.1.3 Wet–Dry Alternating Cycle Test

The wet–dry alternating cycle test is one of the most commonly used accelerating methods. The test was first proposed to study the metal atmospheric corrosion behavior. This test uses a system with a wheel or a lifting device to install specimens on the edge of the wheel. The rotation of the wheel is adjusted so that the specimens can be periodically immersed in the solutions and in atmospheric environment. When the specimens are rotated out of the solution, they are dried at a pre-set temperature [27–30]. Different solutions can be used to simulate different atmospheric environments, for example, distilled water for rural atmospheres, $NaHSO_3$ solution for industrial atmospheres, and NaCl solution for marine atmospheres. Therefore, the wet–dry alternating cycle test method can simulate the corrosion conditions in various natural environments. This method reproduces the three corrosion states of "wetting–humid–drying" on the metal surface in atmospheric corrosion, and the accelerating experimental results have been proven to correlate well with the outdoor field exposure tests.

Li et al. [31] used EIS to compare the deterioration process of organic coatings under immersion and wet–dry cyclic conditions. The results indicated that compared to single immersion, the wet–dry cyclic immersion significantly accelerates the occurrence of corrosion reactions. The deterioration process can be divided into three stages – water permeation of the coating, corrosion of the metal substrate, and development of corrosion and coating failure. Zhao et al. [32] studied the failure process of organic coatings under wet–dry cyclic conditions and established four equivalent circuit models for different stages of coating failure. They analyzed the variation patterns of coating capacitance and resistance. The results showed that the capacitance of the coating gradually increased during the immersion period and decreased during the drying period as the moisture evaporated. The coating resistance decreased during the immersion period and gradually increased during the drying period. This demonstrates the feasibility of wet–dry cycling experiments in simulating the corrosive conditions of the marine splash zone [33, 34].

10.2.1.4 Atmospheric Exposure Test

The outdoor atmospheric exposure test involves fixing samples on exposure frames and conducting field corrosion tests under natural atmospheric conditions [35–37].

This method can accurately reflect the corrosion status of the location and is the most reliable way to obtain the environmental adaptability of materials and equipment. During the exposure process in marine atmospheres, samples are collected and observed at predetermined intervals to analyze corrosion kinetics. Various physical analysis methods can be used to study the composition of corrosion products in the rust layer, as well as the macroscopic and microscopic corrosion morphologies from the surface view and cross-section view. China started to establish a national network of atmospheric exposure corrosion test sites in the late 1950s. There was a pause from the mid-1960s to the late 1970s, but all were restored in the 1980s, initiating long-term, systematic atmospheric corrosion research and yielding numerous results. An atmospheric corrosion regional zoning map was created, based on China's atmospheric conditions, which analyzed the corrosion failure process and patterns of corrosion protection coatings on bridge steel structures in atmospheric corrosion environments, establishing a predictive model for the lifespan of anti-corrosion coatings. Yong et al. [38] studied the failure modes of organic coatings in atmospheric environments, defined the corrosion resistance grades of organic coating systems, and established the electrochemical impedance models. An equivalent relationship was established between the laboratory and the field exposure of organic coating protection systems.

10.2.2 Experimental Testing Methods for Corrosion Protection Coatings

10.2.2.1 Electrochemical Methods

EIS involves applying a small amplitude sinusoidal alternating disturbance signal to the test system and measuring the response of coating systems to obtain an impedance or admittance spectrum. This method is nondestructive and *in situ*, measuring various electrochemical parameters related to coating performance and failure process, such as coating capacitance, resistance, and double-layer capacitance at the coating–metal interface. It became the primary method for studying coating failure processes internationally in the 1980s. The electrochemical noise method measures the spontaneous fluctuations in potential or current over time in an electrochemical system [39, 40]. This nondestructive, *in situ* technique does not require the application of external disturbances to the test system and is particularly useful for studying the behavior of coatings with high resistance or thickness. It has become an important method in recent decades for studying the corrosion rate of metals, the localized corrosion processes, and the screening of corrosion inhibitors. Scanning Kelvin probe technique measures the local corrosion potentials on the sample surface in moist air without direct contact, allowing in situ monitoring and study of the delamination phenomena in coatings [41]. Measuring the differences in surface potentials determines the average delamination speed of the coating and can characterize the effect of metal pretreatment and the performance of the coating itself by the lowest negative electrode potential.

10.2.2.2 Surface Analysis Methods

Scanning electron microscopy (SEM) is the most common instrument for analyzing the metal surface morphology after corrosion experiments. It provides micro-morphological surfaces of the metal material at different stages and can observe the localized corrosion behavior. Atomic force microscopy (AFM), scanning tunneling microscope (STM), and transmission electron microscope (TEM) are also used for microscopic analysis of corrosion morphology. The corrosion products can be analyzed by X-ray photoelectron spectroscopy (XPS) and Auger electron spectroscopy (AES), which can perform semiquantitative analysis of elements in corrosion products. X-ray diffraction (XRD) can perform qualitative analysis of crystalline structure substances in corrosion products. Raman and infrared (IR) spectroscopy is also commonly used for the identification of corrosion products, detecting typical film components on the coating surface [42–46].

10.2.3 Research Progress in Atmospheric Corrosion Monitor Technique

ACM is based on the current signal from a thin liquid film electrochemical cell to reflect the corrosion degree of the atmospheric environment. Mansfield first introduced the concept of ACM in 1976 for detecting and evaluating the corrosion of metals in atmospheric environments. The basic principle of ACM is to use a pair of metals of different potentials to form a galvanic corrosion sensor, with a high-sensitivity ammeter measuring the corrosion current to evaluate the rate or duration of corrosion occurring in the metal couple in the environment.

ACM cells are of two types – bimetallic primary cells and three-electrode electrolytic cells. The bimetallic primary cell consists of different types of metal thin plates (such as Fe/Ag, Zn/Ag) arranged alternately in the form of ABABAB, where A represents the working electrode (like Fe), and B represents the auxiliary electrode (like Ag). Each pair of metals is separated by an insulator (such as polyester film, PTFE film, or resin), and the same materials are connected in parallel and led out. When a conductive liquid film exists between the two metals, the external zero-resistance ammeter can detect the current value between the anode and cathode. Since this current value is directly related to the corrosion current of the active electrode, it can be used for direct evaluation of the anodic material. From then on, the ACM technique has received significant research attention, mostly aimed at providing differences in the environmental corrosion degree through ACM sensors [47–49].

Some researchers have compared the various properties of Fe/Cu and Fe/Zn electrode pairs and concluded that the Fe/Zn cell is superior to the Fe/Cu cell in aspects such as parallelism, reproducibility, and sensitivity of the battery. Xu and coworkers have made improvements to the bimetallic ACM, resulting in a refined ACM that can precisely control the working surface area, with stable performance, long service life, and extensive research into the factors influencing atmospheric corrosion.

The three-electrode electrolytic cell consists of the same type of metal sheets arranged in the [ABRBA] sequence, where R represents the reference electrode. This arrangement is similar to the bimetallic ACM. Gonzalez et al. [50] have improved the three-electrode arrangement to the [(AA)ARA(AA)] configuration. Experimental research indicates that this improved method can effectively prevent induced polarization, especially when applying certain waveform potentials in studies similar to polarization curves. The three-electrode ACM is commonly used to measure electrochemical parameters such as impedance, corrosion current, and Tafel slope. This type of ACM is also a powerful tool for studying the mechanisms of atmospheric corrosion.

Xiao and coworkers [51] demonstrated that replacing traditional gravimetric methods with ACM technology for studying steel corrosion is feasible. This conclusion is confirmed by comparing the outdoor exposure test results with indoor accelerated experiment results. Cao et al. [52] used a Cu/Fe bimetallic ACM sensor to study the corrosion kinetics of iron in different corrosive environments. He demonstrated that the ACM testing results correlated well with those of the gravimetric method. Kainuma et al. [53] used an Ag/Fe electrode-based ACM sensor for long-term outdoor exposure and validated the effectiveness of the evaluation results. Mizuno [14] installed the Fe/Ag bimetallic ACM sensors and Fe, Zn metal coupon samples on six different parts of automobiles. They used the galvanic current generated by the sensors to represent the corrosion rate of those parts and provided data on the corrosion differences in various parts of the vehicle under different environments to guide material selection. The study results showed that the integral value of the galvanic current from the ACM instrument over the measurement period had a good linear correlation with the corrosion rate of carbon steel coupons, with a correlation coefficient as high as 0.913.

Currently, the ACM technique is predominantly used for studying the corrosion degree of atmospheric environments on different metal samples, with relatively few studies focusing on coating systems. Applying coatings to the probe of an ACM instrument and using the instrument for real-time monitoring of the coating's corrosion behavior is not widely reported. This approach could explore the self-healing mechanisms and failure processes of anticorrosion coatings in corrosive environments. It involves accumulating environmental corrosion data for typical corrosion protection coatings, establishing a lifespan assessment system for coatings, and providing theoretical and experimental support for the application of coatings in engineering.

10.3 Study of the Self-healing Behavior of Coatings Using ACM Technique

10.3.1 Materials and Methods

10.3.1.1 Materials

The materials and reagents used in this section are shown in Table 10.1.

10.3.1.2 Preparation of Galvanic Corrosion Sensors

ACM systems consist of galvanic corrosion sensors and monitoring instruments. The galvanic sensor probe is composed of multiple thin metal plates of two different

Table 10.1 Experimental materials and reagents.

Materials	Purity	Company
Zinc phosphate hydration ($Zn_3P_2O_8 \cdot xH_2O$)	Analytical purity	Sinopharm Group
Ethyl alcohol	Analytical purity	Sinopharm Group
Acetone	Analytical purity	Sinopharm Group
Bisphenol A diglycidyl ether (DGEBA)	Analytical purity	Sigma Corporation
Polyetheramine (D230)	Analytical purity	Aladdin, Inc.

types arranged alternately, with each pair of metals separated by an insulator such as polyester film, polytetrafluoroethylene (PTFE) film, or resin to ensure electrical insulation. In this experiment, Q235 steel and graphite are used. Insulation and separation are achieved using glass fiber-reinforced epoxy composite boards. The carbon steel and graphite are then connected in parallel and led out. Each sensor probe in this experiment consists of seven pairs of metal couples. Each carbon steel and graphite electrode has an exposed area of $21 \times 1\ mm^2$. The spacing between the metal plates is nominally controlled by the thickness (0.1 mm) of the epoxy composite material. Figure 10.1 displays a top view and cross-sectional view of the sample. The electrode assembly is sealed with epoxy resin. When an electrolyte film forms on the surface of the sensor, the carbon steel and graphite electrodes are connected to generate an electrochemical corrosion current. The ACM sensor is connected to a microammeter and data logger via wires to monitor and record the current values.

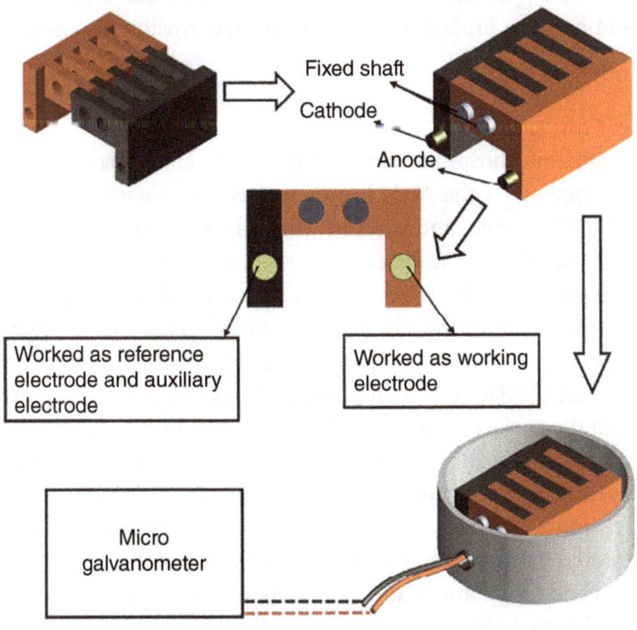

Figure 10.1 Structure diagram of galvanic corrosion sensor.

This methodology outlines the preparation of a typical galvanic corrosion sensor for an ACM, which is fundamental for monitoring atmospheric corrosion in various environments. The sensor's operation relies on the galvanic interaction between dissimilar metals, here Q235 steel and graphite, creating a current flow that can be measured to infer the rate of corrosion. The data gathered from such experiments are crucial for understanding the corrosive processes and can be instrumental in improving protective strategies and material selection in engineering and other applications.

As shown in Figure 10.1, a reference electrode and counter electrode are used along with an electrochemical workstation to test the electrochemical properties of the metal on the sensor surface. When conducting EIS measurements using the sensor probe, one electrode, the Q235 steel end, is used as the working electrode, and the other electrode, the carbon steel parallel connection, serves as both the reference and counter electrodes. The electrochemical experiments involve measuring the open circuit potential (OCP) and galvanic current between the electrodes. Measuring the OCP of the two electrodes serves two purposes – (i) To determine the polarity of the two electrodes and establish the direction of the current. To understand the corrosion process and the flow of electrons between the electrodes, it is important to know which metal acts as the anode and which as the cathode. (ii) To gather other information about the coupled system, such as whether there has been any reversal of polarity. Polarity reversal in a galvanic couple can change the direction of corrosion and may indicate changes in the corrosion environment or the condition of the metals. This method of using EIS and measuring OCP provides valuable insights into the corrosion behaviors of the metals involved in the sensor. It allows for a more nuanced understanding of the galvanic interactions and helps in predicting and mitigating corrosion in practical applications.

10.3.1.3 Preparation of Coatings

Epoxy coatings loaded with zinc phosphate corrosion inhibitors are prepared as the self-healing coating, and applied on the ACM sensor surfaces with a galvanic corrosion couple composed of Q235 steel and graphite. Zinc phosphate is used as the corrosion inhibitor to enhance the self-healing properties of the epoxy coating.

Before preparing the coating, the ACM sensor was first sanded with #150 grit sandpapers until the galvanic couple was completely exposed. This was followed by finer sanding using #240 and #400 grit sandpaper to ensure a smooth surface for the coating. The sanded sensor surface was then cleaned with acetone to remove any oils or contaminants from the surface. Afterward, the sensor was fully immersed in anhydrous ethanol and subjected to ultrasonic cleaning for five min to remove any remaining surface impurities. Once cleaned, the sensor was placed in an oven at 60 °C and dried for three hours. After drying, the current on the sensor was tested, which should have registered between 0 and 0.2 nA. It indicates that the sensor surface was sufficiently clean and free of any conductive contaminants. The meticulous preparation of the clean, dry, and smooth sensor surface is a crucial step to ensure the successful preparation of the self-healing coating on the ACM sensor surface and its subsequent performance in corrosion protection.

(1) Preparation of blank epoxy coating without corrosion inhibitors. DGEBA was first heated at 45–50 °C for 30 minutes to make it transition from a solid to a molten state. A specific amount of the molten DGEBA was then measured, and an epoxy curing agent, polyetheramine D-230, was added. The mixture was heated and stirred at 50 °C until both components were uniformly mixed. The molar ratio between DGEBA and D-230 was controlled at 2 : 1. For the ACM sensors, once the initial current met the requirements (within 0–0.2 nA), the resin mixture was applied to the sensor surface using an 80 μm diameter scraper bar. The coated sensor was then placed in an oven set at 65 °C for a 24 hours curing process to obtain the blank epoxy coating without a corrosion inhibitor.
(2) Preparation of blank epoxy coating with corrosion inhibitors. DGEBA was first heated at 45–50 °C for 30 minutes to make it transition from a solid to a molten state. A certain amount of zinc phosphate hydrate [54, 55] was ground and added to the molten DGEBA. The mixture was stirred for three hours until the zinc phosphate was evenly dispersed. The mixture was heated at 50 °C to avoid the solidifying of DGEBA. Finally, polyetheramine D-230 was added to the molten mixture of DGEBA and zinc phosphate. The mixture was vigorously stirred until it was uniformly mixed. The ratio of DGEBA, D230, and zinc phosphate was maintained at a molar ratio of 2 : 1 and a mass concentration of 15% for zinc phosphate. For the ACM sensors, once the initial current met the requirements (within 0–0.2 nA), the resin mixture was applied to the sensor surface using an 80 μm diameter scraper bar. The sensor is then placed in an oven set at 65 °C for a 24 hour curing process to obtain the self-healing epoxy coating with 15 wt% zinc phosphate. The coatings with 15 wt% zinc phosphate and without corrosion inhibitor are referred to as ZP-15% coating and ZP-0% coating, respectively. Figure 10.2 shows the schematic diagram of coating preparation on the ACM sensor surface.

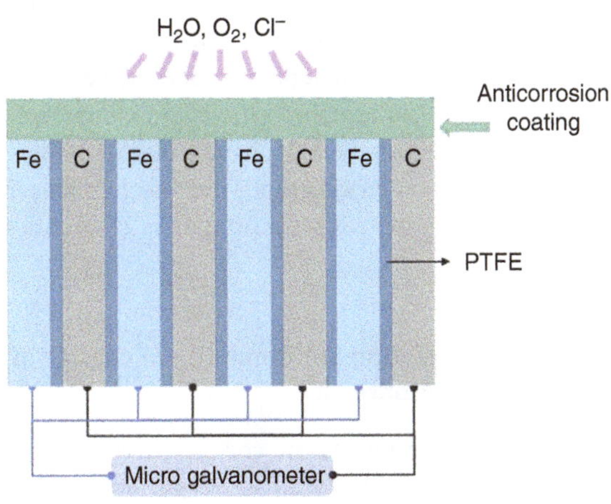

Figure 10.2 Schematic diagram of coating preparation on the ACM sensor surface.

10.3.1.4 Accelerated Corrosion Test

(1) Room temperature immersion environment. The immersion experiment is a well-established method for testing the corrosion resistance of coatings. It includes room temperature immersion and high-temperature immersion. In this experiment, room temperature immersion is used with a 3.5 wt% NaCl solution. The immersion solution is replaced timely during the experiment to maintain constant environmental temperature.

(2) Alternating wet–dry environment. Alternating wet–dry environment is used to explore the anticorrosion performance of coatings. This experiment employs a wet–dry cyclic immersion test chamber, controlling the temperature at room temperature (25 ± 2) °C, with a drying temperature of 40 °C, and a relative humidity (RH) of $60 \pm 5\%$. The test runs on a 60 minutes cycle, meaning a 20 minutes immersion followed by a 40 minutes drying period, using a 3.5 wt% NaCl solution as the experimental liquid.

(3) Atmospheric exposure environment. The atmospheric exposure environment is used to study the anticorrosion performance of coatings. This atmospheric exposure experiment began in August 2020 and ended in September 2020, lasting approximately 30 days, with the location in Beijing. The main environmental factors that cause coating degradation include sunlight, temperature, oxygen, and moisture. Therefore, the temperature, humidity, and precipitation at the sample location were measured during the experiment. This outdoor exposure provides insight into how the coating performs under typical atmospheric conditions that it would encounter in practical applications.

10.3.1.5 Characterization of Coatings

The surface morphology and the microscale morphology of the corrosion products on the coating were observed using a KEYENCE VHX-2000 digital microscope, with a magnification range from 100 to 1000 times. For microscopic observations, a Hitachi SU8100 field emission scanning electron microscope (FE-SEM) was employed to examine the gold-sputtered probe at an accelerating voltage of 15 kV. Additionally, EDS was utilized for point and area analysis of elements at the coating scratch, with a particular focus on key elements such as Fe, O, Zn, and P. The proportion of these elements was then used to obtain related data, which helps in understanding the composition and characteristics of the corrosion products. This detailed analysis aids in comprehensively understanding the protective properties of the coating and the mechanisms of corrosion at a microscopic level. The corrosion-resistant properties of coatings were characterized using EIS measurement.

10.3.2 Self-healing Performance of Coatings in Room Temperature Immersion Environment

This section investigates the corrosion resistance behavior of ZP-0% and ZP-15% coatings applied on the ACM sensor surface in 3.5 wt% NaCl solution at room temperature. By employing OCP values and EIS measurement data, this study compares the variation in corrosion electrochemical behavior of the two coatings during the immersion test. With the aid of a CLSM microscope, the difference in corrosion morphology of the two coatings after immersion experiments was observed. Combined with SEM analysis, this study characterizes the main element distribution at the damaged region of the coating, thereby elucidating the self-healing mechanism of the coating. ACM technique was used to monitor the corrosion currents of coatings in a room temperature immersion environment in real time. The corrosion current data were combined with the corrosion morphology and electrochemical test results to determine the reliability of ACM in monitoring coatings in a room temperature immersion environment.

10.3.2.1 Determination of Corrosion Inhibitor Amount in Coatings

Based on literature references, coating samples with different zinc phosphate contents, specifically 0%, 15%, and 25% were placed in an oven with a Q235 steel substrate. After 24 hours, the coatings were removed and scratched using a scraper. After curing, they were immersed in a 3.5 wt% NaCl solution. The EIS method was used to test the barrier property of coatings with different zinc phosphate contents, and the EIS data were collected after different stages of immersion in the 3.5% NaCl solution at room temperature (25 °C).

Figure 10.3 plots the Nyquist diagram and Bode plots of epoxy coatings with different amounts of zinc phosphate addition. For the coatings containing 15 and 25 wt% zinc phosphate content, the electrolyte solution penetrated the damaged part of the coating, releasing zinc phosphate at the damaged area and forming a film on the surface of the metal substrate. Therefore, the rapid progress of the corrosion reaction was delayed. Compared with the blank epoxy coating, the impedance moduli of the coatings with 15 and 25 wt% zinc phosphate addition were relatively high, indicating good corrosion resistance. However, the impedance modulus of the coating containing 25 wt% zinc phosphate was lower than that of the coating containing 15 wt% zinc phosphate. This was due to the excessive concentration of corrosion inhibitors, which cannot be evenly dispersed in the coating, affecting the overall uniformity and integrity of the coating [56, 57]. Therefore, the addition of zinc phosphate was controlled at 15 wt% to achieve higher barrier property and good self-healing performance. This study provides a critical insight into the optimal concentration for enhancing the self-healing and protective properties of coatings in corrosive environments.

10.3.2.2 Corrosion Morphology on Coating Surfaces

After the immersion experiment, the morphology of coating surfaces was observed using CLSM observation. As depicted in Figure 10.4, at the beginning of immersion, the coatings were free of coating products, and the scratches crossed the Q235 and graphite electrodes. However, as the immersion time progressed, the damaged areas of the ZP-0% coating showed a considerable accumulation of yellow corrosion products. In contrast, the ZP-15% coating showed much fewer corrosion products with uniform distribution at the scratched coating area. No noticeable spread of the corrosion products was observed on the ZP-15% coating, and the edges of the scratch

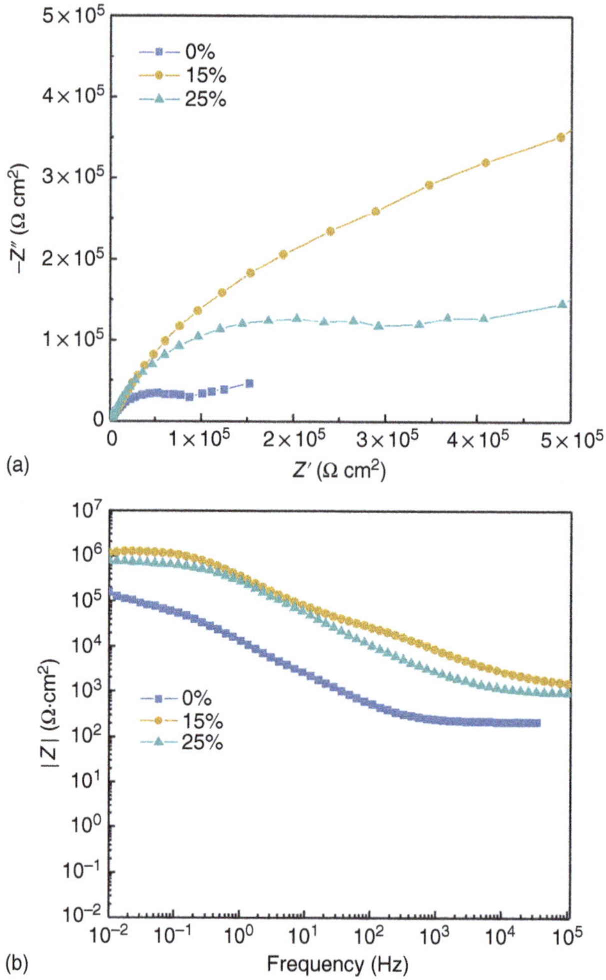

Figure 10.3 (a) Nyquist diagram and (b) Bode plots of epoxy coatings with different amounts of zinc phosphate addition.

remained clearly visible. This suggests that the presence of zinc phosphate corrosion inhibitors in the coating could avoid the occurrence and spread of corrosion in the damaged areas of the coating [58–60]. The addition of zinc phosphate slowed down the corrosion process and limited the extent of coating damage, which is crucial for the longevity and effectiveness of the coating for practical applications.

SEM observation was used to show the morphology and analyze the composition of the two coatings after immersion in 3.5 wt% NaCl solution. The SEM images and EDS spectra of the corrosion products on the ZP-15% coating after 7 days of immersion in 3.5 wt% NaCl solution were illustrated in Figures 10.5 and 10.6. The corrosion products exhibited needle-like structures, which were typically associated with the phosphatization of steel. The EDS results indicate that the main elements at the scratched location were P, O, Zn, and Fe. This also demonstrates the formation of

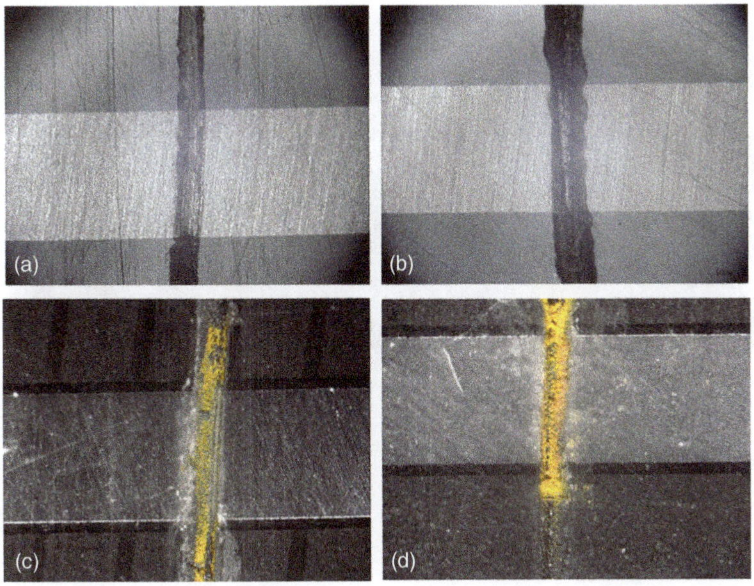

Figure 10.4 Macroscopic corrosion morphologies of ZP-15% coating (a) before immersion and (c) after 7 days of immersion in 3.5 wt% NaCl solution; macroscopic corrosion morphologies of ZP-0% coating (b) before immersion and (d) after 7 days of immersion in 3.5 wt% NaCl solution at room temperature.

a zinc phosphate protective film on the surface of the exposed metal substrate. The presence of such a protective film was indicative of the self-healing properties of the ZP-15% coating, where the zinc phosphate actively contributed to the protection and healing of the damaged area. It mitigated further corrosion and maintained the integrity of the underlying metal. This insight is crucial for the development and improvement of corrosion-resistant and self-healing coatings, particularly in harsh environmental conditions where material durability is of utmost importance.

10.3.2.3 Electrochemical Analysis

ZP-15% and ZP-0% coatings were covered on the surface of ACM sensor probes. Artificial scratches were created on the coating surface of the ACM sensor probes using a blade, and the probes were then immersed in 3.5 wt% NaCl solution for periodic electrochemical testing. The changes in OCP values for the two coatings were monitored. Figure 10.7 illustrates the trend of self-corrosion potential of the coatings. It is observed that at the beginning of immersion, the corrosion potentials of the ZP-0% and ZP-15% coatings were of similar values, maintained at approximately −0.42 V. The OCP values of the ZP-15% coating remained relatively stable during the entire immersion time of four days, maintaining around −0.4 V. In contrast, the ZP-0% coating exhibited significant fluctuations in corrosion potential, dropping from −0.4 to nearly −0.45 V. This noticeable decline is due to the damage in the coating that allowed moisture, oxygen, and other corrosive ions to penetrate the coating and

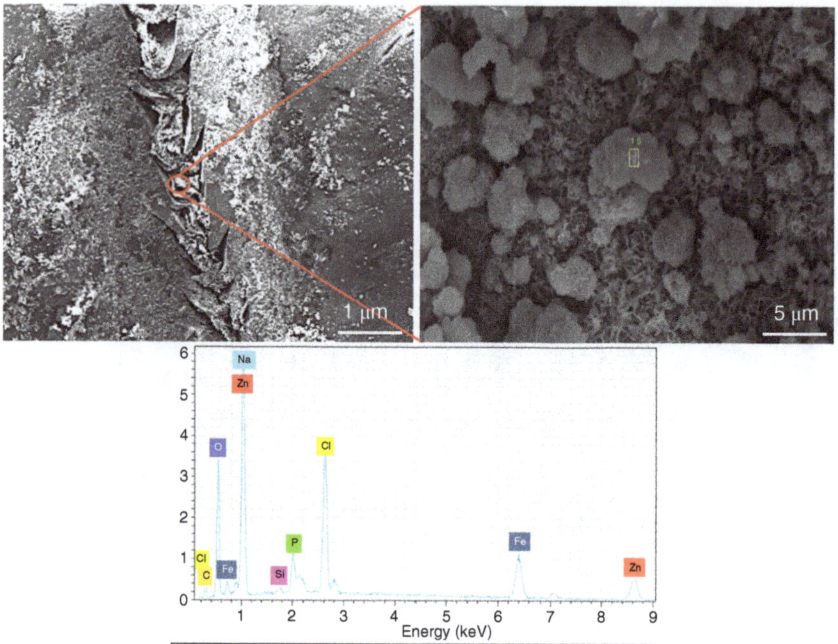

Element	C	O	P	Cl	Fe	Zn
wt%	12.56	31.16	3.43	9.93	7.91	13.00

Figure 10.5 SEM and EDS results of the corrosion products on the ZP-15% coating after seven days of immersion in 3.5 wt% NaCl solution at room temperature.

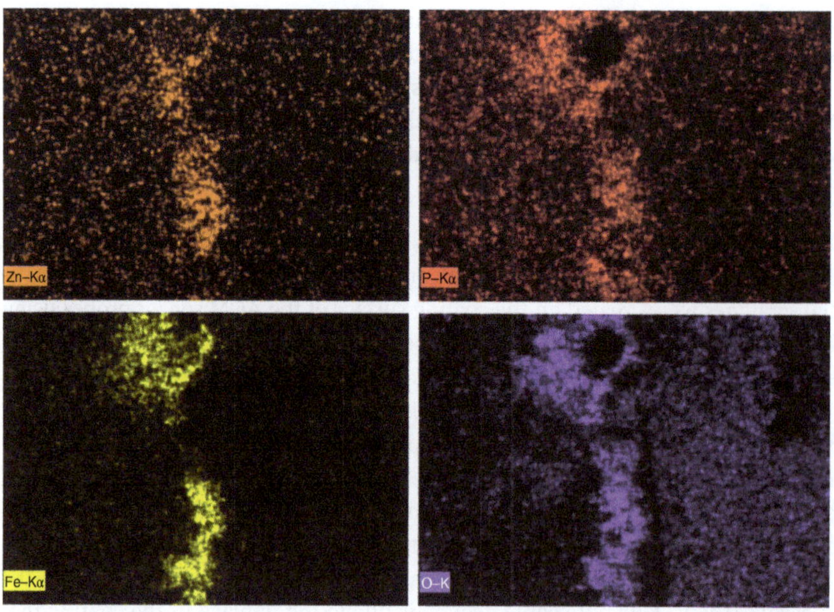

Figure 10.6 EDS mapping results of the corrosion products on the ZP-15% coating after seven days of immersion in 3.5 wt% NaCl solution at room temperature.

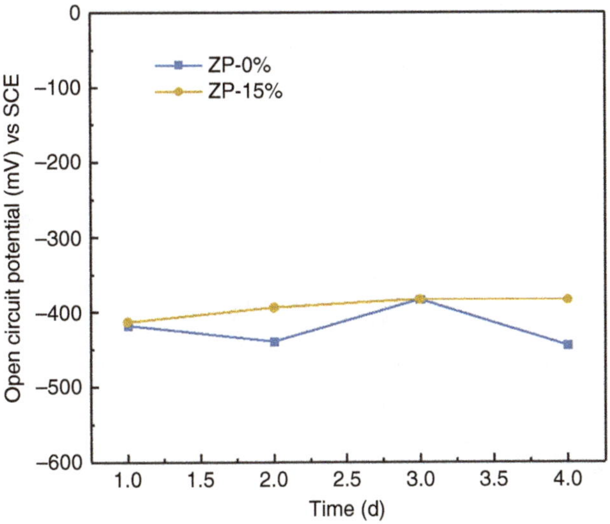

Figure 10.7 OCP values of ZP-0% coating and ZP-15% coating during immersion in 3.5 wt% NaCl solution.

reach the metal interface, leading to corrosion reactions. The continuous infiltration of water, oxygen, and ions opened pathways through the coating, reducing its protective performance [39, 61, 62]. A minor rise in the potential of the ZP-0% coating on the third day might be attributed to the accumulation of corrosion products temporarily elevating the corrosion potential. However, as the solution continued to permeate, the potential resumed its downward trend. On the other hand, the ZP-15% coating maintains its barrier performance over a short period due to the release of the corrosion inhibitor from the coating, which suppressed the corrosion reaction at the damaged site, resulting in slower degradation of the protective performance of the coating. This behavior underscores the effectiveness of zinc phosphate as a corrosion inhibitor in maintaining the integrity and prolonging the durability of the coating under corrosive conditions.

Figure 10.8 shows the EIS measurement data of ZP-0% and ZP-15% coatings after zero, three, five, and seven days of immersion in 3.5 wt% NaCl solution. The assessment of coating corrosion resistance using the magnitude of the low-frequency impedance value at 0.01 Hz is widely applied in coating corrosion resistance evaluation systems. For the ZP-0% coating, the radius of the impedance arc continuously decreased during the immersion process, indicating that the corrosion activity at the metal substrate near the scratch gradually accelerated. In contrast, the impedance arc radius of the ZP-15% coating did not show a significant decrease. In the Bode plot, the $|Z|_{0.01\,\text{Hz}}$ of the ZP-0% coating gradually decreased and dropped to $10^5\,\Omega\,\text{cm}^2$ after one week of immersion, indicating that the protective performance of the coating gradually decreased with the extension of immersion time. Meanwhile, the $|Z|_{0.01\,\text{Hz}}$ value of the ZP-15% coating maintained around $10^6\,\Omega\,\text{cm}^2$ and showed a slight increase. This increase was mainly due to the corrosion-inhibiting action of zinc phosphate. For the phase angle plots, the phase angle widened with the

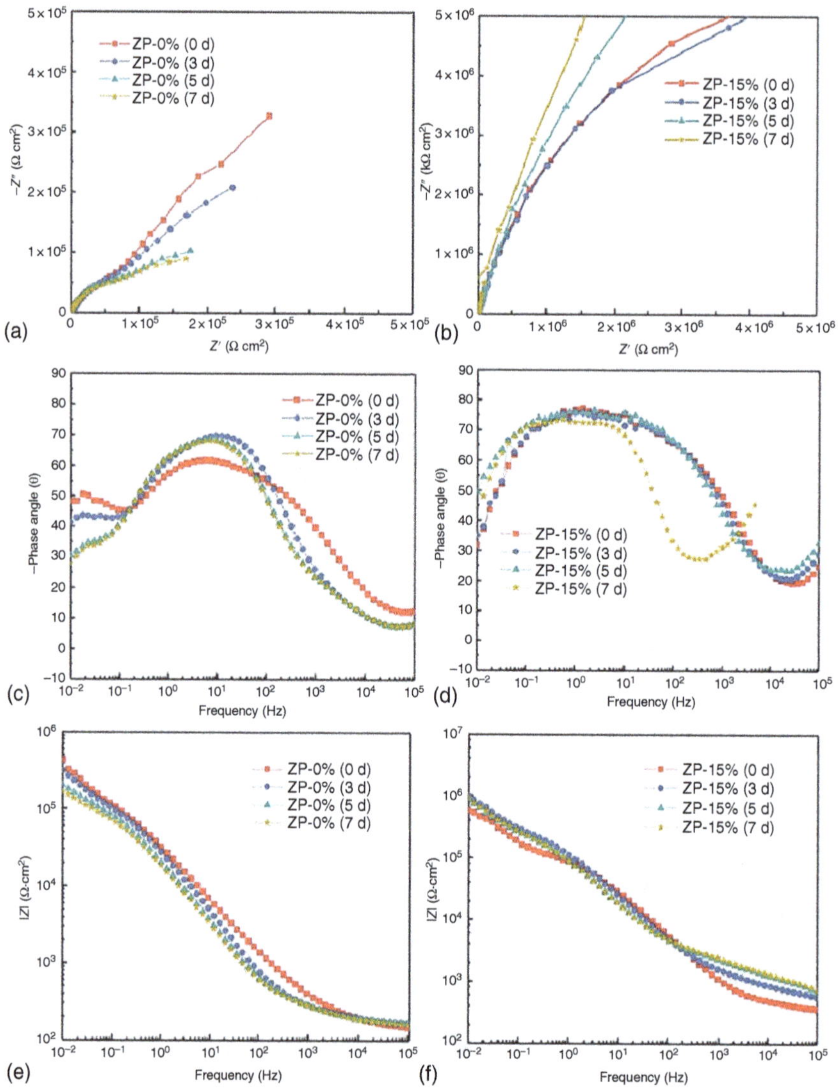

Figure 10.8 Nyquist diagram and Bode plots of (a,c,e) ZP-0% and (b,d,f) ZP-15% coatings after zero, three, five, and seven days of immersion in 3.5 wt% NaCl solution.

addition of zinc phosphate, which was ascribed to the generation of a corrosion protective layer on the metal surface. The results demonstrate the efficiency of zinc phosphate in maintaining and enhancing the coating's corrosion resistance, thereby slowing down the corrosion process and extending the life of the coating and the underlying metal substrate.

The excellent corrosion resistance of the ZP-15% coating is attributed to the formation of phosphates on the steel. At the damaged areas of the coating, the accumulation of phosphate ions exhibits strong chelating abilities, reacting with Fe^{2+}

and Fe^{3+} ions to form a phosphate protective film. This phosphate protective film is insoluble in water and adheres strongly to the substrate metal, effectively preventing further corrosion reactions and achieving a more stable corrosion resistance effect. The protective film acts as a barrier, hindering the diffusion of corrosive agents to the metal substrate and reducing the rate of corrosion. The presence of zinc phosphate in the coating, therefore, significantly enhances the coating's durability and effectiveness, providing a reliable solution for long-term protection of metal surfaces in corrosive environments. This mechanism is a critical aspect of the self-healing properties of ZP-15% coatings, where the coating not only provides an initial protective barrier but also actively responds to damage by forming additional protection, thereby maintaining its integrity and protective capabilities over time.

10.3.2.4 ACM Measurement

The ZP-15% and ZP-0% coatings with artificial scratches were fully immersed in 3.5% NaCl solution, and the change in coating current was monitored using an ACM instrument. The current of the sensor was collected at a frequency of once per minute. The current resolution was 0.1 nA, with a range from 0.1 nA to 50 mA. $I_{\text{ACM-instant}}$ represents the real-time galvanic corrosion current obtained on the ACM sensor, with a one-minute data collection interval. Figure 10.9 presents the complete dataset of instant currents ($I_{\text{ACM-instant}}$) obtained from the ACM sensor throughout the entire immersion process, capturing the instant current values for both ZP-0% and ZP-15% coatings. The experiment was conducted under consistent environmental conditions due to the sensitivity of ACM sensors to environmental changes. To better compare the differences in the corrosiveness of the coatings, the output current values $I_{\text{ACM-instant}}$ were integrated over the test duration to obtain the charge output ($Q_{\text{ACM-instant}}$) using the formula:

$$Q_{\text{ACM-instant}} = \Sigma I_{\text{ACM-instant}} \times 1 \text{ minute} \quad (10.1)$$

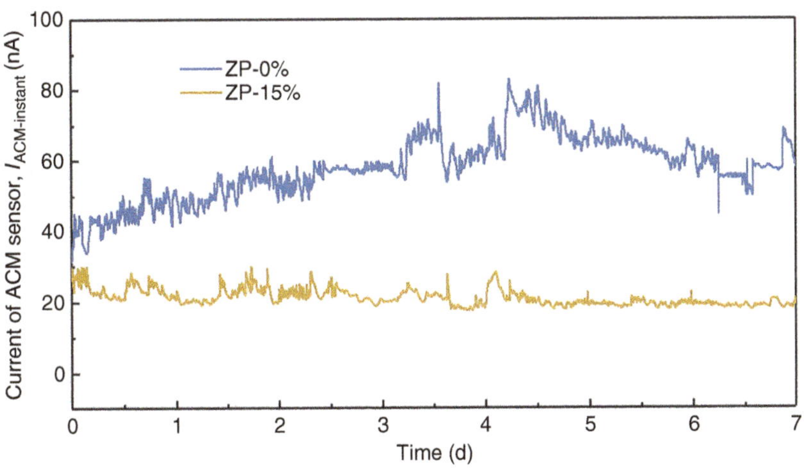

Figure 10.9 Output corrosion current values $I_{\text{ACM-instant}}$ of ZP-15% and ZP-0% coatings.

Here, $I_{ACM\text{-}instant}$ represents the real-time galvanic corrosion current obtained on the ACM sensor, with a data collection interval of one minute. In the initial stages of immersion, the electrolyte solution easily permeates the probe surface through artificial damage, forming a circuit between the cathode and anode and generating a corrosion current. Figure 10.9 shows that at the beginning of the immersion period, both coatings possessed a current value of about 30 nA. As the immersion time extended, the corrosion of the ZP-0% coating gradually intensified, showing a continuous increase in corrosion current, reaching 90 nA after four days. The corrosion current then became stabilized from day 5 to day 7, which might be ascribed to the generation of corrosion products that offered a protective property to some extent. Conversely, the corrosion current for the ZP-15% coating slightly decreased with an increase in the immersion time, maintaining a stable overall state, settling around 20 nA after seven days, which was much lower than that of the ZP-0% coating. The much larger current fluctuation for the ZP-0% coating indicated a higher corrosion activity. To have a better understanding of the overall corrosion behavior on the two coatings, Figure 10.10 shows the corrosion charge curve $Q_{ACM\text{-}instant}$ of the two coatings. The $I_{ACM\text{-}instant}$ and $Q_{ACM\text{-}instant}$ values show similar variation trends. For both ZP-15% and ZP-0% coatings, the $Q_{ACM\text{-}instant}$ values increased gradually with time, and the ZP-15% coating had a much lower $Q_{ACM\text{-}instant}$ at any specific detection time. Both the current and charge monitoring curves indicate that during the immersion process, the ZP-15% coating had superior corrosion resistance performance compared to the ZP-0% coating. These results are consistent with the results from electrochemical tests. It demonstrates that the output from the ACM sensor can reflect the corrosion state of the coating samples in an accurate manner, showing the high reliability of ACM technology in monitoring the corrosion resistance performance of coatings in a room temperature immersion environment [63–65].

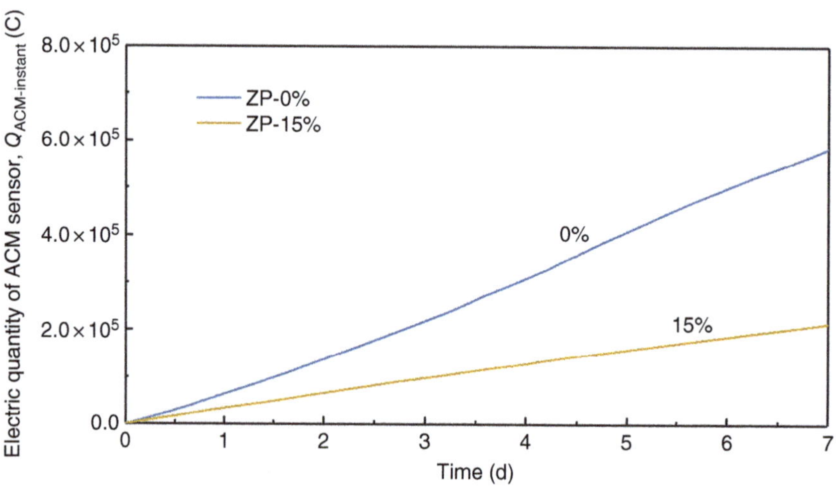

Figure 10.10 Output corrosion charge values $Q_{ACM\text{-}instant}$ of ZP-15% and ZP-0% coatings during immersion test in 3.5 wt% NaCl solution.

10.3.3 Self-healing Performance of Coatings in Alternating Wet–Dry Environment

This section explores the corrosion resistance of ZP-0% and ZP-15% coatings applied to ACM sensor surfaces and exposed to alternating wet–dry environments. Through OCP and EIS assessment results, this study compares the corrosion electrochemical behavior changes of both coatings in the alternating wet–dry environment. Digital cameras, stereo microscopes, and SEM were used to observe the corrosion morphology differences of both coatings after testing. ACM is used to monitor the coatings in real-time under alternating wet–dry conditions, obtaining galvanic corrosion current curves and integrating them to get corrosion charge curves. These are compared with the coating's corrosion morphology and electrochemical test results to determine the reliability of ACM in monitoring coatings in alternating wet–dry environments.

10.3.3.1 Corrosion Morphology on Coating Surfaces

Cameras and CLSM microscope were used to observe and compare the morphology of different coatings after the alternating wet–dry tests. As shown in Figure 10.11, after zero, four, and seven days of alternating wet–dry testing, both the ZP-0% and ZP-15% coatings exhibited a noticeable accumulation of corrosion products in the damaged areas. The ZP-0% coating showed more extensive corrosion product accumulation and lateral spreading from scratch, along with bubbling and more severe deterioration compared to the corrosion behavior during the full immersion

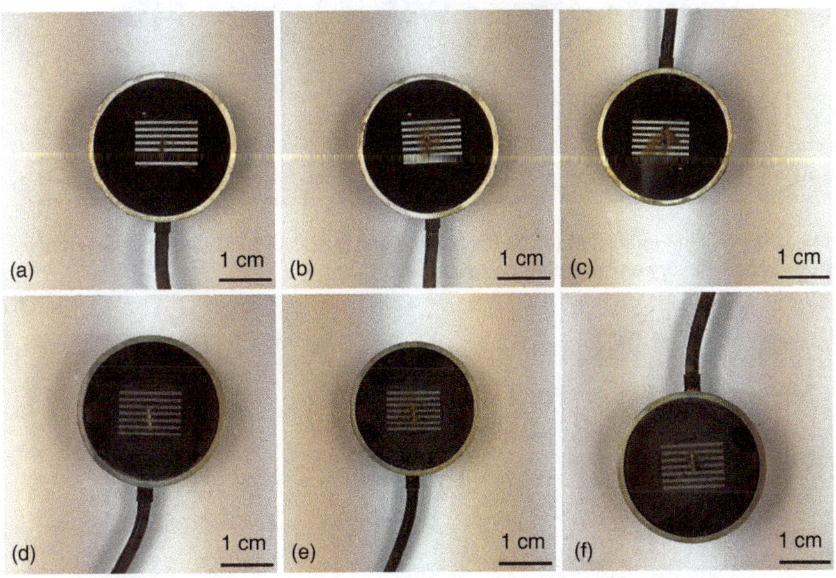

Figure 10.11 Optical images of the ACM sensor surface covered by ZP-0% coating after (a) zero, (b) four, and (c) seven days of alternating wet–dry environment corrosion test; optical images of the ACM sensor surface covered by ZP-15% coating after (d) zero, (e) four, and (f) seven days of alternating wet–dry environment corrosion test.

environment. The corrosion products in the wet–dry environment were sparsely distributed on the surface of the sample. The ZP-15% coating showed uniformly distributed rust at the scratch with a much smaller corroded area than the ZP-0% coating, exhibiting good corrosion resistance related to the formation of phosphate film on the metal substrate surface. The optical images provide us with an intuitive comparison between the corrosion degrees of two coatings.

Further observations of the macroscopic morphology of the damaged areas of the ZP-15% and ZP-0% coatings were made using CLSM observation, as shown in Figure 10.12. The ZP-15% coating showed uniformly distributed corrosion products without significant lateral or longitudinal spreading. The width of the corroded line was approximately 6 mm. In contrast, the ZP-0% coating samples had more loosely distributed corrosion products with apparent corrosion pits, expanding the corroded area to about 10 mm in width. This width was significantly higher than that of the ZP-15% coating. This is also related to the corrosion-inhibiting mechanism of the dense phosphate formed by zinc phosphate.

Figure 10.13 shows the microscale images of the corrosion products at the scratched area of the two coatings. The morphologies of corrosion products on

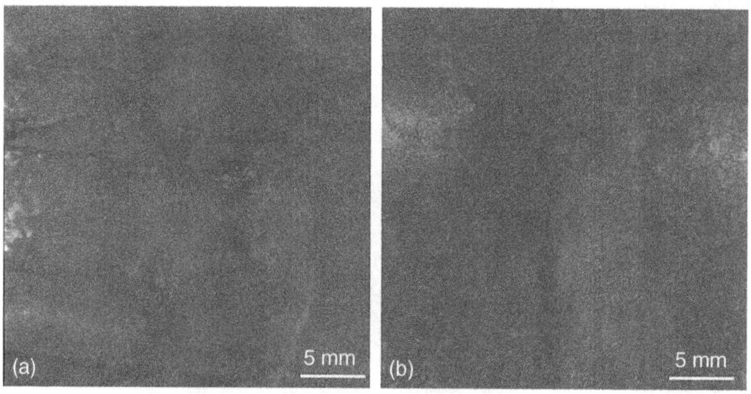

Figure 10.12 CLSM images of (a) ZP-0% coating and (b) ZP-15% coating after seven days of alternating wet–dry environment corrosion test.

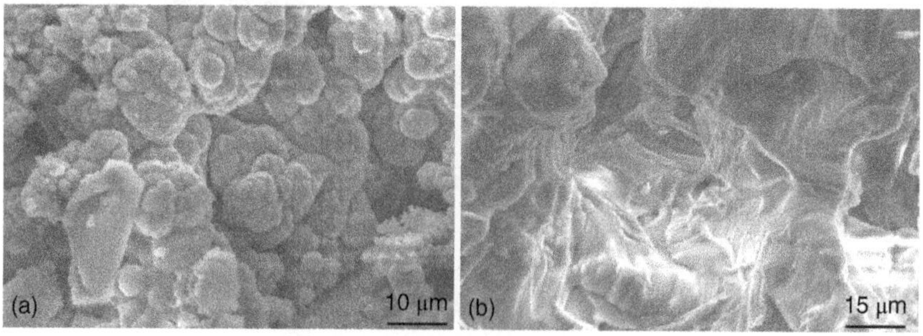

Figure 10.13 SEM images of (a) ZP-15% coating and (b) ZP-0% coating after seven days of alternating wet–dry environment corrosion test.

the two coatings were completely different. The corrosion products in the ZP-15% coating were in particle shape and were densely arranged, whereas those in the ZP-0% coating were more loosely packed and featured more holes in the microscale. The condition of the corrosion products will influence the corrosion process on the underneath metal surface. When the corrosion products were loose and easily detachable, they could not provide reliable protection to the underneath metal substrate, so their impact on corrosion was less significant. On the contrary, when the corrosion products can densely cover the surface of the sensor, they can strengthen the corrosion resistance between the galvanic couples, thereby reducing the corrosion current. The strong hygroscopic nature of the loose corrosion products can increase the wetting time on the substrate surface to some extent [66–68]. This, in turn, prolongs the corrosion effect during the drying period when oxygen is abundant, thereby increasing the total amount of corrosion products.

10.3.3.2 Electrochemical Analysis

The OCP values of the ZP-15% coating and ZP-0% coating are shown in Figure 10.14. The OCP values of the ZP-0% coating showed a clear decrease as the experiment progressed, dropping from approximately −0.42 to −0.58 V. The OCP values of the ZP-15% coating remained relatively stable in the alternating wet–dry environment, maintaining −0.4 V. This indicates that under alternating wet–dry conditions, the ZP-15% coating had superior corrosion resistance compared to the ZP-0% coating. Compared with the OCP values obtained at the room temperature immersion experiment, the OCP values of both coatings show a noticeable decline in the alternating wet–dry environment. It indicates that the alternating environment could provide a more severe corrosion environment that intensifies the failure process of coatings.

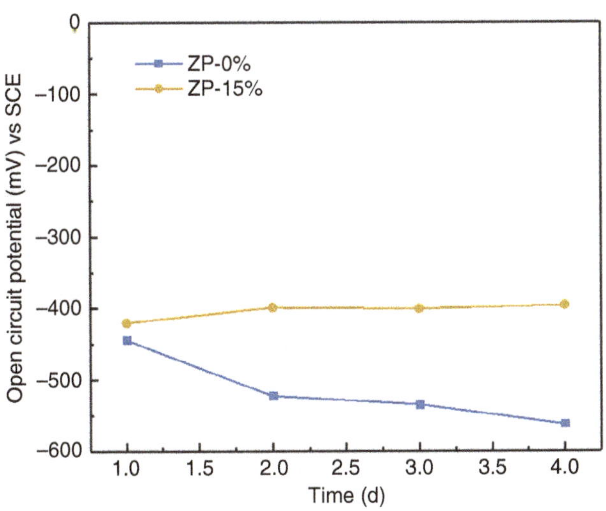

Figure 10.14 OCP values of ZP-0% coating and ZP-15% coating during alternating wet–dry corrosion conditions.

Figure 10.15 displays the EIS data for the ZP-0% and ZP-15% coatings after zero, four, six, and eight days of alternating wet–dry tests. The impedance arc radius of the ZP-0% coating gradually decreased during the alternating wet–dry corrosion test, especially on day 8, indicating a decline in the barrier performance of the coating. The impedance arc radius of the ZP-15% coating remained stable throughout the alternating wet–dry process. The $|Z|_{0.01\,\text{Hz}}$ value in the Bode plots is used to evaluate the corrosion resistance of the two coatings. In the Bode plot, the $|Z|_{0.01\,\text{Hz}}$ of the ZP-0% coating gradually decreased, indicating that the interfacial corrosion reactions of the ZP-0% coating were intense. In contrast, the $|Z|_{0.01\,\text{Hz}}$ of the ZP-15%

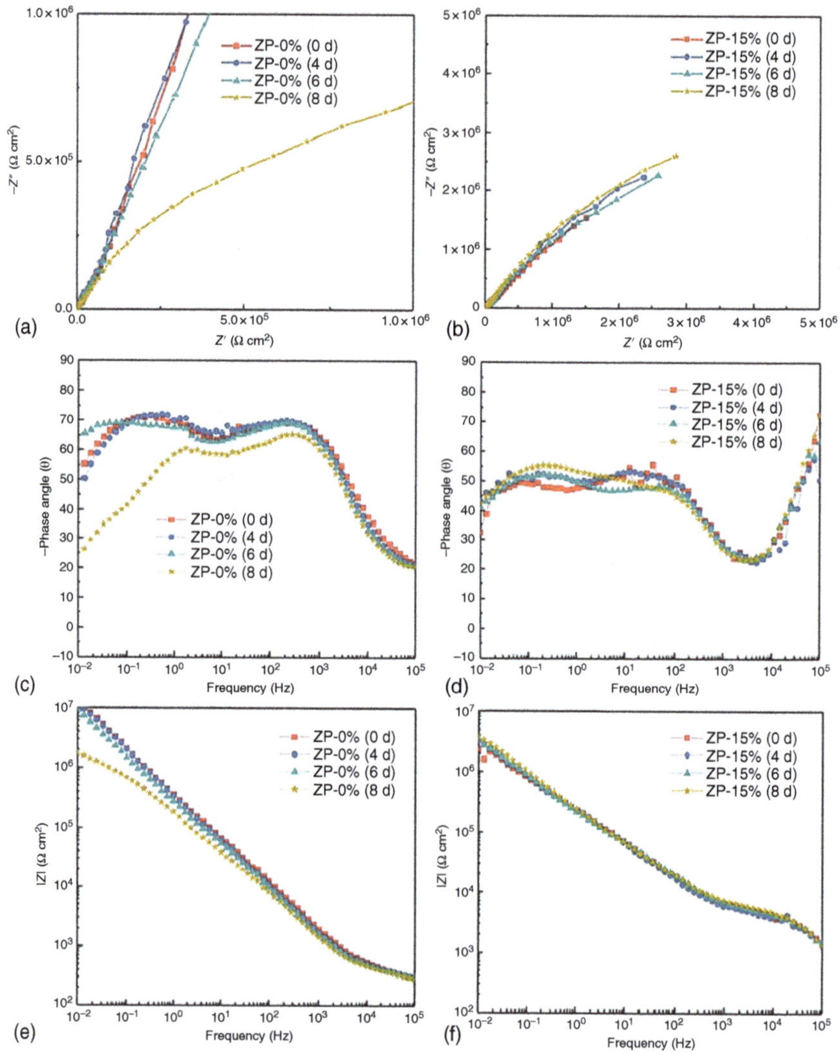

Figure 10.15 Nyquist diagram and Bode plots of (a,c,e) ZP-0% and (b,d,f) ZP-15% coatings after zero, four, six, and eight days of alternating wet–dry corrosion test.

coating with zinc phosphate addition maintains at $10^6\ \Omega\ cm^2$, which was ascribed to the corrosion-inhibiting effect of zinc phosphate that suppressed the charge transfer at the metal–coating interface. From the phase angle plots, the ZP-15% coating illustrated high phase angles at a high frequency. It indicates that the ZP-15% coating had better barrier properties. The results obtained from the alternating wet–dry corrosion test were consistent with the results in the room temperature immersion environment.

10.3.3.3 ACM Measurement

Based on the instantaneous current data from the ACM instrument, the variation curves of $I_{ACM\text{-}instant}$ values for both ZP-0% and ZP-15% coatings were obtained during alternating wet–dry corrosion tests. As shown in Figure 10.16, it is evident that the current of the ZP-0% coating presented an increasing trend as the alternating

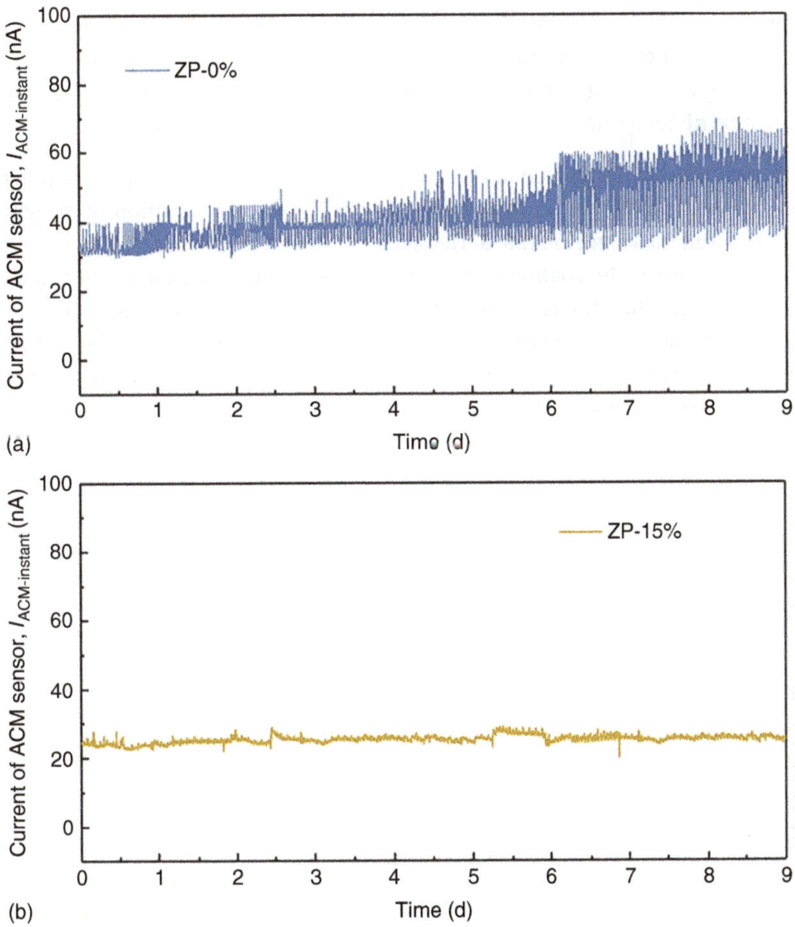

Figure 10.16 Output corrosion current values $I_{ACM\text{-}instant}$ of (a) ZP-0% and (b) ZP-15% coatings during alternating wet–dry corrosion test.

wet–dry cycle progressed, with a larger degree of variation compared to the immersion experiment. The $I_{ACM\text{-}instant}$ values were in the range from 30 to 70 nA, and fluctuated dramatically during the test, which was ascribed to the environmental change at dry period and immersing period. Overall, the increased $I_{ACM\text{-}instant}$ values indicated the intensified corrosion behavior on the ACM sensor covered with the blank epoxy coating. In contrast, the current of the ZP-15% coating remained relatively stable throughout the experiment without significant fluctuation. At the beginning of the test, the $I_{ACM\text{-}instant}$ value was approximately 25 nA, and fluctuated slightly during the dry period and immersing period. After nine days of testing, the $I_{ACM\text{-}instant}$ value was still as low as 25 nA.

In the alternating wet–dry experiments, the variation trends of current changes for ZP-0% and ZP-15% coatings were markedly different. The ZP-15% coating maintained a stable current due to the addition of zinc phosphate corrosion inhibitor, and the current fluctuation was less pronounced throughout the drying and immersing cycles. This is attributed to the action of zinc phosphate and the formation of corrosion products, which suppressed the substrate corrosion. Conversely, the current of ZP-0% coating showed a gradually increasing trend, starting from an initial current of 30 nA and reaching around 70 nA after nine days of the wet–dry cycles. This process can be divided into three specific stages:

(1) Corrosion product formation at the scratch location. As the coating transitions from the immersing period to the drying period, the concentration of oxygen increases and the corrosion rate at the scratch location intensifies. As the drying process continues, the coating loses water on the surface, and corrosion gradually decreases. When the re-submerging process begins, due to the reduction of oxygen concentration, the corrosion at the scratch area continues to slow down [69–71]. This is because the hygroscopic action of the corrosion products at the scratch location and their blocking effect increases the wetting degree of the

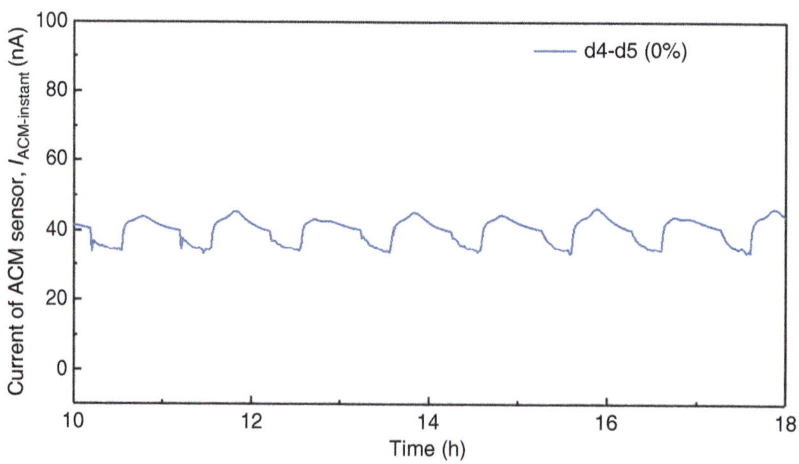

Figure 10.17 Output corrosion current values $I_{ACM\text{-}instant}$ of ZP-0% coating during alternating wet–dry corrosion test at day 4.

substrate surface to some extent. It temporarily blocks the solution from further penetrating into the coating. Simultaneously, the reduction in oxygen concentration causes a continued decrease in corrosion rate at the scratch, preventing a rapid increase in instantaneous current at the beginning of re-submergence. This process is repeated over and over again, causing the corrosion current to increase and decrease periodically. Figure 10.17 provides a detailed corrosion current change of ZP-0% coating during alternating wet–dry corrosion test at day 4 within eight hours.

(2) Gradual expansion of the corrosion area. With the continuation of the alternating wet–dry corrosion cycle, the oxygen concentration increases, and the liquid film becomes thinner due to evaporation in the drying period, whereas the corrosion current still rises rapidly, reaching a maximum before gradually declining as water evaporates. In subsequent submersion, the corrosion current slightly increases at the beginning, then declines, and gradually stabilizes throughout the process. This is due to the continuous penetration of moisture and chloride ions through the pores and defects in the coating, and the accumulation of chloride ions at the coating surface and crevices. More chloride ions enter the coating interior together with the immersing solution, reaching the interface between the coating and the substrate, causing a short-term increase in current at the beginning of the immersion. Figure 10.18 provides a detailed corrosion current change of ZP-0% coating during alternating wet–dry corrosion test at day 6 within eight hours.

(3) Loss of the protective function of coatings. As the alternating wet–dry cycle continues, the evaporation of water during the drying period carries out the soluble salts from the coating, further widening the pathways and allowing oxygen to be promptly transmitted to the substrate surface. Corrosion products then accumulate at the metal–coating interface. The hygroscopic nature of the

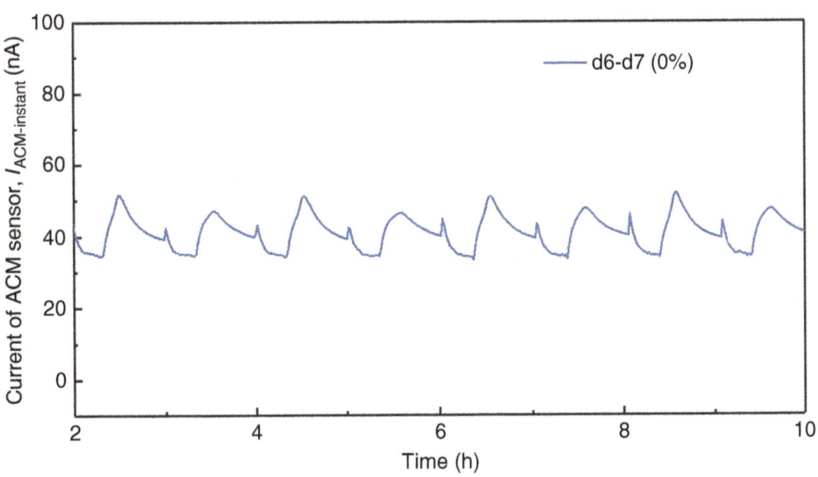

Figure 10.18 Output corrosion current values $I_{ACM\text{-}instant}$ of ZP-0% coating during alternating wet–dry corrosion test at day 6.

formed corrosion products and the osmotic pressure generated between the interface and corrosion products during the immersing period accelerate water penetration to the interface. At the same time, the different contraction and expansion properties of the coating and substrate create internal stress at the interface, generating bubbles and cracks in the coating due to accumulated corrosion products and increased internal stress. During this immersing time, as the electrolyte solution penetrates the layer, the corrosion current rapidly increases, far exceeding the values during the drying period. Subsequently, the decrease in oxygen concentration at the metal interface disturbs the oxygen diffusion process, inhibiting the cathodic corrosion reaction. Figure 10.19 provides a detailed corrosion current change of ZP-0% coating during alternating wet–dry corrosion test at day 8 within eight hours.

Figure 10.20 shows the corrosion charge curves of the ZP-0% and ZP-15% coatings. The Q values of the ZP-0% coating increased significantly, implying that the corrosion of the metal substrate had intensified and continuously occurred. In contrast, the corrosion charge curve of the ZP-15% coating was lower, consistent with its performance in the immersion state. The corrosion current rate for the ZP-0% coating was much higher than that of the ZP-15% coating, indicating that the corrosion-resistant property of the coating with corrosion inhibitors was promoted. After eight hours of alternating wet–dry corrosion test, the corrosion charge value of the ZP-15% coating was approximately 3.2×10^5 °C, which was higher than that obtained during the full immersion test (2.0×10^5 °C). It indicates that the alternating wet–dry corrosion environment can provide more severe corrosion conditions, offering a faster acceleration corrosion environment.

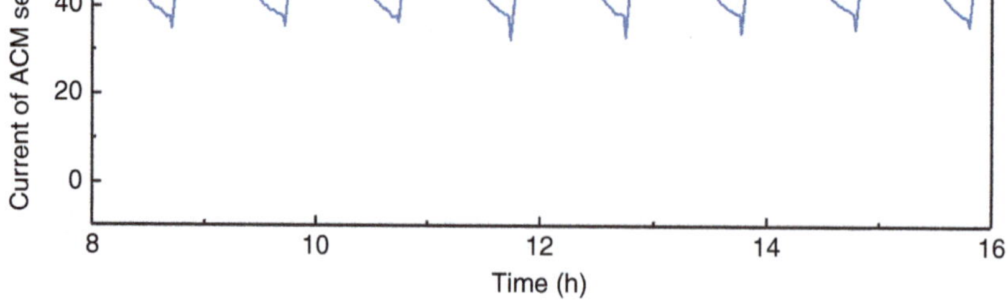

Figure 10.19 Output corrosion current values $I_{ACM\text{-}instant}$ of ZP-0% coatings during alternating wet–dry corrosion test at day 8.

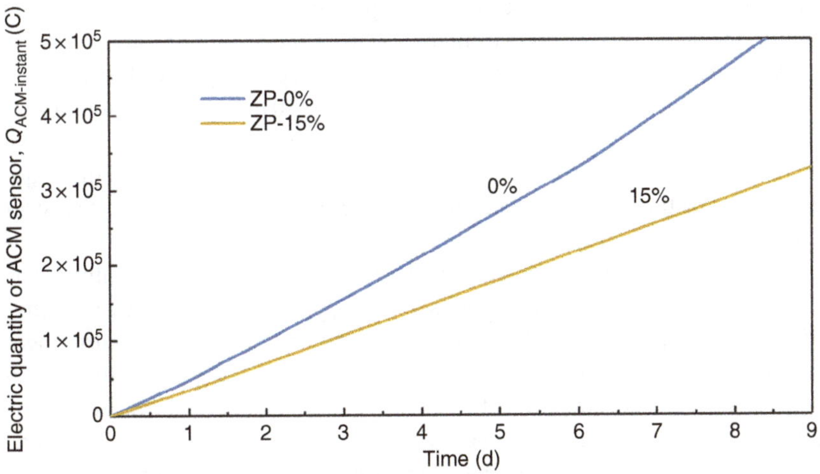

Figure 10.20 Output corrosion charge values $Q_{ACM\text{-}instant}$ of ZP-15% and ZP-0% coatings during alternating wet–dry environment corrosion test.

10.3.4 Self-healing Performance of Coatings in Outdoor Exposure Environment

This experiment explores the protective behavior of coatings in an outdoor atmospheric environment over a 30-day period, with the macroscopic surface corrosion morphology of the samples recorded using a digital camera and real-time monitoring throughout the experiment with an ACM instrument. The on-site images are shown in Figure 10.21.

10.3.4.1 ACM Measurement

The ACM instrument was used for continuous corrosion behavior observation at a Beijing experimental site. The ACM sensor and two parallel samples were installed at a certain distance, more than 1 m above the ground, tilted at 45 degrees southward.

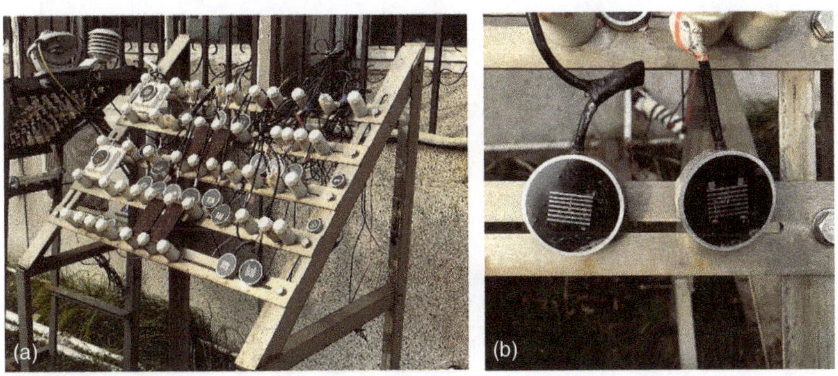

Figure 10.21 The on-site images of ACM sensors exposed to the outdoor environment: (a) placement mode of all ACM sensors, and (b) the enlarged picture of two ACM sensors.

The humidity and temperature sensors were placed beside the ACM sensor, and exposed to the same corrosive environment. During the testing process, the ACM sensor was used to collect the corrosion current in every one minute. The obtained corrosion currents were in the range from 0.1 nA to 50 mA, with a resolution of 0.1 nA. According to the methods specified by the atmospheric site, measurements of the atmospheric RH, temperature, and rainfall in Beijing were taken to get the major environmental factors affecting material atmospheric corrosion, as well as the ACM current and the corrosion charge Q value.

Figure 10.22 shows the corrosion current curves of the two coatings monitored by ACM during the test period. In most cases, the $I_{\text{ACM-instant}}$ values of the two coatings remained lower than 1 nA, which was ascribed to the dry atmospheric environment that was not able to form a continuous and stable thin liquid film on the sensor surface. However, the corrosion current increased sharply to hundreds of nA at some time points, followed by a notable decrease. This phenomenon was clearly observable on both the coatings. The dramatic increase of the corrosion currents did not happen every day but on some specific days. This phenomenon should be explained in combination with the variation of environmental factors.

Figure 10.23 lists the temperature, atmospheric RH, and the amount of rainfall during the outdoor exposure period. The RH value was in the range of 52–85%, and the average temperature was stabilized at 23–27 °C. On dates such as 29 August 2020 and 1 September 2020, the rainfall happened. By combining the meteorological data in Figure 10.23 and the corrosion current data in Figure 10.22, it can be easily concluded that the dramatic increase of the corrosion currents usually coincided with the onset of rainfall, maintaining higher current values during the rainy period. When the rain ended, the surface of the sensor probe went through a wet-to-dry process, where the electrolyte film gradually evaporated and was no longer in a continuous state. It led to a gradual cessation of the electrochemical reaction, thus the $I_{\text{ACM-instant}}$ values dropped. This indicates that the thickness of the electrolyte film on the sensor surface affected the surface corrosion reaction, as seen in the

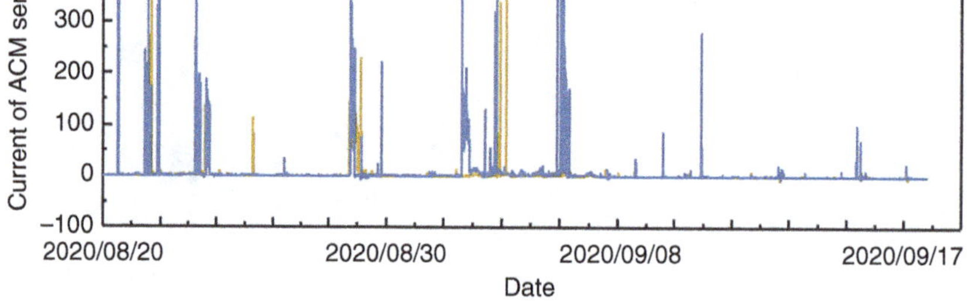

Figure 10.22 The corrosion current curves of the ZP-0% coating and ZP-15% coating monitored by ACM sensors in outdoor exposure conditions.

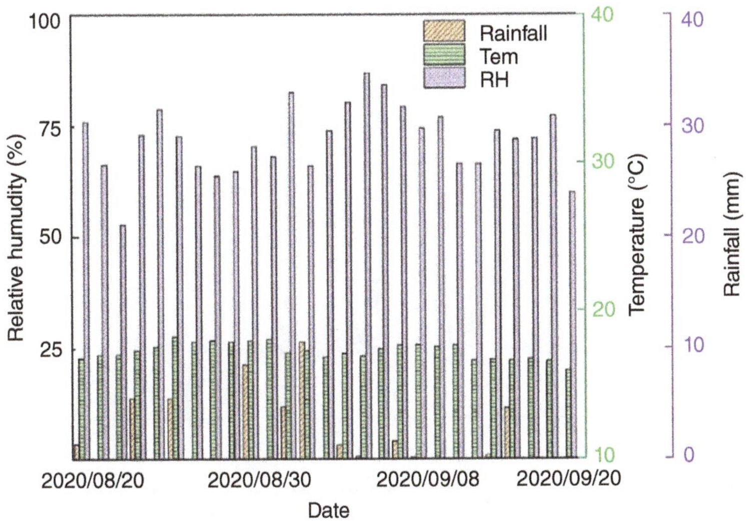

Figure 10.23 Temperature, humidity, and rainfall parameters under outdoor atmospheric exposure.

alternating wet–dry experiment where the probe impeded oxygen transmission due to the excessive thick liquid film during the immersion period, and the electrolyte reaction declined when the surface electrolyte film was too thin to be continuous during the drying phase. Therefore, in an outdoor atmospheric exposure environment, when high humidity and rainfall occur, a continuous electrolyte film forms on the sensor surface along with ample oxygen supply, resulting in higher current peaks during rainfall.

Figure 10.24 displays the integrated corrosion charge curves of the two coatings. The corrosion charge curve of the ZP-0% coating showed a stepwise increase, indicating that the corrosion of the coating was gradually accumulating and

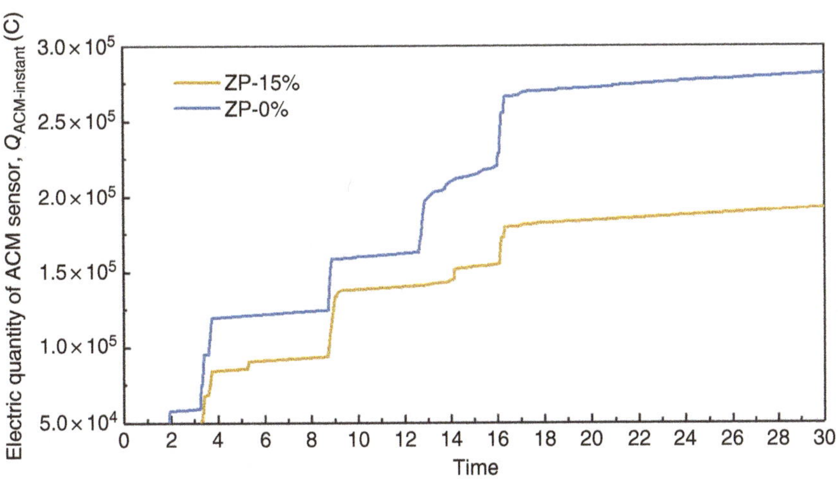

Figure 10.24 The corrosion charge curves of the ZP-0% coating and ZP-15% coating monitored by ACM sensors in outdoor exposure conditions.

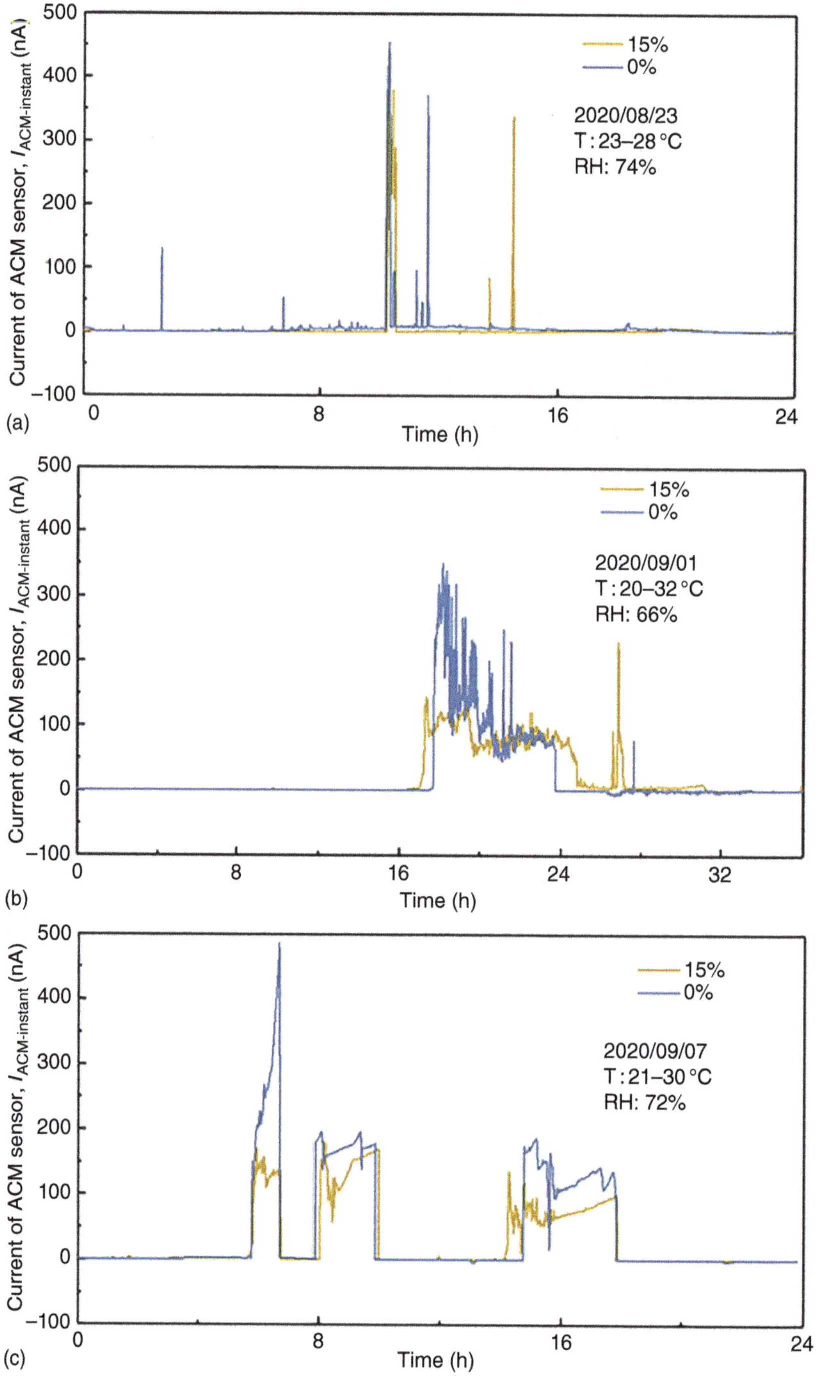

Figure 10.25 Corrosion current curves under an outdoor atmosphere in different time periods: (a) 23 August 2020, (b) 1 September 2020, and (c) 7 September 2020.

Table 10.2 Relationship between ACM corrosion charge (Q) and environmental factors (average atmospheric temperature X_1, average atmospheric RH X_2, rainwater pH X_3, and time X_4).

Coating	Expression	Correlation coefficient
ZP-0%	$Q = -121.8731 + 8.51183X_1 - 0.81399X_2 + 0.38683X_3 + 11.0381X_4$	0.97161
ZP-15%	$Q = -150.72559 + 8.14143X_1 - 0.31245X_2 + 0.50593X_3 + 7.49494X_4$	0.94848

continuously occurring. The sudden increase in the Q values happened on rainy days which dramatically accelerated the corrosion process. The corrosion charge curve of the ZP-15% coating showed a similar trend of change, but with a smaller cumulative value, indicating a suppressed corrosion behavior for the ZP-15% coating. This is consistent with results in room temperature immersion and alternating wet–dry environments, demonstrating superior corrosion resistance and self-healing performance of the coating with inhibitors.

Additionally, Figure 10.25 shows the corrosion current curves under an outdoor atmosphere in different time periods. During the rainfall time, both coatings formed a liquid film on the surface and underwent electrochemical corrosion, the corrosion currents were generated, and the currents fluctuated at different environmental conditions. The ZP-15% coating had lower current values during rainfall, while the ZP-0% coating exhibited a more intense increase in the corrosion current. This is consistent with the results obtained in indoor immersion tests and alternating wet–dry environments, where zinc phosphate acted as a corrosion inhibitor within the coating, demonstrating superior corrosion resistance.

By using the ACM corrosion charge value as the dependent variable and environmental factors as independent variables, a multiple linear regression statistical analysis was performed to obtain the mathematical relationship between the corrosion charge (Q) and environmental factors (average atmospheric temperature X_1, average atmospheric RH X_2, rainwater pH X_3, and time X_4). The established relationship between ACM corrosion charge and environmental factors is shown in Table 10.2, which shows high reliability indicated by the correlation coefficient and F-test.

10.4 Chapter Summary

In this experiment, zinc phosphate was used as a corrosion inhibitor to prepare ZP-15% coating, using the blank ZP-0% coating without corrosion inhibitor as the control sample. The corrosion protection performance of the two types of coatings was studied under room temperature immersion, alternating wet–dry, and outdoor atmospheric exposure conditions. Various techniques such as EIS, SEM, EDS, and CLSM were used to characterize the corrosion morphology and composition of the coatings. Additionally, a galvanic sensor probe made from Q235 steel and graphite was prepared, and the ACM technology was used to monitor the degradation behavior of coatings under various conditions, obtaining reliable and practical applications of ACM technology in coating property monitoring.

Using zinc phosphate as a corrosion-inhibiting additive, 15% mass concentration was selected as the optimal addition amount through pre-experiments. Bisphenol A diglycidyl ether and curing agent polyetheramine D-230 were used in a molar ratio of 2 : 1 to prepare the ZP-0% coating. The ZP-15% coating was similarly prepared with the addition of a 15% mass concentration of zinc phosphate. The two kinds of coatings were covered on the surface of ACM sensors.

The corrosion resistance of the coatings in a room temperature immersion environment was tested. In the immersion experiments, the ZP-15% coating maintained stable electrochemical impedance modulus and OCP, while the ZP-0% coating showed a decrease in impedance modulus and OCP as the experiment proceeded, indicating reduced corrosion resistance. Additionally, the ZP-0% coating showed more corrosion products in the damaged area, while the ZP-15% coating possessed fewer corrosion products with a layer of corrosion inhibitor containing a high content of P and Zn elements. The ACM monitoring current value curve and charge value curve were consistent with the electrochemical test technology and surface morphology results, indicating that ACM technology has high reliability in monitoring the corrosion resistance of coatings in a room temperature immersion environment.

The corrosion resistance of the coatings in an alternating wet–dry cycle environment was tested. The electrochemical impedance spectrum and OCP of the ZP-15% coating were similar to those in the room temperature immersion environment. The ZP-0% coating, however, underwent more severe corrosion reactions, with the accumulation of corrosion products on the metal surface, indicating that the alternating wet–dry environment intensified the corrosion of the coating. The ACM monitoring current value curve and charge value curve were consistent with the electrochemical test technology and surface morphology results, and the different failure stages of the coating were monitored and discussed. The difference between room temperature immersion and alternating wet–dry cycle corrosion was explained in detail, which indicates that the alternating dry period and immersing period could intensify the corrosion reaction. The ACM technology has high reliability in monitoring the corrosion resistance of coatings in an alternating wet–dry environment.

The corrosion resistance of the coatings in an outdoor atmospheric exposure experiment was tested. Real-time monitoring of the corrosion behavior of ZP-0% and ZP-15% coatings using the ACM instrument showed the variation of corrosion behavior of the two types of coatings under rainfall conditions and demonstrated the excellent corrosion resistance of the ZP-15% coating in an outdoor atmospheric exposure environment, consistent with the surface morphology results. Statistical analysis using mathematical methods established a reasonable mathematical relationship between the ACM corrosion charge and atmospheric environmental factors, indicating the good practicality of ACM technology in assessing coatings in outdoor atmospheric exposure environments. Combined with atmospheric environment parameters such as temperature and humidity, rainfall is an important environmental parameter affecting the barrier property of coatings in outdoor atmospheric exposure environments. This method offers a new direction for research on self-healing coatings and enriches the application of ACM in engineering environments. The further development of ACM technology, particularly in its application to coating systems, holds significant importance. It could potentially revolutionize the way we monitor, evaluate, and understand the performance of coatings in real-world environments, leading to better materials and protective strategies in industries reliant on metal structures and components.

Future ACM sensors may incorporate multiparametric sensing technology, enabling real-time monitoring of various environmental parameters related to corrosion. For instance, sensors could integrate chemical, humidity, temperature, and light sensors, providing an all-encompassing environmental monitoring solution. This would aid in more accurate forecasting and assessment of corrosion rates, as well as in analyzing how different environmental factors jointly influence corrosion processes. Nanosensors could offer higher surface area and more sensitive reaction planes, allowing ACM sensors to detect signs of corrosion at an earlier stage. Smart materials, such as shape memory alloys or pH-sensitive polymers, could adjust their physical and chemical properties in response to external environmental changes, thereby enhancing sensor responsiveness and selectivity. The incorporation of wireless communication technology will enable ACM sensors to easily integrate into the Internet of Things (IoT), facilitating wireless data transmission. This would greatly improve the efficiency and real-time nature of monitoring, especially in remote or hazardous environments. Combined with cloud computing platforms, big data analytics can be performed for real-time corrosion trend monitoring and early warning. In recent years, researchers have been developing ACM sensors capable of harvesting energy from the environment, such as converting vibrational, solar, or thermal energy into electrical energy. This self-powering technology can significantly reduce reliance on batteries or external power sources, lower maintenance costs, and increase the feasibility of sensor applications in remote areas. Using the ACM technique, numerous data can be collected during the service life. Therefore, machine learning and artificial intelligence algorithms can process and analyze the vast amounts of data collected from ACM sensors. These algorithms can identify patterns in corrosion development, predict future corrosion behaviors, and provide decision support for maintenance plans and anti-corrosion measures. Such technology can improve the accuracy of predictions and optimize resource allocation. For biomonitoring technologies, biosensors use microorganisms or enzymes to monitor specific corrosion mechanisms. For example, the metabolic activities of certain microbes may be closely related to corrosion processes; by monitoring these biological markers, highly sensitive bio-sensing systems for specific corrosion mechanisms can be developed.

To ensure interoperability among different devices and technologies, an international set of standards and certification processes might be established in the future. This would not only help guarantee the performance of new ACM sensors but also assist users in comparing the performance of various products and selecting the technology that best suits their application needs. From the sustainability and environmental impact, future ACM sensor design will pay greater attention to eco-friendliness and sustainability. Researchers may develop sensors using recyclable, biodegradable materials, or those that minimize environmental impact. Efforts will also be made to reduce energy consumption during sensor manufacturing and operation to lessen their environmental footprint. These directions will collectively drive the evolution of ACM sensors towards more intelligence, automation, and eco-consciousness to address the increasingly severe corrosion challenges on a global scale.

References

1 Sekine, I., Yuasa, M., Hirose, N. et al. (2002). Degradation evaluation of corrosion protective coatings by electrochemical, physicochemical and physical measurements. *Progress in Organic Coatings* 45: 1–13. https://doi.org/10.1016/S0300-9440(02)00113-3.

2 Permeh, S., Lau, K., and Duncan, M. (2020). Degradation of coatings for steel in environments susceptible to corrosion associated with fouling. *Structure and Infrastructure Engineering* 16: 1186–1200. https://doi.org/10.1080/15732479.2019.1694543.

3 Cai, G., Wang, H., Jiang, D. et al. (2018). Degradation of fluorinated polyurethane coating under UVA and salt spray. Part I: corrosion resistance and morphology. *Progress in Organic Coatings* 123: 337–349. https://doi.org/10.1016/j.porgcoat.2018.07.025.

4 Zulkifli, F., Ali, N.A., Yusof, M.S.M. et al. (2017). Henna leaves extract as a corrosion inhibitor in acrylic resin coating. *Progress in Organic Coatings* 105: 310–319. https://doi.org/10.1016/j.porgcoat.2017.01.017.

5 Li, C., Guo, X., and Frankel, G.S. (2021). Corrosion inhibition of AA2024-T3 by a coating containing dual-pH sensitive, corrosion inhibitor loaded microspheres. *Corrosion Science* 192: 109835. https://doi.org/10.1016/j.corsci.2021.109835.

6 Visser, P., Terryn, H., and Mol, J.M.C. (2018). On the importance of irreversibility of corrosion inhibitors for active coating protection of AA2024-T3. *Corrosion Science* 140: 272–285. https://doi.org/10.1016/j.corsci.2018.05.037.

7 Khramov, A.N., Voevodin, N.N., Balbyshev, V.N. et al. (2005). Sol–gel-derived corrosion-protective coatings with controllable release of incorporated organic corrosion inhibitors. *Thin Solid Films* 483: 191–196. https://doi.org/10.1016/j.tsf.2004.12.021.

8 Liu, X., Xiong, J., Lv, Y. et al. (2009). Study on corrosion electrochemical behavior of several different coating systems by EIS. *Progress in Organic Coatings* 64: 497–503. https://doi.org/10.1016/j.porgcoat.2008.08.012.

9 Dong, C.F., Fu, A.Q., Li, X.G. et al. (2008). Localized EIS characterization of corrosion of steel at coating defect under cathodic protection. *Electrochimica Acta* 54: 628–633. https://doi.org/10.1016/j.electacta.2008.07.016.

10 Bakhshandeh, E., Jannesari, A., Ranjbar, Z. et al. (2014). Anti-corrosion hybrid coatings based on epoxy–silica nano-composites: toward relationship between the

morphology and EIS data. *Progress in Organic Coatings* 77: 1169–1183. https://doi.org/10.1016/j.porgcoat.2014.04.005.

11 Amirudin, A. and Thieny, D. (1995). Application of electrochemical impedance spectroscopy to study the degradation of polymer-coated metals. *Progress in Organic Coatings* 26: 1–28. https://doi.org/10.1016/0300-9440(95)00581-1.

12 Ranjbar, Z., Moradian, S., and Attar, M.R.M.Z. (2004). EIS investigation of cataphoretically electrodeposited epoxy coatings having different EEWs. *Progress in Organic Coatings* 51: 87–90. https://doi.org/10.1016/j.porgcoat.2004.02.004.

13 Miszczyk, A. and Darowicki, K. (2001). Accelerated ageing of organic coating systems by thermal treatment. *Corrosion Science* 43: 1337–1343. https://doi.org/10.1016/S0010-938X(00)00156-6.

14 Mizuno, D., Suzuki, S., Fujita, S. et al. (2014). Corrosion monitoring and materials selection for automotive environments by using atmospheric corrosion monitor (ACM) sensor. *Corrosion Science* 83: 217–225. https://doi.org/10.1016/j.corsci.2014.02.020.

15 Pei, Z., Xiao, K., Chen, L. et al. (2022). Investigation of corrosion behaviors on an Fe/Cu-type ACM sensor under various environments. *Metals* 10 (7): 905. https://doi.org/10.3390/met10070905.

16 Li, Z., Fu, D., Li, Y. et al. (2019). Application of an electrical resistance sensor-based automated corrosion monitor in the study of atmospheric corrosion. *Materials* 12 (7): 1065. https://doi.org/10.3390/ma12071065.

17 Ahn, J.-H., Jeong, Y.-S., Kim, I.-T. et al. (2019). A method for estimating time-dependent corrosion depth of carbon and weathering steel using an atmospheric corrosion monitor sensor. *Sensors* 19 (6): 1416. https://doi.org/10.3390/s19061416.

18 Huang, Y., Wang, P., Tan, W. et al. (2022). Photothermal and pH dual-responsive self-healing coating for smart corrosion protection. *Journal of Materials Science and Technology* 107: 34–42. https://doi.org/10.1016/j.jmst.2021.08.044.

19 Ji, X., Wang, W., Zhao, X. et al. (2022). Poly(dimethyl siloxane) anti-corrosion coating with wide pH-responsive and self-healing performance based on core–shell nanofiber containers. *Journal of Materials Science and Technology* 101: 128–145. https://doi.org/10.1016/j.jmst.2021.06.014.

20 Yan, D., Wang, Y., Liu, J. et al. (2020). Self-healing system adapted to different pH environments for active corrosion protection of magnesium alloy. *Journal of Alloys and Compounds* 824: 153918. https://doi.org/10.1016/j.jallcom.2020.153918.

21 Wang, Y., Huang, Z., Yan, Q. et al. (2016). Corrosion behaviors and effects of corrosion products of plasma electrolytic oxidation coated AZ31 magnesium alloy under the salt spray corrosion test. *Applied Surface Science* 378: 435–442. https://doi.org/10.1016/j.apsusc.2016.04.011.

22 Zhao, M.-C., Schmutz, P., Brunner, S. et al. (2009). An exploratory study of the corrosion of Mg alloys during interrupted salt spray testing. *Corrosion Science* 51: 1277–1292. https://doi.org/10.1016/j.corsci.2009.03.014.

23 Huang, M. and Yang, J. (2014). Salt spray and EIS studies on HDI microcapsule-based self-healing anticorrosive coatings. *Progress in Organic Coatings* 77: 168–175. https://doi.org/10.1016/j.porgcoat.2013.09.002.

24 Ng, W.F., Wong, M.H., and Cheng, F.T. (2010). Stearic acid coating on magnesium for enhancing corrosion resistance in Hanks' solution. *Surface and Coatings Technology* 204: 1823–1830. https://doi.org/10.1016/j.surfcoat.2009.11.024.

25 Zhang, D., Qian, H., Wang, L. et al. (2016). Comparison of barrier properties for a superhydrophobic epoxy coating under different simulated corrosion environments. *Corrosion Science* 103: 230–241. https://doi.org/10.1016/j.corsci.2015.11.023.

26 Wang, X., Guo, L., Liu, X. et al. (2024). Enhanced corrosion resistance and cell activity of magnesium alloy by DCPD/$MgHPO_4 \cdot 3H_2O$ coating via one-step chemical conversion. *Surface and Coatings Technology* 476: 130228. https://doi.org/10.1016/j.surfcoat.2023.130228.

27 Wu, J., Cui, J.P., Zheng, Q.J. et al. (2019). Insight into the corrosion evolution of Fe-based amorphous coatings under wet-dry cyclic conditions. *Electrochimica Acta* 319: 966–980. https://doi.org/10.1016/j.electacta.2019.07.058.

28 Park, J.H., Lee, G.D., Ooshige, H. et al. (2003). Monitoring of water uptake in organic coatings under cyclic wet–dry condition. *Corrosion Science* 45: 1881–1894. https://doi.org/10.1016/S0010-938X(03)00024-6.

29 Yadav, A.P., Nishikata, A., and Tsuru, T. (2004). Electrochemical impedance study on galvanized steel corrosion under cyclic wet–dry conditions–influence of time of wetness. *Corrosion Science* 46: 169–181. https://doi.org/10.1016/S0010-938X(03)00130-6.

30 Zhang, W., Chen, X., Yin, P. et al. (2012). EIS study on the deterioration process of organic coatings under immersion and different cyclic wet-dry ratios. *Applied Mechanics and Materials* 161: 58–66. https://doi.org/:10.4028/www.scientific.net/AMM.161.58.

31 Li, Y., Wang, J., Zhang, W. et al. (2010). Comparative studies on the deterioration process of organic coatings under immersed and cyclic wet-dry conditions by EIS. *Journal of Electrochemistry* 16 (4): 393–400. https://doi.org/:10.61558/2993-074X.2059.

32 Zhao, X., Li, W., Hou, B. et al. (2011). Establishment and analysis of equivalent circuit models of electrochemical impedance spectroscopy of organic coating/carbon steel electrode under dry-wet cyclic condition. *Materials Protection* 44: 3. https://doi.org/10.16577/j.cnki.42-1215/tb.2011.03.019.

33 Wang, H.-H. and Du, M. (2017). Corrosion behavior of a low-carbon steel in simulated marine splash zone. *Acta Metallurgica Sinica- English Letters* 30: 585–593. https://doi.org/:10.1007/s40195-017-0535-1.

34 Yu, J., Wang, H., Yu, Y. et al. (2018). Corrosion behavior of X65 pipeline steel: comparison of wet–dry cycle and full immersion. *Corrosion Science* 133: 276–287. https://doi.org/10.1016/j.corsci.2018.01.007.

35 Zhang, T., Zhang, T., He, Y. et al. (2021). Long-term atmospheric aging and corrosion of epoxy primer-coated aluminum alloy in coastal environments. *Coatings* 11 (2): 237. https://doi.org/10.3390/coatings11020237.

36 Vera, R., Cruz, E., Bagnara, M. et al. (2018). Evaluation of anticorrosive coatings on carbon steel in marine environments: accelerated corrosion test and field

exposure. *International Journal of Electrochemical Science* 13: 898–914. https://doi.org/10.20964/2018.01.66.

37 Schmidt, D.P., Shaw, B.A., Sikora, E. et al. (2006). Corrosion protection assessment of sacrificial coating systems as a function of exposure time in a marine environment. *Progress in Organic Coatings* 57: 352–364. https://doi.org/10.1016/j.porgcoat.2006.09.021.

38 Yong, X., Ji, H., Chen, Z., and Ruan, X. (2021). Comparison between accelerated tests and natural exposures for organic coating protective systems based on EIS parameters. *Journal of the Electrochemical Society* 168: 121506.

39 Mabbutt, S., Mills, D.J., and Woodcock, C.P. (2007). Developments of the electrochemical noise method (ENM) for more practical assessment of anti-corrosion coatings. *Progress in Organic Coatings* 59: 192–196. https://doi.org/10.1016/j.porgcoat.2006.09.017.

40 Lv, J., Yue, Q.-X., Ding, R. et al. (2021). The application of electrochemical noise for the study of metal corrosion and organic anticorrosion coatings: a review. *ChemElectroChem* 8: 337–351. https://doi.org/10.1002/celc.202001342.

41 Nazarov, A., Le Bozec, N., and Thierry, D. (2018). Assessment of steel corrosion and deadhesion of epoxy barrier paint by scanning Kelvin probe. *Progress in Organic Coatings* 114: 123–134. https://doi.org/10.1016/j.porgcoat.2017.09.016.

42 Rahimi, E., Rafsanjani-Abbasi, A., Kiani-Rashid, A. et al. (2018). Morphology modification of electrodeposited superhydrophobic nickel coating for enhanced corrosion performance studied by AFM, SEM-EDS and electrochemical measurements. *Colloids and Surfaces A: Physicochemical and Engineering Aspects* 547: 81–94. https://doi.org/10.1016/j.colsurfa.2018.03.045.

43 Gowri, S., Umasankareswari, T., Joseph Rathish, R. et al. (2023). Chapter 8). Atomic force microscopy technique for corrosion measurement. In: *Electrochemical and Analytical Techniques for Sustainable Corrosion Monitoring* (ed. J. Aslam, C. Verma, and C. Mustansar Hussain), 121–140. Elsevier https://doi.org/10.1016/B978-0-443-15783-7.00001-3.

44 Greunz, T., Steinberger, R., Strauss, B. et al. (2018). Reduction of hexavalent chromium embedded in organic insulation and corrosion inhibition layers during X-ray photoelectron spectroscopy (XPS) measurements. *Corrosion Science* 143: 39–45. https://doi.org/10.1016/j.corsci.2018.08.031.

45 Guo, Y., Ali, R., Zhang, X. et al. (2020). Raman and XPS depth profiling technique to investigate the corrosion behavior of FeSiAl alloy in salt spray environment. *Journal of Alloys and Compounds* 834: https://doi.org/10.1016/j.jallcom.2020.155075.

46 Sanchez-Amaya, J.M., Blanco, G., Garcia-Garcia, F.J. et al. (2012). XPS and AES analyses of cerium conversion coatings generated on AA5083 by thermal activation. *Surface & Coatings Technology* 213: 105–116. https://doi.org/10.1016/j.surfcoat.2012.10.027.

47 Wang, J., Ma, L., Huang, Y. et al. (2022). Photothermally activated self-healing protective coating based on the "close and seal" dual-action mechanisms. *Composites Part B: Engineering* 231: https://doi.org/10.1016/j.compositesb.2021.109574.

48 Selvam, M., Saminathan, K., Siva, P. et al. (2016). Corrosion behavior of Mg/graphene composite in aqueous electrolyte. *Materials Chemistry and Physics* 172: 129–136. https://doi.org/10.1016/j.matchemphys.2016.01.051.

49 Popova, K. and Prosek, T. (2022). Corrosion monitoring in atmospheric conditions: a review. *Metals* 12: https://doi.org/10.3390/met12020171.

50 Gonzalez, J., Otero, E., Cabanas, C. et al. (1984). Electrochemical sensors for atmospheric corrosion rates: a new design. *British Corrosion Journal* 19 (2): 89–94. https://doi.org/10.1179/000705984798273399.

51 Cao, X., Xiao, Y., Deng, H. et al. (2009). Evaluation of atmospheric corrosivity by ACM technique. *Materials Science Forum* 610–613: 3–8. https://doi.org/10.4028/www.scientific.net/MSF.610-613.3.

52 Cao, X., Xiao, Y., and Zhang, S. (2008). Atmospheric corrosion monitoring technology was used to study the indoor accelerated corrosion behavior of A3 steel. *Materials Protection* 64–66. 68. https://doi.org/10.16577/j.cnki.42-1215/tb.2008.04.018.

53 Kainuma, S., Sugitani, K., Itoh, Y. et al. (2009). Evaluation method for time-dependent corrosion behavior of carbon steel plate using atmospheric corrosion monitoring sensor. *Key Engineering Materials* 417–418: 417–420. https://doi.org/10.4028/www.scientific.net/KEM.417-418.417.

54 Bastos, A.C., Ferreira, M.G.S., and Simoes, A.M. (2005). Comparative electrochemical studies of zinc chromate and zinc phosphate as corrosion inhibitors for zinc. *Progress in Organic Coatings* 52: 339–350. https://doi.org/10.1016/j.porgcoat.2004.09.009.

55 Shao, Y., Jia, C., Meng, G. et al. (2009). The role of a zinc phosphate pigment in the corrosion of scratched epoxy-coated steel. *Corrosion Science* 51: 371–379. https://doi.org/10.1016/j.corsci.2008.11.015.

56 Ma, L., Wang, J., Zhang, D. et al. (2021). Dual-action self-healing protective coatings with photothermal responsive corrosion inhibitor nanocontainers. *Chemical Engineering Journal* 404: https://doi.org/10.1016/j.cej.2020.127118.

57 Yabuki, A., Kanagaki, M., Nishikawa, C. et al. (2021). Effective release of corrosion inhibitor by cellulose nanofibers and zeolite particles in self-healing coatings for corrosion protection. *Progress in Organic Coatings* 154: https://doi.org/10.1016/j.porgcoat.2021.106194.

58 Xiang, T., Zheng, S., Zhang, M. et al. (2018). Bioinspired slippery zinc phosphate coating for sustainable corrosion protection. *ACS Sustainable Chemistry & Engineering* 6: 10960–10968. https://doi.org/10.1021/acssuschemeng.8b02345.

59 Tian, Y., Huang, H., Wang, H. et al. (2020). Accelerated formation of zinc phosphate coatings with enhanced corrosion resistance on carbon steel by introducing α-zirconium phosphate. *Journal of Alloys and Compounds* 831: 154906. https://doi.org/10.1016/j.jallcom.2020.154906.

60 Wan, H., Song, D., Li, X. et al. (2017). Effect of zinc phosphate on the corrosion behavior of waterborne acrylic coating/metal interface. *Materials* 10: https://doi.org/10.3390/ma10060654.

61 Kim, H., Yarin, A.L., and Lee, M.W. (2020). Self-healing corrosion protection film for marine environment. *Composites Part B: Engineering* 182: https://doi.org/10.1016/j.compositesb.2019.107598.

62 Li, J., Shi, H., Liu, F. et al. (2021). Self-healing epoxy coating based on tung oil-containing microcapsules for corrosion protection. *Progress in Organic Coatings* 156: https://doi.org/10.1016/j.porgcoat.2021.106236.

63 Pei, Z., Zhang, D., Zhi, Y. et al. (2020). Towards understanding and prediction of atmospheric corrosion of an Fe/Cu corrosion sensor via machine learning. *Corrosion Science* 170: https://doi.org/10.1016/j.corsci.2020.108697.

64 Pongsaksawad, W., Palsson, N.S., Khamsuk, P. et al. (2020). Atmospheric corrosion monitoring sensor in corrosion rate prediction of carbon and weathering steels in Thailand. *Materials Transactions* 61: 2348–2356. https://doi.org/10.2320/matertrans.MT-M2020230.

65 To, D., Shinohara, T., and Umezawa, O. (2016). Experimental investigation on the corrosivity of atmosphere through the atmospheric corrosion monitoring (ACM) sensors. *ECS Transactions* 1–10. https://doi.org/10.1149/07529.0001ecst.

66 Wang, P., Zhao, J., Ma, L. et al. (2021). Effect of grain ultra-refinement on microstructure, tensile property, and corrosion behavior of low alloy steel. *Materials Characterization* 179: https://doi.org/10.1016/j.matchar.2021.111385.

67 Wang, P., Cheng, X., Zhang, Z. et al. (2023). Roles of grain refinement in the rust formation and corrosion resistance of weathering steels. *Corrosion Science* 224: https://doi.org/10.1016/j.corsci.2023.111561.

68 Wang, Z., Zhou, Z., Xu, W. et al. (2020). Study on inner corrosion behavior of high strength product oil pipelines. *Engineering Failure Analysis* 115: https://doi.org/10.1016/j.engfailanal.2020.104659.

69 Meng, P.R., Chen, Y., and Liu, Z.L. (2020). Influence of the dissolved oxygen concentration in artificial seawater on the corrosion behavior of Q235 mild steel. *International Journal of Electrochemical Science* 15: 4454–4469. https://doi.org/10.20964/2020.05.45.

70 Zhong, X., Lu, W., Yang, H. et al. (2019). Oxygen corrosion of N80 steel under laboratory conditions simulating high pressure air injection: analysis of corrosion products. *Journal of Petroleum Science and Engineering* 172: 162–170. https://doi.org/10.1016/j.petrol.2018.09.058.

71 Eyu, G.D., Will, G., Dekkers, W. et al. (2016). Effect of dissolved oxygen and immersion time on the corrosion behaviour of mild steel in bicarbonate/chloride solution. *Materials* 9: https://doi.org/10.3390/ma9090748.

11

Future Development Direction of Smart Anticorrosive Coatings

11.1 Introduction

The complex and changeable ecological environment has made creatures in nature subject to harsh selection. After a prolonged era of the survival of the fittest, a wide variety of organisms has evolved their own optimal structures and functions, so that they are perfectly adapted to their surroundings. By studying and imitating the structure and characteristics of organisms, the preparation of biomimetic intelligent materials with integrated structure and function has become one of the frontier hotspots. Biological organisms possess unique morphological structures and strong adaptive capabilities to their environment. Upon external damage, they can autonomously regulate and repair themselves. Inspired by this self-healing ability of biological organisms, the development of intelligent biomimetic materials integrating sensing and actuation capabilities with excellent environmental adaptability has become a popular research direction. Guided by bionics principles, concepts such as self-healing coatings, self-reporting coatings, and superhydrophobic coatings have emerged as intelligent anticorrosion coatings and have garnered widespread attention. The following discussion primarily focuses on the future development and application directions of these intelligent coatings.

11.2 Smart Self-healing Coatings

In recent years, there has been considerable literature reporting on smart self-healing coatings. These smart anticorrosion coatings meet the requirements of reliability and longevity for corrosion protection applications. However, they are still in the stages of theoretical research and preliminary development, with limited industrial applications. Further in-depth research is necessary. First, enhancing the intelligence level of coatings to sensitively detect both external and internal environmental stimuli, responding more swiftly to the damages and environmental changes, will be a key research focus. This improvement aims to maximize the healing efficiency and durability of the coatings. Second, another frontier in the development of self-healing coatings involves integrating different repair mechanisms within the same system. By orchestrating the orderly expression and efficient

Smart Protective Coatings for Corrosion Control, First Edition. Lingwei Ma and Dawei Zhang.

collaboration of various mechanisms, coatings with multiple environmental response capabilities can be developed to meet diverse application needs in varying environments. Third, employing advanced micro and nanoscience technologies to fabricate novel micro- and nanocontainers, and using advanced microscopic characterization methods to understand the release and migration behavior of corrosion inhibitors or healing agents loaded in these containers at the molecular or atomic level represents an important future research direction. In addition, accelerating the development of new methods for preparing intrinsic self-healing coatings and achieving breakthroughs in creating polymer resin materials with both excellent mechanical and self-healing properties is crucial. Fourth, there is a need to simplify the coating preparation process, reduce costs, and select appropriate self-healing structures and agents suited to real-world applications. This will better facilitate their practical use in production and promote the engineering application of self-healing coatings. Finally, to adapt intelligent anticorrosion coatings to harsh environments, more systematic exploration is needed into coatings that combine other functions such as self-cleaning, antibacterial, and self-reporting capabilities.

11.2.1 Application of Self-healing Coatings in Marine Corrosion

The 21st century is the century of the ocean. With the depletion of land resources, various countries and organizations in the world have gradually shifted the focus of economic development from land to the sea. The development of marine resources has become the focus of the current era, and the development of the marine economy has been unprecedentedly strong. Metal materials are the main structural materials for constructing offshore engineering and equipment, such as offshore drilling platforms and ships, and play an important role in supporting offshore operations. However, when metal materials are in service in the marine environment, they will be affected by marine environmental factors, such as humid atmosphere, seawater immersion, tidal erosion, and biological fouling. These marine environmental factors will accelerate the corrosion and destruction of metal materials, resulting in the shortening of the service life or even failure of marine engineering equipment, greatly increasing the cost of equipment maintenance and resulting in the waste of human and material resources. Serious consequences may also cause casualties. In order to slow down the corrosion effect of the marine environment on metal materials, extend the service life of marine engineering and equipment, and avoid safety accidents in marine operations, effective measures must be taken to protect metal materials against corrosion. At present, the coating of organic resin coating on the surface of metal materials is still one of the most effective technical means to prevent metal surface corrosion and improve the service life of metal materials. Neutral sea water contains more dissolved oxygen and various electrolytes, mainly sodium chloride, which makes the marine environment easily cause the corrosion of metal materials. In addition, the metal will also be subjected to ultraviolet aging, biological fouling, quicksand erosion, and other damage, which together constitute a complex marine corrosion environment.

In the application of ship structure anticorrosion technology, spraying anticorrosion coatings on ship hull surfaces is the most effective method used at present. The anticorrosion principle of ship coatings mainly relies on the coating to form a thin film on the metal surface that can be isolated from the external environment, and isolates the substrate surface from the surrounding corrosive media to achieve the effect of anticorrosion. The anticorrosion coating has good isolation and water resistance and can adapt to complex and harsh marine environments. However, due to the influence of environmental factors and the nature of the coating itself, there will be some tiny cracks on the coating surface during service, and these microcracks will gradually expand, resulting in the decline of coating adhesion and anticorrosion ability. The traditional method of repairing anticorrosion coating mainly relies on manual removal of the damaged part of the coating and re-coating; the repair process is complicated, high cost, and cannot achieve the optimal anticorrosion effects. Some coatings contain chemicals that are harmful to the human body and will affect its health. The latest research shows that the global sales share of self-healing coatings is increasing year by year, and the proportion is also increasing year by year. Moreover, due to the harsh service environment of the ship, the performance of the anticorrosion coating is of very high standard. Therefore, the self-healing coating applied in the field of ship anticorrosion will become a research hotspot and has great economic value and practical significance.

Research on self-healing anticorrosion coatings for metals has been ongoing for many years, and while significant progress has been made in the self-healing effectiveness, there are still fundamental issues to be addressed in the development of self-healing coatings for ships. First, the application process of these coatings should be simple, cost effective, and reliable. Second, the operational and application temperatures should range between −30 and 40 °C. In line with green and sustainable development goals, the raw materials used should be as natural as possible. Due to these environmental considerations, current research on self-healing materials for ships predominantly utilizes natural resins modified to serve as eco-friendly coatings.

The harsh, corrosive marine environment poses new requirements for self-healing anticorrosion coatings. (i) Low-temperature self-healing coatings: Current self-healing coatings typically cannot be repaired at room temperature and usually require heating. Self-healing coatings used in ship corrosion prevention generally need a temperature range of 80–160 °C to initiate self-healing. If this temperature is not reached, the coatings cannot sense damage and perform self-healing, failing to achieve the desired anticorrosive effect. Therefore, developing self-healing coatings that operate effectively at −5–40 °C and can sensitively detect damage is a future trend in ship anticorrosion coatings. (ii) Natural material-based self-healing coatings. With increasing environmental awareness, the use of natural materials is expanding to reduce environmental pollution. With their abundance, natural materials, low cost, and easy availability, are becoming ideal industrial raw materials. As significant progress is made in self-healing anticorrosion coatings, it becomes possible to use excellent natural structures to prepare these coatings. (iii) Environmentally friendly self-healing coatings: In response to the green, energy-saving, and eco-friendly development trends in manufacturing, ship coating enterprises should intensify their efforts to develop environmentally friendly self-healing coatings, focusing on core technologies in this field that meet green environmental standards. Future development directions for environmentally friendly self-healing coatings include developing light-induced hydrophobic self-healing coatings to minimize environmental pollution; reducing the use and release of toxic substances in ship coatings; developing self-polishing, antifouling, and anticorrosion coatings that reduce ship drag and fuel consumption; and focusing on maintenance-free, quick healing coating materials. (iv) Marine environmental suitability. Self-healing anticorrosion coatings must be suitable for marine environments. For example, in embedded self-healing coatings, the healing agents released from micro/nanocontainers must be capable of self-healing the barrier or electrochemical properties of the coating in seawater. The marine environment is complex, and a single stimulus-response and healing mechanism may not be suitable for such intricate corrosion environments. Therefore, complex marine environments require self-healing anticorrosion coatings with multiple stimulus responses and repair mechanisms to maintain long-term protective functions under various environmental changes. (v) Harnessing marine environmental energy. Develop materials that can convert energy sources in the marine environment, such as wind, tides, friction, and thermal energy, into the energy needed for the self-healing process.

As research in this field deepens, the limitations of self-healing coatings will gradually be overcome, increasing the variety of self-healing coatings applicable in ship corrosion prevention. In the context of the industry-wide response to green chemistry and low carbon emissions, self-healing coatings will evolve toward low-temperature, natural material-based, environmentally friendly, and super-hydrophobic directions.

11.2.2 Application of Self-healing Coatings in Nuclear Corrosion

The seawater systems in nuclear power plants, providing cooling sources to both primary and secondary circuits, are critical for the safe and stable operation of nuclear power stations. In China, the main anticorrosion measures for equipment and pipelines in seawater systems, such as the circulating water system, circulating water treatment system, key freshwater system, and auxiliary cooling water system of nuclear power plants, involve coating protection. For some equipment and pipelines, like those in the key freshwater system and drum screens, a combination of coatings and cathodic protection is employed. After operating for one fuel cycle, existing coatings generally show varying degrees of damage, leading to increased costs for the power plant. Once the anticorrosion coatings in a nuclear power station degrade and break, the consequences are irreversible. If not detected promptly, this can significantly reduce the coating's protective effect on the metal substrate, causing equipment corrosion and failure. The internal medium of a nuclear power plant's seawater system is primarily seawater. According to the anticorrosion mechanism of intelligent coatings, damage to the coating due to external forces or environmental factors meets the trigger conditions for the self-healing of intelligent anticorrosion coatings. Therefore, it is essential to find a self-healing coating to enhance the anticorrosion performance of the coating, thereby increasing the safety and stability of the seawater system. Intelligent anticorrosion coatings have potential application value and broad market prospects in nuclear power plant seawater systems.

The seawater in the nuclear power plant's seawater system has high salinity and carries a certain amount of sediment. During the operation of the unit, the flowing seawater can impact the coating. Hence, if smart anticorrosion coatings are to replace traditional epoxy coatings in nuclear power plant seawater systems, in addition to addressing industrial mass production issues and reducing manufacturing costs, the following application research needs to be conducted. (i) The resistance of coatings to cathodic disbonding in the seawater environment of nuclear power stations. (ii) The impact resistance and wear resistance of coatings in seawater. (iii) The ability of smart anticorrosion coatings to resist seawater penetration. (iv) Verification of the self-healing ability of smart anticorrosion coatings in seawater environments. Addressing these challenges will be pivotal for the successful integration and performance of smart anticorrosion coatings in such demanding and critical environments.

11.2.3 Application of Self-healing Coatings in Oil Gas Corrosion

The oil and gas industry experiences severe corrosion in various equipment, pipelines, and storage tanks. Data shows pipeline corrosion alone costs the United States up to US$2 billion annually, with the United Kingdom and Germany incurring losses of US$1.7 billion and US$3.3 billion, respectively. Corrosion in the oil and gas sector not only leads to the waste of resources and manpower but also poses severe safety and environmental hazards. In protecting oil and gas storage and transportation pipelines, self-healing anticorrosion coatings in the oil and gas field typically have the following requirements: first, the coatings must have good electrical insulation properties. Second, the coating must possess excellent peel resistance to ensure adhesion between the coating and the pipe material, preventing subsequent delamination. When adding micro/nano fillers into the coating, the impact of filler content on coating adhesion must be fully considered to prevent interface bonding strength reduction due to excessive filler content. Third, the coating and filler system should have strong stability and adapt to high pressure, high temperature, and varying pH conditions. This requires considering the filler system selection under high pressure to prevent deformation or disintegration, the activity of corrosion inhibitors and repair agents under high temperatures to prevent deactivation and decrease in resin shielding, and the effect of pH changes on the coating's corrosion resistance performance.

Here we focus again on the extrinsic self-healing coatings based on the addition of corrosion inhibitors in micro/nanocontainers, as they impose fewer restrictions on resins and have broader applicability. The key to achieving long-lasting self-healing in coatings lies in the strict selection of corrosion inhibitors and the rational design of their carriers. However, evaluating the self-healing effect of inhibitor-based coatings can only be accurately done through long-term corrosion tests in actual or simulated service environments. The preparation process for most smart micro/nanocontainers of corrosion inhibitors is cumbersome, making it challenging for large-scale industrial production. Therefore, it is vital to fully utilize the structural characteristics of some materials (like the sensitivity of layered double hydroxides (LDHs) to environmental Cl^- and metal-organic frameworks (MOFs) to pH) to reduce complex surface modification or modification steps. Designing composite nanocontainers can improve compatibility and uniform dispersion in coatings, enhance the loading efficiency of corrosion inhibitors, and enable diverse stimulus-responsive release behaviors. Challenges still exist in controlling the size of micro/nanocontainers, effectively loading and controllably releasing corrosion inhibitors, and ensuring their uniform distribution within the coating. While combining different nanocontainers can endow the coating with multiple self-healing capabilities and multi-effect protection mechanisms, the complexity of composite design hinders large-scale production. Thus, achieving the compounding of corrosion inhibitors within the same micro/nanocontainer is significant. Although most current research results remain at the laboratory stage, some have been applied in practice and shown promising effects. Therefore, it is foreseeable that self-healing coatings for different industry applications will have a broader prospect.

11.3 Smart Self-reporting Coatings

The use of self-reporting coatings in engineering structures provides a simple, efficient, and intuitive new method for corrosion detection of large-area structures and can realize corrosion detection of structural areas that are difficult to reach by inspectors or equipment. Therefore, self-reporting coatings have attracted more and more attention from corrosion researchers. The development of self-reporting coatings makes it possible to diagnose the early damage of coating in large areas, which can interact with the medium or product in the corrosion process to produce color change or luminescence. When the coating is painted on the surface of the structure, it can be observed whether the coating is damaged or corroded by light irradiation or coloration effect, which can realize the visual diagnosis of the corrosion condition of the engineering structure. Therefore, compared with the ordinary coating, the self-reporting coatings have the advantages of convenient detection, intuitive results, time-saving and labor-saving, and have a good application prospect.

Direct mixing of corrosion indicators and coating matrix may cause premature fluorescence or color signals. To avoid this side effect, indicator molecules can be incorporated into micro- or nanocontainers and then loaded into coatings. At the same time, color-developing agents for pH monitoring (such as phenolphthalein) can observe color changes under visible light, so their application advantages are greater than pH fluorescence agents. Metal ions during corrosion can be monitored by chelating metal ions with fluorescent agents to produce chelation enhanced fluorescence (CHEF) or chelation enhanced fluorescence quenching (CHEQ) effects. Due to the interference of the luminous background effect, the indicator with the fluorescence enhancement effect is better than the fluorescence quenching indicator. Because of the complexity of the photochromic indicator, the microencapsulation effect is used to replace the photoluminescence indicator in the coating microcrack monitoring. More attention is paid to the stability of the microcapsule, the UV shielding performance, and tolerance. At the same time, there are few research on quantitative monitoring of microcracks. A self-reporting and self-healing multifunctional coating can realize the synchronization of monitoring and repair and has certain advantages to realize spontaneous repair by adding corrosion inhibitors, water-sensitive materials, polymer monomers, and other methods. The effect of adding multiple components on the stability of the coating remains to be investigated.

In general, the pH sensing coatings can effectively act on the local corrosion acid-alkalization during the initial stages of corrosion, while the microcrack sensing coatings can provide early warning for coating damage. The metal ion sensing coatings have specific requirements for the metal substrate type, while the chloride ion sensing coating is rarely used at present, and the multifunctional self-sensing coating is a coating with a self-healing function based on the sensing function. Different self-reporting coatings have their own characteristics, and their application conditions are also changed according to the actual needs. The self-reporting coating can not only be used in corrosion monitoring but also in building structure crack monitoring, biological surface monitoring, environmental pollution monitoring, and so on. Although self-reporting coatings have carried out long-term basic and application research, they still face a series of problems to be solved. (i) The stability of the indicator. As the key sensing "element" of self-reporting coatings, the coloration of fluorescence intensity, sensitivity, and stability of indicators are the key points of application in corrosion sensing coating. (ii) Reliability of incorporation effect. Incorporation methods and construction technology have a great influence on the actual effect of coating. The microcapsule or nanocontainer size, distribution uniformity, and loading capacity will affect the coating performance. (iii) Coating transparency problem. When corrosion occurs inside the coating, if the coating does not have good transparency, the coating corrosion cannot be effectively observed. Similarly, if the color of the coating itself is not very different from the color of the indicator, the self-reporting effect of the coating will not be ideal. (iv) In order to facilitate application and manual observation, it is particularly important to develop self-reporting coatings excited under visible light.

11.4 Superhydrophobic Coatings

Due to their exceptional properties of superhydrophobic coatings, they find extensive applications in numerous fields, including corrosion protection, antibacterial, self-cleaning, drag reduction, oil–water separation, and ice prevention. However, during the applications of superhydrophobic coatings, their surfaces are highly susceptible to chemical damage caused by oxidizing agents such as light and ozone, as well as to physical damages resulting from external forces, leading to their failure. Furthermore, as science and technology rapidly advance, materials with only superhydrophobic properties can no longer meet the needs of industrial development. These factors have limited the application scope of superhydrophobic coatings. To address these issues, researchers have been dedicated to integrating self-healing functions with superhydrophobic coatings. When the superhydrophobic coating becomes ineffective, it can spontaneously recover its superhydrophobic properties under certain conditions. With continuous exploration and development, significant progress has been made in the theoretical research and preparation methods of superhydrophobic surfaces and self-healing coatings. It is foreseeable that as science and technology continue to evolve, their contributions will play an increasingly critical role in the field of corrosion protection.

Self-healing superhydrophobic coatings involve incorporating substances with hydrophobic groups and healing factors into the coating matrix to form a three-dimensional network structure. This structure exhibits hydrophobic properties and, upon damage, allows the healing agents to migrate from within the coating to the damaged surface or to reconstruct the rough structure on the surface, thus restoring the superhydrophobicity of coatings. This capability not only enables the reuse of material properties but also extends the material's lifespan and reduces costs. Coatings with self-healing and superhydrophobic functionalities can effectively recover their superhydrophobic surface properties when subjected to certain mechanical or chemical damage, thereby prolonging their useful span. Such coatings find effective applications in metal corrosion protection. Compared to coatings with only superhydrophobic properties, self-healing superhydrophobic coatings offer an extended service life due to their dual functionality, reducing the need for maintenance in terms of both labor and resources. (i) Although research on self-healing superhydrophobic coatings has made progress, there are still few materials capable of industrial-scale production. This limitation is mainly due to the complex preparation process of self-healing superhydrophobic materials, leading to higher production costs and hindering industrial-scale production. (ii) Current methods for preparing self-healing superhydrophobic coatings often simply overlay self-healing properties onto superhydrophobic features without organically integrating the two functionalities. The greater synergistic effects and functional characteristics of this integration remain to be explored. (iii) It is important to focus on enhancing the mechanical strength and durability of self-healing superhydrophobic coatings. (iv) Many superhydrophobic coatings are made from low surface energy fluorinated substances. However, fluorinated compounds are toxic and can harm the environment. Therefore, there is an urgent need to find a low-cost, convenient, and environmentally friendly method for preparing self-healing superhydrophobic coatings to advance their industrial progress. (v) Most research tests the corrosion resistance and electrochemical impedance spectrum of superhydrophobic coatings only in saline or seawater mediums. However, in practical applications, metal materials may face various corrosive environments such as salt spray, humidity, and rain. As most superhydrophobic coatings are not dense enough to prevent water vapor penetration, in some cases, due to water vapor permeation or condensation, the superhydrophobic properties of the coating may be completely lost. Therefore, superhydrophobic coatings currently cannot effectively replace existing metal corrosion protection systems but can be combined with existing coating systems for synergistic protection of the metal substrate. (vi) In addition, the healing mechanisms of some self-healing superhydrophobic coatings are still unclear. A more profound and comprehensive analysis of the interdependence between surface components, structure, wettability, and corrosion resistance at the microscopic scale is necessary. Exploring the universal rules and related theories concerning the impact of substrate micro/nanostructures, roughness, hydrophobicity/hydrophilicity, and chemical stability on superhydrophobic coating materials will provide more theoretical support for preparing durable superhydrophobic surfaces. Only by fundamentally refining the preparation process of self-healing superhydrophobic coatings can achieve significant improvements. (vii) Although some studies have achieved rapid repair at room temperature, most self-healing superhydrophobic coatings still require external stimuli, such as heating or light for repair, making them unsuitable for applications demanding high stability. Continuous development of multifunctional composite self-healing superhydrophobic materials is needed, endowing materials with additional functional properties, such as photo-, electro-, magnetic-, and thermoresponsive features, to further enhance their practicality.

11.5 Application of Atmospheric Corrosion Monitor Technology in Smart Coating Design

Atmospheric corrosion monitor (ACM) technology has been used to reflect the corrosion of materials in the atmospheric environment according to the current signals of thin liquid film electrochemical cells. The concept of ACM was first proposed in 1976 and has been used in the research of metal corrosion detection and evaluation in atmospheric environments. The basic detection principle of ACM is to use two metal pairs with different potentials to form a galvanic corrosion sensor, and use a high-sensitivity galvanometer to detect the corrosion currents generated under the thin liquid film in the atmospheric environment so as to evaluate the corrosion rate of metal pairs in different environments or in different duration. At present, there is much research on atmospheric corrosion of metal materials using ACM, whereas few reports on coating systems using ACM. Therefore, by brushing self-healing coatings on the surface of the ACM sensor probe, the corrosion behavior of the coating is monitored in real time, and the basic environmental parameters, such as temperature and humidity, are equipped with sensors to collect environmental factors and monitor the protection and failure process of the coating in real time. It can solve the problem that the accumulation of traditional electrochemical test data is fragmented and cannot reflect the whole process of self-healing performance and gradual failure of the coating system. Meanwhile, it provides a method for the failure study of anticorrosion coatings in outdoor atmospheric environments, truly reflects the service conditions of the coating, accumulates the environmental corrosion data of the coating, and establishes the life evaluation system of the coating. It provides theoretical support and experimental basis for the application of coating in engineering and provides a new direction for the development of self-healing coatings. Moreover, enriching the application of ACM in an engineering environment is of great significance for the further development of ACM technology.

At present, ACM technology has made great progress, but there are still the following issues to be solved: (i) Data collection lacks enough stability, and data analysis methods need further exploration. The established model for data analysis needs higher accuracy. (ii) The emergence of various new materials and the exploration of various harsh environments have made the atmospheric corrosion behavior of materials more complicated, which put forward new requirements for the ACM online monitoring methods. (iii) The development of modern science and technology provides new ideas and methods for online corrosion monitoring. For example, with the development of image recognition technology, we can directly extract information from the corroded surface of the sample, hoping to conduct qualitative and quantitative analysis of the corrosion situation through a macro photo. (iv) In addition, the online network observation of atmospheric environmental corrosion is the focus of current development. The traditional corrosion monitoring cycle is of a long period, and the corrosion state fluctuation information cannot be obtained in time. The next step should be guided by intelligent anticorrosive management, integrating meteorological data, environmental data, and other acquisition modules to achieve real-time and high-throughput collection and storage. Finally, the data information will be fused to form a network observation platform with the characteristics of "corrosion big data." The corrosion database should be built, and the material life cycle prediction model is established based on various data mining algorithms, and the bridge between the corrosion data and the actual corrosion situation is built.

Index

Smart Protective Coatings for Corrosion Control, First Edition. Lingwei Ma and Dawei Zhang.

www.ingramcontent.com/pod-product-compliance
Lightning Source LLC
Chambersburg PA
CBHW080931240926
51849CB00003BA/429
* 9 7 8 3 5 2 7 3 5 3 8 6 6 *